UNITEXT

La Matematica per il 3+2

Volume 163

The **UNITEXT - La Matematica per il 3+2** series is designed for undergraduate and graduate academic courses, and also includes books addressed to PhD students in mathematics, presented at a sufficiently general and advanced level so that the student or scholar interested in a more specific theme would get the necessary background to explore it.

Originally released in Italian, the series now publishes textbooks in English addressed to students in mathematics worldwide.

Some of the most successful books in the series have evolved through several editions, adapting to the evolution of teaching curricula.

Submissions must include at least 3 sample chapters, a table of contents, and a preface outlining the aims and scope of the book, how the book fits in with the current literature, and which courses the book is suitable for.

For any further information, please contact the Editor at Springer: francesca.bonadei@springer.com

THE SERIES IS INDEXED IN SCOPUS

Lorenzo Brasco

Handbook of Calculus of Variations for Absolute Beginners

Lorenzo Brasco
Dipartimento Matematica e Informatica
Università degli Studi di Ferrara
Ferrara, Italy

ISSN 2038-5714 ISSN 2532-3318 (electronic)
UNITEXT
ISSN 2038-5722 ISSN 2038-5757 (electronic)
La Matematica per il 3+2
ISBN 978-3-031-87163-4 ISBN 978-3-031-87164-1 (eBook)
https://doi.org/10.1007/978-3-031-87164-1

This Springer imprint is published by the registered company Springer Nature Switzerland AG
The registered company address is: Gewerbestrasse 11, 6330 Cham, Switzerland

A Davide e Niccolò, sperando che possano essere fieri del loro babbo

Preface

Foreword

This is not just another book on the Calculus of Variations. We already have several well-written ones by world-leading experts in the field. Without any claim to completeness, we list some of them in the Bibliography at the end of the book. *Ça va sans dire*, such a list has been compiled according to the author's personal taste.

This book is rather a collection of insights we've gained about the Calculus of Variations throughout our careers, which we believe could be useful to any mathematics student approaching the subject for the first time. The content is presented in a way that strongly reflects the author's approach. For this reason, the reader will not find general statements or proofs based on general abstract theories. For example, the use of Functional Analysis is very limited. Also, you will not find here a complete treatment of the most general integral functionals from the Calculus of Variations. Rather, the main focus will be on introducing things from scratch, by means of simple (yet meaningful) explicit examples.

The ultimate scope of the book is to equip students with a survival toolkit, in order to start safely diving into the realm of Calculus of Variations. On the other hand, some experts in the field might simply find the book boring, tedious, largely incomplete, or too easy: in a word, *useless*. Hopefully, they will not judge us too harshly.

Some Classical Problems in Calculus of Variations

We do not attempt here an history of the Calculus of Variations. The interested reader could find more information in the introduction and the notes to the chapters of Giusti's book [6] or in the books by Giaquinta and Hildebrandt [4, 5] and by Goldstine [7]. We can be content to know that the Calculus of Variations, roughly speaking, is a part of Mathematical Analysis that deals with minimization and maximization problems. The prototypical problem involves proving that a certain optimization problem admits a solution, exploring its properties, and possibly determining it explicitly. Often, the functional to be optimized arises from Physics (where it represents some form of energy of a system) or Geometry (where it represents some geometric quantity).

To motivate the reader to engage with the Calculus of Variations, let us present four classical optimization problems that greatly contributed to the birth and development of the field.[1]

Isoperimetric Problem
This is probably one of the most classical problem in Mathematics, it goes back to the Ancient Greeks (see, for example, [64] or [70] for an historical overview of the problem). Let us stick to its original formulation: among all closed curves in the plane having the same length, find the one enclosing the region of the plane having maximal area. This problem usually looks easy to uninitiated people and essentially everybody is immediately tempted to say that the circle should be the solution. This intuition is correct, but the rigorous justification of this fact took several centuries: we will present in Theorem 2.6.1 the classical proof by Hurwitz, given at the beginning of the twentieth century. This has been one of the very first rigorous and complete proofs.

The Brachistocrone Problem
This is another very classical problem, posed for the first time by Johann Bernoulli in the seventeenth century (we refer to [7, Chapter 1] for more historical details on this problem). This can be formulated as follows: in a vertical plane, suppose to have two distinct points **A** and **B** and a mass particle moving in this plane along a curve connecting **A** to **B**. Assuming that the mass particle is only subject to the force of gravity, which is the curve minimizing the total traveling time from **A** to **B**? Here, the intuition might probably fool us and suggest that a segment or an arc of circle should be the solution. Actually, this is not the case: the solution is given by an arc of *cycloid*. We will show in Chap. 2 how to properly settle this problem and obtain the claimed solution, by using some classical methods in the Calculus of Variations.

Gradient Fields with Minimal Energy
This is one of the problems which will guide us throughout the whole book, in order to get acquainted with the celebrated *Direct Method* in the Calculus of Variations, see Chaps. 4 and 6. The problem reads as follows: for an

(continued)

[1] We wish to quote here Giaquinta and Hildebrandt: in the preface of their book [5] they say "*examples are the true lifeblood of the calculus of variations*". We could not agree more.

open bounded set $\Omega \subseteq \mathbb{R}^N$ with smooth boundary (the physically relevant situations are $N = 2$ or $N = 3$) and a given function $U : \partial\Omega \to \mathbb{R}$, we seek for a solution of

$$\inf_{u \in X(\Omega)} \left\{ \int_\Omega |\nabla u|^2 \, dx \, : \, u = U \text{ on } \partial\Omega \right\}. \tag{H}$$

Here $X(\Omega)$ is a suitable "space of functions", which should be carefully defined, in order to guarantee that a solution of this problem exists.

A physical situation where one encounters this kind of problem comes from Electromagnetism: suppose to have a *capacitor* (or *condenser*) consisting of the spatial region enclosed by two spherical plates C_1 and C_2, with the latter contained in the former. We call Ω this region, so that $\partial\Omega = C_1 \cup C_2$ (see Fig. 1). We establish an electric potential difference between the two plates: this is the role of boundary datum U in the formulation above. For example, for simplicity we could say that

$$U \equiv 1 \text{ on } C_2 \qquad \text{and} \qquad U \equiv 0 \text{ on } C_1.$$

This potential difference creates an electric field E in each point of the capacitor Ω: by recalling that the electric field is conservative, we actually must have $E = \nabla v$ for some electric potential v, which coincides with U on the two plates C_1 and C_2. The quantity

$$\int_\Omega |E|^2 \, dx = \int_\Omega |\nabla v|^2 \, dx,$$

is nothing but the electric energy stored in the capacitor (up to some physical constants, which are here neglected). It can be proved that such a potential v is exactly the one solving the problem (H) above. Indeed, by Maxwell's equations we know that

$$\operatorname{div} E = \operatorname{div}(\nabla v) = \frac{\partial^2 v}{\partial x^2} + \frac{\partial^2 v}{\partial y^2} + \frac{\partial^2 v}{\partial z^2} = 0, \qquad \text{in } \Omega,$$

i.e., v is a so-called *harmonic function*. The asserted minimality property of v is then a consequence of the *Dirichlet principle*, which will be presented in Chap. 1. Thus, solving the minimization problem (H) we can compute the electric field E and the value of the energy stored in the condenser. Of course, we assumed for ease of presentation that the plates C_1 and C_2 were spherical, but we can imagine much more general geometric configurations.

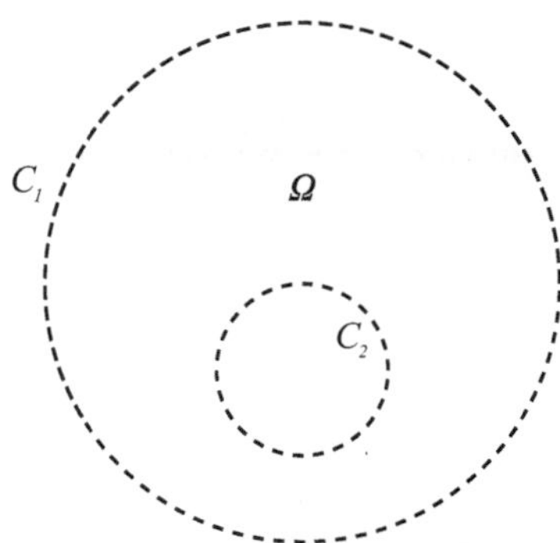

Fig. 1 In dashed line, the two plates C_1 and C_2 of the capacitor Ω

Graphs with Least Area

This is a problem that contributed a lot to the developments of new mathematical tools, like for example the introduction of *functions with bounded variation* (see, for example, [16] or [30]). We suppose to have an open bounded set $\Omega \subseteq \mathbb{R}^2$ with smooth boundary and a function U defined on $\partial\Omega$. We seek for a function v coinciding with U on $\partial\Omega$ and such that its graph

$$\Gamma_v = \Big\{(x, v(x)) \in \mathbb{R}^3 \, : \, x \in \Omega\Big\},$$

has the least possible area. By recalling the expression for the area of the graph of a smooth function, we are thus looking for a minimizer of the following problem

$$\inf_{u \in Y(\Omega)} \left\{ \int_\Omega \sqrt{1 + |\nabla u|^2}\, dx \, : \, u = U \text{ on } \partial\Omega \right\}.$$

Here as well $Y(\Omega)$ is a suitable "space of functions". Classically, this is also called *Plateau's problem in non-parametric form* (see, for example, [53, Chapter IV]).

We will discuss this problem in Chap. 6 and show that, under appropriate assumptions on U and the set Ω, a minimizer actually exists, is unique, and solves (in a suitable sense) the following partial differential equation

$$-\operatorname{div}\left(\frac{\nabla u}{\sqrt{1 + |\nabla u|^2}}\right) = 0, \qquad \text{in } \Omega.$$

A reader with a minimal knowledge of the *geometry of surfaces* may recognize the quantity on the left-hand side as the *sum of the two principal*

(continued)

curvatures of the cartesian surface $\phi(x) = (x, u(x))$, see, for example, [25, Chapter 3, Section 3, Example 5]. For this reason, the differential operator on the left-hand side is also called *mean curvature operator*.

Many more examples could be done: we refer for instance to [1, 9–12] and especially to the list of examples that can be found at the end of [5]. We will stick here to some simple, yet meaningful, ones which will be tackled in this book.

Ferrara, Italy
January 2025

Lorenzo Brasco

Acknowledgments

It is fair to admit that the writing of this book has been quite demanding, in terms of both time and efforts. We feel that a lot of people deserve to be thanked, for having contributed to this work, in a way or another.

Having some excellent teachers has been a good starting point: we thus thank Giuseppe Buttazzo, Guillaume Carlier, Nicola Fusco, and Filippo Santambrogio. We also wish to address a special acknowledgment to Rolando Magnanini, who first introduced us to research in Mathematical Analysis.

We want to sincerely thank all the friends, collaborators, and colleagues who spent some time to revise parts of the book and provide useful suggestions, criticisms, corrections, and improvements: Paolo Baroni, Vladimir Bobkov, Pierre Bousquet, Guillaume Carlier, Erik Lindgren, Michele Miranda, Delio Mugnolo, Roberto Ognibene, Enea Parini, Francesca Prinari, Giulio Tralli. Their help has been invaluable, it would have been impossible to complete the writing of this book without their kind collaboration: *grazie mille!*

We also would like to thank the following people for all the mathematical discussions along the years, which contributed to improve our knowledge of the subject: Guido De Philippis, Yves Dermenjian (*notre amiable collègue de bureau à Marseille*), Giovanni Franzina, Sunra Mosconi, Berardo Ruffini, Marco Squassina, Juan-Luis Vázquez, and Bruno Volzone (who almost forced us to publish this book).

We gratefully acknowledge the kind assistance of Francesca Bonadei at Springer during the production process; we wish to thank her.

Of course, it would be unfair if we do not thank our former Ph.D. students Francesca Bianchi, Francesco Bozzola, and Anna Chiara Zagati: this book has been written, to a large extent, having them in mind.

Last but not least, we wish to express all our heartfelt thanks to Eleonora Cinti: she has been so kind to lovely take care of a final reading of the book. She also provided a lot of comments, criticisms, and encouragements during all the writing process.

As the reader may see, we thanked a lot of people for all the good things (if any) contained in this book. On the contrary, we proudly admit that all the mistakes and the bad things are just our own responsibility.

About This Book

Contents of the Book

The book consists of eight chapters and four appendices. It is time to proceed with an appetizer of what the reader will find in every chapter.

- Chapter 1: we start with some basic results on convex functions. These should be well-known, but we prefer to include them at cost of being pedantic, due to their pivotal importance. Then, we present the classical *Du Bois-Reymond Lemma*, the concept of *Euler-Lagrange equation* (in classical and weak forms), and the *Dirichlet principle*, aimed at introducing the student to the crucial interplay between optimization problems and partial differential equations.
- Chapter 2: this contains some classical one-dimensional variational problems, which can be solved "by hands", thanks to some suitable tricks. This is intended to introduce the students to some classical methods which are important, but usually not treated in those textbooks which are more focused on the so-called *Direct Method* (but we should cite the excellent book [3] by Dacorogna). We point out that we claim no completeness of the presentation in this part: many of the "old school" methods are not discussed.

  In particular, we present the problem of finding curves with minimal kinetic energy; the brachistocrone problem (see above); the one-dimensional harmonic oscillator; and the problem of determining the sharp constant in one-dimensional Poincaré inequalities. As an application of the last problem, we also give Hurwitz's proof of the planar Isoperimetric Inequality, as previously announced.

  In our study of the brachistocrone problem, we prove existence of a solution by means of an *hidden convexity* structure of the problem. This fact is not always highlighted in the references treating the subject: among the places where we have been able to trace back this fact, we mention [2, Chapter 1, Section 3, Example 3] and [5, Chapter 4, Section 2.3, Example 4].

  In the proof by Hurwitz, we tried to introduce some original elements, by replacing the traditional use of Fourier series with a variational argument, based on a convexity trick (the so-called *Picone inequality*). The idea for this variant has been taken from the book [32], that we highly recommend to any reader interested in "inequalities" of any (mathematical) kind.

- Chapter 3: this is a small introductory monograph on Sobolev spaces. These are functional spaces which eventually became an essential ingredient in the modern theory of Calculus of Variations. The contents are quite classical and much of the materials can be found in many books, which we list in the bibliography. For example, we certainly recommend [20, Chapter 9] or [29, Chapter 7]. Here much emphasis is put on the space $W_0^{1,p}$ of functions "vanishing at the boundary", which is usually a bit overlooked, despite its crucial importance in boundary value problems with Dirichlet conditions.
- Chapter 4: this is the main core of the book, here we introduce the student to the *Direct Method* in Sobolev spaces and give some of its applications. As already said, the reader will not find here the most general existence results for minimization problems, but rather a result which is sufficiently general to enclose some interesting problems, comprising the aforementioned problem of the electrostatic condenser. For example, this chapter contains a thorough discussion on some important partial differential equations coming from variational problems, like the *torsional rigidity equation*, the *Lane-Emden equation*, and the *eigenvalue equation for the Dirichlet-Laplacian*.

  We have also decided to add a section on the *Spectral Theorem* about the structure of the spectrum of the Dirichlet-Laplacian, on an open bounded set. Here, we have essentially followed what is done in the beautiful book [12], where an elegant and entirely variational proof is given.
- Chapter 5: this is a counterpart of Chap. 3, with a focus now on Lipschitz continuous functions. It contains all the basic and important results on these functions, including compactness properties, extension results, and differentiability properties. The most sophisticated result presented in this part is the *Rademacher Theorem*, asserting the almost everywhere differentiability of Lipschitz functions.
- Chapter 6: this deals with the Direct Method in the class of Lipschitz functions. This is important because it permits to treat some problems which cannot be handled in Sobolev spaces, but still they are quite interesting. The most prominent example is given by the minimization of the area functional

$$u \mapsto \int \sqrt{1+|\nabla u|^2}\,dx,$$

  that we have already presented. Here we mostly follow the approach of [6, Chapter 1], by making it as elementary as possible. We also expanded a bit the discussion on the so-called *bounded slope condition*, which plays an important role in the theory, by adding some useful sufficient conditions for this to hold.
- Chapter 7: this is the most advanced part of the book. Indeed, this is intended as an introduction to the techniques of the modern Elliptic Regularity Theory. In other words, we present some tools (which eventually became standard in the field) to prove that the minimizers found in Chap. 4, which in principle

are functions differentiable only in a weak sense, are actually more regular. However, here as well, we resist the temptation of any generalization that could obscure the ideas and we simply stick to the case of the Laplace equation, i.e., to minimizers of the problem (H). This is used as a "case study" to expose the main techniques. The reader is then encouraged to apply these techniques to more general situations, through a small list of final problems. Among these problems, we present the proof of the celebrated *De Giorgi-Nash-Moser Theorem*:

- Chapter 8: each of the previous chapters contains a list of problems of varying difficulty. This final chapter collects all the solutions to these problems, except for those of Chap. 7. In this last case, we rather give some hints.

Dulcis in fundo, we included four appendices. The first one, Appendix A, is very simple and contains the expression of the Laplacian operator in polar coordinates in $\mathbb{R}^2$, together with the construction of some explicit harmonic functions (the so-called *circular harmonics*). We found it useful to add this part to the book, since we have the impression that nowadays not all the students are familiar with these important basics.

In Appendix B we recall some compactness results for L^p spaces. We also take the occasion to present an elegant and useful result, the *Mazur Lemma*, which is not always treated in the textbooks.

Appendix C contains a succinct discussion about the so-called *homogeneous Sobolev spaces*, defined through a completion process. These spaces are sometimes important to deal with variational problems on the whole space or, more generally, on unbounded open sets. Despite their importance in the Calculus of Variations, they are a source of misunderstandings even among experts: hopefully this small appendix could contribute to clarify some of their basic aspects.

Finally, in Appendix D we offer a brief treatment of fractional Sobolev spaces, in connection with the theory of *traces*. We limit ourselves to expose the main results and proofs in the meaningful case of the half-space $\mathbb{R}^N \times (0, +\infty)$. This part is put here essentially to give a flavor, for the students who are curious enough, of the fact that $W_0^{1,p}$ should be made of functions that "vanish at the boundary" in some weak sense, at least if the boundary of the set is "good enough". For those who wish to deepen the study of these spaces and of the theory of traces, we recommend the reading of [41, Chapter 6] and [44, Chapters 17 and 18]. The recent monographs [26] and [45] are also recommended.

A Message for the Student

Of course, the expression "absolute beginners" in the title of the book should be taken with a bit of understanding. It is easy to imagine that a person without any prior knowledge in Mathematical Analysis would have some serious difficulties in handling this book. In other words, in spite of our efforts to keep the level of needed prerequisites as low as possible, we assume that the student is familiar with the following facts:

- The contents of the first two years courses in Mathematical Analysis, especially differential and integral Calculus for functions of several variables, geometry of curves and surfaces, the Divergence Theorem.
- Some (very few) basic facts from Topology.
- Some rudiments of the theory of Lebesgue measure and integration. In particular, it is assumed that the reader is familiar with the most famous theorems about interchanging of limit and integral, i.e., Fatou's Lemma, the Monotone Convergence Theorem and the Dominated Convergence Theorem.
- The main properties of L^p spaces. Here, a very good source is given for example by [46], which we strongly recommend. An important property we will use frequently is that "from every sequence converging in L^p norm, we can extract a subsequence converging almost everywhere to the same limit". Some compactness properties of L^p spaces, which should be already known to the reader, have been added in Appendix B, due to their importance. We also expect that the student feels comfortable with the operation of "convolution": if this is not the case, we recommend again [46].

As for the problems that the student will find at the end of each chapter: we made a great effort to provide as much details as possible in their solutions. It was our intention to guide the student, rather than frustrate her/him. For the same reason, we preferred not to put a clear indication on the level of difficulty of the problems: the distinction between "easy" and "hard" is very often quite personal. We can easily imagine a student getting frustrated by the fact that he/she cannot solve a problem that we marked as "easy"... We prefer to avoid these situations.

A Message for the Teacher

It is very unlikely that in a standard course in Calculus of Variations (whatever "standard" means...), the teacher could cover the whole contents of the book. One could imagine to plan different possible courses, according to his personal taste and the level of the students.

Our experience is based on a Master course delivered in the Spring 2020 and 2022, at the University of Ferrara. This was a 48 hours course, aimed at giving a brief introduction to variational methods for minimization problems. It was designed for students having a limited knowledge of Mathematical Analysis, apart from the contents of the courses of the first two years and some basic facts on L^p spaces, as described above.

We used the following planning:

- Chapter 1 (excluded Sect. 1.6)
- Chapter 2 (excluded the *divertissement* of Sect. 2.7)

- Chapter 3 (excluded Sects. 3.9, 3.10, and 3.11)
- Chapter 4 (only Sects. 4.1, 4.2, 4.3, 4.7, and 4.8)

A related, yet slightly different choice could be the following:

- Chapter 1
- Chapter 2
- Chapter 5
- Chapter 6

aimed at giving more emphasis on classical methods. This has the advantage of introducing students to the area functional and, time permitting, to a brief study of *minimal surfaces*, possibly by integrating the material with that of the classical book [30].

Depending on the audience, for example for students already knowing Sobolev spaces and Lipschitz functions, one could also imagine to use the following planning:

- Chapter 4
- Chapter 6
- Chapter 7

Finally, some parts of the book could also be used for some short Ph.D. courses: this is certainly the case of Chap. 7 and of the problems therein contained. We also found useful some parts of Chap. 4 to prepare a Ph.D. course entitled "*Geometry of principal frequencies*", which was focused on geometric estimates for eigenvalues of the Laplacian.

A Message for the Researcher

When compared to all the outstanding books on the subject, the present *handbook* may appear of a very limited utility to researchers working in the area. Nevertheless, we think that some specific parts of the book could be useful to experts, as well. During our career, it happened to us several times to discuss with researchers working in Calculus of Variations (in a broad sense, including for example Spectral Theory and Critical Point Theory), who were "absolute beginners" on some of the topics treated in this book. For example, they did not have any real knowledge in Regularity Theory, apart from having in mind some celebrated statements. In this respect, for example, we think that Chap. 7 could be useful even for a researcher, at least to have a friendly introduction to some basic techniques, which are standard nowadays.

The same can be said for the contents of Chap. 6. We have the impression that they are considered a bit "out of date" and very often researchers, even those working in Regularity Theory, are not familiar with them.[2] Here they could find a simple and brief introduction, which could be useful.

Another source of confusion is sometimes given by Sobolev spaces of functions "vanishing" at the boundary of a set. It is not always clear that these can be defined (of course in a weak sense) without any regularity requirement on the sets and without appealing to the theory of "traces". Moreover, the distinction between spaces obtained as "closure" or "completion" does not seem to be well-understood by everybody. Hopefully, this book could contribute to clarify these aspects.

[2] We must confess that it is only thanks to our friend and collaborator Pierre Bousquet that we understood the importance of these techniques, along the years.

Contents

List of Symbols

We summarize below some of the main notations used throughout the book.

Miscellaneous Notation

$\mathbf{e}_i$	$i-$th unit vector of the canonical basis of $\mathbb{R}^N$
$\langle x, y\rangle$	standard euclidean scalar product on $\mathbb{R}^N$, defined by $\langle x, y\rangle = x_1\, y_1 + \cdots + x_N\, y_N$
$\mathbb{S}^{N-1}$	the $N-$dimensional unit sphere, defined by $\{x \in \mathbb{R}^N\ :\ \lvert x\rvert = 1\}$
$B_R(x_0)$	euclidean ball centered at $x_0 \in \mathbb{R}^N$, having radius $R > 0$
$\lvert E\rvert$	$N-$dimensional Lebesgue measure of the measurable set $E \subseteq \mathbb{R}^N$
ω_N	$\lvert B_1(0)\rvert$, i.e. $N-$dimensional Lebesgue measure of $B_1(0) \subseteq \mathbb{R}^N$
$\delta_{i,j}$	Kronecker delta, given by $\delta_{i,j} = 0$ for $i \neq j$ and $\delta_{i,i} = 1$
$\mathcal{M}_N(\mathbb{R})$	vector space of $N \times N$ matrices with real coefficients
M^{T}	transposed matrix of $M \in \mathcal{M}_N(\mathbb{R})$
$M \cdot x$	if $M \in \mathcal{M}_N(\mathbb{R})$ and $x \in \mathbb{R}^N$, product "matrix times column vector"
$x \cdot M$	if $M \in \mathcal{M}_N(\mathbb{R})$ and $x \in \mathbb{R}^N$, product "row vector times matrix"
Id_N	the $N \times N$ identity matrix
a. e.	almost everywhere (with respect to the Lebesgue measure)
$\Omega' \Subset \Omega$	the open set $\Omega' \subseteq \mathbb{R}^N$ has compact closure $\overline{\Omega'}$ contained in the open set $\Omega \subseteq \mathbb{R}^N$
$\mathrm{dist}(x, E)$	distance of $x \in \mathbb{R}^N$ from $E \subseteq \mathbb{R}^N$, defined by $\mathrm{dist}(x, E) = \inf\limits_{y\in E} \lvert x-y\rvert$
$\mathrm{dist}(E, F)$	distance between two subsets $E, F \subseteq \mathbb{R}^N$, defined by $\mathrm{dist}(E, F) = \inf\limits_{x\in E, y\in F} \lvert x - y\rvert$
$\mathrm{int}_n(\Omega)$	$\{x \in \Omega\ :\ \mathrm{dist}(x, \partial\Omega) > 1/n\}$, with $\Omega \subseteq \mathbb{R}^N$ open set and $n \in \mathbb{N}\setminus\{0\}$
1_E	characteristic function of E, defined by $1_E(x) = 1$ if $x \in E$ and $1_E(x) = 0$ otherwise

$\fint_E u\,dx$ integral average of the locally summable function u over the measurable set E, i.e., $\dfrac{1}{|E|}\displaystyle\int_E u\,dx$

Differential Operators

∇u gradient of the differentiable function of N variables u, defined by $\nabla u = \left(\dfrac{\partial u}{\partial x_1}, \dots, \dfrac{\partial u}{\partial x_N}\right)$

D^2u Hessian matrix of the twice differentiable function of N variables u, defined by $D^2u = \left(\dfrac{\partial^2 u}{\partial x_i\,\partial x_j}\right)_{i,j=1}^N$

$D\phi$ Jacobian matrix of the differentiable vector field of N variables ϕ, defined by $D\phi = \left(\dfrac{\partial \phi_i}{\partial x_j}\right)_{i,j=1}^N$

$\operatorname{div}\phi$ divergence of a differentiable vector field $\phi = (\phi_1, \dots, \phi_N)$, defined by $\operatorname{div}\phi = \displaystyle\sum_{i=1}^N \dfrac{\partial \phi_i}{\partial x_i}$

Δ Laplacian operator, defined for a twice differentiable function of N variables u by $\Delta u = \operatorname{div}(\nabla u) = \displaystyle\sum_{i=1}^N \dfrac{\partial^2 u}{\partial x_i^2}$

Δ_p $p-$Laplacian operator, defined for a twice differentiable function of N variables u by $\Delta_p u = \operatorname{div}(|\nabla u|^{p-2}\,\nabla u) = \displaystyle\sum_{i=1}^N \dfrac{\partial}{\partial x_i}\left(|\nabla u|^{p-2}\,\dfrac{\partial u}{\partial x_i}\right)$

Function Spaces

$C^k(\Omega)$ continuous functions on the open set $\Omega \subseteq \mathbb{R}^N$, which are continuously differentiable k times in Ω (with $k \in \mathbb{N}$)

$C^\infty(\Omega)$ continuous functions on the open set $\Omega \subseteq \mathbb{R}^N$, which are continuously differentiable infinitely many times in Ω

$C_0^k(\Omega)$ compactly supported functions belonging to $C^k(\Omega)$

$C_0^\infty(\Omega)$ compactly supported functions belonging to $C^\infty(\Omega)$

$C^0(\overline{\Omega})$ continuous functions on the closure $\overline{\Omega}$ of the open set $\Omega \subseteq \mathbb{R}^N$

$C_b^0(\overline{\Omega})$ functions of $C^0(\overline{\Omega})$ which are bounded on $\overline{\Omega}$

$C^{0,1}(\overline{\Omega})$ Lipschitz continuous functions belonging to $C_b^0(\overline{\Omega})$

$C_0^{0,1}(\overline{\Omega})$ functions of $C^{0,1}(\overline{\Omega})$ which vanish at the boundary $\partial\Omega$

$C^{0,\alpha}(\overline{\Omega})$	α−Hölder continuous functions belonging to $C^0_b(\overline{\Omega})$ (for $0 < \alpha < 1$)
$C^k(\overline{\Omega})$	functions $u : \overline{\Omega} \to \mathbb{R}$ for which there exist an open set E containing $\overline{\Omega}$ and a function $U \in C^k(E)$ such that $u = U$ on $\overline{\Omega}$
$L^p(\Omega)$	measurable functions $u : \Omega \to \mathbb{R}$ such that $\int_\Omega |u|^p\,dx < +\infty$
$L^\infty(\Omega)$	measurable functions $u : \Omega \to \mathbb{R}$ which are essentially bounded, i.e., such that there exists a constant M for which $|u(x)| \le M$ for almost every $x \in \Omega$
$L^p_{\mathrm{loc}}(\Omega)$	measurable functions $u : \Omega \to \mathbb{R}$ such that $u \in L^p(\Omega')$, for every $\Omega' \Subset \Omega$
$L^p(\Omega; \mathbb{R}^k)$	measurable functions $u : \Omega \to \mathbb{R}^k$ such that $|u| \in L^p(\Omega)$
$W^{1,p}(\Omega)$	for $1 \le p \le \infty$ and an open set $\Omega \subseteq \mathbb{R}^N$, the Sobolev space of $L^p(\Omega)$ functions, whose weak gradient belongs to $L^p(\Omega; \mathbb{R}^N)$
$W^{1,p}_{\mathrm{loc}}(\Omega)$	the local Sobolev space of $L^p_{\mathrm{loc}}(\Omega)$ functions, whose weak gradient belongs $L^p_{\mathrm{loc}}(\Omega; \mathbb{R}^N)$
$W^{1,p}_0(\Omega)$	closure of $C^\infty_0(\Omega)$ in $W^{1,p}(\Omega)$

Norms and Seminorms

$\|u\|_{C^0(\overline{\Omega})}$	for $u \in C^0(\overline{\Omega})$, this is the sup-norm defined by $\sup_{x\in\overline{\Omega}} |u(x)|$
$|u|_{C^{0,\alpha}(\overline{\Omega})}$	for $0 < \alpha \le 1$, this is defined by $\sup_{x\ne y,\ x,y\in\overline{\Omega}} \dfrac{|u(x) - u(y)|}{|x - y|^\alpha}$
$\|u\|_{L^p(\Omega)}$	for $1 \le p < \infty$, this is $\left(\int_\Omega |u|^p\,dx\right)^{1/p}$
$\|u\|_{L^\infty(\Omega)}$	$\inf\left\{M \ge 0 : |u(x)| \le M \text{ for a. e. } x \in \Omega\right\}$
$\|u\|_{L^p(\Omega;\mathbb{R}^k)}$	for $1 \le p \le \infty$, this is the L^p norm of $|u|$
$\|u\|_{W^{1,p}(\Omega)}$	for $1 \le p \le \infty$, this is defined by $\|u\|_{W^{1,p}(\Omega)} = \|u\|_{L^p(\Omega)} + \|\nabla u\|_{L^p(\Omega;\mathbb{R}^N)}$

Chapter 1
Tools

1.1 Introduction

This chapter introduces some foundational material that will be necessary throughout the entire book. Some of this material may already be familiar to the reader, but we have chosen to include it to ensure that the presentation remains accessible to a broad audience.

We begin by recalling some properties of convex functions, starting with the case of a single real variable. We will not delve too deeply into the theory of these functions but will focus on the properties that are needed in the rest of the book. These will be sufficient for the purposes of the text. Among the few results we recall, the most important is the so-called *"above tangent" property*, i.e., the fact that the graph of a convex function always lies above its tangent hyperplanes. This property will play a key role and will be used repeatedly. Readers interested in further expanding their knowledge of Convex Analysis are encouraged to consult references [22, Chapter 3] or [36], for example.

We will also present some useful consequences of the "above tangent" property, including Jensen's inequality, the lower semicontinuity of L^p norms with respect to weak convergence, and Picone's inequality. While the first two results are generally well-known, the third is perhaps less familiar. In fact, Picone's inequality is a very helpful tool and will be used in several places throughout the book.

Once these preliminary tools are introduced, we will turn our attention to the central themes of the book, beginning with some fundamental concepts: the so-called *fundamental lemma of the Calculus of Variations*, a discussion of the *first variation* and the *Euler-Lagrange equation* for integral functionals, and the *Dirichlet principle*, which reveals the close connection between optimization problems and partial differential equations.

L. Brasco, *Handbook of Calculus of Variations for Absolute Beginners*,
La Matematica per il 3+2 163, https://doi.org/10.1007/978-3-031-87164-1_1

1.2 Convex Functions of One Variable

Definition 1.2.1 Let $I \subseteq \mathbb{R}$ be an interval and let $f : I \to \mathbb{R}$ be a function. We say that f is *convex* if we have

$$f(\lambda\, t_0 + (1-\lambda)\, t_1) \le \lambda\, f(t_0) + (1-\lambda)\, f(t_1), \qquad \text{for every } \lambda \in [0,1],\ t_0, t_1 \in I.$$

Moreover, we say that f is *strictly convex* if we have

$$f(\lambda\, t_0 + (1-\lambda)\, t_1) < \lambda\, f(t_0) + (1-\lambda)\, f(t_1),$$
$$\text{for every } \lambda \in (0,1),\ t_0, t_1 \in I \text{ with } t_0 \neq t_1.$$

Lemma 1.2.2 *Let $I \subseteq \mathbb{R}$ be an interval and let $f : I \to \mathbb{R}$ be a convex function. For every $t_0, t_1 \in I$, the function*

$$g(\lambda) = \frac{f(t_0 + \lambda\,(t_1 - t_0)) - f(t_0)}{\lambda}, \qquad \text{for } \lambda \in (0,1],$$

is non-decreasing.

Proof We can assume that $t_0 \neq t_1$, otherwise there is nothing to prove. We take $\lambda_0, \lambda_1 \in (0,1]$ such that $\lambda_0 < \lambda_1$. We then observe that we can write

$$f(t_0 + \lambda_0\,(t_1 - t_0)) = f\left(\frac{\lambda_0}{\lambda_1}\,(t_0 + \lambda_1\,(t_1 - t_0)) + \left(1 - \frac{\lambda_0}{\lambda_1}\right)\, t_0\right).$$

Thus, from the definition of convex function we obtain

$$\begin{aligned} f&\left(\frac{\lambda_0}{\lambda_1}\,(t_0 + \lambda_1\,(t_1 - t_0)) + \left(1 - \frac{\lambda_0}{\lambda_1}\right)\, t_0\right) \\ &\le \frac{\lambda_0}{\lambda_1}\, f(t_0 + \lambda_1\,(t_1 - t_0)) + \left(1 - \frac{\lambda_0}{\lambda_1}\right)\, f(t_0). \end{aligned}$$

By joining these two equations, we have obtained

$$f(t_0 + \lambda_0\,(t_1 - t_0)) \le \frac{\lambda_0}{\lambda_1}\, f(t_0 + \lambda_1\,(t_1 - t_0)) + \left(1 - \frac{\lambda_0}{\lambda_1}\right)\, f(t_0).$$

By subtracting $f(t_0)$ on both sides and dividing by λ_0, we now easily get

$$g(\lambda_0) \le g(\lambda_1),$$

as desired. □

The following crucial property of convex functions will be used several times.

Proposition 1.2.3 ("Above Tangent" Property) *Let $I \subseteq \mathbb{R}$ be an interval and let $f : I \to \mathbb{R}$ be a convex function. If f is differentiable at $t_0 \in I$, then we have*

$$f(t) \geq f(t_0) + f'(t_0)\,(t - t_0), \qquad \text{for every } t \in I.$$

Moreover, if f is strictly convex we have

$$f(t) > f(t_0) + f'(t_0)\,(t - t_0), \qquad \text{for every } t \in I \setminus \{t_0\}.$$

Proof By definition of convex function, we know that for every $\lambda \in (0, 1)$ and every $t \in I$ we have

$$f(\lambda\, t + (1 - \lambda)\, t_0) \leq \lambda\, f(t) + (1 - \lambda)\, f(t_0).$$

This can be rewritten as

$$f(t_0 + \lambda\,(t - t_0)) - f(t_0) \leq \lambda\,(f(t) - f(t_0)),$$

and dividing by $\lambda > 0$, we get

$$\frac{f(t_0 + \lambda\,(t - t_0)) - f(t_0)}{\lambda} \leq f(t) - f(t_0). \tag{1.2.1}$$

If we now take the limit as λ goes to 0 and use the definition of derivative, we end up with

$$f'(t_0)\,(t - t_0) \leq f(t) - f(t_0),$$

as desired.

Finally, if f is strictly convex, then (1.2.1) holds with strict inequality sign for $t \neq t_0$. Moreover, the left-hand side in (1.2.1) is non-decreasing with respect to λ, thanks to Lemma 1.2.2. Thus, we can keep the strict inequality sign after taking the limit as λ goes to 0, as well. □

We now list a couple of remarkable consequences of the previous property.

Lemma 1.2.4 (Picone's Inequality) *Let $I \subseteq \mathbb{R}$ be an interval and let $\varphi, \psi : I \to \mathbb{R}$ be two C^1 functions, such that $\psi(t) > 0$ for every $t \in I$. Then for every $1 < p < \infty$ we have*

$$|\psi'(t)|^{p-2}\,\psi'(t)\left(\frac{|\varphi(t)|^p}{\psi(t)^{p-1}}\right)' \leq |\varphi'(t)|^p, \qquad \text{for every } t \in I. \tag{1.2.2}$$

Proof We first show that the function $t \mapsto |\varphi(t)|^p$ is differentiable on I. Indeed, in the points where $\varphi(t) \neq 0$, this is just the composition of two differentiable functions. On the other hand, if $t_0 \in I$ is such that $\varphi(t_0) = 0$, then since φ is C^1 we have

$$\varphi(t) = \varphi'(t_0)\,(t - t_0) + o(t - t_0), \qquad \text{for } t \to t_0,$$

and thus

$$\begin{aligned}\lim_{h\to 0} \frac{|\varphi(t_0 + h)|^p - |\varphi(t_0)|^p}{h} &= \lim_{h\to 0} \frac{|\varphi'(t_0)\,h + o(h)|^p}{h} \\ &= \lim_{h\to 0} \frac{|h|^p}{h} \left|\varphi'(t_0) + \frac{o(h)}{h}\right|^p = 0.\end{aligned}$$

We now come to the proof of (1.2.2). Thanks to Proposition 1.2.3, if f is a C^1 convex function, we have

$$f(\varphi'(t)) \geq f\left(\varphi(t)\,\frac{\psi'(t)}{\psi(t)}\right) + f'\left(\varphi(t)\,\frac{\psi'(t)}{\psi(t)}\right)\left(\varphi'(t) - \varphi(t)\,\frac{\psi'(t)}{\psi(t)}\right).$$

We use this general fact with the specific choice $f(\tau) = |\tau|^p$. We thus get

$$\begin{aligned}|\varphi'(t)|^p &\geq |\varphi(t)|^p\,\frac{|\psi'(t)|^p}{\psi(t)^p} \\ &\quad + p\,|\varphi(t)|^{p-2}\,\varphi(t)\,\frac{|\psi'(t)|^{p-2}\,\psi'(t)}{\psi(t)^{p-1}}\left(\varphi'(t) - \varphi(t)\,\frac{\psi'(t)}{\psi(t)}\right) \\ &= -(p-1)\,|\varphi(t)|^p\,\frac{|\psi'(t)|^p}{\psi(t)^p} + p\,|\varphi(t)|^{p-2}\,\varphi(t)\,\varphi'(t)\,\frac{|\psi'(t)|^{p-2}\,\psi'(t)}{\psi(t)^{p-1}} \\ &= |\psi'(t)|^{p-2}\,\psi'(t)\left(\frac{|\varphi(t)|^p}{\psi(t)^{p-1}}\right)'.\end{aligned}$$

This is exactly inequality (1.2.2). □

Remark 1.2.5 (Case $p = 2$) In the case $p = 2$, we have a stronger statement. Indeed, in this case we can replace the inequality (1.2.2) with an identity. Namely, we have *Picone's identity*

$$\psi'(t)\left(\frac{|\varphi(t)|^2}{\psi(t)}\right)' = |\varphi'(t)|^2 - \left|\varphi(t)\,\frac{\psi'(t)}{\psi(t)} - \varphi'(t)\right|^2, \qquad \text{for every } t \in I.$$

The proof of this fact is by direct verification. Observe that the second term in the right-hand side has a negative sign, thus by dropping it we get back (1.2.2) for $p = 2$.

The next result is easy, but it deserves to be singled out, due to its importance for this book. It asserts that for a convex function, *critical points are automatically minimum points*.

Lemma 1.2.6 (Important!) *Let $I \subseteq \mathbb{R}$ be an interval and let $f : I \to \mathbb{R}$ be a convex function. If f is differentiable at $t_0 \in I$ and*

$$f'(t_0) = 0,$$

then t_0 is a minimum point for f over I.

Proof Let $t \in I$, by using Proposition 1.2.3 we get

$$f(t) \geq f(t_0) + f'(t_0)\,(t - t_0) = f(t_0).$$

This gives the desired conclusion. □

When f is smooth enough, convexity is equivalent to the fact that the second order derivative is non-negative. This is the content of the next result.

Proposition 1.2.7 *Let $I \subseteq \mathbb{R}$ be an interval and let $f : I \to \mathbb{R}$ be a function of class C^2. Then*

$$f \text{ is convex} \qquad \Longleftrightarrow \qquad f''(t) \geq 0, \qquad \text{for every } t \in I.$$

Proof Let us assume that f is convex. We take $t_0, t_1 \in I$ such that $t_0 < t_1$. By using Proposition 1.2.3, we know that

$$f(t_1) \geq f(t_0) + f'(t_0)\,(t_1 - t_0),$$

and

$$f(t_0) \geq f(t_1) + f'(t_1)\,(t_0 - t_1).$$

By summing up these two inequalities, this gives

$$f(t_1) + f(t_0) \geq f(t_0) + f(t_1) + \Big(f'(t_0) - f'(t_1)\Big)\,(t_1 - t_0),$$

that is

$$\Big(f'(t_0) - f'(t_1)\Big)\,(t_1 - t_0) \leq 0.$$

By recalling that $t_1 > t_0$, this entails that

$$f'(t_0) \leq f'(t_1).$$

In other words, we just showed that f' is a non-decreasing function on I. Since by assumption f' is differentiable and I is an interval, this is equivalent to the fact that

$$f''(t) \geq 0, \qquad \text{for every } t \in I.$$

We now assume that f'' is non-negative on I. We recall that for every $t, s \in I$ there holds

$$f(t) = f(s) + f'(s)\,(t-s) + \int_s^t f''(\tau)\,(t-\tau)\,d\tau.$$

By using the assumption on f'', this in particular implies that

$$f(t) \geq f(s) + f'(s)\,(t-s),$$

for every $t, s \in I$. We use this inequality with

$$t = t_0 \qquad \text{and} \qquad s = \lambda\,t_0 + (1-\lambda)\,t_1,$$

so to get

$$f(t_0) \geq f(\lambda\,t_0 + (1-\lambda)\,t_1) + f'(\lambda\,t_0 + (1-\lambda)\,t_1)\,(1-\lambda)\,(t_0 - t_1). \tag{1.2.3}$$

We can also use the same inequality as above, this time with the choices

$$t = t_1 \qquad \text{and} \qquad s = \lambda\,t_0 + (1-\lambda)\,t_1,$$

so to get

$$f(t_1) \geq f(\lambda\,t_0 + (1-\lambda)\,t_1) - f'(\lambda\,t_0 + (1-\lambda)\,t_1)\,\lambda\,(t_0 - t_1). \tag{1.2.4}$$

We now multiply (1.2.3) by λ and (1.2.4) by $1-\lambda$, then we sum them up. We thus get

$$\begin{aligned}\lambda\,f(t_0) + (1-\lambda)\,f(t_1) &\geq (\lambda + (1-\lambda))\,f(\lambda\,t_0 + (1-\lambda)\,t_1)\\ &= f(\lambda\,t_0 + (1-\lambda)\,t_1),\end{aligned}$$

as desired. □

Remark 1.2.8 By inspecting the previous proof, it is easy to see that the following implication holds true

$$f''(t) > 0, \qquad \text{for every } t \in I \qquad \Longrightarrow \qquad f \text{ is strictly convex.}$$

Lemma 1.2.9 (Young's Inequality) *Let* $1 < p < \infty$*, then for every* $a, b \in \mathbb{R}$ *we have*

$$|a\,b| \le \frac{|a|^p}{p} + \frac{|b|^{p'}}{p'}, \tag{1.2.5}$$

where p' *is the conjugate exponent of* p*, defined by*

$$p' = \frac{p}{p-1}.$$

Proof Without loss of generality, we can suppose that a and b are both positive. Moreover, if $a = 0$ or $b = 0$, then (1.2.5) holds true. Thus, let us assume $a > 0$ and $b > 0$. In order to conclude, we need to prove that

$$a\,b \le \frac{a^p}{p} + \frac{b^{p'}}{p'}.$$

To this aim, we observe that the function

$$f(t) = \frac{t^p}{p}, \qquad \text{for } t \ge 0,$$

is convex. Thus, by using the "above tangent" property of Proposition 1.2.3, we get for every $a, c > 0$

$$\frac{a^p}{p} \ge \frac{c^p}{p} + c^{p-1}\,(a-c) \qquad \text{that is} \qquad \frac{a^p}{p} \ge \left(\frac{1}{p} - 1\right) c^p + c^{p-1}\,a.$$

We now choose $c = b^{1/(p-1)}$, so to get from the last equation

$$\frac{a^p}{p} \ge \left(\frac{1}{p} - 1\right) b^{\frac{p}{p-1}} + b\,a.$$

By recalling the definition of p', we get the desired conclusion. □

Remark 1.2.10 (Generalized Young's Inequality) Very often, a more useful form of Young's inequality is the following one

$$|x\,y| \le \frac{|x|^p}{p}\,\varepsilon + \frac{|y|^{p'}}{p'}\,\varepsilon^{-\frac{1}{p-1}}, \qquad \text{for every } \varepsilon > 0. \tag{1.2.6}$$

This can be obtained by writing

$$x\,y = \left(x\,\varepsilon^{\frac{1}{p}}\right)\left(y\,\varepsilon^{-\frac{1}{p}}\right),$$

and then using (1.2.5) with

$$a = x\,\varepsilon^{\frac{1}{p}} \qquad \text{and} \qquad b = y\,\varepsilon^{-\frac{1}{p}}.$$

In what follows, for every measurable set $E \subseteq \mathbb{R}^N$ such that $0 < |E| < +\infty$, we use the notation

$$⨍_E \varphi(x)\,dx = \frac{1}{|E|}\int_E \varphi(x)\,dx.$$

Proposition 1.2.11 (Jensen's Inequality) *Let $f : \mathbb{R} \to \mathbb{R}$ be a C^1 convex function and let $E \subseteq \mathbb{R}^N$ be a measurable set, such that $0 < |E| < +\infty$. For every summable function $\varphi : E \to \mathbb{R}$ such that $f \circ \varphi$ is still summable, we have*

$$f\left(⨍_E \varphi(x)\,dx\right) \le ⨍_E f(\varphi(x))\,dx.$$

Proof By using the "above tangent" property of Proposition 1.2.3 with

$$t = \varphi(x) \qquad \text{and} \qquad t_0 = ⨍_E \varphi(y)\,dy,$$

we have

$$\begin{aligned} f(\varphi(x)) \ge f&\left(⨍_E \varphi(y)\,dy\right)\\ &+ f'\left(⨍_E \varphi(y)\,dy\right)\left(\varphi(x) - ⨍_E \varphi(y)\,dy\right), \qquad \text{for a. e. } x \in E. \end{aligned}$$

We now integrate this inequality over E and divide by $|E|$, so to get

$$\begin{aligned} ⨍_E f(\varphi(x))\,dx \ge f&\left(⨍_E \varphi(y)\,dy\right)\\ &+ f\left(⨍_E \varphi(y)\,dy\right)\left(⨍_E \varphi(x)\,dx - ⨍_E \varphi(y)\,dy\right), \end{aligned}$$

and the last two terms cancel out. This gives the desired inequality. □

As another application of the "above tangent" property, we recall the following result. This should be well-known to the reader from the basic theory of L^p spaces. However, since this will be used several times, we give a proof.

Proposition 1.2.12 (Weak Lower Semicontinuity of L^p Norms) *Let $1 < p < \infty$ and let $E \subseteq \mathbb{R}^N$ be a measurable set. If the sequence $\{u_n\}_{n\in\mathbb{N}} \subseteq L^p(E)$ is weakly converging to $u \in L^p(E)$, then we have*

$$\liminf_{n\to\infty} \|u_n\|_{L^p(E)} \geq \|u\|_{L^p(E)}.$$

Proof We recall from the theory of L^p spaces that weak convergence means that

$$\lim_{n\to\infty} \int_E (u_n - u)\,\varphi\,dx = 0, \qquad \text{for every } \varphi \in (L^p(E))^* = L^{p'}(E). \tag{1.2.7}$$

We denoted by the symbol $*$ the topological dual space of $L^p(E)$ and we used that for $1 < p < \infty$ this space can be identified with $L^{p'}(E)$ (see for example [46, Theorem 2.14]).

We now observe that the function $f(t) = |t|^p$ is C^1 convex, thus by Proposition 1.2.3 we have

$$|t|^p \geq |s|^p + p\,|s|^{p-2}\,s\,(t-s), \qquad \text{for every } t, s \in \mathbb{R}.$$

We use this inequality with

$$t = u_n(x) \qquad \text{and} \qquad s = u(x),$$

this yields

$$|u_n(x)|^p \geq |u(x)|^p + p\,|u(x)|^{p-2}\,u(x)\,(u_n(x) - u(x)), \qquad \text{for a. e. } x \in E.$$

By integrating this inequality, we get

$$\int_E |u_n|^p\,dx \geq \int_E |u|^p\,dx + p \int_E |u|^{p-2}\,u\,(u_n - u)\,dx.$$

We now observe that $|u|^{p-2}\,u \in L^{p'}(E)$, thus by using (1.2.7) with $\varphi = |u|^{p-2}\,u$, we get

$$\liminf_{n\to\infty} \int_E |u_n|^p\,dx \geq \int_E |u|^p\,dx + p \lim_{n\to\infty} \int_E |u|^{p-2}\,u\,(u_n - u)\,dx = \int_E |u|^p\,dx.$$

By raising to the power $1/p$ and recalling that $t \mapsto t^{1/p}$ is monotone increasing, we get the conclusion. □

1.3 Convex Functions of Several Variables

We now show how to generalize the contents of the previous section to the case of functions of several variables. Given two points $x_0, x_1 \in \mathbb{R}^N$, we recall that the curve

$$\gamma(t) = (1-t)\,x_0 + t\,x_1, \qquad \text{for } t \in [0, 1],$$

parametrizes the segment connecting x_0 and x_1. We will use the symbol

$$\langle z, w\rangle = \sum_{i=1}^{N} z_i\, w_i,$$

to denote the Euclidean scalar product of two vectors $z, w \in \mathbb{R}^N$.

Definition 1.3.1 Let $\Omega \subseteq \mathbb{R}^N$ be a non-empty set. We say that Ω is *convex* if for every $x_0, x_1 \in \Omega$ we have

$$(1-t)\,x_0 + t\,x_1 \in \Omega, \qquad \text{for every } t \in [0,1].$$

Definition 1.3.2 Let $\Omega \subseteq \mathbb{R}^N$ be a convex set and let $f : \Omega \to \mathbb{R}$ be a function. We say that f is *convex* if we have

$$f(\lambda\, x_0 + (1-\lambda)\, x_1) \leq \lambda\, f(x_0) + (1-\lambda)\, f(x_1), \quad \text{for every } \lambda \in [0,1],\ x_0, x_1 \in \Omega.$$

Moreover, we say that f is *strictly convex* if we have

$$f(\lambda\, x_0 + (1-\lambda)\, x_1) < \lambda\, f(x_0) + (1-\lambda)\, f(x_1), \quad \text{for every } \begin{array}{c} \lambda \in (0,1), \\ x_0, x_1 \in \Omega,\ x_0 \neq x_1 \end{array}.$$

Most of the results previously presented in dimension $N = 1$ have their own counterparts in several dimensions. We start with the following one: the proof is omitted, since this is exactly the same as that of Lemma 1.2.2.

Lemma 1.3.3 *Let $\Omega \subseteq \mathbb{R}^N$ be a convex set and let $f : \Omega \to \mathbb{R}$ be a convex function. For every $x_0, x_1 \in \Omega$, the function*

$$g(\lambda) = \frac{f(x_0 + \lambda\,(x_1 - x_0)) - f(x_0)}{\lambda}, \qquad \text{for } \lambda \in (0,1],$$

is non-decreasing.

Proposition 1.3.4 ("Above Tangent" Property—Several Variables) *Let $\Omega \subseteq \mathbb{R}^N$ be a convex set and let $f : \Omega \to \mathbb{R}$ be a convex function. If f is differentiable at $x_0 \in \Omega$ we have*

$$f(x) \geq f(x_0) + \langle \nabla f(x_0), x - x_0\rangle, \qquad \text{for every } x \in \Omega.$$

Moreover, if f is strictly convex we have

$$f(x) > f(x_0) + \langle \nabla f(x_0), x - x_0\rangle, \qquad \text{for every } x \in \Omega \setminus \{x_0\}.$$

Proof By definition of convex function, we know that for every $\lambda \in (0,1)$ and every $x \in \Omega$ we have

$$f(\lambda x + (1-\lambda)\,x_0) \le \lambda\, f(x) + (1-\lambda)\, f(x_0).$$

We observe that for $x = x_0$ there is nothing to prove, thus we can always assume that $x \neq x_0$. This can be rewritten as

$$f(x_0 + \lambda\,(x - x_0)) - f(x_0) \le \lambda\,(f(x) - f(x_0)),$$

and diving by $\lambda > 0$, we get

$$\frac{f(x_0 + \lambda\,(x - x_0)) - f(x_0)}{\lambda} \le f(x) - f(x_0). \tag{1.3.1}$$

We now use that f is differentiable at x_0. We thus have

$$f(x_0 + \lambda\,(x - x_0)) - f(x_0) = \lambda\,\langle \nabla f(x_0), x - x_0\rangle + o(\lambda), \qquad \text{for } \lambda \to 0.$$

By using this fact in (1.3.1), we get

$$\frac{\lambda\,\langle \nabla f(x_0), x - x_0\rangle + o(\lambda)}{\lambda} \le f(x) - f(x_0), \qquad \text{for } \lambda \to 0$$

If we now take the limit as λ goes to 0, we get the desired conclusion.

If f is strictly convex, the second part of the statement follows exactly as in the one-dimensional case (by using Lemma 1.3.3 in place of Lemma 1.2.2). □

Lemma 1.3.5 (Picone's Inequality—Several Variables) *Let $E \subseteq \mathbb{R}^N$ be an open set and let $\varphi, \psi : E \to \mathbb{R}$ be two C^1 functions, such that $\psi(x) > 0$ for every $x \in E$. Then for every $1 < p < \infty$ we have*

$$\left\langle |\nabla \psi(x)|^{p-2}\,\nabla \psi(x), \nabla\left(\frac{|\varphi(x)|^p}{\psi(x)^{p-1}}\right)\right\rangle \le |\nabla \varphi(x)|^p, \qquad \textit{for every } x \in E. \tag{1.3.2}$$

Proof The proof is exactly the same as in the one-dimensional case.

We first show that the function $x \mapsto |\varphi(x)|^p$ is differentiable on E. Indeed, at the points where $\varphi(x) \neq 0$, this is just the composition of two differentiable functions. On the other hand, if $x_0 \in E$ is such that $\varphi(x_0) = 0$, then since φ is C^1 we have

$$\varphi(x_0 + h) = \langle \nabla \varphi(x_0), h\rangle + o(|h|), \qquad \text{for } |h| \to 0.$$

This permits to infer that

$$\begin{aligned}\lim_{|h|\to 0} \frac{|\varphi(x_0+h)|^p - |\varphi(x_0)|^p}{|h|} &= \lim_{h\to 0} \frac{|\langle \nabla \varphi(x_0), h\rangle + o(|h|)|^p}{|h|} \\ &= \lim_{h\to 0} \frac{|h|^p}{|h|} \left|\left\langle \nabla \varphi(x_0), \frac{h}{|h|}\right\rangle + \frac{o(|h|)}{|h|}\right|^p = 0,\end{aligned}$$

thanks to the fact that $p > 1$ and

$$\left\langle \nabla\varphi(x_0), \frac{h}{|h|} \right\rangle \leq |\nabla\varphi(x_0)|, \qquad \text{for every } h \neq 0.$$

The previous computation shows that

$$|\varphi(x_0 + h)|^p = |\varphi(x_0)|^p + o(|h|), \qquad \text{for } |h| \to 0,$$

i.e. $|\varphi|^p$ is differentiable at the point x_0, as well.

We now come to the proof of (1.3.2). Thanks to Proposition 1.3.4, if f is a C^1 convex function on $\mathbb{R}^N$, we have

$$f(\nabla\varphi(x)) \geq f\left(\varphi(x)\,\frac{\nabla\psi(x)}{\psi(x)}\right) + \left\langle \nabla f\left(\varphi(x)\,\frac{\nabla\psi(x)}{\psi(x)}\right), \nabla\varphi(x) - \varphi(x)\,\frac{\nabla\psi(x)}{\psi(x)} \right\rangle.$$

We use this general fact, with the specific choice $f(z) = |z|^p$, for $z \in \mathbb{R}^N$. We thus get

$$\begin{aligned}
|\nabla\varphi(x)|^p &\geq |\varphi(x)|^p\,\frac{|\nabla\psi(x)|^p}{\psi(x)^p} \\
&\quad + p\,|\varphi(x)|^{p-2}\,\varphi(x)\left\langle \frac{|\nabla\psi(x)|^{p-2}\,\nabla\psi(x)}{\psi(x)^{p-1}}, \nabla\varphi(x) - \varphi(x)\,\frac{\nabla\psi(x)}{\psi(x)} \right\rangle \\
&= -(p-1)\,|\varphi(x)|^p\,\frac{|\nabla\psi(x)|^p}{\psi(x)^p} \\
&\quad + p\,|\varphi(x)|^{p-2}\,\varphi(x)\,\frac{\langle |\nabla\psi(x)|^{p-2}\,\nabla\psi(x), \nabla\varphi(x)\rangle}{\psi(x)^{p-1}} \\
&= \left\langle |\nabla\psi(x)|^{p-2}\,\nabla\psi(x), \nabla\left(\frac{|\varphi(x)|^p}{\psi(x)^{p-1}}\right)\right\rangle.
\end{aligned}$$

This is exactly inequality (1.3.2). □

Remark 1.3.6 (Case $p = 2$) As in the one-dimensional case, here as well in the case $p = 2$ we have a stronger statement. Indeed, in this case we can replace the inequality (1.3.2) with an identity. Namely, we have *Picone's identity*

$$\begin{aligned}
\left\langle \nabla\psi(x), \nabla\left(\frac{|\varphi(x)|^2}{\psi(x)}\right)\right\rangle &= |\nabla\varphi(x)|^2 \\
&\quad - \left|\varphi(x)\,\frac{\nabla\psi(x)}{\psi(x)} - \nabla\varphi(x)\right|^2, \qquad \text{for every } x \in E.
\end{aligned}$$

The proof of this fact is by direct verification, again.

Lemma 1.3.7 (Important!) *Let $\Omega \subseteq \mathbb{R}^N$ be a convex set and let $f : \Omega \to \mathbb{R}$ be a convex function. If f is differentiable at $x_0 \in \Omega$ and*

$$\nabla f(x_0) = 0,$$

then x_0 is a minimum point of f over Ω.

Proof Let $x \in \Omega$, by using Proposition 1.3.4 we get

$$f(x) \geq f(x_0) + \langle \nabla f(x_0), x - x_0 \rangle = f(x_0).$$

This gives the desired conclusion. □

We have seen that in dimension $N = 1$, if f is smooth enough, convexity is equivalent to the fact that the second order derivative is non-negative. This remains true in several variables, as well. In what follows, we use the symbol $D^2 f$ to denote the Hessian matrix.

Proposition 1.3.8 *Let $\Omega \subseteq \mathbb{R}^N$ be an open convex set and let $f : \Omega \to \mathbb{R}$ be a function of class C^2. Then*

$$f \text{ is convex} \qquad \Longleftrightarrow \qquad D^2 f(x) \text{ is non-negative definite}, \qquad \text{for every } x \in \Omega.$$

Proof Let us assume that f is convex. We take $x_0, x_1 \in \Omega$ two distinct points. By using Proposition 1.3.4, we know that

$$f(x_1) \geq f(x_0) + \langle \nabla f(x_0), x_1 - x_0 \rangle,$$

and

$$f(x_0) \geq f(x_1) + \langle \nabla f(x_1), x_0 - x_1 \rangle.$$

By summing up these two inequalities, we get

$$f(x_1) + f(x_0) \geq f(x_0) + f(x_1) + \langle \nabla f(x_0) - \nabla f(x_1), x_1 - x_0 \rangle,$$

that is

$$\langle \nabla f(x_1) - \nabla f(x_0), x_1 - x_0 \rangle \geq 0, \qquad \text{for every } x_0, x_1 \in \Omega. \tag{1.3.3}$$

We need to prove that $D^2 f(x)$ is non-negative definite, for every $x \in \Omega$. This property, is equivalent to require that

$$\langle D^2 f(x)\, \xi, \xi \rangle \geq 0, \qquad \text{for every } x \in \Omega,\ \xi \in \mathbb{R}^N. \tag{1.3.4}$$

We fix ξ a unit vector and take $x_1 = x_0 + t\,\xi$ in (1.3.3). This is feasible, since Ω is open and convex and thus

$$x_1 = x_0 + t\,\xi \in \Omega, \qquad \text{for } t \text{ small enough.}$$

By assumption we have that f is of class C^2, thus ∇f is differentiable. We obtain

$$\begin{aligned}
0 &\le \langle \nabla f(x_0 + t\,\xi) - \nabla f(x_0), t\,\xi\rangle \\
&= \sum_{i=1}^{N} \left(\frac{\partial f}{\partial x_i}(x_0 + t\,\xi) - \frac{\partial f}{\partial x_i}(x_0)\right) t\,\xi_i \\
&= t^2 \sum_{i=1}^{N} \left(\sum_{j=1}^{N} \frac{\partial^2 f}{\partial x_i \partial x_j}(x_0)\,\xi_j\right) \xi_i + \sum_{i=1}^{N} o_i(|t|)\,t\,\xi_i \\
&= t^2 \,\langle D^2 f(x_0)\cdot \xi, \xi\rangle + \sum_{i=1}^{N} o_i(|t|)\,t\,\xi_i.
\end{aligned}$$

We now divide by $t^2 \neq 0$ and then let t go to 0, so to obtain

$$0 \le \langle D^2 f(x_0)\cdot \xi, \xi\rangle, \qquad \text{for every } |\xi| = 1.$$

By bilinearity of the scalar product, this is equivalent to (1.3.4) and thus the proof of this implication is concluded.

We now assume that $D^2 f(x)$ is non-negative definite for every $x \in \Omega$, i.e. condition (1.3.4) holds. We take $x, y \in \Omega$ two points and observe that by the Fundamental Theorem of Calculus

$$\begin{aligned}
\nabla f(y) - \nabla f(x) &= \int_0^1 \frac{d}{dt} \nabla f(x + t\,(y-x))\,dt \\
&= \int_0^1 D^2 f(x + t\,(y-x))\cdot (y-x)\,dt.
\end{aligned}$$

If we take the scalar product with the vector $y - x$ and use (1.3.4), we get

$$\langle \nabla f(y) - \nabla f(x), y - x\rangle = \int_0^1 \langle D^2 f(x + t\,(y-x))\cdot (y-x), y-x\rangle\,dt \ge 0. \tag{1.3.5}$$

Observe that we used the convexity of Ω to guarantee that $x + t\,(y-x) \in \Omega$ for every $t \in [0,1]$. In order to prove that f is convex, we take $x_0, x_1 \in \Omega$ and we observe that from (1.3.5) with

$$y = x_0 + t\,(x_1 - x_0) \qquad \text{and} \qquad x = x_0,$$

we have for every $t \in [0, 1]$

$$\langle \nabla f(x_0 + t\,(x_1 - x_0)) - \nabla f(x_0), t\,(x_1 - x_0)\rangle \geq 0,$$

that is, by bilinearity

$$\langle \nabla f(x_0 + t\,(x_1 - x_0)), x_1 - x_0\rangle \geq \langle \nabla f(x_0), x_1 - x_0\rangle. \tag{1.3.6}$$

On the other hand, by the fundamental theorem of Calculus, we know that

$$\begin{aligned} f(x_1) - f(x_0) &= \int_0^1 \frac{d}{dt} f(x_0 + t\,(x_1 - x_0))\,dt \\ &= \int_0^1 \langle \nabla f(x_0 + t\,(x_1 - x_0)), x_1 - x_0\rangle\,dt. \end{aligned}$$

We now use inequality (1.3.6), so to get

$$f(x_1) - f(x_0) \geq \langle \nabla f(x_0), x_1 - x_0\rangle,$$

i.e. f satisfies the "above tangent" property. In order to conclude, we need to show that this property implies that f is[1] convex. From the previous property, we get for every $\lambda \in [0, 1]$ and $x_0, x_1 \in \Omega$

$$f(x_0) - f(\lambda\,x_0 + (1-\lambda)\,x_1) \geq (\lambda - 1)\,\langle \nabla f(\lambda\,x_0 + (1-\lambda)\,x_1), x_1 - x_0\rangle,$$

and also

$$f(x_1) - f(\lambda\,x_0 + (1-\lambda)\,x_1) \geq \lambda\,\langle \nabla f(\lambda\,x_0 + (1-\lambda)\,x_1), x_1 - x_0\rangle.$$

We now multiply the first inequality by λ, the second one by $(1 - \lambda)$ and then sum up. We obtain

$$\lambda\,f(x_0) + (1-\lambda)\,f(x_1) - f(\lambda\,x_0 + (1-\lambda)\,x_1) \geq 0,$$

as desired. □

Remark 1.3.9 By inspecting the previous proof, it is easy to see that the following implication holds true

$$D^2 f(x) \text{ is positive definite, for every } x \in \Omega \qquad \Longrightarrow \qquad f \text{ is strictly convex.}$$

[1] Thus, in light of Proposition 1.3.4, we get that the "above tangent" property is equivalent to the convexity.

Proposition 1.3.10 (Jensen's Inequality—Several Variables) *Let $F : \mathbb{R}^k \to \mathbb{R}$ be a C^1 convex function and let $E \subseteq \mathbb{R}^N$ be a measurable set, such that $0 < |E| < +\infty$. Then for every summable vector-valued function $\varphi : E \to \mathbb{R}^k$ such that $F \circ \varphi$ is still summable, we have*

$$F\left(\fint_E \varphi(x)\,dx\right) \le \fint_E F(\varphi(x))\,dx.$$

Proof The proof is identical to that of Proposition 1.2.11. By using the "above tangent" property of Proposition 1.3.4, for almost every $x \in E$ we have

$$F(\varphi(x)) \ge F\left(\fint_E \varphi(y)\,dy\right) + \left\langle \nabla F\left(\fint_E \varphi(y)\,dy\right), \left(\varphi(x) - \fint_E \varphi(y)\,dy\right)\right\rangle.$$

We now integrate this inequality over E and divide by $|E|$, so to get

$$\begin{aligned}\fint_E F(\varphi(x))\,dx &\ge F\left(\fint_E \varphi(y)\,dy\right)\\ &\quad + \fint_E \left\langle \nabla F\left(\fint_E \varphi(y)\,dy\right), \left(\varphi(x) - \fint_E \varphi(y)\,dy\right)\right\rangle dx.\end{aligned}$$

By observing that

$$\begin{aligned}&\fint_E \left\langle \nabla F\left(\fint_E \varphi(y)\,dy\right), \left(\varphi(x) - \fint_E \varphi(y)\,dy\right)\right\rangle dx\\ &\quad = \fint_E \left\langle \nabla F\left(\fint_E \varphi(y)\,dy\right), \varphi(x)\right\rangle dx\\ &\quad - \fint_E \left\langle \nabla F\left(\fint_E \varphi(y)\,dy\right), \fint_E \varphi(y)\,dy\right\rangle dx\\ &\quad = \left\langle \nabla F\left(\fint_E \varphi(y)\,dy\right), \fint_E \varphi(x)\,dx\right\rangle\\ &\quad - \left\langle \nabla F\left(\fint_E \varphi(y)\,dy\right), \fint_E \varphi(y)\,dy\right\rangle = 0,\end{aligned}$$

we finally get the desired inequality. □

The following result generalizes Proposition 1.2.12 to vector-valued functions. The adaptation of the proof is left to the reader, as an exercise. For $k \in \mathbb{N} \setminus \{0\}$ and $1 < p < \infty$, we use the standard notation

$$\|u\|_{L^p(E;\mathbb{R}^k)} = \left(\int_E |u(x)|^p\,dx\right)^{\frac{1}{p}}, \qquad \text{for } u \in L^p(E;\mathbb{R}^k),$$

where $|u(x)|$ now stands for the modulus of the vector $u(x) \in \mathbb{R}^k$.

Proposition 1.3.11 (Weak Lower Semicontinuity of L^p Norms) *Let $1 < p < \infty$ and let $E \subseteq \mathbb{R}^N$ be a measurable set. If the sequence $\{u_n\}_{n\in\mathbb{N}} \subseteq L^p(E;\mathbb{R}^k)$ is weakly converging to $u \in L^p(E;\mathbb{R}^k)$, then we have*

$$\liminf_{n\to\infty} \|u_n\|_{L^p(E;\mathbb{R}^k)} \ge \|u\|_{L^p(E;\mathbb{R}^k)}.$$

1.4 The Fundamental Lemma of Calculus of Variations

In what follows, given an open set $\Omega \subseteq \mathbb{R}^N$, we will indicate by $C_0^\infty(\Omega)$ the space of infinitely differentiable functions defined on Ω, whose support is a compact set contained in Ω.

The following result will be quite useful, it usually goes under the name of *fundamental lemma of Calculus of Variations*.

Lemma 1.4.1 (Du Bois-Reymond—Classical Version) *Let $\Omega \subseteq \mathbb{R}^N$ be an open set and let $f : \Omega \to \mathbb{R}$ be a continuous function such that*

$$\int_\Omega f\,\varphi\,dx = 0, \qquad \text{for every } \varphi \in C_0^\infty(\Omega).$$

Then f identically vanishes on Ω.

Proof We argue by contradiction and assume that there exists a point $\xi \in \Omega$ such that $f(\xi) > 0$. By continuity of f, there exists a ball $B_r(\xi) \subseteq \Omega$ such that

$$f(x) > 0, \qquad \text{for every } x \in B_r(\xi).$$

We now take the *standard smoothing kernel*

$$\rho(x) = \begin{cases} C_N \exp\left(\dfrac{1}{|x|^2-1}\right), & \text{if } |x| < 1, \\ \\ 0, & \text{otherwise} \end{cases} \tag{1.4.1}$$

which is a non-negative $C_0^\infty(\mathbb{R}^N)$ function (see Fig. 1.1). The constant C_N is given by

$$C_N = \left(\int_{B_1(0)} \exp\left(\frac{1}{|x|^2-1}\right)dx\right)^{-1},$$

so that by construction we have

$$\int_{\mathbb{R}^N} \rho\,dx = 1.$$

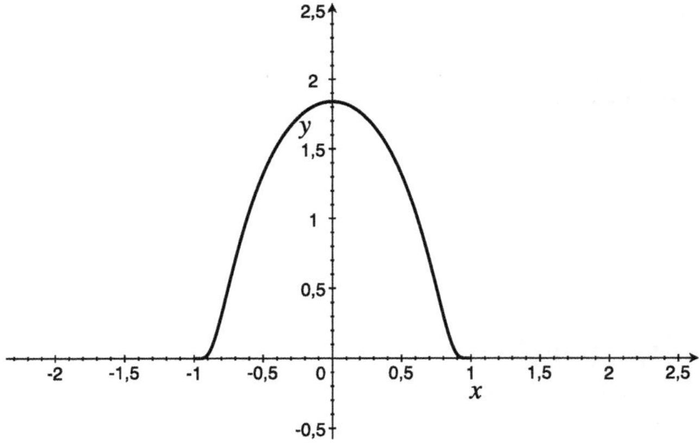

Fig. 1.1 The standard smoothing kernel in dimension $N = 1$

We set

$$\rho_r(x) = \rho\left(\frac{x-\xi}{r}\right),$$

so that

$$\rho_r \in C_0^\infty(\Omega).$$

By assumption, we have

$$0 = \int_\Omega f(x)\,\rho_r(x)\,dx = \int_{B_r(\xi)} f(x)\,\rho_r(x)\,dx.$$

Since both f and ρ_r are strictly positive on $B_r(\xi)$, we get the desired contradiction. □

The previous result can be considerably improved, by appealing to the properties of the Lebesgue integral. To this aim, we first need to recall some further notation. Given two open sets $\Omega' \subseteq \Omega$, in the sequel we will use the symbol

$$\Omega' \Subset \Omega,$$

to say that Ω' has compact closure contained in Ω. In other words, the set $\overline{\Omega'}$ is compact and contained in the open set Ω. This in particular implies that we have

$$\mathrm{dist}(\Omega', \partial\Omega) > 0.$$

For every $1 \le p \le \infty$, we then recall that $f \in L^p_{\mathrm{loc}}(\Omega)$ means that

$$f \in L^p(\Omega'), \qquad \text{for every } \Omega' \Subset \Omega.$$

Lemma 1.4.2 (Du Bois-Reymond—Modern Version) *Let $\Omega \subseteq \mathbb{R}^N$ be an open set and let $f \in L^1_{\mathrm{loc}}(\Omega)$ be a function such that*

$$\int_\Omega f\,\varphi\,dx = 0, \qquad \textit{for every } \varphi \in C_0^\infty(\Omega).$$

Then $f = 0$ almost everywhere in Ω.

Proof We fix $\Omega' \Subset \Omega$ and we consider the L^∞ function

$$g(x) = \begin{cases} \dfrac{f(x)}{|f(x)|}\,1_{\Omega'}(x), & \text{if } f(x) \neq 0, \\ \\ 0, & \text{if } f(x) = 0. \end{cases}$$

Here we denote by 1_E the *characteristic function* of a measurable set E, i.e.

$$1_E(x) = \begin{cases} 1, & \text{if } x \in E, \\ \\ 0, & \text{if } x \notin E. \end{cases}$$

We now consider the sequence of rescaled smoothing kernels

$$\rho_n(x) = n^N\,\rho(n\,x), \qquad \text{for } n \in \mathbb{N} \setminus \{0\},$$

where ρ is the same function as in (1.4.1). Observe that by construction we have

$$\int_{\mathbb{R}^N} \rho_n(x)\,dx = n^N \int_{\mathbb{R}^N} \rho(n\,x)\,dx = \frac{n^N}{n^N} \int_{\mathbb{R}^N} \rho(y)\,dy = 1. \tag{1.4.2}$$

By using the properties of convolutions, we know that there exists $\Omega' \Subset \mathcal{O} \Subset \Omega$ such that[2]

$$g_n = g * \rho_n \in C_0^\infty(\mathcal{O}), \qquad \text{for } n \text{ large enough},$$

[2] More precisely, by observing that ρ_n is supported on the ball $\overline{B_{1/n}(0)}$ and g is supported on $\overline{\Omega'}$, the support of $g * \rho_n$ is contained in

$$\left\{x \in \mathbb{R}^N \,:\, \mathrm{dist}(x, \Omega') \le 1/n\right\},$$

thanks to the properties of convolutions. Thus, by setting $h_0 = \mathrm{dist}(\Omega', \partial\Omega) > 0$ and defining

$$\mathcal{O} = \left\{x \in \Omega \,:\, \mathrm{dist}(x, \Omega') < \frac{h_0}{2}\right\}$$

it is sufficient to choose $n \ge 4/h_0$ in order to guarantee that $g * \rho_n$ has compact support in $\mathcal{O}$.

and

$$\lim_{n\to\infty} \|g_n - g\|_{L^1(\Omega')} = 0.$$

This last property in turn entails that, up to a subsequence, we also have

$$\lim_{n\to\infty} g_n(x) = g(x), \qquad \text{for a. e. } x \in \Omega'. \tag{1.4.3}$$

We also observe that

$$|g_n(x)| = \left| \int_{\Omega'} g(y)\, \rho_n(x-y)\, dy \right| \le \int_{\Omega'} |g(y)|\, \rho_n(x-y)\, dy \le 1,$$

thanks to the fact that $|g| \le 1$ and (1.4.2). In particular, we get

$$|g_n(x)| \le 1_{\mathcal{O}}(x), \qquad \text{for a. e. } x \in \Omega, \tag{1.4.4}$$

by recalling that each g_n is compactly supported in $\mathcal{O}$. By hypothesis we get

$$0 = \int_\Omega f\, g_n\, dx,$$

and both (1.4.3) and (1.4.4) imply that we can use the Dominated Convergence Theorem, so to get

$$\int_\Omega f\, g\, dx = \lim_{n\to\infty} \int_\Omega f\, g_n\, dx = 0.$$

By recalling the definition of g, the last equation gives that

$$0 = \int_\Omega f\, g\, dx = \int_{\{x\in\Omega'\,:\, f(x)\neq 0\}} f(x)\, \frac{f(x)}{|f(x)|}\, dx = \int_{\{x\in\Omega'\,:\, f(x)\neq 0\}} |f(x)|\, dx.$$

This shows that we must have

$$f(x) = 0, \qquad \text{for a. e. } x \in \Omega'.$$

By arbitrariness of $\Omega' \Subset \Omega$, we get the conclusion. □

1.5 The Dirichlet Principle

In this book, we will be interested in minimization problems, usually settled among suitable spaces of functions of one or several variables. The following classical result lies at the basis of the Calculus of Variations. Despite being limited to a very particular situation, it already illustrates quite well the connections between

minimization problems and *differential equations.*

Theorem 1.5.1 (Dirichlet Principle) *Let $\Omega \subseteq \mathbb{R}^N$ be an open bounded set with smooth boundary. Let U be a $C^2(\overline{\Omega})$ function. The following two facts are equivalent:*

(i) $v \in C^2(\overline{\Omega})$ is a solution of

$$\begin{cases} -\Delta v = 0, & in\ \Omega, \\ \quad v = U, & on\ \partial\Omega. \end{cases}$$

(ii) $v \in C^2(\overline{\Omega})$ is a minimizer of the variational problem

$$\inf_{u \in C^2(\overline{\Omega})} \left\{ \frac{1}{2} \int_\Omega |\nabla u|^2 \, dx \ : \ u = U \ on\ \partial\Omega \right\}.$$

Proof We recall that a function which verifies $-\Delta v = 0$ in an open set is called *harmonic*. Let us assume that $v \in C^2(\overline{\Omega})$ is a harmonic function with boundary datum U. We take another function $u \in C^2(\overline{\Omega})$ coinciding with U on the boundary. By using the information on v we get

$$0 = -\int_\Omega \Delta v\,(u - v)\, dx.$$

We observe that

$$-\Delta v\,(u - v) = -\mathrm{div}\,(\nabla v\,(u - v)) + \langle \nabla v, \nabla u - \nabla v \rangle,$$

thus from the previous equation, we get

$$0 = -\int_\Omega \mathrm{div}\,(\nabla v\,(u - v))\, dx + \int_\Omega \langle \nabla v, \nabla u - \nabla v \rangle\, dx.$$

We use the Divergence Theorem on the first integral in the right-hand side, which gives

$$\int_\Omega \mathrm{div}\,(\nabla v\,(u - v))\, dx = \int_{\partial\Omega} \langle \nabla v, \nu_\Omega \rangle\,(u - v)\, d\sigma = 0.$$

Here ν_Ω is the unit outer normal vector of the boundary $\partial\Omega$. In the last identity, we used that $u = v$ on $\partial\Omega$. We thus get

$$0 = \int_\Omega \langle \nabla v, \nabla u - \nabla v \rangle\, dx, \qquad \text{for every admissible } u. \tag{1.5.1}$$

We now use the "above tangent" property (i.e. Proposition 1.3.4) for the convex function

$$f(z) = \frac{1}{2}\,|z|^2,$$

to infer that

$$\frac{1}{2}\,|\nabla u|^2 \geq \frac{1}{2}\,|\nabla v|^2 + \langle \nabla v, \nabla u - \nabla v\rangle.$$

By integrating this pointwise inequality on Ω and using (1.5.1), we then obtain

$$\frac{1}{2}\int_\Omega |\nabla u|^2\,dx \geq \frac{1}{2}\int_\Omega |\nabla v|^2\,dx + \int_\Omega \langle \nabla v, \nabla u - \nabla v\rangle\,dx = \frac{1}{2}\int_\Omega |\nabla v|^2\,dx.$$

This shows that i) $\Longrightarrow$ ii).
We now assume that $v \in C^2(\overline{\Omega})$ is minimal for

$$\inf_{u\in C^2(\overline{\Omega})} \left\{\frac{1}{2}\int_\Omega |\nabla u|^2\,dx\ :\ u = U \text{ on } \partial\Omega\right\},$$

and aim at proving that v is harmonic in Ω. Observe that for every $t \in \mathbb{R}$ and every $\varphi \in C_0^\infty(\Omega)$, we have

$$v + t\,\varphi \in C^2(\overline{\Omega}) \qquad \text{and} \qquad v + t\,\varphi = U \text{ on } \partial\Omega.$$

By minimality of v, we thus get

$$\frac{1}{2}\int_\Omega |\nabla v + t\,\nabla\varphi|^2\,dx \geq \frac{1}{2}\int_\Omega |\nabla v|^2\,dx,$$

that is, by expanding the square on the left-hand side

$$\frac{1}{2}\int_\Omega |\nabla v|^2\,dx + t\int_\Omega \langle \nabla v, \nabla\varphi\rangle\,dx + t^2\,\frac{1}{2}\int_\Omega |\nabla\varphi|^2\,dx \geq \frac{1}{2}\int_\Omega |\nabla v|^2\,dx.$$

By erasing the common factor on both sides, this gives

$$t\int_\Omega \langle \nabla v, \nabla\varphi\rangle\,dx + t^2\,\frac{1}{2}\int_\Omega |\nabla\varphi|^2\,dx \geq 0, \quad \text{for } t \in \mathbb{R} \text{ and } \varphi \in C_0^\infty(\Omega).$$

We can now take $t > 0$, multiply both sides by $1/t$ and then take the limit as t goes to 0. This leads to

$$\int_\Omega \langle \nabla v, \nabla\varphi\rangle\,dx \geq 0, \qquad \text{for every } \varphi \in C_0^\infty(\Omega).$$

By arbitrariness of φ in the previous inequality, we can use it with $-\varphi$ in place of φ. This gives

$$\int_\Omega \langle \nabla v, \nabla \varphi \rangle \, dx \leq 0, \qquad \text{for every } \varphi \in C_0^\infty(\Omega),$$

as well. The last two displays finally imply that we must have

$$\int_\Omega \langle \nabla v, \nabla \varphi \rangle \, dx = 0, \qquad \text{for every } \varphi \in C_0^\infty(\Omega).$$

By using the Divergence Theorem, we then obtain

$$\begin{aligned} 0 = \int_\Omega \langle \nabla v, \nabla \varphi \rangle \, dx &= \int_\Omega \operatorname{div}(\nabla v \, \varphi) - \int_\Omega \Delta v \, \varphi \, dx \\ &= \int_{\partial\Omega} \langle \nabla v, \nu_\Omega \rangle \, \varphi \, d\sigma - \int_\Omega \Delta v \, \varphi \, dx = - \int_\Omega \Delta v \, \varphi \, dx. \end{aligned}$$

We used that the boundary integral vanishes, thanks to the compact support of φ. Up to now, we have shown that if v is minimal, then it verifies

$$- \int_\Omega \Delta v \, \varphi \, dx = 0, \qquad \text{for every } \varphi \in C_0^\infty(\Omega).$$

By applying the Du Bois-Reymond Lemma (Lemma 1.4.1), we obtain that $-\Delta v = 0$ everywhere in Ω. This shows that v is harmonic on Ω. □

Remark 1.5.2 (Uniqueness) We notice that a minimizer of

$$\inf_{u \in C^2(\overline{\Omega})} \left\{ \frac{1}{2} \int_\Omega |\nabla u|^2 \, dx \, : \, u = U \text{ on } \partial\Omega \right\},$$

must be unique, provided it exists. Indeed, observe that the function

$$z \mapsto \frac{1}{2} \, |z|^2,$$

is strictly convex, by Remark 1.3.9. Let us suppose that $v_1, v_2 \in C^2(\overline{\Omega})$ are two minimizers, the function

$$w = \frac{v_1 + v_2}{2},$$

would be a competitor for the minimization problem. By convexity, we would get

$$|\nabla w|^2 \leq \frac{1}{2} \, |\nabla v_1|^2 + \frac{1}{2} \, |\nabla v_2|^2, \tag{1.5.2}$$

which, integrated over Ω, would give

$$\int_\Omega |\nabla w|^2\, dx \le \frac{1}{2}\int_\Omega |\nabla v_1|^2\, dx + \frac{1}{2}\int_\Omega |\nabla v_2|^2\, dx.$$

By recalling that v_1 and v_2 are both minimizers, we get

$$\frac{1}{2}\int_\Omega |\nabla v_1|^2\, dx = \frac{1}{2}\int_\Omega |\nabla v_2|^2\, dx = \inf_{u\in C^2(\overline{\Omega})}\left\{\frac{1}{2}\int_\Omega |\nabla u|^2\, dx\ :\ u = U \text{ on } \partial\Omega\right\},$$

and thus w is a minimizer, as well. This implies that we must have equality for every $x \in \Omega$ in (1.5.2). By strict convexity, this entails that

$$\nabla v_1(x) = \nabla v_2(x), \qquad \text{for every } x \in \Omega,$$

that is

$$\nabla(v_1 - v_2) = (0, \dots, 0) \qquad \text{in } \Omega.$$

Since v_1 and v_2 coincide on $\partial\Omega$, we must have $v_1 = v_2$, as desired.

In light of the previous result, we also get that a C^2 solution of

$$\begin{cases} -\Delta v = 0, & \text{in } \Omega, \\ v = U, & \text{on } \partial\Omega, \end{cases}$$

must be unique, whenever it exists.

Actually, there is nothing specific about harmonic functions and the previous result can be easily generalized to cover much more general cases. Without any attempt of completeteness, we give the following more general statement.

Theorem 1.5.3 (Dirichlet Principle—General Version) *Let $\Omega \subseteq \mathbb{R}^N$ be an open bounded set with smooth boundary. Let U be a $C^2(\overline{\Omega})$ function and $f \in C^0(\overline{\Omega})$. Finally, let $F : \mathbb{R}^N \to \mathbb{R}$ be a convex function of class C^2. Then the following two facts are equivalent:*

(i) $v \in C^2(\overline{\Omega})$ is a solution of

$$\begin{cases} -\mathrm{div}(\nabla F(\nabla v)) = f, & \text{in } \Omega, \\ v = U, & \text{on } \partial\Omega. \end{cases}$$

(ii) $v \in C^2(\overline{\Omega})$ is a minimizer of the variational problem

$$\inf_{u\in C^2(\overline{\Omega})}\left\{\int_\Omega F(\nabla u)\, dx - \int_\Omega f\, u\, dx\ :\ u = U \text{ on } \partial\Omega\right\}.$$

Proof Let us assume that $v \in C^2(\overline{\Omega})$ is a solution of the boundary value problem. We take another function $u \in C^2(\overline{\Omega})$ coinciding with U on the boundary. By using the information on v we get

$$\int_\Omega f\,(u-v)\,dx = -\int_\Omega \mathrm{div}(\nabla F(\nabla v))\,(u-v)\,dx.$$

We observe that

$$-\,\mathrm{div}\Big(\nabla F(\nabla v)\Big)\,(u-v) = -\mathrm{div}\,\Big(\nabla F(\nabla v)\,(u-v)\Big) + \langle \nabla F(\nabla v), \nabla u - \nabla v\rangle,$$

thus from the previous equation, we get

$$\int_\Omega f\,(u-v)\,dx = -\int_\Omega \mathrm{div}\,\Big(\nabla F(\nabla v)\,(u-v)\Big)\,dx + \int_\Omega \langle \nabla F(\nabla v), \nabla u - \nabla v\rangle\,dx.$$

We use the Divergence Theorem for the first integral on the right-hand side, which gives

$$\int_\Omega \mathrm{div}\,\Big(\nabla F(\nabla v)\,(u-v)\Big)\,dx = \int_{\partial\Omega} \langle \nabla F(\nabla v), \nu_\Omega\rangle\,(u-v)\,d\sigma = 0.$$

In the last identity, we used that $u = v$ on $\partial\Omega$. We thus get

$$\int_\Omega f\,(u-v)\,dx = \int_\Omega \langle \nabla F(\nabla v), \nabla u - \nabla v\rangle\,dx, \qquad \text{for every admissible } u. \tag{1.5.3}$$

We now use the "above tangent" property (i.e. Proposition 1.3.4) for the convex function F, to infer that

$$F(\nabla u) \geq F(\nabla v) + \langle \nabla F(\nabla v), \nabla u - \nabla v\rangle.$$

By integrating this pointwise inequality on Ω and using (1.5.3), we then obtain

$$\begin{aligned}\int_\Omega F(\nabla u)\,dx &\geq \int_\Omega F(\nabla v)\,dx + \int_\Omega \langle \nabla F(\nabla v), \nabla u - \nabla v\rangle\,dx\\ &= \int_\Omega F(\nabla v)\,dx - \int_\Omega f\,v\,dx + \int_\Omega f\,u\,dx,\end{aligned}$$

that is

$$\int_\Omega F(\nabla u)\,dx - \int_\Omega f\,u\,dx \geq \int_\Omega F(\nabla v)\,dx - \int_\Omega f\,v\,dx.$$

This shows that $i) \Longrightarrow ii)$.

We now assume that $v \in C^2(\overline{\Omega})$ is minimal for

$$\inf_{u\in C^2(\overline{\Omega})} \left\{ \int_\Omega F(\nabla u)\,dx - \int_\Omega f\,u\,dx \ :\ u = U \text{ on } \partial\Omega \right\},$$

and aim at proving that v is a solution of the boundary value problem. As in the proof of Theorem 1.5.1, for every $t \in \mathbb{R}$ and every $\varphi \in C_0^\infty(\Omega)$, we have

$$v + t\,\varphi \in C^2(\overline{\Omega}) \qquad \text{and} \qquad v + t\,\varphi = U \text{ on } \partial\Omega.$$

By minimality of v, we thus get

$$\int_\Omega F(\nabla v + t\,\nabla\varphi)\,dx - \int_\Omega f\,v\,dx - t\int_\Omega f\,\varphi\,dx \geq \int_\Omega F(\nabla v)\,dx - \int_\Omega f\,v\,dx$$

This implies that the function of one real variable

$$t \mapsto g(t) := \int_\Omega F(\nabla v + t\,\nabla\varphi)\,dx - \int_\Omega f\,v - t\int_\Omega f\,\varphi\,dx,$$

is minimal at $t = 0$. By the classical Fermat Theorem, we must have

$$g'(0) = 0,$$

provided we can prove that g is differentiable at $t = 0$. The verification of this last fact is left to the reader (it is sufficient to adapt the argument of Problem 1.7.16, below), as well as the fact that

$$\begin{aligned}\frac{d}{dt}\left(\int_\Omega F(\nabla v + t\,\nabla\varphi)\,dx - \int_\Omega f\,v\,dx - t\int_\Omega f\,\varphi\,dx\right)_{|t=0}& \\ = \left(\int_\Omega \frac{d}{dt}F(\nabla v + t\,\nabla\varphi)\,dx\right)_{|t=0} - \int_\Omega f\,\varphi\,dx& \\ = \int_\Omega \langle \nabla F(\nabla v), \nabla\varphi\rangle\,dx - \int_\Omega f\,\varphi\,dx.&\end{aligned}$$

The last two displays finally imply that we must have

$$\int_\Omega \langle \nabla F(\nabla v), \nabla\varphi\rangle\,dx = \int_\Omega f\,\varphi\,dx, \qquad \text{for every } \varphi \in C_0^\infty(\Omega).$$

By using the Divergence Theorem, we then obtain

$$\begin{aligned}\int_{\Omega} f\,\varphi\,dx &= \int_{\Omega} \langle \nabla F(\nabla v), \nabla \varphi \rangle\,dx \\ &= \int_{\Omega} \operatorname{div}(\nabla F(\nabla v)\,\varphi)\,dx - \int_{\Omega} \operatorname{div}(\nabla F(\nabla v))\,\varphi\,dx \\ &= \int_{\partial\Omega} \langle \nabla F(\nabla v), \nu_{\Omega} \rangle\,\varphi\,d\sigma - \int_{\Omega} \operatorname{div}(\nabla F(\nabla v))\,\varphi\,dx \\ &= -\int_{\Omega} \operatorname{div}(\nabla F(\nabla v))\,\varphi\,dx.\end{aligned}$$

Up to now, we have shown that if v is minimal, then it verifies

$$\int_{\Omega} \Big[\operatorname{div}(\nabla F(\nabla v)) + f\Big]\,\varphi\,dx = 0, \qquad \text{for every } \varphi \in C_0^{\infty}(\Omega).$$

By applying the Du Bois-Reymond Lemma (Lemma 1.4.1), we obtain that v satisfies $-\operatorname{div}(\nabla F(\nabla v)) = f$ everywhere in Ω. This shows that v is a solution of the claimed partial differential equation. □

Remark 1.5.4 The Dirichlet principle as stated above is quite important from a theoretical point of view. Indeed, it relates existence of solutions for a class of partial differential equations to existence of minimizers for minimization problems. However, from a practical point of view, the previous statements are essentially useless, since proving existence of a minimizer in $C^2(\overline{\Omega})$ is quite a complicated task. We will come back on this point at the beginning of Chap. 3.

Definition 1.5.5 The equation

$$-\operatorname{div}\big(\nabla F(\nabla u)\big) = f, \qquad \text{in } \Omega, \tag{1.5.4}$$

is called *Euler-Lagrange equation* associated to the functional

$$u \mapsto \mathcal{F}(u) = \int_{\Omega} F(\nabla u)\,dx - \int_{\Omega} f\,u\,dx.$$

We say that u is a *weak solution* of (1.5.4) if it verifies

$$\int_{\Omega} \langle \nabla F(\nabla u), \nabla \varphi \rangle\,dx = \int_{\Omega} f\,\varphi\,dx, \qquad \text{for every } \varphi \in C_0^{\infty}(\Omega). \tag{1.5.5}$$

Accordingly, equation (1.5.5) is called *weak formulation* of Eq. (1.5.4).

Remark 1.5.6 (Classical vs. Weak) For the sake of clarity, we explicitly observe that

$$\operatorname{div}(\nabla F(\nabla u)) = \sum_{i,j=1}^{N} a_{i,j}\, \frac{\partial^2 u}{\partial x_i\, \partial x_j}, \qquad \text{where } a_{i,j} = \frac{\partial^2 F}{\partial z_i\, \partial z_j}(\nabla u),$$

i.e. these are the coefficients of the Hessian matrix of the function $z \mapsto F(z)$, computed at the point ∇u.

Then, we observe that in the formulation (1.5.5), only the gradient ∇u appears. In particular, this formulation makes sense for functions which are not necessarily twice differentiable, differently from (1.5.4).

Let us spend some words to clarify the connection between (1.5.5) and (1.5.4). If $u \in C^2(\Omega)$ solves (1.5.4) in classical pointwise sense, then by multiplying the equation by $\varphi \in C_0^\infty(\Omega)$ and integrating, we get

$$-\int_\Omega \operatorname{div}\big(\nabla F(\nabla v)\big)\, \varphi\, dx = \int_\Omega f\, \varphi\, dx.$$

By observing that

$$-\operatorname{div}\big(\nabla F(\nabla u)\big)\, \varphi = -\operatorname{div}\big(\nabla F(\nabla u)\, \varphi\big) + \langle \nabla F(\nabla u), \nabla \varphi \rangle,$$

using the Divergence Theorem and the fact that the vector field $\nabla F(\nabla u)\, \varphi$ is compactly supported in Ω, we obtain

$$\int_\Omega \langle \nabla F(\nabla u)), \nabla \varphi \rangle\, dx = \int_\Omega f\, \varphi\, dx.$$

Thus, a classical solution of (1.5.4) is a weak solution, as well. Observe that these are the same computations we used in the proof of Theorem 1.5.3.

On the other hand, let us suppose that u verifies (1.5.5) and assume in addition that $u \in C^2(\Omega)$. Then we would obtain

$$\begin{aligned}
\int_\Omega f\, \varphi\, dx &= \int_\Omega \langle \nabla F(\nabla v), \nabla \varphi \rangle\, dx \\
&= \int_\Omega \operatorname{div}(\nabla F(\nabla v)\, \varphi)\, dx - \int_\Omega \operatorname{div}(\nabla F(\nabla v))\, \varphi\, dx \\
&= \int_{\partial\Omega} \langle \nabla F(\nabla v), \nu_\Omega \rangle\, \varphi\, d\sigma - \int_\Omega \operatorname{div}(\nabla F(\nabla v))\, \varphi\, dx \\
&= -\int_\Omega \operatorname{div}(\nabla F(\nabla v))\, \varphi\, dx.
\end{aligned}$$

where we used again the Divergence Theorem. In other words, we get

$$\int_\Omega \Big[\operatorname{div}\big(\nabla F(\nabla v)\big) + f\Big]\, \varphi\, dx = 0, \qquad \text{for every } \varphi \in C_0^\infty(\Omega).$$

By the Du Bois-Reymond Lemma (in its classical version, i.e. Lemma 1.4.1), this would imply that

$$\mathrm{div}\big(\nabla F(\nabla v)\big) + f,$$

identically vanishes on Ω, i.e. u is a classical solution of (1.5.4).

Remark 1.5.7 (Important!) We observe that the weak formulation (1.5.5) is formally obtained by considering for every $t \in \mathbb{R}$ and every $\varphi \in C_0^\infty(\Omega)$, the function

$$t \mapsto \mathcal{F}(u + t\,\varphi),$$

computing its derivative at $t = 0$ and imposing that this vanishes. The rigorous justification of this operation usually requires the application of some theorems assuring that one could take the limit under the integral sign (very often, the Dominated Convergence Theorem). We will see some examples of this procedure in Chap. 4. The reader is invited to see Problem 1.7.16 below.

For every fixed function u, the linear functional obtained by performing the operation described in Remark 1.5.7, i.e.

$$\varphi \mapsto \frac{d}{dt}\mathcal{F}(u + t\,\varphi)_{|t=0},$$

is called *first variation of $\mathcal{F}$ computed at u*. Sometimes, this will be denoted by the symbol

$$\delta\mathcal{F}(u)[\varphi].$$

This can be thought as an infinite dimensional analogue of the directional derivative of a function of several variables, with the "test" function φ playing the role of a direction. For the functional

$$\mathcal{F}(u) = \int_\Omega F(\nabla u)\,dx - \int_\Omega f\,u\,dx,$$

this is thus given by

$$\delta\mathcal{F}(u)[\varphi] = \int_\Omega \langle \nabla F(\nabla u), \nabla\varphi\rangle\,dx - \int_\Omega f\,\varphi\,dx.$$

Observe that if u is a weak solution of (1.5.4), by appealing to the previous discussion we can write this fact in compact form

$$\delta\mathcal{F}(u)[\varphi] = 0, \qquad \text{for every } \varphi \in C_0^\infty(\Omega). \tag{1.5.6}$$

Since the first variation can be thought as a directional first derivative of $\mathcal{F}$, it is natural to introduce the following

Definition 1.5.8 A function u is said to be a *critical point* of the functional $\mathcal{F}$, when (1.5.6) is satisfied.

Remark 1.5.9 The reader should not be worried about some lack of accuracy of the previous definitions. For this moment, we just want the reader to familiarize with some general ideas, principles and terminologies that lie at the basis of the Calculus of Variations. In the next chapters, we will be more precise on the functional analytic framework.

1.6 Elliptic Operators

In the previous section, we have seen that there is a close connection between minimization problems and differential equations. We now discuss a class of remarkable differential operators that are very often connected to convex minimization problems.

Definition 1.6.1 Let $A : \mathbb{R}^N \to \mathbb{R}^N$ be a continuous function. We say that the second order differential operator

$$\varphi \mapsto -\mathrm{div}\big(A(\nabla\varphi)\big),$$

is:

- *elliptic* if A satisfies

$$\langle A(z) - A(w), z - w\rangle > 0, \qquad \text{for every } z, w \in \mathbb{R}^N \text{ with } z \neq w;$$

- *strictly elliptic* if there exists a constant $c_A > 0$ such that A satisfies

$$\langle A(z) - A(w), z - w\rangle \geq c_A\,|z - w|^2, \qquad \text{for every } z, w \in \mathbb{R}^N.$$

Proposition 1.6.2 (Convexity vs. Ellipticity) *Let $F : \mathbb{R}^N \to \mathbb{R}$ be a C^1 strictly convex function. Then the second order differential operator*

$$\varphi \mapsto -\mathrm{div}\big(\nabla F(\nabla\varphi)\big),$$

is elliptic. Moreover, if F is C^2 this operator is strictly elliptic if and only if the symmetric matrix D^2F is uniformly positive definite, i.e. if there exists $\lambda > 0$ such that

$$\langle D^2F(z)\cdot\xi, \xi\rangle \geq \lambda\,|\xi|^2, \qquad \textit{for every } z, \xi \in \mathbb{R}^N. \tag{1.6.1}$$

In this case, we have

$$\langle \nabla F(z) - \nabla F(w), z - w \rangle \geq \lambda\, |z - w|^2, \qquad \textit{for every } z, w \in \mathbb{R}^N \tag{1.6.2}$$

Proof The proof is similar to that of Proposition 1.3.8. By using the "above tangent" property of convex functions (i.e. Proposition 1.3.4), we have for every $z, w \in \mathbb{R}^N$

$$F(z) \geq F(w) + \langle \nabla F(w), z - w \rangle,$$

and also

$$F(w) \geq F(z) + \langle \nabla F(z), w - z \rangle.$$

Moreover, since F is strictly convex, the strict inequality sign holds for $z \neq w$. By summing up these two inequalities, we get for every $z \neq w$

$$F(z) + F(w) > F(w) + F(z) + \langle \nabla F(w) - \nabla F(z), z - w \rangle,$$

that is

$$0 > \langle \nabla F(w) - \nabla F(z), z - w \rangle.$$

This proves the ellipticity.

We now prove the second statement. Let us suppose that the operator is strictly elliptic. This means that

$$\langle \nabla F(z) - \nabla F(w), z - w \rangle \geq c\, |z - w|^2, \qquad \text{for every } z, w \in \mathbb{R}^N,$$

for a suitable constant $c > 0$. We need to prove that $D^2 F$ satisfies (1.6.1). We fix ξ a unit vector and take $z = w + t\,\xi$ in the definition of strict ellipticity. We also use that ∇F is differentiable, thanks to the fact that F is of class C^2. We obtain

$$\begin{aligned} C\, t^2 &\leq \langle \nabla F(w + t\,\xi) - \nabla F(w), t\,\xi \rangle \\ &= \sum_{i=1}^{N} \left(\frac{\partial F}{\partial x_i}(w + t\,\xi) - \frac{\partial F}{\partial x_i}(w) \right) t\,\xi_i \\ &= t^2 \sum_{i=1}^{N} \left(\sum_{j=1}^{N} \frac{\partial^2 F}{\partial x_i \partial x_j}(w)\,\xi_j \right) \xi_i + \sum_{i=1}^{N} o_i(|t|)\, t\,\xi_i \\ &= t^2 \langle D^2 F(w) \cdot \xi, \xi \rangle + \sum_{i=1}^{N} o_i(|t|)\, t\,\xi_i. \end{aligned}$$

We divide by $t^2 \neq 0$ and then let t go to 0, so to obtain

$$c \le \langle D^2 F(w) \cdot \xi, \xi \rangle, \qquad \text{for every } |\xi| = 1.$$

By bilinearity of the scalar product, this is equivalent to (1.6.1) and thus the proof of this implication is concluded.

We now assume that $D^2 F$ is uniformly positive definite, i.e. condition (1.6.1) is in force. We take $z, w \in \mathbb{R}^N$ two points and observe that by the fundamental theorem of Calculus

$$\begin{aligned}\nabla F(z) - \nabla F(w) &= \int_0^1 \frac{d}{dt} \nabla F(w + t\,(z - w))\, dt \\ &= \int_0^1 D^2 F(w + t\,(z - w)) \cdot (z - w)\, dt.\end{aligned}$$

If we take the scalar product with the vector $z - w$ and use (1.6.1), we get

$$\begin{aligned}\langle \nabla F(z) - \nabla F(w), z - w \rangle &= \int_0^1 \langle D^2 F(w + t\,(z - w)) \cdot (z - w), z - w \rangle\, dt \\ &\ge \lambda\, |z - w|^2 \int_0^1 dt = \lambda\, |z - w|^2.\end{aligned}$$

This proves that the operator is strictly elliptic and (1.6.2) holds. □

Remark 1.6.3 (Some Remarkable Elliptic Operators) In what follows, we use the notation Id_N for the $N \times N$ identity matrix.

1. By taking the convex function

$$F(z) = \frac{1}{2}\, |z|^2, \qquad z \in \mathbb{R}^N,$$

   the corresponding differential operator is

$$-\operatorname{div}\big(\nabla F(\nabla \varphi)\big) = -\operatorname{div}(\nabla \varphi) = -\Delta \varphi,$$

   i.e. the *Laplacian operator*. We also observe that

$$D^2 F(z) = \mathrm{Id}_N,$$

   thus by Proposition 1.6.2, this is a strictly elliptic operator;
2. by taking the convex function

$$F(z) = \sqrt{1 + |z|^2}, \qquad z \in \mathbb{R}^N,$$

the corresponding differential operator is

$$- \operatorname{div}\big(\nabla F(\nabla\varphi)\big) = -\operatorname{div}\left(\frac{\nabla\varphi}{\sqrt{1+|\nabla\varphi|^2}}\right),$$

which is called the *mean curvature operator*. We also observe that[3]

$$D^2 F(z) = \frac{\mathrm{Id}_N}{\sqrt{1+|z|^2}} - \frac{z\otimes z}{(1+|z|^2)^{\frac{3}{2}}},$$

so that, with simple computations, we have

$$\frac{|\xi|^2}{(1+|z|^2)^{\frac{3}{2}}} \le \langle D^2F(z)\cdot\xi,\xi\rangle \le \frac{|\xi|^2}{\sqrt{1+|z|^2}}.$$

By recalling the definition (1.6.1) of uniform positive definiteness, this shows that D^2F is not uniformly positive definite. Thus, by Proposition 1.6.2, the mean curvature operator is an elliptic operator, but *not* a strictly elliptic one;

3. more generally, we take $\alpha > 0$ and $1 < p < \infty$ and consider the convex function

$$F_\alpha(z) = \frac{1}{p}\,(\alpha + |z|^2)^{\frac{p}{2}}, \qquad z\in\mathbb{R}^N.$$

The corresponding differential operator is

$$- \operatorname{div}\big(\nabla F_\alpha(\nabla\varphi)\big) = -\operatorname{div}\left((\alpha+|\nabla\varphi|^2)^{\frac{p-2}{2}}\,\nabla\varphi\right). \tag{1.6.3}$$

We compute the Hessian matrix of F, which is given by

$$D^2F_\alpha(z) = (\alpha+|z|^2)^{\frac{p-2}{2}}\,\mathrm{Id}_N + (p-2)\,z\otimes z\,(\alpha+|z|^2)^{\frac{p-4}{2}},$$

to that

$$\langle D^2F_\alpha(z)\cdot\xi,\xi\rangle = (\alpha+|z|^2)^{\frac{p-2}{2}}\,|\xi|^2 + (p-2)\,\langle z,\xi\rangle^2\,(\alpha+|z|^2)^{\frac{p-4}{2}}.$$

---

[3] We denote by $z\otimes\xi$ the *tensor product* of two vectors. This is the $N\times N$ matrix whose (i,j) entry is given by $z_i\,\xi_j$. In particular, we get

$$\langle z\otimes z\cdot\xi,\xi\rangle = \sum_{i=1}^N\left(\sum_{j=1}^N z_i\,z_j\,\xi_j\right)\xi_i = \left(\sum_{i=1}^N z_i\,\xi_i\right)\left(\sum_{j=1}^N z_j\,\xi_j\right) = \langle z,\xi\rangle^2.$$

Thus, if $p \geq 2$, we have

$$(\alpha + |z|^2)^{\frac{p-2}{2}} |\xi|^2 \leq \langle D^2 F_\alpha(z) \cdot \xi, \xi \rangle \leq (p-1)\,(\alpha + |z|^2)^{\frac{p-2}{2}} |\xi|^2.$$

By using Proposition 1.6.2, we then get that the operator (1.6.3) is strictly elliptic thanks to the fact that $\alpha > 0$.

Similarly, if $1 < p < 2$, we have

$$(p-1)\,(\alpha + |z|^2)^{\frac{p-2}{2}} |\xi|^2 \leq \langle D^2 F_\alpha(z) \cdot \xi, \xi \rangle \leq (\alpha + |z|^2)^{\frac{p-2}{2}} |\xi|^2.$$

By using Proposition 1.6.2, we get that the operator (1.6.3) is elliptic, but not strictly elliptic;

4. for $\alpha = 0$, the operator (1.6.3) becomes

$$-\operatorname{div}\left(|\nabla \varphi|^{p-2}\, \nabla \varphi\right) =: -\Delta_p \varphi, \tag{1.6.4}$$

which is called *p-Laplacian operator*. It is not difficult to see that the function

$$F(z) = \frac{1}{p}\, |z|^p, \qquad \text{for } z \in \mathbb{R}^N,$$

is still strictly convex. Indeed, by the triangle inequality for the Euclidean norm and the convexity of the power $p > 1$, we have for every $z, w \in \mathbb{R}^N$ and every $\lambda \in [0, 1]$

$$\begin{aligned} F(\lambda\, z + (1-\lambda)\, w) &= \frac{1}{p}\, |\lambda\, z + (1-\lambda)\, w|^p \leq \frac{1}{p}\, (\lambda\, |z| + (1-\lambda)\, |w|)^p \\ &\leq \frac{\lambda}{p}\, |z|^p + \frac{(1-\lambda)}{p}\, |w|^p = \lambda\, F(z) + (1-\lambda)\, F(w). \end{aligned}$$

Moreover, we recall that equality holds in the triangle inequality if and only if $\lambda\, z$ and $(1-\lambda)\, w$ are two linearly dependent vectors, with the same orientation; in the second inequality, for $\lambda \in (0, 1)$ we can have equality if and only if $|z| = |w|$, by the strict convexity of the power $p > 1$. These facts permit to infer that F is strictly convex and thus, by Proposition 1.6.2, the p-Laplacian operator is elliptic.

On the other hand, it is not strictly elliptic for $p \neq 2$: for $p > 2$, it is sufficient to observe that

$$\langle D^2 F(z) \cdot \xi, \xi \rangle \leq (p-1)\, |z|^{p-2}\, |\xi|^2, \qquad \text{for every } z, \xi \in \mathbb{R}^N.$$

Thus, the condition (1.6.1) is not verified and the conclusion follows from Proposition 1.6.2.

In the case $1 < p < 2$, the function F is not C^2, but by supposing that the operator is strictly elliptic one would obtain

$$\langle |z|^{p-2}\, z - |\nabla w|^{p-2}\, w, z - w\rangle \geq c\, |z - w|^2, \qquad \text{for every } z, w \in \mathbb{R}^N,$$

for some $c > 0$. If we take $w = 0$, one would in particular obtain

$$|z|^p \geq c\, |z|^2, \qquad \text{for every } z \in \mathbb{R}^N,$$

which is not true.

1.7 Problems

Problem 1.7.1 Let $0 < \alpha < 1$, prove that we have

$$2^{\alpha-1}\,(t^\alpha + s^\alpha) \leq (t+s)^\alpha \leq t^\alpha + s^\alpha, \qquad \text{for every } t, s \geq 0. \tag{1.7.1}$$

Problem 1.7.2 Let $\alpha > 1$, prove that we have

$$t^\alpha + s^\alpha \leq (t+s)^\alpha \leq 2^{\alpha-1}\,(t^\alpha + s^\alpha), \qquad \text{for every } t, s \geq 0. \tag{1.7.2}$$

Problem 1.7.3 Let $0 < \alpha < 1$, then we have

$$\Big| |t|^{\alpha-1}\, t - |s|^{\alpha-1}\, s \Big| \leq 2^{1-\alpha}\, |t - s|^\alpha, \qquad \text{for every } t, s \in \mathbb{R}.$$

Problem 1.7.4 Let $\alpha \geq 1$, then we have

$$\Big| |t|^{\alpha-1}\, t - |s|^{\alpha-1}\, s \Big| \leq \alpha\, (|t| + |s|)^{\alpha-1}\, |t - s|, \qquad \text{for every } t, s \in \mathbb{R}.$$

Problem 1.7.5 Let $\alpha \geq 1$ and let $G : \mathbb{R} \to \mathbb{R}$ be a C^1 function such that

$$|G'(t)| \leq A\,(|t|^{\alpha-1} + 1), \qquad \text{for every } t \in \mathbb{R},$$

for a constant $A \geq 0$. Prove that

$$|G(t) - G(s)| \leq A\,(|t| + |s|)^{\alpha-1}\, |t - s| + A\, |t - s|.$$

Deduce from this fact that there exists a constant $\widetilde{A} \geq 0$ such that

$$|G(t)| \leq \widetilde{A}\,(|t|^\alpha + 1), \qquad \text{for every } t \in \mathbb{R}.$$

Problem 1.7.6 Let $f : [a, b] \to \mathbb{R}$ be a convex function. Show that f is bounded on $[a, b]$. Also give an example of convex function on (a, b) which is not bounded.

Problem 1.7.7 Let $H : \mathbb{R}^N \to \mathbb{R}$ be a C^2 convex function such that D^2H is uniformly positive definite, i.e. it satisfies (1.6.1). Show that H admits a minimum over $\mathbb{R}^N$ and that there exists a unique minimum point.

Problem 1.7.8 Let $f : [0, +\infty) \to \mathbb{R}$ be a strictly convex non-decreasing function. Show that f is actually increasing. Then prove that the function $F : \mathbb{R}^N \to \mathbb{R}$ defined by

$$F(z) = f(|z|), \qquad \text{for every } z \in \mathbb{R}^N,$$

is strictly convex. Finally, show that if f has two intervals of strict monotonicity, then F is not convex.

Problem 1.7.9 Let $1 < p < \infty$ and let $H : \mathbb{R}^N \to \mathbb{R}$ be a C^1 convex function with the following property: there exist two constants $C_2 \geq C_1 > 0$ such that

$$C_1\,(|z|^p - 1) \leq H(z) \leq C_2\,(|z|^p + 1), \qquad \text{for every } z \in \mathbb{R}^N.$$

Show that there must exist a constant $C_3 > 0$ such that

$$|\nabla H(z)| \leq C_3\,(|z|^{p-1} + 1), \qquad \text{for every } z \in \mathbb{R}^N.$$

Problem 1.7.10 Prove the following generalization of Young's inequality: for every $k \in \mathbb{N}$ such that $k \geq 2$ and every $a_1, \ldots, a_k \geq 0$, we have

$$\prod_{i=1}^{k} a_i \leq \frac{1}{k}\left(\sum_{i=1}^{k} a_i^k\right). \tag{1.7.3}$$

Problem 1.7.11 Let $f \in C^0([a, b])$ be such that

$$\int_a^b f(t)\,\varphi'(t)\,dt = 0, \qquad \text{for every } \varphi \in C^1([a, b]) \text{ such that } \varphi(a) = \varphi(b) = 0.$$

Show that f is constant on $[a, b]$.

Problem 1.7.12 Show that the same conclusion of Problem 1.7.11 holds, provided we only know that

$$\int_a^b f(t)\,\varphi'(t)\,dt = 0, \qquad \text{for every } \varphi \in C_0^\infty((a, b)).$$

Problem 1.7.13 Let $f \in C^0([a, b])$ be such that

$$\int_a^b f(t)\,\varphi(t)\,dt = 0, \qquad \text{for every } \varphi \in C_0^\infty((a, b)) \text{ with } \int_a^b \varphi(t)\,dt = 0.$$

Show that f is constant on $[a, b]$.

Problem 1.7.14 Let $f \in C^0([a, b])$ be such that

$$\int_a^b f(t)\,\varphi''(t)\,dt = 0, \qquad \text{for every } \varphi \in C^2([a, b]) \text{ such that } \begin{array}{l} \varphi(a) = \varphi(b) = 0, \\ \varphi'(a) = \varphi'(b) = 0. \end{array}$$

Show that f is an affine function, i. e. it has the form

$$f(t) = \alpha + \beta\,t,$$

for some $\alpha, \beta \in \mathbb{R}$.

Problem 1.7.15 Prove the following variant of Lemma 1.4.1: let $\Omega \subseteq \mathbb{R}^N$ be an open set and let $f : \Omega \to \mathbb{R}$ be a continuous function such that

$$\int_\Omega f\,\varphi\,dx \geq 0, \qquad \text{for every } \varphi \in C_0^\infty(\Omega) \text{ such that } \varphi \geq 0.$$

Then $f(x) \geq 0$ for every $x \in \Omega$.

Problem 1.7.16 Let $f \in C^0([-1, 1])$, we consider the functional

$$\mathcal{F}(u) = \frac{1}{4}\int_{-1}^1 |u'|^4\,dt - \int_{-1}^1 f\,u\,dt,$$

defined for functions $u \in C^1([-1, 1])$. Compute the first variation of this functional and obtain the corresponding Euler-Lagrange equation, both in weak and classical formulations.

Problem 1.7.17 Find a weak solution $u \in C^1([-1, 1])$ of the Euler-Lagrange equation of Problem 1.7.16, when $f \equiv 1$ and the boundary conditions

$$u(-1) = u(1) = 0,$$

are imposed. Then show that this solution is unique. Is it true that u is twice differentiable?

Problem 1.7.18 We consider the functional

$$\mathcal{F}(u) = \frac{1}{4}\int_{-1}^1 |u'|^4\,dt - \frac{1}{2}\int_{-1}^1 |u|^2\,dt,$$

defined for functions $u \in C^1([-1, 1])$. Compute the first variation of this functional and obtain the corresponding Euler-Lagrange equation, both in weak and classical formulations.

Problem 1.7.19 Let us consider the nonlinear ordinary differential equation

$$-\,t\,u''(t)-u'(t)=|u(t)|^5\,u(t),\qquad \text{for } t\in(0,1).$$

Find a functional $\mathcal{F}$ defined on $C^1([0,1])$ whose Euler-Lagrange equation is given by the equation above.

Problem 1.7.20 Let $1<p<\infty$, find a solution of

$$\inf_{u\in C^1([-1,1])}\left\{\frac{1}{p}\int_{-1}^{1}|u'|^p\,dt-\int_{-1}^{1}u\,dt\ :\ u(-1)=u(1)=0\right\},$$

and show that this is unique. Under which conditions on p such a solution belongs to $C^2([-1,1])$?

Problem 1.7.21 Let $B_1(0)=\{(x,y)\in\mathbb{R}^2\ :\ x^2+y^2<1\}$, we recall that its *moment of inertia* with respect to the axis of rotation passing through its center and orthogonal to the $x\,y$ plane is given by

$$\mathcal{I}=\int_{B_1(0)}(x^2+y^2)\,dx\,dy=\frac{\pi}{2}.$$

Show that

$$\int_{B_1(0)}|\nabla u|^2\,dx\,dy\geq 4\,\mathcal{I},$$

for every $u\in C^1(\overline{B_1(0)})$ which coincides with the function

$$U(x,y)=x^2-y^2,\qquad \text{for } (x,y)\in\partial B_1(0).$$

Problem 1.7.22 (Interpolation in L^p Spaces) Let $E\subseteq\mathbb{R}^N$ be a measurable set and let $f\in L^q(E)\cap L^p(E)$, for $1\leq q<p\leq\infty$. Prove that $f\in L^r(E)$ for every $q<r<p$ and we have

$$\|f\|_{L^r(E)}\leq\|f\|_{L^p(E)}^{\vartheta}\,\|f\|_{L^q(E)}^{1-\vartheta},\tag{1.7.4}$$

where the exponent $\vartheta\in(0,1)$ is given by

$$\vartheta=\begin{cases}\dfrac{p}{r}\,\dfrac{r-q}{p-q}, & \text{if } p<\infty,\\[2ex] \dfrac{r-q}{r}, & \text{if } p=\infty.\end{cases}$$

Problem 1.7.23 Let $E\subseteq\mathbb{R}^N$ be a measurable set and take $k\in\mathbb{N}$ such that $k\geq 2$. Prove that for every $f_i\in L^k(E)$ with $i\in\{1,\ldots,k\}$, we have

$$\int_E \prod_{i=1}^k |f_i|\,dx \le \prod_{i=1}^k \left(\int_E |f_i|^k\,dx\right)^{\frac{1}{k}}.$$

Problem 1.7.24 Let $N \ge 2$ and let $\Phi_1, \dots, \Phi_N \in L^1_{\rm loc}(\mathbb{R}^N)$ be non-negative measurable functions with the following properties:

- Φ_k does not depend on the k-th variable;
- $\Phi_k \in L^{N-1}(\mathbb{R}^{N-1})$.

Show that we have

$$\prod_{k=1}^N \Phi_k \in L^1(\mathbb{R}^N),$$

and the following inequality holds

$$\int_{\mathbb{R}^N} \prod_{k=1}^N \Phi_k\,dx \le \prod_{k=1}^N \left(\int_{\mathbb{R}^{N-1}} \Phi_k^{N-1}\,d\widehat{x}_k\right)^{\frac{1}{N-1}}. \tag{1.7.5}$$

Here we used the notation

$$d\widehat{x}_k := dx_1 \dots dx_{k-1}\,dx_{k+1} \dots dx_N, \qquad \text{for } k \in \{1, \dots, N\}.$$

Problem 1.7.25 For $1 < q \le \infty$, we consider the ℓ^q norm over $\mathbb{R}^2$, defined by

$$\|(x, y)\|_{\ell^q} = \begin{cases} (|x|^q + |y|^q)^{\frac{1}{q}}, & \text{if } 1 < q < \infty, \\ \max\Big\{|x|, |y|\Big\}, & \text{if } q = \infty. \end{cases}$$

Let $1 \le p \le \infty$ and let $E \subseteq \mathbb{R}^N$ be a measurable set. Show that

$$N_{p,q}(u, \phi) := \left\|\Big(\|u\|_{L^p(E;\mathbb{R}^k)}, \|\phi\|_{L^p(E;\mathbb{R}^n)}\Big)\right\|_{\ell^q},$$
$$\text{for } (u, \phi) \in L^p(E; \mathbb{R}^k) \times L^p(E; \mathbb{R}^n),$$

defines a norm on $L^p(E; \mathbb{R}^k) \times L^p(E; \mathbb{R}^n)$. Also show that this is equivalent to the norm

$$\|(u, \phi)\|_{L^p(E;\mathbb{R}^k)\times L^p(E;\mathbb{R}^n)} := \|u\|_{L^p(E;\mathbb{R}^k)} + \|\phi\|_{L^p(E;\mathbb{R}^n)}.$$

Problem 1.7.26 Let $\Omega \subseteq \mathbb{R}^N$ be an open set with $\Omega \ne \mathbb{R}^N$. We define the *distance function* by

$$\mathrm{dist}(x, \partial\Omega) = \inf_{y\in\partial\Omega} |x - y|, \qquad \text{for every } x \in \Omega.$$

Show at first that the infimum is actually a minimum. Then, for every $n \in \mathbb{N} \setminus \{0\}$, we define

$$\mathrm{int}_n(\Omega) := \{x \in \Omega \, : \, \mathrm{dist}(x, \partial\Omega) > 1/n\}.$$

Prove that if $|\Omega| < +\infty$, we have

$$\lim_{n\to\infty} |\Omega \setminus \mathrm{int}_n(\Omega)| = 0.$$

Problem 1.7.27 Let $\Omega \subseteq \mathbb{R}^N$ be an open connected set. Show that there exists an increasing sequence of open bounded *connected* sets $\{\Omega_k\}_{k\in\mathbb{N}}$ such that

$$\Omega = \bigcup_{k\in\mathbb{N}} \Omega_k \qquad \text{and} \qquad \Omega_k \Subset \Omega, \ \text{for every } k \in \mathbb{N}.$$

Problem 1.7.28 Let $E \subseteq \mathbb{R}^N$ be a measurable set. Show that for every $1 \le p < \infty$ the set $L^1(E) \cap L^\infty(E)$ is dense in $L^p(E)$, i.e. for every $u \in L^p(E)$ there exists a sequence $\{u_n\}_{n\in\mathbb{N}} \subseteq L^1(E) \cap L^\infty(E)$ such that

$$\lim_{n\to\infty} \|u - u_n\|_{L^p(E)} = 0.$$

Give an example to show that this is false for $p = \infty$.

Problem 1.7.29 Let $E \subseteq \mathbb{R}^N$ be a measurable set and let $1 < p \ne q < \infty$ be two exponents. We suppose that $\{f_n\}_{n\in\mathbb{N}} \subseteq L^p(E)\cap L^q(E)$ weakly converges in $L^p(E)$ to a function $f \in L^p(E) \cap L^q(E)$. Show that if there exists $M > 0$ such that

$$\|f_n\|_{L^q(E)} \le M, \qquad \text{for every } n \in \mathbb{N},$$

then the whole sequence $\{f_n\}_{n\in\mathbb{N}}$ weakly converges in $L^q(E)$ to f, as well.

Problem 1.7.30 Let $E \subseteq \mathbb{R}^N$ be a measurable set. Let $p_0 \ge 1$ and let $u \in L^p(E)$, for every $p_0 \le p < \infty$. Prove that if

$$\limsup_{p\nearrow\infty} \|u\|_{L^p(E)} \le M < +\infty,$$

then $u \in L^\infty(E)$, with $\|u\|_{L^\infty(E)} \le M$.

Problem 1.7.31 Let $\Omega \subseteq \mathbb{R}^N$ be an open set. Prove that for every continuous function $u : \overline{\Omega} \to \mathbb{R}$ we have

$$\|u\|_{L^\infty(\Omega)} = \sup_{x\in\Omega} |u(x)| = \sup_{x\in\overline{\Omega}} |u(x)|.$$

Chapter 2
Some One-dimensional Variational Problems

2.1 Introduction

In this chapter we consider some classical minimization problems settled over suitable spaces of functions of one real variable. These will be of the form

$$\inf\left\{\int_a^b F(t,u(t),u'(t))\,dt\ :\ u(a)=u_0,\ u(b)=u_1\right\},$$

where u may also be vector-valued, i.e. it could be a curve in the space. However, rather than aiming at giving the most complete theory for a general function F (for which we refer for example to [1, 2] or [3]), we will focus on some particular examples, which are still quite instructive and interesting. In particular, we mostly focus on the *brachistocrone problem* (Sect. 2.3) and on the problem of determining the sharp constant in some one-dimensional *Poincaré inequalities* (Sect. 2.4).

For the first problem, we show that it is possible to prove existence of a (unique) minimizer by detecting a "hidden" convex structure of the problem and then solving the relevant Euler-Lagrange equation. Roughly speaking, we rely on the idea that, for a convex function, criticality is equivalent to minimality.

For the second problem, we are interested in particular in solving the following

$$\inf_{\varphi\in C^1([0,1])\setminus\{0\}}\left\{\frac{\displaystyle\int_0^1|\varphi'(t)|^2\,dt}{\displaystyle\int_0^1|\varphi(t)|^2\,dt}\ :\ \varphi(0)=\varphi(1),\ \int_0^1\varphi(t)\,dt=0\right\}. \tag{2.1.1}$$

The determination of this infimum is tightly connected with the celebrated *isoperimetric inequality* for smooth sets in the plane. This is clarified in Sect. 2.6, where we present the proof (due to Hurwitz) of this result. Indeed, its main ingredient is the exact determination of the infimum (2.1.1): this shows a first intriguing connection

L. Brasco, *Handbook of Calculus of Variations for Absolute Beginners*,
La Matematica per il 3+2 163, https://doi.org/10.1007/978-3-031-87164-1_2

between functional inequalities and geometric problems (we refer the reader to [21] for some comprehensive studies on this subject).

The original proof by Hurwitz used the theory of *Fourier series* in order to determine the value of (2.1.1) (see for example [73]). Here on the contrary, we prefer to use a purely variational point of view, relying on the Euler-Lagrange equation for the problem (2.1.1) and Picone's inequality. With the same approach, we can treat a more general class of Poincaré constants.

The chapter is complemented by other miscellaneous problems, namely: finding curves with minimal kinetic energy connecting two points (Sect. 2.2); a minimization problem connected with the one-dimensional harmonic oscillator (Sect. 2.5); a constrained isoperimetric-like problem for graphs of functions (Sect. 2.7).

2.2 Curves of Minimal Action

We start with a very simple variational problem, i.e. finding the curves of minimal length connecting two given points in the Euclidean space. By recalling that for a regular curve $\gamma : [0, T] \to \mathbb{R}^N$ its length is given by

$$\int_0^T |\gamma'(t)|\, dt,$$

we have the following

Theorem 2.2.1 (Curves of Minimal Length) *Let $x_0, x_1 \in \mathbb{R}^N$, the minimization problem*

$$\min_{\gamma\in C^1([0,T];\mathbb{R}^N)} \left\{ \int_0^T |\gamma'(t)|\, dt \;:\; \gamma(0) = x_0,\; \gamma(T) = x_1 \right\},$$

has infinitely many solutions. Indeed, any orientation preserving reparametrization of the constant speed curve

$$\gamma_{\text{opt}}(t) = \left(1 - \frac{t}{T}\right) x_0 + \frac{t}{T}\, x_1, \qquad \textit{for } t \in [0, T],$$

is a minimizer.

Proof For every admissible curve, we have

$$\int_0^T |\gamma'(t)|\, dt \geq \left| \int_0^T \gamma'(t)\, dt \right| = |\gamma(T) - \gamma(0)| = |x_1 - x_0|, \tag{2.2.1}$$

which shows that

$$\inf_{\gamma\in C^1([0,T];\mathbb{R}^N)}\left\{\int_0^T |\gamma'(t)|\,dt\;:\;\gamma(0)=x_0,\;\gamma(T)=x_1\right\}\geq |x_1-x_0|.$$

On the other hand, by taking the curve $\gamma_{\rm opt}$ we have

$$|\gamma'_{\rm opt}(t)|=\frac{1}{T}\,|x_1-x_0|,$$

so that

$$\begin{aligned}\inf_{\gamma\in C^1([0,T];\mathbb{R}^N)}\left\{\int_0^T |\gamma'(t)|\,dt\;:\;\gamma(0)=x_0,\;\gamma(T)=x_1\right\}&\leq \int_0^T |\gamma'_{\rm opt}(t)|\,dt\\ &=\frac{1}{T}\,T\,|x_0-x_1|.\end{aligned}$$

This shows that the curve $\gamma_{\rm opt}$ is a minimizer of our problem. By observing that the length functional is invariant by time reparametrizations, we get the conclusion. □

If a curve $\gamma : [0,T]\to\mathbb{R}^N$ represents the trajectory of a unit mass particle in the space, the quantity

$$\frac{1}{2}\,|\gamma'(t)|^2,$$

represents the kinetic energy at time t. We now look at the minimization of the total kinetic energy of a curve.

Theorem 2.2.2 (Curves of Minimal Kinetic Energy) *The minimization problem*

$$\min_{\gamma\in C^1([0,T];\mathbb{R}^N)}\left\{\int_0^T |\gamma'(t)|^2\,dt\;:\;\gamma(0)=x_0,\;\gamma(T)=x_1\right\},$$

has a unique solution, given by the constant speed curve

$$\gamma_{\rm opt}(t)=\left(1-\frac{t}{T}\right)x_0+\frac{t}{T}\,x_1,\qquad \text{for } t\in[0,T].$$

Proof By using Jensen's inequality (Proposition 1.3.10), for every admissible curve we have

$$\begin{aligned}\int_0^T |\gamma'(t)|^2\,dt&=T\,\frac{1}{T}\int_0^T |\gamma'(t)|^2\,dt\\ &\geq T\left(\frac{1}{T}\int_0^T |\gamma'(t)|\,dt\right)^2\geq\frac{1}{T}\,|x_0-x_1|^2,\end{aligned}$$

where in the last inequality we used (2.2.1). This shows that

$$\inf_{\gamma \in C^1([0,T];\mathbb{R}^N)} \left\{ \int_0^T |\gamma'(t)|^2 \, dt \,:\, \gamma(0) = x_0,\ \gamma(T) = x_1 \right\} \geq \frac{1}{T} \, |x_0 - x_1|^2.$$

On the other hand, for the curve γ_{opt} we have

$$|\gamma'_{\mathrm{opt}}(t)| = \frac{1}{T} \, |x_0 - x_1|,$$

so that

$$\begin{aligned} \inf_{\gamma \in C^1([0,T];\mathbb{R}^N)} \left\{ \int_0^T |\gamma'(t)|^2 \, dt \,:\, \gamma(0) = x_0,\ \gamma(T) = x_1 \right\} &\leq \int_0^T |\gamma'_{\mathrm{opt}}(t)|^2 \, dt \\ &= \frac{1}{T} \, |x_0 - x_1|^2. \end{aligned}$$

This shows that the curve γ_{opt} is a minimizer of our problem.

Uniqueness now follows from the strict convexity of the function $z \mapsto |z|^2$. Indeed, if γ_1 and γ_2 were two distinct minimizers, we could consider the curve

$$\widetilde{\gamma}(t) = \frac{\gamma_1(t) + \gamma_2(t)}{2}, \qquad \text{for } t \in [0, T],$$

which is still admissible. By minimality of both γ_1 and γ_2, we would get

$$\int_0^T |\widetilde{\gamma}'(t)|^2 \, dt \geq \int_0^T |\gamma_1'(t)|^2 \, dt = \int_0^T |\gamma_2'(t)|^2 \, dt,$$

while by strict convexity, we should obtain

$$\int_0^T |\widetilde{\gamma}'(t)|^2 \, dt = \int_0^T \left| \frac{\gamma_1'(t) + \gamma_2'(t)}{2} \right|^2 dt < \frac{1}{2} \int_0^T |\gamma_1'(t)|^2 \, dt + \frac{1}{2} \int_0^T |\gamma_2'(t)|^2 \, dt.$$

The last two displays are in contrast, thus we get the claimed uniqueness of the mimimizer. This concludes the proof. □

Remark 2.2.3 Observe that the previous two problems have apparently the same solution. However, there is a crucial difference: by minimizing the total kinetic energy, we find a *unique* solution, given by the constant speed parametrization of the geodesic curve connecting x_0 and x_1. Indeed, the total kinetic energy *is not* invariant by time reparametrizations, differently from the length functional.

More generally, we could replace the length functional

$$\gamma \mapsto \int_0^T |\gamma'(t)| \, dt,$$

with a *weighted length functional*, i.e. something of the form

$$\gamma \mapsto \int_0^T g(\gamma(t))\, |\gamma'(t)|\, dt,$$

with $g : \mathbb{R}^N \to [0, +\infty)$ a given continuous function. This is particularly meaningful in connection with the problem of finding geodesics on a Riemannian manifold, see for example [9, Chapter 1, Section 2].

2.3 The Brachistocrone Problem

This is one of the most classical problem in the Calculus of Variations. The formulation of the problem dates back to the seventeenth century and is due to Johann Bernoulli. We present the classical version of this problem: the interested reader may find some generalizations in [69].

We fix two points in $\mathbb{R}^2$, for simplicity we take them to be

$$(0, u_0) \qquad \text{and} \qquad (1, 0),$$

with $u_0 > 0$. We look for the *cartesian curve*

$$\gamma(x) = (x, f(x)), \qquad \text{for } x \in (0, 1),$$

connecting these two points, such that the traveling time of a mass particle moving along such curve under the force of gravity is the least possible.

We recall that the infinitesimal time is given by

$$dt = \frac{ds}{v}, \tag{2.3.1}$$

where ds and v stand for the infinitesimal length and the speed, respectively. The infinitesimal length for a cartesian curve is given by

$$ds = |\gamma'(x)|\, dx = \sqrt{1 + |f'(x)|^2}\, dx. \tag{2.3.2}$$

In order to find the speed, we first recall that a mass falling from height u_0 under the influence of the force of gravity and with initial speed 0, after a time t is located at height

$$\ell = u_0 - \frac{1}{2}\, g\, t^2,$$

where g is the constant acceleration due to the force of gravity. Observe that we automatically have $\ell < u_0$. The speed of such a mass, after a time t is given by

$$v = g\,t.$$

By comparing the last two displays, we thus find the relation

$$v = \sqrt{2\,g\,(u_0 - \ell)}.$$

By observing that for the curve $\gamma(x) = (x, f(x))$, the quantity $f(x)$ represents the vertical displacement, the previous relation implies

$$v = \sqrt{2\,g\,(u_0 - f(x))}. \tag{2.3.3}$$

By using (2.3.2) and (2.3.3) in (2.3.1), we get

$$dt = \frac{\sqrt{1 + |f'(x)|^2}}{\sqrt{2\,g\,(u_0 - f(x))}}\,dx.$$

Thus, the total traveling time is given by

$$\int_0^1 \frac{\sqrt{1 + |f'(x)|^2}}{\sqrt{2\,g\,(u_0 - f(x))}}\,dx,$$

and we are lead to consider the following minimization problem

$$\inf_{f \in C^1((0,1)) \cap C^0([0,1])} \left\{ \int_0^1 \frac{\sqrt{1 + |f'(x)|^2}}{\sqrt{u_0 - f(x)}}\,dx \ : \ \begin{array}{c} f(0) = u_0,\ f(1) = 0 \\ f < u_0 \text{ in } (0,1] \end{array} \right\}. \tag{2.3.4}$$

Observe that we removed the constant $2\,g$, since this is unessential. By operating the change $u = u_0 - f$, the problem (2.3.4) can be equivalently rewritten as

$$\inf_{u \in X([0,1])} \left\{ \int_0^1 \frac{\sqrt{1 + |u'(x)|^2}}{\sqrt{u(x)}}\,dx \ : \ \begin{array}{c} u(0) = 0,\ u(1) = u_0 \\ u > 0 \text{ in } (0,1] \end{array} \right\}, \tag{2.3.5}$$

where $X([0, 1])$ is the following space of functions

$$X([0,1]) := \left\{ C^1((0,1)) \cap C^0([0,1]) \ : \ \int_0^1 \frac{\sqrt{1 + |u'(x)|^2}}{\sqrt{u(x)}}\,dx < +\infty \right\}.$$

Observe that this is not empty, since it contains the function $u(x) = u_0\,x$.

Existence of a solution for the previous problem is not trivial. We start by noticing that a solution is unique, provided it exists. This is the content of the following

Proposition 2.3.1 *Problem* (2.3.5) *admits at most a solution.*

Proof Let us assume that u_1 and u_2 are two distinct solutions of (2.3.5). We then introduce the new functions

$$U_1 = \sqrt{u_1}, \qquad U_2 = \sqrt{u_2}, \qquad V = \frac{U_1 + U_2}{2} = \frac{\sqrt{u_1} + \sqrt{u_2}}{2}, \tag{2.3.6}$$

and set

$$v = V^2 = \left(\frac{\sqrt{u_1} + \sqrt{u_2}}{2}\right)^2.$$

With some elementary, though lengthy, manipulations we can see that v is still admissible in the variational problem (2.3.5), in particular its brachistocrone functional is still finite. Thus, we have

$$\int_0^1 \frac{\sqrt{1 + |v'|^2}}{\sqrt{v}}\,dx \geq \int_0^1 \frac{\sqrt{1 + |u_1'|^2}}{\sqrt{u_1}}\,dx = \int_0^1 \frac{\sqrt{1 + |u_2'|^2}}{\sqrt{u_2}}\,dx. \tag{2.3.7}$$

We now introduce the function of two variables

$$F(x_1, x_2) = \sqrt{\frac{1}{x_1^2} + 4\,x_2^2}, \qquad \text{for } (x_1, x_2) \in (0, +\infty) \times \mathbb{R},$$

and observe *that this is strictly convex*, thanks to Problem 2.8.4 below. By means of the change of variable (2.3.6), we can rewrite

$$\begin{aligned} \int_0^1 \frac{\sqrt{1 + |u_1'|^2}}{\sqrt{u_1}}\,dx &= \int_0^1 \frac{\sqrt{1 + 4\,|U_1|^2\,|U_1'|^2}}{U_1}\,dx \\ &= \int_0^1 \sqrt{\frac{1}{|U_1|^2} + 4\,|U_1'|^2}\,dx = \int_0^1 F(U_1, U_1')\,dx, \end{aligned}$$

and

$$\int_0^1 \frac{\sqrt{1 + |u_2'|^2}}{\sqrt{u_2}}\,dx = \int_0^1 \sqrt{\frac{1}{|U_2|^2} + 4\,|U_2'|^2}\,dx = \int_0^1 F(U_2, U_2')\,dx.$$

Similarly, by definition of v and V, we have

$$\begin{aligned}\int_0^1 \frac{\sqrt{1+|v'|^2}}{\sqrt{v}}\,dx &= \int_0^1 \sqrt{\frac{1}{|V|^2}+4\,|V'|^2}\,dx\\ &= \int_0^1 \sqrt{\frac{1}{\left|\dfrac{U_1+U_2}{2}\right|^2}+4\left|\frac{U_1'+U_2'}{2}\right|^2}\,dx\\ &= \int_0^1 F\left(\frac{U_1+U_2}{2},\frac{U_1'+U_2'}{2}\right)dx.\end{aligned}$$

Thus, from the previous computations and the strict convexity of F, we get

$$\begin{aligned}\int_0^1 \frac{\sqrt{1+|v'|^2}}{\sqrt{v}}\,dx &= \int_0^1 F\left(\frac{U_1+U_2}{2},\frac{U_1'+U_2'}{2}\right)dx\\ &< \frac{1}{2}\int_0^1 F(U_1,U_1')\,dx+\frac{1}{2}\int_0^1 F(U_2,U_2')\,dx\\ &= \frac{1}{2}\int_0^1 \frac{\sqrt{1+|u_1'|^2}}{\sqrt{u_1}}\,dx+\frac{1}{2}\int_0^1 \frac{\sqrt{1+|u_2'|^2}}{\sqrt{u_2}}\,dx.\end{aligned}$$

This contradicts (2.3.7), thus we must have $u_1 = u_2$. □

Remark 2.3.2 We observe that the trick used in the previous proof, i.e. the substitution $U = \sqrt{u}$, is essential in order to detect an *hidden convexity* structure in our problem. On the contrary, it is easy to see that the function

$$G(x_1,x_2) = \frac{\sqrt{1+x_2^2}}{\sqrt{x_1}}, \qquad \text{for } (x_1,x_2)\in(0,+\infty)\times\mathbb{R},$$

which is such that

$$\int_0^1 \frac{\sqrt{1+|u'|^2}}{\sqrt{u}}\,dx = \int_0^1 G(u,u')\,dx,$$

is *not* convex. Indeed, the determinant of its Hessian matrix is given by

$$\det D^2G(x_1,x_2) = \frac{3-x_2^2}{4\,x_1^3\,(1+x_2^2)},$$

which changes sign, according to whether $|x_2| \geq \sqrt{3}$ or $|x_2| < \sqrt{3}$. Thus by Proposition 1.3.8 we get that G is not convex.

The following result gives a sufficient condition to minimality. It is in the same spirit of the *Dirichlet principle* (Theorems 1.5.1 and 1.5.3), since it permits to say that the solution of a suitable differential equation (provided it exists!) is a minimizer of our functional. See also the comment of Remark 2.3.4 below.

Proposition 2.3.3 (Sufficient Conditions I) *Let $u \in C^0([0, 1]) \cap C^2((0, 1))$ be a function such that*

$$u(x) > 0 \quad \text{for every } x \in (0, 1], \quad u(0) = 0, \quad u(1) = u_0 \quad \text{and} \quad \int_0^1 \frac{1}{u}\, dx < +\infty.$$

If it satisfies the differential equation

$$-2\, u''(x)\, u(x) = 1 + |u'(x)|^2, \qquad \text{for every } x \in (0, 1), \tag{2.3.8}$$

then u is the unique minimizer of (2.3.5).

Proof It is sufficient to show that u is a minimizer, since uniqueness has been already proved in Proposition 2.3.1. We first observe that (2.3.8) implies that there exists a constant $C > 0$ such that

$$1 + |u'(x)|^2 = \frac{C}{u(x)}, \qquad \text{for every } x \in (0, 1). \tag{2.3.9}$$

Indeed, we can rewrite (2.3.8) as follows

$$2\, u''\, u + (1 + |u'|^2) = 0, \qquad \text{in } (0, 1).$$

By multiplying both sides by u', we get

$$2\, u''\, u'\, u + u'\, (1 + |u'|^2) = 0, \qquad \text{in } (0, 1).$$

On the left-hand side, we can now recognize a derivative: indeed, the previous equation is the same as

$$\Big(u\, (1 + |u'|^2)\Big)' = 0, \qquad \text{in } (0, 1).$$

The Eq. (2.3.9) now easily follows. By (2.3.9) we get in particular

$$\int_0^1 \frac{\sqrt{1 + |u'|^2}}{\sqrt{u}}\, dx = \int_0^1 \frac{\sqrt{C}}{u}\, dx < +\infty,$$

thus the brachistocrone functional is finite for u.

In order to show that u is a minimizer, let φ be an admissible function in problem (2.3.5): we are going to use the same trick as in Proposition 2.3.1. We set

$$U(x) = \sqrt{u(x)} \qquad \text{and} \qquad \phi(x) = \sqrt{\varphi(x)},$$

then we have

$$\int_0^1 \frac{\sqrt{1+|\varphi'|^2}}{\sqrt{\varphi}}\, dx = \int_0^1 \sqrt{\frac{1}{|\phi|^2} + 4\,|\phi'|^2}\, dx.$$

For simplicity, we keep on indicating by F the strictly convex function of Proposition 2.3.1, i.e.

$$F(x_1, x_2) = \sqrt{\frac{1}{x_1^2} + 4\,x_2^2}, \qquad \text{for } (x_1, x_2) \in (0, +\infty) \times \mathbb{R}.$$

We can then rewrite the previous equation as

$$\int_0^1 \frac{\sqrt{1+|\varphi'|^2}}{\sqrt{\varphi}}\, dx = \int_0^1 F(\phi, \phi')\, dx.$$

From the "above tangent" property of convex functions (i.e. Proposition 1.3.4), we get

$$\begin{aligned} F(\phi, \phi') &\geq F(U, U') + \left\langle \nabla F(U, U'), \begin{bmatrix} \phi - U \\ \phi' - U' \end{bmatrix} \right\rangle \\ &= F(U, U') + \frac{\partial F}{\partial x_1}(U, U')\,(\phi - U) + \frac{\partial F}{\partial x_2}(U, U')\,(\phi' - U'). \end{aligned}$$

By integrating over (0, 1) this inequality, we get

$$\begin{aligned} \int_0^1 \frac{\sqrt{1+|\varphi'|^2}}{\sqrt{\varphi}}\, dx &= \int_0^1 F(\phi, \phi')\, dx \\ &\geq \int_0^1 F(U, U')\, dx \\ &\quad + \int_0^1 \frac{\partial F}{\partial x_1}(U, U')\,(\phi - U)\, dx + \int_0^1 \frac{\partial F}{\partial x_2}(U, U')\,(\phi' - U')\, dx \\ &= \int_0^1 \frac{\sqrt{1+|u'|^2}}{\sqrt{u}}\, dx \\ &\quad + \int_0^1 \frac{\partial F}{\partial x_1}(U, U')\,(\phi - U)\, dx + \int_0^1 \frac{\partial F}{\partial x_2}(U, U')\,(\phi' - U')\, dx. \end{aligned}$$

In order to prove the minimality of u, it is thus sufficient to prove that the sum of the last two integrals vanishes, i.e. that

$$\int_0^1 \left[\frac{\partial F}{\partial x_1}(U, U')\,(\phi - U) + \frac{\partial F}{\partial x_2}(U, U')\,(\phi' - U') \right] dx = 0. \tag{2.3.10}$$

The gradient of F is given by

$$\frac{\partial F}{\partial x_1}(x_1, x_2) = -\frac{1}{x_1^3} \frac{1}{\sqrt{\frac{1}{x_1^2} + 4\,x_2^2}} = -\frac{1}{x_1^3} \frac{1}{F(x_1, x_2)},$$

and

$$\frac{\partial F}{\partial x_2}(x_1, x_2) = \frac{4\,x_2}{\sqrt{\frac{1}{x_1^2} + 4\,x_2^2}} = \frac{4\,x_2}{F(x_1, x_2)}.$$

Then the left-hand side of (2.3.10) becomes

$$\int_0^1 \left[-\frac{1}{U^3} \frac{1}{F(U, U')}\,(\phi - U) + \frac{4\,U'}{F(U, U')}\,(\phi' - U') \right] dx. \tag{2.3.11}$$

We now observe that, by using the definition of F, the relation $U = \sqrt{u}$ and Eq. (2.3.9), we get

$$F(U, U') = \sqrt{\frac{1}{|U|^2} + 4\,|U'|^2} = \frac{\sqrt{1 + 4\,|U|^2\,|U'|^2}}{U} = \frac{1}{U} \frac{\sqrt{C}}{U},$$

thus we can rewrite (2.3.11) as

$$\frac{1}{\sqrt{C}} \int_0^1 \left[-\frac{1}{U}\,(\phi - U) + 4\,U'\,U^2\,(\phi' - U') \right] dx.$$

We now integrate by parts the second term in the integrand: by observing that

$$\phi(0) - U(0) = 0 = \phi(1) - U(1),$$

the previous integral becomes

$$\frac{1}{\sqrt{C}} \int_0^1 \left[-\frac{1}{U} - (4\,U'\,U^2)' \right] (\phi - U)\,dx.$$

By recalling that $U = \sqrt{u}$, we get

$$
\begin{aligned}
-\frac{1}{U} - (4\,U'\,U^2)' &= -\frac{1}{U} - 4\,U''\,U^2 - 8\,(U')^2\,U \\
&= -\frac{1}{\sqrt{u}} - 4\,(\sqrt{u})''\,u - 8\left((\sqrt{u})'\right)^2\sqrt{u} \\
&= -\frac{1}{\sqrt{u}} - 2\left(\frac{u'}{\sqrt{u}}\right)'\,u - 2\left(\frac{u'}{\sqrt{u}}\right)^2\sqrt{u} \\
&= -\frac{1}{\sqrt{u}} - 2\,\frac{u''}{\sqrt{u}}\,u + \frac{(u')^2}{\sqrt{u}} - 2\,\frac{(u')^2}{\sqrt{u}} \\
&= -\frac{(1+(u')^2) + 2\,u''\,u}{\sqrt{u}},
\end{aligned}
$$

and the latter identically vanishes on the interval $(0, 1)$, thanks to the assumption (2.3.8) on u. This finally shows the validity of (2.3.10) and thus the minimality of u. □

Remark 2.3.4 We observe that, with the language of Sect. 1.5 of Chap. 1, Eq. (2.3.8) is nothing but the Euler-Lagrange equation for the functional

$$
\mathcal{F}(u) = \int_0^1 \frac{\sqrt{1+|u'|^2}}{\sqrt{u}}\,dx,
$$

see Problem 2.8.5 below. By recalling that a solution of the Euler-Lagrange equation is a critical point of $\mathcal{F}$, the idea behind the previous result is that, since our problem is convex in a suitable sense, any critical point must be a minimizer.

Actually, by slightly enforcing the assumptions on u, the first order differential Eq. (2.3.9) is still sufficient to give minimality. This is the content of the following

Corollary 2.3.5 (Sufficient Conditions II) *Let $u \in C^0([0,1]) \cap C^2((0,1))$ be a function such that*

$$
u(x) > 0 \quad \textit{for every } x \in (0,1], \quad u(0) = 0, \quad u(1) = u_0 \quad \textit{and} \quad \int_0^1 \frac{1}{u}\,dx < +\infty.
$$

If in addition we suppose that $u'(x) > 0$ for every $x \in (0,1)$ and it satisfies (2.3.9) *for some constant $C > 0$, then u is the unique minimizer of* (2.3.5).

Proof We multiply both sides of (2.3.9) by $u(x)$. This yields

$$
u(x)\,(1 + |u'(x)|^2) = C, \qquad \text{for every } x \in (0,1).
$$

By differentiating, we also get

$$2\,u'(x)\,u''(x)\,u(x) + u'(x)\,(1 + |u'(x)|^2) = 0.$$

If we now simplify the factor $u'(x)$ (which never vanishes on $(0, 1)$, by assumption), we get

$$-\,2\,u''(x)\,u(x) = 1 + |u'(x)|^2, \qquad \text{for every } x \in (0, 1),$$

so that (2.3.8) is satisfied, as well. We then conclude thanks to Proposition 2.3.3. □

We are ready to give an existence result for the brachistocrone problem.

Theorem 2.3.6 *Suppose that*

$$u_0 \geq \frac{2}{\pi},$$

then problem (2.3.5) *admits a unique solution, given by the monotone increasing concave function*

$$u(x) = K^{-1}(x), \qquad \textit{where } K(t) = \int_0^t \sqrt{\frac{\tau}{C_0 - \tau}}\,d\tau, \tag{2.3.12}$$

where the constant $C_0 \geq u_0$ is chosen so that the final condition $u(1) = u_0$ is satisfied. Moreover, the graph of u is an arc of cycloid.

Proof In light of Corollary 2.3.5, it is sufficient to show that the function u belongs to $C^0([0, 1]) \cap C^2((0, 1))$, it has the following properties

$$u(x) > 0 \quad \text{for every } x \in (0, 1], \qquad u(0) = 0, \qquad u(1) = u_0,$$

$$\int_0^1 \frac{1}{u(x)}\,dx < +\infty, \qquad \text{and} \qquad u'(x) > 0, \quad \text{for every } x \in (0, 1),$$

and it solves the differential Eq. (2.3.9). We divide the proof in two parts, for ease of readability.

Step 1: Minimality We first have to verify that u is well-defined, i.e. it is possible to choose $C_0 \geq u_0$ so that the function K defined in (2.3.12) has the following properties

$$K : [0, u_0] \to [0, 1], \qquad \text{with } K(0) = 0,\; K(u_0) = 1,$$

and it is continuous and invertible. Observe that K is a monotone increasing function on the interval $[0, u_0)$, with

$$K'(t) = \sqrt{\frac{t}{C_0 - t}} > 0, \qquad \text{for } t \in (0, u_0) \subseteq (0, C_0),$$

and $K(0) = 0$. Thus, it is continuous and invertible as a function defined on $[0, u_0)$. Moreover, we observe that by Problem 2.8.6 we have

$$K(u_0) = \int_0^{u_0} \sqrt{\frac{\tau}{C_0 - \tau}}\, d\tau = -\sqrt{u_0\,(C_0 - u_0)} + C_0 \arctan\sqrt{\frac{u_0}{C_0 - u_0}}.$$

It is easily seen that the function

$$C \mapsto -\sqrt{u_0\,(C - u_0)} + C \arctan\sqrt{\frac{u_0}{C - u_0}},$$

is continuous on $(u_0, +\infty)$, with

$$\lim_{C \to u_0^+} \left[-\sqrt{u_0\,(C - u_0)} + C \arctan\sqrt{\frac{u_0}{C - u_0}}\right] = u_0\,\frac{\pi}{2},$$

and

$$\lim_{C \to +\infty} \left[-\sqrt{u_0\,(C - u_0)} + C \arctan\sqrt{\frac{u_0}{C - u_0}}\right] = 0.$$

Thus, if $u_0 > 2/\pi$, by the Intermediate Value Theorem it is possible to choose $C_0 > u_0$ such that

$$K(u_0) = \int_0^{u_0} \sqrt{\frac{\tau}{C_0 - \tau}}\, d\tau = 1.$$

In the limit case $u_0 = 2/\pi$, we get the desired property by choosing $C_0 = u_0$, since the above discussion gives

$$\begin{aligned} K(u_0) &= \int_0^{u_0} \sqrt{\frac{\tau}{u_0 - \tau}}\, d\tau \\ &= \lim_{C \to u_0^+} \left[-\sqrt{u_0\,(C - u_0)} + C \arctan\sqrt{\frac{u_0}{C - u_0}}\right] = u_0\,\frac{\pi}{2} = 1. \end{aligned}$$

This finally permits to prove that K has the desired properties. In particular, we have

$$u = K^{-1} \in C^0([0, 1]) \qquad \text{with} \qquad u(0) = K^{-1}(0) = 0,\ u(1) = K^{-1}(1) = u_0.$$

Moreover, it is C^1 on $(0, 1]$, with

$$u'(x) > 0, \quad \text{for } x \in (0, 1) \qquad \text{and} \qquad \lim_{x \to 0^+} u'(x) = +\infty,$$

and by construction $u > 0$ on the interval $(0, 1]$. We finally observe that

$$K''(t) = \frac{C_0}{2\sqrt{t}}(C_0 - t)^{-\frac{3}{2}} > 0, \qquad \text{for } t \in (0, u_0).$$

Thus, K is strictly convex and consequently $u = K^{-1}$ is $C^2((0, 1))$ and strictly concave. We still need to prove that $1/u$ is integrable over $(0, 1)$. To this aim, it is sufficient to prove that

$$u(x) \sim \left(\frac{3\sqrt{C_0}}{2}\right)^{\frac{2}{3}} x^{\frac{2}{3}}, \qquad \text{for } x \to 0^+.$$

Observe that we have

$$\lim_{x\to 0^+} \frac{u(x)}{x^{\frac{2}{3}}} = \lim_{x\to 0^+} \frac{K^{-1}(x)}{x^{\frac{2}{3}}} = \lim_{t\to 0^+} \frac{t}{\big(K(t)\big)^{\frac{2}{3}}}.$$

On the other hand, by using the definition of K and L'Hôpital's rule, we have

$$\lim_{t\to 0^+} \frac{K(t)}{t^{\frac{3}{2}}} = \frac{2}{3}\lim_{t\to 0^+} \frac{1}{\sqrt{t}}\sqrt{\frac{t}{C_0 - t}} = \frac{2}{3\sqrt{C_0}}.$$

By using this information in the limit above, we get

$$\lim_{x\to 0^+} \frac{u(x)}{x^{\frac{2}{3}}} = \lim_{t\to 0^+} \frac{t}{\big(K(t)\big)^{\frac{2}{3}}} = \left(\frac{3\sqrt{C_0}}{2}\right)^{\frac{2}{3}},$$

as claimed.

We can now show that u satisfies (2.3.9). We start by observing that by definition, we have

$$K(u(x)) = x, \qquad \text{for every } x \in [0, 1].$$

By differentiating this identity, we get

$$K'(u(x))\, u'(x) = 1 \qquad \text{that is} \qquad \sqrt{\frac{u(x)}{C_0 - u(x)}}\, u'(x) = 1,$$

thanks to the definition of K. With simple manipulations, from the previous identity we get

$$|u'(x)|^2 = \frac{C_0}{u(x)} - 1 \qquad \text{that is} \qquad 1 + |u'(x)|^2 = \frac{C_0}{u(x)},$$

which shows that u satisfies (2.3.9).

Step 2: Arc of Cycloid We recall at first that an arc of cycloid of radius $R > 0$ can be parametrized by the regular curve

$$\begin{cases} x(t) = R\,(t - \sin t), \\ y(t) = R\,(1 - \cos t), \end{cases} \qquad \text{for } t \in \mathbb{R}.$$

We observe that the function $F(t) = t - \sin t$ is invertible on $\mathbb{R}$ and its inverse function $F^{-1} : \mathbb{R} \to \mathbb{R}$ is infinitely differentiable, except at the points of the form $2\,k\,\pi$, for $k \in \mathbb{Z}$. Thus, the cycloid can be seen as the graph of the function

$$y(x) = R\left(1 - \cos\left(F^{-1}\left(\frac{x}{R}\right)\right)\right),$$

which is infinitely differentiable on each interval of the form

$$(2\,k\,\pi\,R, 2\,(k+1)\,\pi\,R), \qquad \text{for } k \in \mathbb{Z}.$$

We choose $k = 0$ and consider the function

$$v(x) = R\left(1 - \cos\left(F^{-1}\left(\frac{x}{R}\right)\right)\right), \qquad \text{for } x \in [0, 2\,\pi\,R].$$

In order to conclude, we just need to show that it is possible to choose $R \geq 1/\pi$ so that

$$v \equiv u, \qquad \text{in } [0, 1].$$

This in turn will be proved by appealing again to Corollary 2.3.5, i.e. we will show that v satisfies the sufficient conditions for minimality. This will give that v would be another minimizer and thus by uniqueness it must coincide with u.

We first observe that for every $R \geq 1/\pi$ we have

$$[0, 1] \subseteq [0, \pi\,R],$$

and thus by construction $v \in C^0([0, 1]) \cap C^2((0, 1))$. Moreover, we have

$$v(x) > 0, \quad \text{for } x \in (0, 1],$$

and

$$v(0) = 0, \qquad \text{since } F^{-1}(0) = 0.$$

The verification of the boundary condition $v(1) = u_0$ is slightly more delicate and it is precisely here that the choice of the radius R enters: we need to impose that

$$v(1) = R\left(1 - \cos\left(F^{-1}\left(\frac{1}{R}\right)\right)\right) = u_0. \tag{2.3.13}$$

We observe that the function

$$R \mapsto R\left(1 - \cos\left(F^{-1}\left(\frac{1}{R}\right)\right)\right),$$

is continuous and such that

$$\lim_{R\to\left(\frac{1}{\pi}\right)^+} R\left(1 - \cos\left(F^{-1}\left(\frac{1}{R}\right)\right)\right) = \frac{1}{\pi}\left(1 - \cos\left(F^{-1}(\pi)\right)\right) = \frac{2}{\pi},$$

and

$$\lim_{R\to+\infty} R\left(1 - \cos\left(F^{-1}\left(\frac{1}{R}\right)\right)\right) = \lim_{\tau\to 0^+} \frac{1-\cos\tau}{F(\tau)} = +\infty.$$

To compute the last limit, we used the change of variable $R = 1/F(\tau)$. Thus, if $u_0 > 2/\pi$, by the Intermediate Value Theorem it is possible to choose $R > 1/\pi$ such that (2.3.13) holds. In the limit case $u_0 = 2/\pi$, we get the desired property by choosing $R = 1/\pi$.

We have to show that $1/v$ is integrable on $(0, 1)$. By using the change of variable $x = R\,F(\tau) = R\,(\tau - \sin\tau)$, we have

$$\begin{aligned}\int_0^1 \frac{1}{v(x)}\,dx &= \frac{1}{R}\int_0^1 \frac{1}{\left(1 - \cos\left(F^{-1}\left(\frac{x}{R}\right)\right)\right)}\,dx \\ &= \int_0^{F^{-1}\left(\frac{1}{R}\right)} \frac{1-\cos\tau}{1-\cos\tau}\,d\tau = F^{-1}\left(\frac{1}{R}\right) < +\infty,\end{aligned}$$

as desired.

We now compute the first derivative of v: for $x \in (0, 1)$ this is given by

$$v'(x) = R\,\sin\left(F^{-1}\left(\frac{x}{R}\right)\right)\frac{d}{dx}F^{-1}\left(\frac{x}{R}\right) = \frac{\sin\left(F^{-1}\left(\frac{x}{R}\right)\right)}{1 - \cos\left(F^{-1}\left(\frac{x}{R}\right)\right)} > 0.$$

Thus, we get

$$
\begin{aligned}
1+|v'(x)|^2 &= 1+\frac{\sin^2\left(F^{-1}\left(\frac{x}{R}\right)\right)}{\left(1-\cos\left(F^{-1}\left(\frac{x}{R}\right)\right)\right)^2}\\
&= 1+\frac{1-\cos^2\left(F^{-1}\left(\frac{x}{R}\right)\right)}{\left(1-\cos\left(F^{-1}\left(\frac{x}{R}\right)\right)\right)^2}\\
&= 1+\frac{1+\cos\left(F^{-1}\left(\frac{x}{R}\right)\right)}{1-\cos\left(F^{-1}\left(\frac{x}{R}\right)\right)}=\frac{2}{1-\cos\left(F^{-1}\left(\frac{x}{R}\right)\right)}=\frac{2\,R}{v(x)}.
\end{aligned}
$$

This shows that v solves (2.3.9) with $C = 2\,R$. As previously explained, this is enough to conclude the proof. □

Remark 2.3.7 The definition (2.3.12) for the minimizer u may look complicated at first sight, but there is nothing mysterious: this is the expression that one obtains when trying to solve the ordinary differential Eq. (2.3.9). This goes as follows: from (2.3.9) we get

$$
|u'(x)|^2 = \frac{C}{u(x)} - 1 \qquad \text{and thus} \qquad |u'(x)| = \sqrt{\frac{C}{u(x)} - 1}.
$$

If we suppose that $u(x) < C$ and $u'(x) > 0$ (which, a posteriori, is guaranteed by the choice $u_0 \geq 2/\pi$), this can be rephrased as

$$
\sqrt{\frac{u(x)}{C-u(x)}}\,u'(x) = 1, \qquad \text{that is} \qquad \frac{d}{dx}K(u(x)) = 1,
$$

if we recall the expression of K. We point out that, for simplicity, we avoided the direct construction of the solution u in the case

$$
u_0 < \frac{2}{\pi}.
$$

In this case, the construction would be slightly more complicated, since it is no more monotone increasing, but it has two intervals of monotonicity (see Fig. 2.1).

We also refer to [1, Capitolo 9, Sezione 7] for a complete treatment of the brachistocrone problem, by using a different approach.

Remark 2.3.8 We observe in particular that the solution of problem (2.3.5) is not given by the linear function $u(x) = u_0\,x$: indeed, since we know that a solution exists and is unique, in order to disprove the minimality of this linear function it is sufficient to observe that it does not solve the Euler-Lagrange equation (2.3.8). By going back to the original formulation of the brachistocrone problem

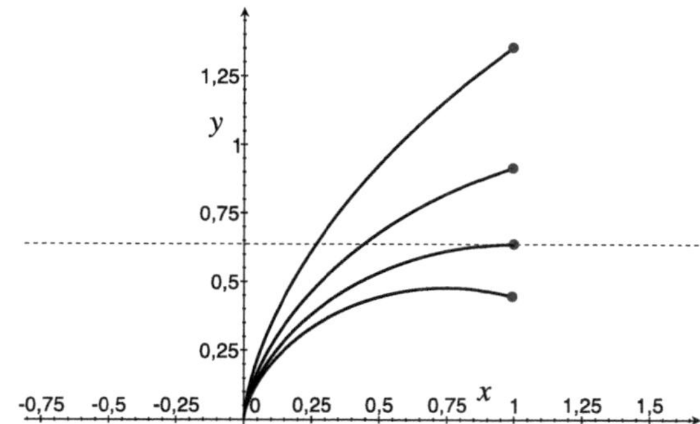

Fig. 2.1 The graphs of the solutions obtained in Theorem 2.3.6, according to different boundary conditions $u_0 > 0$. The dashed line corresponds to the limit case $u_0 = 2/\pi$

$$\inf_{f \in C^1((0,1)) \cap C^0([0,1])} \left\{ \int_0^1 \frac{\sqrt{1+|f'(x)|^2}}{\sqrt{2\,g\,(u_0 - f(x))}}\,dx \;:\; \begin{array}{c} f(0) = u_0,\ f(1) = 0 \\ f < u_0 \text{ in } (0,1] \end{array} \right\},$$

we thus get that the rectilinear path connecting $(0, u_0)$ to $(1, 0)$ is not the one minimizing the total traveling time.

2.4 Sharp Poincaré Constants

Let us suppose that u is a C^1 function on the interval $[0, 1]$, such that $u(0) = 0$. The intuition should suggest that if u' is everywhere small in absolute value, then u is everywhere small in absolute value, as well. Viceversa, if u assume large values somewhere on the interval, correspondingly the derivative u' should be large somewhere.

This intuition is correct and actually there is a quantitative way to express this property: this is the content of the so-called *Poincaré inequalities*. We will see that this kind of results remain true even if the property $u(0) = 0$ is replaced by other suitable conditions, which prevent functions "to assume large values without oscillating too much".

2.4.1 Functions Vanishing at the Endpoints

Lemma 2.4.1 *For every $\varphi \in C^1([0, 1])$ such that $\varphi(0) = 0$, we have the following Poincaré inequality*

$$2 \int_0^1 |\varphi(t)|^2\,dt \le \int_0^1 |\varphi'(t)|^2\,dt.$$

Proof We observe that

$$\varphi(t) = \varphi(t) - \varphi(0) = \int_0^t \varphi'(\tau)\,d\tau, \qquad \text{for every } t \in [0,1].$$

By taking the absolute value, we thus get

$$|\varphi(t)| = \left|\int_0^t \varphi'(\tau)\,d\tau\right| \leq \int_0^t |\varphi'(\tau)|\,d\tau.$$

We now raise both sides to power 2, integrate over [0, 1] and use Jensen's inequality (Proposition 1.2.11). This gives

$$\int_0^1 |\varphi(t)|^2\,dt \leq \int_0^1 \left(\int_0^t |\varphi'(\tau)|\,d\tau\right)^2 dt \leq \int_0^1 t \left(\int_0^t |\varphi'(\tau)|^2\,d\tau\right) dt.$$

If we now observe that

$$\int_0^t |\varphi'(\tau)|^2\,d\tau \leq \int_0^1 |\varphi'(\tau)|^2\,d\tau,$$

we finally obtain

$$\int_0^1 |\varphi(t)|^2\,dt \leq \left(\int_0^1 t\,dt\right)\left(\int_0^1 |\varphi'(\tau)|^2\,d\tau\right) = \frac{1}{2}\left(\int_0^1 |\varphi'(\tau)|^2\,d\tau\right),$$

as desired. □

We are now interested in determining the sharp constant in the previous Poincaré inequality. In other words, we want to determine the largest constant $c > 0$ such that

$$c \int_0^1 |\varphi(t)|^2\,dt \leq \int_0^1 |\varphi'(t)|^2\,dt, \qquad \text{for every } \varphi \in C^1([0,1]) \text{ such that } \varphi(0) = 0.$$

This is the content of the next

Theorem 2.4.2 *Let us set*

$$\sigma([0,1]) = \inf_{\varphi \in C^1([0,1])\setminus\{0\}} \left\{ \frac{\displaystyle\int_0^1 |\varphi'(t)|^2\,dt}{\displaystyle\int_0^1 |\varphi(t)|^2\,dt} : \varphi(0) = 0 \right\}.$$

Then we have

$$\sigma([0,1]) = \left(\frac{\pi}{2}\right)^2,$$

and this sharp constant is uniquely attained by any non-trivial multiple of the function

$$\phi(t) := \sin\left(\frac{\pi}{2}\,t\right), \qquad \textit{for } t \in [0,1].$$

Proof We split the proof in two parts: first we determine $\sigma([0,1])$ and then we show the claimed uniqueness of the minimizers.

Part 1: Determination of the Constant We already know from Lemma 2.4.1 that $\sigma([0,1]) > 0$, more precisely we know that $\sigma([0,1]) \geq 2$. Computing the exact value of the sharp constant in general is quite tricky. We first observe that by definition of sharp constant, we have

$$\sigma([0,1]) \leq \frac{\displaystyle\int_0^1 |\phi'(t)|^2\,dt}{\displaystyle\int_0^1 |\phi(t)|^2\,dt} = \left(\frac{\pi}{2}\right)^2 \frac{\displaystyle\int_0^1 \left|\cos\left(\frac{\pi}{2}\,t\right)\right|^2\,dt}{\displaystyle\int_0^1 \left|\sin\left(\frac{\pi}{2}\,t\right)\right|^2\,dt} = \left(\frac{\pi}{2}\right)^2,$$

where ϕ is the function in the statement. In order to conclude, we need to prove that

$$\left(\frac{\pi}{2}\right)^2 \leq \frac{\displaystyle\int_0^1 |\varphi'(t)|^2\,dt}{\displaystyle\int_0^1 |\varphi(t)|^2\,dt}, \qquad \text{for every } \varphi \in C^1([0,1]) \setminus \{0\} \text{ with } \varphi(0) = 0. \tag{2.4.1}$$

We observe that the function ϕ in the statement is such that

$$\begin{cases} -\phi''(t) = (\pi/2)^2\,\phi(t), \text{ for } t \in [0,1], \\ \phi(0) = 0, \\ \phi'(1) = 0. \end{cases}$$

We multiply this equation by $\varphi \in C^1([0,1])$ and integrate over $[0,1]$, so to get

$$-\int_0^1 \phi''(t)\,\varphi(t)\,dt = \left(\frac{\pi}{2}\right)^2 \int_0^1 \phi(t)\,\varphi(t)\,dt.$$

We now use an integration by parts on the first integral: we have

$$-\int_0^1 \phi''(t)\,\varphi(t)\,dt = -\Big[\phi'(t)\,\varphi(t)\Big]_0^1 + \int_0^1 \phi'(t)\,\varphi'(t)\,dt.$$

Thus, by using the boundary conditions of both φ and ϕ', we get

$$\left(\frac{\pi}{2}\right)^2 \int_0^1 \phi(t)\,\varphi(t)\,dt = \int_0^1 \phi'(t)\,\varphi'(t)\,dt, \tag{2.4.2}$$

which holds for every C^1 function φ, such that $\varphi(0) = 0$. This can be seen as the weak form of the differential equation satisfied by ϕ.

We take ψ any admissible function and use (2.4.2) with the choice

$$\varphi = \frac{\psi^2}{\phi}.$$

By using that $\phi > 0$ on $(0, 1]$, that $\phi'(0) \neq 0$ and a Taylor expansion centered at $t = 0$, it is not difficult to see that the latter is indeed a C^1 function on $[0, 1]$, such that $\varphi(0) = 0$. We thus obtain

$$\left(\frac{\pi}{2}\right)^2 \int_0^1 |\psi(t)|^2\,dt = \left(\frac{\pi}{2}\right)^2 \int_0^1 \phi(t)\,\frac{|\psi(t)|^2}{\phi(t)}\,dt = \int_0^1 \phi'(t)\left(\frac{|\psi(t)|^2}{\phi(t)}\right)'\,dt. \tag{2.4.3}$$

We can now use *Picone's inequality* for $p = 2$ (see Lemma 1.2.4), so to get

$$\left(\frac{\pi}{2}\right)^2 \int_0^1 |\psi(t)|^2\,dt \le \int_0^1 |\psi'(t)|^2\,dt.$$

This proves (2.4.1) and thus we have determined the value of $\sigma([0, 1])$.

Part 2: Uniqueness of the Minimizers We are left with proving that the only minimizers of our variational problem are given by

$$v(t) = C\,\phi(t), \qquad \text{with } C \in \mathbb{R},\ \phi(t) = \sin\left(\frac{\pi}{2}\,t\right).$$

Let us suppose that $v \in C^1([0, 1]) \setminus \{0\}$ is a minimizer, i.e. v solves

$$\inf_{\varphi\in C^1([0,1])\setminus\{0\}} \left\{ \frac{\displaystyle\int_0^1 |\varphi'(t)|^2\,dt}{\displaystyle\int_0^1 |\varphi(t)|^2\,dt} \,:\, \varphi(0) = 0 \right\} = \sigma([0, 1]) = \left(\frac{\pi}{2}\right)^2.$$

In particular, we have

$$\frac{\displaystyle\int_0^1 |v'(t)|^2\,dt}{\displaystyle\int_0^1 |v(t)|^2\,dt} = \left(\frac{\pi}{2}\right)^2. \tag{2.4.4}$$

By using (2.4.3) with $\psi = v$, we get

$$\left(\frac{\pi}{2}\right)^2 \int_0^1 |v(t)|^2\, dt = \int_0^1 \phi'(t) \left(\frac{|v(t)|^2}{\phi(t)}\right)' dt.$$

We now use *Picone's identity* (see Remark 1.2.5), so to get

$$\left(\frac{\pi}{2}\right)^2 \int_0^1 |v(t)|^2\, dt = \int_0^1 |v'(t)|^2\, dt - \int_0^1 \left|v(t)\,\frac{\phi'(t)}{\phi(t)} - v'(t)\right|^2 dt.$$

On account of (2.4.4), we get from the previous identity

$$\int_0^1 \left|v(t)\,\frac{\phi'(t)}{\phi(t)} - v'(t)\right|^2 dt = 0.$$

This in turns implies

$$v'(t) - v(t)\,\frac{\phi'(t)}{\phi(t)} = 0, \qquad \text{for } t \in (0, 1].$$

By multiplying the previous equation by $1/\phi(t)$, we get

$$\frac{v'(t)\,\phi(t) - v(v)\,\phi'(t)}{|\phi(t)|^2} = 0, \qquad \text{for } t \in (0, 1],$$

that is

$$\left(\frac{v(t)}{\phi(t)}\right)' = 0, \qquad \text{for } t \in (0, 1].$$

This implies that there exists a constant $C \neq 0$ such that

$$\frac{v(t)}{\phi(t)} = C, \qquad \text{for } t \in (0, 1].$$

This gives the desired conclusion. □

Remark 2.4.3 (How Did We Find the Optimal Function ϕ?) The guess of the optimal function ϕ in the previous proof may appear "a kind of magic", at first sight. We now explain how one could obtain it. By definition of $\sigma([0, 1])$, we have

$$\frac{\displaystyle\int_0^1 |\varphi'(t)|^2\,dt}{\displaystyle\int_0^1 |\varphi(t)|^2\,dt} \geq \sigma([0,1]), \qquad \text{for } \varphi \in C^1([0,1]) \setminus \{0\} \text{ such that } \varphi(0) = 0,$$

which implies that

$$\int_0^1 |\varphi'(t)|^2\,dt - \sigma([0,1]) \int_0^1 |\varphi(t)|^2\,dt \geq 0, \text{ for } \varphi \in C^1([0,1]) \text{ such that } \varphi(0) = 0.$$

On the other hand, if $\sigma([0,1])$ admits a minimizer v, we have

$$\frac{\displaystyle\int_0^1 |v'(t)|^2\,dt}{\displaystyle\int_0^1 |v(t)|^2\,dt} = \sigma([0,1]) \quad \text{that is} \quad \int_0^1 |v'(t)|^2\,dt - \sigma([0,1]) \int_0^1 |v(t)|^2\,dt = 0.$$

This simple argument shows that v is a minimizer of the functional

$$\mathcal{F}(\varphi) := \int_0^1 |\varphi'(t)|^2\,dt - \sigma([0,1]) \int_0^1 |\varphi(t)|^2\,dt,$$

as well, among C^1 functions vanishing at $t = 0$. This implies that for every $\varphi \in C^1([0,1])$ such that $\varphi(0) = 0$, we have

$$\mathcal{F}(v + \varepsilon\,\varphi) \geq \mathcal{F}(v), \qquad \text{for every } \varepsilon \in \mathbb{R}.$$

In other words, the function of one real variable

$$g(\varepsilon) = \mathcal{F}(v + \varepsilon\,\varphi),$$

attains its minimum at $\varepsilon = 0$. Thus, we must have

$$g'(0) = 0.$$

By recalling the definition of first variation of a functional (see Sect. 1.5 of Chap. 1), this entails that

$$\delta\mathcal{F}(v)[\varphi] = 0, \qquad \text{for every } \varphi \in C^1([0,1]) \text{ such that } \varphi(0) = 0,$$

i.e. v must be a critical point of $\mathcal{F}$. Let us compute this first variation: by expanding the squares, we have

$$
\begin{aligned}
g(\varepsilon) = \mathcal{F}(v + \varepsilon\,\varphi) = & \int_0^1 |v'(t)|^2\,dt + 2\,\varepsilon \int_0^1 v'(t)\,\varphi'(t)\,dt + \varepsilon^2 \int_0^1 |\varphi'(t)|^2\,dt \\
& - \sigma([0,1]) \int_0^1 |v(t)|^2\,dt - 2\,\sigma([0,1])\,\varepsilon \int_0^1 v(t)\,\varphi(t)\,dt \\
& - \sigma([0,1])\,\varepsilon^2 \int_0^1 |\varphi(t)|^2\,dt.
\end{aligned}
$$

By taking the derivative with respect to ε and computing it at $\varepsilon = 0$, we get

$$
0 = \delta\mathcal{F}(v)[\varphi] = 2 \int_0^1 \Big[v'(t)\,\varphi'(t) - \sigma([0,1])\,v(t)\,\varphi(t)\Big]\,dt. \tag{2.4.5}
$$

With the language already introduced in Chap. 1, this is the *Euler-Lagrange equation in weak form* associated to the functional $\mathcal{F}$. It expresses the optimality condition that needs to be satisfied by any minimizer v, provided it exists.

Let us study it a bit further, by obtaining the equation in classical form: supposing for a moment that v exists and is $C^2([0,1])$, an integration by parts in the first integral in (2.4.5) would lead to

$$
0 = \Big[v'(t)\,\varphi(t)\Big]_0^1 - \int_0^1 \Big[v''(t) + \sigma([0,1])\,v(t)\Big]\,\varphi(t)\,dt,
$$

for every $\varphi \in C^1([0,1])$ with $\varphi(0) = 0$. This gives

$$
\int_0^1 \Big[v''(t) + \sigma([0,1])\,v(t)\Big]\,\varphi(t)\,dt = v'(1)\,\varphi(1). \tag{2.4.6}
$$

In particular, this entails that

$$
\int_0^1 \Big[v''(t) + \sigma([0,1])\,v(t)\Big]\,\varphi(t)\,dt = 0, \qquad \text{for every } \varphi \in C_0^\infty((0,1)),
$$

and by the usual *Du Bois-Reymond Lemma* (see Lemma 1.4.1), we finally get that any minimizer v of class C^2 must satistify the ordinary differential equation

$$
-\,v''(t) = \sigma([0,1])\,v(t), \qquad \text{on } (0,1).
$$

This is the *Euler-Lagrange equation in classical form*, associated to our functional $\mathcal{F}$. In addition to the boundary condition

$$
v(0) = 0,
$$

taken at the beginning, from (2.4.6) we also get the following one

$$v'(1) = 0.$$

Indeed, once we obtained that

$$v''(t) + \sigma([0, 1])\, v(t) = 0, \qquad \text{for every } t \in (0, 1),$$

we can insert this information in (2.4.6) and get

$$0 = v'(1)\, \varphi(1), \qquad \text{for every } \varphi \in C^1([0, 1]) \text{ such that } \varphi(0) = 0.$$

By taking for example $\varphi(t) = t$ in the previous equation, we obtain the claimed boundary condition.

Observe that the parameter $\sigma([0, 1])$ entering in the functional $\mathcal{F}$ and in its Euler-Lagrange equation is not given, but actually *it is exactly what we need to determine!* However, it is now easy to see that any solution of

$$\begin{cases} -v''(t) = \sigma([0, 1])\, v(t), \ \text{for } t \in [0, 1], \\ v(0) = 0, \\ v'(1) = 0, \end{cases}$$

has the form

$$v(t) = A \sin\left(\sqrt{\sigma([0, 1])}\, t\right) + B \cos\left(\sqrt{\sigma([0, 1])}\, t\right),$$

with

$$B = 0 \qquad \text{and} \qquad A \cos\left(\sqrt{\sigma([0, 1])}\right) = 0. \tag{2.4.7}$$

This already explains why the optimal function ϕ in the previous result must be a sine function. Moreover, the second condition in (2.4.7) permits to identify the value of the optimal constant $\sigma([0, 1])$: indeed, if we are seeking for nontrivial solutions, we must take $A \neq 0$ and thus

$$\cos\left(\sqrt{\sigma([0, 1])}\right) = 0.$$

This implies that we must have

$$\sqrt{\sigma([0, 1])} = (2k + 1)\, \frac{\pi}{2} \qquad \text{i.e.} \qquad \sigma([0, 1]) = (2k + 1)^2 \left(\frac{\pi}{2}\right)^2.$$

The smallest possible value is obtained by taking $k = 0$, so that

$$\sigma([0, 1]) = \left(\frac{\pi}{2}\right)^2 \qquad \text{and} \qquad v(t) = A \sin\left(\frac{\pi}{2}\, t\right).$$

Corollary 2.4.4 *For two real numbers $a < b$, let us set*

$$\sigma([a,b]) = \inf_{\varphi \in C^1([a,b])\setminus\{0\}} \left\{ \frac{\displaystyle\int_a^b |\varphi'(t)|^2\,dt}{\displaystyle\int_a^b |\varphi(t)|^2\,dt} \,:\, \varphi(a) = 0 \right\}.$$

Then we have

$$\sigma([a,b]) = \left(\frac{\pi}{2}\right)^2 \frac{1}{(b-a)^2},$$

and this sharp constant is uniquely attained by any non-trivial multiple of the function

$$\phi(t) := \sin\left(\frac{\pi}{2\,(b-a)}\,(t-a)\right), \qquad \textit{for } t \in [a,b].$$

Proof We could repeat verbatim the previous proof, but there is a quicker way, just by using a scaling. We take $\varphi \in C^1([a,b]) \setminus \{0\}$ such that $\varphi(a) = 0$, then we define

$$\varphi_{a,b}(t) = \varphi\,(t\,(b-a) + a)\,, \qquad t \in [0,1].$$

This is admissible for the variational problem which defines $\sigma([0,1])$, thus we get

$$\begin{aligned}
\sigma([0,1]) &\le \inf_{\varphi \in C^1([a,b])} \frac{\displaystyle\int_0^1 |\varphi_{a,b}'(t)|^2\,dt}{\displaystyle\int_0^1 |\varphi_{a,b}(t)|^2\,dt} \\
&= \inf_{\varphi \in C^1([a,b])} \frac{(b-a)^2 \displaystyle\int_0^1 |\varphi'(t\,(b-a)+a)|^2\,dt}{\displaystyle\int_0^1 |\varphi(t\,(b-a)+a)|^2\,dt} \\
&= \inf_{\varphi \in C^1([a,b])} \frac{(b-a)^2 \displaystyle\int_a^b |\varphi'(\tau)|^2\,d\tau}{\displaystyle\int_a^b |\varphi(\tau)|^2\,d\tau} = (b-a)^2\,\sigma([a,b]).
\end{aligned}$$

In a similar way, we can take $\psi \in C^1([0,1]) \setminus \{0\}$ such that $\psi(0) = 0$ and define

$$\psi_{a,b}(t) = \psi\left(\frac{t-a}{b-a}\right), \qquad \text{for } t \in [a,b].$$

As before, we get

$$\sigma([a,b]) \le \inf_{\psi\in C^1([0,1])} \frac{\displaystyle\int_a^b |\psi'_{a,b}(t)|^2\,dt}{\displaystyle\int_a^b |\psi_{a,b}(t)|^2\,dt}$$

$$= \inf_{\psi\in C^1([0,1])} \frac{1}{(b-a)^2} \frac{\displaystyle\int_a^b |\psi'((t-a)/(b-a))|^2\,dt}{\displaystyle\int_a^b |\psi((t-a)/(b-a))|^2\,dt}$$

$$= \inf_{\psi\in C^1([0,1])} \frac{1}{(b-a)^2} \frac{\displaystyle\int_0^1 |\psi'(\tau)|^2\,d\tau}{\displaystyle\int_0^1 |\psi(\tau)|^2\,d\tau} = \frac{1}{(b-a)^2}\,\sigma([0,1]).$$

This finally gives

$$\sigma([a,b]) = \frac{1}{(b-a)^2}\,\sigma([0,1]) = \left(\frac{\pi}{2}\right)^2 \frac{1}{(b-a)^2},$$

as desired. □

Remark 2.4.5 It is not difficult to see that the previous sharp constant is unchanged if admissible functions vanish on the right endpoint b, rather than on the left one a. It is sufficient to observe that if $\varphi \in C^1([a,b])$ is such that $\varphi(b) = 0$, then

$$\widetilde{\varphi}(t) = \varphi((b-t)+a), \qquad \text{for } t \in [a,b],$$

is such that $\widetilde{\varphi}(a) = \varphi(b) = 0$ and

$$\int_a^b |\widetilde{\varphi}(t)|^2\,dt = \int_a^b |\varphi(t)|^2\,dt, \qquad \int_a^b |\widetilde{\varphi}'(t)|^2\,dt = \int_a^b |\varphi'(t)|^2\,dt.$$

Thus, we still have

$$\inf_{\varphi\in C^1([a,b])\setminus\{0\}} \left\{ \frac{\displaystyle\int_a^b |\varphi'(t)|^2\,dt}{\displaystyle\int_a^b |\varphi(t)|^2\,dt} \,:\, \varphi(b) = 0 \right\} = \left(\frac{\pi}{2}\right)^2 \frac{1}{(b-a)^2},$$

and this sharp constant is uniquely attained by any non-trivial multiple of the function

$$\phi(t) = \sin\left(\frac{\pi}{2\,(b-a)}\,(b-t)\right).$$

Let us now consider the following subset of functions

$$\mathcal{A} = \Big\{\varphi \in C^1([0,1])\ :\ \varphi(0) = \varphi(1) = 0\Big\}.$$

In light of the previous discussion, we have

$$\left(\frac{\pi}{2}\right)^2 \int_0^1 |u(t)|^2\,dt \le \int_0^1 |u'(t)|^2\,dt, \qquad \text{for every } \varphi \in \mathcal{A}.$$

This shows that if we set

$$\lambda([0,1]) = \inf_{\varphi \in C^1([0,1])\setminus\{0\}} \left\{ \frac{\displaystyle\int_0^1 |\varphi'(t)|^2\,dt}{\displaystyle\int_0^1 |\varphi(t)|^2\,dt}\ :\ \varphi(0) = \varphi(1) = 0 \right\},$$

then we have

$$\lambda([0,1]) \ge \left(\frac{\pi}{2}\right)^2 = \sigma([0,1]),$$

and in principle the strict inequality sign could hold, since the infimum on the left-hand side is performed on a narrower class.

Actually, this is exactly what happens. More precisely, we have the following

Theorem 2.4.6 *Let us set*

$$\lambda([0,1]) = \inf_{\varphi \in C^1([0,1])\setminus\{0\}} \left\{ \frac{\displaystyle\int_0^1 |\varphi'(t)|^2\,dt}{\displaystyle\int_0^1 |\varphi(t)|^2\,dt}\ :\ \varphi(0) = \varphi(1) = 0 \right\}.$$

Then we have

$$\lambda([0,1]) = \pi^2,$$

and this sharp constant is uniquely attained by any non-trivial multiple of the function

$$\phi(t) := \sin(\pi\, t), \qquad \textit{for } t \in [0,1].$$

Proof We first observe that

$$\frac{\displaystyle\int_0^1 |\phi'(t)|^2\,dt}{\displaystyle\int_0^1 |\phi(t)|^2\,dt} = \pi^2 \frac{\displaystyle\int_0^1 |\cos(\pi\, t)|^2\,dt}{\displaystyle\int_0^1 |\sin(\pi\, t)|^2\,dt} = \pi^2,$$

thus we have $\lambda([0,1]) \leq \pi^2$. In order to prove the reverse estimate, we could use a proof similar to that of Theorem 2.4.2, but we prefer to use a trick showing that we can actually reduce to the case of the constant $\sigma([0,1])$.

For every $\varphi \in C^1([0,1]) \setminus \{0\}$ such that $\varphi(0) = \varphi(1) = 0$, we get in particular that φ is admissible for the variational problem which defines $\sigma([0,1/2])$. Thus by Corollary 2.4.4 we get

$$\int_0^{\frac{1}{2}} |\varphi'(t)|^2\,dt \geq \sigma\left(\left[0, \frac{1}{2}\right]\right) \int_0^{\frac{1}{2}} |\varphi(t)|^2\,dt = \pi^2 \int_0^{\frac{1}{2}} |\varphi(t)|^2\,dt. \tag{2.4.8}$$

Similarly, we can observe that φ is C^1 on the interval $[1/2, 1]$ and vanishes on the right endpoint, as well. Thus, by Corollary 2.4.4 and Remark 2.4.5, we also have

$$\int_{\frac{1}{2}}^1 |\varphi'(t)|^2\,dt \geq \pi^2 \int_{\frac{1}{2}}^1 |\varphi(t)|^2\,dt. \tag{2.4.9}$$

By summing up (2.4.8) and (2.4.9), we get

$$\int_0^1 |\varphi'(t)|^2\,dt \geq \pi^2 \int_0^1 |\varphi(t)|^2\,dt,$$

which proves that $\lambda([0,1]) \geq \pi^2$, as well. This gives the value of $\lambda([0,1])$ and the optimality of ϕ.

In order to prove that each optimizer is necessarily a multiple of ϕ, we observe that if $v \in C^1([0,1]) \setminus \{0\}$ is a minimizer, then (2.4.8) and (2.4.9) must both hold as identities. By appealing to Corollary 2.4.4 and Remark 2.4.5, we thus get that there exist two constants $C_1, C_2 \neq 0$ such that

$$v(t) = C_1\,\sin(\pi\, t), \qquad \text{for } t \in \left[0, \frac{1}{2}\right],$$

and

$$v(t) = C_2\,\sin(\pi\,(1-t)) = C_2\,\sin(\pi\, t), \qquad \text{for } t \in \left[\frac{1}{2}, 1\right].$$

Since v is a C^1 function, we must have $C_1 = C_2$ and thus v is a multiple of ϕ. □

As before, with a simple scaling argument, we get a similar result for a generic interval, not necessarily of length 1. The proof is left to the reader.

Corollary 2.4.7 *Given two real numbers $a < b$, let us set*

$$\lambda([a,b]) = \inf_{\varphi \in C^1([a,b])\setminus\{0\}} \left\{ \frac{\displaystyle\int_a^b |\varphi'(t)|^2\,dt}{\displaystyle\int_a^b |\varphi(t)|^2\,dt} \,:\, \varphi(a) = \varphi(b) = 0 \right\}.$$

Then we have

$$\lambda([a,b]) = \frac{\pi^2}{(b-a)^2},$$

and this sharp constant is uniquely attained by any non-trivial multiple of the function

$$\phi(t) := \sin\left(\frac{\pi}{b-a}\,(t-a)\right), \qquad \textit{for } t \in [a,b].$$

2.4.2 Functions with Zero Average

We now consider the case of functions which do not necessarily vanish at the endpoints of the interval, but have zero average. More precisely, we start with the following

Lemma 2.4.8 *For every C^1 function $\varphi : [0,1] \to \mathbb{R}$ such that*

$$\int_0^1 \varphi(t)\,dt = 0,$$

we have

$$3 \int_0^1 |\varphi(t)|^2\,dt \le \int_0^1 |\varphi'(t)|^2\,dt.$$

Proof We take $t, s \in [0,1]$, then we have

$$\varphi(t) - \varphi(s) = \int_s^t \varphi'(\tau)\,d\tau.$$

We integrate over $s \in [0, 1]$ and use that $\int_0^1 \varphi(s)\,ds = 0$. This gives

$$\varphi(t) = \int_0^1 \int_s^t \varphi'(\tau)\,d\tau\,ds,$$

and we split the last integral as follows

$$\begin{aligned}\varphi(t) &= \int_0^t \int_s^t \varphi'(\tau)\,d\tau\,ds + \int_t^1 \int_s^t \varphi'(\tau)\,d\tau\,ds \\ &= \int_0^t \int_s^t \varphi'(\tau)\,d\tau\,ds - \int_t^1 \int_t^s \varphi'(\tau)\,d\tau\,ds.\end{aligned}$$

By taking the absolute value, we obtain in particular

$$\begin{aligned}|\varphi(t)| &\le \left|\int_0^t \int_s^t \varphi'(\tau)\,d\tau\,ds\right| + \left|\int_t^1 \int_t^s \varphi'(\tau)\,d\tau\,ds\right| \\ &\le \int_0^t \int_s^t |\varphi'(\tau)|\,d\tau\,ds + \int_t^1 \int_t^s |\varphi'(\tau)|\,d\tau\,ds.\end{aligned}$$

By introducing the notation (see Fig. 2.2)

$$\Sigma_t = \Big\{(s,\tau)\ :\ 0 \le s \le t,\ s \le \tau \le t\Big\} \cup \Big\{(s,\tau)\ :\ t \le s \le 1,\ t \le \tau \le s\Big\},$$

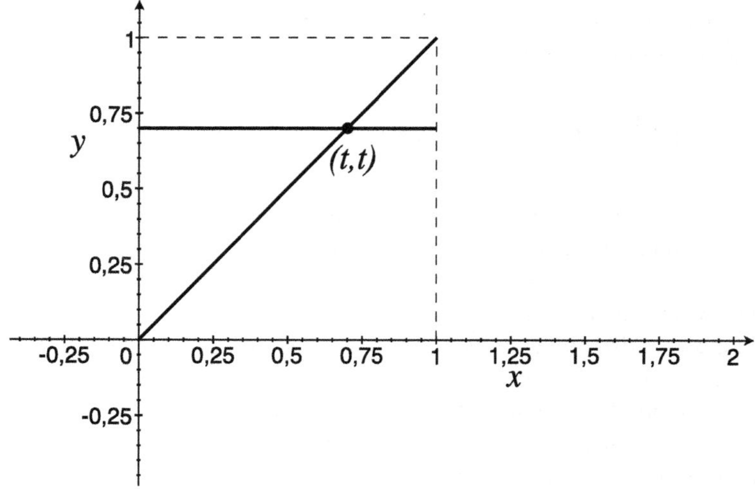

Fig. 2.2 The set Σ_t in the proof of Lemma 2.4.8 is given by the two triangles having the common vertex at (t, t)

the previous inequality can be rewritten as

$$|\varphi(t)| \leq \iint_{\Sigma_t} |\varphi'(\tau)|\, d\tau\, ds.$$

We take the square on both sides and use Jensen's inequality, so to obtain

$$|\varphi(t)|^2 \leq \left(\iint_{\Sigma_t} |\varphi'(\tau)|\, d\tau\, ds \right)^2 \leq |\Sigma_t| \iint_{\Sigma_t} |\varphi'(\tau)|^2\, d\tau\, ds. \tag{2.4.10}$$

We now observe that

$$|\Sigma_t| = \frac{t^2 + (1-t)^2}{2},$$

and

$$\iint_{\Sigma_t} |\varphi'(\tau)|^2\, d\tau\, ds \leq \int_0^1 \int_0^1 |\varphi'(\tau)|^2\, d\tau\, ds = \int_0^1 |\varphi'(\tau)|^2\, d\tau.$$

By inserting these two informations in (2.4.10) and integrating over $t \in [0, 1]$, we get the desired conclusion. □

Again, we are interested in obtaining the sharp constant in the previous Poincaré inequality. This is the content of the next

Theorem 2.4.9 *Let us set*

$$\mu([0,1]) = \inf_{\varphi \in C^1([0,1]) \setminus \{0\}} \left\{ \frac{\displaystyle\int_0^1 |\varphi'(t)|^2\, dt}{\displaystyle\int_0^1 |\varphi(t)|^2\, dt} \; : \; \int_0^1 \varphi(t)\, dt = 0 \right\}.$$

Then we have

$$\mu([0,1]) = \lambda([0,1]) = \pi^2,$$

and this sharp constant is uniquely attained by any non-trivial multiple of the function

$$\phi(t) := \cos(\pi\, t), \qquad \textit{for } t \in [0,1].$$

Proof We already know from Lemma 2.4.8 that $\mu([0,1]) > 0$, more precisely we have that $\mu([0,1]) \geq 3$. In order to compute the exact value of $\mu([0,1])$, we will use a trick which permits to reduce the problem to the constant $\sigma([0,1/2])$ of

Corollary 2.4.4. As before, we first observe that by definition of sharp constant, we have

$$\mu([0,1]) \le \frac{\displaystyle\int_0^1 |\phi'(t)|^2\,dt}{\displaystyle\int_0^1 |\phi(t)|^2\,dt} = \pi^2\,\frac{\displaystyle\int_0^1 |\sin(\pi\,t)|^2\,dt}{\displaystyle\int_0^1 |\cos(\pi\,t)|^2\,dt} = \pi^2.$$

In order to conclude, we need to prove that

$$\pi^2 \le \frac{\displaystyle\int_0^1 |\varphi'(t)|^2\,dt}{\displaystyle\int_0^1 |\varphi(t)|^2\,dt}, \qquad \text{for every } \varphi \in C^1([0,1]) \setminus \{0\} \text{ with } \int_0^1 \varphi(t)\,dt = 0. \tag{2.4.11}$$

We take $\varphi \in C^1([0,1])$, then we set

$$\widetilde{\varphi}(t) = \varphi(t) - \varphi\left(\frac{1}{2}\right), \qquad \text{for } t \in [0,1].$$

Observe that this can be regarded as a C^1 function on the intervals $[0,1/2]$ and $[1/2,1]$, vanishing at one of the endpoints. Thus, by Corollary 2.4.4 and Remark 2.4.5, we have

$$\begin{aligned}\pi^2 \int_0^{\frac{1}{2}} |\widetilde{\varphi}(t)|^2\,dt &= \sigma\left(\left[0,\frac{1}{2}\right]\right) \int_0^{\frac{1}{2}} |\widetilde{\varphi}(t)|^2\,dt \\ &\le \int_0^{\frac{1}{2}} |\widetilde{\varphi}'(t)|^2\,dt = \int_0^{\frac{1}{2}} |\varphi'(t)|^2\,dt,\end{aligned}$$

and

$$\pi^2 \int_{\frac{1}{2}}^1 |\widetilde{\varphi}(t)|^2\,dt = \sigma\left(\left[\frac{1}{2},1\right]\right) \int_{\frac{1}{2}}^1 |\widetilde{\varphi}(t)|^2\,dt \le \int_{\frac{1}{2}}^1 |\widetilde{\varphi}'(t)|^2\,dt = \int_{\frac{1}{2}}^1 |\varphi'(t)|^2\,dt.$$

By summing up these two inequalities and recalling the definition of $\widetilde{\varphi}$, we obtain

$$\pi^2 \int_0^1 \left|\varphi(t) - \varphi\left(\frac{1}{2}\right)\right|^2 dt \le \int_0^1 |\varphi'(t)|^2\,dt.$$

This is valid for every $\varphi \in C^1([0,1])$. We now further assume that $\int_0^1 \varphi(t)\,dt = 0$ and expand the square on the left-hand side: this yields

$$\pi^2 \int_0^1 |\varphi(t)|^2\,dt + \pi^2 \left|\varphi\left(\frac{1}{2}\right)\right|^2 \le \int_0^1 |\varphi'(t)|^2\,dt. \tag{2.4.12}$$

This in particular implies (2.4.11), as desired.

We are left with proving uniqueness of the minimizers. Let us suppose that ψ is another minimizer, thus we have

$$\pi^2 \int_0^1 |\psi(t)|^2\,dt = \int_0^1 |\psi'(t)|^2\,dt.$$

By comparing with (2.4.12), this implies that $\psi(1/2) = 0$. The proof above now gives that the restrictions of ψ to $[0, 1/2]$ and $[1/2, 1]$ must be minimizers of the two problems

$$\inf_{\varphi\in C^1([0,1/2])\setminus\{0\}} \left\{ \frac{\displaystyle\int_0^{\frac{1}{2}} |\varphi'(t)|^2\,dt}{\displaystyle\int_0^{\frac{1}{2}} |\varphi(t)|^2\,dt} \ :\ \varphi\left(\frac{1}{2}\right) = 0 \right\}$$

and

$$\inf_{\varphi\in C^1([1/2,1])\setminus\{0\}} \left\{ \frac{\displaystyle\int_{\frac{1}{2}}^{1} |\varphi'(t)|^2\,dt}{\displaystyle\int_{\frac{1}{2}}^{1} |\varphi(t)|^2\,dt} \ :\ \varphi\left(\frac{1}{2}\right) = 0 \right\},$$

respectively. By Corollary 2.4.4 and Remark 2.4.5, we then get that there exist two constants C_1, C_2 such that

$$\psi(t) = C_1 \sin\left(\pi\,\left(\frac{1}{2} - t\right)\right) = C_1 \cos(\pi\,t), \qquad \text{for } t \in \left[0, \frac{1}{2}\right],$$

and

$$\psi(t) = C_2 \sin\left(\pi\,\left(t - \frac{1}{2}\right)\right) = -C_2 \cos(\pi\,t), \qquad \text{for } t \in \left[\frac{1}{2}, 1\right].$$

Finally, since ψ is C^1 on the interval $[0, 1]$, we must have $C_2 = -C_1$. □

As usual, we can get a similar result for a general interval, by a simple scaling argument. The reader is invited to write down the details, as a useful exercise.

Corollary 2.4.10 *For two real numbers $a < b$, let us set*

$$\mu([a,b]) = \inf_{\varphi \in C^1([a,b]) \setminus \{0\}} \left\{ \frac{\displaystyle\int_a^b |\varphi'(t)|^2\, dt}{\displaystyle\int_a^b |\varphi(t)|^2\, dt} \ : \ \int_a^b \varphi(t)\, dt = 0 \right\}.$$

Then we have

$$\mu([a,b]) = \frac{\pi^2}{(b-a)^2},$$

and this sharp constant is uniquely attained by any non-trivial multiple of the function

$$\phi(t) = \cos\left(\frac{\pi}{b-a}\,(t-a)\right).$$

2.4.3 Periodic Functions with Zero Average

Finally, in addition to having zero average, we further impose that the functions on $[0, 1]$ are *periodic*, i.e. they assume the same values at the endpoints of the interval. We are thus interested in determining the sharp constant

$$\inf_{\varphi \in C^1([0,1]) \setminus \{0\}} \left\{ \frac{\displaystyle\int_0^1 |\varphi'(t)|^2\, dt}{\displaystyle\int_0^1 |\varphi(t)|^2\, dt} \ : \ \varphi(0) = \varphi(1),\ \int_0^1 \varphi(t)\, dt = 0 \right\}.$$

As anticipated in the Introduction to this chapter, the exact determination of such a constant is tightly connected with the celebrated *isoperimetric problem* in the plane.

Theorem 2.4.11 *Let us set*

$$\Lambda([0,1]) = \inf_{\varphi \in C^1([0,1]) \setminus \{0\}} \left\{ \frac{\displaystyle\int_0^1 |\varphi'(t)|^2\, dt}{\displaystyle\int_0^1 |\varphi(t)|^2\, dt} \ : \ \varphi(0) = \varphi(1),\ \int_0^1 \varphi(t)\, dt = 0 \right\}.$$

Then we have

$$\Lambda([0,1]) = 4\,\pi^2,$$

and this sharp constant is uniquely attained by any function of the form

$$C\,\phi(t-\alpha), \qquad \textit{with } C \in \mathbb{R}\setminus\{0\},\ \alpha \in \mathbb{R},$$

where

$$\phi(t) := \sin(2\,\pi\,t), \qquad \textit{for } t \in [0,1].$$

Proof We already know from Theorem 2.4.9 that the infimum above is not 0, since $\Lambda([0,1]) \geq \mu([0,1]) = \pi^2$. We also observe that, if ϕ is the function in the statement, then

$$\int_0^1 \phi(t)\,dt = 0 \qquad \text{and} \qquad \phi(0) = \phi(1) = 0,$$

thus by definition

$$\Lambda([0,1]) \leq \frac{\displaystyle\int_0^1 |\phi'(t)|^2\,dt}{\displaystyle\int_0^1 |\phi(t)|^2\,dt} = 4\,\pi^2\,\frac{\displaystyle\int_0^1 |\cos(2\,\pi\,t)|^2\,dt}{\displaystyle\int_0^1 |\sin(2\,\pi\,t)|^2\,dt} = 4\,\pi^2.$$

In order to conclude, we need to prove that for every $\varphi \in C^1([0,1])$ such that $\varphi(0) = \varphi(1)$ and $\int_0^1 \varphi\,dt = 0$, we have

$$4\,\pi^2 \leq \frac{\displaystyle\int_0^1 |\varphi'(t)|^2\,dt}{\displaystyle\int_0^1 |\varphi(t)|^2\,dt}.$$

Let us take an admissible function φ and observe that the function

$$z(t) = \varphi\left(t + \frac{1}{2}\right) - \varphi(t), \qquad \text{for } t \in \left[0, \frac{1}{2}\right],$$

is continuous and such that

$$z(0) = \varphi\left(\frac{1}{2}\right) - \varphi(0) = -\left(\varphi(1) - \varphi\left(\frac{1}{2}\right)\right) = -z\left(\frac{1}{2}\right).$$

Observe that we used that $\varphi(0) = \varphi(1)$, by assumption. By the Intermediate Values Theorem, this implies that there exists $t_0 \in [0, 1/2)$ such that

$$\varphi\left(t_0 + \frac{1}{2}\right) - \varphi(t_0) = z(t_0) = 0.$$

We now take $\psi(t) = \sin(2\pi(t - t_0))$ and consider the new function

$$V(t) = \begin{cases} \dfrac{(\varphi(t) - \varphi(t_0))^2}{\psi(t)}, & \text{for } t \in [0,1] \setminus \left\{t_0, t_0 + \dfrac{1}{2}\right\}, \\ 0, & \text{for } t = t_0 \text{ or } t = t_0 + \dfrac{1}{2}. \end{cases}$$

We observe that the function V is continuous. Indeed, by definition

$$\psi'(t_0) = 2\pi \cos(0) = 2\pi \neq 0.$$

Thus, by a simple Taylor expansion we have (recall that φ is C^1)

$$\begin{aligned} V(t) &= \frac{(\varphi'(t_0)\,(t - t_0))^2 + o((t - t_0)^2)}{\psi'(t_0)\,(t - t_0) + o(t - t_0)} \\ &= \frac{(\varphi'(t_0))^2\,(t - t_0) + o(t - t_0)}{\psi'(t_0) + o(1)} \\ &= \Big((\varphi'(t_0))^2\,(t - t_0) + o(t - t_0)\Big)\,\frac{1 + o(1)}{\psi'(t_0)} \\ &= \frac{(\varphi'(t_0))^2}{\psi'(t_0)}\,(t - t_0) + o(t - t_0), \qquad \text{for } t \to t_0. \end{aligned}$$

Similarly, we have

$$V(t) = \frac{\left(\varphi'\left(t_0 + \frac{1}{2}\right)\right)^2}{\psi'\left(t_0 + \frac{1}{2}\right)}\left(t - t_0 - \frac{1}{2}\right) + o\left(t - t_0 - \frac{1}{2}\right), \qquad \text{for } t \to t_0 + \frac{1}{2}.$$

This shows that

$$\lim_{t \to t_0} V(t) = 0 = V(t_0) \qquad \text{and} \qquad \lim_{t \to t_0 + \frac{1}{2}} V(t) = 0 = V\left(t_0 + \frac{1}{2}\right).$$

The previous expansions also show that V is differentiable at both $t = t_0$ and $t = t_0 + 1/2$, with

$$V'(t_0) = \frac{(\varphi'(t_0))^2}{\psi'(t_0)} = \frac{(\varphi'(t_0))^2}{2\pi},$$

$$V'\left(t_0 + \frac{1}{2}\right) = \frac{\left(\varphi'\left(t_0 + \frac{1}{2}\right)\right)^2}{\psi'\left(t_0 + \frac{1}{2}\right)} = -\frac{\left(\varphi'\left(t_0 + \frac{1}{2}\right)\right)^2}{2\pi}.$$

Moreover, it is not difficult to see that V' is continuous at these two points, thus $V \in C^1([0, 1])$. Finally, we have

$$V(0) = \frac{(\varphi(0) - \varphi(t_0))^2}{\psi(0)} = \frac{(\varphi(1) - \varphi(t_0))^2}{\psi(1)} = V(1), \tag{2.4.13}$$

thanks to the properties of φ and the 1-periodicity of ψ.

The proof now runs similarly to that of Theorem 2.4.2. We notice that ψ solves

$$-\psi''(t) = 4\pi^2\, \psi(t), \qquad \text{for } t \in [0, 1].$$

We multiply this equation by V, integrate over $[0, 1]$ and perform an integration by parts: this gives

$$4\pi^2 \int_0^1 \psi(t)\, V(t)\, dt = \Big[-\psi'(t)\, V(t)\Big]_0^1 + \int_0^1 \psi'(t)\, V'(t)\, dt = \int_0^1 \psi'(t)\, V'(t)\, dt.$$

In the second identity, we used (2.4.13) and the fact that

$$\psi'(1) = 2\pi\, \cos(2\pi\,(1 - t_0)) = 2\pi\, \cos(-2\pi\, t_0) = \psi'(0).$$

By recalling the definition of V and the fact that $\psi(t) \neq 0$ for $t \neq t_0$, this is the same as

$$\begin{aligned}
4\pi^2 \int_0^1 (\varphi(t) - \varphi(t_0))^2\, dt &= \int_0^1 \psi'(t) \left(\frac{(\varphi(t) - \varphi(t_0))^2}{\psi(t)}\right)' dt \\
&= \int_0^{t_0} \psi'(t) \left(\frac{(\varphi(t) - \varphi(t_0))^2}{\psi(t)}\right)' dt \\
&\quad + \int_{t_0}^{t_0+\frac{1}{2}} \psi'(t) \left(\frac{(\varphi(t) - \varphi(t_0))^2}{\psi(t)}\right)' dt \\
&\quad + \int_{t_0+\frac{1}{2}}^1 \psi'(t) \left(\frac{(\varphi(t) - \varphi(t_0))^2}{\psi(t)}\right)' dt.
\end{aligned}$$

We can now use Picone's inequality (see Lemma 1.2.4) on each integral in the right-hand side[1] and obtain

$$4\pi^2 \int_0^1 (\varphi(t) - \varphi(t_0))^2 \, dt \leq \int_0^1 |\varphi'(t)|^2 \, dt. \tag{2.4.14}$$

Now the proof is almost over: we still have to use that $\int_0^1 \varphi(t)\, dt = 0$. By expanding the square in the left-hand side of (2.4.14) and using such a property, we get

$$4\pi^2 \int_0^1 |\varphi(t)|^2 \, dt + 4\pi^2 \, |\varphi(t_0)|^2 \leq \int_0^1 |\varphi'(t)|^2 \, dt.$$

This entails in particular that

$$4\pi^2 \int_0^1 |\varphi(t)|^2 \, dt \leq \int_0^1 |\varphi'(t)|^2 \, dt,$$

as desired. The uniqueness statement is left as an exercise for the reader. □

Remark 2.4.12 The inequality

$$\Lambda([0,1]) \int_0^1 |\varphi(t)|^2 \, dt \leq \int_0^1 |\varphi'(t)|^2 \, dt,$$

valid for every $\varphi \in C^1([0,1])$ such that

$$\varphi(0) = \varphi(1), \qquad \int_0^1 \varphi(t)\, dt = 0,$$

is sometimes called *Wirtinger's inequality*. The argument giving the exact value of $\Lambda([0,1])$ presented above is inspired by that of Hans Lewy, which can be found in the classical book by Hardy, Littlewood and Pólya, see [32, Theorem 258, page 185]. However, our use of Picone's inequality appears to be original.

In many textbooks, Wirtinger's inequality is proved by appealing to the theory of Fourier series: the main tool in this case is *Plancherel's identity* (see Problem 2.8.11 below). For yet another different proof, based on a discretization of the integrals, the reader can see [12, Teorema 2.4.1].

[1] Observe that on each interval, the function ψ has constant sign. In particular, on $(t_0, t_0 + 1/2)$ this is positive and we can apply directly Lemma 1.2.4. On the intervals $(0, t_0)$ and $(t_0 + 1/2, 1)$ we have $\psi < 0$ and we can simply observe that

$$\psi'(t) \left(\frac{(\varphi(t) - \varphi(t_0))^2}{\psi(t)} \right)' = (-\psi)'(t) \left(\frac{(\varphi(t) - \varphi(t_0))^2}{-\psi(t)} \right)'.$$

Thus, we can apply Lemma 1.2.4 with $-\psi$ in place of ψ.

As usual, Wirtinger's inequality for a general interval can be obtained by a scaling argument.

Corollary 2.4.13 *For two real numbers $a < b$, let us set*

$$\Lambda([a,b]) = \inf_{\varphi\in C^1([a,b])\setminus\{0\}} \left\{ \frac{\displaystyle\int_a^b |\varphi'(t)|^2\,dt}{\displaystyle\int_a^b |\varphi(t)|^2\,dt} \,:\, \varphi(a)=\varphi(b),\ \int_a^b \varphi(t)\,dt = 0 \right\}.$$

Then we have

$$\Lambda([a,b]) = \frac{4\pi^2}{(b-a)^2},$$

and this sharp constant is uniquely attained by any function of the form

$$C\,\phi(t-\alpha), \qquad \text{with } C \in \mathbb{R}\setminus\{0\},\ \alpha\in\mathbb{R},$$

where

$$\phi(t) := \sin\left(\frac{2\pi}{b-a}\,t\right), \qquad \text{for } t\in[a,b].$$

2.5 The Harmonic Oscillator

Let us consider a unit mass particle, constrained on the real line for simplicity, and attached through a spring to a fixed point O. Without loss of generality, we can suppose that O coincides with the origin of the real line. By the so-called *Hooke law*, we know that this particle experiences a force $\mathbf{F}$ directed towards the fixed point O, whenever displaced from the origin. In other words, this force tends to restore the particle to the position O. More precisely, if we indicate by x the position of the particle on the real line, *Hooke's law* asserts that the particle undergoes the action of the conservative force

$$\mathbf{F} = -k\,x\,\mathbf{e}_1,$$

where $k > 0$ is a given constant, which only depends on the material composing the spring. By using this equation and *Newton's second law*, we can then obtain the equation of motion of the mass particle: if $x(t)$ represents the position at time t, we must have

$$\text{force} = \text{mass} \times \text{acceleration},$$

that is (recall that we took the mass to be 1, for simplicity)

$$x''(t) = -k\,x(t). \tag{2.5.1}$$

By finding all the solutions of the previous ordinary differential equation, we thus get that the particle periodically oscillates, according to

$$x(t) = A\,\cos(\sqrt{k}\,t) + B\,\sin(\sqrt{k}\,t), \tag{2.5.2}$$

for some $A, B \in \mathbb{R}$. This represents the general solution of the equation of motion (2.5.1), which has to be coupled with some initial conditions on the position and the speed of the particle, i.e.

$$x(0) = x_0, \qquad x'(0) = v_0.$$

Observe that the period of oscillation is given by

$$\mathrm{t} = \frac{2\pi}{\sqrt{k}},$$

i.e. it is inversely proportional to the square root of the constitutive constant k. This in particular implies that the "stronger" is the spring (i.e. the higher is the value of k), the less is the period of oscillation, i. e. more rapidly the mass oscillates.

In what follows, we will discuss whether the oscillatory solution $x(t)$ obtained above could also be obtained as the solution of a minimization problem, where the quantity to be minimized is some physically relevant "energy". In order to do this, we observe that the *kinetic energy* at time t for the mass particle is given by

$$K(t) = \frac{1}{2}\,|x'(t)|^2,$$

while the *potential energy* at time t is given by

$$U(t) = -\frac{k}{2}\,|x(t)|^2.$$

Consequently, the *mechanical energy* at time t is given by

$$E(t) = K(t) + U(t) = \frac{1}{2}\,|x'(t)|^2 - \frac{k}{2}\,|x(t)|^2.$$

Then one could consider the *total mechanical energy* over the time interval $[0, T]$

$$\int_0^T E(t)\,dt,$$

for a suitable $T > 0$, and ask whether the solution in (2.5.2) can be obtained as a minimizer of this total energy. It turns out that *this is not always true*, but this is the case for oscillations such that

$$x(0) = 0,$$

provided that the length T of the time interval is suitably chosen. This is the content of the next result, which will be just an application of Theorem 2.4.2 and Corollary 2.4.4.

Theorem 2.5.1 *Let $T > 0$ and $k > 0$, we set*

$$T_0 := \frac{\pi}{2\sqrt{k}}.$$

The following variational problem

$$\inf_{x \in C^1([0,T])} \left\{ \int_0^T \left[\frac{|x'(t)|^2}{2} - k\,\frac{|x(t)|^2}{2} \right] dt \,:\, x(0) = 0 \right\},$$

is such that:

- *if $T < T_0$, it has a unique solution, given by $u(t) \equiv 0$;*
- *if $T = T_0$, it has infinitely many solutions, given by any multiple of the function*

$$\sin(\sqrt{k}\,t);$$

- *if $T > T_0$, it does not have any solution and the infimum above is $-\infty$.*

Proof Let us set for simplicity

$$\mathcal{F}(x) = \int_0^T \left[\frac{|x'(t)|^2}{2} - k\,\frac{|x(t)|^2}{2} \right] dt, \qquad \text{for every } x \in C^1([0,T]).$$

Then, for every admissible function x, by Corollary 2.4.4 we have

$$\begin{aligned}
\mathcal{F}(x) &= \int_0^T \left[\frac{|x'(t)|^2}{2} - k\,\frac{|x(t)|^2}{2} \right] dt \\
&\geq \frac{\pi^2}{4\,T^2} \int_0^T \frac{|x(t)|^2}{2}\,dt - k \int_0^T \frac{|x(t)|^2}{2}\,dt \\
&= \left(\frac{\pi^2}{4\,T^2} - k \right) \int_0^T \frac{|x(t)|^2}{2}\,dt \\
&= k \left(\left(\frac{T_0}{T} \right)^2 - 1 \right) \int_0^T \frac{|x(t)|^2}{2}\,dt.
\end{aligned} \tag{2.5.3}$$

Observe that we also used the definition of T_0. This simple estimate implies that if $T < T_0$, then the value of $\mathcal{F}(x)$ is always non-negative and vanishes if and only if

$$\int_0^T |x(t)|^2\, dt = 0,$$

i.e. if and only if x identically vanishes on $[0, T]$. This proves the first point of the statement.

We now assume that $T = T_0$. Still by Corollary 2.4.4, we know that if we set

$$\phi_C(t) = C\, \sin\left(\frac{\pi}{2\,T}\, t\right), \qquad \text{for } C \in \mathbb{R},$$

then

$$\int_0^T |\phi_C'(t)|^2\, dt = \frac{\pi^2}{4\,T^2} \int_0^T |\phi_C(t)|^2\, dt = k\, \frac{T_0^2}{T^2} \int_0^T |\phi_C(t)|^2\, dt.$$

By recalling that $T = T_0$, this implies that

$$\mathcal{F}(\phi_C) = 0,$$

while by (2.5.3) we still have

$$\mathcal{F}(x) \geq 0, \qquad \text{for every } x \in C^1([0, 1]) \text{ with } x(0) = 0.$$

We thus get that the infimum of $\mathcal{F}$ is 0 and it is uniquely attained by any function ϕ_C.

Finally, for the case $T > T_0$, we take again

$$\phi_C(t) = C\, \sin\left(\frac{\pi}{2\,T}\, t\right), \qquad \text{for } C \in \mathbb{R}.$$

We have

$$\begin{aligned} \mathcal{F}(\phi_C) = \int_0^T \left[\frac{|\phi_C'(t)|^2}{2} - k\, \frac{|\phi_C(t)|^2}{2}\right] dt &= \left(\frac{\pi^2}{4\,T^2} - k\right) \int_0^T \frac{|\phi_C(t)|^2}{2}\, dt \\ &= C^2 \left(\frac{\pi^2}{4\,T^2} - k\right) \int_0^T \frac{|\phi_1(t)|^2}{2}\, dt \\ &= C^2\, k \left(\frac{T_0^2}{T^2} - 1\right) \int_0^T \frac{|\phi_1(t)|^2}{2}\, dt. \end{aligned}$$

This implies that, when $T > T_0$, we have

$$\lim_{|C|\to+\infty} \mathcal{F}(\phi_C) = \lim_{|C|\to+\infty} C^2\, k \left(\frac{T_0^2}{T^2} - 1\right) \int_0^T \frac{|\phi_1(t)|^2}{2}\, dt = -\infty.$$

This concludes the proof. □

2.6 The Classical Isoperimetric Inequality in the Plane

We now prove the celebrated isoperimetric property of disks, among sets in the plane. This asserts that among smooth simply connected subsets of the plane with given perimeter, disks maximize the area. The result can be obtained with a clever combination of the Divergence Theorem and Theorem 2.4.11. The final outcome is quite remarkable.

Theorem 2.6.1 (Hurwitz) *Let $\Omega \subseteq \mathbb{R}^2$ be an open bounded set, whose boundary $\partial\Omega$ coincides with the image of a closed simple C^1 curve. We indicate by $L(\Omega) > 0$ the length of $\partial\Omega$. Then the area of Ω is smaller than or equal to the area of the disk having perimeter $L(\Omega)$. In other words, we have the* isoperimetric inequality

$$|\Omega| \le \frac{\Big(L(\Omega)\Big)^2}{4\,\pi}, \tag{2.6.1}$$

and we have equality when Ω is a disk.

Proof We first prove (2.6.1) and then show how this implies the claimed property of disks.

For notational simplicity, we will write L in place of $L(\Omega)$. By the Divergence Theorem, we know that for every C^1 vector field $\mathbf{F} : \overline{\Omega} \to \mathbb{R}^2$ we have

$$\int_\Omega \operatorname{div} \mathbf{F}\, dx\, dy = \int_{\partial\Omega} \langle \mathbf{F}, \nu_\Omega\rangle\, d\ell(x, y),$$

where ν_Ω is the unit outer normal vector. We use this formula with the choice

$$\mathbf{F}(x, y) = (x - x_1, y - y_1),$$

where (x_1, y_1) is an arbitrary point. This gives

$$\begin{aligned} 2\,|\Omega| = 2 \int_\Omega dx\, dy &= \int_\Omega \operatorname{div} \mathbf{F}(x, y)\, dx\, dy \\ &= \int_{\partial\Omega} \left\langle \begin{bmatrix} x - x_1 \\ y - y_1 \end{bmatrix}, \nu_\Omega \right\rangle d\ell(x, y). \end{aligned}$$

In the last integral, we can further use Cauchy-Schwarz inequality and Hölder's inequality, so to get

$$\begin{aligned}
&\int_{\partial\Omega} \left\langle \begin{bmatrix} x - x_1 \\ y - y_1 \end{bmatrix}, \nu_\Omega \right\rangle d\ell(x, y) \\
&\leq \int_{\partial\Omega} \sqrt{(x - x_1)^2 + (y - y_1)^2}\, d\ell(x, y) \\
&\leq \left(\int_{\partial\Omega} ((x - x_1)^2 + (y - y_1)^2)\, d\ell(x, y) \right)^{\frac{1}{2}} \left(\int_{\partial\Omega} d\,\ell \right)^{\frac{1}{2}}.
\end{aligned}$$

Thus, up to now, we obtained

$$2\,|\Omega| \leq \left(\int_{\partial\Omega} ((x - x_1)^2 + (y - y_1)^2)\, d\ell(x, y) \right)^{\frac{1}{2}} \sqrt{L}. \tag{2.6.2}$$

We now take a parametrization $\gamma : [0, 1] \to \mathbb{R}^2$ of the boundary $\partial\Omega$. Without loss of generality, up to a reparametrization, we can take it such that

$$|\gamma'(t)| = L, \qquad \text{for } t \in [0, 1].$$

Then the integral becomes

$$\begin{aligned}
&\int_{\partial\Omega} ((x - x_1)^2 + (y - y_1)^2)\, d\ell(x, y) \\
&= L \int_0^1 |\gamma_1(t) - x_1|^2\, dt + L \int_0^1 |\gamma_2(t) - y_1|^2\, dt.
\end{aligned} \tag{2.6.3}$$

We now make a suitable choice of (x_1, y_1): we set

$$x_1 = \int_0^1 \gamma_1(\tau)\, d\tau \qquad \text{and} \qquad y_1 = \int_0^1 \gamma_2(\tau)\, d\tau.$$

Then we observe that the two functions

$$\widetilde{\gamma}_i(t) = \gamma_i(t) - \int_0^1 \gamma_i(\tau)\, d\tau, \qquad \text{for } i = 1, 2,$$

are such that

$$\int_0^1 \widetilde{\gamma}_i(t)\, dt = 0 \qquad \text{and} \qquad \widetilde{\gamma}_i(0) = \widetilde{\gamma}_i(1), \qquad \text{for } i = 1, 2.$$

We can thus apply Theorem 2.4.11 and infer that

$$\int_0^1 |\gamma_i(t) - x_i|^2\, dt = \int_0^1 |\widetilde{\gamma}_i(t)|^2\, dt \le \frac{1}{4\,\pi^2} \int_0^1 |\widetilde{\gamma}_i'(t)|^2\, dt = \frac{1}{4\,\pi^2} \int_0^1 |\gamma_i'(t)|^2\, dt.$$

In light of (2.6.3), we thus obtain

$$\int_{\partial\Omega} ((x - x_1)^2 + (y - y_1)^2)\, d\ell(x, y) \le \frac{L}{4\,\pi^2} \int_0^1 \left[|\gamma_1'(t)|^2 + |\gamma_2'(t)|^2\right] dt = \frac{L^3}{4\,\pi^2}.$$

We spend this information into (2.6.2), so to get

$$2\,|\Omega| \le \frac{L^2}{2\,\pi},$$

which is exactly the desired estimate (2.6.1).

We now show that disks have the claimed maximality property, among sets with given perimeter. For every Ω as in the statement, we have from (2.6.1)

$$\frac{|\Omega|}{\Big(L(\Omega)\Big)^2} \le \frac{1}{4\,\pi}.$$

On the other hand, for a disk $D \subseteq \mathbb{R}^2$, we have

$$L(D) = 2\,\pi\, R \qquad \text{and} \qquad |D| = \pi\, R^2.$$

Thus, we get

$$\frac{|D|}{\Big(L(D)\Big)^2} = \frac{1}{4\,\pi}.$$

If we take D such that $L(D) = L(\Omega)$, from

$$\frac{|\Omega|}{\Big(L(\Omega)\Big)^2} \le \frac{1}{4\,\pi} = \frac{|D|}{\Big(L(D)\Big)^2},$$

we get in particular that $|\Omega| \le |D|$, as desired. □

Remark 2.6.2 In light of the proof above, another way to write (2.6.1) is the following

$$\frac{|\Omega|}{\Big(L(\Omega)\Big)^2} \le \frac{|D|}{\Big(L(D)\Big)^2}. \tag{2.6.4}$$

where $D \subseteq \mathbb{R}^2$ is any open disk. Then, from (2.6.4) we can also assert that

"*among sets in the plane with given area, a disk minimizes the perimeter*"

This is a sort of equivalent "dual" formulation of the classical isoperimetric statement.

Remark 2.6.3 (Equality Cases) With a little extra work, it would not be difficult to show that equality in (2.6.1) can hold *only if* Ω is a disk. It would be sufficient to trace back the equality cases in the three inequalities used in the proof above: Cauchy-Schwarz, Hölder and Wirtinger.

2.7 A Further Isoperimetric-type Problem

We suppose for simplicity to work on the interval [0, 1]. We consider the following geometric problem: among all $C^1([0,1])$ non-negative functions φ such that $\varphi(0) = \varphi(1) = 0$ and such that the subgraph of φ has given area, find the one whose perimeter of the enclosed region is minimal. This problem sounds quite similar to the Isoperimetric Problem treated in Sect. 2.6, but observe that we have an additional constraint given by the fact that the regions considered have to be subgraphs of functions of one real variable.

In an analytic framework, the previous problem can be reformulated as follows

$$\inf_{\varphi \in C^1([0,1])} \left\{ \int_0^1 \sqrt{1 + |\varphi'(t)|^2}\, dt \; : \; \begin{array}{c} \varphi(0) = \varphi(1) = 0, \; \varphi \geq 0 \\ \text{and } \int_0^1 \varphi(t)\, dt = C \end{array} \right\}, \tag{2.7.1}$$

where the constant $C > 0$ represents the given area (see Fig. 2.3). Then we have the following

Theorem 2.7.1 *With the previous notation, we have:*

- *if* $0 < C < \pi/8$, *the infimum* (2.7.1) *is uniquely attained by the function*

$$u(t) = y_0 + \sqrt{y_0^2 + \frac{1}{4} - \left(t - \frac{1}{2}\right)^2}, \qquad t \in [0, 1], \tag{2.7.2}$$

whose graph is an arc of circle. Here y_0 *is the unique real number* $y_0 < 0$ *such that*

$$\int_0^1 u(t)\, dt = C.$$

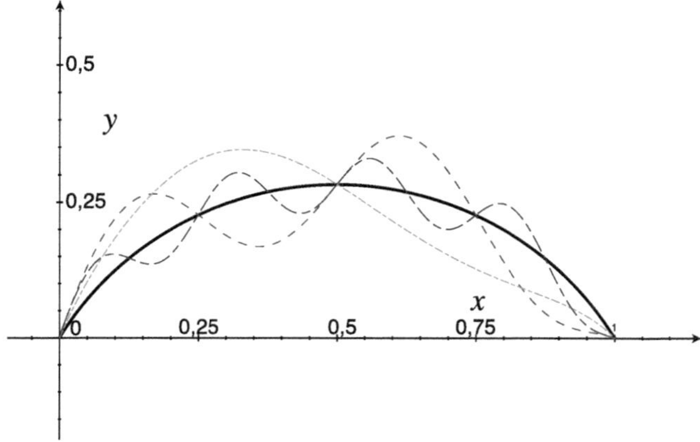

Fig. 2.3 The function u attaining the infimum in (2.7.1) (bold black line) and some admissible functions (dotted lines)

The value of the infimum (2.7.1) *is then given by*

$$2\sqrt{y_0^2+\frac{1}{4}}\arcsin\left(\frac{1}{2\sqrt{y_0^2+\frac{1}{4}}}\right); \tag{2.7.3}$$

- *if* $C = \pi/8$, *the infimum above is* $\pi/2$, *but it is not attained. In other words, the variational problem* (2.7.1) *does not have a solution in* $C^1([0, 1])$.

Proof We treat the two cases separately.

1. Case $0 < C < \pi/8$ We first observe that $\pi/8$ corresponds to the area of the half-circle centered at $(1/2, 0)$, with radius $1/2$. The requirement that $C < \pi/8$ implies that the real number y_0 in (2.7.2) is $y_0 < 0$. Thus, the function u defined by (2.7.2) belongs to $C^1([0, 1])$.

We now set for simplicity

$$F(t) = \sqrt{1+t^2},$$

and observe that this is a C^2 convex function. In particular, from the “above tangent” property (see Proposition 1.2.3), for every admissible function φ we have

$$F(\varphi'(t)) \ge F(u'(t)) + F'(u'(t))\,(\varphi'(t) - u'(t)), \qquad \text{for } t \in [0, 1].$$

We integrate this pointwise inequality over $[0, 1]$. This gives

$$\int_0^1 \sqrt{1+|\varphi'(t)|^2}\,dt \ge \int_0^1 \sqrt{1+|u'(t)|^2}\,dt + \int_0^1 F'(u'(t))\,(\varphi'(t) - u'(t))\,dt.$$

In order to conclude, it would be sufficient to prove that

$$\int_0^1 F'(u'(t))\,(\varphi'(t) - u'(t))\,dt = 0.$$

Before going further, we notice that this integral identity can be seen as the Euler-Lagrange equation of our variational problem. We observe that for every $t \in [0, 1]$

$$F'(u'(t)) = \frac{u'(t)}{\sqrt{1 + |u'(t)|^2}} \qquad \text{and} \qquad u'(t) = -\frac{t - \dfrac{1}{2}}{\sqrt{y_0^2 + \dfrac{1}{4} - \left(t - \dfrac{1}{2}\right)^2}}.$$

In particular, by direct computation, we get

$$F'(u'(t)) = \frac{1}{\sqrt{y_0^2 + \dfrac{1}{4}}}\left(\frac{1}{2} - t\right), \qquad \text{for } t \in [0, 1]. \tag{2.7.4}$$

We now use this expression and perform an integration by parts as follows

$$\begin{aligned}
\int_0^1 F'(u'(t))\,(\varphi'(t) - u'(t))\,dt &= \frac{1}{\sqrt{y_0^2 + \dfrac{1}{4}}}\int_0^1 \left(\frac{1}{2} - t\right)(\varphi'(t) - u'(t))\,dt \\
&= \frac{1}{\sqrt{y_0^2 + \dfrac{1}{4}}}\left[\left(\frac{1}{2} - t\right)(\varphi(t) - u(t))\right]_0^1 \\
&+ \frac{1}{\sqrt{y_0^2 + \dfrac{1}{4}}}\int_0^1 (\varphi(t) - u(t))\,dt \\
&= \frac{1}{\sqrt{y_0^2 + \dfrac{1}{4}}}\int_0^1 (\varphi(t) - u(t))\,dt.
\end{aligned}$$

We used that both φ and u vanish at the endpoints of the interval. If we now recall that

$$\int_0^1 \varphi(t)\,dt = C = \int_0^1 u(t)\,dt,$$

from the previous computations we get

$$\int_0^1 F'(u'(t))\,(\varphi'(t) - u'(t))\,dt = 0,$$

as desired. The minimal value (2.7.3) can now be obtained by simply computing the integral

$$\int_0^1 \sqrt{1 + |u'(t)|^2}\,dt,$$

and this is left as an exercise for the reader.

We still have to prove that u is the unique minimizer: for this fact, we can observe that we are minimizing a strictly convex functional, over a convex set of admissible functions. Indeed, the function F is C^2 and such that

$$F''(t) = \frac{1}{(1+t^2)^{\frac{3}{2}}} > 0, \qquad \text{for every } t \in \mathbb{R}.$$

Thus, F is strictly convex, thanks to Remark 1.2.8. If we assume that $v \in C^1([0, 1])$ is another minimizer, we then consider the new function

$$U(t) = \frac{u(t) + v(t)}{2},$$

which is still admissible for our variational problem. By convexity, we have

$$\begin{aligned}\sqrt{1 + |U'(t)|^2} &= F(U'(t)) = F\left(\frac{u'(t) + v'(t)}{2}\right)\\ &\leq \frac{F(u'(t)) + F(v'(t))}{2} = \frac{\sqrt{1 + |u'(t)|^2} + \sqrt{1 + |v'(t)|^2}}{2},\end{aligned}$$

and the inequality sign is strict, whenever $u'(t) \neq v'(t)$. Since u and v do not coincide, the set $\{t \in [0, 1] \,:\, u'(t) \neq v'(t)\}$ must have positive measure. Thus, by integrating over $[0, 1]$ we would get

$$\int_0^1 \sqrt{1 + |U'(t)|^2}\,dt < \frac{1}{2}\int_0^1 \sqrt{1 + |u'(t)|^2}\,dt + \frac{1}{2}\int_0^1 \sqrt{1 + |v'(t)|^2}\,dt.$$

Since both u and v are minimizers, the last two integrals coincide with the minimum of our variational problem, i.e. we get

$$\begin{aligned}&\int_0^1 \sqrt{1 + |U'(t)|^2}\,dt\\ &\quad < \min_{\varphi \in C^1([0,1])} \left\{ \int_0^1 \sqrt{1 + |\varphi'(t)|^2}\,dt \,:\, \begin{array}{c} \varphi(0) = \varphi(1) = 0,\ \varphi \geq 0 \\ \text{and } \int_0^1 \varphi(t)\,dt = C \end{array} \right\}.\end{aligned}$$

This gives the desired contradiction and thus the uniqueness of u.

2. *Case* $C = \pi/8$ In this case, the function u coincides with the function whose graph is the half-circle, centered at $(1/2, 0)$ with radius $1/2$. Thus, this is given by

$$u(t) = \sqrt{\frac{1}{4} - \left(t - \frac{1}{2}\right)^2},$$

which corresponds to take $y_0 = 0$ in (2.7.2). Observe that $u \notin C^1([0, 1])$ but only $u \in C^1((0, 1))$, since

$$\lim_{t \to 0^+} u'(t) = +\infty \qquad \text{and} \qquad \lim_{t \to 1^-} u'(t) = -\infty. \tag{2.7.5}$$

However, for every admissible $\varphi \in C^1([0, 1])$, we still have

$$F(\varphi'(t)) \geq F(u'(t)) + F'(u'(t))\,(\varphi'(t) - u'(t)), \qquad \text{for } t \in (0, 1),$$

where we now excluded the endpoints of the interval. By recalling the expression of F, this is the same as

$$\sqrt{1 + |\varphi'(t)|^2} \geq \sqrt{1 + |u'(t)|^2} + 2\left(\frac{1}{2} - t\right)(\varphi'(t) - u'(t)), \qquad \text{for } t \in (0, 1), \tag{2.7.6}$$

where we used (2.7.4), with $y_0 = 0$. We now observe that $u'(t) \in L^1([0, 1])$. Indeed, this derivative is given by

$$u'(t) = -\frac{t - \dfrac{1}{2}}{\sqrt{\dfrac{1}{4} - \left(t - \dfrac{1}{2}\right)^2}},$$

thus we have

$$u'(t) \sim \frac{1}{2}\frac{1}{\sqrt{t}}, \text{ for } t \to 0^+ \qquad \text{and} \qquad u'(t) \sim -\frac{1}{2}\frac{1}{\sqrt{1-t}}, \text{ for } t \to 1^-.$$

We can integrate the pointwise inequality (2.7.6) on the interval and get as before

$$\int_0^1 \sqrt{1 + |\varphi'(t)|^2}\, dt \geq \int_0^1 \sqrt{1 + |u'(t)|^2}\, dt + 2\int_0^1 \left(\frac{1}{2} - t\right)(\varphi'(t) - u'(t))\, dt.$$

We can again integrate by parts as in the previous case and get

$$\int_0^1 \left(\frac{1}{2} - t\right)(\varphi'(t) - u'(t))\, dt = 0.$$

This in turn implies that

$$\int_0^1 \sqrt{1+|\varphi'(t)|^2}\,dt \geq \int_0^1 \sqrt{1+|u'(t)|^2}\,dt,$$

for every admissible function φ. An explicit computation shows that the value of the last integral is $\pi/2$, thus we have

$$\int_0^1 \sqrt{1+|\varphi'(t)|^2}\,dt \geq \frac{\pi}{2}.$$

In order to show that $\pi/2$ is the infimum (2.7.1) of our variational problem and it is not attained, we now have to show that:

(a) actually, we have

$$\int_0^1 \sqrt{1+|\varphi'(t)|^2}\,dt > \frac{\pi}{2},$$

for every admissible function φ;

(b) there exists a sequence $\{\varphi_n\}_{n\in\mathbb{N}} \subseteq C^1([0,1])$ of admissible functions, such that

$$\lim_{n\to\infty} \int_0^1 \sqrt{1+|\varphi_n'(t)|^2}\,dt = \frac{\pi}{2}.$$

In order to prove point a), we still exploit the strict convexity of F. Let us assume that there exists $\varphi \in C^1([0,1])$ such that

$$\int_0^1 \sqrt{1+|\varphi'(t)|^2}\,dt = \frac{\pi}{2} = \int_0^1 \sqrt{1+|u'(t)|^2}\,dt. \tag{2.7.7}$$

By recalling Remark 1.2.8, by strict convexity of F we can infer

$$F(\varphi'(t)) \geq F(u'(t)) + F'(u'(t))\,(\varphi'(t)-u'(t)), \qquad \text{for } t \in (0,1)$$

and the inequality sign is strict, whenever $\varphi'(t) \neq u'(t)$. The set where this happens has positive measure, by recalling that $\varphi \in C^1([0,1])$, while u satisfies (2.7.5). Thus, when we integrate the previous inequality we get

$$\int_0^1 \sqrt{1+|\varphi'(t)|^2}\,dt > \int_0^1 \sqrt{1+|u'(t)|^2}\,dt + 2\int_0^1 \left(\frac{1}{2}-t\right)(\varphi'(t)-u'(t))\,dt.$$

By recalling that

$$\int_0^1 \left(\frac{1}{2} - t\right) (\varphi'(t) - u'(t))\, dt = 0,$$

we thus get a contradiction with (2.7.7).

We are only left with proving point (b): we take $\{y_n\}_{n\in\mathbb{N}}$ a sequence of negative numbers, such that

$$\lim_{n\to\infty} y_n = 0.$$

Accordingly, we take the functions

$$u_n(t) = y_n + \sqrt{y_n^2 + \frac{1}{4} - \left(t - \frac{1}{2}\right)^2},$$

which are of class $C^1([0, 1])$ (thanks to the fact that $y_n < 0$) and vanish at $t = 0$ and $t = 1$. By construction, we have

$$\int_0^1 u_n(t)\, dt = C_n < \frac{\pi}{8} \qquad \text{with} \qquad \lim_{n\to\infty} C_n = \frac{\pi}{8}.$$

We then choose the sequence of admissible functions $\{\varphi_n\}_{n\in\mathbb{N}}$ given by

$$\varphi_n(t) = \frac{\pi}{8} \frac{u_n(t)}{C_n}.$$

We have

$$\begin{aligned}
\int_0^1 \sqrt{1 + |\varphi_n'(t)|^2}\, dt &= \int_0^1 \sqrt{1 + \left(\frac{\pi/8}{C_n}\right)^2 |u_n'(t)|^2}\, dt \\
&< \left(\frac{\pi/8}{C_n}\right) \int_0^1 \sqrt{1 + |u_n'(t)|^2}\, dt \\
&= \left(\frac{\pi/8}{C_n}\right) 2\sqrt{y_n^2 + \frac{1}{4}}\, \arcsin\left(\frac{1}{2\sqrt{y_n^2 + \frac{1}{4}}}\right).
\end{aligned}$$

By taking the limit as n goes to ∞ and recalling that y_n converges to 0, while C_n converges to $\pi/8$, we get

$$\frac{\pi}{2} \le \liminf_{n\to\infty} \int_0^1 \sqrt{1+|\varphi_n'(t)|^2}\,dt \le \limsup_{n\to\infty} \int_0^1 \sqrt{1+|\varphi_n'(t)|^2}\,dt$$

$$\le \lim_{n\to\infty} \left(\frac{\pi/8}{C_n}\right) 2\sqrt{y_n^2+\frac{1}{4}}\,\arcsin\left(\frac{1}{2\sqrt{y_n^2+\frac{1}{4}}}\right) = \frac{\pi}{2},$$

as desired. This finally concludes the proof. □

2.8 Problems

Problem 2.8.1 Find the Euler-Lagrange equation for the problem of Theorem 2.2.2.

Problem 2.8.2 Let $F : \mathbb{R}^N \to [0, +\infty)$ be a strictly convex function. Show that the minimization problem

$$\min_{\gamma\in C^1([0,T];\mathbb{R}^N)} \left\{ \int_0^T F(\gamma'(t))\,dt \,:\, \gamma(0) = x_0,\ \gamma(T) = x_1 \right\},$$

has a unique solution, given by the constant speed curve

$$\gamma_{\mathrm{opt}}(t) = \left(1 - \frac{t}{T}\right) x_0 + \frac{t}{T} x_1, \qquad \text{for } t \in [0, T].$$

Problem 2.8.3 Let us consider the 2×2 symmetric matrix

$$A(t) = \begin{bmatrix} t & 1 \\ 1 & 2 \end{bmatrix}, \qquad \text{for } t \in [0, 1].$$

Consider the functional defined on $C^1([0, 1]; \mathbb{R}^2)$ by

$$\mathcal{F}(\gamma) = \frac{1}{2} \int_0^1 \langle A(t) \cdot \gamma'(t), \gamma'(t)\rangle\,dt.$$

Then:

1. compute the first variation of $\mathcal{F}$ and its Euler-Lagrange equation, both in weak and classical form;
2. find the general solution to this equation;
3. show that for every $\alpha \neq 0$, the problem

$$\inf_{\gamma\in C^1([0,1];\mathbb{R}^2)}\Big\{\mathcal{F}(\gamma)\,:\,\gamma(0)=(0,0),\ \gamma(1)=(\alpha,1)\Big\},\tag{2.8.1}$$

does not admit a solution.

Problem 2.8.4 Show that the function

$$F(x_1,x_2)=\sqrt{\frac{1}{x_1^2}+4\,x_2^2},\qquad \text{for } (x_1,x_2)\in(0,+\infty)\times\mathbb{R},$$

is strictly convex.

Problem 2.8.5 By proceeding informally, show that the Euler-Lagrange equation of the brachistocrone functional

$$\mathcal{F}(u)=\int_0^1\frac{\sqrt{1+|u'(x)|^2}}{\sqrt{u(x)}}\,dx,$$

is given by the following ordinary differential equation

$$-\,2\,u''(x)\,u(x)=1+|u'(x)|^2,\qquad \text{for } x\in(0,1).$$

Problem 2.8.6 Compute the integral

$$\int_0^T\sqrt{\frac{\tau}{C-\tau}}\,d\tau,$$

for every $C\geq T>0$.

Problem 2.8.7 Show that for every $\alpha\neq 0$ we have

$$\inf_{\varphi\in C^1([0,1])\setminus\{0\}}\left\{\frac{\displaystyle\int_0^1|\varphi'|^2\,dt}{\displaystyle\int_0^1|\varphi|^2\,dt}\,:\,\varphi(0)=\alpha\right\}=0.$$

Problem 2.8.8 Show that

$$\inf_{\varphi\in C^1([0,1])\setminus\{0\}}\left\{\frac{\|\varphi'\|_{L^\infty([0,1])}}{\|\varphi\|_{L^\infty([0,1])}}\,:\,\varphi(0)=0\right\}=1,$$

and that the infimum is attained by any linear function.

Problem 2.8.9 Show that

$$\inf_{\varphi \in C^1([0,1])\setminus\{0\}} \left\{ \frac{\displaystyle\int_0^1 |\varphi'|\,dt}{\displaystyle\int_0^1 |\varphi|\,dt} \,:\, \varphi(0) = 0 \right\} = 1,$$

and that the infimum is not attained.

Problem 2.8.10 Show that for every $1 \leq p < \infty$ we have

$$\inf_{\varphi \in C_0^\infty(\mathbb{R})\setminus\{0\}} \frac{\displaystyle\int_{\mathbb{R}} |\varphi'|^p\,dt}{\displaystyle\int_{\mathbb{R}} |\varphi|^p\,dt} = 0.$$

Problem 2.8.11 Prove again the Wirtinger inequality of Theorem 2.4.11, by using Fourier series.

Problem 2.8.12 Prove that we have the following Poincaré inequality

$$\pi^2 \inf_{c \in \mathbb{R}} \int_0^1 |\varphi - c|^2\,dt \leq \int_0^1 |\varphi'|^2\,dt, \qquad \text{for every } \varphi \in C^1([0,1]).$$

Show that the constant π^2 is optimal.

Problem 2.8.13 Let $\alpha > 0$, we define

$$\sup_{\varphi \in C^1([-1,1])\setminus\{0\}} \left\{ \frac{\left|\displaystyle\int_{-1}^1 \varphi\,dt\right|^\alpha}{\displaystyle\int_{-1}^1 |\varphi'|^2\,dt} \,:\, \varphi(-1) = \varphi(1) = 0 \right\}.$$

Show that for $\alpha \neq 2$ this supremum is $+\infty$. Then take $\alpha = 2$ and call $\mathcal{T}([-1,1])$ the previous supremum:

1. show that $\mathcal{T}([-1,1]) < +\infty$, by proving that

$$\mathcal{T}([-1,1]) \leq \frac{2}{\lambda([-1,1])},$$

   where λ is the Poincaré constant of Corollary 2.4.7;
2. prove that

$$\mathcal{T}([-1,1]) = \sup_{\varphi \in C^1([-1,1])} \left\{ 2 \int_{-1}^1 \varphi\,dt - \int_{-1}^1 |\varphi'|^2\,dt \,:\, \varphi(-1) = \varphi(1) = 0 \right\};$$

3. by using the previous point, compute the value of $\mathcal{T}([-1,1])$, by finding a function which attains the supremum.

Problem 2.8.14 Use the previous problem to compute the quantity

$$\mathcal{T}([a,b]) := \sup_{\varphi \in C^1([a,b])\setminus\{0\}} \left\{ \frac{\left(\displaystyle\int_a^b \varphi\, dt\right)^2}{\displaystyle\int_a^b |\varphi'|^2\, dt} : \varphi(a) = \varphi(b) = 0 \right\},$$

for every $a < b$.

Problem 2.8.15 Let $B_1(0) = \{(x, y) \in \mathbb{R}^2 : x^2 + y^2 < 1\}$ be the open disk centered at $(0, 0)$, with radius 1.

1. Prove that for every $u \in C^1(\overline{B_1(0)})$ such that $u = 0$ on $\partial B_1(0)$, we have the following *weighted* Poincaré inequality

$$\iint_{B_1(0)} \frac{1}{\sqrt{x^2+y^2}} |\nabla u|^2\, dx\, dy \geq \left(\frac{\pi}{2}\right)^2 \iint_{B_1(0)} \frac{1}{\sqrt{x^2+y^2}} |u|^2\, dx\, dy;$$

2. prove that equality holds if and only if

$$u(x, y) = C \cos\left(\frac{\pi}{2}\sqrt{x^2+y^2}\right),$$

for some constant $C \in \mathbb{R}$.

Problem 2.8.16 Let $\alpha \neq 2$. With the notation of Theorem 2.6.1, show that there can not exist any constant $C > 0$ such that the following inequality

$$|\Omega| \leq C\left(L(\Omega)\right)^\alpha,$$

holds for every smooth simply connected open bounded set $\Omega \subseteq \mathbb{R}^2$.

Problem 2.8.17 Let $\Omega \subseteq \mathbb{R}^2$ be an open bounded set, whose boundary $\partial\Omega$ coincides with the image of a closed and simple C^1 curve. We define the *moment of inertia of $\partial\Omega$ with respect to a point $(\overline{x}, \overline{y}) \in \mathbb{R}^2$* as the quantity

$$\mathcal{I}_{\partial\Omega}(\overline{x}, \overline{y}) = \int_{\partial\Omega} \left((x-\overline{x})^2 + (y-\overline{y})^2\right) d\ell(x, y).$$

1. Show that there exists a unique point which attains the following infimum

$$\inf_{(\overline{x},\overline{y})\in\mathbb{R}^2} \mathcal{I}_{\partial\Omega}(\overline{x}, \overline{y}),$$

and characterize this point in terms of $\partial\Omega$;

2. by calling $\mathcal{I}(\partial\Omega)$ the previous infimum,[2] show the following isoperimetric-type inequality

$$\frac{\mathcal{I}(\partial\Omega)}{\Big(L(\Omega)\Big)^3} \leq \frac{\mathcal{I}(\partial D)}{\Big(L(D)\Big)^3},$$

where $D \subseteq \mathbb{R}^2$ is any open disk and L still denotes the perimeter of a set. In other words, show that disks maximize the quantity $\mathcal{I}(\partial\Omega)$ among smooth simply connected open bounded sets in the plane, with given perimeter.

Problem 2.8.18 Let $F : [0,1] \times \mathbb{R} \times [0,+\infty) \to \mathbb{R}$ be a continuous function, such that for every $(t,z) \in [0,1] \times [0,+\infty)$ the function

$$u \mapsto F(t,u,z) \quad \text{is non-increasing on } \mathbb{R}.$$

Consider the minimization problem

$$\inf_{u \in C^1([0,1])} \left\{ \int_0^1 F(t,u(t),|u'(t)|)\,dt \,:\, u(0) = 0 \right\},$$

and suppose that this admits a unique minimizer v. Show that v must be a non-decreasing function.

[2] This is also called *polar moment of inertia* of $\partial\Omega$.

Chapter 3
Sobolev Spaces

3.1 Introduction: Why Do We Need Sobolev Spaces?

Let us suppose that we want to prove existence of the solution to the following boundary value problem for the Laplace equation

$$\begin{cases} -\Delta u = 0, & \text{in } \Omega, \\ u = g, & \text{on } \partial\Omega. \end{cases}$$

Here $\Omega \subseteq \mathbb{R}^N$ is a smooth bounded open set and g is a given smooth function. By Theorem 1.5.1 (i.e. the Dirichlet principle), we know that this is the same as seeking a solution of

$$\inf_{u \in C^2(\overline{\Omega})} \left\{ \frac{1}{2} \int_\Omega |\nabla u|^2\, dx \,:\, u = g \text{ on } \partial\Omega \right\}.$$

We thus have reduced the initial question to proving existence of a minimizer for a suitable *functional*, i.e.

$$\mathcal{F}(u) = \frac{1}{2} \int_\Omega |\nabla u|^2\, dx.$$

How to prove existence of a minimizer? A possible strategy can be guessed by recalling the very first result about existence of minimizers, usually encountered in the first year course in Mathematical Analysis: the *Weierstrass Theorem.*

Unfortunately, we can not directly apply Weierstrass Theorem as usually presented in first year courses: this usually concerns continuous functions defined on *compact* subsets of the finite dimensional space $\mathbb{R}^N$. On the contrary, our functional $\mathcal{F}$ is considered over the space of functions

L. Brasco, *Handbook of Calculus of Variations for Absolute Beginners*,
La Matematica per il 3+2 163, https://doi.org/10.1007/978-3-031-87164-1_3

$$C^2_g(\overline{\Omega}) := \Big\{u \in C^2(\overline{\Omega})\ :\ u = g \text{ on } \partial\Omega\Big\},$$

which is infinite dimensional. However, we can try to mimick the proof of the Weierstrass Theorem and see if this can be amended, in order to fit into our framework.

- The **first step** is to take a *minimizing sequence*, that is a sequence of admissible functions $\{u_n\}_{n\in\mathbb{N}}$ such that

$$\mathcal{F}(u_n) \le \mathfrak{m} + \frac{1}{n+1}, \qquad \text{for every } n \in \mathbb{N}.$$

  Here $\mathfrak{m}$ is defined to be

$$\mathfrak{m} = \inf_{u\in C^2(\overline{\Omega})} \left\{\frac{1}{2}\int_\Omega |\nabla u|^2\,dx\ :\ u = g \text{ on } \partial\Omega\right\} < +\infty,$$

  thus the existence of such a sequence is just a plain consequence of the definition of infimum.
- The **second step** in the classical Weierstrass Theorem is to use that a sequence in a compact set is actually converging, up to a subsequence.
- Then, the **third step** uses the lower semicontinuity of the function to infer that

$$\mathcal{F}(u) \le \liminf_{n\to\infty} \mathcal{F}(u_n) \le \lim_{n\to\infty}\left(\mathfrak{m} + \frac{1}{n+1}\right) = \mathfrak{m}.$$

  This would show that u is the desired minimizer.

However, the **second step** badly fails to be true in our setting: it is not true that the space

$$C^2_g(\overline{\Omega}) := \Big\{u \in C^2(\overline{\Omega})\ :\ u = g \text{ on } \partial\Omega\Big\},$$

is sequentially compact. As a simple counterexample, it is sufficient to take

$$\overline{\Omega} = [-1, 1], \qquad g = 1,$$

and the sequence

$$u_n(t) = 1 + \sqrt{\frac{1}{n} + t^2} - \sqrt{\frac{1}{n} + 1}, \qquad \text{for } n \in \mathbb{N}\setminus\{0\}, \tag{3.1.1}$$

see Fig. 3.1. Indeed, observe that $u_n \in C^\infty([-1, 1])$ and as n goes to ∞, we have uniform convergence to the function

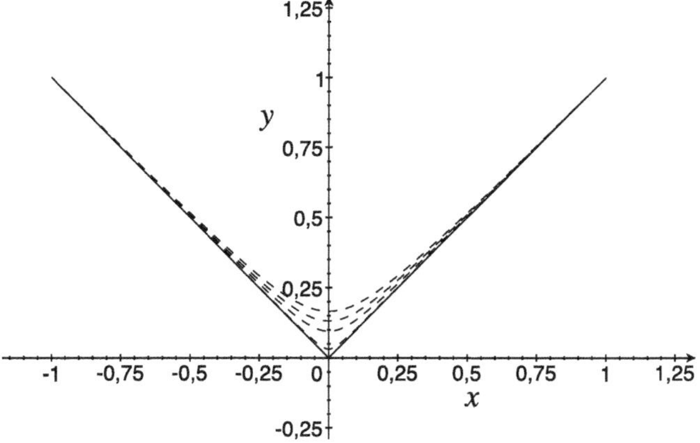

Fig. 3.1 The sequence of smooth functions u_n defined by (3.1.1)

$$u(t) = |t| \notin C^2([-1, 1]).$$

We also observe that our sequence has the further property that

$$\mathcal{F}(u_n) = \frac{1}{2}\int_{-1}^{1} |u_n'(t)|^2\,dt \le C, \qquad \text{for every } n \in \mathbb{N} \setminus \{0\},$$

since by direct computation we have

$$\frac{1}{2}\int_{-1}^{1} |u_n'(t)|^2\,dt = \frac{1}{2}\int_{-1}^{1} \frac{t^2}{\dfrac{1}{n} + t^2}\,dt \le 1.$$

The previous example seems to suggest that apparently there is no hope to generalize this kind of proof in our setting. However, this is not completely true: there is a problem of *topology* here. Working on the space $C^2(\overline{\Omega})$, endowed with its norm topology, is too restrictive in order to apply the Weierstrass-type argument to the problem

$$\inf\left\{\frac{1}{2}\int_{\Omega} |\nabla u|^2\,dx \,:\, u = g\right\},$$

since minimizing sequences may fail to be compact with respect to this topology.

We should then ask ourselves: is there any form of compactness that we can infer for a minimizing sequence? We need to go back to the **first step**: recall that for every minimizing sequence $\{u_n\}_{n\in\mathbb{N}}$, we can infer in particular that

$$\mathcal{F}(u_n) = \int_\Omega |\nabla u_n|^2\,dx \le C, \qquad \text{for every } n \in \mathbb{N}.$$

This implies that $\{\nabla u_n\}_{n\in\mathbb{N}}$ is a bounded sequence in $L^2(\Omega;\mathbb{R}^N)$. If we now want to mimick the **second step**, we could apply the Banach-Alaoglu Theorem (see Theorem B.1.1) and infer that actually we have a suitable form of sequential compactness: this is weak compactness *of the gradients* in $L^2(\Omega;\mathbb{R}^N)$. We thus get that there exists a vector field $\phi \in L^2(\Omega;\mathbb{R}^N)$ such that, up to a subsequence, we have

$$\lim_{n\to\infty} \int_\Omega \langle \nabla u_n, \Psi\rangle\,dx = \int_\Omega \langle \phi, \Psi\rangle\,dx, \qquad \text{for every } \Psi \in L^2(\Omega;\mathbb{R}^N),$$

by weak convergence in L^2. In principle, ϕ is just a vector field in $L^2(\Omega;\mathbb{R}^N)$: let us suppose for a moment that one could prove that

$$\phi = \nabla v,$$

for some function v such that $v = g$ on $\partial\Omega$. Then we could perform the **third step**: by using that $\{u_n\}_{n\in\mathbb{N}}$ is a minimizing sequence and the lower semicontinuity of the L^2 norm with respect to the weak convergence (see Proposition 1.3.11), we would get

$$\mathfrak{m} = \liminf_{n\to\infty} \frac{1}{2}\int_\Omega |\nabla u_n|^2\,dx \ge \frac{1}{2}\int_\Omega |\nabla v|^2\,dx.$$

Thus, v would be the desired minimizer.
As it should be clear from the previous discussion, this function v in principle does not belong to $C^2(\overline{\Omega})$, but only to a *larger space* $X(\Omega)$. This larger space $X(\Omega)$ is still mysterious for the moment: it should be a space of functions with the following property, crucially exploited in the previous argument

"the weak limit of a sequence of gradients is still a gradient" (3.1.2)

As we will see, it turns out that *Sobolev spaces*, introduced and studied in this chapter, exactly have this property.

Important
The previous method to prove existence of a minimizer, i.e.

- taking a minimizing sequence
- infer compactness

(continued)

- use lower semicontinuity

is called *the Direct Method in the Calculus of Variations*. This is the core of the modern theory of Calculus of Variations. We will explore it in more details in Chap. 4.

This chapter has to be intended as a gentle introduction to the theory of Sobolev spaces. For the reader who will be willing to study them in much more details, we suggest the classical books [13, 14, 41, 44] or [62]. Should the reader fall in love with the subject, we absolutely recommend the monographs [49, 50].

3.2 Weak Derivatives

Let $\Omega \subseteq \mathbb{R}^N$ be an open set. We recall that $u \in L^p_{\mathrm{loc}}(\Omega)$ means that

$$u \in L^p(\Omega'), \qquad \text{for every } \Omega' \Subset \Omega.$$

Definition 3.2.1 Let $1 \le p \le \infty$ and let $u \in L^1_{\mathrm{loc}}(\Omega)$. We say that u has a *weak partial derivative in* $L^p_{\mathrm{loc}}(\Omega)$ *with respect to the k-th variable*, if there exists $g_k \in L^p_{\mathrm{loc}}(\Omega)$ such that

$$\int_\Omega u\,\frac{\partial \varphi}{\partial x_k}\,dx = -\int_\Omega g_k\,\varphi\,dx, \qquad \text{for every } \varphi \in C^\infty_0(\Omega). \tag{3.2.1}$$

When this happens, we use the notation[1]

$$g_k = \frac{\partial u}{\partial x_k}.$$

If u has a weak partial derivative in $L^p_{\mathrm{loc}}(\Omega)$ with respect to every variable, we say that u has a *weak gradient in* $L^p_{\mathrm{loc}}(\Omega;\mathbb{R}^N)$ and we will use the notation

$$\nabla u = \left(\frac{\partial u}{\partial x_1},\dots,\frac{\partial u}{\partial x_N}\right).$$

The following simple result shows that the previous definition is well-posed.

[1] Observe that we are using the same symbol to denote both the classical and weak derivatives, as it is customary.

Lemma 3.2.2 *Let $1 \leq p \leq \infty$ and let $u \in L^1_{\mathrm{loc}}(\Omega)$. If $g_k, h_k \in L^p_{\mathrm{loc}}(\Omega)$ are two weak partial derivatives of u with respect to the k-th variable, then*

$$g_k = h_k, \qquad \textit{a. e. in } \Omega.$$

Proof By using the definition, we have

$$\int_\Omega u \, \frac{\partial \varphi}{\partial x_k} \, dx = - \int_\Omega g_k \, \varphi \, dx, \qquad \text{for every } \varphi \in C^\infty_0(\Omega),$$

and also

$$\int_\Omega u \, \frac{\partial \varphi}{\partial x_k} \, dx = - \int_\Omega h_k \, \varphi \, dx, \qquad \text{for every } \varphi \in C^\infty_0(\Omega).$$

By subtracting the two equations, we thus get

$$\int_\Omega (g_k - h_k) \, \varphi \, dx = 0, \qquad \text{for every } \varphi \in C^\infty_0(\Omega).$$

We now get the desired conclusion by appealing to the Du Bois-Reymond Lemma (in its modern version, see Lemma 1.4.2). □

We can verify that the concept of *weak derivative* is a genuine extension of the classical definition of derivative. In other words, classical derivatives are weak ones, as well. The verification of this fact is essentially based on the "integration by parts" formula. This is the content of the next result.

Proposition 3.2.3 *Let $\Omega \subseteq \mathbb{R}^N$ be an open set and let $u \in C^1(\Omega)$. Its classical gradient coincides with the weak one almost everywhere and it belongs to $L^\infty_{\mathrm{loc}}(\Omega; \mathbb{R}^N)$.*

Proof By hypothesis, each partial derivative of u is a continuous function on Ω and thus

$$\frac{\partial u}{\partial x_k} \in L^\infty_{\mathrm{loc}}(\Omega), \qquad \text{for every } k \in \{1, \dots, N\}.$$

We need to show that for every $k \in \{1, \dots, N\}$ we have

$$\int_\Omega u \, \frac{\partial \varphi}{\partial x_k} \, dx = - \int_\Omega \frac{\partial u}{\partial x_k} \, \varphi \, dx, \qquad \text{for every } \varphi \in C^\infty_0(\Omega).$$

To this aim, given $\varphi \in C^\infty_0(\Omega)$, we observe that the function $u \, \varphi$ is C^1 and compactly supported on Ω. This in particular implies that the support of $u \, \varphi$ is contained in the hyper-cube

$$Q_L = (-L, L)^N,$$

for L large enough. Thus, by extending $u\,\varphi$ by zero outside Ω, we have $u\,\varphi \in C^1_0(Q_L)$. We will use the notation $x = (x_1, x') \in \mathbb{R} \times \mathbb{R}^{N-1}$ and set

$$Q'_L = (-L, L)^{N-1} \qquad \text{so that} \qquad Q_L = (-L, L) \times Q'_L.$$

By integrating over Q_L, we get

$$\begin{aligned}\int_{Q_L} \frac{\partial}{\partial x_1}(u\,\varphi)\,dx &= \int_{Q'_L} \left(\int_{-L}^{L} \frac{\partial}{\partial x_1}(u\,\varphi)\,dx_1\right) dx' \\ &= \int_{Q'_L} \left[u(L, x')\,\varphi(L, x') - u(-L, x')\,\varphi(-L, x')\right] dx' = 0.\end{aligned}$$

We used that $\varphi(L, x') = \varphi(-L, x') = 0$, thanks to the compact support of φ. On the other hand, by recalling that $u\,\varphi$ is compactly supported both in Ω and Q_L, we get

$$\int_\Omega \frac{\partial}{\partial x_1}(u\,\varphi)\,dx = \int_{Q_L} \frac{\partial}{\partial x_1}(u\,\varphi)\,dx.$$

Thus, in particular, by using the last two equations we obtain

$$\int_\Omega \frac{\partial u}{\partial x_1}\,\varphi\,dx + \int_\Omega \frac{\partial \varphi}{\partial x_1}\,u\,dx = 0,$$

that is

$$\int_\Omega \frac{\partial u}{\partial x_1}\,\varphi\,dx = -\int_\Omega \frac{\partial \varphi}{\partial x_1}\,u\,dx.$$

This proves the statement for $k = 1$. By repeating the argument for the other partial derivatives, we conclude. □

Remark 3.2.4 Let us consider the function of one variable

$$f(x) = |x|, \qquad \text{for } x \in (-1, 1).$$

This function is differentiable in classical sense only in $(-1, 0) \cup (0, 1)$ and it is not differentiable in 0. The classical derivative is given by

$$f'(x) = \begin{cases} 1, & \text{for } 0 < x < 1, \\ -1, & \text{for } -1 < x < 0. \end{cases}$$

On the other hand, it is not difficult to see that f has a weak derivative in $L^\infty((-1, 1))$ and this weak derivative coincides almost everywhere with the classical one f'. We refer to Problem 3.12.3 below.

Remark 3.2.5 (Important! Derivatives a. e. vs Weak Derivatives) In general, it is not true that if u is differentiable almost everywhere and $\nabla u \in L^p_{\mathrm{loc}}$, then it has a *weak* gradient in L^p_{loc}. For example, we take the piecewise constant function

$$u(x) = 1 \text{ for a.e. } x \in (0, 1), \qquad u(x) = 0 \text{ for a.e. } x \in (-1, 0),$$

considered on the open set $\Omega = (-1, 1)$. Let us suppose that u has a weak derivative in $L^p_{\mathrm{loc}}((-1, 1))$, i.e. there exists $g \in L^p_{\mathrm{loc}}((-1, 1))$ such that

$$\int_{-1}^{1} g(x)\,\varphi(x)\,dx = -\int_{-1}^{1} \varphi'(x)\,u(x)\,dx, \qquad \text{for every } \varphi \in C_0^\infty((-1, 1)).$$

By using the definition of u, this is the same as

$$\int_{-1}^{1} g(x)\,\varphi(x)\,dx = -\int_{-1}^{1} \varphi'(x)\,u(x)\,dx = -\int_{0}^{1} \varphi'(x)\,dx = \varphi(0), \tag{3.2.2}$$

for every $\varphi \in C_0^\infty((-1, 1))$. In particular, by taking $\varphi \in C_0^\infty((0, 1))$, we get from the previous identity

$$\int_{0}^{1} g(x)\,\varphi(x)\,dx = \int_{-1}^{1} g(x)\,\varphi(x)\,dx = 0.$$

By the Du Bois-Reymond Lemma (see Lemma 1.4.2), this implies that

$$g(x) = 0 \qquad \text{for a.e. } x \in (0, 1).$$

By using the same argument with test functions $\varphi \in C_0^\infty((-1, 0))$, we also get

$$g(x) = 0 \quad \text{for a.e. } x \in (-1, 0).$$

In conclusion, we get that $g \in L^p_{\mathrm{loc}}((-1, 1))$ vanishes almost everywhere. By using this into (3.2.2), we get

$$0 = \varphi(0), \qquad \text{for every } \varphi \in C_0^\infty((-1, 1)).$$

The latter is not always true, thus we get a contradiction.

We now need an *approximation* result: it will be quite useful. It asserts that the weak gradient can always be approximated in L^p by a sequence of classical gradients, when $p < \infty$. We fix at first some notation: for an open set $\Omega \subseteq \mathbb{R}^N$ with $\Omega \neq \mathbb{R}^N$, we define

$$r_\Omega := \sup_{x \in \Omega} \operatorname{dist}(x, \partial\Omega). \tag{3.2.3}$$

Then, as in Problem 1.7.26, we set

$$\mathrm{int}_n(\Omega) := \left\{x \in \Omega \, : \, \mathrm{dist}(x, \partial\Omega) > \frac{1}{n}\right\},$$

which is a non-empty open set, for every $n \in \mathbb{N}$ such that $n > 1/r_\Omega$. When $r_\Omega = +\infty$, the quantity $1/r_\Omega$ has to be simply interpreted as 0. In the particular case $\Omega = \mathbb{R}^N$, the sets $\mathrm{int}_n(\Omega)$ are intended to coincide with the whole $\mathbb{R}^N$.

Proposition 3.2.6 *Let $1 \le p < \infty$ and let $\Omega \subseteq \mathbb{R}^N$ be an open set. If $u \in L^p_{\mathrm{loc}}(\Omega)$ has a weak gradient in $L^p_{\mathrm{loc}}(\Omega; \mathbb{R}^N)$, for every $n > 1/r_\Omega$ we define the function*

$$\begin{aligned} u_n(x) := u * \rho_n(x) &= \int_{\mathbb{R}^N} u(y)\, \rho_n(x-y)\, dy \\ &= \int_{B_{\frac{1}{n}}(x)} u(y)\, \rho_n(x-y)\, dy, \qquad \textit{for } x \in \mathrm{int}_n(\Omega), \end{aligned}$$

where

$$\rho_n(x) = n^N\, \rho(n\, x),$$

and ρ is the standard smoothing kernel defined in (1.4.1). *Then:*

1. *for every $n > 1/r_\Omega$, we have $u_n \in C^\infty(\mathrm{int}_n(\Omega))$;*
2. *for every $n > 1/r_\Omega$, the gradient of u_n is given by*

$$\nabla u_n(x) = (\nabla u) * \rho_n(x), \qquad \textit{for a. e. } x \in \mathrm{int}_n(\Omega); \tag{3.2.4}$$

3. *for every $\Omega' \Subset \Omega$, we have*

$$\lim_{n\to\infty} \|u_n - u\|_{L^p(\Omega')} = \lim_{n\to\infty} \|\nabla u_n - \nabla u\|_{L^p(\Omega';\mathbb{R}^N)} = 0.$$

Proof The fact that $u_n \in C^\infty(\mathrm{int}_n(\Omega))$ follows from the regularity of ρ_n and the properties of convolutions (see for example [46, Theorem 2.16] and the subsequent remarks). Thus, we have in particular

$$u_n \in L^p_{\mathrm{loc}}(\mathrm{int}_n(\Omega)) \qquad \text{and} \qquad \nabla u_n \in L^p_{\mathrm{loc}}(\mathrm{int}_n(\Omega)),$$

thanks to Proposition 3.2.3. We now prove that the weak gradient is given by (3.2.4). We already know that

$$\nabla u_n(x) = u * (\nabla \rho_n)(x), \qquad \text{for } x \in \mathrm{int}_n(\Omega),$$

see again [46, Theorem 2.16], for example. In particular, for $k \in \{1, \dots, N\}$ we have

$$\begin{aligned}\frac{\partial u_n}{\partial x_k}(x) = u * \left(\frac{\partial \rho_n}{\partial x_k}\right)(x) &= \int_{B_{\frac{1}{n}}(x)} u(y)\,\frac{\partial \rho_n}{\partial x_k}(x-y)\,dy\\ &= -\int_{B_{\frac{1}{n}}(x)} u(y)\,\frac{\partial \rho_n}{\partial y_k}(x-y)\,dy\\ &= -\int_{B_r(x)} u(y)\,\frac{\partial \rho_n}{\partial y_k}(x-y)\,dy,\end{aligned}$$

where $r > 0$ is a radius such that $1/n < r < \mathrm{dist}(x, \partial\Omega)$. The choice of r entails that we have $B_{1/n}(x) \Subset B_r(x) \Subset \mathrm{int}_n(\Omega)$. By observing that $y \mapsto \rho_n(x-y)$ is a function in $C_0^\infty(B_r(x))$ and using the definition of weak derivative, we get

$$\begin{aligned}-\int_{B_r(x)} u(y)\,\frac{\partial \rho_n}{\partial y_k}(x-y)\,dy &= \int_{B_r(x)} \frac{\partial u}{\partial y_k}(y)\,\rho_n(x-y)\,dy\\ &= \int_{B_{\frac{1}{n}}(x)} \frac{\partial u}{\partial y_k}(y)\,\rho_n(x-y)\,dy.\end{aligned}$$

By arbitrariness of $k \in \{1, \dots, N\}$, this proves (3.2.4).

The last point of the statement is again a consequence of the properties of convolutions. As before, we refer the reader to [46, Chapter 2] for the missing details. □

The previous result will permit to show that many of the usual properties of functions which are differentiable in classical sense, actually hold for weakly differentiable functions, as well. Roughly speaking, it will be sufficient to argue by approximation.

We start with the following result: it asserts that the composition of a weakly differentiable function with a smooth invertible map is still weakly differentiable. This comes with the natural *Chain Rule formula* for computing the weak gradient of this composition. For clarity, we briefly recall some notation: we denote by $\mathcal{M}_N(\mathbb{R})$ the vector space of $N \times N$ matrices with real coefficients. If $M \in \mathcal{M}_N(\mathbb{R})$ and $x \in \mathbb{R}^N$, we indicate by

$$x \cdot M,$$

the product "row vector times matrix". In other words, if $M = (m_{ij})_{i,j=1}^N$, this product gives the vector of $\mathbb{R}^N$ whose components are

$$x \cdot M = \left(\sum_{i=1}^N x_i\, m_{i1}, \dots, \sum_{i=1}^N x_i\, m_{iN}\right).$$

Theorem 3.2.7 (Change of Variables) *Let $\Omega, \mathcal{O} \subseteq \mathbb{R}^N$ be two open sets and let $\Phi : \mathcal{O} \to \Omega$ be a C^1 invertible map, such that its inverse function Φ^{-1} is still C^1. Let $u \in L^p_{\mathrm{loc}}(\Omega)$ be a function having a weak gradient in $L^p_{\mathrm{loc}}(\Omega; \mathbb{R}^N)$, for $1 \le p \le \infty$. Then, the composition $u_\Phi := u \circ \Phi$ is such that*

$$u_\Phi \in L^p_{\mathrm{loc}}(\mathcal{O}), \ \nabla u_\Phi \in L^p_{\mathrm{loc}}(\mathcal{O}; \mathbb{R}^N),$$

and the weak gradient is given by

$$\nabla u_\Phi(x) = \nabla u(\Phi(x)) \cdot D\Phi(x), \qquad \text{for a. e. } x \in \mathcal{O}. \tag{3.2.5}$$

Proof We first notice that the Jacobian matrix

$$D\Phi = \left(\frac{\partial \Phi_i}{\partial x_j}\right)_{i,j=1}^N,$$

have entries in $L^\infty_{\mathrm{loc}}(\mathcal{O})$, where we denoted by Φ_i the i-th component of Φ. We also observe that

$$\det D\Phi(x) > 0, \qquad \text{for every } x \in \mathcal{O},$$

thanks to the fact that Φ^{-1} is C^1 (see for example [19, Chapter II, Section 6]). Thus, by using the continuity of $D\Phi$, we have in particular that for every $\mathcal{O}' \Subset \mathcal{O}$ there exists $c_{\mathcal{O}'} > 0$ such that

$$\det D\Phi(x) \ge c_{\mathcal{O}'}, \qquad \text{for every } x \in \mathcal{O}',$$

This in turn implies that

$$\det D\Phi^{-1} \in L^\infty_{\mathrm{loc}}(\Omega). \tag{3.2.6}$$

By using (3.2.6), the property $u_\Phi \in L^p_{\mathrm{loc}}(\mathcal{O})$ follows by making the change of variables $y = \Phi(x)$. Let us focus on the weak gradient. Observe that if (3.2.5) holds true, then we get that $\nabla u_\Phi \in L^p_{\mathrm{loc}}(\mathcal{O}; \mathbb{R}^N)$. Indeed, for every $\mathcal{O}' \Subset \mathcal{O}$, by using the change of variables $y = \Phi(x)$ and (3.2.6), we have for $1 \le p < \infty$

$$\begin{aligned}
\int_{\mathcal{O}'} |\nabla u_\Phi(x)|^p \, dx &= \int_{\mathcal{O}'} |\nabla u(\Phi(x)) \cdot D\Phi(x)|^p \, dx \\
&= \int_{\Phi(\mathcal{O}')} |\nabla u(y) \cdot D\Phi(\Phi^{-1}(y))|^p \, |\det D\Phi^{-1}(y)| \, dy \\
&\le \| \det D\Phi^{-1} \|_{L^\infty(\Phi(\mathcal{O}'))} \int_{\Phi(\mathcal{O}')} |\nabla u(y) \cdot D\Phi(\Phi^{-1}(y))|^p \, dy.
\end{aligned}$$

The latter is finite, since $\Phi(\mathcal{O}') \Subset \Omega$ and by Cauchy-Schwarz inequality we have[2]

$$|\nabla u \cdot D\Phi| = \sqrt{\sum_{j=1}^{N}\left(\sum_{i=1}^{N}\frac{\partial u}{\partial y_i}\,\frac{\partial \Phi_i}{\partial x_j}\right)^2} \le |\nabla u|\,\sqrt{\sum_{i,j=1}^{N}\left|\frac{\partial \Phi_i}{\partial x_j}\right|^2} \in L^p_{\mathrm{loc}}(\Omega).$$

The case $p = \infty$ follows from the previous estimate, as well.

We need to prove that u_Φ has a weak gradient, given by (3.2.5), i.e. we have to show that for $\varphi \in C_0^\infty(\mathcal{O})$ and $k \in \{1, \dots, N\}$, we have

$$\int_{\mathcal{O}} u(\Phi(x))\,\frac{\partial \varphi}{\partial x_k}(x)\,dx = -\sum_{i=1}^{N}\int_{\mathcal{O}}\frac{\partial u}{\partial y_i}(\Phi(x))\,\frac{\partial \Phi_i}{\partial x_k}(x)\,\varphi(x)\,dx. \tag{3.2.7}$$

At this aim, we proceed by approximation: we take the sequence $\{u_n\}_{n\in\mathbb{N}}$ of Proposition 3.2.6 and an open set $\Omega' \Subset \Omega$ such that $\mathcal{O}' := \Phi^{-1}(\Omega')$ contains the support of φ. We observe that

$$\int_{\mathcal{O}} u(\Phi(x))\,\frac{\partial \varphi}{\partial x_k}(x)\,dx = \lim_{n\to\infty}\int_{\mathcal{O}} u_n(\Phi(x))\,\frac{\partial \varphi}{\partial x_k}(x)\,dx.$$

Indeed, for n large enough we have $\Omega' \Subset \mathrm{int}_n(\Omega)$ and thus

$$\begin{aligned}
&\left|\int_{\mathcal{O}} u(\Phi(x))\,\frac{\partial \varphi}{\partial x_k}(x)\,dx - \int_{\mathcal{O}} u_n(\Phi(x))\,\frac{\partial \varphi}{\partial x_k}(x)\,dx\right| \\
&\qquad \le \|\nabla\varphi\|_{L^\infty(\mathcal{O}';\mathbb{R}^N)}\int_{\mathcal{O}'}|u(\Phi(x)) - u_n(\Phi(x))|\,dx \\
&\qquad \le \|\nabla\varphi\|_{L^\infty(\mathcal{O}';\mathbb{R}^N)}\int_{\Omega'}|u(y) - u_n(y)|\,|\det D\Phi^{-1}(y)|\,dy \\
&\qquad \le \|\nabla\varphi\|_{L^\infty(\mathcal{O}';\mathbb{R}^N)}\,\|\det D\Phi^{-1}\|_{L^\infty(\Omega')} \\
&\qquad \times \int_{\Omega'}|u(y) - u_n(y)|\,dy,
\end{aligned}$$

again by (3.2.6). The last integral converges to 0 as n goes to 0, by recalling the properties of $\{u_n\}_{n\in\mathbb{N}}$. With a similar argument, we get that

$$\lim_{n\to\infty}\sum_{i=1}^{N}\int_{\mathcal{O}}\frac{\partial u_n}{\partial y_i}(\Phi(x))\,\frac{\partial \Phi_i}{\partial x_k}(x)\,\varphi(x)\,dx = \sum_{i=1}^{N}\int_{\mathcal{O}}\frac{\partial u}{\partial y_i}(\Phi(x))\,\frac{\partial \Phi_i}{\partial x_k}(x)\,\varphi(x)\,dx.$$

[2] We use Cauchy-Schwarz inequality for the vectors ∇u and $\partial\Phi/\partial x_j$.

In light of these limits, in order to establish (3.2.7) it is sufficient to prove it for u_n in place of u, for n large enough. We observe that $u_n \circ \Phi \in C^1(\mathcal{O}_n)$, where $\mathcal{O}_n := \Phi^{-1}(\mathrm{int}_n(\Omega))$, and by Proposition 3.2.3 its weak gradient coincides with the classical one. This is given by the usual Chain Rule formula

$$\nabla u_n(\Phi(x)) \cdot D\Phi(x) = \left(\sum_{i=1}^{N} \frac{\partial u_n}{\partial y_i}(\Phi(x))\,\frac{\partial \Phi_i}{\partial x_1}(x), \dots, \sum_{i=1}^{N} \frac{\partial u_n}{\partial y_i}(\Phi(x))\,\frac{\partial \Phi_i}{\partial x_N}(x)\right).$$

Moreover, we know that this belongs to $L^\infty(\mathcal{O}')$, since $\mathcal{O}' = \Phi^{-1}(\Omega') \Subset \Phi^{-1}(\mathrm{int}_n(\Omega))$, for n large enough. In particular, we get

$$\int_{\mathcal{O}} u_n \circ \Phi\,\frac{\partial \varphi}{\partial x_k}\,dx = -\int_{\mathcal{O}} \frac{\partial}{\partial x_k}(u_n \circ \Phi)\,\varphi\,dx,$$

which is exactly (3.2.7) for u_n, thanks to the formula above for the gradient of $u_n \circ \Phi$. As explained above, this is enough to conclude the proof. □

Remark 3.2.8 It is clear that we can give a *global* character to the previous result, provided we enforce its assumptions as follows:

- $\Phi : \mathcal{O} \to \Omega$ is a C^1 invertible map with a C^1 inverse map Φ^{-1}, the Jacobian matrix

$$D\Phi = \left(\frac{\partial \Phi_i}{\partial x_j}\right)_{i,j=1}^{N},$$

  have entries in $L^\infty(\mathcal{O})$ and $\det D\Phi^{-1} \in L^\infty(\Omega)$;
- $u \in L^p(\Omega)$ and $\nabla u \in L^p(\Omega; \mathbb{R}^N)$.

In this case the composition $u_\Phi := u \circ \Phi$ is such that

$$u_\Phi \in L^p(\mathcal{O}), \ \nabla u_\Phi \in L^p(\mathcal{O}; \mathbb{R}^N).$$

The next result extends to weakly differentiable functions a classical fact about functions having a vanishing gradient.

Proposition 3.2.9 *Let $\Omega \subseteq \mathbb{R}^N$ be an open connected set. Let us suppose that $u \in L^1_{\mathrm{loc}}(\Omega)$ has a weak gradient in $L^1_{\mathrm{loc}}(\Omega; \mathbb{R}^N)$ such that*

$$\nabla u(x) = (0, \dots, 0), \qquad \text{for a. e. } x \in \Omega.$$

Then there exists a constant C such that

$$u(x) = C, \qquad \text{for a. e. } x \in \Omega.$$

Proof We take again the sequence $\{u_n\}_{n\in\mathbb{N}}$ of Proposition 3.2.6 and the sequence $\{\Omega_k\}_{k\in\mathbb{N}}$ of open *connected* subsets of Ω given by Problem 1.7.27. We recall that they have the following properties

$$\Omega = \bigcup_{k\in\mathbb{N}} \Omega_k, \qquad \Omega_k \subseteq \Omega_{k+1} \qquad \text{and} \qquad \Omega_k \Subset \Omega, \text{ for every } k \in \mathbb{N}.$$

We fix $k \in \mathbb{N}$, then there exists $\overline{n}_k \in \mathbb{N}$ such that $\Omega_k \Subset \mathrm{int}_n(\Omega)$, for every $n \geq \overline{n}_k$. Thus, by the properties of u_n we have

$$\nabla u_n(x) = (\nabla u) * \rho_n(x) = (0, \dots, 0), \qquad \text{for } x \in \Omega_k \text{ and } n \geq \overline{n}_k.$$

Since u_n is a smooth function on Ω_k, from the previous property we get that u_n is constant on the connected set Ω_k, for $n \geq \overline{n}_k$. Let us call $C_n(k)$ such a constant. For every $n, m \geq \overline{n}_k$ we then have

$$\begin{aligned} |C_n(k) - C_m(k)| &= \frac{1}{|\Omega_k|} \int_{\Omega_k} |u_n - u_m|\, dx \\ &\leq \frac{1}{|\Omega_k|} \int_{\Omega_k} |u_n - u|\, dx + \frac{1}{|\Omega_k|} \int_{\Omega_k} |u_m - u|\, dx. \end{aligned}$$

Since by Proposition 3.2.6 we have

$$\lim_{n\to\infty} \|u_n - u\|_{L^1(\Omega_k)} = 0,$$

we get that $\{C_n(k)\}_{n\in\mathbb{N}}$ is a Cauchy sequence of real numbers, thus it converges to $C(k) \in \mathbb{R}$. We then obtain

$$\begin{aligned} \int_{\Omega_k} |u - C(k)|\, dx &\leq \int_{\Omega_k} |u - u_n|\, dx + \int_{\Omega_k} |u_n - C(k)|\, dx \\ &= \int_{\Omega_k} |u - u_n|\, dx + |\Omega_k|\, |C_n(k) - C(k)|, \qquad \text{for every } n \geq \overline{n}_k. \end{aligned}$$

By taking the limit as n goes to ∞, we get from the previous estimate

$$\int_{\Omega_k} |u - C(k)|\, dx = 0.$$

This entails that u coincides with the constant $C(k)$ almost everywhere on Ω_k, for every $k \in \mathbb{N}$. Since $\Omega_k \subseteq \Omega_{k+1}$, we actually get that $C(k)$ does not depend on $k \in \mathbb{N}$. This implies that

$$u = C, \qquad \text{for a. e. } x \in \Omega_k, \text{ for every } k \in \mathbb{N}.$$

By recalling that $\{\Omega_k\}_{k\in\mathbb{N}}$ is an exhaustion of Ω, we get the conclusion. □

3.3 Sobolev Spaces

Definition 3.3.1 Let $1 \leq p \leq \infty$ and let $\Omega \subseteq \mathbb{R}^N$ be an open set. We define the *Sobolev space* $W^{1,p}(\Omega)$ by

$$W^{1,p}(\Omega) = \Big\{u \in L^p(\Omega) \ : \ \nabla u \in L^p(\Omega; \mathbb{R}^N)\Big\}.$$

The "local" version $W^{1,p}_{\mathrm{loc}}(\Omega)$ is defined accordingly, as

$$W^{1,p}_{\mathrm{loc}}(\Omega) = \Big\{u \in L^p_{\mathrm{loc}}(\Omega) \ : \ \nabla u \in L^p_{\mathrm{loc}}(\Omega; \mathbb{R}^N)\Big\}.$$

By the linearity of the weak gradient and the fact that L^p spaces are vector spaces, we immediately obtain that $W^{1,p}(\Omega)$ and $W^{1,p}_{\mathrm{loc}}(\Omega)$ are vector spaces over $\mathbb{R}$, as well.

Theorem 3.3.2 *Let $1 \leq p \leq \infty$ and let $\Omega \subseteq \mathbb{R}^N$ be an open set. We endow $W^{1,p}(\Omega)$ with the norm*

$$\|u\|_{W^{1,p}(\Omega)} := \|u\|_{L^p(\Omega)} + \|\nabla u\|_{L^p(\Omega;\mathbb{R}^N)}. \tag{3.3.1}$$

Then $W^{1,p}(\Omega)$ is a Banach space, i.e. for every Cauchy sequence $\{u_n\}_{n\in\mathbb{N}} \subseteq W^{1,p}(\Omega)$ there exists $u \in W^{1,p}(\Omega)$ such that

$$\lim_{n\to\infty} \|u_n - u\|_{W^{1,p}(\Omega)} = 0.$$

Proof It is not difficult to see that (3.3.1) actually defines a norm. We verify that $W^{1,p}(\Omega)$ is a Banach space. We take $\{u_n\}_{n\in\mathbb{N}} \subseteq W^{1,p}(\Omega)$ a Cauchy sequence. This means that for every $\varepsilon > 0$, there exists $n_\varepsilon \in \mathbb{N}$ such that

$$\|u_n - u_m\|_{W^{1,p}(\Omega)} < \varepsilon, \qquad \text{for every } n, m \geq n_\varepsilon.$$

Thus, thanks to the definition (3.3.1), we have that $\{u_n\}_{n\in\mathbb{N}}$ is a Cauchy sequence in $L^p(\Omega)$ and $\{\nabla u_n\}_{n\in\mathbb{N}}$ is a Cauchy sequence in $L^p(\Omega; \mathbb{R}^N)$. By using that L^p spaces are Banach spaces (see for example [46, Theorem 2.7]), we get that there exist $u \in L^p(\Omega)$ and $\phi \in L^p(\Omega; \mathbb{R}^N)$ such that

$$\lim_{n\to\infty} \|u_n - u\|_{L^p(\Omega)} = 0 \qquad \text{and} \qquad \lim_{n\to\infty} \|\nabla u_n - \phi\|_{L^p(\Omega;\mathbb{R}^N)} = 0.$$

In order to conclude, we need to show that $u \in W^{1,p}(\Omega)$ and $\nabla u = \phi$. For every $\varphi \in C^\infty_0(\Omega)$ and every $k \in \{1, \ldots, N\}$, by Hölder's inequality we have

$$\left| \int_\Omega \phi_k \, \varphi \, dx - \int_\Omega \frac{\partial u_n}{\partial x_k} \, \varphi \, dx \right|$$
$$\leq \int_\Omega |\varphi| \left| \phi_k - \frac{\partial u_n}{\partial x_k} \right| dx \leq \|\varphi\|_{L^{p'}(\Omega)} \left\| \phi_k - \frac{\partial u_n}{\partial x_k} \right\|_{L^p(\Omega)}.$$

Thus, by using the convergence claimed above, we get

$$\lim_{n\to\infty} \int_\Omega \frac{\partial u_n}{\partial x_k} \, \varphi \, dx = \int_\Omega \phi_k \, \varphi \, dx.$$

Analogously, we also get

$$\lim_{n\to\infty} \int_\Omega u_n \, \frac{\partial \varphi}{\partial x_k} \, dx = \int_\Omega u \, \frac{\partial \varphi}{\partial x_k} \, dx.$$

Thus, by using the definition of weak derivative, we obtain

$$\begin{aligned} \int_\Omega \phi_k \, \varphi \, dx &= \lim_{n\to\infty} \int_\Omega \frac{\partial u_n}{\partial x_k} \, \varphi \, dx \\ &= - \lim_{n\to\infty} \int_\Omega u_n \, \frac{\partial \varphi}{\partial x_k} \, dx = - \int_\Omega u \, \frac{\partial \varphi}{\partial x_k} \, dx. \end{aligned}$$

This shows that u has a weak gradient in L^p, coinciding with ϕ. We thus have proven that every Cauchy sequence in $W^{1,p}(\Omega)$ converges. □

Remark 3.3.3 (An Equivalent Norm) For $1 < p < \infty$, sometimes it is useful to use the following modified norm

$$u \mapsto \left(\|u\|^p_{L^p(\Omega)} + \|\nabla u\|^p_{L^p(\Omega;\mathbb{R}^N)} \right)^{\frac{1}{p}}, \qquad \text{for every } u \in W^{1,p}(\Omega). \tag{3.3.2}$$

By Problem 1.7.25, we know that this is indeed a norm, which is equivalent to (3.3.1). The modified norm (3.3.2) is particularly interesting in the case $p = 2$: indeed, in this case we see that it comes from the *scalar product*

$$\big[u, v\big]_{W^{1,2}(\Omega)} := \int_\Omega u \, v \, dx + \int_\Omega \langle \nabla u, \nabla v \rangle \, dx, \qquad \text{for every } u, v \in W^{1,2}(\Omega).$$

Accordingly, when endowed with this scalar product and the associated norm (3.3.2), the space $W^{1,2}(\Omega)$ is a *Hilbert space*. We refer the reader to [22], [24, Chpater VI], [35, Chapter IV] or [54] for an introduction to the general theory of Banach and Hilbert spaces.

Lemma 3.3.4 *Let* $1 \leq p \leq \infty$ *and let* Ω *be an open set. Let* $f \in L^p(\Omega)$ *and* $g \in W^{1,p}(\Omega)$ *be two functions such that*

$$f = g, \qquad \textit{a. e. in } \Omega.$$

Then $f \in W^{1,p}(\Omega)$ *and*

$$\nabla f = \nabla g, \qquad \textit{a. e. in } \Omega,$$

as well.

Proof For every $\varphi \in C_0^\infty(\Omega)$ and every $k \in \{1, \dots, N\}$, by definition of weak partial derivative we have

$$\int_\Omega \frac{\partial g}{\partial x_k}\, \varphi\, dx = -\int_\Omega g\, \frac{\partial \varphi}{\partial x_k}\, dx.$$

If we now use the assumption on f, we get from the previous identity

$$\int_\Omega \frac{\partial g}{\partial x_k}\, \varphi\, dx = -\int_\Omega f\, \frac{\partial \varphi}{\partial x_k}\, dx.$$

Thanks to the definition, this means that f has weak partial derivative in $L^p(\Omega)$ with respect to the k-th variable and there holds

$$\frac{\partial f}{\partial x_k} = \frac{\partial g}{\partial x_k}.$$

By arbitrariness of $k \in \{1, \dots, N\}$, we get the conclusion. □

Definition 3.3.5 Let $1 \le p < \infty$ and let $\Omega \subseteq \mathbb{R}^N$ be an open set. We say that a sequence $\{u_n\}_{n\in\mathbb{N}} \subseteq W^{1,p}(\Omega)$ *weakly converges* to $u \in W^{1,p}(\Omega)$ if

$$\lim_{n\to\infty} \int_\Omega u_n\, \varphi\, dx = \int_\Omega u\, \varphi \qquad \text{and} \qquad \lim_{n\to\infty} \int_\Omega \langle \nabla u_n, \phi\rangle\, dx = \int_\Omega \langle \nabla u, \phi\rangle\, dx,$$

for every $\varphi \in L^{p'}(\Omega)$ and $\phi \in L^{p'}(\Omega; \mathbb{R}^N)$.

The following compactness result will be useful.

Theorem 3.3.6 ("Banach-Alaoglu Meet Sobolev") *Let* $1 < p < \infty$ *and let* $\Omega \subseteq \mathbb{R}^N$ *be an open set. For every* $M > 0$, *the set*

$$\mathcal{B}_M = \Big\{u \in W^{1,p}(\Omega)\, :\, \|u\|_{W^{1,p}(\Omega)} \le M\Big\},$$

is sequentially weakly compact, i.e. for every $\{u_n\}_{n\in\mathbb{N}} \subseteq \mathcal{B}_M$, *there exists a subsequence weakly converging in* $W^{1,p}(\Omega)$ *to a function* $u \in \mathcal{B}_M$.

Proof It is sufficient to use the relevant properties of L^p spaces, together with the definition of Sobolev space. Indeed, if

$$\|u_n\|_{L^p(\Omega)} + \|\nabla u_n\|_{L^p(\Omega;\mathbb{R}^N)} = \|u_n\|_{W^{1,p}(\Omega)} \le M,$$

this implies that the sequence $\{(u_n, \nabla u_n)\}_{n\in\mathbb{N}}$ is bounded in $L^p(\Omega; \mathbb{R}^{N+1})$. Indeed, observe that

$$\begin{aligned}\|(u_n, \nabla u_n)\|_{L^p(\Omega;\mathbb{R}^{N+1})} &= \left(\int_\Omega (|u_n|^2 + |\nabla u_n|^2)^{\frac{p}{2}}\, dx\right)^{\frac{1}{p}} \\ &\le \left(\int_\Omega (|u_n| + |\nabla u_n|)^p\, dx\right)^{\frac{1}{p}} \\ &\le \|u_n\|_{L^p(\Omega)} + \|\nabla u_n\|_{L^p(\Omega;\mathbb{R}^N)},\end{aligned}$$

thanks to the fact that $\sqrt{a^2+b^2} \le |a| + |b|$ (by Problem 1.7.1 with $\alpha = 1/2$) and to Minkowski's inequality.

By using the Banach-Alaoglu Theorem for L^p spaces (see Theorem B.1.1), we can infer weak convergence (up to a subsequence) to an element $(u, \phi) \in L^p(\Omega; \mathbb{R}^{N+1})$. It is not difficult to see that we must have

$$\phi = (\phi_1, \dots, \phi_N) = \left(\frac{\partial u}{\partial x_1}, \dots, \frac{\partial u}{\partial x_N}\right). \tag{3.3.3}$$

Indeed, for every $\varphi \in C_0^\infty(\Omega)$ and $k \in \{1, \dots, N\}$, by the weak convergence and the definition of weak derivative we get

$$\begin{aligned}\int_\Omega \phi_k\, \varphi\, dx &= \lim_{n\to\infty} \int_\Omega \frac{\partial u_n}{\partial x_k}\, \varphi\, dx \\ &= -\lim_{n\to\infty} \int_\Omega u_n\, \frac{\partial \varphi}{\partial x_k}\, dx = -\int_\Omega u\, \frac{\partial \varphi}{\partial x_k}\, dx.\end{aligned}$$

This shows (3.3.3) and thus $u \in W^{1,p}(\Omega)$. We are only left with proving that we still have

$$\|u\|_{W^{1,p}(\Omega)} \le M.$$

For this, it is sufficient to use the bound on the norm of u_n, together with the lower semicontinuity of the L^p norm, with respect to the weak convergence (see Propositions 1.2.12 and 1.3.11). □

Remark 3.3.7 The previous proof shows that $W^{1,p}(\Omega)$ has exactly the remarkable property announced in (3.1.2), i.e. the weak limit of a family of gradients is still a gradient. Of course, here “gradients” must be understood in the weak sense.

Remark 3.3.8 (The Cases $p = 1$ and $p = \infty$) The conclusion of Theorem 3.3.6 *does not hold* for the limit case $p = 1$. We refer to Problem 3.12.4 below for a counter-example.

On the contrary, in the case $p = \infty$ the result would hold, provided one replaces the *weak convergence* with the so-called $*$-*weak convergence* (see for example [20] for the definition and more details). We prefer however to exclude this case, since it will not be needed for the scopes of this book.

3.4 Some Operations on Sobolev Functions

Proposition 3.4.1 (Chain Rule—C^1 Case) *Let $f : \mathbb{R} \to \mathbb{R}$ be a C^1 function with bounded derivative. Let $1 \le p \le \infty$ and let $\Omega \subseteq \mathbb{R}^N$ be an open set with finite measure. Then for every $u \in W^{1,p}(\Omega)$, we have*

$$f \circ u \in W^{1,p}(\Omega),$$

as well. Moreover, its weak gradient is given by the formula

$$\nabla(f \circ u) = f'(u)\, \nabla u. \tag{3.4.1}$$

Proof We first observe that for every $t, s \in \mathbb{R}$ we have

$$|f(t) - f(s)| = \left| \int_s^t f'(\tau)\, d\tau \right|.$$

By using that f' is bounded, if we set $L = \sup_{\mathbb{R}} |f'|$ we obtain in particular

$$|f(t) - f(s)| \le L\, |t - s|, \qquad \text{for } t, s \in \mathbb{R}. \tag{3.4.2}$$

By taking $s = 0$ and $t = u(x)$, we get in particular that

$$|f(u(x))| \le |f(u(x)) - f(0)| + |f(0)| \le L\, |u(x)| + |f(0)|, \qquad \text{for a. e. } x \in \Omega,$$

which implies that

$$\|f \circ u\|_{L^p(\Omega)} \le L\, \|u\|_{L^p(\Omega)} + |\Omega|^{\frac{1}{p}}\, |f(0)| < +\infty,$$

thanks to the fact that Ω has finite measure. This shows that $f \circ u \in L^p(\Omega)$.

We now have to show that this function has a weak gradient in $L^p(\Omega; \mathbb{R}^N)$, given by (3.4.1). We take $\varphi \in C_0^\infty(\Omega)$ and consider the sequence $\{u_n\}_{n\in\mathbb{N}}$ of Proposition 3.2.6. Since φ is compactly supported in Ω, there exists $\Omega' \Subset \Omega$ such that Ω' contains the support of φ. We observe that $\Omega' \Subset \mathrm{int}_n(\Omega)$, for n large enough.

For these functions, by Proposition 3.2.6 we have $f \circ u_n \in C^1(\mathrm{int}_n(\Omega))$ and the classical gradient is given by

$$\nabla f \circ u_n = f'(u_n)\, \nabla u_n.$$

By Proposition 3.2.3, we get that this is the weak gradient, as well, and it belongs to $L^p(\Omega'; \mathbb{R}^N)$. Thus, we obtain for every $k \in \{1, \dots, N\}$

$$\int_\Omega \frac{\partial \varphi}{\partial x_k}\, f \circ u_n \, dx = -\int_\Omega \varphi \, \frac{\partial}{\partial x_k}(f \circ u_n)\, dx = -\int_\Omega \varphi \, f'(u_n)\, \frac{\partial u_n}{\partial x_k}\, dx. \tag{3.4.3}$$

We observe that $W^{1,p}(\Omega') \subseteq W^{1,1}(\Omega')$ (see Problem 3.12.6 below), thus from Proposition 3.2.6 we also have

$$\lim_{n\to\infty} \|u_n - u\|_{L^1(\Omega')} = 0 \qquad \text{and} \qquad \lim_{n\to\infty} \|\nabla u_n - \nabla u\|_{L^1(\Omega';\mathbb{R}^N)} = 0. \tag{3.4.4}$$

Up to take a subsequence, we can further suppose that u_n converges almost everywhere to u in Ω'. From the first information in (3.4.4), by using estimate (3.4.2) and the fact that the support of φ is contained in Ω', we get

$$\begin{aligned}
\lim_{n\to\infty} \left| \int_\Omega \frac{\partial \varphi}{\partial x_k}\, f \circ u_n \, dx - \int_\Omega \frac{\partial \varphi}{\partial x_k}\, f \circ u \, dx \right| & \\
&\le \|\nabla \varphi\|_{L^\infty(\Omega';\mathbb{R}^N)} \lim_{n\to\infty} \int_{\Omega'} |f(u_n) - f(u)|\, dx \\
&\le \|\nabla \varphi\|_{L^\infty(\Omega';\mathbb{R}^N)}\, L \lim_{n\to\infty} \int_{\Omega'} |u_n - u|\, dx = 0.
\end{aligned}$$

As for the right-hand side of (3.4.3), we have

$$\begin{aligned}
\left| \int_\Omega \varphi\, f'(u_n)\, \frac{\partial u_n}{\partial x_k}\, dx - \int_\Omega \varphi\, f'(u)\, \frac{\partial u}{\partial x_k}\, dx \right| & \\
&\le \|\varphi\|_{L^\infty(\Omega')} \int_{\Omega'} \left| f'(u_n)\, \frac{\partial u_n}{\partial x_k} - f'(u)\, \frac{\partial u}{\partial x_k} \right| dx \\
&\le \|\varphi\|_{L^\infty(\Omega')} \int_{\Omega'} |f'(u_n)| \left| \frac{\partial u_n}{\partial x_k} - \frac{\partial u}{\partial x_k} \right| dx \\
&+ \|\varphi\|_{L^\infty(\Omega')} \int_{\Omega'} |f'(u_n) - f'(u)| \left| \frac{\partial u}{\partial x_k} \right| dx.
\end{aligned}$$

By using that f' is bounded, from (3.4.4) we get that

$$\lim_{n\to\infty} \int_{\Omega'} |f'(u_n)| \left| \frac{\partial u_n}{\partial x_k} - \frac{\partial u}{\partial x_k} \right| dx = 0.$$

Moreover, by using that f' is bounded and continuous and that u_n converges almost everywhere to u in Ω', we can apply the Dominated Convergence Theorem in the second integral and get

$$\lim_{n\to\infty} \int_{\Omega'} |f'(u_n) - f'(u)| \left| \frac{\partial u}{\partial x_k} \right| dx = 0,$$

as well. By taking the limit on both sides of (3.4.3), we then get

$$\int_\Omega \frac{\partial \varphi}{\partial x_k} f \circ u \, dx = - \int_\Omega \varphi \, f'(u) \, \frac{\partial u}{\partial x_k} \, dx.$$

Since $\varphi \in C_0^\infty(\Omega)$ and $k \in \{1, \ldots, N\}$ are arbitrary, this shows that $f \circ u$ has a weak gradient in $L^p(\Omega; \mathbb{R}^N)$ and this is given by

$$\nabla(f \circ u) = \left(f'(u) \, \frac{\partial u}{\partial x_1}, \ldots, f'(u) \, \frac{\partial u}{\partial x_N} \right).$$

This concludes the proof. □

Remark 3.4.2 The previous result continues to hold even for an open set $\Omega \subseteq \mathbb{R}^N$ such that $|\Omega| = +\infty$, provided that we add the further requirement $f(0) = 0$.

We also notice that formula (3.4.1) still holds if $u \in W^{1,p}_{\mathrm{loc}}(\Omega)$: in this case, we have $f \circ u \in W^{1,p}_{\mathrm{loc}}(\Omega)$, as well. This immediately follows from the proof above.

The following result contains a particular Chain Rule formula, for compositions with some remarkable functions, which are not C^1.

Proposition 3.4.3 *Let $1 \le p \le \infty$ and let $\Omega \subseteq \mathbb{R}^N$ be an open set. For every $u \in W^{1,p}(\Omega)$ we set*

$$u_+(x) = \max\{u(x), 0\} \qquad \textit{and} \qquad u_-(x) = \min\{u(x), 0\},$$

then we still have $u_+, u_- \in W^{1,p}(\Omega)$. Moreover, for almost every $x \in \Omega$, their weak gradients are given by

$$\nabla u_+(x) = \begin{cases} \nabla u(x), & \textit{if } u(x) > 0, \\ 0, & \textit{if } u(x) \le 0, \end{cases} \qquad \textit{and} \qquad \nabla u_-(x) = \begin{cases} \nabla u(x), & \textit{if } u(x) < 0, \\ 0, & \textit{if } u(x) \ge 0. \end{cases}$$

Proof We will prove the result for u_+ only, by leaving to the reader the case of u_-. The fact that $u_+ \in L^p(\Omega)$ follows from the fact that $u_+ \le |u|$. For every $\varepsilon > 0$, we consider the C^1 function

$$f_\varepsilon(t) = \begin{cases} \sqrt{\varepsilon^2 + t^2} - \varepsilon, & \text{if } t \ge 0, \\ 0, & \text{if } t < 0, \end{cases}$$

whose derivative is given by

$$f_\varepsilon'(t) = \begin{cases} \dfrac{t}{\sqrt{\varepsilon^2+t^2}}, & \text{if } t \geq 0, \\ 0, & \text{if } t < 0. \end{cases}$$

Then we observe that

$$|f_\varepsilon(u(x)) - u_+(x)| \leq 2\,\varepsilon, \qquad \text{for a. e. } x \in \Omega. \tag{3.4.5}$$

Indeed, by using the definition we have

$$|f_\varepsilon(u(x)) - u_+(x)| = 0, \qquad \text{if } u(x) < 0,$$

while for $u(x) \geq 0$ we get

$$\begin{aligned} |f_\varepsilon(u(x)) - u_+(x)| &= \left|\sqrt{\varepsilon^2 + u(x)^2} - \varepsilon - u(x)\right| \\ &\leq \left|\sqrt{\varepsilon^2 + u(x)^2} - u(x)\right| + \varepsilon \\ &= \sqrt{\varepsilon^2 + u(x)^2} - u(x) + \varepsilon \leq \varepsilon + u(x) - u(x) + \varepsilon = 2\,\varepsilon. \end{aligned}$$

In the last inequality, we used that $\sqrt{a+b} \leq \sqrt{a} + \sqrt{b}$ for every $a, b \geq 0$, thanks to Problem 1.7.1. This implies (3.4.5).

Since f_ε is C^1 with bounded derivative and such that $f_\varepsilon(0) = 0$, from Proposition 3.4.1 and Remark 3.4.2 we can infer that $f_\varepsilon \circ u \in W^{1,p}(\Omega)$. Moreover, the weak gradient is given by

$$\nabla f_\varepsilon \circ u(x) = f_\varepsilon'(u(x))\, \nabla u(x), \qquad \text{for a. e. } x \in \Omega.$$

By using the definition of weak derivative, for every $\varphi \in C_0^\infty(\Omega)$ and $k \in \{1, \ldots, N\}$ we have

$$\int_\Omega \frac{\partial \varphi}{\partial x_k}\, f_\varepsilon(u)\, dx = -\int_\Omega \varphi\, \frac{\partial (f_\varepsilon \circ u)}{\partial x_k}\, dx = -\int_\Omega \varphi\, f_\varepsilon'(u)\, \frac{\partial u}{\partial x_k}\, dx. \tag{3.4.6}$$

We now want to take this identity to the limit. Observe that by construction

$$|f_\varepsilon'(u(x))| \leq 1, \qquad \text{for a. e. } x \in \Omega,$$

and for almost every $x \in \Omega$

$$\lim_{\varepsilon \to 0^+} f_\varepsilon'(u(x)) = \begin{cases} 1, & \text{if } u(x) > 0, \\ 0 & \text{if } u(x) \leq 0. \end{cases}$$

Thus, we can pass to the limit as ε goes to 0 in (3.4.6), by using (3.4.5) in the left-hand side and the Dominated Convergence Theorem in the right-hand side (recall that φ has compact support). This gives

$$\int_\Omega \frac{\partial \varphi}{\partial x_k}\, u_+\, dx = -\int_\Omega \varphi\, \frac{\partial u}{\partial x_k}\, 1_{\{u>0\}}\, dx, \qquad \text{for every } \varphi \in C_0^\infty(\Omega).$$

By recalling the definition of weak derivative, this shows that u_+ has a weak gradient in $L^p(\Omega; \mathbb{R}^N)$, whose components are given by

$$\frac{\partial u_+}{\partial x_k}(x) = \frac{\partial u}{\partial x_k}(x)\, 1_{\{u>0\}}(x), \qquad \text{for } k \in \{1, \dots, N\},$$

as desired. □

As an important consequence of the previous result, we have the following

Corollary 3.4.4 *Let $1 \le p \le \infty$ and let $\Omega \subseteq \mathbb{R}^N$ be an open set. For every pair $u, v \in W^{1,p}(\Omega)$, we define*

$$W = \max\{u, v\} \qquad \textit{and} \qquad w = \min\{u, v\}.$$

Then $W, w \in W^{1,p}(\Omega)$, as well. Their weak gradients are given by

$$\nabla W(x) = \begin{cases} \nabla u(x), & \textit{if } u(x) > v(x), \\ \nabla v(x), & \textit{if } v(x) \ge u(x), \end{cases} \qquad \textit{and} \qquad \nabla w(x) = \begin{cases} \nabla u(x), & \textit{if } u(x) < v(x), \\ \nabla v(x), & \textit{if } u(x) \ge v(x). \end{cases}$$

Proof We observe that for every $a, b \in \mathbb{R}$ we have

$$\max\{a, b\} = (a - b)_+ + b.$$

This implies that

$$W(x) = (u(x) - v(x))_+ + v(x), \qquad \text{for } x \in \Omega.$$

It is now sufficient to use Proposition 3.4.3 for the function $u - v$ and the fact that $W^{1,p}(\Omega)$ is a vector space. The proof for w is similar and it is left to the reader. □

Remark 3.4.5 It is clear that

$$\max\{u, v\} = \max\{v, u\} \qquad \text{and} \qquad \min\{v, u\} = \min\{u, v\}.$$

In particular, we have

$$\nabla \max\{u, v\} = \nabla \max\{v, u\}, \qquad \text{a. e. in } \Omega.$$

By using the formula of Corollary 3.4.4, we get that for almost every $x \in \Omega$

$$\begin{cases} \nabla u(x), & \text{if } u(x) > v(x) \\ \nabla v(x), & \text{if } v(x) \geq u(x) \end{cases} = \begin{cases} \nabla v(x), & \text{if } v(x) > u(x) \\ \nabla u(x), & \text{if } u(x) \geq v(x). \end{cases}$$

This in particular implies that the following remarkable identity holds

$$\nabla u(x) = \nabla v(x), \qquad \text{for a. e. } x \in \{y \in \Omega \,:\, u(y) = v(y)\}. \tag{3.4.7}$$

Corollary 3.4.6 (Chain Rule—The Absolute Value) *Let $1 \leq p \leq \infty$ and let $\Omega \subseteq \mathbb{R}^N$ be an open set. For every $u \in W^{1,p}(\Omega)$, we have $|u| \in W^{1,p}(\Omega)$, as well. Moreover, it holds*

$$\big|\nabla |u(x)|\big| = \big|\nabla u(x)\big|, \qquad \text{for a. e. } x \in \Omega.$$

Proof Observe that

$$|u(x)| = \max\{u(x), -u(x)\}.$$

By applying Corollary 3.4.4, we have for almost every $x \in \Omega$

$$\nabla |u(x)| = \begin{cases} \nabla u(x), & \text{if } u(x) > 0, \\ -\nabla u(x), & \text{if } u(x) \leq 0. \end{cases}$$

By taking the modulus on both sides, we get the desired conclusion. □

Proposition 3.4.7 (Leibniz Rule) *Let $1 \leq p \leq \infty$ and let $\Omega \subseteq \mathbb{R}^N$ be an open set. If $u \in W^{1,p}(\Omega)$ and $\eta \in C^1(\Omega) \cap W^{1,\infty}(\Omega)$, then we have $\eta\, u \in W^{1,p}(\Omega)$, as well. Moreover, its weak gradient is given by*

$$\nabla(\eta\, u) = u\, \nabla\eta + \eta\, \nabla u. \tag{3.4.8}$$

Proof We notice at first that we have $\eta\, u \in L^p(\Omega)$, since

$$|\eta(x)\, u(x)| \leq \|\eta\|_{L^\infty(\Omega)}\, |u(x)|, \qquad \text{for a. e. } x \in \Omega.$$

This implies that

$$\|\eta\, u\|_{L^p(\Omega)} \leq \|\eta\|_{L^\infty(\Omega)}\, \|u\|_{L^p(\Omega)} < +\infty.$$

We now have to show that $\eta\, u$ has a weak gradient in L^p, given by formula (3.4.8). To this aim, we first observe that

$$u\, \nabla\eta + \eta\, \nabla u \in L^p(\Omega; \mathbb{R}^N),$$

since by Minkowski's inequality

$$\begin{aligned}\|u\,\nabla\eta+\eta\,\nabla u\|_{L^p(\Omega;\mathbb{R}^N)} &\le \|u\,\nabla\eta\|_{L^p(\Omega;\mathbb{R}^N)}+\|\eta\,\nabla u\|_{L^p(\Omega;\mathbb{R}^N)}\\ &\le \|\nabla\eta\|_{L^\infty(\Omega;\mathbb{R}^N)}\,\|u\|_{L^p(\Omega)}\\ &+\|\eta\|_{L^\infty(\Omega)}\,\|\nabla u\|_{L^p(\Omega;\mathbb{R}^N)}<+\infty.\end{aligned}$$

We now take $\varphi \in C_0^\infty(\Omega)$ and consider again the sequence $\{u_n\}_{n\in\mathbb{N}}$ of Proposition 3.2.6. If we take $\Omega' \Subset \Omega$ containing the support of φ, the product $\eta\, u_n$ belongs to $C^1(\Omega')$, for n large enough. By the classical Leibniz rule we have

$$\nabla(\eta\, u_n)=u_n\,\nabla\eta+\eta\,\nabla u_n.$$

Since φ has compact support contained in Ω', we have

$$\varphi\,\frac{\partial(\eta\, u_n)}{\partial x_k}=\varphi\left(u_n\,\frac{\partial\eta}{\partial x_k}+\eta\,\frac{\partial u_n}{\partial x_k}\right)\in L^1(\Omega'),$$

and also

$$\frac{\partial\varphi}{\partial x_k}\,\eta\, u_n\in L^1(\Omega').$$

By Proposition 3.2.3, we get for $k\in\{1,\dots,N\}$

$$\begin{aligned}\int_\Omega \frac{\partial\varphi}{\partial x_k}\,\eta\,u_n\,dx &= \int_{\Omega'}\frac{\partial\varphi}{\partial x_k}\,\eta\,u_n\,dx\\ &=-\int_{\Omega'}\varphi\,\frac{\partial(\eta\,u_n)}{\partial x_k}=-\int_{\Omega'}\varphi\left(u_n\,\frac{\partial\eta}{\partial x_k}+\eta\,\frac{\partial u_n}{\partial x_k}\right)dx.\end{aligned}$$

We now recall that the sequence $\{u_n\}_{n\in\mathbb{N}}$ converges strongly in $W^{1,1}(\Omega')$ to u, by Proposition 3.2.6. Thus, we can pass to the limit as n goes to ∞, so to obtain

$$\begin{aligned}\int_\Omega\frac{\partial\varphi}{\partial x_k}\,\eta\,u\,dx &=-\int_{\Omega'}\varphi\left(u\,\frac{\partial\eta}{\partial x_k}+\eta\,\frac{\partial u}{\partial x_k}\right)dx\\ &=-\int_{\Omega}\varphi\left(u\,\frac{\partial\eta}{\partial x_k}+\eta\,\frac{\partial u}{\partial x_k}\right)dx,\qquad \text{for } k\in\{1,\dots,N\}.\end{aligned}$$

By appealing to the definition of weak derivative, this shows that $\eta\,u$ has a weak gradient given by (3.4.8). Moreover, this lies in $L^p(\Omega;\mathbb{R}^N)$, as already verified. This concludes the proof. □

Remark 3.4.8 We refer to Problems 3.12.18 and 3.12.19 for some useful variants of the previous result.

3.5 Poincaré-type Inequalities

In this section we consider Poincaré-type inequalities, i.e. inequalities of the form

$$\int_\Omega |\varphi|^p\, dx \le C \int_\Omega |\nabla\varphi|^p\, dx, \qquad \text{for every } \varphi \in C_0^\infty(\Omega),$$

for an open set $\Omega \subseteq \mathbb{R}^N$, thus generalizing to dimensions larger than 1 the result of Theorem 2.4.6. Differently from Chap. 2, we admit a general exponent p, not necessarily equal to 2. We focus on giving *sufficient* conditions ensuring the validity of this inequality, without discussing the optimality[3] of the relevant constant obtained.

We need at first the following

Definition 3.5.1 We say that an open set $\Omega \subseteq \mathbb{R}^N$ is *bounded in direction* $\omega \in \mathbb{S}^{N-1}$ if

$$\inf_{x\in\Omega}\langle x, \omega\rangle > -\infty \qquad \text{and} \qquad \sup_{x\in\Omega}\langle x, \omega\rangle < +\infty.$$

In this case, we call *width of* Ω *in direction* ω the quantity

$$\mathrm{width}(\Omega; \omega) := \sup_{x\in\Omega}\langle x, \omega\rangle - \inf_{x\in\Omega}\langle x, \omega\rangle.$$

Observe that if we set $b = \sup_{x\in\Omega}\langle x, \omega\rangle$ and $a = \inf_{x\in\Omega}\langle x, \omega\rangle$, we have that

$$\Omega \subseteq \Big\{x \in \mathbb{R}^N \,:\, a \le \langle x, \omega\rangle \le b\Big\},$$

i.e. Ω is contained in a *slab*: this is an infinite region enclosed by two parallel hyperplanes.

Proposition 3.5.2 (Poincaré Inequality) *Let* $1 \le p \le \infty$ *and let* $\Omega \subseteq \mathbb{R}^N$ *be an open set bounded in direction* $\omega \in \mathbb{S}^{N-1}$. *Then for every* $\varphi \in C_0^\infty(\Omega)$ *we have*

$$\frac{1}{\mathrm{width}(\Omega; \omega)}\, \|\varphi\|_{L^p(\Omega)} \le \|\nabla\varphi\|_{L^p(\Omega;\mathbb{R}^N)}. \tag{3.5.1}$$

Proof The proof is quite similar to that of Lemma 2.4.1 for the one-dimensional case. Let us set $w_\Omega = \mathrm{width}(\Omega; \omega)$, up to a rigid motion we can suppose that $\omega = \mathbf{e}_1 = (1, 0, \ldots, 0)$ and

$$\Omega \subseteq [0, w_\Omega] \times \mathbb{R}^{N-1} =: \Sigma.$$

[3] We will partly come back on the question of the sharp constant in Chap. 4.

We take $\varphi \in C_0^\infty(\Omega)$ and extend it by 0 outside Ω. By the fundamental theorem of Calculus, we have for every $x = (x_1, x') \in [0, w_\Omega] \times \mathbb{R}^{N-1}$

$$|\varphi(x)| = |\varphi(x_1, x')| = \left|\varphi(x_1, x') - \varphi\left(0, x'\right)\right| = \left|\int_0^{x_1} \frac{\partial \varphi}{\partial x_1}(t, x')\, dt\right|$$
$$\leq \int_0^{x_1} \left|\frac{\partial \varphi}{\partial x_1}(t, x')\right| dt \qquad (3.5.2)$$
$$\leq \int_0^{w_\Omega} \left|\frac{\partial \varphi}{\partial x_1}(t, x')\right| dt.$$

From this estimate, the case $p = \infty$ is readily obtained. Indeed, we get in particular for every $x \in \Sigma$

$$|\varphi(x)| \leq w_\Omega \left\|\frac{\partial \varphi}{\partial x_1}\right\|_{L^\infty(\Sigma)} \leq w_\Omega \|\nabla\varphi\|_{L^\infty(\Sigma;\mathbb{R}^N)} = w_\Omega \|\nabla\varphi\|_{L^\infty(\Omega;\mathbb{R}^N)}.$$

In the second inequality we used that

$$\left|\frac{\partial \varphi}{\partial x_1}\right| \leq |\nabla\varphi|. \qquad (3.5.3)$$

We now consider the case $1 \leq p < \infty$. We raise (3.5.2) to the power p and use Jensen's inequality (Proposition 1.2.11), so to get

$$|\varphi(x)|^p \leq w_\Omega^{p-1} \int_0^{w_\Omega} \left|\frac{\partial \varphi}{\partial x_1}(t, x')\right|^p dt.$$

We integrate with respect to x_1 on the interval $[0, w_\Omega]$. This yields

$$\int_0^{w_\Omega} |\varphi(x_1, x')|^p\, dx_1 \leq w_\Omega^{p-1} \int_0^{w_\Omega} \int_0^{w_\Omega} \left|\frac{\partial \varphi}{\partial x_1}(t, x')\right|^p dt\, dx_1$$
$$= w_\Omega^p \int_0^{w_\Omega} \left|\frac{\partial \varphi}{\partial x_1}(t, x')\right|^p dt.$$

We can integrate with respect to the remaining group of variables, so to get

$$\int_\Sigma |\varphi|^p\, dx \leq w_\Omega^p \int_\Sigma \left|\frac{\partial \varphi}{\partial x_1}\right|^p dx.$$

By recalling that φ has compact support in $\Omega \subseteq \Sigma$ and using (3.5.3), we get the conclusion for $1 \leq p < \infty$, as well. □

Remark 3.5.3 A closer inspection of the previous proof immediately reveals that the following stronger inequality holds

$$\frac{1}{\text{width}(\Omega;\omega)}\,\|\varphi\|_{L^p(\Omega)} \le \left\|\frac{\partial\varphi}{\partial\omega}\right\|_{L^p(\Omega)}, \qquad \text{for every } \varphi \in C_0^\infty(\Omega),$$

under the assumptions of Proposition 3.5.2.

In what follows, for an open bounded set $\Omega \subseteq \mathbb{R}^N$, we indicate its *diameter* by

$$\text{diam}(\Omega) = \sup_{x,y\in\Omega} |x-y|.$$

Corollary 3.5.4 *Let $1 \le p \le \infty$ and let $\Omega \subseteq \mathbb{R}^N$ be an open bounded set. Then for every $\varphi \in C_0^\infty(\Omega)$ we have*

$$\frac{1}{\text{diam}(\Omega)}\,\|\varphi\|_{L^p(\Omega)} \le \|\nabla\varphi\|_{L^p(\Omega;\mathbb{R}^N)}. \tag{3.5.4}$$

Proof It is enough to prove that, for an open bounded set, we have

$$\text{width}(\Omega;\omega) \le \text{diam}(\Omega), \qquad \text{for every } \omega \in \mathbb{S}^{N-1}. \tag{3.5.5}$$

Then (3.5.4) would follow directly from (3.5.1). In order to prove (3.5.5), we fix $\omega \in \mathbb{S}^{N-1}$ and call

$$a_\omega = \inf_{x\in\Omega}\langle x,\omega\rangle \qquad \text{and} \qquad b_\omega = \sup_{x\in\Omega}\langle x,\omega\rangle.$$

Since Ω is bounded, it is easy to see that these quantities are finite. Thus, for every $\varepsilon > 0$ there exist $x_\varepsilon, y_\varepsilon \in \Omega$ such that

$$\langle x_\varepsilon,\omega\rangle > b_\omega - \frac{\varepsilon}{2} \qquad \text{and} \qquad \langle y_\varepsilon,\omega\rangle < a_\omega + \frac{\varepsilon}{2}.$$

In particular, we get

$$b_\omega - a_\omega - \varepsilon < \langle x_\varepsilon,\omega\rangle - \langle y_\varepsilon,\omega\rangle = \langle x_\varepsilon - y_\varepsilon,\omega\rangle \le |x_\varepsilon - y_\varepsilon|.$$

The last term $|x_\varepsilon - y_\varepsilon|$ is not larger than $\text{diam}(\Omega)$, by definition of diameter. Thus, by recalling the definition of width, we obtain

$$\text{width}(\Omega;\omega) - \varepsilon < \text{diam}(\Omega).$$

By arbitrariness of both $\varepsilon > 0$ and $\omega \in \mathbb{S}^{N-1}$, we get (3.5.5). □

Remark 3.5.5 (A Necessary Condition) For a general open set, Poincaré inequality does not hold. For example, for $\Omega = \mathbb{R}^N$ it is easily seen that it can not be valid: it is sufficient to use a scaling argument as in the solution of Problem 2.8.10. More generally, the same argument shows that Poincaré inequality fails to hold for every open set containing arbitrarily large balls. Thus, a necessary condition for its validity is that

$$\sup\{r > 0 \,:\, \exists x_0 \in \Omega \text{ such that } B_r(x_0) \subseteq \Omega\} < +\infty.$$

The next result is a sort of "weighted" Poincaré inequality, for compactly supported functions on the punctured space $\mathbb{R}^N \setminus \{0\}$.

Theorem 3.5.6 (Hardy's Inequality) *Let $1 \le p < \infty$ be such that $p \neq N$. Then, for every $\varphi \in C_0^\infty(\mathbb{R}^N \setminus \{0\})$ we have*

$$\left|\frac{N-p}{p}\right|^p \int_{\mathbb{R}^N} \frac{|\varphi|^p}{|x|^p}\,dx \le \int_{\mathbb{R}^N} |\nabla\varphi|^p\,dx. \tag{3.5.6}$$

Moreover, in the case $1 \le p < N$, the inequality still holds if $\varphi \in C_0^\infty(\mathbb{R}^N)$.

Proof We first prove the inequality (3.5.6) for $1 \le p < \infty$ such that $p \neq N$. We take $\alpha \neq 0$ and consider the function $V_\alpha(x) = |x|^\alpha$, which is C^∞ in $\mathbb{R}^N \setminus \{0\}$. Its gradient is given by

$$\nabla V_\alpha(x) = \alpha\,|x|^{\alpha-2}\,x, \qquad \text{for } x \neq 0.$$

We then compute the p-Laplacian operator for the function V_α (recall (1.6.4))

$$\begin{aligned}
-\Delta_p V_\alpha &= -\mathrm{div}(|\nabla V_\alpha|^{p-2}\,\nabla V_\alpha)\\
&= -|\alpha|^{p-2}\,\alpha\,\mathrm{div}(|x|^{\alpha\,(p-1)-p}\,x)\\
&= -|\alpha|^{p-2}\,\alpha\left(|x|^{\alpha\,(p-1)-p}\mathrm{div}x + \langle x, \nabla|x|^{\alpha\,(p-1)-p}\rangle\right)\\
&= -|\alpha|^{p-2}\,\alpha\,(N+\alpha\,(p-1)-p)\,|x|^{\alpha\,(p-1)-p}.
\end{aligned}$$

By recalling the definition of V_α, the latter can be rewritten as

$$-\Delta_p V_\alpha(x) = -|\alpha|^{p-2}\,\alpha\,(N+\alpha\,(p-1)-p)\,\frac{(V_\alpha(x))^{p-1}}{|x|^p}, \qquad \text{for every } x \neq 0. \tag{3.5.7}$$

We restrict for the moment to the case $p > 1$ and take a function $\varphi \in C_0^\infty(\mathbb{R}^N \setminus \{0\})$. We multiply (3.5.7) by[4] $|\varphi|^p/V_\alpha^{p-1}$ and integrate over $\mathbb{R}^N \setminus \{0\}$. We get

$$-\alpha\,|\alpha|^{p-2}\,(N+\alpha\,(p-1)-p)\int_{\mathbb{R}^N}\frac{|\varphi|^p}{|x|^p}\,dx = -\int_{\mathbb{R}^N}\mathrm{div}(|\nabla V_\alpha|^{p-2}\,\nabla V_\alpha)\,\frac{|\varphi|^p}{V_\alpha^{p-1}}\,dx.$$

[4] Observe that this function belongs to $C_0^1(\mathbb{R}^N \setminus \{0\})$, since $p > 1$.

We can now write the rightmost integral as follows

$$-\int_{\mathbb{R}^N} \operatorname{div}(|\nabla V_\alpha|^{p-2}\,\nabla V_\alpha)\,\frac{|\varphi|^p}{V_\alpha^{p-1}}\,dx = -\int_{\mathbb{R}^N} \operatorname{div}\left(|\nabla V_\alpha|^{p-2}\,\nabla V_\alpha\,\frac{|\varphi|^p}{V_\alpha^{p-1}}\right)dx$$
$$+\int_{\mathbb{R}^N}\left\langle |\nabla V_\alpha|^{p-2}\,\nabla V_\alpha, \nabla\left(\frac{|\varphi|^p}{V_\alpha^{p-1}}\right)\right\rangle dx,$$

and then use the Divergence Theorem on the first integral on the right-hand side. By using that φ is compactly supported, we get that such an integral vanishes. Thus, we obtain

$$-\alpha\,|\alpha|^{p-2}\,(N+\alpha\,(p-1)-p)\int_{\mathbb{R}^N}\frac{|\varphi|^p}{|x|^p}\,dx$$
$$=\int_{\mathbb{R}^N}\left\langle |\nabla V_\alpha|^{p-2}\,\nabla V_\alpha, \nabla\left(\frac{|\varphi|^p}{V_\alpha^{p-1}}\right)\right\rangle dx.$$

On the right-hand side we apply Picone's inequality (see Lemma 1.3.5), so to obtain

$$-\alpha\,|\alpha|^{p-2}\,(N+\alpha\,(p-1)-p)\int_{\mathbb{R}^N}\frac{|\varphi|^p}{|x|^p}\,dx \le \int_{\mathbb{R}^N}|\nabla\varphi|^p\,dx.$$

Observe that the right-hand side is independent of α, while in the left-hand side we can choose any $\alpha \neq 0$. We make the following choice[5]

$$\alpha = \frac{p-N}{p},$$

which is feasible, since $p \neq N$. We then obtain

$$\left|\frac{N-p}{p}\right|^p \int_{\mathbb{R}^N}\frac{|\varphi|^p}{|x|^p}\,dx \le \int_{\mathbb{R}^N}|\nabla\varphi|^p\,dx, \quad \text{for every } \varphi \in C_0^\infty(\mathbb{R}^N\setminus\{0\}). \tag{3.5.8}$$

This proves the desired inequality for $p \neq N$ and $p > 1$.

The case $p = 1 < N$ can be handled similarly: for every $\varepsilon > 0$ we introduce the non-negative function

$$f_\varepsilon(t) = \sqrt{\varepsilon^2 + t^2} - \varepsilon, \qquad \text{for } t \in \mathbb{R}.$$

[5] Observe that this is the value of α making the quantity

$$-\alpha\,|\alpha|^{p-2}\,(N+\alpha\,(p-1)-p),$$

as large as possible.

We take again $\varphi \in C_0^\infty(\mathbb{R}^N \setminus \{0\})$, multiply (3.5.7) by $f_\varepsilon(\varphi)$ and then integrate it again over $\mathbb{R}^N$. Observe that $f_\varepsilon(\varphi)$ is compactly supported in $\mathbb{R}^N \setminus \{0\}$. This time we get

$$-\frac{\alpha}{|\alpha|}\,(N-1)\int_{\mathbb{R}^N} \frac{f_\varepsilon(\varphi)}{|x|}\,dx = -\int_{\mathbb{R}^N} \operatorname{div}\left(\frac{\nabla V_\alpha}{|\nabla V_\alpha|}\right) f_\varepsilon(\varphi)\,dx,$$

for every $\alpha \neq 0$. In the right-hand side, we proceed as before and use the Divergence Theorem. We obtain

$$-\frac{\alpha}{|\alpha|}\,(N-1)\int_{\mathbb{R}^N} \frac{f_\varepsilon(\varphi)}{|x|}\,dx = \int_{\mathbb{R}^N} \left\langle \frac{\nabla V_\alpha}{|\nabla V_\alpha|}, \nabla\varphi \right\rangle f_\varepsilon'(\varphi)\,dx.$$

By using the Cauchy-Schwarz inequality and the fact that

$$|f_\varepsilon'(t)| = \frac{|t|}{\sqrt{\varepsilon^2+t^2}} \le 1, \qquad \text{for every } t \in \mathbb{R},$$

we obtain in particular

$$-\frac{\alpha}{|\alpha|}\,(N-1)\int_{\mathbb{R}^N} \frac{f_\varepsilon(\varphi)}{|x|}\,dx \le \int_{\mathbb{R}^N} |\nabla\varphi|\,dx.$$

By choosing $\alpha = 1-N$, taking the limit as ε goes to 0 and using Fatou's Lemma, we get the desired conclusion in the case $p = 1 < N$, as well.
In order to conclude, we need to show that we can "fill the hole", if $p < N$. We take $\varphi \in C_0^\infty(\mathbb{R}^N)$ and we take $\zeta \in C^\infty(\mathbb{R}^N)$ such that

$$0 \le \zeta \le 1, \quad \zeta(x) = 1 \text{ for } x \in \mathbb{R}^N \setminus B_2(0), \quad \zeta(x) = 0 \text{ for } x \in B_1(0).$$

We also define the rescaled function

$$\zeta_n(x) = \zeta(n\,x),$$

and write

$$\varphi = \varphi\,\zeta_n + (1-\zeta_n)\,\varphi.$$

We then have by Minkowski's inequality

$$\left(\int_{\mathbb{R}^N} \frac{|\varphi|^p}{|x|^p}\,dx\right)^{\frac{1}{p}} \le \left(\int_{\mathbb{R}^N} \frac{|\varphi\,\zeta_n|^p}{|x|^p}\,dx\right)^{\frac{1}{p}} + \left(\int_{\mathbb{R}^N} \frac{|(1-\zeta_n)\,\varphi|^p}{|x|^p}\,dx\right)^{\frac{1}{p}}.$$

Observe that the function $\varphi\,\zeta_n \in C_0^\infty(\mathbb{R}^N \setminus \{0\})$, i.e. it vanishes around the origin, due to the presence of ζ_n which is identically zero in $B_{1/n}(0)$. We can then apply (3.5.8) to this function and get

$$\left(\int_{\mathbb{R}^N}\frac{|\varphi|^p}{|x|^p}\,dx\right)^{\frac{1}{p}}\le\frac{p}{N-p}\left(\int_{\mathbb{R}^N}|\nabla(\varphi\,\zeta_n)|^p\,dx\right)^{\frac{1}{p}}+\left(\int_{\mathbb{R}^N}\frac{|(1-\zeta_n)\,\varphi|^p}{|x|^p}\,dx\right)^{\frac{1}{p}}.$$

We need to estimate the last two integrals. By construction, we have that

$$\zeta_n(x)=1 \text{ for } x\in\mathbb{R}^N\setminus B_{2/n}(0), \tag{3.5.9}$$

thus we get

$$\int_{\mathbb{R}^N}\frac{|(1-\zeta_n)\,\varphi|^p}{|x|^p}\,dx=\int_{B_{2/n}(0)}\frac{|(1-\zeta_n)\,\varphi|^p}{|x|^p}\,dx\le\|\varphi\|^p_{L^\infty(\mathbb{R}^N)}\int_{B_{2/n}(0)}\frac{1}{|x|^p}\,dx.$$

The last integral can be explicitly computed by using spherical coordinates. Namely, we have

$$\int_{B_{2/n}(0)}\frac{1}{|x|^p}\,dx=N\,\omega_N\int_0^{\frac{2}{n}}\varrho^{N-1-p}\,d\varrho=\frac{N\,\omega_N}{N-p}\left(\frac{2}{n}\right)^{N-p}.$$

By recalling that we are considering the case $N>p$, we can then infer

$$\lim_{n\to\infty}\left(\int_{\mathbb{R}^N}\frac{|(1-\zeta_n)\,\varphi|^p}{|x|^p}\,dx\right)^{\frac{1}{p}}=0.$$

Thus, we arrive at

$$\left(\int_{\mathbb{R}^N}\frac{|\varphi|^p}{|x|^p}\,dx\right)^{\frac{1}{p}}\le\frac{p}{N-p}\limsup_{n\to\infty}\left(\int_{\mathbb{R}^N}|\nabla(\varphi\,\zeta_n)|^p\,dx\right)^{\frac{1}{p}}. \tag{3.5.10}$$

We compute the gradient of the product in the last integral and then use Minkowski's inequality. This yields

$$\begin{aligned}\left(\int_{\mathbb{R}^N}|\nabla(\varphi\,\zeta_n)|^p\,dx\right)^{\frac{1}{p}}&\le\left(\int_{\mathbb{R}^N}|\nabla\varphi|^p\,|\zeta_n|^p\,dx\right)^{\frac{1}{p}}\\&\quad+\left(\int_{\mathbb{R}^N}|\varphi|^p\,|\nabla\zeta_n|^p\,dx\right)^{\frac{1}{p}}.\end{aligned} \tag{3.5.11}$$

For the second integral on the right-hand side, by recalling (3.5.9), we have

$$\int_{\mathbb{R}^N}|\varphi|^p\,|\nabla\zeta_n|^p\,dx=\int_{B_{2/n}(0)}|\varphi|^p\,|\nabla\zeta_n|^p\,dx\le\|\varphi\|^p_{L^\infty(\mathbb{R}^N)}\int_{B_{2/n}(0)}|\nabla\zeta_n|^p\,dx.$$

We use the definition of ζ_n and a change of variable, so to get

$$\int_{B_{2/n}(0)} |\nabla \zeta_n|^p \, dx = n^p \int_{B_{2/n}(0)} |\nabla \zeta(n\,x)|^p \, dx = n^{p-N} \int_{B_2(0)} |\nabla \zeta(y)|^p \, dy.$$

By using again that $N > p$, the last two centered formulas show that

$$\lim_{n\to\infty} \int_{\mathbb{R}^N} |\varphi|^p \, |\nabla \zeta_n|^p \, dx = 0.$$

From (3.5.11) and the fact that $|\zeta_n| \le 1$, we then obtain

$$\limsup_{n\to\infty} \left(\int_{\mathbb{R}^N} |\nabla(\varphi\, \zeta_n)|^p \, dx \right)^{\frac{1}{p}} \le \left(\int_{\mathbb{R}^N} |\nabla \varphi|^p \, dx \right)^{\frac{1}{p}}.$$

By inserting this information into (3.5.10), we finally get Hardy's inequality for every $\varphi \in C_0^\infty(\mathbb{R}^N)$ and $1 \le p < N$. □

Remark 3.5.7 It is not difficult to see that in the case $p > N$ we can not "fill the hole" in the previous result, i.e. inequality (3.5.6) *can not hold* for functions in $C_0^\infty(\mathbb{R}^N)$. Indeed, by taking $\varphi \in C_0^\infty(\mathbb{R}^N)$ such that $\varphi \ge 1$ on the ball $B_1(0)$, we have

$$\int_{\mathbb{R}^N} \frac{|\varphi|^p}{|x|^p} \, dx \ge \int_{B_1(0)} \frac{1}{|x|^p} \, dx = N\, \omega_N \int_0^1 \frac{1}{\varrho^p} \, \varrho^{N-1} \, d\varrho = +\infty,$$

for $p > N$, while

$$\int_{\mathbb{R}^N} |\nabla \varphi|^p \, dx < +\infty.$$

We also remark that the constant appearing (3.5.6) can be shown to be the best possible: this requires some extra work, we refer the interested reader to [81].

Remark 3.5.8 (The Case $p = \infty$) The previous result still holds in the case $p = \infty$, in the following form

$$\left\| \frac{\varphi}{|x|} \right\|_{L^\infty(\mathbb{R}^N)} \le \|\nabla \varphi\|_{L^\infty(\mathbb{R}^N;\mathbb{R}^N)}, \qquad \text{for every } \varphi \in C_0^\infty(\mathbb{R}^N \setminus \{0\}). \tag{3.5.12}$$

The proof simply follows from the Fundamental Theorem of Calculus. Indeed, for every $\varphi \in C_0^\infty(\mathbb{R}^N \setminus \{0\})$ and $x \ne 0$, we have

$$
\begin{aligned}
|\varphi(x)| = |\varphi(x) - \varphi(0)| &= \left| \int_0^1 \frac{d}{dt}\varphi(t\,x)\,dt \right| = \left| \int_0^1 \langle \nabla\varphi(t\,x), x\rangle\,dt \right| \\
&\leq \int_0^1 |\langle \nabla\varphi(t\,x), x\rangle|\,dt \\
&\leq |x| \int_0^1 |\nabla\varphi(t\,x)|\,dt \leq |x|\,\|\nabla\varphi\|_{L^\infty(\mathbb{R}^N;\mathbb{R}^N)}.
\end{aligned}
$$

If we divide by $|x|$ and take the supremum (recall that φ vanishes in a neighborhood of the origin), we obtain (3.5.12).

In the previous result, one could think of the weight

$$
\frac{1}{|x|^p}, \qquad x \neq 0,
$$

appearing in (3.5.6) as the power p of the distance of a point $x \in \mathbb{R}^N \setminus \{0\}$ from the boundary of this set, which consists of the singleton $\{0\}$. The next result is in the same spirit: it is a particular case of Hardy's inequality for open sets (see for example [23, Chapter 1, Section 4] or [52] for some studies on the subject).

Theorem 3.5.9 (Hardy's Inequality for a Ball) *Let $1 < p < \infty$. For every $\varphi \in C_0^\infty(B_R(x_0))$ we have*

$$
\left(\frac{p-1}{p}\right)^p \int_{B_R(x_0)} \frac{|\varphi|^p}{(R - |x - x_0|)^p}\,dx \leq \int_{B_R(x_0)} |\nabla\varphi|^p\,dx. \tag{3.5.13}
$$

Proof Observe that the function

$$
R - |x - x_0|, \qquad \text{for } x \in B_R(x_0),
$$

coincides with the distance from the boundary $\partial B_R(x_0)$ of a point $x \in B_R(x_0)$. Without loss of generality, we can suppose that $R = 1$ and $x_0 = 0$. We indicate by B the open ball centered at the origin, with radius 1. We then set

$$
U(x) = 1 - |x|, \qquad \text{for } x \in B.
$$

We will prove (3.5.13) by using an argument similar to that of Theorem 3.5.6. The idea of the proof is quite simple, but we will have to cope with some technical issues linked to a lack of regularity of the function U at $x = 0$.

We start by noticing that U belongs to $C^\infty(B \setminus \{0\})$ and its gradient is given by

$$
\nabla U(x) = -\frac{x}{|x|}, \qquad \text{for } x \in B \setminus \{0\}. \tag{3.5.14}
$$

Then we get that the Laplacian of U is given by

$$
\begin{aligned}
-\Delta U = -\mathrm{div}\,(\nabla U) &= \mathrm{div}\left(\frac{x}{|x|}\right) \\
&= \frac{\mathrm{div}\,x}{|x|} + \langle x, \nabla |x|^{-1}\rangle = \frac{N-1}{|x|}, \qquad \text{for } x \in B \setminus \{0\}.
\end{aligned}
$$

In particular, it is a non-negative quantity. We then take $\varphi \in C_0^1(B)$ such that $\varphi \geq 0$, for every $0 < \varepsilon < 1$ we have

$$
0 \leq \int_{B\setminus B_\varepsilon} -\Delta U\, \varphi\, dx = -\int_{B\setminus B_\varepsilon} \mathrm{div}\,(\nabla U\, \varphi)\, dx + \int_{B\setminus B_\varepsilon} \langle \nabla U, \nabla \varphi\rangle\, dx,
$$

where $B_\varepsilon = \{x \in \mathbb{R}^N : |x| < \varepsilon\}$. By using the Divergence Theorem and the fact that φ is compactly supported in B, we then obtain

$$
0 \leq \int_{\partial B_\varepsilon} \langle \nabla U, \nu_{B_\varepsilon}\rangle\, \varphi\, d\sigma + \int_{B\setminus B_\varepsilon} \langle \nabla U, \nabla\varphi\rangle\, dx,
$$

where ν_{B_ε} is the outer unit normal vector of ∂B_ε. Observe that this is given by $\nu_{B_\varepsilon}(x) = x/|x|$. Thus, by recalling (3.5.14), the last equation can be rewritten as

$$
\begin{aligned}
0 &\leq \int_{\partial B_\varepsilon} \left\langle -\frac{x}{|x|}, \frac{x}{|x|}\right\rangle \varphi\, d\sigma + \int_{B\setminus B_\varepsilon} \left\langle -\frac{x}{|x|}, \nabla\varphi\right\rangle dx \\
&= -\int_{\partial B_\varepsilon} \varphi\, d\sigma + \int_{B\setminus B_\varepsilon} \left\langle -\frac{x}{|x|}, \nabla\varphi\right\rangle dx.
\end{aligned}
$$

In particular, since $\varphi \geq 0$ this gives that

$$
0 \leq \int_{B\setminus B_\varepsilon} \left\langle -\frac{x}{|x|}, \nabla\varphi\right\rangle dx.
$$

This is valid for every $\varphi \in C_0^1(B)$ such that $\varphi \geq 0$ and every $0 < \varepsilon < 1$. We now wish to take the limit as ε goes to 0: by the Cauchy-Schwarz inequality

$$
\left|\left\langle \frac{x}{|x|}, \nabla\varphi\right\rangle\right| \leq |\nabla\varphi| \in L^1(B),
$$

thus we can apply the Dominated Convergence Theorem. In conclusion, we get

$$
0 \leq \int_B \left\langle -\frac{x}{|x|}, \nabla\varphi\right\rangle dx, \qquad \text{for every } \varphi \in C_0^1(B) \text{ such that } \varphi \geq 0.
$$

We then take $\psi \in C_0^\infty(B)$ and use the previous inequality with the choice

$$\varphi = \frac{|\psi|^p}{U_n^{p-1}}, \qquad \text{where } U_n(x) = \sqrt{1+\frac{1}{n}} - \sqrt{|x|^2+\frac{1}{n}},$$

for $n \in \mathbb{N} \setminus \{0\}$. Observe that this function is non-negative and it belongs to $C_0^1(B)$, thanks to the properties of U_n and ψ. We get

$$\begin{aligned}
0 \le \int_B \left\langle -\frac{x}{|x|}, \nabla \left(\frac{|\psi|^p}{U_n^{p-1}}\right)\right\rangle dx &= -(p-1)\int_B \left\langle -\frac{x}{|x|}, \frac{\nabla U_n}{U_n^p}\right\rangle |\psi|^p\, dx \\
&\quad + p \int_B \left\langle -\frac{x}{|x|}, \nabla \psi\right\rangle \frac{|\psi|^{p-2}\,\psi}{U_n^{p-1}}\, dx \\
&\le -(p-1)\int_B \left\langle -\frac{x}{|x|}, \frac{\nabla U_n}{U_n^p}\right\rangle |\psi|^p\, dx \\
&\quad + p \int_B |\nabla \psi|\, \frac{|\psi|^{p-1}}{U_n^{p-1}}\, dx.
\end{aligned}$$

By rearranging the terms, we get

$$\int_B \left\langle -\frac{x}{|x|}, \frac{\nabla U_n}{U_n^p}\right\rangle |\psi|^p\, dx \le \frac{p}{p-1}\int_B |\nabla \psi|\, \frac{|\psi|^{p-1}}{U_n^{p-1}}\, dx.$$

We now observe that

$$\left\langle -\frac{x}{|x|}, \nabla U_n\right\rangle = \left\langle -\frac{x}{|x|}, -\frac{x}{\sqrt{|x|^2+1/n}}\right\rangle = \frac{|x|}{\sqrt{|x|^2+1/n}}.$$

We thus have obtained

$$\int_B \frac{|x|}{\sqrt{|x|^2+1/n}}\, \frac{|\psi|^p}{U_n^p}\, dx \le \frac{p}{p-1}\int_B |\nabla \psi|\, \frac{|\psi|^{p-1}}{U_n^{p-1}}\, dx.$$

On the right-hand side, we can apply the generalized Young inequality (1.2.6), so to get

$$\int_B \frac{|x|}{\sqrt{|x|^2+1/n}}\, \frac{|\psi|^p}{U_n^p}\, dx \le \frac{\varepsilon}{p-1}\int_B |\nabla \psi|^p\, dx + \varepsilon^{-\frac{1}{p-1}} \int_B \frac{|\psi|^p}{U_n^p}\, dx,$$

for every $\varepsilon > 0$ and $n \in \mathbb{N} \setminus \{0\}$. We now pass to the limit as n goes to ∞: on the left-hand side it is sufficient to use Fatou's Lemma, while on the right-hand side we can apply the Monotone Convergence Theorem. Indeed, it is not difficult to see that the sequence $\{1/U_n^p\}_{n\in\mathbb{N}}$ is monotonically *decreasing* converging[6] to $1/U^p$ and for $n = 1$ we have

$$\int_B \frac{|\psi|^p}{U_1^p}\,dx < +\infty,$$

thanks to the compact support of ψ. Thus, by taking the limit as n goes to ∞, for every $\varepsilon > 0$ we obtain

$$\int_B \frac{|\psi|^p}{(1-|x|)^p}\,dx \le \frac{\varepsilon}{p-1}\int_B |\nabla\psi|^p\,dx + \varepsilon^{-\frac{1}{p-1}}\int_B \frac{|\psi|^p}{(1-|x|)^p}\,dx,$$

that is

$$(p-1)\left(1-\varepsilon^{-\frac{1}{p-1}}\right)\frac{1}{\varepsilon}\int_B \frac{|\psi|^p}{(1-|x|)^p}\,dx \le \int_B |\nabla\psi|^p\,dx.$$

By observing that the quantity in the left-hand side is maximal for $\varepsilon = (p/(p-1))^{p-1}$, we get

$$\left(\frac{p-1}{p}\right)^p \int_B \frac{|\psi|^p}{(1-|x|)^p}\,dx \le \int_B |\nabla\psi|^p\,dx,$$

as desired. □

Remark 3.5.10 Also in this case, the constant obtained in (3.5.13) can be shown to be the best possible (we refer again to [81]). We notice that such a constant vanishes for $p = 1$. This is not by chance: actually, one can prove that for $p = 1$ the result of the previous theorem can not hold.

Remark 3.5.11 (The Case $p = \infty$) Also the previous result still holds in the limit case $p = \infty$, in the following form

$$\left\|\frac{\varphi}{R-|x-x_0|}\right\|_{L^\infty(B_R(x_0))} \le \|\nabla\varphi\|_{L^\infty(B_R(x_0);\mathbb{R}^N)}, \qquad \text{for every } \varphi \in C_0^\infty(B_R(x_0)).$$

[6] It is sufficient to notice that for every $x \in B$, the quantity

$$h(\delta) = \sqrt{1+\delta} - \sqrt{|x|^2+\delta},$$

is monotone decreasing. Indeed, we have

$$h'(\delta) = \frac{1}{2}\left(\frac{1}{\sqrt{1+\delta}} - \frac{1}{\sqrt{|x|^2+\delta}}\right) < 0.$$

Thus, in particular $h(1/n)$ is monotone increasing and $1/h(1/n)$ is monotone decreasing.

The proof is again based on a straightforward application of the Fundamental Theorem of Calculus. Indeed, let $\varphi \in C_0^\infty(B_R(x_0))$ and $x \in B_R(x_0)$. Let us indicate by $\overline{x}$ a point such that[7]

$$\overline{x} \in \partial B_R(x_0) \qquad \text{and} \qquad |x - \overline{x}| = R - |x - x_0|.$$

Thus, we have

$$\begin{aligned}
|\varphi(x)| = |\varphi(x) - \varphi(\overline{x})| &= \left| \int_0^1 \frac{d}{dt} \varphi(t\,x + (1-t)\,\overline{x})\,dt \right| \\
&= \left| \int_0^1 \langle \nabla\varphi(t\,x + (1-t)\,\overline{x}), x - \overline{x}\rangle\,dt \right| \\
&\le |x - \overline{x}| \int_0^1 |\nabla\varphi(t\,x + (1-t)\,\overline{x})|\,dt \\
&\le |x - \overline{x}|\, \|\nabla\varphi\|_{L^\infty(\mathbb{R}^N;\mathbb{R}^N)} \\
&= (R - |x - x_0|)\, \|\nabla\varphi\|_{L^\infty(\mathbb{R}^N;\mathbb{R}^N)}.
\end{aligned}$$

This ie enough to get the claimed inequality.

3.6 Sobolev-type Inequalities

We collect in this section some integral inequalities for smooth functions with compact support. Differently from the Poincaré inequality of the previous section, we show that the L^p norm of the gradient controls a L^q norm of the function, for a *larger* exponent $q > p$. As we will show in Sect. 3.8, these inequalities have an important impact on the summability properties of functions belonging to Sobolev spaces.

We start with the following remarkable result, originally due to Sergej L'vovič Sobolev [87].

Theorem 3.6.1 (Sobolev's Inequality) *Let* $1 \le p < N$, *we set*

$$p^* = \frac{N\,p}{N - p}.$$

---

[7] If $x = x_0$, we can choose as $\overline{x}$ any point of the boundary. Otherwise, we have

$$\overline{x} = x_0 + R\,\frac{x - x_0}{|x - x_0|}.$$

Then for every $\varphi \in C_0^\infty(\mathbb{R}^N)$ we have

$$\|\varphi\|_{L^{p^*}(\mathbb{R}^N)} \le \mathcal{S}\,\|\nabla\varphi\|_{L^p(\mathbb{R}^N;\mathbb{R}^N)}, \tag{3.6.1}$$

for a constant $\mathcal{S} > 0$ depending on N and p only.

Proof For every $\varphi \in C_0^\infty(\mathbb{R}^N)$ and $\alpha \ge 1$, we consider the function $x \mapsto |\varphi(x)|^{\alpha-1}\varphi(x)$. Thanks to the fact that $\alpha \ge 1$, this is a C^1 function with compact support. Thus, for every $k \in \{1,\dots,N\}$, we have

$$\begin{aligned}
|\varphi(x)|^\alpha = \left||\varphi(x)|^{\alpha-1}\varphi(x)\right| &= \left|\int_{-\infty}^{x_k} \frac{\partial|\varphi|^{\alpha-1}\varphi}{\partial x_k}(x_1,\dots,x_{k-1},t,x_{k+1},\dots,x_N)\,dt\right| \\
&= \alpha\left|\int_{-\infty}^{x_k} \frac{\partial\varphi}{\partial x_k}\,|\varphi|^{\alpha-1}\,dt\right| \\
&\le \alpha\int_{-\infty}^{+\infty}\left|\frac{\partial\varphi}{\partial x_k}\right|\,|\varphi|^{\alpha-1}\,dt.
\end{aligned}$$

In particular, by using the previous estimate with $k \in \{1,\dots,N\}$, we have

$$|\varphi(x)|^{N\alpha} = \underbrace{|\varphi(x)|^\alpha\cdot\ldots\cdot|\varphi(x)|^\alpha}_{N\text{ times}} \le \alpha^N\prod_{k=1}^N\int_{-\infty}^{+\infty}\left|\frac{\partial\varphi}{\partial x_k}\right|\,|\varphi|^{\alpha-1}\,dt.$$

We raise to the power $1/(N-1)$ and integrate over $\mathbb{R}^N$, so to get

$$\int_{\mathbb{R}^N}|\varphi|^{\frac{N\alpha}{N-1}}\,dx \le \alpha^{\frac{N}{N-1}}\int_{\mathbb{R}^N}\prod_{k=1}^N\left(\int_{-\infty}^{+\infty}\left|\frac{\partial\varphi}{\partial x_k}\right|\,|\varphi|^{\alpha-1}\,dt\right)^{\frac{1}{N-1}}dx. \tag{3.6.2}$$

We now set

$$\Phi_k = \left(\int_{-\infty}^{+\infty}\left|\frac{\partial\varphi}{\partial x_k}\right|\,|\varphi|^{\alpha-1}\,dt\right)^{\frac{1}{N-1}}, \qquad \text{for } k \in \{1,\dots,N\},$$

and observe that, by its very definition, we have that Φ_k is non-negative, it does not depend on the variable x_k and it belongs to $L^{N-1}(\mathbb{R}^{N-1})$. By applying Problem 1.7.24, we get from (3.6.2)

$$\int_{\mathbb{R}^N}|\varphi|^{\frac{N\alpha}{N-1}}\,dx \le \alpha^{\frac{N}{N-1}}\int_{\mathbb{R}^N}\prod_{k=1}^N\Phi_k\,dx \le \alpha^{\frac{N}{N-1}}\prod_{k=1}^N\left(\int_{\mathbb{R}^{N-1}}\Phi_k^{N-1}\,d\widehat{x}_k\right)^{\frac{1}{N-1}}.$$

We used the notation

$$d\widehat{x_k} := dx_1 \dots dx_{k-1}\, dx_{k+1} \dots dx_N, \qquad \text{for } k \in \{1, \dots, N\}.$$

By recalling the definition of Φ_k, this is the same as

$$\begin{aligned} \int_{\mathbb{R}^N} |\varphi|^{\frac{N\alpha}{N-1}}\, dx &\le \alpha^{\frac{N}{N-1}} \prod_{k=1}^{N} \left(\int_{\mathbb{R}^{N-1}} \int_{-\infty}^{+\infty} \left| \frac{\partial \varphi}{\partial x_k} \right| |\varphi|^{\alpha-1}\, dt\, d\widehat{x_k} \right)^{\frac{1}{N-1}} \\ &= \alpha^{\frac{N}{N-1}} \prod_{k=1}^{N} \left(\int_{\mathbb{R}^N} \left| \frac{\partial \varphi}{\partial x_k} \right| |\varphi|^{\alpha-1}\, dx \right)^{\frac{1}{N-1}}. \end{aligned} \tag{3.6.3}$$

In the case $p = 1$, we can take $\alpha = 1$ and obtain

$$\int_{\mathbb{R}^N} |\varphi|^{\frac{N}{N-1}}\, dx \le \prod_{k=1}^{N} \left(\int_{\mathbb{R}^N} \left| \frac{\partial \varphi}{\partial x_k} \right| dx \right)^{\frac{1}{N-1}},$$

which is sufficient to obtain the claimed inequality (3.6.1), in light of the fact that

$$\left| \frac{\partial \varphi}{\partial x_k} \right| \le |\nabla \varphi|, \qquad \text{for } k \in \{1, \dots, N\}. \tag{3.6.4}$$

In the case $1 < p < N$, we can apply Hölder's inequality in each of the integrals on the right-hand side of (3.6.3), so to get

$$\begin{aligned} \int_{\mathbb{R}^N} |\varphi|^{\frac{N\alpha}{N-1}}\, dx \le \alpha^{\frac{N}{N-1}} &\left(\int_{\mathbb{R}^N} |\varphi|^{(\alpha-1)\frac{p}{p-1}}\, dx \right)^{\frac{N}{N-1}\frac{p-1}{p}} \\ &\times \prod_{k=1}^{N} \left(\int_{\mathbb{R}^N} \left| \frac{\partial \varphi}{\partial x_k} \right|^p dx \right)^{\frac{1}{p(N-1)}}. \end{aligned} \tag{3.6.5}$$

We now choose $\alpha > 1$ such that the power on φ is the same on both sides, i.e. we impose

$$\frac{N\alpha}{N-1} = (\alpha - 1)\, \frac{p}{p-1}.$$

With simple computations, we obtain the choice

$$\alpha = \frac{p\,(N-1)}{N-p}.$$

Observe that such a choice is feasible, thanks to the fact that

$$\frac{p\,(N-1)}{N-p} > 1 \qquad \Longleftrightarrow \qquad 1 < p < N.$$

Thus, from (3.6.5) we get

$$\int_{\mathbb{R}^N} |\varphi|^{p^*}\,dx \le \left(\frac{p\,(N-1)}{N-p}\right)^{\frac{N}{N-1}} \prod_{k=1}^{N}\left(\int_{\mathbb{R}^N}\left|\frac{\partial\varphi}{\partial x_k}\right|^p dx\right)^{\frac{1}{p\,(N-1)}}$$
$$\times\left(\int_{\mathbb{R}^N}|\varphi|^{p^*}\,dx\right)^{\frac{N}{N-1}\,\frac{p-1}{p}}.$$

We raise to the power $(N-1)/N$ and then simplify the L^{p^*} norm of φ, so to get

$$\left(\int_{\mathbb{R}^N}|\varphi|^{p^*}\,dx\right)^{\frac{1}{p^*}} \le \frac{p\,(N-1)}{N-p}\prod_{k=1}^{N}\left(\int_{\mathbb{R}^N}\left|\frac{\partial\varphi}{\partial x_k}\right|^p dx\right)^{\frac{1}{N\,p}}.$$

If we use (3.6.4), we finally get the desired conclusion. □

Remark 3.6.2 (Dimensional Analysis) There is no mystery in the appearing of the exponent p^*, which is usually called *critical Sobolev exponent*. Indeed, this is the unique exponent such that the two norms

$$\|\nabla\varphi\|_{L^p(\mathbb{R}^N;\mathbb{R}^N)} \qquad \text{and} \qquad \|\varphi\|_{L^{p^*}(\mathbb{R}^N)},$$

have the same scaling. More precisely, let us take $\varphi \in C_0^\infty(\mathbb{R}^N)$ and consider $\varphi_\lambda(x) = \varphi(\lambda\,x)$ for some $\lambda > 0$. Then we have

$$\left(\int_{\mathbb{R}^N}|\nabla\varphi_\lambda(x)|^p\,dx\right)^{\frac{1}{p}} = \lambda\left(\int_{\mathbb{R}^N}|\nabla\varphi(\lambda\,x)|^p\,dx\right)^{\frac{1}{p}}$$
$$= \lambda^{1-\frac{N}{p}}\left(\int_{\mathbb{R}^N}|\nabla\varphi(y)|^p\,dy\right)^{\frac{1}{p}},$$

and

$$\left(\int_{\mathbb{R}^N}|\varphi_\lambda|^{p^*}\,dx\right)^{\frac{1}{p^*}} = \left(\int_{\mathbb{R}^N}|\varphi(\lambda\,x)|^{p^*}\,dx\right)^{\frac{1}{p^*}} = \lambda^{-\frac{N}{p^*}}\left(\int_{\mathbb{R}^N}|\varphi|^{p^*}\,dy\right)^{\frac{1}{p^*}}.$$

Then a necessary condition for the validity of inequality (3.6.1) is that

$$\lambda^{-\frac{N}{p^*}} = \lambda^{1-\frac{N}{p}} \qquad \text{i.e.} \qquad -\frac{N}{p^*} = 1 - \frac{N}{p}.$$

With simple computations, we see that the last condition is equivalent to require that

$$\frac{N-p}{N\,p} = \frac{1}{p^*}.$$

We also refer to Problem 3.12.10 below.

Remark 3.6.3 The previous elementary proof of Sobolev's inequality is due to Gagliardo ([75]) and Nirenberg ([86]), independently. It provides the following explicit expression

$$\mathcal{S} = \frac{p\,(N-1)}{N-p},$$

for the constant appearing in (3.6.1). However, this is not the best possible.

The determination of the sharp constant in Sobolev's inequality is much more delicate, with respect to the case of Hardy's inequality: it goes beyond the scopes of this book. The interested reader could consult [63, 90] for the classical proof based on *symmetrization techniques* or [48, Theorem 9.3] and [59, Theorem 6.21] for a proof based on *Optimal Transport theory*. Here, we only point out that such an optimal constant $\mathcal{T}_{N,p}$ has the following asymptotic behaviour

$$\mathcal{T}_{N,p} \sim C_N \left(\frac{1}{N-p}\right)^{\frac{p-1}{p}}, \qquad \text{as } p \nearrow N,$$

for a dimensional constant $C_N > 0$ (see for example [90, equation (2)]).

For the limit case $p = N$, we have the following

Theorem 3.6.4 (Ladyzhenskaya's Inequality) *Let $N \geq 2$, for every $\varphi \in C_0^\infty(\mathbb{R}^N)$ and every $N < q < \infty$ we have*

$$\|\varphi\|_{L^q(\mathbb{R}^N)} \leq \mathcal{L}\, \|\varphi\|_{L^N(\mathbb{R}^N)}^{\frac{N}{q}}\, \|\nabla\varphi\|_{L^N(\mathbb{R}^N;\mathbb{R}^N)}^{1-\frac{N}{q}}, \tag{3.6.6}$$

for a constant $\mathcal{L} > 0$ depending on N and q only.

Proof We go back to (3.6.5) with $p = N$. By further using (3.6.4), we get

$$\int_{\mathbb{R}^N} |\varphi|^{\frac{N\,\alpha}{N-1}}\, dx \leq \alpha^{\frac{N}{N-1}} \left(\int_{\mathbb{R}^N} |\nabla\varphi|^N\, dx\right)^{\frac{1}{N-1}} \left(\int_{\mathbb{R}^N} |\varphi|^{(\alpha-1)\,\frac{N}{N-1}}\, dx\right). \tag{3.6.7}$$

In particular, by choosing $\alpha = N$ and raising to the power $(N-1)/N^2$, we get

$$\left(\int_{\mathbb{R}^N} |\varphi|^{\frac{N^2}{N-1}}\,dx\right)^{\frac{N-1}{N^2}} \le N^{\frac{1}{N}} \left(\int_{\mathbb{R}^N} |\nabla\varphi|^N\,dx\right)^{\frac{1}{N^2}} \left(\int_{\mathbb{R}^N} |\varphi|^N\,dx\right)^{\frac{N-1}{N^2}}. \tag{3.6.8}$$

This is exactly (3.6.6) for the particular case $q = N^2/(N-1)$.

In order to deal with the general case, we now have to distinguish two possibilities.

1. *Case* $N < q < N^2/(N-1)$. In this case, it is enough to use a simple interpolation argument in L^p spaces, i.e. we use that (see Problem 1.7.22)

$$\|\varphi\|_{L^q(\mathbb{R}^N)} \le \|\varphi\|^{\vartheta}_{L^{\frac{N^2}{N-1}}(\mathbb{R}^N)} \|\varphi\|^{1-\vartheta}_{L^N(\mathbb{R}^N)},$$

where the exponent $\vartheta \in (0,1)$ is given by

$$\vartheta = N - \frac{N^2}{q}.$$

By using (3.6.8) in the right-hand side of this estimate, we get

$$\|\varphi\|_{L^q(\mathbb{R}^N)} \le N^{\frac{\vartheta}{N}} \left(\int_{\mathbb{R}^N} |\nabla\varphi|^N\,dx\right)^{\frac{\vartheta}{N^2}} \left(\int_{\mathbb{R}^N} |\varphi|^N\,dx\right)^{\frac{N-1}{N^2}\vartheta + \frac{1-\vartheta}{N}}.$$

By observing that

$$\frac{\vartheta}{N^2} = \frac{1}{N} - \frac{1}{q} \qquad \text{and} \qquad \frac{N-1}{N^2}\vartheta + \frac{1-\vartheta}{N} = \frac{1}{q},$$

we get the desired inequality (3.6.6) in this case, as well.

2. *Case* $q > N^2/(N-1)$. This is slightly more tricky. By choosing

$$\alpha = \frac{N-1}{N}\,q,$$

in (3.6.7), we get

$$\int_{\mathbb{R}^N} |\varphi|^q\,dx \le \left(\frac{N-1}{N}\,q\right)^{\frac{N}{N-1}} \left(\int_{\mathbb{R}^N} |\nabla\varphi|^N\,dx\right)^{\frac{1}{N-1}} \left(\int_{\mathbb{R}^N} |\varphi|^{q-\frac{N}{N-1}}\,dx\right). \tag{3.6.9}$$

We observe that in this case

$$N < q - \frac{N}{N-1} < q,$$

thus we can still use an interpolation inequality for L^p norms, this time on the right-hand side of (3.6.9). Namely, we use that

$$\int_{\mathbb{R}^N} |\varphi|^{q-\frac{N}{N-1}}\,dx \leq \left(\|\varphi\|_{L^q(\mathbb{R}^N)}\right)^{\vartheta\left(q-\frac{N}{N-1}\right)} \left(\|\varphi\|_{L^N(\mathbb{R}^N)}\right)^{(1-\vartheta)\left(q-\frac{N}{N-1}\right)},$$

where $\vartheta \in (0, 1)$ is now given by

$$\vartheta = \frac{q - \dfrac{N^2}{N-1}}{q - \dfrac{N}{N-1}}\,\frac{q}{q-N}.$$

The combination of the two inequalities yields

$$\begin{aligned}\int_{\mathbb{R}^N} |\varphi|^q\,dx &\leq \left(\frac{N-1}{N}\,q\right)^{\frac{N}{N-1}} \left(\int_{\mathbb{R}^N} |\nabla\varphi|^N\,dx\right)^{\frac{1}{N-1}} \\ &\quad\times \left(\|\varphi\|_{L^q(\mathbb{R}^N)}\right)^{\vartheta\left(q-\frac{N}{N-1}\right)} \left(\|\varphi\|_{L^N(\mathbb{R}^N)}\right)^{(1-\vartheta)\left(q-\frac{N}{N-1}\right)}.\end{aligned}$$

We observe that

$$\vartheta\left(q - \frac{N}{N-1}\right) = \frac{q}{q-N}\left(q - \frac{N^2}{N-1}\right) = q - \frac{q}{q-N}\,\frac{N}{N-1},$$

and thus

$$(1-\vartheta)\left(q - \frac{N}{N-1}\right) = q - \frac{N}{N-1} - \left(q - \frac{q}{q-N}\,\frac{N}{N-1}\right) = \frac{N}{N-1}\,\frac{N}{q-N}.$$

After a simplification, we get

$$\begin{aligned}\left(\|\varphi\|_{L^q(\mathbb{R}^N)}\right)^{\frac{N}{N-1}\,\frac{q}{q-N}} &\leq \left(\frac{N-1}{N}\,q\right)^{\frac{N}{N-1}} \left(\int_{\mathbb{R}^N} |\nabla\varphi|^N\,dx\right)^{\frac{1}{N-1}} \\ &\quad\times \left(\|\varphi\|_{L^N(\mathbb{R}^N)}\right)^{\frac{N}{N-1}\,\frac{N}{q-N}}.\end{aligned}$$

By raising both sides to the power $(N-1)\,(q-N)/(N\,q)$, we get the desired conclusion. □

Remark 3.6.5 There is a small abuse of terminology in the previous result. The original Ladyzhenskaya inequality was proved by Olga Ladyzhenskaya only for the two-dimensional case $N = 2$ and for the exponent $q = 4$, i.e.

$$\|\varphi\|_{L^4(\mathbb{R}^2)} \le \mathcal{L}\, \|\varphi\|_{L^2(\mathbb{R}^2)}^{\frac{1}{2}}\, \|\nabla\varphi\|_{L^2(\mathbb{R}^2;\mathbb{R}^2)}^{\frac{1}{2}},$$

see [42, Lemma 1]. Observe that this corresponds to (3.6.8) with $N = 2$. However, the proof for the general case (3.6.6) is essentially the same. For completeness, we notice that the previous proof provides the following value for the constant

$$\mathcal{L} = \begin{cases} N^{\left(1-\frac{N}{q}\right)}, & \text{if } N < q \le \dfrac{N^2}{N-1}, \\ \\ \left(\dfrac{N-1}{N}\, q\right)^{\left(1-\frac{N}{q}\right)}, & \text{if } \dfrac{N^2}{N-1} < q < \infty. \end{cases}$$

However, in general this is not the sharp constant: to the best of our knowledge, the value of the optimal constant in (3.6.6) is still not known, already in the original case $N = 2$ and $q = 4$ (we refer to [76, Lemma 1] for an estimate of the sharp constant in this case).

We mention that the inequality (3.6.6) is a particular case of a larger family of interpolation inequalities, which nowadays are usually called *Gagliardo-Nirenberg inequalities* (see for example [44, Chapter 12, Section 5]).

The reader may notice that the two previous statements do not permit to consider the limit case $p = N = 1$. The latter is somehow exceptional, it deserves a separated statement.

Theorem 3.6.6 (One-dimensional Sobolev-Ladyzhenskaya Inequalities) *For every $\varphi \in C_0^\infty(\mathbb{R})$ and every $1 < q < \infty$, we have*

$$\|\varphi\|_{L^\infty(\mathbb{R})} \le \frac{1}{2}\, \|\varphi'\|_{L^1(\mathbb{R})}, \tag{3.6.10}$$

and

$$\|\varphi\|_{L^q(\mathbb{R})} \le \left(\frac{1}{2}\right)^{\frac{1}{q'}} \|\varphi'\|_{L^1(\mathbb{R})}^{\frac{1}{q'}}\, \|\varphi\|_{L^1(\mathbb{R})}^{\frac{1}{q}}.$$

Proof By the Fundamental Theorem of Calculus, for every $x \in \mathbb{R}$ we have

$$|\varphi(x)| = \left|\int_{-\infty}^{x} \varphi'(t)\, dt\right| \le \int_{-\infty}^{x} |\varphi'(t)|\, dt.$$

Similarly, we also have

$$|\varphi(x)| = \left|\int_{x}^{+\infty} \varphi'(t)\, dt\right| \le \int_{x}^{+\infty} |\varphi'(t)|\, dt.$$

By summing up these two inequalities and taking the supremum over x, we get (3.6.10).

We now take $1 < q < \infty$ and observe that

$$\left(\int_{\mathbb{R}} |\varphi|^q \, dx\right)^{\frac{1}{q}} = \left(\int_{\mathbb{R}} |\varphi|^{q-1} \, |\varphi| \, dx\right)^{\frac{1}{q}} \leq \|\varphi\|_{L^\infty(\mathbb{R})}^{\frac{q-1}{q}} \left(\int_{\mathbb{R}} |\varphi| \, dx\right)^{\frac{1}{q}}.$$

By applying (3.6.10) in the right-hand side to estimate the L^∞ norm, we get the second inequality, as well. □

Remark 3.6.7 It can be shown that the constant $1/2$ in (3.6.10) is sharp, see for example [67, Remark 5.6].

Finally, we consider the case $p > N$. In what follows, for $0 < \alpha \leq 1$ we use the notation

$$|\varphi|_{C^{0,\alpha}(\mathbb{R}^N)} = \sup_{x \neq y} \frac{|\varphi(x) - \varphi(y)|}{|x - y|^\alpha}, \qquad \text{for every } \varphi \in C_0^\infty(\mathbb{R}^N).$$

We then have the following result, which is usually named after Charles Bradfield Morrey Jr.

Theorem 3.6.8 (Morrey's Inequalities) *Let $N < p < \infty$, then for every $\varphi \in C_0^\infty(\mathbb{R}^N)$ we have*

$$|\varphi|_{C^{0,\alpha}(\mathbb{R}^N)} \leq \mathcal{M}_1 \, \|\nabla \varphi\|_{L^p(\mathbb{R}^N;\mathbb{R}^N)}, \qquad \text{with } \alpha = 1 - \frac{N}{p}, \tag{3.6.11}$$

and

$$\|\varphi\|_{L^\infty(\mathbb{R}^N)} \leq \mathcal{M}_2 \, \|\varphi\|_{L^p(\mathbb{R}^N)}^{1-\frac{N}{p}} \, \|\nabla \varphi\|_{L^p(\mathbb{R}^N;\mathbb{R}^N)}^{\frac{N}{p}}, \tag{3.6.12}$$

for two constants $\mathcal{M}_1, \mathcal{M}_2 > 0$ depending on N and p only.

Proof We recall that the volume of a ball having radius 1 is denoted by the dimensional constant ω_N. Then by scaling we obtain

$$|B_R(x_0)| = R^N \, \omega_N. \tag{3.6.13}$$

Let us fix two distinct points $x, y \in \mathbb{R}^N$ and set $\delta = |x - y|$. We consider the intersection of the two balls

$$B_\delta(x) \cap B_\delta(y),$$

which is non-empty by construction.

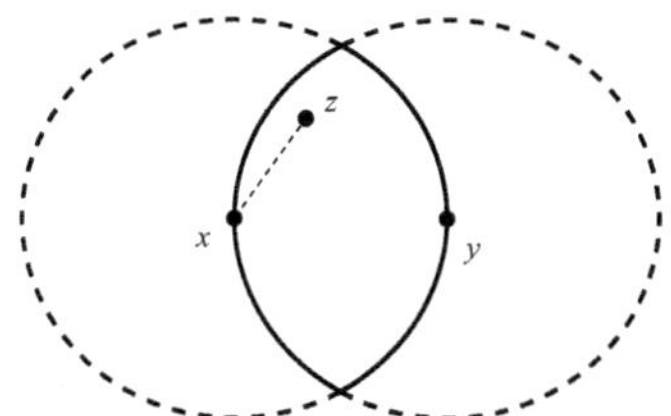

Fig. 3.2 The construction for the proof of Morrey's inequalities: for every point z in the intersection of the two balls, we can write the difference $\varphi(x) - \varphi(z)$ as an integral of its gradient over the segment joining the two points

Then for every $z \in B_\delta(x) \cap B_\delta(y)$, from the Fundamental Theorem of Calculus and Cauchy-Schwarz inequality we have (see Fig. 3.2)

$$\begin{aligned}
|\varphi(x) - \varphi(z)| &\le \left| \int_0^1 \frac{d}{dt} \varphi(t\,x + (1-t)\,z)\,dt \right| \\
&= \left| \int_0^1 \langle \nabla\varphi(t\,x + (1-t)\,z), x - z \rangle\,dt \right| \\
&\le |x - z| \int_0^1 |\nabla\varphi(t\,x + (1-t)\,z)|\,dt \\
&\le \delta \int_0^1 |\nabla\varphi(t\,x + (1-t)\,z)|\,dt.
\end{aligned}$$

We integrate with respect to $z \in B_\delta(x) \cap B_\delta(y)$, so to obtain

$$\begin{aligned}
\int_{B_\delta(x)\cap B_\delta(y)} |\varphi(x) - \varphi(z)|\,dz &\le \delta \int_{B_\delta(x)\cap B_\delta(y)} \int_0^1 |\nabla\varphi(t\,x + (1-t)\,z)|\,dt\,dz \\
&\le \delta \int_{B_\delta(x)} \int_0^1 |\nabla\varphi(t\,x + (1-t)\,z)|\,dt\,dz \\
&= \delta \int_0^1 \left(\int_{B_\delta(x)} |\nabla\varphi(t\,x + (1-t)\,z)|\,dz \right) dt.
\end{aligned} \tag{3.6.14}$$

For every $t \in [0, 1)$, in the integral

$$\int_{B_\delta(x)} |\nabla\varphi(t\,x + (1-t)\,z)|\,dz,$$

we perform the change of variables

$$w = t\,x + (1-t)\,z \in B_{(1-t)\,\delta}(x).$$

By using that $dw = (1-t)^N\,dz$, we obtain

$$\begin{aligned}
&\int_{B_\delta(x)} |\nabla\varphi(t\,x + (1-t)\,z)|\,dz \\
&= (1-t)^{-N} \int_{B_{(1-t)\,\delta}(x)} |\nabla\varphi(w)|\,dw \\
&\le (1-t)^{-N}\,|B_{(1-t)\,\delta}(x)|^{1-\frac{1}{p}} \left(\int_{B_{(1-t)\,\delta}(x)} |\nabla\varphi|^p\,dw\right)^{\frac{1}{p}} \\
&\le (1-t)^{-\frac{N}{p}} \left(\omega_N\,\delta^N\right)^{\frac{p-1}{p}} \left(\int_{\mathbb{R}^N} |\nabla\varphi|^p\,dw\right)^{\frac{1}{p}}.
\end{aligned}$$

Observe that we used (3.6.13), for the volume of the ball $B_{(1-t)\,\delta}(x)$. By using this estimate in (3.6.14), we get[8]

$$\begin{aligned}
&\int_{B_\delta(x)\cap B_\delta(y)} |\varphi(x) - \varphi(z)|\,dz \\
&\le \omega_N^{\frac{p-1}{p}}\,\delta^{1+N-\frac{N}{p}} \left(\int_{\mathbb{R}^N} |\nabla\varphi|^p\,dw\right)^{\frac{1}{p}} \left(\int_0^1 (1-t)^{-\frac{N}{p}}\,dt\right) \\
&= \omega_N^{\frac{p-1}{p}}\,\frac{p}{p-N}\,\delta^{1+N-\frac{N}{p}} \left(\int_{\mathbb{R}^N} |\nabla\varphi|^p\,dw\right)^{\frac{1}{p}}.
\end{aligned}$$

With exactly the same proof, we get the same estimate for

$$\int_{B_\delta(x)\cap B_\delta(y)} |\varphi(y) - \varphi(z)|\,dz.$$

From the triangle inequality, we thus get

[8] Observe that the integral

$$\int_0^1 (1-t)^{-\frac{N}{p}}\,dt,$$

is finite, thanks to the fact that $p > N$. It is here that the assumption $p > N$ comes into play.

$$\begin{aligned} |\varphi(x) - \varphi(y)|\,|B_\delta(x) \cap B_\delta(y)| &= \int_{B_\delta(x)\cap B_\delta(y)} |\varphi(x) - \varphi(y)|\,dz \\ &\le \int_{B_\delta(x)\cap B_\delta(y)} |\varphi(x) - \varphi(z)|\,dz \\ &+ \int_{B_\delta(x)\cap B_\delta(y)} |\varphi(y) - \varphi(z)|\,dz \\ &\le 2\,\omega_N^{\frac{p-1}{p}}\,\frac{p}{p-N}\,\delta^{1+N-\frac{N}{p}} \left(\int_{\mathbb{R}^N} |\nabla\varphi|^p\,dw\right)^{\frac{1}{p}}. \end{aligned}$$

By observing that

$$|B_\delta(x) \cap B_\delta(y)| \ge \left|B_{\frac{\delta}{2}}\left(\frac{x+y}{2}\right)\right| = \omega_N \left(\frac{\delta}{2}\right)^N,$$

and recalling that

$$\delta = |x - y|,$$

we now obtain the desired estimate (3.6.11), thanks to the arbitrariness of x and y. Let us now prove (3.6.12). From (3.6.11), for every $x, y \in \mathbb{R}^N$ we have

$$|\varphi(x) - \varphi(y)| \le \mathcal{M}_1\,|x - y|^\alpha \left(\int_{\mathbb{R}^N} |\nabla\varphi|^p\,dx\right)^{\frac{1}{p}}.$$

Thus, by using the triangle inequality

$$\begin{aligned} |\varphi(x)| &\le |\varphi(x) - \varphi(y)| + |\varphi(y)| \\ &\le \mathcal{M}_1\,|x - y|^\alpha \left(\int_{\mathbb{R}^N} |\nabla\varphi|^p\,dx\right)^{\frac{1}{p}} + |\varphi(y)|. \end{aligned}$$

For every $x \in \mathbb{R}^N$, we integrate the previous inequality with respect to $y \in B_1(x)$. This yields

$$\omega_N\,|\varphi(x)| \le \mathcal{M}_1 \int_{B_1(x)} |x - y|^\alpha\,dy \left(\int_{\mathbb{R}^N} |\nabla\varphi|^p\,dx\right)^{\frac{1}{p}} + \int_{B_1(x)} |\varphi(y)|\,dy. \tag{3.6.15}$$

Observe that by using spherical coordinates centered at x

$$y = x + \varrho\,\omega, \qquad \text{for } \varrho \in [0, 1),\ \omega \in \mathbb{S}^{N-1},$$

we have[9]

$$\int_{B_1(x)} |x-y|^\alpha \, dy = N\,\omega_N \int_0^1 \varrho^{\alpha+N-1}\, d\varrho = \frac{N\,\omega_N}{N+\alpha},$$

while by Hölder's inequality

$$\int_{B_1(x)} |\varphi|\, dy \le |B_1(x)|^{1-\frac{1}{p}} \left(\int_{B_1(x)} |\varphi|^p\, dy\right)^{\frac{1}{p}} \le \omega_N^{\frac{p-1}{p}} \left(\int_{\mathbb{R}^N} |\varphi|^p\, dy\right)^{\frac{1}{p}}.$$

By using these facts in (3.6.15), we finally get

$$|\varphi(x)| \le \mathcal{M}_1\, \frac{N\,\omega_N}{N+\alpha} \left(\int_{\mathbb{R}^N} |\nabla\varphi|^p\, dx\right)^{\frac{1}{p}} + \omega_N^{\frac{p-1}{p}} \left(\int_{\mathbb{R}^N} |\varphi(y)|^p\, dy\right)^{\frac{1}{p}}.$$

By arbitrariness of $x \in \mathbb{R}^N$, this proves

$$\|\varphi\|_{L^\infty(\mathbb{R}^N)} \le \mathcal{M}_1\, \frac{N\,\omega_N}{N+\alpha} \left(\int_{\mathbb{R}^N} |\nabla\varphi|^p\, dx\right)^{\frac{1}{p}} + \omega_N^{\frac{p-1}{p}} \left(\int_{\mathbb{R}^N} |\varphi|^p\, dx\right)^{\frac{1}{p}}. \tag{3.6.16}$$

In order to arrive at the desired estimate (3.6.12), we use a *scaling argument*. Namely, we take $\varphi \in C_0^\infty(\mathbb{R}^N)$ and apply (3.6.16) to the function $\varphi_\lambda(x) = \varphi(\lambda\, x)$, where $\lambda > 0$. This gives

$$\|\varphi_\lambda\|_{L^\infty(\mathbb{R}^N)} \le \mathcal{M}_1\, \frac{N\,\omega_N}{N+\alpha}\, \lambda \left(\int_{\mathbb{R}^N} |\nabla\varphi(\lambda\, x)|^p\, dx\right)^{\frac{1}{p}} + \omega_N^{\frac{p-1}{p}} \left(\int_{\mathbb{R}^N} |\varphi(\lambda\, x)|^p\, dx\right)^{\frac{1}{p}}.$$

By observing that

$$\|\varphi_\lambda\|_{L^\infty(\mathbb{R}^N)} = \|\varphi\|_{L^\infty(\mathbb{R}^N)},$$

and making the change of variables $\lambda\, x = y$, we thus obtain

[9] We use that $\mathbb{S}^{N-1} = \partial B_1$, thus its $(N-1)$-dimensional surface measure can be computed by the Divergence Theorem

$$|\mathbb{S}^{N-1}| = \int_{\partial B_1(0)} d\sigma(x) = \int_{\partial B_1(0)} \langle x, \nu_{B_1}\rangle\, d\sigma(x)$$
$$= \int_{B_1(0)} \operatorname{div} x\, dx = N\,|B_1(0)| = N\,\omega_N.$$

$$\|\varphi\|_{L^\infty(\mathbb{R}^N)} \leq \mathcal{M}_1 \frac{N\,\omega_N}{N+\alpha}\,\lambda^{1-\frac{N}{p}} \left(\int_{\mathbb{R}^N} |\nabla\varphi|^p\,dy\right)^{\frac{1}{p}} + \omega_N^{\frac{p-1}{p}}\,\lambda^{-\frac{N}{p}} \left(\int_{\mathbb{R}^N} |\varphi|^p\,dy\right)^{\frac{1}{p}},$$

which holds for every $\lambda > 0$. In particular, we obtain

$$\begin{aligned}\|\varphi\|_{L^\infty(\mathbb{R}^N)} \leq \inf_{\lambda>0} \Bigg[& \mathcal{M}_1 \frac{N\,\omega_N}{N+\alpha}\,\lambda^{1-\frac{N}{p}} \left(\int_{\mathbb{R}^N} |\nabla\varphi|^p\,dy\right)^{\frac{1}{p}} \\ & + \omega_N^{\frac{p-1}{p}}\,\lambda^{-\frac{N}{p}} \left(\int_{\mathbb{R}^N} |\varphi|^p\,dy\right)^{\frac{1}{p}} \Bigg].\end{aligned}$$

The last infimum can be explicitly computed, by studying the function of one variable

$$h(\lambda) = \lambda^{1-\frac{N}{p}}\,A + \lambda^{-\frac{N}{p}}\,B, \qquad \text{for } \lambda > 0.$$

Thanks to the fact that $p > N$, the last function is minimal for

$$\lambda = \frac{B}{A}\,\frac{N}{p-N}.$$

By using this fact with

$$A = \mathcal{M}_1 \frac{N\,\omega_N}{N+\alpha} \left(\int_{\mathbb{R}^N} |\nabla\varphi|^p\,dy\right)^{\frac{1}{p}} \qquad \text{and} \qquad B = \omega_N^{\frac{p-1}{p}} \left(\int_{\mathbb{R}^N} |\varphi|^p\,dy\right)^{\frac{1}{p}},$$

and computing the minimum above, we finally get the desired estimate (3.6.12). □

Remark 3.6.9 A closer inspection of the proof of the previous result reveals that we never used that φ has compact support. Thus, inequality (3.6.11) still holds for every function $\varphi \in C^1(\mathbb{R}^N)$, such that

$$\|\nabla\varphi\|_{L^p(\mathbb{R}^N;\mathbb{R}^N)} < +\infty,$$

while (3.6.12) still holds for functions $\varphi \in C^1(\mathbb{R}^N) \cap W^{1,p}(\mathbb{R}^N)$. For $N < p < \infty$, the constants found in the previous proof are given by

$$\mathcal{M}_1 = \frac{2^{N+1}}{\omega_N^{1/p}}\,\frac{p}{p-N},$$

and

$$\mathcal{M}_2 = \left(\mathcal{M}_1 \frac{N\,\omega_N}{N+\alpha}\right)^{\frac{N}{p}} \omega_N^{\frac{(p-1)\,(p-N)}{p^2}} \frac{p}{p-N} \left(\frac{p-N}{N}\right)^{\frac{N}{p}}.$$

Once again, these constants are not the sharpest possible: we refer to [80] and [91, Remark 2.5] for a discussion on this issue.

Remark 3.6.10 (The Case $p = \infty$) The previous result holds also in the case $p=\infty$. The proof is straightforward in this case. Indeed, (3.6.12) simply reduces to an identity with $\mathcal{M}_2 = 1$, by interpreting N/p as 0. As for the first inequality (3.6.11), this holds with $\alpha = 1$ and $\mathcal{M}_1 = 1$: for every $x, y \in \mathbb{R}^N$ we can directly infer

$$|\varphi(x) - \varphi(y)| \leq |x-y| \int_0^1 |\nabla\varphi(t\,x + (1-t)\,y)|\,dt \leq |x-y|\,\|\nabla\varphi\|_{L^\infty(\mathbb{R}^N;\mathbb{R}^N)}.$$

This in particular implies that

$$|\varphi|_{C^{0,1}(\mathbb{R}^N)} \leq \|\nabla\varphi\|_{L^\infty(\mathbb{R}^N;\mathbb{R}^N)}. \tag{3.6.17}$$

3.7 The Space $W_0^{1,p}$

The following subspace of $W^{1,p}$ will play an important role: it will be natural space to prescribe a boundary datum in minimization problems.

Definition 3.7.1 Let $1 \leq p \leq \infty$ and let $\Omega \subseteq \mathbb{R}^N$ be an open set. We define $W_0^{1,p}(\Omega)$ as the closure of $C_0^\infty(\Omega)$ in the Sobolev space $W^{1,p}(\Omega)$. In particular, by construction we have $W_0^{1,p}(\Omega) \subseteq W^{1,p}(\Omega)$.

Remark 3.7.2 Roughly speaking, we can think of the space $W_0^{1,p}(\Omega)$ as the subspace of $W^{1,p}(\Omega)$ made of functions "vanishing at the boundary" in a suitable sense. When $\partial\Omega$ is sufficiently regular, this can be made more precise by using the so-called *trace theory*: see Appendix D for a brief introduction to such a theory and [44, Chapter 18] for a comprehensive study.

Remark 3.7.3 It is useful to observe that, by taking the closure of $C_0^1(\Omega)$ in the Sobolev space $W^{1,p}(\Omega)$, one still gets the same space $W_0^{1,p}(\Omega)$. Indeed, since we have

$$C_0^\infty(\Omega) \subseteq C_0^1(\Omega) \subseteq W^{1,p}(\Omega),$$

it is clear that $W_0^{1,p}(\Omega)$ is contained in the closure of $C_0^1(\Omega)$. On the other hand, for every $\varepsilon > 0$ and every $\varphi \in C_0^1(\Omega)$, there exists $\varphi_\varepsilon \in C_0^\infty(\Omega)$ such that

$$\|\varphi - \varphi_\varepsilon\|_{W^{1,p}(\Omega)} < \varepsilon.$$

This can be constructed by using convolutions. We now take u belonging to the closure of $C_0^1(\Omega)$, thus there exists $\{\varphi_n\}_{n\in\mathbb{N}} \subseteq C_0^1(\Omega)$ such that

$$\lim_{n\to\infty} \|u - \varphi_n\|_{W^{1,p}(\Omega)} = 0.$$

For every $n \in \mathbb{N} \setminus \{0\}$, we take $v_n \in C_0^\infty(\Omega)$ such that

$$\|v_n - \varphi_n\|_{W^{1,p}(\Omega)} < \frac{1}{n}.$$

We thus obtain

$$\lim_{n\to\infty} \|u - v_n\|_{W^{1,p}(\Omega)} \le \lim_{n\to\infty} \|u - \varphi_n\|_{W^{1,p}(\Omega)} + \lim_{n\to\infty} \|v_n - \varphi_n\|_{W^{1,p}(\Omega)} = 0,$$

which shows that $u \in W_0^{1,p}(\Omega)$.

Theorem 3.7.4 *Let $1 \le p \le \infty$ and let $\Omega \subseteq \mathbb{R}^N$ be an open set. Then $W_0^{1,p}(\Omega)$ is a Banach subspace of $W^{1,p}(\Omega)$, i.e. for every Cauchy sequence $\{u_n\}_{n\in\mathbb{N}} \subseteq W_0^{1,p}(\Omega)$ there exists $u \in W_0^{1,p}(\Omega)$ such that*

$$\lim_{n\to\infty} \|u_n - u\|_{W^{1,p}(\Omega)} = 0.$$

For $1 < p < \infty$, this space is also weakly closed.

Proof We first prove that $W_0^{1,p}(\Omega)$ is a vector space. We take $u, v \in W_0^{1,p}(\Omega)$ and $\alpha, \beta \in \mathbb{R}$. We want to prove that

$$\alpha\, u + \beta\, v \in W_0^{1,p}(\Omega).$$

By construction, we know that there exists $\{u_n\}_{n\in\mathbb{N}}, \{v_n\}_{n\in\mathbb{N}} \subseteq C_0^\infty(\Omega)$ such that

$$\lim_{n\to\infty} \|u_n - u\|_{W^{1,p}(\Omega)} = 0 = \lim_{n\to\infty} \|v_n - v\|_{W^{1,p}(\Omega)}.$$

By defining the function

$$\phi_n = \alpha\, u_n + \beta\, v_n \in C_0^\infty(\Omega),$$

we have that

$$\lim_{n\to\infty} \|\phi_n - (\alpha\, u + \beta\, v)\|_{W^{1,p}(\Omega)} \le |\alpha| \lim_{n\to\infty} \|u_n - u\|_{W^{1,p}(\Omega)} + |\beta| \lim_{n\to\infty} \|v_n - v\|_{W^{1,p}(\Omega)} = 0.$$

This shows that $\alpha\, u + \beta\, v$ belongs to the closure, as desired. The fact that $W_0^{1,p}(\Omega)$ is a Banach space follows from the fact that it is a closed subspace of the Banach space $W^{1,p}(\Omega)$ (*write down the details as an exercise*).

We now take $1 < p < \infty$ and show that $W_0^{1,p}(\Omega)$ is weakly closed, as well. Let us take a sequence $\{u_n\}_{n\in\mathbb{N}} \subseteq W_0^{1,p}(\Omega)$ weakly converging in $W^{1,p}(\Omega)$ to a function $u \in W^{1,p}(\Omega)$. We need to show that $u \in W_0^{1,p}(\Omega)$. To this aim, we will use Mazur's Lemma (see Theorem B.3.2). By considering the sequence

$$\{\phi_n\}_{n\in\mathbb{N}} \subseteq L^p(\Omega;\mathbb{R}^{N+1}), \qquad \text{with } \phi_n = (u_n, \nabla u_n),$$

this is weakly converging to $\phi = (u, \nabla u)$ in $L^p(\Omega;\mathbb{R}^{N+1})$, thanks to the weak convergence of $\{u_n\}_{n\in\mathbb{N}}$. By Theorem B.3.2 and Remark B.3.3, we can construct a new sequence $\{\Phi_n\}_{n\in\mathbb{N}}$ such that

$$\Phi_n \in \operatorname{conv}\Big(\{\phi_0,\dots,\phi_n\}\Big), \qquad \text{for every } n \in \mathbb{N},$$

and

$$\lim_{n\to\infty} \|\Phi_n - \phi\|_{L^p(\Omega;\mathbb{R}^{N+1})} = 0. \tag{3.7.1}$$

Here we denoted by $\operatorname{conv}(\{\phi_0,\dots,\phi_n\})$ the convex hull of the set $\{\phi_0,\dots,\phi_n\}$ (see Definition B.3.1). By construction each Φ_n has the form

$$\Phi_n = \left(\sum_{i=0}^{n} \lambda_{i,n}\, u_i, \sum_{i=0}^{n} \lambda_{i,n}\, \nabla u_i\right), \qquad \text{for some } \lambda_{i,n} \ge 0 \text{ such that } \sum_{i=0}^{n} \lambda_{i,n} = 1.$$

We can then define

$$v_n = \sum_{i=0}^{n} \lambda_{i,n}\, u_i, \qquad \text{for } n \in \mathbb{N},$$

and observe that $v_n \in W_0^{1,p}(\Omega)$, since each $u_i \in W_0^{1,p}(\Omega)$ and the latter is a vector space. By noticing that $\Phi_n = (v_n, \nabla v_n)$, from (3.7.1) we thus get

$$\lim_{n\to\infty} \|v_n - u\|_{W^{1,p}(\Omega)} = 0.$$

This shows that u is the limit in the norm topology of a sequence $\{v_n\}_{n\in\mathbb{N}} \subseteq W_0^{1,p}(\Omega)$. Since the space $W_0^{1,p}(\Omega)$ is closed in the norm topology by construction, we finally get

$$u \in W_0^{1,p}(\Omega),$$

as desired. □

Remark 3.7.5 In the previous statement about weak closedness, we did not discussed for simplicity the extremal cases $p = 1$ and $p = \infty$. This is not restrictive for our scopes, since in the next chapter we will need this property only for the case $1 < p < \infty$.

In what follows, we will use Definition 3.5.1.

Theorem 3.7.6 (Poincaré Inequality for $W_0^{1,p}(\Omega)$) *Let $1 \leq p \leq \infty$ and let $\Omega \subseteq \mathbb{R}^N$ be an open set bounded in direction $\omega \in \mathbb{S}^{N-1}$. Then we have*

$$\frac{1}{\mathrm{width}(\Omega;\,\omega)}\,\|\varphi\|_{L^p(\Omega)} \leq \|\nabla\varphi\|_{L^p(\Omega;\mathbb{R}^N)}, \qquad \textit{for every } \varphi \in W_0^{1,p}(\Omega).$$

In particular

$$\varphi \mapsto \|\nabla\varphi\|_{L^p(\Omega;\mathbb{R}^N)},$$

is a norm on $W_0^{1,p}(\Omega)$ and this is equivalent to the one of $W^{1,p}(\Omega)$.

Proof Let $\varphi \in W_0^{1,p}(\Omega)$, by definition there exists a sequence $\{\varphi_n\}_{n\in\mathbb{N}} \subseteq C_0^\infty(\Omega)$ such that

$$\lim_{n\to\infty} \|\varphi_n - \varphi\|_{L^p(\Omega)} = \lim_{n\to\infty} \|\nabla\varphi_n - \nabla\varphi\|_{L^p(\Omega;\mathbb{R}^N)} = 0. \tag{3.7.2}$$

For each φ_n, by Proposition 3.5.2 we have

$$\frac{1}{\mathrm{width}(\Omega;\,\omega)}\,\|\varphi_n\|_{L^p(\Omega)} \leq \|\nabla\varphi_n\|_{L^p(\Omega;\mathbb{R}^N)}.$$

By (3.7.2), we can pass to the limit in the previous inequality and get the validity of Poincaré inequality for $\varphi \in W_0^{1,p}(\Omega)$, as well.

Let us now show the equivalence of the norms. The fact that

$$\|\nabla\varphi\|_{L^p(\Omega;\mathbb{R}^N)} \leq \|\varphi\|_{W^{1,p}(\Omega)},$$

is a direct consequence of the definition of the norm of $W^{1,p}(\Omega)$. On the other hand, by Poincaré inequality, we have

$$\|\varphi\|_{W^{1,p}(\Omega)} = \|\varphi\|_{L^p(\Omega)} + \|\nabla\varphi\|_{L^p(\Omega;\mathbb{R}^N)} \leq (\mathrm{width}(\Omega;\,\omega) + 1)\,\|\nabla\varphi\|_{L^p(\Omega;\mathbb{R}^N)}.$$

This concludes the proof. □

Remark 3.7.7 For an open bounded set $\Omega \subseteq \mathbb{R}^N$, by recalling (3.5.5), we also get from Theorem 3.7.6

$$\frac{1}{\operatorname{diam}(\Omega)}\,\|\varphi\|_{L^p(\Omega)} \le \|\nabla\varphi\|_{L^p(\Omega;\mathbb{R}^N)}, \qquad \text{for every } \varphi \in W_0^{1,p}(\Omega).$$

The next result will be important, in order to prove compactness of some embeddings for Sobolev spaces. For a given vector $h \in \mathbb{R}^N$, we will use the following notation

$$\mathrm{T}_h\varphi(x) := \varphi(x+h), \qquad \text{for every } x \in \mathbb{R}^N,$$

for a measurable function φ.

Lemma 3.7.8 (Control on the Translations) *Let* $1 \le p \le \infty$*, for every* $\varphi \in W_0^{1,p}(\mathbb{R}^N)$ *and every vector* $h \in \mathbb{R}^N$*, we have*

$$\|\mathrm{T}_h\varphi - \varphi\|_{L^p(\mathbb{R}^N)} \le |h|\,\|\nabla\varphi\|_{L^p(\mathbb{R}^N;\mathbb{R}^N)}.$$

Proof By density, it is enough to prove the inequality for $\varphi \in C_0^\infty(\mathbb{R}^N)$. For $p = \infty$, we observe that the result immediately follows from the extremal Morrey inequality (3.6.17). We thus take $1 \le p < \infty$: by the Fundamental Theorem of Calculus and Cauchy-Schwarz inequality, we have

$$\begin{aligned}|\varphi(x+h)-\varphi(x)| = \left|\int_0^1 \frac{d}{dt}\varphi(x+t\,h)\,dt\right| &= \left|\int_0^1 \langle\nabla\varphi(x+t\,h), h\rangle\,dt\right| \\ &\le |h|\int_0^1 |\nabla\varphi(x+t\,h)|\,dt.\end{aligned}$$

We raise the previous inequality to the power p and get

$$|\varphi(x+h)-\varphi(x)|^p \le |h|^p\left(\int_0^1 |\nabla\varphi(x+t\,h)|\,dt\right)^p \le |h|^p\int_0^1 |\nabla\varphi(x+t\,h)|^p\,dt,$$

where we also used Jensen's inequality (Proposition 1.2.11) in the last inequality. We now integrate with respect to $x \in \mathbb{R}^N$, so to get

$$\begin{aligned}\int_{\mathbb{R}^N} |\varphi(x+h)-\varphi(x)|^p\,dx &\le |h|^p\int_{\mathbb{R}^N}\left(\int_0^1 |\nabla\varphi(x+t\,h)|^p\,dt\right)dx \\ &= |h|^p\int_0^1\left(\int_{\mathbb{R}^N} |\nabla\varphi(x+t\,h)|^p\,dx\right)dt,\end{aligned}$$

where we exchanged the order of integration, by the Fubini-Tonelli Theorem. It is only left to observe that

$$\int_{\mathbb{R}^N} |\nabla\varphi(x+t\,h)|^p\,dx = \int_{\mathbb{R}^N} |\nabla\varphi(y)|^p\,dy, \qquad \text{for every } h \in \mathbb{R}^N \text{ and } t \in [0,1].$$

Thus we get

$$\int_{\mathbb{R}^N} |\varphi(x+h) - \varphi(x)|^p \, dx \le |h|^p \int_{\mathbb{R}^N} |\nabla \varphi|^p \, dx,$$

as desired. □

Functions belonging to $W_0^{1,p}(\Omega)$ enjoy a natural extension property, contained in the following

Lemma 3.7.9 (Extension by Zero) *Let $1 \le p \le \infty$ and let $\Omega \subsetneq E \subseteq \mathbb{R}^N$ be two open sets. We define the "extension by zero" operator*

$$\mathcal{Z} : W_0^{1,p}(\Omega) \to L^p(E),$$

through

$$\mathcal{Z}[u](x) = \begin{cases} u(x), & \text{if } x \in \Omega, \\ 0, & \text{if } x \in E \setminus \Omega. \end{cases}$$

Then this is a linear operator such that

$$\mathcal{Z}[u] \in W_0^{1,p}(E), \qquad \text{for every } u \in W_0^{1,p}(\Omega), \tag{3.7.3}$$

and

$$\big\| \mathcal{Z}[u] \big\|_{W^{1,p}(E)} = \|u\|_{W^{1,p}(\Omega)}, \qquad \text{for every } u \in W_0^{1,p}(\Omega). \tag{3.7.4}$$

Moreover, for every $u \in W_0^{1,p}(\Omega)$ the weak gradient of $\mathcal{Z}[u]$ is given by

$$\nabla \mathcal{Z}[u](x) = \begin{cases} \nabla u(x), & \text{if } x \in \Omega, \\ (0, \dots, 0), & \text{if } x \in E \setminus \Omega. \end{cases}$$

Proof The linearity of $\mathcal{Z}$ is obvious. Let us show the property (3.7.3). By definition of $W_0^{1,p}(\Omega)$, we know that for every $u \in W_0^{1,p}(\Omega)$ there exists a sequence $\{u_n\}_{n\in\mathbb{N}} \subseteq C_0^\infty(\Omega)$ such that

$$\lim_{n\to\infty} \|u_n - u\|_{W^{1,p}(\Omega)} = 0. \tag{3.7.5}$$

In particular, we have that $\{u_n\}_{n\in\mathbb{N}} \subseteq C_0^\infty(\Omega)$ is a Cauchy sequence in $W^{1,p}(\Omega)$. Moreover, we clearly have $\mathcal{Z}[u_n] \in C_0^\infty(E)$ and

$$\big\| \mathcal{Z}[u_n] \big\|_{W^{1,p}(E)} = \|u_n\|_{W^{1,p}(\Omega)}, \qquad \text{for every } n \in \mathbb{N}, \tag{3.7.6}$$

thanks to the fact that u_n has compact support in Ω. We thus get, by using also the linearity, that

$$\left\| \mathcal{Z}[u_n] - \mathcal{Z}[u_m] \right\|_{W^{1,p}(E)} = \|u_n - u_m\|_{W^{1,p}(\Omega)}, \qquad \text{for every } n, m \in \mathbb{N},$$

which shows that $\{\mathcal{Z}[u_n]\}_{n\in\mathbb{N}} \subseteq C_0^\infty(E)$ is a Cauchy sequence in $W_0^{1,p}(E)$. The latter being a Banach space (recall Theorem 3.7.4), we get that $\{\mathcal{Z}[u_n]\}_{n\in\mathbb{N}}$ converges in $W^{1,p}(E)$ to a function $U \in W_0^{1,p}(E)$. Since by construction we have $\mathcal{Z}[u] \in L^p(E)$, in order to conclude that $\mathcal{Z}[u] \in W_0^{1,p}(E)$ it is enough to prove that

$$U = \mathcal{Z}[u], \qquad \text{a. e. in } E, \tag{3.7.7}$$

thanks to Lemma 3.3.4. To this aim, we observe that $\mathcal{Z}[u_n] = u_n$ and $\mathcal{Z}[u] = u$ on Ω, by construction. By using this fact, we get

$$\begin{aligned} \left\| U - \mathcal{Z}[u] \right\|_{W^{1,p}(\Omega)} &= \|U - u\|_{W^{1,p}(\Omega)} \\ &\le \left\| U - \mathcal{Z}[u_n] \right\|_{W^{1,p}(\Omega)} + \left\| \mathcal{Z}[u_n] - u \right\|_{W^{1,p}(\Omega)} \\ &= \left\| U - \mathcal{Z}[u_n] \right\|_{W^{1,p}(\Omega)} + \|u_n - u\|_{W^{1,p}(\Omega)}, \end{aligned}$$

and both quantities on the right-hand side converge to 0, as n goes to ∞. This shows that

$$U = \mathcal{Z}[u], \qquad \text{a. e. in } \Omega.$$

On the other hand, U is the strong L^p limit of functions which vanish almost everywhere on $E \setminus \Omega$, thus

$$\|U\|_{L^p(E\setminus\Omega)} = \lim_{n\to\infty} \left\| \mathcal{Z}[u_n] \right\|_{L^p(E\setminus\Omega)} = 0.$$

This shows that U vanishes almost everywhere on $E \setminus \Omega$, exactly as $\mathcal{Z}[u]$. We thus get (3.7.7) and $\mathcal{Z}[u] \in W_0^{1,p}(E)$, as explained above.

The proof of (3.7.4) is now very simple, it is sufficient to pass to the limit in (3.7.6).

Finally, we come to the computation of the weak gradient of $\mathcal{Z}[u]$. For every $\varphi \in C_0^\infty(E)$ and every $k \in \{1, \dots, N\}$, we have

$$\int_E \frac{\partial \varphi}{\partial x_k}\, \mathcal{Z}[u]\, dx = \int_\Omega \frac{\partial \varphi}{\partial x_k}\, u\, dx, \tag{3.7.8}$$

by definition of $\mathcal{Z}[u]$. We consider again the sequence $\{u_n\}_{n\in\mathbb{N}} \subseteq C_0^\infty(\Omega)$ such that (3.7.5) holds. By using the definition of weak derivative on Ω for φ and recalling that its classical gradient coincides with the weak one (Proposition 3.2.3), we have

$$\int_\Omega \frac{\partial \varphi}{\partial x_k}\, u_n\, dx = -\int_\Omega \varphi\, \frac{\partial u_n}{\partial x_k}\, dx.$$

By using the strong convergence (3.7.5), we can take the limit as n goes to ∞ and get

$$\int_\Omega \frac{\partial \varphi}{\partial x_k}\, u\, dx = -\int_\Omega \varphi\, \frac{\partial u}{\partial x_k}\, dx.$$

By recalling (3.7.8), this finally shows that

$$\int_E \frac{\partial \varphi}{\partial x_k}\, \mathcal{Z}[u]\, dx = -\int_\Omega \varphi\, \frac{\partial u}{\partial x_k}\, dx, \qquad \text{for every } \varphi \in C_0^\infty(E).$$

The definition of weak partial derivative implies that

$$\frac{\partial \mathcal{Z}[u]}{\partial x_k} = \frac{\partial u}{\partial x_k}\, 1_\Omega, \qquad \text{for } k \in \{1, \dots, N\},$$

and thus we conclude the proof. □

In particular, the previous result guarantees that every function $u \in W_0^{1,p}(\Omega)$ can be naturally extended to a Sobolev function defined in the whole $\mathbb{R}^N$, i.e. as an element of $W_0^{1,p}(\mathbb{R}^N)$. Actually, one can prove that

$$W_0^{1,p}(\mathbb{R}^N) = W^{1,p}(\mathbb{R}^N), \tag{3.7.9}$$

provided that $1 \le p < \infty$. This is the content of the next result, that we record for completeness.

Proposition 3.7.10 *Let* $1 \le p < \infty$*, then the identity* (3.7.9) *holds. On the contrary, we have*

$$W_0^{1,\infty}(\mathbb{R}^N) \subseteq W^{1,\infty}(\mathbb{R}^N) \qquad \text{and} \qquad W_0^{1,\infty}(\mathbb{R}^N) \neq W^{1,\infty}(\mathbb{R}^N).$$

Proof For every $1 \le p \le \infty$, the inclusion

$$W_0^{1,p}(\mathbb{R}^N) \subseteq W^{1,p}(\mathbb{R}^N),$$

is straightforward, by the definition of $W_0^{1,p}(\mathbb{R}^N)$. To show the reverse inclusion for $1 \le p < \infty$, we need to show that for every $u \in W^{1,p}(\mathbb{R}^N)$ there exists a sequence $v_n \in C_0^\infty(\mathbb{R}^N)$ such that

$$\lim_{n\to\infty} \|v_n - u\|_{W^{1,p}(\mathbb{R}^N)} = 0.$$

To this aim, we first take the sequence $\{u_n\}_{n\in\mathbb{N}}$ of Proposition 3.2.6, with $\Omega = \mathbb{R}^N$, i.e.

$$u_n(x) = u * \rho_n(x) = \int_{B_{\frac{1}{n}}(0)} u(x-y)\,\rho_n(y)\,dy, \qquad \text{for } x \in \mathbb{R}^N.$$

By the properties of convolutions (see for example [46, Theorem 2.16]), we know that $u_n \in C^\infty(\mathbb{R}^N) \cap L^p(\mathbb{R}^N)$ and that

$$\lim_{n\to\infty} \|u_n - u\|_{L^p(\mathbb{R}^N)} = 0. \tag{3.7.10}$$

By Proposition 3.2.6 we also know that

$$\nabla u_n = (\nabla u) * \rho_n.$$

Thus, we actually have $u_n \in C^\infty(\mathbb{R}^N) \cap W^{1,p}(\mathbb{R}^N)$ and, again by using the properties of convolutions, we get

$$\lim_{n\to\infty} \|\nabla u - \nabla u_n\|_{L^p(\mathbb{R}^N;\mathbb{R}^N)} = 0,$$

as well. Up to now, we have shown that we can approximate a $W^{1,p}(\mathbb{R}^N)$ function by a sequence of $C^\infty(\mathbb{R}^N) \cap W^{1,p}(\mathbb{R}^N)$ functions. We still have to show that we can take the approximating functions with compact support. We take a *cut-off function* $\eta_n \in C_0^\infty(\mathbb{R}^N)$ such that

$$0 \le \eta_n \le 1, \qquad \eta_n = 1 \text{ on } B_n(0), \qquad \eta_n = 0 \text{ on } \mathbb{R}^N \setminus B_{n+1}(0),$$

and

$$|\nabla \eta_n| \le C,$$

for some $C > 0$ independent of n. We set $v_n = u_n\,\eta_n \in C_0^\infty(\mathbb{R}^N)$ and observe that by the triangle inequality

$$\|v_n - u\|_{W^{1,p}(\mathbb{R}^N)} \le \|v_n - u_n\|_{W^{1,p}(\mathbb{R}^N)} + \|u_n - u\|_{W^{1,p}(\mathbb{R}^N)}.$$

In order to conclude the proof, we need to show that

$$\lim_{n\to\infty} \|u_n - v_n\|_{W^{1,p}(\mathbb{R}^N)} = 0.$$

We have

$$\left(\int_{\mathbb{R}^N} |u_n - v_n|^p \, dx\right)^{\frac{1}{p}} = \left(\int_{\mathbb{R}^N} |1-\eta_n|^p \, |u_n|^p \, dx\right)^{\frac{1}{p}}$$
$$\leq \left(\int_{\mathbb{R}^N \setminus B_n(0)} |u_n|^p \, dx\right)^{\frac{1}{p}}$$
$$\leq \left(\int_{\mathbb{R}^N \setminus B_n(0)} |u_n - u|^p \, dx\right)^{\frac{1}{p}} + \left(\int_{\mathbb{R}^N \setminus B_n(0)} |u|^p \, dx\right)^{\frac{1}{p}}.$$

The last two integrals converges to 0 as n goes to ∞: the first one, by (3.7.10); the second one, by the Dominated Convergence Theorem.

As for the gradients, by using the properties of η_n we have

$$\left(\int_{\mathbb{R}^N} |\nabla u_n - \nabla v_n|^p \, dx\right)^{\frac{1}{p}}$$
$$= \left(\int_{\mathbb{R}^N} |\nabla u_n \, \eta_n + u_n \, \nabla \eta_n - \nabla u_n|^p \, dx\right)^{\frac{1}{p}}$$
$$\leq \left(\int_{\mathbb{R}^N} |1-\eta_n|^p \, |\nabla u_n|^p \, dx\right)^{\frac{1}{p}} + \left(\int_{\mathbb{R}^N} |u_n|^p \, |\nabla \eta_n|^p \, dx\right)^{\frac{1}{p}}$$
$$\leq \left(\int_{\mathbb{R}^N \setminus B_n(0)} |\nabla u_n|^p \, dx\right)^{\frac{1}{p}} + C\left(\int_{B_{n+1}(0)\setminus B_n(0)} |u_n|^p \, dx\right)^{\frac{1}{p}}.$$

The last two integrals converge to 0 as n goes to ∞, thanks to an argument similar to the one used above. We leave the details to the reader.

Finally, we need to show that $W_0^{1,\infty}(\mathbb{R}^N) \neq W^{1,\infty}(\mathbb{R}^N)$. We take the function $u \equiv 1$. We have $u \in L^\infty(\mathbb{R}^N)$ and moreover, by Proposition 3.2.3, its weak gradient identically vanishes. Thus, $\nabla u \in L^\infty(\mathbb{R}^N; \mathbb{R}^N)$ and this shows that $u \in W^{1,\infty}(\mathbb{R}^N)$. We argue by contradiction and suppose that $u \in W_0^{1,\infty}(\mathbb{R}^N)$. This implies that there exists a sequence $\{u_n\}_{n\in\mathbb{N}} \subseteq C_0^\infty(\mathbb{R}^N)$ such that

$$\lim_{n\to\infty} \|u_n - 1\|_{L^\infty(\mathbb{R}^N)} = 0 = \lim_{n\to\infty} \|\nabla u_n\|_{L^\infty(\mathbb{R}^N;\mathbb{R}^N)}.$$

However, we clearly have

$$\|u_n - 1\|_{L^\infty(\mathbb{R}^N)} \geq 1, \qquad \text{for every } n \in \mathbb{N},$$

because u_n is compactly supported. This gives a contradiction. □

We can show that the only constant function belonging to $W_0^{1,p}(\Omega)$ is the null one. Observe that we *do not* need to assume that Ω is connected. This result enforces the idea that functions in $W_0^{1,p}(\Omega)$ are "vanishing at the boundary", in some sense (recall Remark 3.7.2).

Proposition 3.7.11 *Let $1 \le p \le \infty$ and let $\Omega \subseteq \mathbb{R}^N$ be an open set. If $u \in W_0^{1,p}(\Omega)$ is such that*

$$\nabla u = (0, \dots, 0), \qquad a.\ e.\ in\ \Omega,$$

then u vanishes almost everywhere in Ω.

Proof We can consider the extension by 0 of u, i.e.

$$\mathcal{Z}[u] = \begin{cases} u, \text{ on } \Omega, \\ 0, \text{ on } \mathbb{R}^N \setminus \Omega. \end{cases}$$

By Lemma 3.7.9 with $E = \mathbb{R}^N$, we know that $\mathcal{Z}[u] \in W_0^{1,p}(\mathbb{R}^N)$. Moreover, still by Lemma 3.7.9 we know that the weak gradient of $\mathcal{Z}[u]$ is given by

$$\nabla \mathcal{Z}[u](x) = \begin{cases} \nabla u(x), \text{ if } x \in \Omega, \\ (0, \dots, 0), \text{ otherwise}, \end{cases} = (0, \dots, 0).$$

By Proposition 3.2.9, we get that $\mathcal{Z}[u]$ must be constant on $\mathbb{R}^N$ (the latter being a connected set, obviously). This in turn implies that there exists a constant $C \in \mathbb{R}$ such that

$$\mathcal{Z}[u] \equiv C, \qquad \text{a. e. on } \mathbb{R}^N.$$

Since $\mathcal{Z}[u]$ in particular belongs to $L^p(\mathbb{R}^N)$, for $1 \le p < \infty$ the only possibility is that $C = 0$. By definition of $\mathcal{Z}[u]$, this implies that we have $u = 0$ almost everywhere in Ω, as desired.

For the case $p = \infty$, the argument is slightly different: by the same argument used at the end of the proof Proposition 3.7.10, we see that if $\mathcal{Z}[u] \equiv C \neq 0$, then it could not belong to $W_0^{1,\infty}(\mathbb{R}^N)$. □

Finally, we conclude this section by particularizing to the space $W_0^{1,p}(\Omega)$ the results of Sect. 3.4, about compositions with some suitable functions.

Proposition 3.7.12 *Let $f : \mathbb{R} \to \mathbb{R}$ be a C^1 function with bounded derivative, such that $f(0) = 0$. Let $1 \le p \le \infty$ and let $\Omega \subseteq \mathbb{R}^N$ be an open set. Then for every $u \in W_0^{1,p}(\Omega)$, we have*

$$f \circ u \in W_0^{1,p}(\Omega),$$

as well.

Proof Let f be a function as in the statement and let $u \in W_0^{1,p}(\Omega)$. We already know that $f \circ u$ belongs to $W^{1,p}(\Omega)$, thanks to Proposition 3.4.1 and Remark 3.4.2. We have to show that it actually belongs to $W_0^{1,p}(\Omega)$. By definition, we know that there exists $\{u_n\}_{n\in\mathbb{N}} \subseteq C_0^\infty(\Omega)$ such that

$$\lim_{n\to\infty} \|u_n - u\|_{W^{1,p}(\Omega)} = 0.$$

Up to extract a subsequence, we can further suppose that u_n converges almost everywhere to u. We observe that $f \circ u_n \in C_0^1(\Omega)$: we claim that

$$\lim_{n\to\infty} \|f(u_n) - f(u)\|_{W^{1,p}(\Omega)} = 0. \tag{3.7.11}$$

By recalling Remark 3.7.3, this would be sufficient to establish that $f \circ u \in W_0^{1,p}(\Omega)$. To verify (3.7.11), we notice that by using (3.4.2) we have

$$\lim_{n\to\infty} \|f(u_n) - f(u)\|_{L^p(\Omega)} \le L \lim_{n\to\infty} \|u_n - u\|_{L^p(\Omega)} = 0,$$

where $L = \sup_{\mathbb{R}} |f'|$. Moreover, by recalling that $\nabla(f \circ u) = f'(u)\,\nabla u$ thanks to Proposition 3.4.1, we also have

$$\begin{aligned}\|\nabla(f \circ u_n) - \nabla(f \circ u)\|_{L^p(\Omega;\mathbb{R}^N)} &= \|f'(u_n)\,\nabla u_n - f'(u)\,\nabla u\|_{L^p(\Omega;\mathbb{R}^N)} \\ &\le \|f'(u_n)\,(\nabla u_n - \nabla u)\|_{L^p(\Omega;\mathbb{R}^N)} \\ &\quad + \|(f'(u_n) - f'(u))\,\nabla u\|_{L^p(\Omega;\mathbb{R}^N)}.\end{aligned}$$

The last two terms converge to 0 as n goes to ∞. Indeed, for the first one, it is sufficient to use that f' is bounded, thus

$$\|f'(u_n)\,(\nabla u_n - \nabla u)\|_{L^p(\Omega;\mathbb{R}^N)} \le L\,\|\nabla u_n - \nabla u\|_{L^p(\Omega;\mathbb{R}^N)},$$

and the latter converges to 0, as n goes to ∞.

For the second one, in the case $1 \le p < \infty$ one can observe that

$$\lim_{n\to\infty} |f'(u_n(x)) - f'(u(x))|^p\,|\nabla u(x)|^p = 0 \qquad \text{for a. e. } x \in \Omega,$$

thanks to the fact that u_n converges almost everywhere and that f' is continuous; moreover, we have

$$|f'(u_n) - f'(u)|^p\,|\nabla u|^p \le 2^p\,L^p\,|\nabla u|^p \in L^1(\Omega).$$

Thus, we can use the Dominated Convergence Theorem.

For the case $p = \infty$, we proceed as follows: we observe that, since $\{u_n\}_{n\in\mathbb{N}} \subseteq C_0^\infty(\Omega)$ converges in $L^\infty(\Omega)$ to u, it is in particular a Cauchy sequence in $L^\infty(\Omega)$. On the other hand, we have (see Problem 1.7.31)

$$\|u_n\|_{L^\infty(\Omega)} = \sup_{x\in\overline{\Omega}} |u_n(x)|.$$

This implies that $\{u_n\}_{n\in\mathbb{N}} \subseteq C_0^\infty(\Omega)$ is also a Cauchy sequence in the space

$$C_b^0(\overline{\Omega}) = \Big\{\varphi : \overline{\Omega} \to \mathbb{R} \,:\, \varphi \text{ is continuous and bounded on } \overline{\Omega}\Big\},$$

which is a Banach space, when endowed with the norm

$$\varphi \mapsto \sup_{x\in\overline{\Omega}} |\varphi(x)|,$$

(see [35, Theorem 7.9] or Problem 5.5.9 below). Thus, this sequence converges uniformly on $\overline{\Omega}$ to a continuous and bounded function: by uniqueness of the limit, it must coincide with u. Thus, we have $u \in C_b^0(\overline{\Omega})$. Finally, this permits to infer that

$$\begin{aligned}\lim_{n\to\infty} &\|(f'(u_n) - f'(u))\,\nabla u\|_{L^\infty(\Omega;\mathbb{R}^N)} \\ &\le \|\nabla u\|_{L^\infty(\Omega;\mathbb{R}^N)} \lim_{n\to\infty} \sup_{x\in\overline{\Omega}} |f'(u_n(x)) - f'(u(x))| = 0,\end{aligned}$$

thanks to the continuity of f' and the uniform convergence of $\{u_n\}_{n\in\mathbb{N}}$. □

Proposition 3.7.13 *Let $1 \le p < \infty$ and let $\Omega \subseteq \mathbb{R}^N$ be an open set. Then, for every $u \in W_0^{1,p}(\Omega)$ the following functions still belong to $W_0^{1,p}(\Omega)$:*

1. u_+ and u_-, where as before

$$u_+(x) = \max\{u(x), 0\} \qquad \textit{and} \qquad u_-(x) = \min\{u(x), 0\};$$

2. the absolute value $|u|$.

Proof We know that these functions belong to $W^{1,p}(\Omega)$, thanks to Proposition 3.4.3 and Corollary 3.4.6.

1. For every $\varepsilon > 0$, we consider again the C^1 function with bounded derivative

$$f_\varepsilon(t) = \begin{cases} \sqrt{\varepsilon^2 + t^2} - \varepsilon, & \text{if } t \ge 0, \\ 0, & \text{if } t < 0. \end{cases}$$

By noticing that $f_\varepsilon(0) = 0$, we get that $f_\varepsilon \circ u \in W_0^{1,p}(\Omega)$, thanks to Proposition 3.7.12. Thus, if we can show that

$$\lim_{\varepsilon \to 0^+} \|f_\varepsilon \circ u - u_+\|_{W^{1,p}(\Omega)} = 0, \tag{3.7.12}$$

this would imply that $u_+ \in W_0^{1,p}(\Omega)$, since this space is closed by construction. We observe that for $t \geq 0$, we have

$$f_\varepsilon(t) = \sqrt{\varepsilon^2 + t^2} - \varepsilon \leq \varepsilon + t - \varepsilon = t,$$

thanks to Problem 1.7.1 with $\alpha = 1/2$. In particular, this gives

$$|f_\varepsilon(u) - u_+|^p = (u_+ - f_\varepsilon(u))^p \leq u_+^p \in L^1(\Omega).$$

By observing that

$$\lim_{\varepsilon \to 0^+} f_\varepsilon(u(x)) = u_+(x), \qquad \text{for a. e. } x \in \Omega,$$

we can apply the Dominated Convergence Theorem and get

$$\lim_{\varepsilon \to 0^+} \int_\Omega |f_\varepsilon \circ u - u_+|^p \, dx = 0.$$

As for the gradients, we have

$$\begin{aligned}\int_\Omega |\nabla(f_\varepsilon \circ u) - \nabla u_+|^p \, dx &= \int_\Omega |f_\varepsilon'(u) \, \nabla u - \nabla u_+|^p \, dx \\ &= \int_{\{u>0\}} \left| \frac{u}{\sqrt{\varepsilon^2 + u}} \, \nabla u - \nabla u \right|^p \, dx \\ &= \int_{\{u>0\}} |\nabla u|^p \left| \frac{u}{\sqrt{\varepsilon^2 + u}} - 1 \right|^p \, dx,\end{aligned}$$

where we used Proposition 3.4.3 for the gradient of u_+. By using again the Dominated Convergence Theorem, we get again that the last integral converges to 0, as ε goes to 0. In conclusion, we have obtained (3.7.12).

For u_-, one can simply observe that

$$u_-(x) = \min\{u(x), 0\} = -\max\{-u(x), 0\} = -(-u)_+(x),$$

and then apply the previous conclusion, with $-u$ in place of u.

2. It is sufficient to observe that $|u| = u_+ - u_-$, then use point 1 and the fact that $W_0^{1,p}(\Omega)$ is a vector space.

This concludes the proof. □

Remark 3.7.14 We briefly notice that Proposition 3.7.13 fails to be true for the space $W_0^{1,\infty}(\Omega)$. Indeed, one can prove that $W_0^{1,\infty}(\Omega) \subseteq C^1(\Omega)$: taking the positive/negative part or the absolute value of a C^1 function in general gives a function which is no more in C^1.

3.8 Embedding Theorems for $W_0^{1,p}$

In what follows, given two normed vector spaces $(X, \|\cdot\|_X)$ and $(Y, \|\cdot\|_Y)$, we will use the symbol

$$Y \hookrightarrow X,$$

to indicate that the following properties hold at the same time:

(i) $Y \subseteq X$;
(ii) the identity map

$$\begin{aligned} i : Y &\to X \\ y &\mapsto y \end{aligned}$$

is a linear and *continuous* operator, i.e. there exists a constant $C > 0$ such that

$$\|y\|_X \leq C \, \|y\|_Y, \qquad \text{for every } y \in Y.$$

In this case, we will say that *Y is continuously embedded in X*.
We will also say that the embedding $Y \hookrightarrow X$ is *compact* if the following additional property holds: from any sequence $\{y_n\}_{n\in\mathbb{N}} \subseteq Y$ such that

$$\|y_n\|_Y \leq C, \qquad \text{for every } n \in \mathbb{N},$$

we can extract a subsequence converging in X.

We recall that for an open set $\Omega \subseteq \mathbb{R}^N$ we indicate

$$C^0_{\mathrm{b}}(\overline{\Omega}) = \Big\{\varphi : \overline{\Omega} \to \mathbb{R} \, : \, \varphi \text{ is continuous and bounded on } \overline{\Omega}\Big\}.$$

Then, for an exponent $0 < \alpha < 1$, we denote by $C^{0,\alpha}(\overline{\Omega})$ the space

$$C^{0,\alpha}(\overline{\Omega}) = \Big\{\varphi \in C^0_{\mathrm{b}}(\overline{\Omega}) \, : \, |\varphi|_{C^{0,\alpha}(\overline{\Omega})} < +\infty\Big\},$$

where

$$|\varphi|_{C^{0,\alpha}(\overline{\Omega})} = \sup_{x \neq y,\ x,y \in \overline{\Omega}} \frac{|\varphi(x) - \varphi(y)|}{|x - y|^\alpha}.$$

In other words, this is the space of continuous and bounded functions over $\overline{\Omega}$, which are *Hölder continuous*. Observe that this is a Banach space (see for example Problem 5.5.10), when endowed with the norm[10]

$$\|\varphi\|_{C^{0,\alpha}(\overline{\Omega})} := \sup_{\overline{\Omega}} |\varphi| + |\varphi|_{C^{0,\alpha}(\overline{\Omega})} = \|\varphi\|_{L^\infty(\Omega)} + |\varphi|_{C^{0,\alpha}(\overline{\Omega})}.$$

Then we have the following result: observe that this is valid for every open set, without any further assumption.

Theorem 3.8.1 (Sobolev Embeddings) *Let $\Omega \subseteq \mathbb{R}^N$ be an open set, then:*

(i) *if $1 \le p < N$, for every*

$$p < q \le p^* = \frac{Np}{N-p},$$

we have the continuous embedding

$$W_0^{1,p}(\Omega) \hookrightarrow L^q(\Omega).$$

More precisely, for every $u \in W_0^{1,p}(\Omega)$ we have

$$\|u\|_{L^q(\Omega)} \le \mathcal{S}^\theta \, \|u\|_{L^p(\Omega)}^{1-\theta} \, \|\nabla u\|_{L^p(\Omega;\mathbb{R}^N)}^\theta, \qquad \text{with } \theta = \frac{N}{p} - \frac{N}{q}, \tag{3.8.1}$$

where $\mathcal{S} = \mathcal{S}(N, p) > 0$ is the same constant as in Theorem 3.6.1;

(ii) *if $p = N$, for every $N < q < \infty$ we have the continuous embedding*

$$W_0^{1,N}(\Omega) \hookrightarrow L^q(\Omega).$$

More precisely, for every $u \in W_0^{1,N}(\Omega)$ we have

$$\|u\|_{L^q(\Omega)} \le \mathcal{L} \, \|u\|_{L^N(\Omega)}^{\frac{N}{q}} \, \|\nabla u\|_{L^N(\Omega;\mathbb{R}^N)}^{1-\frac{N}{q}},$$

where $\mathcal{L} = \mathcal{L}(N, q) > 0$ is the same constant as in Theorem 3.6.4;

(iii) *if $N < p < \infty$, for every $p < q \le \infty$ we have the continuous embeddings*

[10] As observed in the proof of Proposition 3.7.12, we use that for continuous functions on $\overline{\Omega}$, we have $\|u\|_{L^\infty(\Omega)} = \sup_{\overline{\Omega}} |u|$, see Problem 1.7.31.

$$W_0^{1,p}(\Omega) \hookrightarrow L^q(\Omega) \qquad \textit{and} \qquad W_0^{1,p}(\Omega) \hookrightarrow C^{0,\alpha}(\overline{\Omega}), \ \textit{with}\ \alpha = 1 - \frac{N}{p}.$$

More precisely, for every $u \in W_0^{1,p}(\Omega)$ *we have*

$$\|u\|_{L^q(\Omega)} \le \mathcal{M}_2^\theta\, \|u\|^{1-\theta}_{L^p(\Omega)}\, \|\nabla u\|^{\theta}_{L^p(\Omega;\mathbb{R}^N)}, \qquad \textit{with}\ \theta = \frac{N}{p} - \frac{N}{q},$$

and

$$|u|_{C^{0,\alpha}(\overline{\Omega})} \le \mathcal{M}_1\, \|\nabla u\|_{L^p(\Omega;\mathbb{R}^N)},$$

where $\mathcal{M}_i = \mathcal{M}_i(N, p) > 0$, *for* $i \in \{1, 2\}$, *are the same constants as in Theorem 3.6.8.*

Proof It is sufficient to use the definition of $W_0^{1,p}(\Omega)$ and the fact that $C_0^\infty(\Omega) \subseteq C_0^\infty(\mathbb{R}^N)$, together with: Sobolev's inequality (Theorem 3.6.1) if $1 \le p < N$; Ladyzhenskaya's inequality (Theorems 3.6.4 and 3.6.6) if $p = N$ and Morrey's inequalities (Theorem 3.6.8) if $N < p < \infty$.

We give the details for the case (i) and leave the other cases for the reader. We first prove the embedding for the limit exponent $q = p^*$. Let $u \in W_0^{1,p}(\Omega)$, by definition there exists a sequence $\{u_n\}_{n\in\mathbb{N}} \subseteq C_0^\infty(\Omega)$ such that

$$\lim_{n\to\infty} \|u_n - u\|_{W^{1,p}(\Omega)} = 0.$$

This implies in particular that $\{\nabla u_n\}_{n\in\mathbb{N}} \subseteq L^p(\Omega; \mathbb{R}^N)$ is a Cauchy sequence. By using Theorem 3.6.1 for the function $u_n - u_m \in C_0^\infty(\Omega)$, we get

$$\|u_n - u_m\|_{L^{p^*}(\Omega)} \le \mathcal{S}\, \|\nabla u_n - \nabla u_m\|_{L^p(\Omega;\mathbb{R}^N)}, \qquad \text{for } n, m \in \mathbb{N},$$

which shows that $\{u_n\}_{n\in\mathbb{N}}$ is a Cauchy sequence in $L^{p^*}(\Omega)$. The latter being a Banach space, we get that this sequence converges. By uniqueness of the limit, u_n is converging to u in $L^{p^*}(\Omega)$. We can now pass to the limit as n goes to ∞ in

$$\|u_n\|_{L^{p^*}(\Omega)} \le \mathcal{S}\, \|\nabla u_n\|_{L^p(\Omega;\mathbb{R}^N)},$$

and get (3.8.1) for $u \in W_0^{1,p}(\Omega)$, in the limit case $q = p^*$ (so that $\theta = 1$). For $p < q < p^*$, we can now simply use interpolation in L^p spaces: indeed, by Problem 1.7.22 we get

$$\|u\|_{L^q(\Omega)} \le \|u\|^{1-\theta}_{L^p(\Omega)}\, \|u\|^{\theta}_{L^{p^*}(\Omega)}, \qquad \text{for every } u \in W_0^{1,p}(\Omega),$$

where θ is as in (3.8.1). By using Sobolev's inequality to control the L^{p^*} norm, we get the claimed inequality. Finally, observe that by using Young's inequality (Lemma 1.2.9) with conjugate exponents $1/\theta$ and $1/(1-\theta)$, from (3.8.1) we also get

$$\|u\|_{L^q(\Omega)} \le \mathcal{S}^\theta \left[(1-\theta)\,\|u\|_{L^p(\Omega)} + \theta\,\|\nabla u\|_{L^p(\Omega;\mathbb{R}^N)}\right] \le \mathcal{S}^\theta\,\|u\|_{W^{1,p}(\Omega)},$$

thus proving the continuity of the embedding, i.e. $W_0^{1,p}(\Omega) \hookrightarrow L^q(\Omega)$. □

Remark 3.8.2 (The Case $p = N$) In the borderline case $p = N$, the previous result gives that $W_0^{1,N}(\Omega)$ is "almost" embedded in $L^\infty(\Omega)$. However, it can be shown that for $N \ge 2$ we have

$$W_0^{1,N}(\Omega) \not\hookrightarrow L^\infty(\Omega),$$

see Problem 3.12.22 below. The one-dimensional case $N = 1$ is exceptional: in this case, it is true that $W_0^{1,1}(\Omega)$ continuously embeds into $L^\infty(\Omega)$ and even in $C_b^0(\overline{\Omega})$, as a consequence of Theorem 3.6.6. See Problem 3.12.23 below for more details.

Theorem 3.8.3 (Rellich-Kondrašov) *Let $1 \le p < \infty$ and let $\Omega \subseteq \mathbb{R}^N$ be an open bounded set. Then we have the compact embedding*

$$W_0^{1,p}(\Omega) \hookrightarrow L^p(\Omega),$$

i.e., from any sequence $\{u_n\}_{n\in\mathbb{N}} \subseteq W_0^{1,p}(\Omega)$ such that

$$\|u_n\|_{W^{1,p}(\Omega)} \le M, \qquad \text{for every } n \in \mathbb{N},$$

we can extract a subsequence which converges strongly in $L^p(\Omega)$. Moreover, if $1 < p < \infty$ the limit function still belongs to $W_0^{1,p}(\Omega)$.

Proof The fact that

$$W_0^{1,p}(\Omega) \hookrightarrow L^p(\Omega),$$

is straightforward, from the definition of $W_0^{1,p}(\Omega)$. In order to prove that this embedding is compact, we want to apply the Riesz-Fréchet-Kolmogorov Theorem, see Theorem B.2.1 below. We take a sequence $\{u_n\}_{n\in\mathbb{N}} \subseteq W_0^{1,p}(\Omega)$ as in the statement. Thanks to Lemma 3.7.9 we can extend these functions by zero outside Ω and consider them as elements of $W_0^{1,p}(\mathbb{R}^N)$. This sequence in particular is bounded in $L^p(\mathbb{R}^N)$. Thus, hypothesis (H1) of Theorem B.2.1 is satisfied.

We also observe that hypothesis (H2) of the same theorem comes for free, since each u_n has been extended by zero outside Ω, the latter being bounded. As for hypothesis (H3), by Lemma 3.7.8 we have

$$\int_{\mathbb{R}^N} |u_n(x+h)-u_n(x)|^p \, dx \leq |h|^p \int_{\mathbb{R}^N} |\nabla u_n|^p \, dx \leq M^p \, |h|^p, \text{ for every } n \in \mathbb{N}.$$

This shows that

$$\lim_{|h|\to 0} \sup_{n\in\mathbb{N}} \int_{\mathbb{R}^N} |u_n(x+h) - u_n(x)|^p \, dx \leq \lim_{|h|\to 0} M^p \, |h|^p = 0.$$

We can thus apply Theorem B.2.1 and get that there exists $u \in L^p(\mathbb{R}^N)$ such that, up to subsequences, we have

$$\lim_{n\to\infty} \|u_n - u\|_{L^p(\mathbb{R}^N)} = 0.$$

To conclude, we need to show that $u \in W_0^{1,p}(\Omega)$, under the further restriction that $1 < p < \infty$. For this, we observe that thanks to Theorem 3.3.6, the sequence $\{u_n\}_{n\in\mathbb{N}}$ weakly converges in $W^{1,p}(\Omega)$, again up to subsequences. By uniqueness of the limit, such a limit must coincide with u. Finally, by using that $W_0^{1,p}(\Omega)$ is weakly closed (see Theorem 3.7.4), we get that $u \in W_0^{1,p}(\Omega)$, as desired. □

Remark 3.8.4 We recall from Theorem 3.7.6 that on an open *bounded* set $\Omega \subseteq \mathbb{R}^N$, the two norms

$$\|u\|_{W^{1,p}(\Omega)} \qquad \text{and} \qquad \|\nabla u\|_{L^p(\Omega;\mathbb{R}^N)}, \qquad \text{for } u \in W_0^{1,p}(\Omega),$$

are equivalent. Thus, the conclusion of Theorem 3.8.3 still holds for a sequence $\{u_n\}_{n\in\mathbb{N}} \subseteq W_0^{1,p}(\Omega)$ such that

$$\|\nabla u_n\|_{L^p(\Omega;\mathbb{R}^N)} \leq M, \qquad \text{for every } n \in \mathbb{N}.$$

By joining Theorems 3.8.3 and 3.8.1, we get the following improvement of Theorem 3.8.3.

Corollary 3.8.5 (Rellich-Kondrašov—Extended Version) *Let $1 \leq p < \infty$ and let $\Omega \subseteq \mathbb{R}^N$ be an open bounded set. Then for every*

$$1 \leq q \begin{cases} < p^*, \text{ if } 1 \leq p < N, \\ < \infty, \text{ if } p = N, \\ \leq \infty, \text{ if } p > N, \end{cases}$$

we have the compact embedding

$$W_0^{1,p}(\Omega) \hookrightarrow L^q(\Omega).$$

In other words, from any sequence $\{u_n\}_{n\in\mathbb{N}} \subseteq W_0^{1,p}(\Omega)$ *such that*

$$\|u_n\|_{W^{1,p}(\Omega)} \leq M, \qquad \textit{for every } n \in \mathbb{N},$$

we can extract a subsequence which converges strongly in $L^q(\Omega)$, *for every* q *as above. Moreover, if* $1 < p < \infty$ *the limit function still belongs to* $W_0^{1,p}(\Omega)$.

Proof We have to distinguish various cases:

- if $1 \leq q < p$, we can use Hölder's inequality to obtain that

$$\int_\Omega |u|^q\,dx \leq |\Omega|^{1-\frac{q}{p}} \left(\int_\Omega |u|^p\,dx\right)^{\frac{q}{p}}, \qquad \text{for every } u \in L^p(\Omega), \tag{3.8.2}$$

  which shows that we have the continuous embedding $W_0^{1,p}(\Omega) \hookrightarrow L^q(\Omega)$. Moreover, from Theorem 3.8.3 we know that from any sequence $\{u_n\}_{n\in\mathbb{N}} \subseteq W_0^{1,p}(\Omega)$ such that

$$\|u_n\|_{W^{1,p}(\Omega)} \leq M, \qquad \text{for every } n \in \mathbb{N},$$

  we can extract a subsequence $\{u_{n_k}\}_{k\in\mathbb{N}}$ which converges strongly in $L^p(\Omega)$ to $u \in W_0^{1,p}(\Omega)$. From (3.8.2), we get

$$\lim_{k\to\infty} \int_\Omega |u_{n_k} - u|^q\,dx \leq |\Omega|^{1-\frac{q}{p}} \lim_{k\to\infty} \left(\int_\Omega |u_{n_k} - u|^p\,dx\right)^{\frac{q}{p}} = 0,$$

  thus we have strong convergence in $L^q(\Omega)$, as well. This shows that the embedding in $L^q(\Omega)$ is compact;
- if $p < q < p^*$ and $p < N$, we already know from Theorem 3.8.1 that the embedding $W_0^{1,p}(\Omega) \hookrightarrow L^q(\Omega)$ is continuous. We analyze the compactness of this embedding: again from Theorem 3.8.3, we know that from any sequence $\{u_n\}_{n\in\mathbb{N}} \subseteq W_0^{1,p}(\Omega)$ such that

$$\|u_n\|_{W^{1,p}(\Omega)} \leq M, \qquad \text{for every } n \in \mathbb{N},$$

  we can extract a subsequence $\{u_{n_k}\}_{k\in\mathbb{N}}$ which converges strongly in $L^p(\Omega)$. By using (3.8.1) for the function $u_{n_k} - u$, we have

$$\begin{aligned}
\|u_{n_k} - u\|_{L^q(\Omega)} &\leq \mathcal{S}^\theta\, \|\nabla u_{n_k} - \nabla u\|^\theta_{L^p(\Omega;\mathbb{R}^N)}\, \|u_{n_k} - u\|^{1-\theta}_{L^p(\Omega)} \\
&\leq \mathcal{S}^\theta \left(\|\nabla u_{n_k}\|_{L^p(\Omega;\mathbb{R}^N)} + \|\nabla u\|_{L^p(\Omega;\mathbb{R}^N)}\right)^\theta \|u_{n_k} - u\|^{1-\theta}_{L^p(\Omega)} \\
&\leq \mathcal{S}^\theta \left(M + \|\nabla u\|_{L^p(\Omega;\mathbb{R}^N)}\right)^\theta \|u_{n_k} - u\|^{1-\theta}_{L^p(\Omega)}.
\end{aligned}$$

This permits to infer that $\{u_{n_k}\}_{k\in\mathbb{N}}$ strongly converges in $L^q(\Omega)$, as well. As before, this shows that the embedding in $L^q(\Omega)$ is compact;

- if $p \geq N$ and $q > p$ as in the statement: this is similar to the previous case, the details are left to the reader.

The proof is concluded. □

Remark 3.8.6 In the case $p > N$, actually one could be more precise and obtain that also the embedding

$$W_0^{1,p}(\Omega) \hookrightarrow C^0(\overline{\Omega}),$$

is compact. Even more sophisticated, one could prove the same conclusion for the embedding

$$W_0^{1,p}(\Omega) \hookrightarrow C^{0,\beta}(\overline{\Omega}), \qquad \text{for every } 0 < \beta < 1 - \frac{N}{p}.$$

We skip the proof of this fact.

The following fact is sometimes useful: we will use it in Chap. 4. We stress that the main point of interest is the convergence of the *whole sequence*, without extracting subsequences.

Lemma 3.8.7 *Let* $1 < p < \infty$ *and let* $\Omega \subseteq \mathbb{R}^N$ *be an open bounded set. Let us suppose that* $\{u_n\}_{n\in\mathbb{N}} \subseteq W_0^{1,p}(\Omega)$ *weakly converges in* $W^{1,p}(\Omega)$ *to some function* $u \in W_0^{1,p}(\Omega)$*. Then for every*

$$1 \leq q \begin{cases} < p^*, \text{ if } 1 \leq p < N, \\ < \infty, \text{ if } p = N, \\ \leq \infty, \text{ if } p > N, \end{cases}$$

we have

$$\lim_{n\to\infty} \|u_n - u\|_{L^q(\Omega)} = 0,$$

as well.

Proof We first notice that by the so-called *uniform boundedness principle* (see Proposition B.1.2), there exists $M > 0$ such that

$$\|u_n\|_{W^{1,p}(\Omega)} \leq M, \qquad \text{for every } n \in \mathbb{N}.$$

We now take a subsequence $\{u_{n_k}\}_{k\in\mathbb{N}} \subseteq \{u_n\}_{n\in\mathbb{N}}$. The previous uniform bound and Corollary 3.8.5 imply that we can extract a further subsequence $\{u_{n_{k_m}}\}_{m\in\mathbb{N}} \subseteq \{u_{n_k}\}_{k\in\mathbb{N}}$ which converges strongly in $L^q(\Omega)$ to a function $v \in W_0^{1,p}(\Omega)$, for every q as in the statement. It is not difficult to see that $v = u$: indeed, for every $\varphi \in C_0^\infty(\Omega)$, by using the weak convergence in $L^p(\Omega)$ we have

$$\int_\Omega u\,\varphi\,dx = \lim_{m\to\infty}\int_\Omega u_{n_{k_m}}\,\varphi\,dx = \int_\Omega v\,\varphi\,dx.$$

From Lemma 1.4.2, we get that v and u coincides almost everywhere.

In conclusion, we proved that from every subsequence $\{u_{n_k}\}_{k\in\mathbb{N}} \subseteq \{u_n\}_{n\in\mathbb{N}}$ we can extract a further subsequence converging in $L^q(\Omega)$ *to the same limit* u. This shows the claimed convergence of the whole sequence $\{u_n\}_{n\in\mathbb{N}}$. □

3.9 Embedding Theorems for $W^{1,p}$

We have seen in the previous section that functions in $W_0^{1,p}(\Omega)$ enjoy better summability properties, for every open set $\Omega \subseteq \mathbb{R}^N$. This was the content of Theorem 3.8.1. In the case of the Sobolev space $W^{1,p}(\Omega)$, we immediately find counterexamples to this property, if the boundary of Ω is not smooth enough.

Example In $\mathbb{R}^2$, we consider the open bounded set Ω given by (see Fig. 3.3)

$$\Omega = \left\{(x, y) \in \mathbb{R}^2 \,:\, x \in (0,1),\ -e^{\frac{1}{x-1}} < y < e^{\frac{1}{x-1}}\right\},$$

and the $C^1(\Omega)$ function

$$u(x, y) = (x-1)^2\, e^{\frac{1}{2(1-x)}}, \qquad \text{for every } (x, y) \in \Omega.$$

It is not difficult to see that $u \in L^2(\Omega)$ and $\nabla u \in L^2(\Omega;\mathbb{R}^2)$. Indeed, we have

$$\begin{aligned}\iint_\Omega |u|^2\,dx\,dy &= \int_0^1 (x-1)^4\, e^{\frac{1}{1-x}} \left(\int_{-e^{\frac{1}{x-1}}}^{e^{\frac{1}{x-1}}} dy\right) dx \\ &= 2\int_0^1 (x-1)^4\,dx < +\infty,\end{aligned}$$

and

(continued)

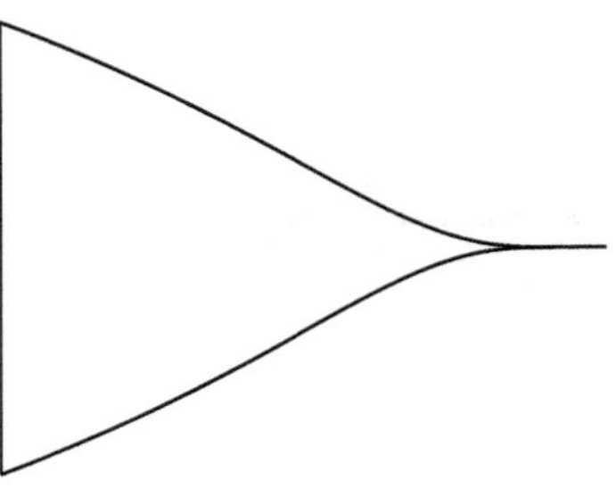

Fig. 3.3 The set Ω of Example

$$
\begin{aligned}
\iint_{\Omega} |\nabla u|^2 \, dx\, dy &= \iint_{\Omega} \left|\frac{\partial u}{\partial x}\right|^2 dx\, dy \\
&= \int_0^1 2\, e^{\frac{1}{x-1}} \left[2\,(x-1)\, e^{\frac{1}{2\,(1-x)}} + \frac{1}{2}\, e^{\frac{1}{2\,(1-x)}}\right]^2 dx \\
&= 2 \int_0^1 \left[2\,(x-1) + \frac{1}{2}\right]^2 dx < +\infty.
\end{aligned}
$$

On the other hand, we have that

$$
u \notin L^q(\Omega), \qquad \text{for every } q > 2,
$$

since

$$
\begin{aligned}
\iint_{\Omega} |u|^q \, dx\, dy &= \int_0^1 (x-1)^{2q}\, e^{\frac{q}{2\,(1-x)}} \left(\int_{-e^{\frac{1}{x-1}}}^{e^{\frac{1}{x-1}}} dy\right) dx \\
&= 2 \int_0^1 (x-1)^{2q}\, e^{\frac{q-2}{2}\,\frac{1}{1-x}}\, dx = +\infty.
\end{aligned}
$$

We will see in this section that embedding results similar to those for $W_0^{1,p}(\Omega)$ hold also for $W^{1,p}(\Omega)$, provided $\partial\Omega$ *is smooth enough*. The cornerstone of these results is the following

Proposition 3.9.1 (Extension Operator) *Let $1 \le p < \infty$ and let $\Omega \subseteq \mathbb{R}^N$ be an open bounded set, with C^1 boundary. Then there exists an open bounded set $\mathcal{O} \subseteq \mathbb{R}^N$ such that $\Omega \Subset \mathcal{O}$ and a linear operator*

$$
E : W^{1,p}(\Omega) \to W_0^{1,p}(\mathcal{O}),
$$

such that:

- $E[u](x) = u(x)$, *for a. e.* $x \in \Omega$*;*
- E *is continuous, i.e. there exists a constant* $C = C(N, p, \Omega) > 0$ *such that*

$$\left\| E[u] \right\|_{W^{1,p}(\mathcal{O})} \leq C \, \|u\|_{W^{1,p}(\Omega)}, \qquad \textit{for every } u \in W^{1,p}(\Omega).$$

Proof We prove the result in the particular case of $\Omega = B_R(x_0)$, i.e. a ball of radius R. We will show that in this case we can construct an extension operator by simply composing functions with a suitable "change of variables" and then multiplying by a cut-off function. The idea of the proof is quite simple, but fixing all the details will require some efforts. The general case is more involved, we refer to [6, Theorem 3.10] or [29, Theorem 7.25], for example.

We can suppose that the ball is centered at the origin and that the radius is $R = 1$, without loss of generality. We thus consider $\Omega = B_1(0)$, in what follows. We will see that in this case we can choose $\mathcal{O} = B_2(0)$, for example. We divide the proof in 4 steps.

Step 1: A Change of Variables We introduce the *inversion with respect to* $\mathbb{S}^{N-1}$, i.e. the mapping

$$\begin{aligned} \mathcal{H} : \mathbb{R}^N \setminus \{0\} &\to \mathbb{R}^N \setminus \{0\}. \\ x \quad &\mapsto \quad \frac{x}{|x|^2} \end{aligned}$$

It is easily seen that this is a C^1 invertible mapping, with

$$\mathcal{H}^{-1}(y) = \frac{y}{|y|^2} = \mathcal{H}(y), \qquad \text{for } y \in \mathbb{R}^N \setminus \{0\}, \tag{3.9.1}$$

and

$$\mathcal{H}(\lambda\, x) = \frac{1}{\lambda}\, \mathcal{H}(x), \qquad \text{for } x \in \mathbb{R}^N \setminus \{0\} \text{ and } \lambda \neq 0.$$

We denote by $D\mathcal{H}$ the Jacobian matrix of $\mathcal{H}$: for every $i, j \in \{1, \dots, N\}$ its entries are given by

$$\frac{\partial \mathcal{H}_i}{\partial x_j}(x) = \frac{\delta_{i,j}}{|x|^2} - 2\, \frac{x_i\, x_j}{|x|^4} = \frac{\partial \mathcal{H}_j}{\partial x_i}(x), \qquad \text{for } x \in \mathbb{R}^N \setminus \{0\},$$

where $\mathcal{H}_i(x)$ is the i-th component of $\mathcal{H}(x)$. In particular, the matrix $D\mathcal{H}$ is symmetric. Observe that from the previous formula we get

$$\frac{\partial \mathcal{H}_i}{\partial x_j}(\mathcal{H}(x)) = \frac{\delta_{i,j}}{|\mathcal{H}(x)|^2} - 2\,\frac{\mathcal{H}_i(x)\,\mathcal{H}_j(x)}{|\mathcal{H}(x)|^4} = \delta_{i,j}\,|x|^2 - 2\,x_i\,x_j. \tag{3.9.2}$$

This in particular implies the following identity

$$\Big(D\mathcal{H}(\mathcal{H}(x))\Big)^2 = |x|^4\,\mathrm{Id}_N, \qquad \text{for every } x \in \mathbb{R}^N \setminus \{0\}. \tag{3.9.3}$$

Indeed, thanks to (3.9.2) we have that the $(i,\,j)$ entry of this matrix is given by

$$\begin{aligned} m_{i,j} &= \sum_{k=1}^{N} \Big(\delta_{i,k}\,|x|^2 - 2\,x_i\,x_k\Big)\Big(\delta_{k,j}\,|x|^2 - 2\,x_k\,x_j\Big) \\ &= \delta_{i,j}\,|x|^4 - 2\,|x|^2\,x_i\,x_j - 2\,|x|^2\,x_j\,x_i + 4\,x_i\,x_j\sum_{k=1}^{N} x_k^2 = \delta_{i,j}\,|x|^4. \end{aligned}$$

In turn, from (3.9.3) we get the following identity

$$|\xi \cdot D\mathcal{H}(\mathcal{H}(x))|^2 = |x|^4\,|\xi|^2, \qquad \text{for every } x \in \mathbb{R}^N \setminus \{0\},\ \xi \in \mathbb{R}^N. \tag{3.9.4}$$

It is sufficient to observe that, by the symmetry of $D\mathcal{H}(\mathcal{H}(x))$, we get

$$\begin{aligned} |\xi \cdot D\mathcal{H}(\mathcal{H}(x))|^2 &= \langle \xi \cdot D\mathcal{H}(\mathcal{H}(x)), \xi \cdot D\mathcal{H}(\mathcal{H}(x))\rangle \\ &= \Big\langle \xi, \xi \cdot \Big(D\mathcal{H}(\mathcal{H}(x))\Big)^2\Big\rangle = |x|^4\,\langle \xi, \xi\rangle. \end{aligned}$$

We can also prove that

$$|\det D\mathcal{H}(x)| = \frac{1}{|x|^{2N}}, \qquad \text{for } x \in \mathbb{R}^N \setminus \{0\}. \tag{3.9.5}$$

To this aim, we first observe that for every $x_0 \in \mathbb{R}^N \setminus \{0\}$ and $R_0 < |x_0|$, it holds

$$\mathcal{H}(B_{R_0}(x_0)) = B_{R_1}(x_1), \qquad \text{where } R_1 = \frac{R_0}{|x_0|^2 - R_0^2} \text{ and } x_1 = \frac{x_0}{|x_0|^2 - R_0^2}.$$

Indeed, for every $x \neq 0$ we have

$$
\begin{aligned}
|x-x_0|^2 < R_0^2 &\iff |x_0|^2 - 2\langle x_0, x\rangle + |x|^2 < R_0^2 \\
&\iff (|x_0|^2 - R_0^2) - 2\langle x_0, x\rangle + |x|^2 < 0 \\
&\iff 1 - \frac{2\langle x_0, x\rangle}{|x_0|^2 - R_0^2} + \frac{|x|^2}{|x_0|^2 - R_0^2} < 0 \\
&\iff \frac{1}{|x|^2} - \frac{2\langle x_0, x\rangle}{|x|^2(|x_0|^2 - R_0^2)} + \frac{1}{|x_0|^2 - R_0^2} < 0 \\
&\iff \frac{1}{|x|^2} - \frac{2\langle x_0, x\rangle}{|x|^2(|x_0|^2 - R_0^2)} + \frac{|x_0|^2}{(|x_0|^2 - R_0^2)^2} < \frac{R_0^2}{(|x_0|^2 - R_0^2)^2} \\
&\iff \left|\frac{x}{|x|^2} - \frac{x_0}{|x_0|^2 - R_0^2}\right|^2 < \frac{R_0^2}{(|x_0|^2 - R_0^2)^2} \\
&\iff |\mathcal{H}(x) - x_1|^2 < R_1^2,
\end{aligned}
$$

which proves the previous claim. By using this property, we can now prove (3.9.5): let $x \in \mathbb{R}^N \setminus \{0\}$ and let $0 < \varepsilon < |x|$, by the change of variable formula we obtain

$$
\left|B_{r_\varepsilon}(x_\varepsilon)\right| = |\mathcal{H}(B_\varepsilon(x))| = \int_{B_\varepsilon(x)} |\det D\mathcal{H}(y)|\, dy,
$$

where

$$
r_\varepsilon = \frac{\varepsilon}{|x|^2 - \varepsilon^2} \qquad \text{and} \qquad x_\varepsilon = \frac{x}{|x|^2 - \varepsilon^2}.
$$

In particular, we get

$$
\frac{1}{|B_\varepsilon(x)|} \int_{B_\varepsilon(x)} |\det D\mathcal{H}(y)|\, dy = \frac{|B_{r_\varepsilon}(x_\varepsilon)|}{|B_\varepsilon(x)|} = \left(\frac{r_\varepsilon}{\varepsilon}\right)^N.
$$

By using the definition of r_ε, this is the same as

$$
\frac{1}{|B_\varepsilon(x)|} \int_{B_\varepsilon(x)} |\det D\mathcal{H}(y)|\, dy = \left(\frac{1}{|x|^2 - \varepsilon^2}\right)^N.
$$

By taking the limit as ε goes to 0 and using the Mean Value Theorem and the continuity of $D\mathcal{H}$, we finally get (3.9.5).

For later reference, we also observe that $\mathcal{H}$ transforms annular regions centered at the origin into annular regions centered at the origin. More precisely, we have

$$\mathcal{H}(B_R(0) \setminus B_r(0)) = \overline{B_{\frac{1}{r}}(0)} \setminus \overline{B_{\frac{1}{R}}(0)}. \tag{3.9.6}$$

The verification of this fact is left to the reader.

Step 2: Extension by Inversion For every $u \in W^{1,p}(B_1(0))$, we define

$$u_{\mathcal{H}}(x) = \begin{cases} u(x), \text{ if } x \in B_1(0), \\ u(\mathcal{H}(x)), \text{ if } x \in B_2(0) \setminus B_1(0). \end{cases}$$

We first observe that this is well-defined, since by (3.9.6)

$$\mathcal{H}(B_2(0) \setminus B_1(0)) = \overline{B_1(0)} \setminus \overline{B_{\frac{1}{2}}(0)} \subseteq \overline{B_1(0)}.$$

We claim that $u_{\mathcal{H}} \in W^{1,p}(B_2(0))$. By definition, we have

$$\begin{aligned} \int_{B_2(0)} |u_{\mathcal{H}}|^p \, dx &= \int_{B_1(0)} |u|^p \, dx + \int_{B_2(0) \setminus B_1(0)} |u(\mathcal{H}(x))|^p \, dx \\ &= \int_{B_1(0)} |u|^p \, dx + \int_{B_1(0) \setminus B_{\frac{1}{2}}(0)} |u(y)|^p \, |\det D\mathcal{H}(y)| \, dy, \end{aligned}$$

where we used the change of variable $\mathcal{H}(x) = y$, together with the property (3.9.1). By further using (3.9.5), we then obtain

$$\begin{aligned} \int_{B_2(0)} |u_{\mathcal{H}}|^p \, dx &= \int_{B_1(0)} |u|^p \, dx + \int_{B_1(0) \setminus B_{\frac{1}{2}}(0)} |u(x)|^p \, \frac{1}{|x|^{2N}} \, dx \\ &\leq \int_{B_1(0)} |u|^p \, dx + 4^N \int_{B_1(0) \setminus B_{\frac{1}{2}}(0)} |u|^p \, dx \\ &\leq (4^N + 1) \int_{B_1(0)} |u|^p \, dx, \end{aligned} \tag{3.9.7}$$

which shows that $u_{\mathcal{H}} \in L^p(B_2(0))$.

Step 3: Computing the Weak Gradient We now have to show that $u_{\mathcal{H}}$ has a weak gradient in $L^p(B_2(0))$. We introduce the vector field

$$\phi_{\mathcal{H}}(x) = \begin{cases} \nabla u(x), \text{ if } x \in B_1(0), \\ \nabla u(\mathcal{H}(x)) \cdot D\mathcal{H}(x), \text{ if } x \in B_2(0) \setminus B_1(0). \end{cases}$$

The notation $\nabla u \cdot D\mathcal{H}$ means as usual the vector given by

$$\nabla u \cdot D\mathcal{H} = \left(\sum_{i=1}^{N} \frac{\partial u}{\partial y_i} \frac{\partial \mathcal{H}_i}{\partial x_1}, \dots, \sum_{i=1}^{N} \frac{\partial u}{\partial y_i} \frac{\partial \mathcal{H}_i}{\partial x_N} \right).$$

We claim that

$$\phi_{\mathcal{H}} = \nabla u_{\mathcal{H}}, \qquad \text{a. e. in } B_2(0).$$

Observe that we already know that $u_{\mathcal{H}}$ is a Sobolev function, when restricted to the two open sets $B_1(0)$ and $B_2(0) \setminus \overline{B_1(0)}$. In the second case, this follows from Theorem 3.2.7 and Remark 3.2.8. The delicate point is proving that this is still true on the whole set $B_2(0)$, i.e. that $\phi_{\mathcal{H}}$ verifies the integration by parts formula (3.2.1) with test functions compactly supported in $B_2(0)$, and not only in $B_1(0)$ or $B_2(0) \setminus \overline{B_1(0)}$, separately. The proof will be a technically refined version of the computations needed in solving Problem 3.12.5 below, the idea being exactly the same.

For every $0 < \varepsilon \le 1/4$, we need a special cut-off function: we consider a function of one variable $\psi_\varepsilon \in C^\infty([0,1])$ such that

$$0 \le \psi_\varepsilon \le 1, \qquad \psi_\varepsilon \equiv 1 \text{ on } [0, 1 - 2\varepsilon], \qquad \psi_\varepsilon \equiv 0 \text{ on } [1 - \varepsilon, 1),$$

and

$$|\psi_\varepsilon'(t)| \le \frac{C}{\varepsilon}, \qquad \text{for } t \in [0,1].$$

We then set

$$\eta_\varepsilon(x) = \psi_\varepsilon(|x|) \in C_0^\infty(B_1(0)), \tag{3.9.8}$$

and finally define

$$\Psi_\varepsilon(x) = \begin{cases} \eta_\varepsilon(x), \text{ if } x \in B_1(0), \\ \eta_\varepsilon(\mathcal{H}(x)), \text{ if } x \in B_2(0) \setminus B_1(0). \end{cases}$$

Observe that by construction we have that $\Psi_\varepsilon \in C^\infty(B_2(0))$, it is radially symmetric, it satisfies $0 \le \Psi_\varepsilon \le 1$ and moreover

$$\Psi_\varepsilon \equiv 0 \text{ on the annular set } A_\varepsilon := B_{\frac{1}{1-\varepsilon}}(0) \setminus B_{1-\varepsilon}(0). \tag{3.9.9}$$

$$\Psi_\varepsilon \equiv 1 \text{ on } B_{1-2\varepsilon}(0) \cup \left(B_2(0) \setminus B_{\frac{1}{1-2\varepsilon}}(0) \right). \tag{3.9.10}$$

In particular, we have that $\Psi_\varepsilon(x)$ converges to 1 as ε goes to 0, for almost every $x \in B_2(0)$ (see Fig. 3.4).

Fig. 3.4 The cut-off function Ψ_ε in the proof of Proposition 3.9.1: it coincides with 1 on a large part of $B_2(0)$ and it "jumps" to zero around the boundary $\partial B_1(0)$

For every $\varphi \in C_0^\infty(B_2(0))$ and every $k \in \{1, \ldots, N\}$, we have

$$\int_{B_2(0)} u_{\mathcal{H}} \frac{\partial \varphi}{\partial x_k} \, dx = \lim_{\varepsilon \to 0} \int_{B_2(0)} u_{\mathcal{H}} \frac{\partial \varphi}{\partial x_k} \Psi_\varepsilon \, dx, \tag{3.9.11}$$

where we used the properties of Ψ_ε and the Dominated Convergence Theorem. We rewrite the last integral as follows

$$\begin{aligned}
\int_{B_2(0)} u_{\mathcal{H}} \frac{\partial \varphi}{\partial x_k} \Psi_\varepsilon \, dx &= \int_{B_2(0)} u_{\mathcal{H}} \frac{\partial}{\partial x_k} (\varphi \, \Psi_\varepsilon) \, dx - \int_{B_2(0)} u_{\mathcal{H}} \frac{\partial \Psi_\varepsilon}{\partial x_k} \varphi \, dx \\
&= \int_{B_1(0)} u \frac{\partial}{\partial x_k} (\varphi \, \Psi_\varepsilon) \, dx + \int_{B_2(0) \setminus B_1(0)} u_{\mathcal{H}} \frac{\partial}{\partial x_k} (\varphi \, \Psi_\varepsilon) \, dx \\
&\quad - \int_{B_{\frac{1}{1-2\varepsilon}}(0) \setminus B_{\frac{1}{1-\varepsilon}}(0)} u_{\mathcal{H}} \frac{\partial \Psi_\varepsilon}{\partial x_k} \varphi \, dx \\
&\quad - \int_{B_{1-\varepsilon}(0) \setminus B_{1-2\varepsilon}(0)} u \frac{\partial \Psi_\varepsilon}{\partial x_k} \varphi \, dx,
\end{aligned} \tag{3.9.12}$$

In the second equality, we used (3.9.9), (3.9.10) and the fact that $u_{\mathcal{H}} = u$ on $B_1(0)$. We analyze separately these four integrals. We start with the first integral, which is the simplest one. We observe that

$$\Psi_\varepsilon \, \varphi \in C_0^\infty(B_1(0)),$$

by construction. Thus, we can use the definition of weak derivative for u and obtain

$$\int_{B_1(0)} u \frac{\partial}{\partial x_k} (\varphi \, \Psi_\varepsilon) \, dx = -\int_{B_1(0)} \frac{\partial u}{\partial x_k} \varphi \, \Psi_\varepsilon \, dx = -\int_{B_1(0)} (\phi_{\mathcal{H}})_k \, \varphi \, \Psi_\varepsilon \, dx.$$

We indicated by $(\phi_{\mathcal{H}})_k$ the k-th component of the vector field $\phi_{\mathcal{H}}$. By using the Dominated Convergence Theorem, we thus get

$$\lim_{\varepsilon \to 0^+} \int_{B_1(0)} u \frac{\partial}{\partial x_k} (\varphi \, \Psi_\varepsilon) \, dx = -\int_{B_1(0)} (\phi_{\mathcal{H}})_k \, \varphi \, dx.$$

For the second integral in the right-hand side of (3.9.12), we observe that

$$\Psi_\varepsilon\, \varphi \in C_0^\infty(B_2(0) \setminus B_1(0)).$$

Thus, we can use the definition of weak derivative and Theorem 3.2.7 for $u_{\mathcal{H}} = u\circ\mathcal{H}$ on $B_2(0) \setminus \overline{B_1(0)}$, to get

$$\begin{aligned}
\int_{B_2(0)\setminus B_1(0)} u_{\mathcal{H}}\, \frac{\partial}{\partial x_k}\, (\varphi\, \Psi_\varepsilon)\, dx &= -\int_{B_2(0)\setminus B_1(0)} \frac{\partial u_{\mathcal{H}}}{\partial x_k}\, \varphi\, \Psi_\varepsilon\, dx \\
&= -\int_{B_2(0)\setminus B_1(0)} \sum_{i=1}^{N} \frac{\partial u}{\partial y_i}\, \frac{\partial \mathcal{H}_i}{\partial x_k}\, \varphi\, \Psi_\varepsilon\, dx \\
&= -\int_{B_2(0)\setminus B_1(0)} (\phi_{\mathcal{H}})_k\, \varphi\, \Psi_\varepsilon\, dx.
\end{aligned}$$

We indicated again by $(\phi_{\mathcal{H}})_k$ the $k-$th component of the vector field $\phi_{\mathcal{H}}$. By using the Dominated Convergence Theorem, we thus get

$$\lim_{\varepsilon\to 0^+} \int_{B_2(0)\setminus B_1(0)} u_{\mathcal{H}}\, \frac{\partial}{\partial x_k}\, (\varphi\, \Psi_\varepsilon)\, dx = -\int_{B_2(0)\setminus B_1(0)} (\phi_{\mathcal{H}})_k\, \varphi\, dx.$$

By recalling (3.9.11) and (3.9.12), up to now we have obtained that

$$\int_{B_2(0)} u_{\mathcal{H}}\, \frac{\partial \varphi}{\partial x_k}\, dx = -\int_{B_2(0)} (\phi_{\mathcal{H}})_k\, \varphi\, dx - \lim_{\varepsilon\to 0} \mathcal{I}(\varepsilon), \tag{3.9.13}$$

where we set

$$\mathcal{I}(\varepsilon) = \int_{B_{\frac{1}{1-2\varepsilon}}(0)\setminus B_{\frac{1}{1-\varepsilon}}(0)} u_{\mathcal{H}}\, \frac{\partial \Psi_\varepsilon}{\partial x_k}\, \varphi\, dx + \int_{B_{1-\varepsilon}(0)\setminus B_{1-2\varepsilon}(0)} u\, \frac{\partial \Psi_\varepsilon}{\partial x_k}\, \varphi\, dx.$$

Thus, in order to conclude we need to show that $\mathcal{I}(\varepsilon)$ converges to 0, as ε goes to 0. By using the definitions of both $u_{\mathcal{H}}$ and Ψ_ε and the change of variables $y = \mathcal{H}(x)$, we get

$$\begin{aligned}
\mathcal{I}(\varepsilon) &= \int_{B_{\frac{1}{1-2\varepsilon}}(0)\setminus B_{\frac{1}{1-\varepsilon}}(0)} u(\mathcal{H}(x)) \sum_{i=1}^{N} \frac{\partial \eta_\varepsilon}{\partial y_i}(\mathcal{H}(x))\, \frac{\partial \mathcal{H}_i}{\partial x_k}(x)\, \varphi(x)\, dx \\
&\quad + \int_{B_{1-\varepsilon}(0)\setminus B_{1-2\varepsilon}(0)} u\, \frac{\partial \eta_\varepsilon}{\partial x_k}\, \varphi\, dx \\
&= \int_{B_{1-\varepsilon}(0)\setminus B_{1-2\varepsilon}(0)} u(y) \sum_{i=1}^{N} \frac{\partial \eta_\varepsilon}{\partial y_i}(y)\, \frac{\partial \mathcal{H}_i}{\partial x_k}(\mathcal{H}(y))\, \varphi(\mathcal{H}(y))\, |\det D\mathcal{H}(y)|\, dy
\end{aligned}$$

$$+\int_{B_{1-\varepsilon}(0)\setminus B_{1-2\varepsilon}(0)} u\,\frac{\partial \eta_\varepsilon}{\partial x_k}\,\varphi\,dx.$$

We also used that $x = \mathcal{H}^{-1}(y) = \mathcal{H}(y)$ by (3.9.1). By recalling (3.9.2) and (3.9.8), we have

$$\begin{aligned}\sum_{i=1}^{N} \frac{\partial \eta_\varepsilon}{\partial y_i}(y)\,\frac{\partial \mathcal{H}_i}{\partial x_k}(\mathcal{H}(y)) &= \sum_{i=1}^{N} \frac{\partial \eta_\varepsilon}{\partial y_i}(y)\,(\delta_{i,k}\,|y|^2 - 2\,y_i\,y_k)\\ &= |y|^2\,\frac{\partial \eta_\varepsilon}{\partial y_k}(y) - 2\,\langle \nabla \eta_\varepsilon(y), y\rangle\, y_k\\ &= |y|^2\,\left\langle \psi_\varepsilon'(|y|)\,\frac{y}{|y|}, \mathbf{e}_k\right\rangle - 2\,\left\langle \psi_\varepsilon'(|y|)\,\frac{y}{|y|}, y\right\rangle\, y_k\\ &= -|y|\,\psi_\varepsilon'(|y|)\,y_k.\end{aligned}$$

We can then obtain

$$\begin{aligned}\mathcal{I}(\varepsilon) = &-\int_{B_{1-\varepsilon}(0)\setminus B_{1-2\varepsilon}(0)} u(x)\,|x|\,\psi_\varepsilon'(|x|)\,x_k\,\varphi(\mathcal{H}(x))\,|\det D\mathcal{H}(x)|\,dx\\ &+\int_{B_{1-\varepsilon}(0)\setminus B_{1-2\varepsilon}(0)} u(x)\,\psi_\varepsilon'(|x|)\,\frac{x_k}{|x|}\,\varphi(x)\,dx.\end{aligned}$$

Thus, we can also write

$$\mathcal{I}(\varepsilon) = \int_{B_{1-\varepsilon}(0)\setminus B_{1-2\varepsilon}(0)} u(x)\,\psi_\varepsilon'(|x|)\,x_k\,J(x)\,dx,$$

where

$$J(x) = -|x|\,\varphi(\mathcal{H}(x))\,|\det D\mathcal{H}(x)| + \frac{\varphi(x)}{|x|} = -|x|^{1-2N}\,\varphi(\mathcal{H}(x)) + \frac{\varphi(x)}{|x|},$$

thanks to (3.9.5). By means of tedious, yet elementary, estimates one can prove that for $0 < \varepsilon \leq 1/4$, we have[11]

[11] By the triangle inequality, we have

$$\begin{aligned}|J(x)| &\leq \frac{|\varphi(x) - \varphi(\mathcal{H}(x))|}{|x|} + |\varphi(\mathcal{H}(x))|\,\left||x|^{1-2N} - |x|^{-1}\right|\\ &\leq \frac{\|\nabla\varphi\|_{L^\infty(B_2(0);\mathbb{R}^N)}}{|x|}\,|x - \mathcal{H}(x)| + \frac{\|\varphi\|_{L^\infty(B_2(0))}}{|x|}\,\left(\left||x|^{2-2N} - 1\right|\right).\end{aligned}$$

Starting from this estimate, one can get the claimed inequality. The reader should try to write down the details as an exercise.

$$|J(x)| \le C\,\varepsilon, \qquad \text{for } x \in B_{1-\varepsilon}(0) \setminus B_{1-2\varepsilon}(0),$$

for a constant C depending on N and the L^∞ norm of both φ and $\nabla\varphi$, but not on ε. By using this fact, we can now easily conclude

$$|\mathcal{I}(\varepsilon)| \le \widetilde{C} \int_{B_{1-\varepsilon}(0)\setminus B_{1-2\varepsilon}(0)} |u|\,|x_k|\,dx,$$

thanks to the properties of ψ_ε', where the constant $\widetilde{C}$ does not depend on ε. This finally gives the claimed convergence to zero, by using that $u \in L^p(B_1(0))$ and that

$$\lim_{\varepsilon\to 0} |B_{1-\varepsilon}(0) \setminus B_{1-2\varepsilon}(0)| = 0.$$

In conclusion, from (3.9.13) we have obtained that for every $\varphi \in C_0^\infty(B_2(0))$ and every $k \in \{1, \dots, N\}$, we have

$$\int_{B_2(0)} u_{\mathcal{H}} \frac{\partial \varphi}{\partial x_k}\,dx = -\int_{B_2(0)} (\phi_{\mathcal{H}})_k\,\varphi\,dx,$$

which finally shows that $u_{\mathcal{H}}$ has a weak gradient in $B_2(0)$, given by

$$\nabla u_{\mathcal{H}} = \phi_{\mathcal{H}} = \begin{cases} \nabla u(x), \text{ if } x \in B_1(0), \\ \nabla u(\mathcal{H}(x)) \cdot D\mathcal{H}(x), \text{ if } x \in B_2(0) \setminus B_1(0). \end{cases}$$

We also have that $\nabla u_{\mathcal{H}} \in L^p(B_2(0); \mathbb{R}^N)$. Indeed, by using the change of variables $\mathcal{H}(x) = y$ and recalling (3.9.4), (3.9.5), we get

$$\begin{aligned} \int_{B_2(0)} |\nabla u_{\mathcal{H}}|^p\,dx &= \int_{B_1(0)} |\nabla u|^p\,dx \\ &\quad + \int_{B_2(0)\setminus B_1(0)} |\nabla u(\mathcal{H}(x)) \cdot D\mathcal{H}(x)|^p\,dx \\ &= \int_{B_1(0)} |\nabla u|^p\,dx \\ &\quad + \int_{B_1(0)\setminus B_{\frac{1}{2}}(0)} |\nabla u(y) \cdot D\mathcal{H}(\mathcal{H}(y))|^p\,|\det D\mathcal{H}(y)|\,dy \\ &= \int_{B_1(0)} |\nabla u|^p\,dx + \int_{B_1(0)\setminus B_{\frac{1}{2}}(0)} |y|^{2p}\,|\nabla u(y)|^p \frac{1}{|y|^{2N}}\,dy \\ &\le \left(1 + \max\{1, 4^{N-p}\}\right) \int_{B_1(0)} |\nabla u|^p\,dx. \end{aligned} \tag{3.9.14}$$

Step 4: Construction of the Operator E We are finally ready to define our extension operator. We take $\Theta \in C_0^\infty(B_2(0))$ a fixed cut-off function, such that

$$0 \le \Theta \le 1, \qquad \Theta \equiv 1 \text{ on } B_1(0), \qquad |\nabla \Theta(x)| \le C.$$

Then we define the extension operator

$$E : W^{1,p}(B_1(0)) \to W_0^{1,p}(B_2(0)),$$

given by

$$E[u](x) = u_{\mathcal{H}}(x)\,\Theta(x), \qquad \text{for every } u \in W^{1,p}(B_1(0)).$$

We observe that $E[u] \in W_0^{1,p}(B_2(0))$ by Problem 3.12.18 below. Moreover, the linearity of E is straightforward by its definition. The property

$$E[u](x) = u(x), \qquad \text{for } x \in B_1(0),$$

is evident, as well. It is just a consequence of the fact that

$$\Theta(x) = 1, \text{ for } x \in B_1(0) \qquad \text{and} \qquad u_{\mathcal{H}}(x) = u(x), \text{ for } x \in B_1(0).$$

We are only left with verifying that E is a continuous operator: by using the Leibniz rule (i.e. Proposition 3.4.7), the properties of Θ and (3.9.7), (3.9.14), we readily get the desired conclusion. □

Remark 3.9.2 The C^1 assumption in the statement of the previous result is not optimal. Apart for the aforementioned results [6, Theorem 3.10] and [29, Theorem 7.25], we refer the reader to [13, Chapter IV] and [50, Chapter 1, Section 5] for a thorough discussion on extension operators in Sobolev spaces.

By using the extension operator and the embedding results for $W_0^{1,p}$, we can now easily get similar results for the space $W^{1,p}$.

Theorem 3.9.3 (Continuous Embeddings) *Let $\Omega \subseteq \mathbb{R}^N$ be an open bounded set, with C^1 boundary. Then:*

1. *if $1 \le p < N$, we have the continuous embedding*

$$W^{1,p}(\Omega) \hookrightarrow L^{p^*}(\Omega), \qquad \text{with } p^* = \frac{N\,p}{N-p};$$

2. *if $p = N$, for every $1 \le q < \infty$ we have the continuous embedding*

$$W^{1,N}(\Omega) \hookrightarrow L^q(\Omega);$$

3. if $N < p < \infty$, we have the continuous embedding

$$W^{1,p}(\Omega) \hookrightarrow C^{0,\alpha}(\overline{\Omega}), \qquad \textit{with } \alpha = 1 - \frac{N}{p}.$$

Proof It is sufficient to use the extension operator $E : W^{1,p}(\Omega) \to W_0^{1,p}(\mathcal{O})$ of Proposition 3.9.1 and then apply Theorem 3.8.1 to the function $E[u]$, for every $u \in W^{1,p}(\Omega)$. We leave the details to the reader. □

Theorem 3.9.4 (Compact Embeddings) *Let $\Omega \subseteq \mathbb{R}^N$ be an open bounded set, with C^1 boundary. Then for every $1 \le p < \infty$ and*

$$1 \le q \begin{cases} < p^*, \textit{ if } 1 \le p < N, \\ < \infty, \textit{ if } p = N, \\ \le \infty, \textit{ if } p > N, \end{cases}$$

the embedding

$$W^{1,p}(\Omega) \hookrightarrow L^q(\Omega),$$

is compact. In other words, from any sequence $\{u_n\}_{n\in\mathbb{N}} \subseteq W^{1,p}(\Omega)$ such that

$$\|u_n\|_{W^{1,p}(\Omega)} \le M, \qquad \textit{for every } n \in \mathbb{N},$$

we can extract a subsequence which converges strongly in $L^q(\Omega)$, for every q as above. Moreover, if $1 < p < \infty$ the limit function still belongs to $W^{1,p}(\Omega)$.

Proof We take a bounded sequence $\{u_n\}_{n\in\mathbb{N}} \subseteq W^{1,p}(\Omega)$, then by Proposition 3.9.1 the sequence $\{E[u_n]\}_{n\in\mathbb{N}} \subseteq W_0^{1,p}(\mathcal{O})$ is bounded, as well, thanks to the fact that

$$\left\| E[u_n] \right\|_{W^{1,p}(\mathcal{O})} \le C \, \|u_n\|_{W^{1,p}(\Omega)}, \qquad \text{for every } n \in \mathbb{N}.$$

By Corollary 3.8.5, we know that the embedding $W_0^{1,p}(\mathcal{O}) \hookrightarrow L^q(\mathcal{O})$ is compact. Thus, up to a subsequence, we have that $\{E[u_n]\}_{n\in\mathbb{N}}$ converges strongly in $L^q(\mathcal{O})$ to a function U, for every q as in the statement. By using that $E[u_n]$ is an extension of u_n, we get

$$\lim_{n\to\infty} \left\| u_n - U \right\|_{L^q(\Omega)} = \lim_{n\to\infty} \left\| E[u_n] - U \right\|_{L^q(\Omega)} \le \lim_{n\to\infty} \left\| E[u_n] - U \right\|_{L^q(\mathcal{O})} = 0.$$

This shows that $\{u_n\}_{n\in\mathbb{N}}$ converges strongly in $L^q(\Omega)$, up to subsequences. The desired property is then proven. □

Remark 3.9.5 With a very minor modification of the proof, we see that the conclusion of Lemma 3.8.7 still holds with $W^{1,p}(\Omega)$ in place of $W_0^{1,p}(\Omega)$, provided Ω has a C^1 boundary. The details are left to the reader.

3.10 Some Further Poincaré-type Inequalities

In what follows, for every $u \in L^1_{loc}(\mathbb{R}^N)$ and every measurable set $E \subseteq \mathbb{R}^N$ with positive finite measure, we recall the notation

$$\fint_E u\,dx = \frac{1}{|E|}\int_E u\,dx,$$

i.e. the integral average of u over E.

Proposition 3.10.1 *Let $1 < p < \infty$ and let $\Omega \subseteq \mathbb{R}^N$ be an open bounded connected set, with C^1 boundary. Then there exists a constant $C_{N,p,\Omega} > 0$ depending on N, p and Ω only, such that*

$$\int_\Omega \left|u - \fint_E u\,dy\right|^p dx \le C_{N,p,\Omega}\int_\Omega |\nabla u|^p\,dx, \qquad \textit{for every } u \in W^{1,p}(\Omega).$$

Proof We argue by contradiction and assume that for every $n \geq 1$, there exists a non-constant function $u_n \in W^{1,p}(\Omega)$ such that

$$\int_\Omega |u_n - \fint_\Omega u_n\,dy|^p\,dx \ge n\int_\Omega |\nabla u_n|^p\,dx.$$

We now set

$$v_n = \frac{u_n - \fint_\Omega u_n\,dy}{\left\|u_n - \fint_\Omega u_n\,dy\right\|_{L^p(\Omega)}},$$

then from the previous inequality we obtain

$$\int_\Omega |\nabla v_n|^p\,dx \le \frac{1}{n} \qquad \text{and} \qquad \int_\Omega |v_n|^p\,dx = 1. \tag{3.10.1}$$

Moreover, by construction we also have

$$\fint_\Omega v_n\,dy = 0. \tag{3.10.2}$$

The two conditions (3.10.1) imply that $\{v_n\}_{n\in\mathbb{N}}$ is a bounded sequence in $W^{1,p}(\Omega)$. By using Theorem 3.3.6, we can infer weak convergence in $W^{1,p}(\Omega)$ (up to a subsequence) to a function $v \in W^{1,p}(\Omega)$. Moreover, thanks to Theorem 3.9.4 and Remark 3.9.5, we have strong convergence in $L^p(\Omega)$, as well. Thus, such a limit function must satisfy

$$\int_\Omega |v|^p\,dx = \lim_{n\to\infty}\int_\Omega |v_n|^p\,dx = 1,$$

and (by lower semicontinuity of the L^p norm, see Proposition 1.3.11)

$$\int_\Omega |\nabla v|^p\,dx \le \liminf_{n\to\infty}\int_\Omega |\nabla v_n|^p\,dx \le \lim_{n\to\infty}\frac{1}{n} = 0.$$

By Proposition 3.2.9, we get that v must be a constant function, with unit $L^p(\Omega)$ norm. On the other hand, by using (3.10.2), we also have

$$\left|\fint_\Omega v\,dx\right| = \left|\fint_\Omega (v - v_n)\,dx\right| \le \fint_\Omega |v - v_n|\,dx \le \left(\fint_\Omega |v - v_n|^p\,dx\right)^{\frac{1}{p}},$$

thanks to Hölder's inequality. Thus, by taking the limit as n goes to ∞, we get $\int_\Omega v\,dx = 0$. The latter contradicts the fact that v is a non-trivial constant (due to the normalization on the L^p norm). This concludes the proof. □

Remark 3.10.2 The previous proof exploits a contradiction argument and does not provide an explicit expression for the constant $C_{N,p,\Omega}$ (compare with the constructive proofs of Sect. 3.5). In the case of convex sets, it would be possible to use a more direct argument and get an explicit constant C, depending on geometric features of Ω: the interested reader is referred to [29, Chapter 7, Section 8] or [46, Chapter 8, Exercise 6].

We also point out that the connectedness assumption in Proposition 3.10.1 can not be removed: take for example Ω to be the union of two disjoint open balls B_1 and B_2. For the piecewise constant function

$$u(x) = \begin{cases} |B_1|^{-1}, & \text{if } x \in B_1,\\ -|B_2|^{-1}, & \text{if } x \in B_2,\end{cases}$$

we have

$$\int_\Omega u\,dx = \frac{1}{|B_1|}\int_{B_1} dx - \frac{1}{|B_2|}\int_{B_2} dx = 0,$$

and

$$\int_\Omega |u|^p\,dx = \int_{B_1} |u|^p\,dx + \int_{B_2} |u|^p\,dx = \frac{1}{|B_1|^{p-1}} + \frac{1}{|B_2|^{p-1}},$$

while

$$\int_\Omega |\nabla u|^p\,dx = 0.$$

Thus, the Poincaré inequality of Proposition 3.10.1 can not hold for a set like this.

When the open set coincides with a ball, we can be more precise about the dependence of the constant C in the previous result.

Corollary 3.10.3 *Let $1 < p < \infty$ and let $B_R(x_0) \subseteq \mathbb{R}^N$ be the N-dimensional open ball of radius R, centered at x_0. There exists a constant $\mu_{N,p} > 0$ depending only on N and p such that*

$$\int_{B_R(x_0)} \left| u - \fint_{B_R(x_0)} u\,dy \right|^p dx \le \frac{R^p}{\mu_{N,p}} \int_{B_R(x_0)} |\nabla u|^p\,dx,$$

for every $u \in W^{1,p}(B_R(x_0))$.

Proof Without loss of generality, we can assume that x_0 coincides with the origin, then we simply write B_R. For every $u \in W^{1,p}(B_R)$, we define

$$u_R(x) = u(R\,x), \qquad \text{for every } x \in B_1.$$

We observe that $u_R \in W^{1,p}(B_1)$ by Theorem 3.2.7 and Remark 3.2.8. Thus, by Proposition 3.10.1 we have

$$\int_{B_1} \left| u_R - \fint_{B_1} u_R\,dy \right|^p dx \le C_{N,p,B_1} \int_{B_1} |\nabla u_R|^p\,dx,$$

where the constant C_{N,p,B_1} depends only on N, p and B_1, the latter being a universal fixed set. We then simply call this constant $1/\mu_{N,p}$. We now recall the definition of u_R, so that the previous inequality becomes

$$\int_{B_1} \left| u(R\,x) - \fint_{B_1} u_R\,dy \right|^p dx \le \frac{R^p}{\mu_{N,p}} \int_{B_1} |\nabla u(R\,x)|^p\,dx.$$

The change of variable $R\,x = z$ gives

$$R^{-N} \int_{B_R} \left| u - \fint_{B_1} u_R\,dy \right|^p dz \le \frac{1}{\mu_{N,p}}\,R^{p-N} \int_{B_R} |\nabla u|^p\,dz.$$

By multiplying both sides by R^N and observing that

$$\fint_{B_1} u_R\,dy = \frac{1}{|B_1|} \int_{B_1} u(R\,y)\,dy = \frac{1}{R^N\,|B_1|} \int_{B_R} u(z)\,dz = \fint_{B_R} u\,dz,$$

we obtain the desired conclusion. □

In the next result, we prove that it is still possible to have a Poincaré inequality, provided that functions vanish in a "sufficiently large" set. This kind of result, which is interesting in itself, will be needed somewhere in Chap. 7.

Proposition 3.10.4 *Let $1 < p < \infty$ and let $\Omega \subseteq \mathbb{R}^N$ be an open bounded connected set, with C^1 boundary. Let $u \in W^{1,p}(\Omega)$ be such that*

$$\big|\{x \in \Omega : u(x) = 0\}\big| > 0.$$

Then we have

$$\int_\Omega |u|^p\, dx \le 2^p\, C_{N,p,\Omega} \frac{|\Omega|}{\big|\{x \in \Omega : u(x) = 0\}\big|} \int_\Omega |\nabla u|^p\, dx,$$

where $C_{N,p,\Omega} > 0$ is the same constant as in Proposition 3.10.1.

Proof We can suppose that

$$\big|\{x \in \Omega : u(x) = 0\}\big| < |\Omega|,$$

otherwise there is nothing to prove. We then have

$$\begin{aligned}\big|\{x \in \Omega : u(x) = 0\}\big| \left|\fint_\Omega u\, dy\right|^p &= \int_{\{x\in\Omega : u(x)=0\}} \left|\fint_\Omega u\, dy\right|^p dx\\ &= \int_{\{x\in\Omega : u(x)=0\}} \left|u - \fint_\Omega u\, dy\right|^p dx\\ &\le \int_\Omega \left|u - \fint_\Omega u\, dy\right|^p dx.\end{aligned}$$

This gives

$$\left|\fint_\Omega u\, dy\right|^p \le \frac{1}{\big|\{x \in \Omega : u(x) = 0\}\big|} \int_\Omega \left|u - \fint_\Omega u\, dy\right|^p dx. \qquad (3.10.3)$$

On the other hand, by the triangle inequality and Problem 1.7.2 we have

$$\begin{aligned}\int_\Omega |u|^p\, dx &\le 2^{p-1} \int_\Omega \left|u - \fint_\Omega u\, dy\right|^p dx + 2^{p-1} \int_\Omega \left|\fint_\Omega u\, dy\right|^p dx\\ &\le 2^{p-1} \left(1 + \frac{|\Omega|}{\big|\{x \in \Omega : u(x) = 0\}\big|}\right) \int_\Omega \left|u - \fint_\Omega u\, dy\right|^p dx,\end{aligned}$$

where in the second inequality we used (3.10.3). By using Proposition 3.10.1 to estimate the last integral, we then get the desired conclusion. □

Once again, in the case of a ball, we can be slightly more precise about the dependence of the constant in the inequality. From Proposition 3.10.4 and recalling the form of the constant in Corollary 3.10.3, we get the following

Corollary 3.10.5 *Let $1 < p < \infty$ and let $u \in W^{1,p}(B_R(x_0))$ be such that*

$$\Big|\{x \in B_R(x_0) \,:\, u(x) = 0\}\Big| > 0.$$

Then we have

$$\int_{B_R(x_0)} |u|^p \, dx \le 2^p \, \frac{R^p}{\mu_{N,p}} \left(\frac{|B_R(x_0)|}{\Big|\{x \in B_R(x_0) \,:\, u(x) = 0\}\Big|} \right) \int_{B_R(x_0)} |\nabla u|^p \, dx,$$

where $\mu_{N,p} > 0$ is the same constant as in Corollary 3.10.3.

3.11 Finite Differences and Sobolev Spaces

Given a measurable function φ and a vector $h \in \mathbb{R}^N$, we still use the notation

$$\mathrm{T}_h\varphi(x) = \varphi(x+h), \qquad \text{for } x \in \mathbb{R}^N,$$

as in Lemma 3.7.8. The following result is a sort of "localized" version of such a result.

Proposition 3.11.1 *Let $1 \le p \le \infty$ and let $\Omega \subseteq \mathbb{R}^N$ be an open set such that $\Omega \ne \mathbb{R}^N$. If $u \in W^{1,p}_{\mathrm{loc}}(\Omega)$, then for every $B_r(x_0) \Subset \Omega$ we have*

$$\begin{aligned} &\sup_{0<|h|<h_0} \left\| \frac{\mathrm{T}_h u - u}{|h|} \right\|_{L^p(B_r(x_0))} \\ &\le C \left(\|\nabla u\|_{L^p(B_{r+h_0}(x_0);\mathbb{R}^N)} + \frac{1}{h_0} \|u\|_{L^p(B_{r+3\,h_0}(x_0))} \right), \end{aligned}$$

where

$$h_0 = \frac{1}{4}\,\mathrm{dist}(B_r(x_0), \partial\Omega),$$

and $C > 0$ depends on N and p, only.

Proof Observe that $B_{r+3\,h_0}(x_0) \Subset \Omega$, thanks to the definition of h_0. We take $\eta \in C^\infty_0(B_{r+h_0}(x_0))$ to be a non-negative cut-off function, such that

$$0 \le \eta \le 1, \qquad \eta \equiv 1 \text{ on } B_r(x_0), \qquad |\nabla \eta(x)| \le \frac{C}{h_0}.$$

Then $u\,\eta \in W_0^{1,p}(\Omega)$, thanks to Problem 3.12.18. By keeping in mind Lemma 3.7.9, we can consider $u\,\eta$ as a function of $W_0^{1,p}(\mathbb{R}^N)$, by extending it by zero outside Ω. Then by Lemma 3.7.8, we get that

$$\|\mathrm{T}_h(u\,\eta) - u\,\eta\|_{L^p(\mathbb{R}^N)} \le |h|\,\|\nabla(u\,\eta)\|_{L^p(\mathbb{R}^N;\mathbb{R}^N)} = |h|\,\|\eta\,\nabla u + u\,\nabla\eta\|_{L^p(\mathbb{R}^N;\mathbb{R}^N)}.$$

We observe that

$$\|\eta\,\nabla u + u\,\nabla\eta\|_{L^p(\mathbb{R}^N;\mathbb{R}^N)} \le \|\eta\,\nabla u\|_{L^p(\mathbb{R}^N;\mathbb{R}^N)} + \|u\,\nabla\eta\|_{L^p(\mathbb{R}^N;\mathbb{R}^N)},$$

and using the properties of η, we get

$$\|\mathrm{T}_h(u\,\eta) - u\,\eta\|_{L^p(\mathbb{R}^N)} \le C\,|h|\,\|\nabla u\|_{L^p(B_{r+h_0}(x_0);\mathbb{R}^N)} + \frac{C}{h_0}\,|h|\,\|u\|_{L^p(B_{r+h_0}(x_0))}. \tag{3.11.1}$$

In order to conclude, we observe that[12]

$$\mathrm{T}_h(u\,\eta) - u\,\eta = (\mathrm{T}_h u - u)\,\eta + \mathrm{T}_h u\,(\mathrm{T}_h\eta - \eta).$$

Thus, from Minkowski's inequality we get

$$\begin{aligned}\|(\mathrm{T}_h u - u)\,\eta\|_{L^p(\mathbb{R}^N)} &= \left\|\big(\mathrm{T}_h(u\,\eta) - u\,\eta\big) - \mathrm{T}_h u\,(\mathrm{T}_h\eta - \eta)\right\|_{L^p(\mathbb{R}^N)}\\ &\le \|\mathrm{T}_h(u\,\eta) - u\,\eta\|_{L^p(\mathbb{R}^N)} + \|\mathrm{T}_h u\,(\mathrm{T}_h\eta - \eta)\|_{L^p(\mathbb{R}^N)},\end{aligned}$$

that is

$$\begin{aligned}\|\mathrm{T}_h(u\,\eta) - u\,\eta\|_{L^p(\mathbb{R}^N)} &\ge \|(\mathrm{T}_h u - u)\,\eta\|_{L^p(\mathbb{R}^N)} - \|\mathrm{T}_h u\,(\mathrm{T}_h\eta - \eta)\|_{L^p(\mathbb{R}^N)}\\ &\ge \|\mathrm{T}_h u - u\|_{L^p(B_r(x_0))} - \|\mathrm{T}_h u\,(\mathrm{T}_h\eta - \eta)\|_{L^p(\mathbb{R}^N)},\end{aligned}$$

where we used that $\eta \equiv 1$ on $B_r(x_0)$. By using this estimate in (3.11.1), we get

$$\begin{aligned}\|\mathrm{T}_h u - u\|_{L^p(B_r)} \le C\,|h|\,\|\nabla u\|_{L^p(B_{r+h_0}(x_0);\mathbb{R}^N)} &+ \frac{C}{h_0}\,|h|\,\|u\|_{L^p(B_{r+h_0}(x_0))}\\ &+ \|\mathrm{T}_h u\,(\mathrm{T}_h\eta - \eta)\|_{L^p(\mathbb{R}^N)}.\end{aligned} \tag{3.11.2}$$

We notice that $\eta_h - \eta$ has compact support in $B_{r+2h_0}(x_0)$ for $0 < |h| < h_0$ and it holds

[12] We notice that this is a sort of "discrete" version of Leibniz rule. The proof is by direct verification.

$$|\mathrm{T}_h\eta(x) - \eta(x)| = \left|\int_0^1 \frac{d}{dt}\eta(x+t\,h)\,dt\right| \le |h| \int_0^1 |\nabla\eta(x+t\,h)|\,dt$$
$$\le \|\nabla\eta\|_{L^\infty(\mathbb{R}^N;\mathbb{R}^N)}\,|h| \le \frac{C}{h_0}\,|h|.$$

By using this fact, we thus obtain for every $0 < |h| < h_0$

$$\|\mathrm{T}_h u\,(\mathrm{T}_h\eta - \eta)\|_{L^p(\mathbb{R}^N)} \le \frac{C}{h_0}\,|h|\,\|\mathrm{T}_h u\|_{L^p(B_{r+2\,h_0}(x_0))} \le \frac{C}{h_0}\,|h|\,\|u\|_{L^p(B_{r+3\,h_0}(x_0))}.$$

We insert this estimate in (3.11.2) and divide by $|h|$, so to get the desired conclusion. □

The previous result can be reverted, in a sense. Both results are useful for proving some regularity estimates for Sobolev minimizers of an integral functional, see Sect. 7.6 of Chap. 7. In the next result, we will denote by $\mathbf{e}_i$ the i-th unit vector of the canonical orthonormal basis of $\mathbb{R}^N$, as usual.

Proposition 3.11.2 *Let $1 < p < \infty$ and let $u \in L^p(B_R(x_0))$. If for a ball $B_r(x_0) \Subset B_R(x_0)$ there exists a constant $C > 0$ such that*

$$\sup_{0<|h|<h_0} \int_{B_r(x_0)} \left|\frac{\mathrm{T}_{h\,\mathbf{e}_i} u - u}{h}\right|^p dx \le C, \qquad \textit{for some } 0 < h_0 < R - r,$$

then we have that $\partial u/\partial x_i \in L^p(B_r(x_0))$. Moreover, there exists a sequence $\{\tau_n\}_{n\in\mathbb{N}} \subseteq (-h_0, h_0) \setminus \{0\}$ converging to 0, such that

$$\frac{\mathrm{T}_{\tau_n\mathbf{e}_i} u - u}{\tau_n} \qquad \textit{weakly converges in } L^p(B_r(x_0)) \textit{ to} \qquad \frac{\partial u}{\partial x_i}.$$

Proof Thanks to our assumption, we have that

$$\left\{\frac{\mathrm{T}_{h\,\mathbf{e}_i} u - u}{h}\right\}_{0<|h|<h_0} \subseteq L^p(B_r(x_0)),$$

is a bounded family of functions. By Banach-Alaoglu Theorem B.1.1, there exists a sequence $\{\tau_n\}_{n\in\mathbb{N}} \subseteq (-h_0, h_0) \setminus \{0\}$ converging to 0, such that

$$\frac{\mathrm{T}_{\tau_n\mathbf{e}_i} u - u}{\tau_n} \quad \text{weakly converges in } L^p(B_r(x_0)) \text{ to} \quad g_i \in L^p(B_r(x_0)).$$

We claim that

$$g_i = \frac{\partial u}{\partial x_i}. \tag{3.11.3}$$

Let $\varphi \in C_0^\infty(B_r(x_0))$, then for every $0 < |h| < h_0$ we have

$$\begin{aligned}
&\int_{B_r(x_0)} \frac{\mathrm{T}_{h\,\mathbf{e}_i} u(x) - u(x)}{h}\,\varphi(x)\,dx \\
&= \int_{B_r(x_0)} \frac{u(x + h\,\mathbf{e}_i)}{h}\,\varphi(x)\,dx - \int_{B_r(x_0)} \frac{u(x)}{h}\,\varphi(x)\,dx \\
&= \int_{B_r(x_0+h\,\mathbf{e}_i)} \frac{u(y)}{h}\,\varphi(y - h\,\mathbf{e}_i)\,dy - \int_{B_r(x_0)} \frac{u(y)}{h}\,\varphi(y)\,dy \\
&= \int_{B_{r+h_0}(x_0)} u(y)\,\frac{\mathrm{T}_{-h\,\mathbf{e}_i}\varphi(y) - \varphi(y)}{h}\,dy.
\end{aligned} \tag{3.11.4}$$

In the last identity we used that

$$B_r(x_0) \subseteq B_{r+h_0}(x_0) \qquad \text{and} \qquad B_r(x_0 + h\,\mathbf{e}_i) \subseteq B_{r+h_0}(x_0), \ \text{for } 0 < |h| < h_0,$$

together with the fact that φ has compact support in $B_r(x_0)$ and $\varphi_{-h\,\mathbf{e}_i}$ has compact support in $B_r(x_0 + h\,\mathbf{e}_i)$. We now observe that (recall that φ is C^∞)

$$\lim_{h\to 0} \frac{\mathrm{T}_{-h\,\mathbf{e}_i}\varphi(y) - \varphi(y)}{h} = -\frac{\partial \varphi}{\partial x_i}(y), \qquad \text{for every } y \in B_{r+h_0}(x_0).$$

Moreover, by using the Fundamental Theorem of Calculus, we see that

$$\left|\frac{\mathrm{T}_{-h\,\mathbf{e}_i}\varphi(y) - \varphi(y)}{h}\right| = \left|-\frac{1}{h}\int_0^1 \langle \nabla\varphi(y - t\,h\,\mathbf{e}_i), h\,\mathbf{e}_i\rangle\,dt\right| \le \|\nabla\varphi\|_{L^\infty(\mathbb{R}^N;\mathbb{R}^N)}.$$

We can then use (3.11.4) with $h = \tau_n$, apply the Dominated Convergence Theorem and pass to the limit as n goes to ∞ in

$$\int_{B_r(x_0)} \frac{\mathrm{T}_{\tau_n \mathbf{e}_i} u(x) - u(x)}{\tau_n}\,\varphi(x)\,dx = \int_{B_{r+h_0}(x_0)} u(y)\,\frac{\mathrm{T}_{-\tau_n \mathbf{e}_i}\varphi(y) - \varphi(y)}{\tau_n}\,dy.$$

This gives

$$\int_{B_r(x_0)} g_i\,\varphi\,dx = -\int_{B_{r+h_0}(x_0)} u\,\frac{\partial\varphi}{\partial x_i}\,dx,$$

which shows the validity of (3.11.3), by recalling that φ is compactly supported in $B_r(x_0)$. □

Remark 3.11.3 In the previous result, we excluded the extremal cases $p = 1$ and $p = \infty$, for simplicity. In the first case, the result would fail to be true, because bounded sequences in L^1 do not converge weakly in L^1, in general. In order to extend this result for $p = 1$, one should rather work with *functions of bounded variation* (see [16, 30] or [50, Chapter 9]), in place of Sobolev functions. On the contrary, in the case $p = \infty$ the statement of Proposition 3.11.2 would still hold, provided we replace the weak convergence with the $*$-weak convergence.

3.12 Problems

Problem 3.12.1 Show that if $1/2 < \alpha \le 1$ the function

$$\psi(x) = |x|^\alpha, \qquad \text{for } x \in (-1, 1),$$

has a weak derivative in $L^2((-1, 1))$, given by the function

$$x \mapsto \alpha\, |x|^{\alpha-2}\, x.$$

On the contrary, show that for $\alpha = 1/2$ such a function does not belong to $W^{1,2}((-1, 1))$.

Problem 3.12.2 Show that if $0 < \alpha \le 1$ the function

$$\psi(x, y) = (x^2 + y^2)^{\frac{\alpha}{2}}, \qquad \text{for } (x, y) \in B_1(0),$$

has a weak gradient in $L^2(B_1(0))$. Here $B_1(0) \subseteq \mathbb{R}^2$ is the unit disk, centered at the origin.

Problem 3.12.3 Let $\psi : [a, b] \to \mathbb{R}$ be a continuous function, which is piecewise C^1 in the following sense: there exists a partition of the interval

$$a = t_0 < t_1 < \cdots < t_{n-1} < t_n = b,$$

such that for every $i = 0, \dots, n-1$, we have $\psi \in C^1([t_i, t_{i+1}])$. Prove that

$$\psi \in W^{1,p}\big((a, b)\big), \qquad \text{for every } 1 \le p \le \infty,$$

and that the weak derivative of ψ coincides almost everywhere with the classical one.

Problem 3.12.4 Let us consider the sequence $\{\psi_n\}_{n\in\mathbb{N}} \subseteq W^{1,1}((0, 1))$ defined by

$$\psi_n(t) = \min\{n\, t,\ 1\}, \qquad \text{for } t \in (0, 1).$$

Show that there exists $M > 0$ such that

$$\|\psi_n\|_{W^{1,1}((0,1))} \le M, \qquad \text{for every } n \in \mathbb{N},$$

but the conclusion of the Theorem 3.3.6 does not hold: this sequence can not contain any subsequence which is weakly converging in $W^{1,1}((-1,1))$.

Problem 3.12.5 Let $u \in W^{1,1}((0,1))$ and consider its "extension by even reflection" $\widetilde{u}$ given by

$$\widetilde{u}(x) = \begin{cases} u(x), & \text{for a. e. } x \in [0,1], \\ u(-x), & \text{for a. e. } x \in [-1,0]. \end{cases}$$

Show that $\widetilde{u} \in W^{1,1}((-1,1))$ and compute its weak derivative.

Problem 3.12.6 Let $1 \le q < p \le \infty$ and let $\Omega \subseteq \mathbb{R}^N$ be an open set with finite measure. Show that we have

$$W^{1,p}(\Omega) \subseteq W^{1,q}(\Omega).$$

Then show with an example that this inclusion fails to be true, if Ω has not finite measure.

Problem 3.12.7 (A Variant of the Chain Rule) Prove the following variant of Proposition 3.4.1: let $\Omega \subseteq \mathbb{R}^N$ be an open set and let $1 < p < \infty$, for every $u \in W^{1,p}(\Omega)$ we have

$$|u|^p \in W^{1,1}(\Omega).$$

Moreover, its weak gradient is given by the formula

$$\nabla |u|^p = p\, |u|^{p-2}\, u\, \nabla u.$$

Problem 3.12.8 (Hardy's Inequality for the Half-space) Let us set

$$\mathbb{H}^1_+ = (0,+\infty), \qquad \mathbb{H}^N_+ = \mathbb{R}^{N-1} \times (0,+\infty) \text{ for } N \ge 2.$$

Let $1 < p < \infty$, show that for every $\varphi \in W^{1,p}_0(\mathbb{H}^N_+)$ we have

$$\left(\frac{p-1}{p}\right)^p \int_{\mathbb{H}^N_+} \frac{|\varphi|^p}{x_N^p}\, dx \le \int_{\mathbb{H}^N_+} |\nabla \varphi|^p\, dx,$$

where x_N is the N-th coordinate of a point x.

Problem 3.12.9 Let $N \ge 3$ and let us suppose that for some $1 \le q < \infty$ and $\alpha > 0$, there exists a constant $C > 0$ such that

$$\left(\int_{\mathbb{R}^N} |\varphi|^q\, dx\right)^{\frac{\alpha}{q}} \le C \int_{\mathbb{R}^N} |\nabla \varphi|^2\, dx, \qquad \text{for every } \varphi \in C^\infty_0(\mathbb{R}^N).$$

Show that we must necessarily have

$$\alpha = 2 \qquad \text{and} \qquad q = \frac{2N}{N-2}.$$

Problem 3.12.10 Show that a Sobolev inequality of the type

$$\|\varphi\|^p_{L^q(\mathbb{R}^N)} \le C \int_{\mathbb{R}^N} |\nabla\varphi|^p \, dx, \qquad \text{for every } \varphi \in C_0^\infty(\mathbb{R}^N),$$

can not hold for $C > 0$, when $p > N$ and $1 \le q \le \infty$.

Problem 3.12.11 (Poincaré Inequality for Sets with Finite Measure) Let $1 \le p \le \infty$ and let $\Omega \subseteq \mathbb{R}^N$ be an open set with finite measure. Show that we have the Poincaré inequality

$$\|\varphi\|_{L^p(\Omega)} \le C\, |\Omega|^{\frac{1}{N}}\, \|\nabla\varphi\|_{L^p(\Omega;\mathbb{R}^N)}, \qquad \text{for every } \varphi \in W_0^{1,p}(\Omega),$$

for a constant $C = C(N, p) > 0$.

Problem 3.12.12 Show that the function $u(x) = 1 - |x|$ belongs to $W_0^{1,p}((-1,1))$ for every $1 \le p < \infty$.

Problem 3.12.13 Show that the function $u(x) = 1 - |x|$ does not belong to $W_0^{1,\infty}((-1,1))$.

Problem 3.12.14 Generalize the previous two problems to the N-dimensional case, with $N \ge 2$: show that the function $u(x) = 1 - |x|$ belongs to $W_0^{1,p}(B_1(0))$, for every $1 \le p < \infty$. Here as usual $B_1(0)$ is the N-dimensional ball of radius 1, centered at the origin. Also show that on the contrary

$$u \notin W_0^{1,\infty}(B_1(0)).$$

Problem 3.12.15 Let $1 \le p < \infty$ and let $\Omega \subseteq \mathbb{R}^N$ be an open bounded set. Prove that

$$\Big\{u \in C^0(\overline{\Omega}) \cap C^1(\Omega) \,:\, u = 0 \text{ on } \partial\Omega,\ \nabla u \in L^\infty(\Omega;\mathbb{R}^N)\Big\} \subseteq W_0^{1,p}(\Omega).$$

Problem 3.12.16 Let $1 \le p < \infty$ and let $\Omega \subseteq \mathbb{R}^N$ be an open bounded set. Prove that

$$X_0^{1,p}(\Omega) := \Big\{u \in C^0(\overline{\Omega}) \cap W^{1,p}(\Omega) \,:\, u = 0 \text{ on } \partial\Omega\Big\} \subseteq W_0^{1,p}(\Omega).$$

Problem 3.12.17 Show that the conclusions of Problems 3.12.15 and 3.12.16 still hold for a general open set $\Omega \subsetneq \mathbb{R}^N$, not necessarily bounded.

Problem 3.12.18 (A Variant of Leibniz Rule I) Let $1 \leq p < \infty$ and let $\Omega \subseteq \mathbb{R}^N$ be an open set. Prove that if $\eta \in C^1_0(\Omega)$ and $u \in W^{1,p}_{\rm loc}(\Omega)$, then

$$\eta\, u \in W^{1,p}_0(\Omega).$$

Moreover, its weak gradient is given by

$$\nabla(\eta\, u) = u\,\nabla\eta + \eta\,\nabla u.$$

Problem 3.12.19 (A Variant of Leibniz Rule II) Let $1 \leq p < \infty$ and let $\Omega \subseteq \mathbb{R}^N$ be an open set. Prove that if $\eta \in C^1(\Omega) \cap W^{1,\infty}(\Omega)$ and $u \in W^{1,p}_0(\Omega)$, then

$$\eta\, u \in W^{1,p}_0(\Omega).$$

Problem 3.12.20 Let $1 < p < \infty$ and let $B_R(x_0) \subseteq \mathbb{R}^N$ be the ball of radius $R > 0$, centered at x_0. Show that Hardy's inequality (3.5.13) still holds for functions in $W^{1,p}_0(B_R(x_0))$. Use this fact to show that for a function $u \in W^{1,p}(B_R(x_0))$ we have

$$u \in W^{1,p}_0(B_R(x_0)) \qquad \Longleftrightarrow \qquad \int_{B_R(x_0)} \frac{|u|^p}{(R - |x - x_0|)^p}\, dx < +\infty. \tag{3.12.1}$$

Problem 3.12.21 (Rellich-Kondrašov for Sets with Finite Measure) Show that the statement of Rellich-Kondrašov Theorem 3.8.3 still holds for an open set $\Omega \subseteq \mathbb{R}^N$ with finite measure.

Problem 3.12.22 Let $N \geq 2$ and let $B_1(0) \subseteq \mathbb{R}^N$ be the N-dimensional open ball centered at the origin, with radius 1. Show that

$$W^{1,N}_0(B_1(0)) \not\hookrightarrow L^\infty(B_1(0)).$$

Problem 3.12.23 (One-dimensional Sobolev Embeddings) Extend the statement of Theorems 3.8.1 in the case $N = p = 1$ and $\Omega = (a, b) \subseteq \mathbb{R}$, as follows: if we set[13]

$$C^0_{\rm b}([a, b]) = \Big\{u : [a, b] \to \mathbb{R} \,:\, u \text{ is continuous and bounded on } [a, b]\Big\},$$

endowed with the norm

$$\|u\|_{C^0_{\rm b}([a,b])} := \sup_{x\in[a,b]} |u(x)|,$$

[13] We remark that the interval (a, b) is not necessarily bounded.

then we have the continuous embedding

$$W_0^{1,1}((a,b)) \hookrightarrow C_b^0([a,b]).$$

Also show that if (a,b) is bounded, then the embedding

$$W_0^{1,1}((a,b)) \hookrightarrow L^q((a,b)),$$

is compact for every $1 \le q < \infty$, but it is not compact for $q = \infty$.

Problem 3.12.24 Let $1 \le p \le \infty$ and let $\Omega \subseteq \mathbb{R}^N$ be an open set. Let us suppose that there exists $\Phi : \mathbb{R}^N \to \mathbb{R}^N$ an affine invertible map such that $\Phi(\Omega) = \Omega$, i.e.

$$\Phi(x) = x \cdot A + \mathbf{b}, \qquad \text{for every } x \in \mathbb{R}^N,$$

for some $\mathbf{b} \in \mathbb{R}^N$ and some $N \times N$ matrix $A \in \mathcal{M}_N(\mathbb{R})$ with $\det A \neq 0$. Show that if $u \in W_0^{1,p}(\Omega)$, the new function defined by

$$u_\Phi(x) := u(\Phi(x)), \qquad \text{for a. e. } x \in \Omega,$$

still belongs to $W_0^{1,p}(\Omega)$.

Problem 3.12.25 (Picone's Inequality for Sobolev Functions) Let $\Omega \subseteq \mathbb{R}^N$ be an open set and let $1 < p < \infty$. Show that for every $\varepsilon > 0$, every $\eta \in C_0^1(\Omega)$ and every $u \in W_{\mathrm{loc}}^{1,p}(\Omega)$ such that $u \ge 0$ almost everywhere in Ω, we have

$$\frac{|\eta|^p}{(\varepsilon + u)^{p-1}} \in W_0^{1,p}(\Omega),$$

and

$$\left\langle |\nabla u|^{p-2}\, \nabla u, \nabla \left(\frac{|\eta|^p}{(\varepsilon + u)^{p-1}} \right) \right\rangle \le |\nabla \eta|^p, \qquad \text{a. e. in } \Omega.$$

Problem 3.12.26 Let $\Omega \subseteq \mathbb{R}^N$ be an open set and let $1 \le p \le \infty$. Let $u \in W_0^{1,p}(\Omega)$ be such that $u \ge 0$ almost everywhere in Ω. Show that there exists a sequence $\{u_n\}_{n\in\mathbb{N}} \subseteq C_0^\infty(\Omega)$ such that

$$\lim_{n\to\infty} \|u_n - u\|_{W^{1,p}(\Omega)} = 0,$$

and

$$u_n(x) \ge 0, \qquad \text{for every } x \in \Omega \text{ and every } n \in \mathbb{N}.$$

Problem 3.12.27 Let $N \ge 2$ and let B be an open N-dimensional ball. Show that if $u \in L^1(B)$ is such that $\nabla u \in L^q(B;\mathbb{R}^N)$ for some $1 < q < \infty$, then $u \in W^{1,q}(B)$.

Chapter 4
The Direct Method in Sobolev Spaces

4.1 A Model Case: Harmonic Functions

We go back to the problem mentioned at the beginning of Sect. 3.1, i.e. finding a solution of

$$\begin{cases} -\Delta u = 0, & \text{in } \Omega, \\ u = g, & \text{on } \partial\Omega, \end{cases} \tag{4.1.1}$$

through the solution of the following minimization problem

$$\inf_u \left\{ \frac{1}{2} \int_\Omega |\nabla u|^2 \, dx \ : \ u = g \text{ on } \partial\Omega \right\}.$$

Here $\Omega \subseteq \mathbb{R}^N$ is an open bounded set. As explained there, we use the *Direct Method*, i.e. a generalization of the idea at the basis of the Weierstrass Theorem. We first have to declare the space of functions where the infimum is seeked. It turns out that a Sobolev space will do the job, i.e. we take $g \in W^{1,2}(\Omega)$ and we consider

$$\inf_{u \in W^{1,2}(\Omega)} \left\{ \frac{1}{2} \int_\Omega |\nabla u|^2 \, dx \ : \ u - g \in W^{1,2}_0(\Omega) \right\}.$$

Observe that, according to Remark 3.7.2, the condition

$$u - g \in W^{1,2}_0(\Omega),$$

can be interpreted as a weak surrogate of the requirement $u = g$ on $\partial\Omega$.

We then take a minimizing sequence $\{u_n\}_{n \in \mathbb{N}}$ of admissible functions. As already observed at the beginning of Chap. 3 we have

L. Brasco, *Handbook of Calculus of Variations for Absolute Beginners*,
La Matematica per il 3+2 163, https://doi.org/10.1007/978-3-031-87164-1_4

$$\int_{\Omega} |\nabla u_n|^2 \, dx \le C, \qquad \text{for every } n \in \mathbb{N}.$$

Moreover, by using the Poincaré inequality (see Remark 3.7.7) for the function $u_n - g \in W_0^{1,2}(\Omega)$, we have

$$\begin{aligned}
\|u_n\|_{L^2(\Omega)} &\le \|u_n - g\|_{L^2(\Omega)} + \|g\|_{L^2(\Omega)} \\
&\le \operatorname{diam}(\Omega) \, \|\nabla u_n - \nabla g\|_{L^2(\Omega;\mathbb{R}^N)} + \|g\|_{L^2(\Omega)} \\
&\le \operatorname{diam}(\Omega) \sqrt{C} + \operatorname{diam}(\Omega) \, \|\nabla g\|_{L^2(\Omega;\mathbb{R}^N)} + \|g\|_{L^2(\Omega)}.
\end{aligned}$$

This shows that $\{u_n\}_{n\in\mathbb{N}}$ is a bounded sequence in $W^{1,2}(\Omega)$. Thus, by the Banach-Alaoglu Theorem 3.3.6, we can infer weak convergence (up to a subsequence) to a function $v \in W^{1,2}(\Omega)$. By using that $\{u_n\}_{n\in\mathbb{N}}$ is a minimizing sequence and recalling that the L^2 norm is lower semicontinuous with respect to the weak convergence (recall Proposition 1.3.11), we have

$$\begin{aligned}
\inf_{u\in W^{1,2}(\Omega)} \left\{ \frac{1}{2} \int_{\Omega} |\nabla u|^2 \, dx \; : \; u - g \in W_0^{1,2}(\Omega) \right\} &= \liminf_{n\to\infty} \frac{1}{2} \int_{\Omega} |\nabla u_n|^2 \, dx \\
&\ge \frac{1}{2} \int_{\Omega} |\nabla v|^2 \, dx.
\end{aligned}$$

We are only left with observing that $u_n - g \in W_0^{1,2}(\Omega)$ weakly converges to $v - g$ and the space $W_0^{1,2}(\Omega)$ is weakly closed, thanks to Theorem 3.7.4. This shows that

$$v - g \in W_0^{1,2}(\Omega),$$

thus the limit function v is admissible for the minimization problem. By the previous discussion, we finally conclude that

$$\inf_{u\in W^{1,2}(\Omega)} \left\{ \frac{1}{2} \int_{\Omega} |\nabla u|^2 \, dx \; : \; u - g \in W_0^{1,2}(\Omega) \right\} = \frac{1}{2} \int_{\Omega} |\nabla v|^2 \, dx,$$

i.e. v is the desired minimizer!

If we now go back to our initial task of proving the existence of a solution to (4.1.1), it is not clear whether we succeeded... or not! Indeed, the function v previously obtained is just a Sobolev function: for the moment we can not guarantee that $v \in C^2(\Omega)$ and that v is harmonic in the usual sense. However, we can reproduce the proof of the Dirichlet Principle (i.e. Theorem 1.5.1) and show that the minimizer v must be a critical point for the functional

$$\mathcal{F}(u) = \frac{1}{2} \int_{\Omega} |\nabla u|^2 \, dx, \tag{4.1.2}$$

i.e. v must solve (in weak sense) the associated Euler-Lagrange equation. Indeed, for every $t \in \mathbb{R}$ and every $\varphi \in C_0^\infty(\Omega)$, by minimality of v we have

$$\mathcal{F}(v + t\,\varphi) \geq \mathcal{F}(v).$$

This implies that the function

$$\begin{aligned}\mathcal{F}(v + t\,\varphi) &= \frac{1}{2}\int_\Omega |\nabla v + t\,\nabla\varphi|^2\,dx\\ &= \frac{1}{2}\int_\Omega |\nabla v|^2\,dx + t\int_\Omega \langle \nabla v, \nabla\varphi\rangle\,dx + \frac{t^2}{2}\int_\Omega |\nabla\varphi|^2\,dx,\end{aligned}$$

is minimal for $t = 0$. By the classical Fermat Theorem in one real variable, it must thus result that the first variation vanishes

$$\delta\mathcal{F}(v)[\varphi] = \frac{d}{dt}\mathcal{F}(v + t\,\varphi)_{|t=0} = 0.$$

This is the same as

$$\int_\Omega \langle \nabla v, \nabla\varphi\rangle\,dx = 0, \qquad \text{for every } \varphi \in C_0^\infty(\Omega).$$

By recalling Definition 1.5.5, we thus can say that the minimizer v is at least a *weak solution* of the Laplace equation.

If one could prove that the minimizer v is a C^2 function, by recalling Remark 1.5.6 we would get that v is actually a true harmonic function! The task of proving that a Sobolev minimizer is actually more regular (for example C^2 or even more) is exactly the aim of the so-called *Regularity Theory*. We will give a flavor of this theory in Chap. 7: see for example Corollary 7.7.2 below.

In this chapter we will see how to extend the above reasonings to prove existence of minimizers for functionals more general than (4.1.2): accordingly, we will get existence of weak solutions to some boundary value problems for quite general second order partial differential equations of elliptic type (recall Sect. 1.6).

4.2 A Lower Semicontinuity Result

If we aim at generalizing the example of the previous section, the first step is to prove lower semicontinuity of a certain class of integral functionals with respect to the weak convergence in Sobolev spaces. We will not give the most general class of lower semicontinuous functionals on Sobolev spaces: we refer to [3, Chapter 3] and [6, Chapter 4] for the general theory. Here, we will rather stick to a specific class of functionals, which is however general enough to encompass many important examples. This will be enough for our purposes. In what follows, for $1 \le p < N$ we still use the notation

$$p^* = \frac{N\,p}{N-p},$$

as in Theorem 3.6.1.

Theorem 4.2.1 *Let $\Omega \subseteq \mathbb{R}^N$ be an open bounded set. Let $1 < p < \infty$ and $q \ge 1$ be two exponents such that*

$$\begin{cases} q < p^*, & \text{if } p < N, \\ q < \infty, & \text{if } p \ge N. \end{cases}$$

We also take:

- *a C^1 convex function $H : \mathbb{R}^N \to \mathbb{R}$ such that*

$$C_1\,(|z|^p - 1) \le H(z) \le C_2\,(|z|^p + 1), \qquad \text{for every } z \in \mathbb{R}^N,$$

  for two constants $C_2 \ge C_1 > 0$;
- *a C^1 function $G : \mathbb{R} \to \mathbb{R}$ such that*

$$|G'(t)| \le A\,(|t|^{q-1} + 1), \qquad \text{for every } t \in \mathbb{R},$$

  for a constant $A \ge 0$;
- *a function $f \in L^\gamma(\Omega)$ with*

$$\begin{cases} \gamma \ge \left(\dfrac{p^*}{q}\right)', & \text{if } p < N, \\ \gamma > 1, & \text{if } p = N, \\ \gamma \ge 1, & \text{if } p > N. \end{cases}$$

Then the functional $\mathcal{F} : W_0^{1,p}(\Omega) \to \mathbb{R}$ defined by

$$\mathcal{F}(\varphi) = \int_\Omega H(\nabla\varphi)\,dx - \int_\Omega f\,G(\varphi)\,dx, \qquad \text{for every } \varphi \in W_0^{1,p}(\Omega), \tag{4.2.1}$$

is lower semicontinuous with respect to the weak convergence in $W^{1,p}(\Omega)$. In other words, if $\{\varphi_n\}_{n\in\mathbb{N}} \subseteq W_0^{1,p}(\Omega)$ weakly converges to $\varphi \in W_0^{1,p}(\Omega)$, then we have

$$\liminf_{n\to\infty} \mathcal{F}(\varphi_n) \ge \mathcal{F}(\varphi).$$

Proof We first show that the functional $\mathcal{F}$ is well-defined on $W_0^{1,p}(\Omega)$. Indeed, the assumption on H assures that

$$\left|\int_\Omega H(\nabla\varphi)\,dx\right| \le C_2 \int_\Omega |\nabla\varphi|^p\,dx + C_2\,|\Omega| < +\infty, \qquad \text{for every } \varphi \in W_0^{1,p}(\Omega).$$

The finiteness of the term containing G is slightly more elaborated. We suppose for simplicity that $p < N$. The other cases can be treated in a similar way and the details are left to the reader, as a useful exercise. In this case, by Theorem 3.8.1 we have the continuous embedding

$$W_0^{1,p}(\Omega) \hookrightarrow L^{p^*}(\Omega).$$

Thus, for every $\varphi \in W_0^{1,p}(\Omega)$ we have $\varphi \in L^{p^*}(\Omega)$, as well. Thanks to the assumption on G, we know that there exists $\widetilde{A} \ge 0$ such that

$$|G(t)| \le \widetilde{A}\,(|t|^q + 1), \qquad \text{for every } t \in \mathbb{R}, \tag{4.2.2}$$

see Problem 1.7.5. This entails that for every $\varphi \in W_0^{1,p}(\Omega)$ we have

$$\left|\int_\Omega f\,G(\varphi)\,dx\right| \le \int_\Omega |f|\,|G(\varphi)|\,dx \le \widetilde{A}\int_\Omega |f|\,|\varphi|^q\,dx + \widetilde{A}\int_\Omega |f|\,dx.$$

In particular, by using Hölder's inequality with exponents p^*/q and $(p^*/q)'$ we have

$$\left|\int_\Omega f\,G(\varphi)\,dx\right| \le \widetilde{A}\,\|f\|_{L^{(p^*/q)'}(\Omega)}\left(\|\varphi\|^q_{L^{p^*}(\Omega)} + |\Omega|^{\frac{q}{p^*}}\right) < +\infty. \tag{4.2.3}$$

Observe that we used that $f \in L^\gamma(\Omega) \subseteq L^{(p^*/q)'}(\Omega)$, thanks to the choice of γ and the boundedness of Ω. We have finally obtained that the functional $\mathcal{F}$ is well-defined.

Let $\{\varphi_n\}_{n\in\mathbb{N}} \subseteq W_0^{1,p}(\Omega)$ be a sequence weakly converging to a function $\varphi \in W_0^{1,p}(\Omega)$. Observe that, by Proposition B.1.2, we know that this sequence is bounded in $W^{1,p}(\Omega)$. We are going to show that

$$\liminf_{n\to\infty}\int_\Omega H(\nabla\varphi_n)\,dx \ge \int_\Omega H(\nabla\varphi)\,dx, \tag{4.2.4}$$

and

$$\lim_{n\to\infty}\int_\Omega f\,G(\varphi_n)\,dx = \int_\Omega f\,G(\varphi)\,dx. \tag{4.2.5}$$

This will be sufficient to establish the claimed semicontinuity property.

In order to prove (4.2.4), the convexity of H will play a crucial role. Indeed, by the "above tangent" property of convex functions (i.e. Proposition 1.3.4), we have

$$H(\nabla\varphi_n(x)) \geq H(\nabla\varphi(x)) + \langle \nabla H(\nabla\varphi(x)), \nabla\varphi_n(x) - \nabla\varphi(x)\rangle, \qquad \text{for a. e. } x \in \Omega.$$

By integrating this inequality over Ω, we get

$$\int_\Omega H(\nabla\varphi_n)\,dx \geq \int_\Omega H(\nabla\varphi)\,dx + \int_\Omega \langle \nabla H(\nabla\varphi), \nabla\varphi_n - \nabla\varphi\rangle\,dx. \tag{4.2.6}$$

We now observe that by Problem 1.7.9, the assumptions on H guarantee that there exists $C_3 > 0$ such that

$$|\nabla H(z)| \leq C_3\,(|z|^{p-1} + 1), \qquad \text{for every } z \in \mathbb{R}^N.$$

This implies that

$$\begin{aligned}\int_\Omega |\nabla H(\nabla\varphi)|^{p'}\,dx &\leq C_3^{p'} \int_\Omega (1 + |\nabla\varphi|^{p-1})^{p'}\,dx \\ &\leq C_3^{p'}\,2^{p'-1}\,|\Omega| + C_3^{p'}\,2^{p'-1} \int_\Omega |\nabla\varphi|^p\,dx < +\infty,\end{aligned}$$

i.e. $\nabla H(\nabla\varphi) \in L^{p'}(\Omega;\mathbb{R}^N)$. Observe that we used Problem 1.7.2 with $\alpha = p'$, together with the fact that $(p-1)\,p' = p$. By the weak convergence, we know that

$$\lim_{n\to\infty} \int_\Omega \langle \phi, \nabla\varphi_n - \nabla\varphi\rangle\,dx = 0, \qquad \text{for every } \phi \in L^{p'}(\Omega;\mathbb{R}^N).$$

Thus, in particular by taking $\phi = \nabla H(\nabla\varphi)$ we can infer

$$\lim_{n\to\infty} \int_\Omega \langle \nabla H(\nabla\varphi), \nabla\varphi_n - \nabla\varphi\rangle\,dx = 0.$$

By taking the $\liminf$ as n goes to ∞ in (4.2.6) and using the previous information, we then get

$$\begin{aligned}\liminf_{n\to\infty} \int_\Omega H(\nabla\varphi_n)\,dx &\geq \int_\Omega H(\nabla\varphi)\,dx \\ &\quad + \liminf_{n\to\infty} \int_\Omega \langle \nabla H(\nabla\varphi), \nabla\varphi_n - \nabla\varphi\rangle\,dx = \int_\Omega H(\nabla\varphi)\,dx.\end{aligned}$$

This establishes the lower semicontinuity of the gradient term, i.e. (4.2.4).

In order to prove (4.2.5), we will restrict again to the case $p < N$ and leave the case $p \geq N$ to the reader. By Lemma 3.8.7, we know that

$$\lim_{n\to\infty} \|\varphi_n - \varphi\|_{L^r(\Omega)} = 0, \qquad \text{for every } 1 \le r < p^*. \tag{4.2.7}$$

By Problem 1.7.28, we can take a sequence $\{f_k\}_{k\in\mathbb{N}} \subseteq L^1(\Omega) \cap L^\infty(\Omega)$ such that

$$\lim_{k\to\infty} \|f_k - f\|_{L^{(p^*/q)'}(\Omega)} = 0. \tag{4.2.8}$$

By some standard algebraic manipulations, we get

$$\begin{aligned}
\left|\int_\Omega f\, G(\varphi_n)\, dx - \int_\Omega f\, G(\varphi)\, dx\right| &\le \left|\int_\Omega (f - f_k)\Big(G(\varphi_n) - G(\varphi)\Big)\, dx\right| \\
&+ \left|\int_\Omega f_k \Big(G(\varphi_n) - G(\varphi)\Big)\, dx\right| \\
&\le \int_\Omega |f - f_k| \Big|G(\varphi_n) - G(\varphi)\Big|\, dx \\
&+ \int_\Omega |f_k| \Big|G(\varphi_n) - G(\varphi)\Big|\, dx \\
&\le \int_\Omega |f - f_k| \Big(|G(\varphi_n)| + |G(\varphi)|\Big)\, dx \\
&+ \int_\Omega |f_k| \Big|G(\varphi_n) - G(\varphi)\Big|\, dx.
\end{aligned}$$

We now set

$$\mathcal{I}_{n,k} := \int_\Omega |f - f_k| \Big(|G(\varphi_n)| + |G(\varphi)|\Big)\, dx \qquad \mathcal{J}_{n,k} =: \int_\Omega |f_k| \Big|G(\varphi_n) - G(\varphi)\Big|\, dx,$$

thus the previous estimate rewrites as follows

$$\left|\int_\Omega f\, G(\varphi_n)\, dx - \int_\Omega f\, G(\varphi)\, dx\right| \le \mathcal{I}_{n,k} + \mathcal{J}_{n,k}. \tag{4.2.9}$$

Let us discuss the two terms in the right-hand side of (4.2.9) separately. For $\mathcal{I}_{n,k}$, we may notice that from the growth estimate (4.2.2) on G, we get

$$|G(\varphi_n)| + |G(\varphi)| \le \widetilde{A}\,(|\varphi_n|^q + |\varphi|^q + 2) \le \widetilde{A}\,(|\varphi_n| + |\varphi|)^q + 2\,\widetilde{A},$$

where we also used the inequality of Problem 1.7.2. This shows that

$$\mathcal{I}_{n,k} \le \widetilde{A} \int_\Omega |f - f_k|\,(|\varphi_n| + |\varphi|)^q\, dx + 2\,\widetilde{A} \int_\Omega |f - f_k|\, dx.$$

We can apply again Hölder's inequality with exponents p^*/q and $(p^*/q)'$, so to get

$$\mathcal{I}_{n,k} \le \widetilde{A}\, \|f - f_k\|_{L^{(p^*/q)'}(\Omega)} \left(\Big\| |\varphi_n| + |\varphi| \Big\|^q_{L^{p^*}(\Omega)} + 2\, |\Omega|^{\frac{q}{p^*}} \right).$$

Observe that the right-hand side is uniformly bounded in n, thanks to Theorem 3.8.1 and the boundedness of the sequence $\{\varphi_n\}_{n\in\mathbb{N}}$. Thus, we get

$$\mathcal{I}_{n,k} \le C\, \|f - f_k\|_{L^{(p^*/q)'}(\Omega)}, \qquad \text{for every } k, n \in \mathbb{N},$$

with C independent of both n and k. By spending this information into (4.2.9), we get

$$\left| \int_\Omega f\, G(\varphi_n)\, dx - \int_\Omega f\, G(\varphi)\, dx \right| \le C\, \|f - f_k\|_{L^{(p^*/q)'}(\Omega)} + \mathcal{J}_{n,k}. \tag{4.2.10}$$

We now handle the term $\mathcal{J}_{n,k}$. We start by observing that the assumption on G entails that

$$|G(t) - G(s)| \le A\, (|t| + |s|)^{q-1}\, |t - s| + A\, |t - s|,$$

see Problem 1.7.5. This yields

$$\mathcal{J}_{n,k} = \int_\Omega |f_k|\, \Big|G(\varphi_n) - G(\varphi)\Big|\, dx \le A \int_\Omega |f_k| \Big(|\varphi_n| + |\varphi|\Big)^{q-1} |\varphi_n - \varphi|\, dx + A \int_\Omega |f_k|\, |\varphi_n - \varphi|\, dx.$$

The last two integrals converge to 0 as n goes to ∞, for every fixed $k \in \mathbb{N}$. Indeed, we have (recall that each f_k is in $L^1(\Omega) \cap L^\infty(\Omega)$)

$$\begin{aligned} \int_\Omega |f_k| \Big(|\varphi_n| + |\varphi|\Big)^{q-1} |\varphi_n - \varphi|\, dx &\le \|f_k\|_{L^\infty(\Omega)} \int_\Omega \Big(|\varphi_n|+|\varphi|\Big)^{q-1} |\varphi_n - \varphi|\, dx \\ &\le \|f_k\|_{L^\infty(\Omega)} \left(\int_\Omega \Big(|\varphi_n| + |\varphi|\Big)^q dx \right)^{\frac{q-1}{q}} \|\varphi_n - \varphi\|_{L^q(\Omega)}, \end{aligned}$$

and

$$\int_\Omega |f_k|\, |\varphi_n - \varphi|\, dx \le \|f_k\|_{L^\infty(\Omega)} \int_\Omega |\varphi_n - \varphi|\, dx.$$

By recalling (4.2.7), we thus have that the last two terms converge to 0 as n goes to ∞, so that

$$\lim_{n\to\infty} \mathcal{J}_{n,k} = 0, \qquad \text{for every } k \in \mathbb{N},$$

as claimed. We now go back to (4.2.10) and take the limit as n goes to ∞. The above discussion gives

$$\begin{aligned}\limsup_{n\to\infty} \left| \int_\Omega f\, G(\varphi_n)\, dx - \int_\Omega f\, G(\varphi)\, dx \right| &\le C\, \|f - f_k\|_{L^{(p^*/q)'}(\Omega)} + \lim_{n\to\infty} \mathcal{J}_{n,k} \\ &= C\, \|f - f_k\|_{L^{(p^*/q)'}(\Omega)}.\end{aligned}$$

It is only left to observe that the left-hand side is independent of k, while by (4.2.8) the right-hand side converges to 0 as k goes to ∞. Thus, we finally get

$$\lim_{n\to\infty} \left| \int_\Omega f\, G(\varphi_n)\, dx - \int_\Omega f\, G(\varphi)\, dx \right| = 0,$$

which is the desired continuity property (4.2.5). □

Remark 4.2.2 The above proof of (4.2.5) could appear a little bit trickier than one could have expected: this is due to the fact that we admitted the borderline case $\gamma = (p^*/q)'$. In the case $\gamma > (p^*/q)'$ the proof is easier: it is sufficient to observe that

$$\begin{aligned}&\left| \int_\Omega f\, G(\varphi_n)\, dx - \int_\Omega f\, G(\varphi)\, dx \right| \\ &\le \int_\Omega |f|\, |G(\varphi_n) - G(\varphi)|\, dx \\ &\le A \int_\Omega |f|\, (|\varphi_n| + |\varphi|)^{q-1}\, |\varphi_n - \varphi|\, dx \\ &+ A \int_\Omega |f|\, |\varphi_n - \varphi|\, dx \\ &\le A\, \|f\|_{L^\gamma(\Omega)} \left(\int_\Omega (|\varphi_n| + |\varphi|)^{(q-1)\,\gamma'}\, |\varphi_n - \varphi|^{\gamma'}\, dx \right)^{\frac{1}{\gamma'}} \\ &+ A\, \|f\|_{L^\gamma(\Omega)}\, \|\varphi_n - \varphi\|_{L^{\gamma'}(\Omega)}.\end{aligned}$$

The last term converges to 0, as n goes to ∞, thanks to (4.2.7). For the other term, by Hölder's inequality with exponents q and q' we have

$$\begin{aligned}\int_\Omega (|\varphi_n| + |\varphi|)^{(q-1)\,\gamma'}\, |\varphi_n - \varphi|^{\gamma'}\, dx &\le \left(\int_\Omega (|\varphi_n| + |\varphi|)^{q\,\gamma'}\, dx \right)^{\frac{q-1}{q}} \\ &\times \left(\int_\Omega |\varphi_n - \varphi|^{q\,\gamma'}\, dx \right)^{\frac{1}{q}}.\end{aligned}$$

The assumption $\gamma > (p^*/q)'$ implies that $q\,\gamma' < p^*$ and thus the last term converges to 0, as n goes to ∞, again thanks to (4.2.7).

Remark 4.2.3 Theorem 4.2.1 continues to hold more generally for weakly converging sequences $\{\varphi_n\}_{n\in\mathbb{N}} \subseteq W^{1,p}(\Omega)$, provided that Ω has C^1 boundary. The proof is the same as before: the C^1 assumption permits to apply the embedding results of Theorems 3.9.3 and 3.9.4 (see also Remark 3.9.5), needed to prove (4.2.5). We leave the details to the reader.

4.3 An Existence Result in Sobolev Spaces

We give two existence results for the minimization of the functional $\mathcal{F}$ in (4.2.1), among $W^{1,p}$ functions with given "boundary values". We first consider the case

$$u = 0, \qquad \text{on } \partial\Omega,$$

which formally corresponds to functions in $W^{1,p}_0(\Omega)$. Then we consider the case of a more general boundary datum.

Theorem 4.3.1 (Minimization in $W^{1,p}_0$) *Under the same assumptions of Theorem 4.2.1, we further suppose that $1 \le q < p$. Then the minimization problem*

$$\inf_{u\in W^{1,p}_0(\Omega)} \left\{ \int_\Omega H(\nabla u)\,dx - \int_\Omega f\,G(u)\,dx \right\},$$

does admit (at least) a solution. Moreover, any minimizer v of the previous problem is a weak solution of the boundary value problem

$$\begin{cases} -\mathrm{div}(\nabla H(\nabla v)) = f\,G'(v), & \text{in } \Omega,\\ \qquad\qquad v = 0, & \text{on } \partial\Omega, \end{cases} \tag{4.3.1}$$

that is

$$\int_\Omega \langle \nabla H(\nabla v), \nabla\varphi\rangle\,dx = \int_\Omega f\,G'(v)\,\varphi\,dx, \quad \text{for every } \varphi \in C^\infty_0(\Omega).$$

Proof We divide the proof in various steps. For simplicity, we deal with the case $1 < p < N$, by leaving to the reader the details for the case $p \ge N$.

1. *The infimum is not $-\infty$.* We take $u \in W^{1,p}_0(\Omega)$ to be admissible in our variational problem. By the estimate (4.2.3) from the proof of Theorem 4.2.1, we have in particular

$$- \int_\Omega f\, G(u)\, dx \geq -\widetilde{A}\, \|f\|_{L^{(p^*/q)'}(\Omega)} \left(\|u\|^q_{L^{p^*}(\Omega)} + |\Omega|^{\frac{q}{p^*}} \right). \tag{4.3.2}$$

By Sobolev inequality (i.e. Theorem 3.8.1 for $1 < p < N$ and $q = p^*$), we have

$$\|u\|_{L^{p^*}(\Omega)} \leq \mathcal{S}\, \|\nabla u\|_{L^p(\Omega;\mathbb{R}^N)}.$$

We use this estimate in (4.3.2), so to get

$$- \int_\Omega f\, G(u)\, dx \geq -\widetilde{A}\, \|f\|_{L^{(p^*/q)'}(\Omega)} \left(\mathcal{S}^q\, \|\nabla u\|^q_{L^p(\Omega;\mathbb{R}^N)} + |\Omega|^{\frac{q}{p^*}} \right).$$

We now apply the generalized Young inequality (1.2.6), with conjugate exponents[1]

$$\frac{p}{q} \qquad \text{and} \qquad \frac{p}{p-q},$$

which gives

$$\begin{aligned}\widetilde{A}\, \mathcal{S}^q\, \|f\|_{L^{(p^*/q)'}(\Omega)}\, \|\nabla u\|^q_{L^p(\Omega;\mathbb{R}^N)} &\leq \varepsilon\, \frac{q}{p}\, \|\nabla u\|^p_{L^p(\Omega;\mathbb{R}^N)} \\ &\quad + \varepsilon^{-\frac{q}{p-q}}\, \frac{p-q}{p}\, (\widetilde{A}\, \mathcal{S}^q)^{\frac{p}{p-q}}\, \|f\|^{\frac{p}{p-q}}_{L^{(p^*/q)'}(\Omega)},\end{aligned}$$

for $\varepsilon > 0$ arbitrary. This finally gives the following lower bound

$$\begin{aligned}- \int_\Omega f\, G(u)\, dx &\geq -\varepsilon\, \frac{q}{p}\, \|\nabla u\|^p_{L^p(\Omega;\mathbb{R}^N)} \\ &\quad - \varepsilon^{-\frac{q}{p-q}}\, \frac{p-q}{p}\, (\widetilde{A}\, \mathcal{S}^q)^{\frac{p}{p-q}}\, \|f\|^{\frac{p}{p-q}}_{L^{(p^*/q)'}(\Omega)} \\ &\quad - \widetilde{A}\, \|f\|_{L^{(p^*/q)'}(\Omega)}\, |\Omega|^{\frac{q}{p^*}}.\end{aligned} \tag{4.3.3}$$

We can now prove that the infimum of our variational problem is not $-\infty$. Indeed, by using (4.3.3) and the lower bound on H, we have

$$\begin{aligned}\int_\Omega H(\nabla u)\, dx - \int_\Omega f\, G(u)\, dx &\geq \left(C_1 - \frac{q}{p}\, \varepsilon \right) \int_\Omega |\nabla u|^p\, dx - C_1\, |\Omega| \\ &\quad - \varepsilon^{-\frac{q}{p-q}}\, \frac{p-q}{p}\, (\widetilde{A}\, \mathcal{S}^q)^{\frac{p}{p-q}}\, \|f\|^{\frac{p}{p-q}}_{L^{(p^*/q)'}(\Omega)} \\ &\quad - \widetilde{A}\, \|f\|_{L^{(p^*/q)'}(\Omega)}\, |\Omega|^{\frac{q}{p^*}},\end{aligned}$$

[1] Here we crucially exploit that $q < p$.

for an arbitrary $\varepsilon > 0$. For example, we can make the choice

$$\varepsilon = \frac{p}{q}\frac{C_1}{2} > 0,$$

and get

$$\int_\Omega H(\nabla u)\,dx - \int_\Omega f\,G(u)\,dx \geq \frac{C_1}{2}\int_\Omega |\nabla u|^p\,dx - M, \tag{4.3.4}$$

where $M > 0$ is a constant which only depends on the data of the problem: the $L^{(p^*/q)'}$ norm of f, the measure of Ω, the exponents p and q, the constants $\widetilde{A}, \mathcal{S}, C_1$. The estimate (4.3.4) clearly shows that our functional is uniformly bounded from below, on the admissible class of functions we are considering. Thus, we obtain

$$\mathfrak{m} := \inf_{u\in W^{1,p}_0(\Omega)} \left\{\int_\Omega H(\nabla u)\,dx - \int_\Omega f\,G(u)\,dx\right\} > -\infty,$$

as desired.

2. *Existence of a minimizer.* With the notation above, for every $n \in \mathbb{N}$ we take $u_n \in W^{1,p}_0(\Omega)$ such that

$$\int_\Omega H(\nabla u_n)\,dx - \int_\Omega f\,G(u_n)\,dx \leq \mathfrak{m} + \frac{1}{n+1}, \qquad \text{for every } n \in \mathbb{N}. \tag{4.3.5}$$

Such a function exists by definition of infimum. By using this estimate in conjunction with (4.3.4), we obtain

$$\int_\Omega |\nabla u_n|^p\,dx \leq \frac{2}{C_1}\left(M + \mathfrak{m} + \frac{1}{n+1}\right), \qquad \text{for every } n \in \mathbb{N}.$$

Thanks to Poincaré inequality (see Theorem 3.7.6 and Remark 3.7.7), this permits to infer that $\{u_n\}_{n\in\mathbb{N}}$ is a bounded sequence in $W^{1,p}_0(\Omega)$. By Theorem 3.3.6, this in turn implies that there exists $v \in W^{1,p}(\Omega)$ such that, up to a subsequence, we have that $\{u_n\}_{n\in\mathbb{N}}$ weakly converges in $W^{1,p}(\Omega)$ to v. Moreover, by recalling that the space $W^{1,p}_0(\Omega)$ is weakly closed (see Theorem 3.7.4), we have $v \in W^{1,p}_0(\Omega)$, as well. By using the semicontinuity result of Theorem 4.2.1 and Eq. (4.3.5), we get

$$\begin{aligned}\mathfrak{m} &\geq \liminf_{n\to\infty}\left[\int_\Omega H(\nabla u_n)\,dx - \int_\Omega f\,G(u_n)\,dx\right]\\ &\geq \int_\Omega H(\nabla v)\,dx - \int_\Omega f\,G(v)\,dx \geq \mathfrak{m}.\end{aligned}$$

Observe that the last inequality comes from the definition of $\mathfrak{m}$. This shows that $v \in W_0^{1,p}(\Omega)$ is a minimizer.

3. *Euler-Lagrange equation.* We still have to show that a minimizer v is a weak solution of (4.3.1), i.e. by recalling Definition 1.5.5, it satisfies

$$\int_\Omega \langle \nabla H(\nabla v), \nabla \varphi \rangle \, dx = \int_\Omega f\, G'(v)\, \varphi \, dx, \qquad \text{for every } \varphi \in C_0^\infty(\Omega).$$

We take $\varphi \in C_0^\infty(\Omega) \subseteq W_0^{1,p}(\Omega)$, then for every $t \in \mathbb{R}$ the function $v + t\,\varphi$ is still admissible in our minimization problem. Consequently, we have

$$\int_\Omega H(\nabla v + t\,\nabla \varphi)\, dx - \int_\Omega f\, G(v + t\,\varphi)\, dx \geq \int_\Omega H(\nabla v)\, dx + \int_\Omega f\, G(v)\, dx.$$

This means that the function of one real variable

$$g(t) = \int_\Omega H(\nabla v + t\,\nabla \varphi)\, dx - \int_\Omega f\, G(v + t\,\varphi)\, dx, \tag{4.3.6}$$

is minimal at $t = 0$. If we can show that g is differentiable at $t = 0$, then by Fermat's Theorem we would get $g'(0) = 0$. Let us show that g is differentiable at $t = 0$: we have

$$\begin{aligned} \frac{g(t) - g(0)}{t} &= \int_\Omega \frac{H(\nabla v + t\,\nabla \varphi) - H(\nabla v)}{t}\, dx \\ &\quad - \int_\Omega f\, \frac{G(v + t\,\varphi) - G(v)}{t}\, dx. \end{aligned} \tag{4.3.7}$$

By the Fundamental Theorem of Calculus, we have

$$\begin{aligned} H(\nabla v + t\,\nabla \varphi) - H(\nabla v) &= \int_0^t \frac{d}{d\tau} H(\nabla v + \tau\,\nabla \varphi)\, d\tau \\ &= \int_0^t \langle \nabla H(\nabla v + \tau\,\nabla \varphi), \nabla \varphi \rangle \, d\tau. \end{aligned}$$

Thus, for every $0 < t < 1$ we can infer

$$\begin{aligned} \left| \frac{H(\nabla v + t\,\nabla \varphi) - H(\nabla v)}{t} \right| &\leq \frac{1}{t} \int_0^t |\nabla H(\nabla v + \tau\,\nabla \varphi)|\, |\nabla \varphi|\, d\tau \\ &\leq \frac{C_3}{t}\, |\nabla \varphi| \int_0^t (|\nabla v + \tau\,\nabla \varphi|^{p-1} + 1)\, d\tau, \end{aligned}$$

where we used the assumption on H and Problem 1.7.9 to deduce the last estimate. Moreover, by using the triangle inequality and Problems 1.7.1 and 1.7.2, we know that

$$|\nabla v+\tau\,\nabla\varphi|^{p-1}\leq(|\nabla v|+\tau\,|\nabla\varphi|)^{p-1}\leq\max\{2^{p-2},1\}\,(|\nabla v|^{p-1}+\tau^{p-1}\,|\nabla\varphi|^{p-1}).$$

Thus, we get for $0<t<1$

$$\begin{aligned}\left|\frac{H(\nabla v+t\,\nabla\varphi)-H(\nabla v)}{t}\right|&\leq\frac{C}{t}\,|\nabla\varphi|\int_0^t(|\nabla v|^{p-1}+\tau^{p-1}\,|\nabla\varphi|^{p-1}+1)\,d\tau\\&=C\,|\nabla\varphi|\,(|\nabla v|^{p-1}+1)+\frac{C}{p}\,t^{p-1}\,|\nabla\varphi|^p\\&\leq C\,|\nabla\varphi|\,(|\nabla v|^{p-1}+1)+\frac{C}{p}\,|\nabla\varphi|^p,\end{aligned}$$

where $C>0$ is a constant, possibly different from C_3. It is not difficult to see that the last function is in $L^1(\Omega)$. Moreover, we have

$$\begin{aligned}\lim_{t\to0^+}\frac{H(\nabla v+t\,\nabla\varphi)-H(\nabla v)}{t}&=\frac{d}{dt}H(\nabla v+t\,\nabla\varphi)_{|t=0}\\&=\langle\nabla H(\nabla v),\nabla\varphi\rangle,\qquad\text{a. e. in }\Omega.\end{aligned}$$

Thus, we can apply the Dominated Convergence Theorem and obtain that

$$\lim_{t\to0^+}\int_\Omega\frac{H(\nabla v+t\,\nabla\varphi)-H(\nabla v)}{t}\,dx=\int_\Omega\langle\nabla H(\nabla v),\nabla\varphi\rangle\,dx,$$

for every $\varphi\in C_0^\infty(\Omega)$. We also observe that

$$\begin{aligned}&\lim_{t\to0^-}\int_\Omega\frac{H(\nabla v+t\,\nabla\varphi)-H(\nabla v)}{t}\,dx\\&=-\lim_{\tau\to0^+}\int_\Omega\frac{H(\nabla v+\tau\,(-\nabla\varphi))-H(\nabla v)}{\tau}\,dx,\end{aligned}$$

and thus using the previous result with $-\varphi$ in place of φ, we also get

$$\lim_{t\to0^-}\int_\Omega\frac{H(\nabla v+t\,\nabla\varphi)-H(\nabla v)}{t}\,dx=\int_\Omega\langle\nabla H(\nabla v),\nabla\varphi\rangle\,dx.$$

For the second term in (4.3.7) we proceed similarly. We have

$$G(v+t\,\varphi)-G(v)=\int_0^t\frac{d}{d\tau}G(v+\tau\,\varphi)\,d\tau=\left(\int_0^t G'(v+\tau\,\varphi)\,d\tau\right)\varphi.$$

Then for every $0 < t < 1$ we get

$$\left| f\, \frac{G(v + t\,\varphi) - G(v)}{t} \right| \leq \frac{|f|\,|\varphi|\,A}{t} \int_0^t (|v + \tau\,\varphi|^{q-1} + 1)\, d\tau,$$

thanks to the assumption on G. By using the properties of power functions as before, we see that we can apply again the Dominated Convergence Theorem. This gives

$$\lim_{t \to 0^+} \int_\Omega f\, \frac{G(v + t\,\varphi) - G(v)}{t}\, dx = \int_\Omega f\, G'(v)\, \varphi\, dx.$$

Moreover, with the same trick as above, we can also get that

$$\lim_{t \to 0^-} \int_\Omega f\, \frac{G(v + t\,\varphi) - G(v)}{t}\, dx = \int_\Omega f\, G'(v)\, \varphi\, dx.$$

We thus have shown that g defined by (4.3.6) is differentiable at $t = 0$. The above discussion then gives

$$0 = g'(0) = \int_\Omega \langle \nabla H(\nabla v), \nabla \varphi \rangle\, dx - \int_\Omega f\, G'(v)\, \varphi\, dx,$$

which is valid for every $\varphi \in C_0^\infty(\Omega)$. The proof is over. □

Remark 4.3.2 The restriction $1 \leq q < p$ in the previous theorem can not be removed, in general. Indeed, it may happen that for $q \geq p$ the functional

$$\mathcal{F}(u) = \int_\Omega H(\nabla u)\, dx - \int_\Omega f\, G(u)\, dx,$$

in unbounded from below on $W_0^{1,p}(\Omega)$. In this case, we would have

$$\inf_{u \in W_0^{1,p}(\Omega)} \left\{ \int_\Omega H(\nabla u)\, dx - \int_\Omega f\, G(u)\, dx \right\} = -\infty,$$

and no minimizers exist. We have already encountered this phenomenon in Chap. 2, in the case of the harmonic oscillator (see Theorem 2.5.1, case $T > T_0$): in that case, we have

$$N = 1, \quad \Omega = (0, T), \quad H(z) = \frac{|z|^2}{2}, \quad G(t) = \frac{|t|^2}{2}, \quad f \equiv k > \left(\frac{\pi}{2}\right)^2 \frac{1}{T^2}.$$

See also the proof of Theorem 4.9.1 for an example of this phenomenon in the case $q > p$.

Theorem 4.3.3 (Minimization in $W^{1,p}(\Omega)$) *We assume the same hypotheses as in Theorem 4.3.1. We further assume that $\Omega \subseteq \mathbb{R}^N$ has a C^1 boundary. Then for every $U \in W^{1,p}(\Omega)$ the minimization problem*

$$\inf_{u\in W^{1,p}(\Omega)} \left\{ \int_\Omega H(\nabla u)\, dx - \int_\Omega f\, G(u)\, dx \,:\, u - U \in W_0^{1,p}(\Omega) \right\},$$

does admit (at least) a minimizer. Moreover, any minimizer v is a weak solution of the boundary value problem

$$\begin{cases} -\mathrm{div}(\nabla H(\nabla v)) = f\, G'(v), & \text{in } \Omega, \\ \qquad\qquad\qquad\quad v = U, & \text{on } \partial\Omega, \end{cases}$$

that is $v - U \in W_0^{1,p}(\Omega)$ and

$$\int_\Omega \langle \nabla H(\nabla v), \nabla\varphi\rangle\, dx = \int_\Omega f\, G'(v)\, \varphi\, dx, \qquad \text{for every } \varphi \in C_0^\infty(\Omega).$$

Proof We reproduce almost verbatim the proof of Theorem 4.3.1, though some modifications are needed. We divide again the proof in three steps and only deal with the case $1 < p < N$.

1. *The infimum is not $-\infty$.* We take $u \in W^{1,p}(\Omega)$ to be admissible in our variational problem. From (4.2.3) we still have

$$-\int_\Omega f\, G(u)\, dx \geq -\widetilde{A}\, \|f\|_{L^{(p^*/q)'}(\Omega)} \left(\|u\|^q_{L^{p^*}(\Omega)} + |\Omega|^{\frac{q}{p^*}} \right). \tag{4.3.8}$$

By Minkowski's and Sobolev's inequalities for the function $u - U \in W_0^{1,p}(\Omega)$, we have

$$\begin{aligned} \|u\|_{L^{p^*}(\Omega)} &\leq \|u - U\|_{L^{p^*}(\Omega)} + \|U\|_{L^{p^*}(\Omega)} \\ &\leq \mathcal{S}\, \|\nabla u - \nabla U\|_{L^p(\Omega;\mathbb{R}^N)} + \|U\|_{L^{p^*}(\Omega)} \\ &\leq \mathcal{S}\, \|\nabla u\|_{L^p(\Omega;\mathbb{R}^N)} + \mathcal{S}\, \|\nabla U\|_{L^p(\Omega;\mathbb{R}^N)} + \|U\|_{L^{p^*}(\Omega)}. \end{aligned}$$

By raising to the power q, from Problem 1.7.2 we also get

$$\begin{aligned} \|u\|^q_{L^{p^*}(\Omega)} &\leq \left(\mathcal{S}\, \|\nabla u\|_{L^p(\Omega;\mathbb{R}^N)} + \mathcal{S}\, \|\nabla U\|_{L^p(\Omega;\mathbb{R}^N)} + \|U\|_{L^{p^*}(\Omega)} \right)^q \\ &\leq 2^{q-1}\, \mathcal{S}^q\, \|\nabla u\|^q_{L^p(\Omega;\mathbb{R}^N)} + 2^{q-1} \left(\mathcal{S}\, \|\nabla U\|_{L^p(\Omega;\mathbb{R}^N)} + \|U\|_{L^{p^*}(\Omega)} \right)^q. \end{aligned}$$

Let us write

$$\Theta := 2^{q-1} \left(\mathcal{S} \, \|\nabla U\|_{L^p(\Omega;\mathbb{R}^N)} + \|U\|_{L^{p^*}(\Omega)} \right)^q,$$

and observe that this quantity only depends on the data of the problem. The previous estimate can be rewritten as

$$\|u\|^q_{L^{p^*}(\Omega)} \leq 2^{q-1} \, \mathcal{S}^q \, \|\nabla u\|^q_{L^p(\Omega)} + \Theta.$$

By using this estimate in (4.3.8), we then get

$$- \int_\Omega f \, G(u) \, dx \geq -\widetilde{A} \, \|f\|_{L^{(p^*/q)'}(\Omega)} \left(2^{q-1} \, \mathcal{S}^q \, \|\nabla u\|^q_{L^p(\Omega;\mathbb{R}^N)} + \Theta + |\Omega|^{\frac{q}{p^*}} \right).$$

We can now proceed exactly as in the proof of Theorem 4.3.1 and get again

$$\int_\Omega H(\nabla u) \, dx - \int_\Omega f \, G(u) \, dx \geq \frac{C_1}{2} \int_\Omega |\nabla u|^p \, dx - M, \tag{4.3.9}$$

where $M > 0$ is a constant, different from before, which only depends on the data of the problem (this time it also depends on the boundary datum U through the quantity Θ above). The estimate (4.3.9) clearly shows that our functional is uniformly bounded from below, on the admissible class of functions we are considering. Thus, we get again

$$\mathfrak{m} := \inf_{u \in W^{1,p}(\Omega)} \left\{ \int_\Omega H(\nabla u) \, dx - \int_\Omega f \, G(u) \, dx \, : \, u - U \in W^{1,p}_0(\Omega) \right\} > -\infty,$$

as claimed.

2. *Existence of a minimizer.* For every $n \in \mathbb{N}$ we take $u_n \in W^{1,p}(\Omega)$ admissible, such that

$$\int_\Omega H(\nabla u_n) \, dx - \int_\Omega f \, G(u_n) \, dx \leq \mathfrak{m} + \frac{1}{n+1}, \qquad \text{for every } n \in \mathbb{N}. \tag{4.3.10}$$

Such a function exists by definition of infimum. This uniform bound on the functional, in conjunction with the estimate (4.3.9), gives that

$$\int_\Omega |\nabla u_n|^p \, dx \leq \frac{2}{C_1} \left(M + \mathfrak{m} + \frac{1}{n+1} \right), \qquad \text{for every } n \in \mathbb{N}. \tag{4.3.11}$$

We claim that this uniform bound permits to infer that $\{u_n\}_{n \in \mathbb{N}}$ is a bounded sequence in $W^{1,p}(\Omega)$. Observe that now we can not directly appeal to the Poincaré inequality, as in the proof of Theorem 4.3.1, because $u_n \notin W^{1,p}_0(\Omega)$. However, we can use the same trick applied at the beginning of this chapter. We have

$$\begin{aligned}\|u_n\|_{L^p(\Omega)} &\leq \|u_n - U\|_{L^p(\Omega)} + \|U\|_{L^p(\Omega)}\\&\leq \mathrm{diam}(\Omega)\,\|\nabla u_n - \nabla U\|_{L^p(\Omega;\mathbb{R}^N)} + \|U\|_{L^p(\Omega)}\\&\leq \mathrm{diam}(\Omega)\,\|\nabla u_n\|_{L^p(\Omega;\mathbb{R}^N)} + \mathrm{diam}(\Omega)\,\|\nabla U\|_{L^p(\Omega;\mathbb{R}^N)} + \|U\|_{L^p(\Omega)},\end{aligned}$$

where in the second inequality we used the Remark 3.7.7, applied to the function $u_n - U \in W_0^{1,p}(\Omega)$. By using (4.3.11), we can infer a uniform bound on the L^p norms of $\{u_n\}_{n\in\mathbb{N}}$, as well. Thus, in conclusion, the minimizing sequence $\{u_n\}_{n\in\mathbb{N}}$ is a bounded sequence in $W^{1,p}(\Omega)$.

By Theorem 3.3.6, this implies that there exists $v \in W^{1,p}(\Omega)$ such that, up to a subsequence, we have that $\{u_n\}_{n\in\mathbb{N}}$ weakly converges in $W^{1,p}(\Omega)$ to v. By using the semicontinuity result of Theorem 4.2.1 (recall Remark 4.2.3) and (4.3.10), we get

$$\begin{aligned}\mathfrak{m} &\geq \liminf_{n\to\infty}\left[\int_\Omega H(\nabla u_n)\,dx - \int_\Omega f\,G(u_n)\,dx\right]\\&\geq \int_\Omega H(\nabla v)\,dx - \int_\Omega f\,G(v)\,dx.\end{aligned}$$

In order to conclude that v is the desired minimizer, we are left to show that it is still admissible for the variational problem, i.e. we need to show that

$$v - U \in W_0^{1,p}(\Omega).$$

It is sufficient to observe that from the weak convergence of $\{u_n\}_{n\in\mathbb{N}}$, we get that $\{u_n - U\}_{n\in\mathbb{N}} \subseteq W_0^{1,p}(\Omega)$ weakly converges to $\{v - U\}_{n\in\mathbb{N}}$. Moreover, the space $W_0^{1,p}(\Omega)$ is weakly closed (see Theorem 3.7.4). Thus, $v - U \in W_0^{1,p}(\Omega)$, as desired.

3. *Euler-Lagrange equation.* This part of the proof is exactly the same as in Theorem 4.3.1 and thus we omit it. □

We conclude this section with a statement in the spirit of the Dirichlet principle (recall Theorems 1.5.1 and 1.5.3), which is valid in the setting of the previous existence theorems. This will be useful in the sequel.

Proposition 4.3.4 *Under the assumptions of Theorem 4.3.1, we further suppose that $G(t) = t$, for every $t \in \mathbb{R}$. Then, every weak solution $v \in W_0^{1,p}(\Omega)$ of the boundary value problem*

$$\begin{cases}-\mathrm{div}(\nabla H(\nabla v)) = f, & \text{in } \Omega,\\ v = 0, & \text{on } \partial\Omega,\end{cases}$$

is a solution of

$$\inf_{u\in W_0^{1,p}(\Omega)} \left\{\int_\Omega H(\nabla u)\,dx - \int_\Omega f\,u\,dx\right\}.$$

Proof By assumption, we have

$$\int_\Omega \langle \nabla H(\nabla v), \nabla\varphi\rangle\,dx = \int_\Omega f\,\varphi\,dx, \qquad \text{for every } \varphi \in C_0^\infty(\Omega).$$

By a density argument, we see that we can even admit test functions $\varphi \in W_0^{1,p}(\Omega)$. We observe that, for almost every $x \in \Omega$, the function

$$(u, z) \mapsto H(z) - f(x)\,u,$$

is convex. Thus, from the "above tangent" property (see Proposition 1.3.4), for every $u \in W_0^{1,p}(\Omega)$ and almost every $x \in \Omega$ we get

$$\begin{aligned} H(\nabla u(x)) - f(x)\,u(x) \geq{}& H(\nabla v(x)) - f(x)\,v(x) \\ &+ \langle \nabla H(\nabla v(x)), \nabla u(x) - \nabla v(x)\rangle \\ &- f(x)\,(u(x) - v(x)). \end{aligned}$$

If we now integrate this pointwise inequality over Ω and use the weak formulation of the equation with $\varphi = u - v$, we get

$$\int_\Omega H(\nabla u)\,dx - \int_\Omega f\,u\,dx \geq \int_\Omega H(\nabla v)\,dx - \int_\Omega f\,v\,dx,$$

as desired. □

With exactly the same proof, we can obtain the same result also for general boundary values. We omit the details.

Proposition 4.3.5 *Under the assumptions of Theorem 4.3.3, we further suppose that* $G(t) = t$ *for every* $t \in \mathbb{R}$. *Then, every weak solution* $v \in W^{1,p}(\Omega)$ *of the boundary value problem*

$$\begin{cases} -\mathrm{div}(\nabla H(\nabla v)) = f, & \text{in } \Omega, \\ v = U, & \text{on } \partial\Omega, \end{cases}$$

is a solution of

$$\inf_{u\in W^{1,p}(\Omega)} \left\{\int_\Omega H(\nabla u)\,dx - \int_\Omega f\,u\,dx \;:\; u - U \in W_0^{1,p}(\Omega)\right\}.$$

4.4 *Intermezzo*: Superharmonic Functions

In this section, we briefly investigate the *Minimum Principle* for functions u which satisfies

$$-\Delta_p u \geq 0,$$

in a weak sense. Here $1 < p < \infty$ and Δ_p is the p-Laplacian operator, already encountered in Chap. 1, see Eq. (1.6.4). We recall that for $p = 2$ it reduces to the usual Laplacian operator.

Definition 4.4.1 Let $\Omega \subseteq \mathbb{R}^N$ be an open set and let $u \in W^{1,p}_{\mathrm{loc}}(\Omega)$, for some $1 < p < \infty$. We say that u is *weakly* p*-superharmonic in* Ω if

$$\int_\Omega \langle |\nabla u|^{p-2}\, \nabla u, \nabla \varphi\rangle \, dx \geq 0, \qquad \text{for every } \varphi \in C^\infty_0(\Omega) \text{ with } \varphi \geq 0. \tag{4.4.1}$$

By using that $|\nabla u|^{p-2}\, \nabla u \in L^{p'}(\Omega'; \mathbb{R}^N)$ for every $\Omega' \Subset \Omega$ and Problem 3.12.26, it is easily seen that (4.4.1) continues to hold for every $\varphi \in W^{1,p}_0(\Omega')$ such that $\varphi \geq 0$ and every $\Omega' \Subset \Omega$.

In the particular case $p = 2$, we will simply say that u is *weakly superharmonic.*

Remark 4.4.2 If $u \in C^2(\Omega)$ is superharmonic in classical sense, i.e. if

$$-\Delta u(x) \geq 0, \qquad \text{for every } x \in \Omega,$$

it is not difficult to see that u is weakly superharmonic. On the other hand, a weakly superharmonic function which is C^2, would be superharmonic in classical sense, as well (see Problem 4.10.22 below). The same observation applies to the case $p \neq 2$.

It is a well-known fact that a superharmonic function $u \in C^2(\Omega) \cap C^0(\overline{\Omega})$ admits its minimum on the boundary (see for example [29, Theorem 2.3]). We are going to show that this property remains true, in a suitable sense, for weakly p-superharmonic functions, as well. We recall the notation

$$u_+ = \max\{u, 0\} \qquad \text{and} \qquad u_- = \min\{u, 0\},$$

and refer to Remark 4.4.4 below for a comment on the next result.

Lemma 4.4.3 (Measure-theoretic Minimum Principle) *Let* $1 < p < \infty$ *and let* $\Omega \subseteq \mathbb{R}^N$ *be an open set. Let* $u \in W^{1,p}(\Omega)$ *be a weakly* p*-superharmonic function such that*

$$u_- \in W^{1,p}_0(\Omega).$$

Then we have

$$u(x) \geq 0, \qquad \textit{for a. e. } x \in \Omega.$$

Proof We first observe that, since u belongs to $W^{1,p}(\Omega)$ and not only to $W^{1,p}_{\mathrm{loc}}(\Omega)$, by density (4.4.1) still holds for non-negative functions $\varphi \in W^{1,p}_0(\Omega)$. It is sufficient to use Problem 3.12.26.

By the "above tangent" property of Proposition 1.3.4 for the function $f(z) = |z|^p$ and using (4.4.1) with $\varphi = u_+ - u$, we get

$$\begin{aligned}\int_\Omega |\nabla u_+|^p\,dx &\geq \int_\Omega |\nabla u|^p\,dx + p\int_\Omega \langle |\nabla u|^{p-2}\,\nabla u, \nabla u_+ - \nabla u\rangle\,dx \\ &\geq \int_\Omega |\nabla u|^p\,dx.\end{aligned}$$

Observe that the choice $\varphi = u_+ - u$ is feasible, since $u = u_+ + u_-$ and thus we have $u_+ - u = -u_- \in W^{1,p}_0(\Omega)$ by assumption. Moreover, we clearly have $u_+ - u \geq 0$. By using that (see Proposition 3.4.3)

$$\nabla u_+ = \nabla u \cdot 1_{\{u>0\}}, \qquad \nabla u_- = \nabla u \cdot 1_{\{u<0\}},$$

and recalling that ∇u vanishes almost everywhere on $\{u = 0\}$ (see (3.4.7)), the previous estimate gives

$$\int_\Omega |\nabla u_-|^p\,dx = 0 \qquad \text{that is} \qquad \nabla u_- = (0,\ldots,0) \quad \text{a. e. in } \Omega.$$

However, the assumption $u_- \in W^{1,p}_0(\Omega)$ entails that we must have $u_- = 0$ almost everywhere in Ω, thanks to Proposition 3.7.11. This finally implies that $u = u_+$ almost everywhere and thus the claimed property follows. □

Remark 4.4.4 If we keep in mind that $\varphi \in W^{1,p}_0(\Omega)$ is a weak surrogate for the condition $\varphi = 0$ on $\partial\Omega$ (recall Remark 3.7.2), the assumption $u_- \in W^{1,p}_0(\Omega)$ in the previous result can be thought as a weak formulation of the fact that $u \geq 0$ on $\partial\Omega$. Accordingly, Lemma 4.4.3 guarantees that a weakly p-superharmonic function with this property is non-negative on the whole open set.

The following result enforces the previous one. This can be seen as a generalization of the strong minimum principle for functions which are superharmonic in classical sense (see [29, Theorem 2.2]).

Lemma 4.4.5 (Measure-theoretic Strong Minimum Principle) *Let $1 < p < \infty$ and let $\Omega \subseteq \mathbb{R}^N$ be an open connected set. Let $u \in W^{1,p}_{\mathrm{loc}}(\Omega)$ be a weakly p-superharmonic function in Ω. If $u \geq 0$ almost everywhere in Ω, then:*

- *either u vanishes almost everywhere in Ω;*
- *or we have $u > 0$ almost everywhere in Ω.*

Proof Let us suppose that u does not vanish almost everywhere in Ω. We divide the proof in two parts.

Part 1 Here we prove the following fact: if $B_R(x_0) \Subset \Omega$ is such that

$$|\{x \in B_R(x_0) \,:\, u(x) = 0\}| > 0, \tag{4.4.2}$$

then u must vanish almost everywhere in the whole ball $B_R(x_0)$.

Since $B_R(x_0)$ is compactly contained in Ω, there exists $R' > R$ such that we still have $B_{R'}(x_0) \Subset \Omega$. We choose a function $\eta \in C_0^\infty(B_{R'}(x_0))$ and use in (4.4.1) the test function

$$\varphi = \frac{|\eta|^p}{(\varepsilon + u)^{p-1}},$$

where $\varepsilon > 0$. Observe that this is a feasible test function, since $\varphi \in W_0^{1,p}(B_{R'}(x_0))$ by Problem 3.12.25. We thus have

$$\begin{aligned} 0 \le \int_{B_{R'}(x_0)} &\left\langle |\nabla u|^{p-2}\,\nabla u, \nabla\left(\frac{|\eta|^p}{(\varepsilon+u)^{p-1}}\right)\right\rangle dx \\ &= -(p-1)\int_{B_{R'}(x_0)} \left|\frac{\nabla u}{\varepsilon+u}\right|^p |\eta|^p\,dx \\ &+ p\int_{B_{R'}(x_0)} \left\langle \frac{|\nabla u|^{p-2}\,\nabla u}{(\varepsilon+u)^{p-1}}, \nabla\eta\right\rangle |\eta|^{p-2}\,\eta\,dx \\ &\le -(p-1)\int_{B_{R'}(x_0)} \left|\frac{\nabla u}{\varepsilon+u}\right|^p |\eta|^p\,dx \\ &+ p\int_{B_{R'}(x_0)} \left|\frac{\nabla u}{\varepsilon+u}\right|^{p-1} |\nabla\eta|\,|\eta|^{p-1}\,dx. \end{aligned}$$

By rearranging the terms, this is the same as

$$\int_{B_{R'}(x_0)} \left|\frac{\nabla u}{\varepsilon+u}\right|^p |\eta|^p\,dx \le \frac{p}{p-1}\int_{B_{R'}(x_0)} \left|\frac{\nabla u}{\varepsilon+u}\right|^{p-1} |\nabla\eta|\,|\eta|^{p-1}\,dx.$$

We use the generalized Young inequality (1.2.6) on the right-hand side, so that for every $\delta > 0$

$$\begin{aligned} \int_{B_{R'}(x_0)} \left|\frac{\nabla u}{\varepsilon+u}\right|^p |\eta|^p\,dx &\le \delta\int_{B_{R'}(x_0)} \left|\frac{\nabla u}{\varepsilon+u}\right|^p |\eta|^p\,dx \\ &+ \frac{\delta^{1-p}}{(p-1)}\int_{B_{R'}(x_0)} |\nabla\eta|^p\,dx. \end{aligned}$$

By taking $\delta = (p-1)/p$, we can absorb the term containing ∇u on the right-hand side. This yields

$$\int_{B_{R'}(x_0)} \left|\frac{\nabla u}{\varepsilon + u}\right|^p |\eta|^p\,dx \le \left(\frac{p}{p-1}\right)^p \int_{B_{R'}(x_0)} |\nabla \eta|^p\,dx. \tag{4.4.3}$$

This is valid for every $\eta \in C_0^\infty(B_{R'}(x_0))$. We now choose η with the following properties:

$$0 \le \eta \le 1, \qquad \eta \equiv 1 \text{ on } B_R(x_0) \qquad \text{and} \qquad |\nabla \eta| \le \frac{C}{R'-R}.$$

By inserting this function in (4.4.3), we get in particular

$$\begin{aligned}\int_{B_R(x_0)} \left|\frac{\nabla u}{\varepsilon + u}\right|^p dx &\le \left(\frac{p}{p-1}\right)^p \int_{B_{R'}(x_0)} |\nabla \eta|^p dx\\ &\le \left(\frac{p}{p-1}\right)^p \frac{C^p}{(R'-R)^p}\,|B_{R'}(x_0)|,\end{aligned}$$

for every $\varepsilon > 0$. We observe that if we set

$$f_\varepsilon(t) = \begin{cases} \log\left(1 + \dfrac{t}{\varepsilon}\right), & \text{if } t \ge 0,\\[2ex] \dfrac{t}{\varepsilon}, & \text{if } t < 0,\end{cases}$$

this is a C^1 function with bounded derivative, for every fixed $\varepsilon > 0$. Thus, $f_\varepsilon(u) \in W^{1,p}_{\rm loc}(\Omega)$ thanks to Proposition 3.4.1 and Remark 3.4.2. Moreover, by recalling that $u \ge 0$ almost everywhere, we have

$$\nabla f_\varepsilon(u) = \nabla \log\left(1 + \frac{u}{\varepsilon}\right) = \frac{\nabla u}{\varepsilon + u}.$$

Thus, up to now we obtained

$$\int_{B_R(x_0)} \left|\nabla \log\left(1 + \frac{u}{\varepsilon}\right)\right|^p dx \le \left(\frac{p}{p-1}\right)^p \frac{C^p}{(R'-R)^p}\,|B_{R'}(x_0)|.$$

Observe that

$$\left\{x \in B_R(x_0)\,:\, \log\left(1 + \frac{u}{\varepsilon}\right) = 0\right\} = \{x \in B_R(x_0)\,:\, u(x) = 0\},$$

and this set has positive measure, by assumption (4.4.2). We can then apply the Poincaré inequality of Corollary 3.10.5 to the function $f_\varepsilon(u) \in W^{1,p}(B_R(x_0))$, so to get

$$\int_{B_R(x_0)} \left|\log\left(1+\frac{u}{\varepsilon}\right)\right|^p\,dx \le C',$$

where $C' > 0$ does not depend on $\varepsilon > 0$. By taking the limit at ε goes to 0, such a uniform control implies that we must have $u = 0$ almost everywhere in $B_R(x_0)$.

Part 2 Let us suppose that

$$|\{x \in \Omega\,:\, u(x) = 0\}| > 0,$$

then there exists two balls $B_r(x_0) \Subset \Omega$ and $B_R(y_0) \Subset \Omega$ such that[2]

$$|\{x \in B_r(x_0)\,:\, u(x) = 0\}| > 0 \qquad \text{and} \qquad |\{x \in B_R(y_0)\,:\, u(x) > 0\}| > 0.$$

The second fact holds true since we are assuming that u does not identically vanishes. Since Ω is open and connected, it is arcwise connected, as well (see for example [60, Corollary 27.10]). This implies that there exists a continuous curve $\gamma : [0,1] \to \Omega$ such that

$$\gamma(0) = x_0 \qquad \text{and} \qquad \gamma(1) = y_0.$$

The image of γ is a compact subset of Ω, thus it can be covered by a finite number of open balls $B_1, \dots, B_k \Subset \Omega$: they can be chosen so that

$$B_i \cap B_{i+1} \neq \emptyset, \qquad \text{for } i = 1, \dots, k-1,$$

and

$$B_1 = B_r(x_0), \qquad B_k = B_R(y_0),$$

see Fig. 4.1. We now start from the ball $B_1 = B_r(x_0)$ and iteratively use the result of *Part 1*. We thus get that u vanishes almost everywhere on every ball B_i and finally on $B_k = B_R(y_0)$. This last fact gives the desired contradiction. Thus, if u is not identically zero, we must have $u > 0$ almost everywhere in Ω. □

---

[2] In light of *Part 1*, we can suppose that these two balls are disjoint.

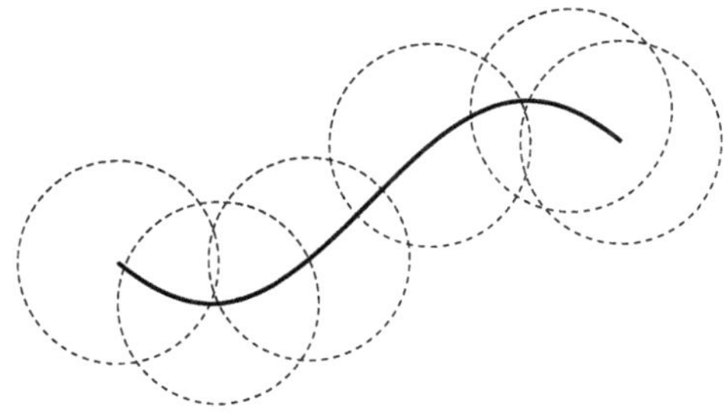

Fig. 4.1 The image of γ and the chain of balls $B_1, \ldots, B_k$ covering it, in the proof of Lemma 4.4.5

4.5 Example: Torsional Rigidity

Besides the example of harmonic functions, already encountered in Sect. 4.1, an interesting application of the existence results of Sect. 4.2 is given by the following

Proposition 4.5.1 (Torsional Rigidity) *Let $\Omega \subseteq \mathbb{R}^N$ be an open bounded connected set. Then there exists a unique weak solution $v \in W_0^{1,2}(\Omega)$ of the problem*

$$\begin{cases} -\Delta u = 1, & in\ \Omega, \\ \quad u = 0, & on\ \partial\Omega. \end{cases} \tag{4.5.1}$$

Moreover, such a solution is strictly positive almost everywhere in Ω.

Proof We divide the proof in three parts.

Existence A weak solution v can be obtained by minimizing

$$\inf_{u \in W_0^{1,2}(\Omega)} \left\{ \frac{1}{2} \int_\Omega |\nabla u|^2 \, dx - \int_\Omega u \, dx \right\}. \tag{4.5.2}$$

For such a functional we can directly apply Theorem 4.3.1 with $p = 2$, $q = 1$ and

$$H(z) = \frac{1}{2} |z|^2, \qquad f \equiv 1, \qquad G(t) = t.$$

Uniqueness Let $w \in W_0^{1,2}(\Omega)$ be another weak solution of (4.5.1), i.e. we have

$$\int_\Omega \langle \nabla w, \nabla \varphi \rangle \, dx = \int_\Omega \varphi \, dx, \qquad \text{for every } \varphi \in C_0^\infty(\Omega). \tag{4.5.3}$$

According to Proposition 4.3.4, w is another minimizer of (4.5.2). In order to prove that $w = v$, it is sufficient to observe that the function

$$F(u, z) = \frac{1}{2} |z|^2 - u, \qquad \text{for } (u, z) \in \mathbb{R} \times \mathbb{R}^N,$$

is convex in the variables (u, z) and strictly convex in the variable z. Since both v and w are minimizers of (4.5.2), we have

$$\begin{aligned}\int_\Omega F(v, \nabla v)\, dx &= \int_\Omega F(w, \nabla w)\, dx \\ &\le \int_\Omega F(u, \nabla u)\, dx, \qquad \text{for every } u \in W_0^{1,2}(\Omega).\end{aligned} \tag{4.5.4}$$

We consider the convex combination

$$U = \frac{v+w}{2} \in W_0^{1,2}(\Omega).$$

By convexity, we get that

$$F(U, \nabla U) \le \frac{1}{2} F(v, \nabla v)\, dx + \frac{1}{2} F(w, \nabla w), \qquad \text{a. e. on } \Omega. \tag{4.5.5}$$

If we integrate over Ω, we thus obtain

$$\int_\Omega F(U, \nabla U)\, dx \le \frac{1}{2} \int_\Omega F(v, \nabla v)\, dx + \frac{1}{2} \int_\Omega F(w, \nabla w)\, dx.$$

By joining this with (4.5.4), we get that it must result

$$\int_\Omega F(U, \nabla U)\, dx = \frac{1}{2} \int_\Omega F(v, \nabla v)\, dx + \frac{1}{2} \int_\Omega F(w, \nabla w)\, dx.$$

The latter implies that we must have equality almost everywhere in (4.5.5). Hence the strict convexity of F in the gradient variable implies that

$$\nabla v = \nabla w, \qquad \text{a. e. on } \Omega,$$

that is $v - w \in W_0^{1,2}(\Omega)$ is such that $\nabla(v - w) = (0, \dots, 0)$ almost everywhere on Ω. Finally, we can apply Proposition 3.7.11 to infer that v and w coincides almost everywhere on Ω. This gives the desired uniqueness result.

Positivity In order to prove that v is strictly positive, it is sufficient to observe that from (4.5.3) we get

$$\int_\Omega \langle \nabla v, \nabla \varphi \rangle\, dx \ge 0, \qquad \text{for every } \varphi \in C_0^\infty(\Omega) \text{ such that } \varphi \ge 0,$$

so that $v \in W_0^{1,2}(\Omega)$ is weakly superharmonic. Moreover, we have $v_- \in W_0^{1,2}(\Omega)$, as well, thanks to Proposition 3.7.13. Thus, by Lemma 4.4.3 we get that $v \ge 0$ almost everywhere in Ω. By Lemma 4.4.5, in order to get that $v > 0$ almost everywhere, it is sufficient to exclude that v identically vanishes. However, if this were the case, from the weak formulation (4.5.3) we would obtain

$$0 = \int_\Omega \varphi\, dx, \qquad \text{for every } \varphi \in C_0^\infty(\Omega),$$

which is not possible. □

Remark 4.5.2 The solution $v \in W_0^{1,2}(\Omega)$ of problem (4.5.1) is called *torsion function of* Ω. Correspondingly, the integral quantity

$$\int_\Omega |\nabla v|^2\, dx,$$

is called *torsional rigidity of* Ω. We observe that by taking as a test function in (4.5.3) the solution v itself, we have that the torsional rigidity can also be written as

$$\int_\Omega |\nabla v|^2\, dx = \int_\Omega v\, dx.$$

For the physical motivations behind this terminology for simply connected sets in dimension $N = 2$, we refer to [43, Chapter II, Section 16]. In this case, the two-dimensional open set Ω represents the cross section of a beam, which undergoes a twisting.

4.6 Example: The Sublinear Lane-Emden Equation

As a further application of the general existence result of Theorem 4.3.1, we can show existence of a solution for the so-called *Lane-Emden equation*, in the sublinear case. This equation finds many applications, for example it is linked to stationary solutions of the *porous medium equation*, see Problem 4.10.26 below.

Proposition 4.6.1 *Let* $1 < q < 2$ *and let* $\Omega \subseteq \mathbb{R}^N$ *be an open bounded connected set. Then there exists a unique nontrivial weak solution* $v \in W_0^{1,2}(\Omega)$ *of the problem*

$$\begin{cases} -\Delta u = |u|^{q-1}, & \text{in } \Omega, \\ \quad u = 0, & \text{on } \partial\Omega. \end{cases}$$

Moreover, such a solution is strictly positive almost everywhere in Ω.

Proof As before, we divide the proof in three parts.

Existence The existence of a solution v follows directly from Theorem 4.3.1, by making the choices

$$H(z) = \frac{1}{2}\,|z|^2, \qquad f \equiv 1, \qquad G(t) = \frac{1}{q}\,|t|^{q-1}\,t.$$

In other words, v is obtained by solving

$$\inf_{u \in W_0^{1,2}(\Omega)} \left\{ \frac{1}{2} \int_\Omega |\nabla u|^2\,dx - \frac{1}{q} \int_\Omega |u|^{q-1}\,u\,dx \right\}. \tag{4.6.1}$$

We just need to prove that this solution v is not trivial. Let us argue by contradiction and assume that $v \equiv 0$: since v has been obtained by solving (4.6.1), this would mean that

$$\frac{1}{2} \int_\Omega |\nabla u|^2\,dx - \frac{1}{q} \int_\Omega |u|^{q-1}\,u\,dx \geq 0, \qquad \text{for every } u \in W_0^{1,2}(\Omega),$$

that is

$$\frac{1}{2} \int_\Omega |\nabla u|^2\,dx \geq \frac{1}{q} \int_\Omega |u|^{q-1}\,u\,dx, \qquad \text{for every } u \in W_0^{1,2}(\Omega).$$

We now take $u \in W_0^{1,2}(\Omega)$ such that

$$\int_\Omega |u|^{q-1}\,u\,dx > 0,$$

and we use the previous inequality for $t\,u$ with $t > 0$ arbitrary. This gives

$$\frac{t^2}{2} \int_\Omega |\nabla u|^2\,dx \geq \frac{t^q}{q} \int_\Omega |u|^{q-1}\,u\,dx, \qquad \text{for every } t > 0.$$

By multiplying both sides by t^{-2} and then taking the limit as t goes to 0, we would get a contradiction from the previous estimate (thanks to the fact that $q < 2$). Thus, v is nontrivial.

Positivity For this, we first need to prove that every weak solution of our problem is non-negative on Ω. Let $v \in W_0^{1,2}(\Omega)$ be a weak solution, i.e.

$$\int_\Omega \langle \nabla v, \nabla \varphi \rangle\,dx = \int_\Omega |v|^{q-1}\,\varphi\,dx, \qquad \text{for every } \varphi \in C_0^\infty(\Omega).$$

By taking $\varphi \geq 0$, we easily get from this formulation that v is weakly superharmonic. By observing that $v_- \in W_0^{1,2}(\Omega)$ thanks to Proposition 3.7.13, we get $v \geq 0$ by Lemma 4.4.3. Moreover, if v is nontrivial, then we can apply Lemma 4.4.5 and infer that it must result $v > 0$ almost everywhere on Ω.

Uniqueness The uniqueness is slightly more complicated than in Proposition 4.5.1. Indeed, observe that now the function

$$F(u, z) = \frac{1}{2} |z|^2 - \frac{1}{q} |u|^{q-1} u, \qquad \text{for } (u, z) \in \mathbb{R} \times \mathbb{R}^N,$$

is *not* convex. In order to prove uniqueness of the solution, we use a trick based on Picone's inequality (1.3.2): this is a technical refinement of an idea by Brezis and Oswald (see [68]). Let us suppose that $v_1, v_2 \in W_0^{1,2}(\Omega)$ are two nontrivial weak solutions, by the previous part of the proof we know that $v_1 > 0$ and $v_2 > 0$. We have

$$\int_\Omega \langle \nabla v_1, \nabla \varphi \rangle \, dx = \int_\Omega v_1^{q-1} \, \varphi \, dx, \tag{4.6.2}$$

and

$$\int_\Omega \langle \nabla v_2, \nabla \varphi \rangle \, dx = \int_\Omega v_2^{q-1} \, \varphi \, dx, \tag{4.6.3}$$

for every $\varphi \in C_0^\infty(\Omega)$. As usual, by a density argument we can admit test functions $\varphi \in W_0^{1,2}(\Omega)$. By definition of $W_0^{1,2}(\Omega)$, there exists two sequences $\{v_{1,n}\}_{n\in\mathbb{N}}, \{v_{2,n}\}_{n\in\mathbb{N}} \subseteq C_0^\infty(\Omega)$ such that

$$\lim_{n\to\infty} \|v_{i,n} - v_i\|_{W^{1,2}(\Omega)} = 0, \qquad i = 1, 2.$$

For every $\varepsilon > 0$ and $n \in \mathbb{N}$, we take the test function

$$\varphi = \frac{(v_{2,n})^2}{v_1 + \varepsilon} - v_1,$$

in (4.6.2) and the test function

$$\varphi = \frac{(v_{1,n})^2}{v_2 + \varepsilon} - v_2,$$

in (4.6.3). Thanks to Problem 3.12.25, both functions belong to $W_0^{1,2}(\Omega)$.

By summing up the resulting identities, we get

$$\begin{aligned}\int_\Omega v_1^{q-1}\frac{(v_{2,n})^2}{v_1+\varepsilon}\,dx - \int_\Omega v_1^q\,dx + \int_\Omega v_2^{q-1}\frac{(v_{1,n})^2}{v_2+\varepsilon}\,dx - \int_\Omega v_2^q\,dx \\ = \int_\Omega \left\langle \nabla v_1, \nabla\frac{(v_{2,n})^2}{v_1+\varepsilon}\right\rangle dx - \int_\Omega |\nabla v_1|^2\,dx \\ + \int_\Omega \left\langle \nabla v_2, \nabla\frac{(v_{1,n})^2}{v_2+\varepsilon}\right\rangle dx - \int_\Omega |\nabla v_2|^2\,dx.\end{aligned}$$

We now use Picone's inequality for Sobolev functions (see Problem 3.12.25) in the right-hand side of this identity, so to obtain

$$\begin{aligned}\int_\Omega v_1^{q-1}\frac{(v_{2,n})^2}{v_1+\varepsilon}\,dx - \int_\Omega v_1^q\,dx + \int_\Omega v_2^{q-1}\frac{(v_{1,n})^2}{v_2+\varepsilon}\,dx - \int_\Omega v_2^q\,dx \\ \le \int_\Omega |\nabla v_{2,n}|^2\,dx - \int_\Omega |\nabla v_1|^2\,dx \\ + \int_\Omega |\nabla v_{1,n}|^2\,dx - \int_\Omega |\nabla v_2|^2\,dx.\end{aligned} \tag{4.6.4}$$

We claim that

$$\lim_{n\to\infty}\int_\Omega v_1^{q-1}\frac{(v_{2,n})^2}{v_1+\varepsilon}\,dx = \int_\Omega v_1^{q-1}\frac{v_2^2}{v_1+\varepsilon}\,dx,$$

and

$$\lim_{n\to\infty}\int_\Omega v_2^{q-1}\frac{(v_{1,n})^2}{v_2+\varepsilon}\,dx = \int_\Omega v_2^{q-1}\frac{v_1^2}{v_2+\varepsilon}\,dx.$$

Indeed, by observing that[3] $v_1^{q-1}/(\varepsilon+v_1) \le \varepsilon^{q-2}$, we get

---

[3] It is sufficient to use that for every $a, \varepsilon > 0$ we have

$$\frac{a^{q-1}}{a+\varepsilon} < \frac{(a+\varepsilon)^{q-1}}{a+\varepsilon} = (a+\varepsilon)^{q-2} < \varepsilon^{q-2},$$

thanks to the fact that $1 < q < 2$.

$$\begin{aligned}
&\left| \int_\Omega v_1^{q-1} \frac{(v_{2,n})^2}{v_1+\varepsilon}\,dx - \int_\Omega v_1^{q-1} \frac{v_2^2}{v_1+\varepsilon}\,dx \right| \\
&\le \varepsilon^{q-2} \int_\Omega \left| (v_{2,n})^2 - v_2^2 \right| dx \\
&= \varepsilon^{q-2} \int_\Omega \left| (v_{2,n}+v_2)\,(v_{2,n}-v_2) \right| dx \\
&\le \varepsilon^{q-2}\, \|v_{2,n}+v_2\|_{L^2(\Omega)}\, \|v_{2,n}-v\|_{L^2(\Omega)}.
\end{aligned}$$

By recalling the properties of $\{v_{2,n}\}_{n\in\mathbb{N}}$, we get that the last term converges to 0, as n goes to ∞. With the same argument, we also get the limit for the integral containing $v_{1,n}$.

Thus, by taking the limit as n goes to ∞ in (4.6.4), we get

$$\int_\Omega v_1^{q-1} \frac{v_2^2}{v_1+\varepsilon}\,dx - \int_\Omega v_1^q\,dx + \int_\Omega v_2^{q-1} \frac{v_1^2}{v_2+\varepsilon}\,dx - \int_\Omega v_2^q\,dx \le 0.$$

By using Fatou's Lemma, we can take the limit as ε goes to 0, so to obtain

$$\int_\Omega v_1^{q-2}\,v_2^2\,dx - \int_\Omega v_1^q\,dx + \int_\Omega v_2^{q-2}\,v_1^2\,dx - \int_\Omega v_2^q\,dx \le 0,$$

thanks to the fact that $v_1, v_2 > 0$ almost everywhere in Ω. The previous inequality can be rewritten as

$$\int_\Omega \left(v_1^{q-2} - v_2^{q-2}\right)(v_2^2 - v_1^2)\,dx \le 0. \tag{4.6.5}$$

We now observe that

$$(a^{q-2} - b^{q-2})\,(b^2 - a^2) > 0, \qquad \text{for every } a, b > 0 \text{ with } a \ne b,$$

thanks to the fact that $q-2<0$. By using this observation in (4.6.5), we thus must have

$$v_1 = v_2, \qquad \text{a. e. in } \Omega.$$

This gives the desired uniqueness property. □

4.7 Example: The First Eigenvalue of the Dirichlet-Laplacian

We give at first the definition of eigenvalue of the Laplacian, with homogeneous Dirichlet boundary conditions.

Definition 4.7.1 Let $\Omega \subseteq \mathbb{R}^N$ be an open set. We say that a real number $\lambda \in \mathbb{R}$ is an *eigenvalue of the Dirichlet-Laplacian on* Ω if there exists a *nontrivial* weak solution $v \in W_0^{1,2}(\Omega)$ to the boundary value problem

$$\begin{cases} -\Delta v = \lambda\, v, & \text{in } \Omega, \\ \quad v = 0, & \text{on } \partial\Omega. \end{cases}$$

Such a solution is called *eigenfunction of the Dirichlet-Laplacian on* Ω, associated to λ. In other words, by definition an eigenfunction v satisfies

$$\int_\Omega \langle \nabla v, \nabla \varphi \rangle \, dx = \lambda \int_\Omega v\, \varphi \, dx, \qquad \text{for every } \varphi \in C_0^\infty(\Omega). \tag{4.7.1}$$

By density of $C_0^\infty(\Omega)$ in $W_0^{1,2}(\Omega)$, in (4.7.1) we can enlarge the class of test functions to the whole $W_0^{1,2}(\Omega)$.

We will also set

$$\mathrm{Eigen}(\Omega) = \Big\{ \lambda \in \mathbb{R} \, : \, \lambda \text{ is an eigenvalue of the Dirichlet-Laplacian on } \Omega \Big\},$$

i.e. the whole collection of eigenvalues of the Dirichlet-Laplacian on Ω. In the next section, we will see that if Ω is bounded the eigenvalues and eigenfunctions have a structure similar to that of the case of a symmetric matrix (see Theorem 4.8.1 below). For more general open sets, understanding the spectral properties of their Dirichlet-Laplacian is quite a difficult task: this is the scope of the so-called *Spectral Theory* of differential operators. The interested reader is referred for example to [18, 23, 37] or [50, Chapter 18], for more details.

We start by making the following observation: if $\lambda \in \mathrm{Eigen}(\Omega)$ is an eigenvalue with eigenfunction v, by inserting $\varphi = v$ as a test function in (4.7.1), we get

$$\int_\Omega |\nabla v|^2 \, dx = \lambda \int_\Omega |v|^2 \, dx.$$

This shows at first that we always have $\lambda > 0$. Moreover, if we set

$$\lambda_1(\Omega) := \inf_{u \in W_0^{1,2}(\Omega) \setminus \{0\}} \frac{\displaystyle\int_\Omega |\nabla u|^2 \, dx}{\displaystyle\int_\Omega |u|^2 \, dx}, \tag{4.7.2}$$

we clearly have that

$$\lambda = \frac{\displaystyle\int_\Omega |\nabla v|^2\,dx}{\displaystyle\int_\Omega |v|^2\,dx} \geq \lambda_1(\Omega). \tag{4.7.3}$$

This shows that

$$\mathrm{Eigen}(\Omega) \subseteq [\lambda_1(\Omega), +\infty).$$

Remark 4.7.2 Observe that, by its very definition, the value $\lambda_1(\Omega)$ coincides with the sharp constant for the Poincaré inequality

$$c \int_\Omega |u|^2\,dx \leq \int_\Omega |\nabla u|^2\,dx, \qquad \text{for every } u \in W_0^{1,2}(\Omega),$$

that we have already encountered in Chaps. 2 and 3. In particular, from Theorem 3.7.6, we get the following geometric lower bound

$$\left(\frac{1}{\mathrm{width}(\Omega;\omega)}\right)^2 \leq \lambda_1(\Omega),$$

if Ω is bounded in the direction $\omega \in \mathbb{S}^{N-1}$ (we refer to Definition 3.5.1). We also recall that, in the case $N = 1$ and when $\Omega = (a, b)$, we have already determined such a sharp constant in Theorem 2.4.6 and Corollary 2.4.7, where we discovered that this is given by

$$\frac{\pi^2}{(b-a)^2}.$$

We are interested in this section in studying the minimization problem (4.7.2). We start with the main result of this section.

Theorem 4.7.3 *Let $\Omega \subseteq \mathbb{R}^N$ be an open bounded set, then the variational problem* (4.7.2) *does admit a solution. Moreover, the following facts hold:*

(1) $\lambda_1(\Omega) \in \mathrm{Eigen}(\Omega)$ *and thus, in light of* (4.7.3)*, it is the smallest possible eigenvalue. Correspondingly, every minimizer $v \in W_0^{1,2}(\Omega) \setminus \{0\}$ is an eigenfunction associated to $\lambda_1(\Omega)$;*
(2) if there exists an eigenvalue $\lambda \in \mathrm{Eigen}(\Omega)$ with eigenfunction $v \in W_0^{1,2}(\Omega)\setminus\{0\}$ such that $v > 0$ almost everywhere in Ω, then we must have $\lambda = \lambda_1(\Omega)$ and v is a solution of (4.7.2).

Proof We divide the proof in various parts, for ease of readability.

Existence The minimization problem (4.7.2) is not of the type considered in this chapter. However, we can easily adapt the Direct Method to this case, as well. We first observe that problem (4.7.2) is equivalent to the following one

$$\inf_{u\in W_0^{1,2}(\Omega)} \left\{ \int_\Omega |\nabla u|^2\, dx \,:\, \int_\Omega |u|^2\, dx = 1 \right\}. \tag{4.7.4}$$

This is similar to the problems previously encountered, but there is an important difference here: this is a *constrained* problem. In other words, our admissible functions must satisfy a constraint on the L^2 norm.

We can still prove existence of a minimizer by using the Direct Method: we take a minimizing sequence $\{u_n\}_{n\in\mathbb{N}} \subseteq W_0^{1,2}(\Omega)$ for problem (4.7.4), i.e.

$$\lim_{n\to\infty} \int_\Omega |\nabla u_n|^2\, dx = \lambda_1(\Omega) \qquad \text{and} \qquad \int_\Omega |u_n|^2\, dx = 1, \quad \text{for every } n \in \mathbb{N}.$$

These properties implies in particular that such a sequence is bounded in $W_0^{1,2}(\Omega)$. By Theorem 3.3.6 and Lemma 3.8.7, we know that (up to a subsequence) $\{u_n\}_{n\in\mathbb{N}}$ converges weakly in $W^{1,2}(\Omega)$ and strongly in $L^2(\Omega)$ to a function $v \in W_0^{1,2}(\Omega)$. We recall that v belongs to $W_0^{1,2}(\Omega)$ thanks to the fact that the latter is weakly closed (see Theorem 3.7.4).

By the strong L^2 convergence, we get that

$$1 = \lim_{n\to\infty} \int_\Omega |u_n|^2\, dx = \int_\Omega |v|^2\, dx,$$

thus v is still admissible in (4.7.4). Moreover, by the lower semicontinuity of the L^2 norm with respect to the weak convergence (recall Proposition 1.3.11), we have

$$\int_\Omega |\nabla v|^2\, dx \le \lim_{n\to\infty} \int_\Omega |\nabla u_n|^2\, dx = \lambda_1(\Omega).$$

This finally gives that v is a solution of (4.7.4) and thus of (4.7.2), as well.

Proof of Point (1) We have to prove that a minimizer of (4.7.2) $v \in W_0^{1,2}(\Omega) \setminus \{0\}$ verifies the optimality condition

$$\int_\Omega \langle \nabla v, \nabla \varphi \rangle\, dx = \lambda_1(\Omega) \int_\Omega v\, \varphi\, dx, \qquad \text{for every } \varphi \in W_0^{1,2}(\Omega). \tag{4.7.5}$$

This would give that $\lambda_1(\Omega)$ is an eigenvalue and thus, in light of (4.7.3), it would be the first one.

There are various methods for proving (4.7.5), for example one could use a suitable infinite-dimensional version of the *Lagrange's Multipliers Rule*, applied to the constrained problem (4.7.4) (see for example [61, Chapter 4]). We prefer here to use an elementary argument, which is the same we already used in Remark 2.4.3. By definition of $\lambda_1(\Omega)$, we have

$$\frac{\displaystyle\int_\Omega |\nabla u|^2\,dx}{\displaystyle\int_\Omega |u|^2\,dx} \geq \lambda_1(\Omega), \qquad \text{for every } u \in W_0^{1,2}(\Omega)\setminus\{0\},$$

which is the same as

$$\mathcal{F}(u) := \int_\Omega |\nabla u|^2\,dx - \lambda_1(\Omega)\int_\Omega |u|^2\,dx \geq 0, \qquad \text{for every } u \in W_0^{1,2}(\Omega).$$

Moreover, for the minimizer v we have

$$\frac{\displaystyle\int_\Omega |\nabla v|^2\,dx}{\displaystyle\int_\Omega |v|^2\,dx} = \lambda_1(\Omega),$$

and thus

$$\mathcal{F}(u) \geq 0 \quad \text{for every } u \in W_0^{1,2}(\Omega) \qquad \text{and} \qquad \mathcal{F}(v) = 0.$$

This argument shows that v is a minimizer of the functional $\mathcal{F}$. It is now sufficient to compute the first variation of this functional, in order to conclude. Indeed, by the minimality of v, for every $t \in \mathbb{R}$ and $\varphi \in W_0^{1,2}(\Omega)$, we have

$$\mathcal{F}(v + t\,\varphi) \geq \mathcal{F}(v).$$

Thus, if we set $g(t) = \mathcal{F}(v + t\,\varphi)$, we must have $g'(0) = 0$. We now compute this derivative: thanks to the fact that $\mathcal{F}$ only contains L^2 norms squared, the computations are very simple and does not rely on the use of the Dominated Convergence Theorem (compare with the proof of Theorem 4.3.1). Indeed, we have for every $t \in \mathbb{R}$

$$\begin{aligned} g(t) = \mathcal{F}(v + t\,\varphi) &= \int_\Omega |\nabla(v + t\,\varphi)|^2\,dx - \lambda_1(\Omega)\int_\Omega |v + t\,\varphi|^2\,dx \\ &= \int_\Omega |\nabla v|^2\,dx - \lambda_1(\Omega)\int_\Omega |v|^2\,dx \\ &+ 2\,t\int_\Omega \langle \nabla v, \nabla\varphi\rangle\,dx - 2\,\lambda_1(\Omega)\,t\int_\Omega v\,\varphi\,dx \\ &+ t^2\int_\Omega |\nabla\varphi|^2\,dx - \lambda_1(\Omega)\,t^2\int_\Omega |\varphi|^2\,dx. \end{aligned}$$

Thus, we immediately get

$$0 = g'(0) = 2 \int_\Omega \langle \nabla v, \nabla \varphi \rangle \, dx - 2\,\lambda_1(\Omega) \int_\Omega v\, \varphi \, dx.$$

By arbitrariness of $\varphi \in W_0^{1,2}(\Omega)$, this gives that v satisfies (4.7.5).

Proof of Point (2) We now suppose that $v \in W_0^{1,2}(\Omega)$ is such that $v > 0$ almost everywhere and satisfies (4.7.1), for some real number λ. We take $\eta \in C_0^\infty(\Omega)$ and $\varepsilon > 0$ and insert in (4.7.1) the test function

$$\varphi = \frac{\eta^2}{v + \varepsilon} \in W_0^{1,2}(\Omega).$$

We observe that this function belongs to $W_0^{1,2}(\Omega)$, thanks to Problem 3.12.25. We thus obtain

$$\lambda \int_\Omega v \, \frac{\eta^2}{v+\varepsilon} \, dx = \int_\Omega \left\langle \nabla v, \nabla \left(\frac{\eta^2}{v+\varepsilon} \right) \right\rangle dx.$$

We can use Picone's inequality for Sobolev functions (i.e. Problem 3.12.25) on the right-hand side, so to get

$$\lambda \int_\Omega v \, \frac{\eta^2}{v+\varepsilon} \, dx \le \int_\Omega |\nabla \eta|^2 \, dx, \qquad \text{for every } \eta \in C_0^\infty(\Omega).$$

By taking the limit as ε goes to 0 on the left-hand side, using the Monotone Convergence Theorem (or Fatou's Lemma) and the fact that $v > 0$ almost everywhere in[4] Ω, we get

$$\lambda \int_\Omega \eta^2 \, dx \le \int_\Omega |\nabla \eta|^2 \, dx, \qquad \text{for every } \eta \in C_0^\infty(\Omega).$$

This entails

$$\lambda \le \inf_{\eta \in C_0^\infty(\Omega)\setminus\{0\}} \frac{\displaystyle\int_\Omega |\nabla \eta|^2 \, dx}{\displaystyle\int_\Omega |\eta|^2 \, dx}.$$

[4] We stress that this assumption is crucial to get that

$$\lim_{\varepsilon \to 0^+} \frac{v(x)}{v(x)+\varepsilon} = 1, \qquad \text{for a. e. } x \in \Omega.$$

The weaker assumption "$v \ge 0$ almost everywhere in Ω" would not be sufficient.

By density of $C_0^\infty(\Omega)$ in $W_0^{1,2}(\Omega)$, we have

$$\inf_{\eta\in C_0^\infty(\Omega)\setminus\{0\}} \frac{\displaystyle\int_\Omega |\nabla \eta|^2\,dx}{\displaystyle\int_\Omega |\eta|^2\,dx} = \inf_{\eta\in W_0^{1,2}(\Omega)\setminus\{0\}} \frac{\displaystyle\int_\Omega |\nabla \eta|^2\,dx}{\displaystyle\int_\Omega |\eta|^2\,dx} = \lambda_1(\Omega),$$

and thus we obtain $\lambda \le \lambda_1(\Omega)$. The reverse inequality follows from (4.7.3). This gives that $\lambda = \lambda_1(\Omega)$ and thus v is a minimizer. □

Remark 4.7.4 (First Eigenvalue of a Rectangle) The previous result can be used to compute the first eigenvalue in some particular cases. Let $\Omega = (0,\ell)\times(0,L)$ be the open rectangle with sides length ℓ and L. We can show that

$$\lambda_1(\Omega) = \frac{\pi^2}{\ell^2} + \frac{\pi^2}{L^2}. \tag{4.7.6}$$

Indeed, if we consider the function

$$v(x,y) = \sin\left(\frac{\pi}{\ell}\,x\right)\,\sin\left(\frac{\pi}{L}\,y\right), \qquad \text{for } (x,y)\in(0,\ell)\times(0,L),$$

we see that

$$v \in \left\{u \in C^0(\overline{\Omega})\cap C^1(\Omega)\ :\ u = 0 \text{ on } \partial\Omega,\ \nabla u \in L^\infty(\Omega;\mathbb{R}^2)\right\}.$$

Thus, $v \in W_0^{1,2}(\Omega)$ thanks to Problem 3.12.15. Moreover, a direct computation shows that v is a classical (and thus weak) solution of

$$-\Delta v = \left(\frac{\pi^2}{\ell^2} + \frac{\pi^2}{L^2}\right)\,v, \qquad \text{in } \Omega.$$

By further noticing that $v > 0$ on Ω, we get by point (2) of Theorem 4.7.3 that $\lambda_1(\Omega)$ is given by formula (4.7.6).

The next result gives some additional informations on first eigenfunctions, in the case of *connected* sets.

Proposition 4.7.5 *Let $\Omega \subseteq \mathbb{R}^N$ be an open bounded connected set. Then every eigenfunction $v \in W_0^{1,2}(\Omega)\setminus\{0\}$ associated to $\lambda_1(\Omega)$ has the following properties:*

- *it must have constant sign in Ω;*
- *it satisfies $v \ne 0$ almost everywhere in Ω.*

Proof We argue by contradiction and suppose that v changes sign. This implies that

$$v_+ = \max\{v, 0\} \qquad \text{and} \qquad v_- = \min\{v, 0\},$$

are both nontrivial. We recall that $v_+, v_- \in W_0^{1,2}(\Omega)$ by Proposition 3.7.13, then we can use them as test functions in (4.7.1). On account of Proposition 3.4.3, this gives

$$\int_\Omega |\nabla v_+|^2 \, dx = \lambda_1(\Omega) \int_\Omega |v_+|^2 \, dx,$$

and

$$\int_\Omega |\nabla v_-|^2 \, dx = \lambda_1(\Omega) \int_\Omega |v_-|^2 \, dx,$$

that is

$$\lambda_1(\Omega) = \frac{\displaystyle\int_\Omega |\nabla v_+|^2 \, dx}{\displaystyle\int_\Omega |v_+|^2 \, dx} = \frac{\displaystyle\int_\Omega |\nabla v_-|^2 \, dx}{\displaystyle\int_\Omega |v_-|^2 \, dx}.$$

This shows that both v_+ and v_- are still minimizers of (4.7.2). Thus, by Theorem 4.7.3 we have that they both solve (4.7.1) with $\lambda = \lambda_1(\Omega)$. In particular, by using that $v_+ \geq 0$ and testing with a non-negative function $\varphi \in C_0^\infty(\Omega)$, we get

$$\int_\Omega \langle \nabla v_+, \nabla \varphi \rangle \, dx \geq 0.$$

This shows that v_+ is weakly superharmonic and thus, by Lemma 4.4.5, we must have $v_+ > 0$ almost everywhere on Ω. This in turn implies that v_- must vanish almost everywhere, thus giving a contradiction. In conclusion, we get that either $v = v_+$ or $v = v_-$, so that v has constant sign. Moreover, it must result $v \neq 0$ almost everywhere, again thanks to the strong minimum principle of Lemma 4.4.5. □

4.8 Example: The Spectrum of the Dirichlet-Laplacian

By using the Direct Method and the compactness of the embedding

$$W_0^{1,2}(\Omega) \hookrightarrow L^2(\Omega),$$

for an open bounded set $\Omega \subseteq \mathbb{R}^N$, it is possible to give a variational characterization of the spectrum of the Dirichlet-Laplacian on Ω.

Theorem 4.8.1 (Spectral Theorem for the Dirichlet-Laplacian) *Let $\Omega \subseteq \mathbb{R}^N$ be an open bounded set. For every $k \in \mathbb{N} \setminus \{0\}$, we inductively define the number $\lambda_k(\Omega)$ as follows:*

- `step 1` *for $k = 1$, we set*

$$\lambda_1(\Omega) = \inf_{u \in W_0^{1,2}(\Omega)} \left\{ \int_\Omega |\nabla u|^2 \, dx \, : \, \int_\Omega |u|^2 \, dx = 1 \right\};$$

- `step k + 1` *for $k \geq 1$, we set*

$$\lambda_{k+1}(\Omega) = \inf_{u \in W_0^{1,2}(\Omega)} \left\{ \int_\Omega |\nabla u|^2 \, dx \, : \, \begin{array}{c} \int_\Omega |u|^2 \, dx = 1 \text{ and} \\ \int_\Omega u \, v_j \, dx = 0 \text{ for } j \in \{1, \dots, k\} \end{array} \right\},$$

where each v_j is a minimizer for $\lambda_j(\Omega)$, with $j \in \{1, \dots, k\}$. Then we have:

(1) each $\lambda_k(\Omega)$ is a minimum and

$$\lambda_k(\Omega) \leq \lambda_{k+1}(\Omega), \qquad \text{for every } k \in \mathbb{N} \setminus \{0\}.$$

Moreover, each $\lambda_k(\Omega)$ is an eigenvalue of the Dirichlet-Laplacian on Ω and a minimizer v_k is an associated eigenfunction;

(2) the sequence $\{v_k\}_{k \in \mathbb{N} \setminus \{0\}}$ is such that

$$\int_\Omega |v_n|^2 \, dx = 1, \quad \int_\Omega v_n \, v_m \, dx = 0 = \int_\Omega \langle \nabla v_n, \nabla v_m \rangle \, dx,$$

for every $n \neq m$;

(3) the sequence $\{\lambda_k(\Omega)\}_{k \in \mathbb{N} \setminus \{0\}}$ diverges to $+\infty$, as k goes to ∞;

(4) for every $u \in W_0^{1,2}(\Omega)$, we have

$$u = \sum_{k=1}^{\infty} \widehat{u}(k) \, v_k \qquad \text{and} \qquad \nabla u = \sum_{k=1}^{\infty} \widehat{u}(k) \, \nabla v_k,$$

where the convergence of the series has to be understood in the L^2 sense and the coefficients $\{\widehat{u}(k)\}_{k \in \mathbb{N} \setminus \{0\}}$ are given by

$$\widehat{u}(k) = \int_\Omega u \, v_k \, dx, \qquad \text{for every } k \in \mathbb{N} \setminus \{0\}.$$

Moreover, it holds

$$\int_\Omega |u|^2\,dx = \sum_{k=1}^\infty |\widehat{u}(k)|^2 \quad \textit{and} \quad \int_\Omega |\nabla u|^2\,dx = \sum_{k=1}^\infty \lambda_k(\Omega)\,|\widehat{u}(k)|^2; \tag{4.8.1}$$

(5) for every $u, \psi \in W_0^{1,2}(\Omega)$, we have

$$\int_\Omega u\,\psi\,dx = \sum_{k=1}^\infty \widehat{u}(k)\,\widehat{\psi}(k) \quad \textit{and} \quad \int_\Omega \langle \nabla u, \nabla\psi\rangle\,dx = \sum_{k=1}^\infty \lambda_k(\Omega)\,\widehat{u}(k)\,\widehat{\psi}(k);$$

(6) finally, we have

$$\mathrm{Eigen}(\Omega) = \Big\{\lambda_k(\Omega)\Big\}_{k\in\mathbb{N}\setminus\{0\}}.$$

Proof We prove each point separately.

(1) We proceed by induction: for $k = 1$, the statement is true by Theorem 4.7.3. We now assume that the statement is true for k and prove it for $k + 1$: in other words, we suppose that each $\lambda_1(\Omega), \dots, \lambda_k(\Omega)$ is well-defined, it is attained at some function $v_1, \dots, v_k$ and it is an eigenvalue with eigenfunction $v_1, \dots, v_k$.
We consider the minimization problem defining $\lambda_{k+1}(\Omega)$ and take a minimizing sequence $\{u_n\}_{n\in\mathbb{N}}$, as in the proof of Theorem 4.7.3, i.e.

$$\lim_{n\to\infty} \int_\Omega |\nabla u_n|^2\,dx = \lambda_{k+1}(\Omega), \qquad \int_\Omega |u_n|^2\,dx = 1, \quad \text{for every } n \in \mathbb{N},$$

and

$$\int_\Omega u_n\,v_j\,dx = 0, \qquad \text{for } j \in \{1, \dots, k\} \text{ and every } n \in \mathbb{N}.$$

Exactly as in the proof of Theorem 4.7.3, we can assure that $\{u_n\}_{n\in\mathbb{N}}$ converges weakly in $W^{1,2}(\Omega)$ and strongly in $L^2(\Omega)$, to a limit function $v \in W_0^{1,2}(\Omega)$, up to a subsequence.
By the strong L^2 convergence, we get that

$$1 = \lim_{n\to\infty} \int_\Omega |u_n|^2\,dx = \int_\Omega |v|^2\,dx,$$

and

$$\int_\Omega v\,v_j\,dx = \lim_{n\to\infty} \int_\Omega u_n\,v_j\,dx = 0, \qquad \text{for } j \in \{1, \dots, k\},$$

thus v is still admissible in the minimization problem which defines $\lambda_{k+1}(\Omega)$. Moreover, by the lower semicontinuity of the L^2 norm with respect to weak convergence, we have

$$\int_\Omega |\nabla v|^2\,dx \le \lim_{n\to\infty} \int_\Omega |\nabla u_n|^2\,dx = \lambda_{k+1}(\Omega).$$

This finally gives that $\lambda_{k+1}(\Omega)$ is a minimum, with minimizer v.
We still need to show that v is an eigenfunction, with eigenvalue $\lambda_{k+1}(\Omega)$. To this aim, we use the same homogeneity trick as in the proof of Theorem 4.7.3. By definition of $\lambda_{k+1}(\Omega)$, we have

$$\frac{\displaystyle\int_\Omega |\nabla u|^2\,dx}{\displaystyle\int_\Omega |u|^2\,dx} \ge \lambda_{k+1}(\Omega), \qquad \begin{array}{l}\text{for every } u \in W^{1,2}_0(\Omega)\\ \text{such that } \displaystyle\int_\Omega u\,v_j\,dx = 0 \text{ for } j \in \{1,\dots,k\}.\end{array}$$

If we define

$$\mathcal{F}(u) = \int_\Omega |\nabla u|^2\,dx - \lambda_{k+1}(\Omega) \int_\Omega |u|^2\,dx,$$

then we can infer that

$$\mathcal{F}(u) \ge 0, \qquad \begin{array}{l}\text{for every } u \in W^{1,2}_0(\Omega)\\ \text{such that } \displaystyle\int_\Omega u\,v_j\,dx = 0 \text{ for } j \in \{1,\dots,k\},\end{array}$$

and $\mathcal{F}(v) = 0$. Thus, v minimizes $\mathcal{F}$, as well, among $W^{1,2}_0(\Omega)$ functions such that

$$\int_\Omega u\,v_j\,dx = 0, \qquad \text{for } j \in \{1,\dots,k\}. \tag{4.8.2}$$

For every $\varphi \in W^{1,2}_0(\Omega)$ which satisfies (4.8.2), we set

$$g(t) := \mathcal{F}(v + t\,\varphi), \qquad \text{for } t \in \mathbb{R}.$$

Thus, by minimality we get

$$0 = g'(0) = 2 \int_\Omega \langle \nabla v, \nabla\varphi\rangle\,dx - 2\,\lambda_{k+1}(\Omega) \int_\Omega v\,\varphi\,dx. \tag{4.8.3}$$

This is still not enough to conclude that $\lambda_{k+1}(\Omega)$ is an eigenvalue, since *we are restricting test functions to those which satisfy the additional condition* (4.8.2). In order to remove this restriction, we now take an arbitrary $\varphi \in W^{1,2}_0(\Omega)$ and define

$$\widetilde{\varphi} = \varphi - \sum_{j=1}^{k} \left(\int_\Omega \varphi\, v_j\, dx \right) v_j.$$

It is not difficult to see that $\widetilde{\varphi} \in W^{1,2}_0(\Omega)$ satisfies (4.8.2), thus we can infer from (4.8.3)

$$\int_\Omega \langle \nabla v, \nabla \widetilde{\varphi} \rangle\, dx = \lambda_{k+1}(\Omega) \int_\Omega v\, \widetilde{\varphi}\, dx.$$

We observe that

$$\int_\Omega v\, \widetilde{\varphi}\, dx = \int_\Omega v \left(\varphi - \sum_{j=1}^{k} \left(\int_\Omega \varphi\, v_j\, dy \right) v_j \right) dx = \int_\Omega v\, \varphi\, dx,$$

since by construction v satisfies (4.8.2). Moreover, we have

$$\begin{aligned} \int_\Omega \langle \nabla v, \nabla \widetilde{\varphi} \rangle\, dx &= \int_\Omega \langle \nabla v, \nabla \varphi \rangle\, dx \\ &\quad - \sum_{j=1}^{k} \left(\int_\Omega \varphi\, v_j\, dy \right) \int_\Omega \langle \nabla v, \nabla v_j \rangle\, dx \\ &= \int_\Omega \langle \nabla v, \nabla \varphi \rangle\, dx. \end{aligned}$$

Observe that in the last identity we used the equations for $v_1, \ldots, v_k$ tested with v: these give

$$\int_\Omega \langle \nabla v_j, \nabla v \rangle\, dx = \lambda_j(\Omega) \int_\Omega v_j\, v\, dx = 0, \qquad \text{for } j \in \{1, \ldots, k\}.$$

The previous discussion finally gives that

$$\int_\Omega \langle \nabla v, \nabla \varphi \rangle\, dx = \lambda_{k+1}(\Omega) \int_\Omega v\, \varphi\, dx, \qquad \text{for every } \varphi \in W^{1,2}_0(\Omega).$$

Thus, $\lambda_{k+1}(\Omega)$ is an eigenvalue and v is a relevant eigenfunction. This concludes the proof of the first point.

(2) The fact that

$$\int_\Omega v_n\, v_m\, dx = 0, \qquad \text{for every } n \neq m,$$

follows directly from the construction. The condition on the gradients follows from the eigenvalue equation and the previous relation: indeed, by taking $n \neq m$ and using v_m as a test function in the equation for v_n, we have

$$\int_\Omega \langle \nabla v_n, \nabla v_m \rangle \, dx = \lambda_n(\Omega) \int_\Omega v_n \, v_m \, dx = 0.$$

(3) Let us argue by contradiction and suppose that there exists $M > 0$ such that

$$\lambda_k(\Omega) \leq M, \qquad \text{for every } k \in \mathbb{N} \setminus \{0\}.$$

Since by construction we have

$$\int_\Omega |\nabla v_k|^2 \, dx = \lambda_k(\Omega), \qquad \text{for every } k \in \mathbb{N} \setminus \{0\},$$

we would obtain that $\{v_k\}_{k\in\mathbb{N}\setminus\{0\}}$ is bounded in $W_0^{1,2}(\Omega)$. By Theorem 3.8.3, we would get that this sequence converges strongly in $L^2(\Omega)$, up to a subsequence. However, this is not possible, since by point (3) we have for $n \neq m$

$$\int_\Omega |v_n - v_m|^2 \, dx = \int_\Omega |v_n|^2 \, dx + \int_\Omega |v_m|^2 \, dx = 2.$$

Thus, the positive sequence $\{\lambda_k(\Omega)\}_{k\in\mathbb{N}\setminus\{0\}}$ must diverge.

(4) Let us take $u \in W_0^{1,2}(\Omega)$ and set

$$u_n = \sum_{k=1}^{n} \widehat{u}(k) \, v_k, \qquad \text{with } \widehat{u}(k) = \int_\Omega u \, v_k \, dx.$$

We compute

$$\begin{aligned} \int_\Omega |u - u_n|^2 &= \int_\Omega |u|^2 \, dx - 2 \int_\Omega u \, u_n \, dx + \int_\Omega |u_n|^2 \, dx \\ &= \int_\Omega |u|^2 \, dx - 2 \sum_{k=1}^{n} \widehat{u}(k) \int_\Omega u \, v_k \, dx + \sum_{k=1}^{n} |\widehat{u}(k)|^2 \\ &= \int_\Omega |u|^2 \, dx - \sum_{k=1}^{n} |\widehat{u}(k)|^2. \end{aligned} \tag{4.8.4}$$

Observe that we used point (2) to infer that

$$\int_\Omega |u_n|^2 \, dx = \sum_{k=1}^{n} |\widehat{u}(k)|^2.$$

Similarly, still by point (2) we have

$$\int_\Omega |\nabla u_n|^2\,dx = \sum_{k=1}^n \lambda_k(\Omega)\,|\widehat{u}(k)|^2,$$

while by using the equation for each v_k, tested with $\varphi = u$, we have

$$\int_\Omega \langle \nabla u, \nabla u_n\rangle = \sum_{k=1}^n \widehat{u}(k)\int_\Omega \langle \nabla u, \nabla v_k\rangle\,dx = \sum_{k=1}^n \lambda_k(\Omega)\,|\widehat{u}(k)|^2.$$

These two identities permit to infer that

$$\begin{aligned}\int_\Omega |\nabla u - \nabla u_n|^2 &= \int_\Omega |\nabla u|^2\,dx - 2\int_\Omega \langle \nabla u, \nabla u_n\rangle\,dx + \int_\Omega |\nabla u_n|^2\,dx \\ &= \int_\Omega |\nabla u|^2\,dx - \sum_{k=1}^n \lambda_k(\Omega)\,|\widehat{u}(k)|^2.\end{aligned} \tag{4.8.5}$$

Since by construction we have

$$\int_\Omega (u - u_n)\,v_k\,dx = 0, \qquad \text{for } k = 1, \dots, n,$$

the definition of $\lambda_{n+1}(\Omega)$ entails that we have

$$\int_\Omega |u - u_n|^2 \le \frac{1}{\lambda_{n+1}(\Omega)}\int_\Omega |\nabla u - \nabla u_n|^2\,dx \le \frac{1}{\lambda_{n+1}(\Omega)}\int_\Omega |\nabla u|^2\,dx.$$

We also used (4.8.5) in the last estimate. By taking the limit as n goes to ∞ and using point (3), we get

$$\lim_{n\to\infty}\int_\Omega |u - u_n|^2\,dx = 0,$$

as desired. Moreover, by using this fact in (4.8.4), we get the first identity in (4.8.1), as well.

We still need to prove that the gradient admits the same type of decomposition. We observe at first that from (4.8.5), we have in particular that

$$\sum_{k=1}^n \lambda_k(\Omega)\,|\widehat{u}(k)|^2 \le \int_\Omega |\nabla u|^2\,dx, \qquad \text{for every } n \in \mathbb{N}\setminus\{0\},$$

and thus

$$\sum_{k=1}^{\infty} \lambda_k(\Omega)\, |\widehat{u}(k)|^2 < +\infty.$$

Then we notice that $\{u - u_n\}_{n\in\mathbb{N}\setminus\{0\}}$ is a Cauchy sequence in $W_0^{1,2}(\Omega)$: indeed, we have[5] for every $m > n \geq 1$

$$\int_\Omega |\nabla(u-u_n)-\nabla(u-u_m)|^2\, dx = \int_\Omega |\nabla u_n - \nabla u_m|^2\, dx = \sum_{k=n+1}^{m} \lambda_k(\Omega)\, |\widehat{u}(k)|^2,$$

and the claim follows from the convergence of the series of the terms

$$\left\{\lambda_k(\Omega)\, |\widehat{u}(k)|^2\right\}_{k\in\mathbb{N}\setminus\{0\}},$$

that we obtained above. Since $W_0^{1,2}(\Omega)$ is a Banach space (recall Theorem 3.7.4), we get that $\{u - u_n\}_{n\in\mathbb{N}}$ converges strongly in $W_0^{1,2}(\Omega)$. We have already proved that its strong L^2 limit is the null function, thus we conclude that

$$\lim_{n\to\infty} \int_\Omega |\nabla u - \nabla u_n|^2\, dx = 0,$$

as well. The second identity in (4.8.1) now follows by using this information in (4.8.5).

(5) Let $u, \psi \in W_0^{1,2}(\Omega)$, from the previous point we know that if we set

$$u_n = \sum_{k=1}^{n} \widehat{u}(k)\, v_k \qquad \text{and} \qquad \psi_n = \sum_{k=1}^{n} \widehat{\psi}(k)\, v_k,$$

then we have

$$\lim_{n\to\infty} \|u_n - u\|_{W^{1,2}(\Omega)} = \lim_{n\to\infty} \|\psi_n - \psi\|_{W^{1,2}(\Omega)} = 0.$$

This permits to infer that

[5] Recall that by Theorem 3.7.6, the quantity

$$u \mapsto \|\nabla u\|_{L^2(\Omega;\mathbb{R}^N)},$$

is a norm on $W_0^{1,2}(\Omega)$ equivalent to the standard one, if Ω is bounded.

$$\int_\Omega u\,\psi\,dx = \lim_{n\to\infty}\int_\Omega u_n\,\psi_n\,dx = \lim_{n\to\infty}\int_\Omega\left(\sum_{k=1}^n \widehat{u}(k)\,v_k\right)\left(\sum_{j=1}^n \widehat{\psi}(j)\,v_j\right)dx$$
$$= \lim_{n\to\infty}\sum_{k,j=1}^n\left(\widehat{u}(k)\,\widehat{\psi}(j)\int_\Omega v_k\,v_j\,dx\right)$$
$$= \lim_{n\to\infty}\sum_{k=1}^n \widehat{u}(k)\,\widehat{\psi}(k) = \sum_{k=1}^\infty \widehat{u}(k)\,\widehat{\psi}(k),$$

where we used point (2). The identity for the product of the gradients can be proved in exactly the same way and it is left to the reader.

(6) Let us suppose that $\lambda \in \mathrm{Eigen}(\Omega)$. Accordingly, we take an associated eigenfunction $v \in W_0^{1,2}(\Omega)$. In particular, by definition this is a non-trivial function. We argue by contradiction and suppose that

$$\lambda \neq \lambda_k(\Omega), \qquad \text{for every } k \in \mathbb{N}\setminus\{0\}.$$

By testing the equation for v with v_k, we get

$$\int_\Omega \langle \nabla v, \nabla v_k\rangle\,dx = \lambda\int_\Omega v\,v_k\,dx.$$

We can also test the equation for v_k with v, so to get

$$\int_\Omega \langle \nabla v_k, \nabla v\rangle\,dx = \lambda_k(\Omega)\int_\Omega v\,v_k\,dx.$$

By symmetry of the scalar product, the last two equations imply that we must have

$$\widehat{v}(k) = \int_\Omega v\,v_k\,dx = 0, \qquad \text{for every } k \in \mathbb{N}\setminus\{0\}.$$

From point (4) applied to v we get in particular that $v \equiv 0$, which gives the desired contradiction.

The proof is now complete. □

The previous result shows in particular that every function of $W_0^{1,2}(\Omega)$ can be written as an infinite linear combination of the eigenfunctions $\{v_k\}_{k\in\mathbb{N}\setminus\{0\}}$. Such a linear combination has to be intended as a series which converges in $L^2(\Omega)$, and even in $W_0^{1,2}(\Omega)$. With a little extra work, this property can be extended to any element of $L^2(\Omega)$, by a density argument. This is the content of the following

Proposition 4.8.2 *Let $\Omega \subseteq \mathbb{R}^N$ be an open bounded set. We still indicate by $\{v_k\}_{k\in\mathbb{N}\setminus\{0\}}$ the eigenfunctions obtained in Theorem 4.8.1. Then for every $u \in L^2(\Omega)$ we have*

$$u = \sum_{k=1}^{\infty} \widehat{u}(k)\, v_k, \qquad \text{with } \widehat{u}(k) = \int_\Omega u\, v_k\, dx, \qquad \text{for every } k \in \mathbb{N}\setminus\{0\},$$

where the convergence of the series has to be understood in the L^2 sense. Moreover, we have

$$\int_\Omega |u|^2\, dx = \sum_{k=1}^{\infty} |\widehat{u}(k)|^2. \tag{4.8.6}$$

Finally, if $\psi \in L^2(\Omega)$, we have

$$\int_\Omega u\, \psi\, dx = \sum_{k=1}^{\infty} \widehat{u}(k)\, \widehat{\psi}(k).$$

Proof By density of $C_0^\infty(\Omega)$ in $L^2(\Omega)$ (see [6, Corollary 2.1] or [46, Theorem 2.16 & Lemma 2.19]), we know that there exists a sequence $\{u_n\}_{n\in\mathbb{N}} \subseteq C_0^\infty(\Omega)$ such that

$$\lim_{n\to\infty} \|u_n - u\|_{L^2(\Omega)} = 0.$$

By Theorem 4.8.1, we know that

$$u_n = \sum_{k=1}^{\infty} \widehat{u_n}(k)\, v_k \qquad \text{and} \qquad \int_\Omega |u_n|^2\, dx = \sum_{k=1}^{\infty} |\widehat{u_n}(k)|^2,$$

while thanks to the strong L^2 convergence, we have

$$\lim_{n\to\infty} \widehat{u_n}(k) = \widehat{u}(k), \qquad \text{for every } k \in \mathbb{N}\setminus\{0\}.$$

We now fix a $k \in \mathbb{N}\setminus\{0\}$, for every $n \in \mathbb{N}$ by using the triangle inequality we have

$$\left| \left(\sum_{j=1}^{k} |\widehat{u}(j)|^2\right)^{\frac{1}{2}} - \left(\sum_{j=1}^{k} |\widehat{u_n}(j)|^2\right)^{\frac{1}{2}} \right| \le \left(\sum_{j=1}^{k} |\widehat{u_n}(j) - \widehat{u}(j)|^2\right)^{\frac{1}{2}}. \tag{4.8.7}$$

Moreover, by using the convergence of the coefficients and (4.8.1) applied to $u_n - u_m$, we also have

$$\begin{aligned}\sum_{j=1}^{k} |\widehat{u_n}(j) - \widehat{u}(j)|^2 &= \lim_{m\to\infty} \sum_{j=1}^{k} |\widehat{u_n}(j) - \widehat{u_m}(j)|^2 \\ &\le \lim_{m\to\infty} \sum_{j=1}^{\infty} |\widehat{u_n}(j) - \widehat{u_m}(k)|^2 \\ &= \lim_{m\to\infty} \int_\Omega |u_n - u_m|^2\,dx = \int_\Omega |u_n - u|^2\,dx.\end{aligned} \tag{4.8.8}$$

By using that

$$\sum_{j=1}^{\infty} |\widehat{u_n}(j)|^2 = \int_\Omega |u_n|^2\,dx < +\infty,$$

and the estimate (4.8.8), from (4.8.7) we obtain in particular that

$$\sum_{j=1}^{\infty} |\widehat{u}(j)|^2 < +\infty,$$

as well. We can combine (4.8.7) and (4.8.8) and take the limit as k goes to ∞ in (4.8.7): we obtain

$$\left| \left(\sum_{j=1}^{\infty} |\widehat{u}(j)|^2\right)^{\frac{1}{2}} - \left(\sum_{j=1}^{\infty} |\widehat{u_n}(j)|^2\right)^{\frac{1}{2}} \right| \le \left(\int_\Omega |u_n - u|^2\,dx\right)^{\frac{1}{2}},$$

By virtue of (4.8.1), this is the same as

$$\left| \left(\sum_{j=1}^{\infty} |\widehat{u}(j)|^2\right)^{\frac{1}{2}} - \left(\int_\Omega |u_n|^2\,dx\right)^{\frac{1}{2}} \right| \le \left(\int_\Omega |u_n - u|^2\,dx\right)^{\frac{1}{2}}.$$

If we now take the limit as n goes to ∞ and use that the last term goes to 0, we obtain

$$\sum_{j=1}^{\infty} |\widehat{u}(j)|^2 = \lim_{n\to\infty} \int_\Omega |u_n|^2\,dx = \int_\Omega |u|^2\,dx.$$

This proves the validity of (4.8.6). In order to prove that u can be written as a series, we observe that for every $k \in \mathbb{N} \setminus \{0\}$, we have as in (4.8.4)

$$\int_\Omega \left| u - \sum_{j=1}^{k} \widehat{u}(j)\, v_j \right|^2 dx = \int_\Omega |u|^2\, dx - \sum_{j=1}^{k} |\widehat{u}(j)|^2.$$

If we now take the limit as k goes to ∞ and use (4.8.6), we get the desired conclusion.

Finally, the proof of the last statement can be done exactly in the same way as in point (5) of Theorem 4.8.1. This is left to the reader. □

Remark 4.8.3 With the language of Hilbert spaces, we just proved that the family of eigenfunctions $\{v_k\}_{k\in\mathbb{N}\setminus\{0\}}$ gives an *orthonormal basis* of $L^2(\Omega)$. The latter is indeed a Hilbert space, when endowed with the scalar product

$$\left[u, \psi\right]_{L^2(\Omega)} := \int_\Omega u\, \psi\, dx.$$

Observe that with this notation, we have

$$\widehat{u}(k) = \left[u, v_k\right]_{L^2(\Omega)}, \qquad \text{for every } k \in \mathbb{N} \setminus \{0\}.$$

Similary, Theorem 4.8.1 gives that the family

$$\left\{ \frac{v_k}{\sqrt{\lambda_k(\Omega)}} \right\}_{k\in\mathbb{N}\setminus\{0\}},$$

is an orthonormal basis of the Hilbert space $W_0^{1,2}(\Omega)$, equipped with the scalar product

$$\left[u, \psi\right]_{W_0^{1,2}(\Omega)} := \int_\Omega \langle \nabla u, \nabla \psi \rangle\, dx \qquad \text{for every } u, \psi \in W_0^{1,2}(\Omega).$$

Theorem 4.8.1 and Proposition 4.8.2 can be used to prove existence of a weak solution to the *Poisson equation*, with homogeneous Dirichlet boundary conditions. Namely, we have the following

Corollary 4.8.4 *Let $\Omega \subseteq \mathbb{R}^N$ be an open bounded set and $f \in L^2(\Omega)$. With the notation of Theorem 4.8.1, the function*

$$v = \sum_{k=1}^{\infty} \frac{\widehat{f}(k)}{\lambda_k(\Omega)}\, v_k \in W_0^{1,2}(\Omega),$$

is the unique weak solution to the following problem

$$\begin{cases} -\Delta u = f, & \text{in } \Omega, \\ \quad u = 0, & \text{on } \partial\Omega. \end{cases}$$

Proof We first observe that v is well-defined and belongs to $W_0^{1,2}(\Omega)$, thanks to Problem 4.10.23 below. It is sufficient to check that

$$\sum_{k=1}^{\infty}\left|\frac{\widehat{f}(k)}{\lambda_k(\Omega)}\right|^2 \lambda_k(\Omega)=\sum_{k=1}^{\infty}\frac{|\widehat{f}(k)|^2}{\lambda_k(\Omega)}<+\infty.$$

This condition follows from the fact that (see Proposition 4.8.2)

$$\sum_{k=1}^{\infty}|\widehat{f}(k)|^2=\|f\|_{L^2(\Omega)}^2<+\infty,$$

and the fact that the sequence $\{\lambda_k(\Omega)\}_{k\in\mathbb{N}\setminus\{0\}}$ is non-decreasing. We verify that v is a weak solution: for every $\varphi\in W_0^{1,2}(\Omega)$ we have

$$\int_\Omega\langle\nabla v,\nabla\varphi\rangle\,dx=\sum_{k=1}^{\infty}\lambda_k(\Omega)\,\widehat{v}(k)\,\widehat{\varphi}(k)=\sum_{k=1}^{\infty}\widehat{f}(k)\,\widehat{\varphi}(k)=\int_\Omega f\,\varphi\,dx.$$

We used Theorem 4.8.1, Proposition 4.8.2 and the fact that

$$\widehat{v}(k)=\frac{\widehat{f}(k)}{\lambda_k(\Omega)},\qquad\text{for every } k\in\mathbb{N}\setminus\{0\}.$$

Thus, v is a weak solution of the claimed equation.

In order to prove uniqueness, we can proceed as follows: let us suppose that $w\in W_0^{1,2}(\Omega)$ is another weak solution. By testing the weak formulation with $\varphi=v_k$, we would obtain

$$\int_\Omega\langle\nabla w,\nabla v_k\rangle\,dx=\int_\Omega f\,v_k\,dx=\widehat{f}(k),\qquad\text{for every } k\in\mathbb{N}\setminus\{0\}.$$

On the other hand, by using the equation for v_k tested with w, we obtain

$$\int_\Omega\langle\nabla w,\nabla v_k\rangle\,dx=\lambda_k(\Omega)\int_\Omega w\,v_k\,dx=\lambda_k(\Omega)\,\widehat{w}(k),\qquad\text{for every } k\in\mathbb{N}\setminus\{0\}.$$

By joining the last two equations, we get

$$\widehat{w}(k)=\frac{\widehat{f}(k)}{\lambda_k(\Omega)}\qquad\text{and thus}\qquad w=\sum_{k=1}^{\infty}\frac{\widehat{f}(k)}{\lambda_k(\Omega)}\,v_k=v,$$

as desired. □

Remark 4.8.5 As a final remark, we point out that in this section the boundedness assumption on Ω has been taken only to assure the compactness of the embedding $W_0^{1,2}(\Omega) \hookrightarrow L^2(\Omega)$. Actually, it is not difficult to see that Theorem 4.8.1 and Corollary 4.8.2 remain valid for every open set having such an embedding property, with exactly the same proofs. This is the case for example of open sets with finite measure, thanks to Problem 3.12.21. For more general conditions on the open set ensuring the compactness of the embedding $W_0^{1,2}(\Omega) \hookrightarrow L^2(\Omega)$, we refer the experienced reader to [50, Chapter 15].

4.9 Example: The Superlinear Lane-Emden Equation

Proposition 4.9.1 *Let $q > 2$ be such that*

$$q < \begin{cases} \dfrac{2\,N}{N-2}, & \textit{if } N \geq 3, \\ \infty, & \textit{if } N \in \{1,2\}, \end{cases}$$

and let $\Omega \subseteq \mathbb{R}^N$ be an open bounded set. Then there exists at least a nontrivial weak solution $v \in W_0^{1,2}(\Omega)$ of the problem

$$\begin{cases} -\Delta u = |u|^{q-1}, & \textit{in } \Omega, \\ u = 0, & \textit{on } \partial\Omega. \end{cases} \tag{4.9.1}$$

Moreover, every weak solution of (4.9.1) *is strictly positive on Ω.*

Proof One could try to mimick the proof of Proposition 4.6.1 and obtain existence of a nontrivial solution by minimizing

$$\inf_{u \in W_0^{1,2}(\Omega)} \left\{ \frac{1}{2} \int_\Omega |\nabla u|^2\, dx - \frac{1}{q} \int_\Omega |u|^{q-1}\, u\, dx \right\}.$$

However, it is easy to see that *this method now is bound to fail*, due to the fact that $q > 2$. Indeed, observe that by taking a function $\psi \in W_0^{1,2}(\Omega)$ such that

$$\int_\Omega |\psi|^{q-1}\, \psi\, dx > 0, \tag{4.9.2}$$

the function $t\,\psi$ is still admissible for the variational problem, for every $t > 0$. Thus we get

$$\inf_{u\in W_0^{1,2}(\Omega)} \left\{\frac{1}{2}\int_\Omega |\nabla u|^2\,dx - \frac{1}{q}\int_\Omega |u|^{q-1}\,u\,dx\right\}$$
$$\le \frac{t^2}{2}\int_\Omega |\nabla \psi|^2\,dx - \frac{t^q}{q}\int_\Omega |\psi|^{q-1}\,\psi\,dx, \qquad \text{for every } t>0.$$

By letting t go to $+\infty$, using that $t^2 = o(t^q)$ and (4.9.2), one obtains

$$\inf_{u\in W_0^{1,2}(\Omega)} \left\{\frac{1}{2}\int_\Omega |\nabla u|^2\,dx - \frac{1}{q}\int_\Omega |u|^{q-1}\,u\,dx\right\} = -\infty,$$

i.e. the functional is now unbounded from below and no minimizers can exist.

We need to change our strategy: we rather try to mimick the proof of Theorem 4.7.3 and consider a *constrained minimization problem*. We first recall that, for q as in the statement, by combining Theorem 3.8.1 and Poincaré inequality (see Theorem 3.7.6) we have

$$\int_\Omega |\nabla u|^2\,dx \ge c\left(\int_\Omega |u|^q\,dx\right)^{\frac{2}{q}}, \qquad \text{for every } u\in W_0^{1,2}(\Omega),$$

for a constant $c = c(N, q, \Omega) > 0$. We seek the sharp constant in this inequality: at the end, we will show how this problem is connected with our primary target.

We thus set

$$\lambda_{2,q}(\Omega) = \inf_{u\in W_0^{1,2}(\Omega)\setminus\{0\}} \frac{\displaystyle\int_\Omega |\nabla u|^2\,dx}{\left(\displaystyle\int_\Omega |u|^q\,dx\right)^{\frac{2}{q}}}, \tag{4.9.3}$$

and show that this infimum is indeed attained. To this aim, we can copy almost verbatim the existence part in the proof of Theorem 4.7.3. We first observe that problem (4.9.3) is equivalent to the following one

$$\inf_{u\in W_0^{1,2}(\Omega)} \left\{\int_\Omega |\nabla u|^2\,dx \,:\, \int_\Omega |u|^q\,dx = 1\right\}. \tag{4.9.4}$$

We use the Direct Method: we take a minimizing sequence $\{u_n\}_{n\in\mathbb{N}} \subseteq W_0^{1,2}(\Omega)$ for problem (4.9.3), i.e.

$$\lim_{n\to\infty}\int_\Omega |\nabla u_n|^2\,dx = \lambda_{2,q}(\Omega) \qquad \text{and} \qquad \int_\Omega |u_n|^q\,dx = 1, \quad \text{for every } n\in\mathbb{N}.$$

These properties imply that the sequence is bounded in $W_0^{1,2}(\Omega)$. By Theorem 3.3.6 and Lemma 3.8.7, we know that $\{u_n\}_{n\in\mathbb{N}}$ converges weakly in $W^{1,2}(\Omega)$ and strongly in $L^q(\Omega)$ (here we use the assumption on the exponent q) to a limit function $v \in W_0^{1,2}(\Omega)$, up to a subsequence.

By the strong L^q convergence, we get that

$$1 = \lim_{n\to\infty} \int_\Omega |u_n|^q \, dx = \int_\Omega |v|^q \, dx,$$

thus v is still admissible in (4.9.4). Moreover, by the lower semicontinuity of the L^2 norm with respect to the weak convergence, we have

$$\int_\Omega |\nabla v|^2 \, dx \le \liminf_{n\to\infty} \int_\Omega |\nabla u_n|^2 \, dx = \lambda_{2,q}(\Omega).$$

Thus, v is a solution of (4.9.4) and of (4.9.3), as well.

We derive the Euler-Lagrange equation for problem (4.9.3). We use the same trick as in the proof of Theorem 4.7.3. By definition of $\lambda_{2,q}(\Omega)$, we have

$$\frac{\displaystyle\int_\Omega |\nabla u|^2 \, dx}{\left(\displaystyle\int_\Omega |u|^q \, dx\right)^{\frac{2}{q}}} \ge \lambda_{2,q}(\Omega), \qquad \text{for every } u \in W_0^{1,2}(\Omega) \setminus \{0\},$$

which is the same as

$$\mathcal{F}_q(u) := \int_\Omega |\nabla u|^2 \, dx - \lambda_{2,q}(\Omega) \left(\int_\Omega |u|^q \, dx\right)^{\frac{2}{q}} \ge 0, \qquad \text{for every } u \in W_0^{1,2}(\Omega).$$

Moreover, for the minimizer v we have

$$\frac{\displaystyle\int_\Omega |\nabla v|^2 \, dx}{\left(\displaystyle\int_\Omega |v|^q \, dx\right)^{\frac{2}{q}}} = \lambda_{2,q}(\Omega),$$

and thus

$$\mathcal{F}_q(u) \ge 0 \quad \text{for every } u \in W_0^{1,2}(\Omega) \qquad \text{and} \qquad \mathcal{F}_q(v) = 0.$$

This argument shows that v is a minimizer of the functional $\mathcal{F}_q$, as well. It is now sufficient to compute the first variation of this functional, in order to conclude. For every $\varphi \in C_0^\infty(\Omega)$, we must have

$$\delta\mathcal{F}_q(v)[\varphi] = \frac{d}{dt}\mathcal{F}_q(v + t\,\varphi)_{|t=0} = 0,$$

provided that $t \mapsto \mathcal{F}_q(v + t\,\varphi)$ is differentiable at $t = 0$. We have[6]

$$\delta\mathcal{F}_q(v)[\varphi] = 2\int_\Omega \langle \nabla v, \nabla\varphi\rangle\,dx - \frac{2}{q}\,\lambda_{2,q}(\Omega)\left(\int_\Omega |v|^q\,dx\right)^{\frac{2}{q}-1} q\int_\Omega |v|^{q-2}\,v\,\varphi\,dx.$$

We recall that v is a solution of the constrained problem (4.9.4), thus in particular we have $\int_\Omega |v|^q\,dx = 1$. From the previous computation of the first variation, we thus get

$$\int_\Omega \langle \nabla v, \nabla\varphi\rangle\,dx - \lambda_{2,q}(\Omega)\int_\Omega |v|^{q-2}\,v\,\varphi\,dx = 0, \qquad \text{for every } \varphi \in C_0^\infty(\Omega).$$

This shows that $v \in W_0^{1,2}(\Omega)$ is a weak solution of

$$-\Delta v = \lambda_{2,q}(\Omega)\,|v|^{q-2}\,v, \qquad \text{in } \Omega.$$

We are not far from the conclusion: at first, we observe that we can suppose that $v \geq 0$ in Ω. Indeed, if this were not the case, it would be sufficient to replace it with $|v|$, which is still admissible in (4.9.4) (thanks to Proposition 3.7.13) and satisfies

$$\int_\Omega |\nabla v|^2\,dx = \int_\Omega |\nabla |v||^2\,dx,$$

thanks to Corollary 3.4.6. This shows that $|v|$ is still a minimizer. We then set

$$w(x) = \lambda_{2,q}(\Omega)^{\frac{1}{q-2}}\,v(x),$$

and observe that for every $\varphi \in C_0^\infty(\Omega)$

---

[6] Observe that the function

$$t \mapsto \left(\int_\Omega |v + t\,\varphi|^q\,dx\right)^{\frac{2}{q}},$$

can be written as the composition of the functions $\psi(t) = \int_\Omega |v + t\,\varphi|^q\,dx$ and $\Psi(t) = t^{2/q}$. The function ψ is differentiable at $t = 0$, with $\psi(0) > 0$. The function Ψ is differentiable outside the origin. Thus, the function

$$\left(\int_\Omega |v + t\,\varphi|^q\,dx\right)^{\frac{2}{q}} = \Psi(\psi(t)),$$

is differentiable at $t = 0$.

$$\int_\Omega \langle \nabla w, \nabla \varphi \rangle\, dx = \lambda_{2,q}(\Omega)^{\frac{1}{q-2}} \int_\Omega \langle \nabla v, \nabla \varphi \rangle\, dx$$
$$= \lambda_{2,q}(\Omega)^{\frac{1}{q-2}}\, \lambda_{2,q}(\Omega) \int_\Omega |v|^{q-2}\, v\, \varphi\, dx$$
$$= \lambda_{2,q}(\Omega)^{\frac{q-1}{q-2}} \int_\Omega |v|^{q-2}\, v\, \varphi\, dx = \int_\Omega |w|^{q-2}\, w\, \varphi\, dx.$$

In other words, $w \in W_0^{1,2}(\Omega)$ is a weak solution of

$$-\Delta w = |w|^{q-2}\, w, \qquad \text{in } \Omega.$$

Moreover, since $w \geq 0$, then $|w|^{q-2}\, w = |w|^{q-1} = w^{q-1}$ and thus w is a weak solution of the equation in the statement. We observe that by construction it is not trivial, since

$$\int_\Omega |w|^q\, dx = \lambda_{2,q}(\Omega)^{\frac{q}{q-2}} \int_\Omega |v|^q\, dx = \lambda_{2,q}(\Omega)^{\frac{q}{q-2}} > 0.$$

The fact that $w > 0$ in Ω now follows from the strong minimum principle of Lemma 4.4.5. □

4.10 Problems

Problem 4.10.1 Under the assumptions of Theorem 4.3.3, we further suppose that:

- H is strictly convex and radially symmetric;
- $\Omega \subseteq \mathbb{R}^N$ is either a ball or a spherical shell, i.e. it has one of the following forms

$$\{x \in \mathbb{R}^N : |x| < R\} \qquad \text{or} \qquad \{x \in \mathbb{R}^N : R_1 < |x| < R_2\};$$

- f and U are radially symmetric, as well.

Prove that the solution of

$$\min_{u \in W^{1,p}(\Omega)} \left\{ \int_\Omega H(\nabla u)\, dx - \int_\Omega f\, u\, dx \, : \, u - U \in W_0^{1,p}(\Omega) \right\},$$

is unique and radially symmetric.

Problem 4.10.2 Let $\Omega \subseteq \mathbb{R}^N$ be an open set with finite measure, possibly unbounded. For $g \in W^{1,2}(\Omega)$, show the existence of a weak solution to

$$\begin{cases} -\Delta u = 0, & \text{in } \Omega, \\ u = g, & \text{on } \partial\Omega. \end{cases}$$

Also prove that this solution is unique.

Problem 4.10.3 Let $0 < r < R$ and let

$$A_{r,R} = \{x \in \mathbb{R}^2 \,:\, r < |x| < R\}.$$

We take $g \in W^{1,2}(A_{r,R})$ the function

$$g(x) = \frac{R - |x|}{R - r}.$$

Compute explicitly the unique solution of

$$\min_{u \in W^{1,2}(A_{r,R})} \left\{ \frac{1}{2} \int_{A_{r,R}} |\nabla u|^2 \, dx \,:\, u - g \in W_0^{1,2}(A_{r,R}) \right\}.$$

Problem 4.10.4 Compute explicitly the torsion function of a disk of radius R (see Remark 4.5.2 for the definition of torsion function).

Problem 4.10.5 Let $0 < r < R$ and let

$$A_{r,R} = \{x \in \mathbb{R}^2 \,:\, r < |x| < R\}.$$

Compute explicitly the torsion function of $A_{r,R}$.

Problem 4.10.6 Let $B_1(0) \subseteq \mathbb{R}^2$ be the two-dimensional open disk centered at the origin, with radius 1. Show that the equation

$$-\Delta u = \frac{1}{\sqrt{x^2 + y^2}}, \qquad \text{in } B_1(0),$$

admits a weak solution $v \in W_0^{1,2}(B_1(0))$. Then, prove that such a solution is unique and compute it.

Problem 4.10.7 Let $B_1(0) \subseteq \mathbb{R}^2$ be the two-dimensional open disk centered at the origin, with radius 1. Show that for every $\alpha > -2$ the equation

$$-\Delta u = (x^2 + y^2)^{\frac{\alpha}{2}}, \qquad \text{in } B_1(0),$$

admits a unique weak solution $v \in W_0^{1,2}(B_1(0))$. Compute such a solution in the special case $\alpha = 2$.

Problem 4.10.8 Let $B_1(0) \subseteq \mathbb{R}^2$ be the two-dimensional open disk centered at the origin, with radius 1. Let $a \in L^\infty(B_1(0))$ be such that

$$\|a\|_{L^\infty(B_1(0))} < \lambda_1(B_1(0)),$$

where $\lambda_1(B_1(0))$ is the first eigenvalue of the Dirichlet-Laplacian on $B_1(0)$, introduced in Theorem 4.7.3. Show that for every $f \in L^q(B_1(0))$ with $q > 1$, the equation

$$-\Delta u = a\,u + f, \qquad \text{in } B_1(0),$$

admits a unique weak solution $v \in W_0^{1,2}(B_1(0))$.

Problem 4.10.9 Let $\Omega \subseteq \mathbb{R}^N$ be an open bounded set. Prove that for every $f \in L^2(\Omega)$ there exists a unique weak solution $v \in W_0^{1,2}(\Omega)$ of the problem

$$\begin{cases} -\Delta u = f, & \text{in } \Omega, \\ u = 0, & \text{on } \partial\Omega. \end{cases}$$

Show that for v we have the estimate

$$\int_\Omega |\nabla v|^2\,dx \le \left(\text{diam}(\Omega)\right)^2 \int_\Omega |f|^2\,dx.$$

Problem 4.10.10 Show that there exists a constant $C > 0$, depending on the dimension N only, such that for every open set $\Omega \subseteq \mathbb{R}^N$ with finite measure, we have

$$C\,|\Omega|^{-\frac{2}{N}} \le \lambda_1(\Omega),$$

where $\lambda_1(\Omega)$ is the first eigenvalue of the Dirichlet-Laplacian on Ω, introduced in Theorem 4.7.3.

Problem 4.10.11 Let $B_1(0) \subseteq \mathbb{R}^3$ be the three-dimensional open ball centered at the origin, with radius 1. Show that for every $f \in L^{6/5}(B_1(0))$ the equation

$$-\Delta u = f, \qquad \text{in } B_1(0),$$

admits a unique weak solution $v \in W_0^{1,2}(B_1(0))$. Prove that if $f \ge 0$ almost everywhere in $B_1(0)$ and $f \not\equiv 0$, then we have $v > 0$ almost everywhere in $B_1(0)$.

Problem 4.10.12 Let $f \in L^q(\mathbb{R}^2)$ for some $1 < q \le 2$ and let $V \in L^2(\mathbb{R}^2)$ be a function such that

$$V(x) \ge c > 0, \qquad \text{for a. e. } x \in \mathbb{R}^2,$$

for some constant c. Show that there exists a unique weak solution $v \in W^{1,2}(\mathbb{R}^2)$ to the equation

$$-\Delta u + V\,u = f \qquad \text{in } \mathbb{R}^2.$$

Problem 4.10.13 Let $\Omega \subseteq \mathbb{R}^N$ be an open bounded set and let $a \in L^\infty(\Omega)$ be such that

$$a(x) \geq \overline{a} > 0, \qquad \text{for a. e. } x \in \Omega,$$

for some constant $\overline{a}$. Show that the functional

$$u \mapsto \frac{1}{2} \int_\Omega a\,|\nabla u|^2\,dx,$$

is weakly lower semicontinuous on $W_0^{1,2}(\Omega)$. Then, show that for every $f \in L^2(\Omega)$, the equation

$$-\operatorname{div}(a\,\nabla u) = f, \qquad \text{in } \Omega,$$

admits a unique weak solution in $W_0^{1,2}(\Omega)$.

Problem 4.10.14 Take the following unbounded open subset of $\mathbb{R}^2$

$$\Omega = \Big(\mathbb{R} \times (-1,1)\Big) \cup \Big((-1,1) \times \mathbb{R}\Big).$$

Show that $\lambda_1(\Omega) > 0$. Then use this fact to show that for every $f \in L^2(\Omega)$ there exists a unique weak solution in $W_0^{1,2}(\Omega)$ to the equation

$$-\Delta u = f, \qquad \text{in } \Omega.$$

Problem 4.10.15 Generalize Proposition 4.5.1 to $1 < p < \infty$: show that on an open bounded connected set $\Omega \subseteq \mathbb{R}^N$ there exists a unique weak solution $w \in W_0^{1,p}(\Omega)$ of the problem

$$\begin{cases} -\Delta_p u = 1, & \text{in } \Omega, \\ u = 0, & \text{on } \partial\Omega, \end{cases}$$

where Δ_p is the p-Laplacian operator, defined in (1.6.4). Moreover, such a solution is strictly positive almost everywhere in Ω. The function w is called *p-torsion function of* Ω.

Problem 4.10.16 Compute the p-torsion function of an N-dimensional open ball of radius R.

Problem 4.10.17 Generalize Theorem 4.7.3 to $1 < p < \infty$: show that on an open bounded set $\Omega \subseteq \mathbb{R}^N$ the variational problem

$$\lambda_{1,p}(\Omega) = \inf_{u \in W_0^{1,p}(\Omega)\setminus\{0\}} \frac{\displaystyle\int_\Omega |\nabla u|^p \, dx}{\displaystyle\int_\Omega |u|^p \, dx}, \tag{4.10.1}$$

admits a solution. Show that any minimizer v of this problem is a weak solution of the boundary value problem

$$\begin{cases} -\Delta_p u = \lambda_{1,p}(\Omega)\, |u|^{p-2}\, u, & \text{in } \Omega, \\ u = 0, & \text{on } \partial\Omega, \end{cases}$$

where Δ_p is the p-Laplacian operator.

Problem 4.10.18 (Comparison Principle) Let $1 < p < \infty$ and let $\Omega \subseteq \mathbb{R}^N$ be an open set. Let $u_1, u_2 \in W^{1,p}(\Omega)$ be weak solutions of the equation

$$-\Delta_p u_i = f_i, \qquad \text{in } \Omega,\ i = 1, 2,$$

with $f_1, f_2 \in L^{p'}(\Omega)$. Prove that if

$$f_2 \geq f_1 \quad \text{a.e. in } \Omega \qquad \text{and} \qquad (u_1 - u_2)_+ \in W_0^{1,p}(\Omega),$$

then we have $u_2 \geq u_1$ almost everywhere in Ω.

Problem 4.10.19 Let $1 < p < \infty$ and let $B_R \subseteq \mathbb{R}^N$ be an N-dimensional open ball, with radius R. Prove that for every $f \in L^{p'}(B_R)$ there exists a unique weak solution $v \in W_0^{1,p}(B_R)$ of the problem

$$\begin{cases} -\Delta_p u = f, & \text{in } B_R, \\ u = 0, & \text{on } \partial B_R. \end{cases}$$

Show that if we additionally assume that $f \in L^\infty(B_R)$, then we have $v \in L^\infty(B_R)$, as well. We also have the estimate

$$\|v\|_{L^\infty(B_R)} \leq \frac{p-1}{p} \left(\frac{1}{N}\right)^{\frac{1}{p-1}} R^{\frac{p}{p-1}} \, \|f\|_{L^\infty(B_R)}^{\frac{1}{p-1}}.$$

Problem 4.10.20 Let $B_1(0) \subseteq \mathbb{R}^3$ be the three-dimensional open ball centered at the origin, with radius 1. Let $f \in L^2(B_1(0))$ be a given function. Show that for every $2 \leq p < \infty$ there exists a unique weak solution $v \in W_0^{1,p}(B_1(0))$ to the equation

$$-\Delta_p u = f, \qquad \text{in } B_1(0).$$

What can be said for $1 < p < 2$?

Problem 4.10.21 Let $\Omega \subseteq \mathbb{R}^2$ be an open bounded set, let $p > 2$ and $f \in L^1(\Omega)$. Show that there exists a weak solution $v \in W_0^{1,p}(\Omega)$ of the equation

$$-\frac{\partial}{\partial x}\left(\left|\frac{\partial u}{\partial x}\right|^{p-2}\frac{\partial u}{\partial x}\right)-\frac{\partial}{\partial y}\left(\left|\frac{\partial u}{\partial y}\right|^{p-2}\frac{\partial u}{\partial y}\right)=f, \qquad \text{in } \Omega.$$

Also show that such a solution belongs to $L^\infty(\Omega)$. Finally, show the validity of the following estimate on v

$$\|v\|_{L^\infty(\Omega)} \leq C\,\Big(\mathrm{diam}(\Omega)\Big)^{\frac{p-2}{p-1}}\,\|f\|_{L^1(\Omega)}^{\frac{1}{p-1}},$$

for a constant $C > 0$ depending only on p.

Problem 4.10.22 Prove the assertions of Remark 4.4.2.

Problem 4.10.23 With the notation of the Spectral Theorem (i.e. Theorem 4.8.1), let $\{\alpha_k\}_{k\in\mathbb{N}} \subseteq \mathbb{R}$ be a sequence of real numbers such that

$$\sum_{k=1}^{\infty} \lambda_k(\Omega)\,|\alpha_k|^2 < +\infty.$$

Show that the series

$$\sum_{k=1}^{\infty} \alpha_k\, v_k,$$

defines a function S belonging to $W_0^{1,2}(\Omega)$. Also show that we have

$$\widehat{S}(k) := \int_\Omega S\, v_k = \alpha_k, \qquad \text{for every } k \in \mathbb{N}\setminus\{0\}.$$

Problem 4.10.24 (Heat Equation) Let $\Omega \subseteq \mathbb{R}^N$ be an open bounded connected set. We consider the *heat equation*

$$\frac{\partial u}{\partial t} = \Delta u, \qquad \text{in } \Omega\times(0,+\infty).$$

Here Δ is the Laplacian with respect to the spatial variable $x \in \Omega$. This equation describes the evolution in time of the temperature u of the heat conductor Ω. It has to be coupled with a boundary condition, for example

$$u(x,t) = g(x,t), \qquad \text{for } (x,t) \in \partial\Omega\times(0,+\infty),$$

and an initial condition

$$u(x,0) = u_0(x), \qquad \text{for } x \in \Omega.$$

We refer to [27, Chapter 2, Section 3] for more details on the heat equation.

A *standing heat wave* is a non-trivial solution of the heat equation u of the form

$$u(x,t) = X(x)\,T(t),$$

i.e. a solution with a separated variables structure.

Use the separation of variables technique and the Spectral Theorem (i.e. Theorem 4.8.1) to find all possible solutions of the form $u(x,t) = T(t)\,X(x)$ of the heat equation, with boundary condition

$$u(x,t) = 0, \qquad \text{for } (x,t) \in \partial\Omega \times (0,+\infty).$$

Problem 4.10.25 (Vibrations of a Drum) Let $\Omega \subseteq \mathbb{R}^2$ be an open bounded connected set. We suppose that Ω represents the membrane of a drum, fixed along its boundary $\partial\Omega$. The vertical displacements of Ω during vibrations are described by a function

$$u(x,t), \qquad \text{with } (x,t) \in \Omega \times (0,+\infty),$$

which solves the *wave equation*

$$\frac{\partial^2 u}{\partial t^2} = \Delta u, \qquad \text{in } \Omega \times (0,+\infty).$$

The fact that Ω is fixed along the boundary is encoded in the boundary condition

$$u(x,t) = 0, \qquad \text{for } (x,t) \in \partial\Omega \times (0,+\infty).$$

We refer to [27, Chapter 2, Section 4] for more details on the wave equation.

A *standing wave* is a non-trivial solution of the wave equation u of the form

$$u(x,t) = X(x)\,T(t),$$

i.e. a solution with a separated variables structure.

Use the separation of variables technique and the Spectral Theorem (i.e. Theorem 4.8.1) to find all possible standing waves of the membrane Ω. Which is the frequency of vibration of each of these solutions?

Problem 4.10.26 (Porous Medium Equation) Let $\Omega \subseteq \mathbb{R}^N$ be an open bounded connected set and let $m > 1$. We consider the *porous medium equation*

$$\frac{\partial u}{\partial t} = \Delta(u^m), \qquad \text{in } \Omega \times (0, +\infty),$$

where u is a non-negative function.

This is a nonlinear evolution equation, which arises in many mathematical models from Physics and Biology. We refer to [58] for a thorough study of this equation. Just like the heat equation (to which this reduces for $m = 1$), this equation has to be coupled with a boundary condition, for example

$$u(x, t) = g(x, t), \qquad \text{for } (x, t) \in \partial\Omega \times (0, +\infty),$$

and an initial condition

$$u(x, 0) = u_0(x), \qquad \text{for } x \in \Omega.$$

Use the separation of variables technique and Proposition 4.6.1 to find a positive solution of the form $u(x, t) = X(x)\,T(t)$ to the porous medium equation, with boundary condition

$$u(x, t) = 0, \qquad \text{for } (x, t) \in \partial\Omega \times (0, +\infty).$$

Chapter 5
Lipschitz Functions

5.1 Definitions and Basic Properties

In this chapter, we discuss another important class of functions which is quite useful in minimization problems: the class of Lipschitz continuous functions. We start with their definition and then explore some of their basic properties, which will be needed in the Chap. 6. For further properties of Lipschitz functions, we refer the reader to [33].

Definition 5.1.1 Let $\Omega \subseteq \mathbb{R}^N$ be a non-empty set and let $f : \Omega \to \mathbb{R}$ be a function. Let $A \subseteq \Omega$, we say that f is *Lipschitz continuous on* A if there exists a constant $L > 0$ such that

$$|f(x) - f(y)| \leq L\,|x - y|, \qquad \text{for every } x, y \in A.$$

If f is Lipschitz on A, we define its *Lipschitz constant on* A as

$$|f|_{C^{0,1}(A)} := \sup_{x,y\in A,\, x\neq y} \frac{|f(x) - f(y)|}{|x - y|}.$$

When $A = \Omega$, we will say that f is *Lipschitz continuous* or simply *Lipschitz*.

Remark 5.1.2 It should be clear from its definition that a Lipschitz function $f : \Omega \to \mathbb{R}$ is continuous at each accumulation point $x \in \Omega$. Moreover, it can always be continuously extended to the set of accumulation points of Ω. Indeed, if $x \in \mathbb{R}^N \setminus \Omega$ is an accumulation point of Ω, then there exists a sequence $\{x_n\}_{n\in\mathbb{N}} \subseteq \Omega$ such that

$$\lim_{n\to\infty} x_n = x.$$

L. Brasco, *Handbook of Calculus of Variations for Absolute Beginners*,
La Matematica per il 3+2 163, https://doi.org/10.1007/978-3-031-87164-1_5

In particular, it is a Cauchy sequence in $\mathbb{R}^N$. For every $n, m \in \mathbb{N}$, by the Lipschitz property, we have

$$|f(x_n) - f(x_m)| \le L\,|x_n - x_m|,$$

which shows that $\{f(x_n)\}_{n\in\mathbb{N}} \subseteq \mathbb{R}$ is a Cauchy sequence, as well. By completeness of $\mathbb{R}$, we can infer convergence of this sequence and we can set

$$f(x) := \lim_{n\to\infty} f(x_n).$$

It is not difficult to see that such a limit does not depend on the particular sequence $\{x_n\}_{n\in\mathbb{N}}$ chosen.

In particular, if $\Omega \subseteq \mathbb{R}^N$ is open, then f can be continuously extended to $\overline{\Omega}$: it is sufficient to recall that, in this case, $\overline{\Omega}$ consists of all the accumulation points of Ω. Moreover, in this case, we have

$$|f|_{C^{0,1}(\overline{\Omega})} = |f|_{C^{0,1}(\Omega)},$$

see Problem 5.5.1 below.

In the next simple result, we highlight an important class of functions which are Lipschitz continuous.

Lemma 5.1.3 *Let $\Omega \subseteq \mathbb{R}^N$ be an open convex set. Let $f : \Omega \to \mathbb{R}$ be a differentiable function, such that*

$$M := \sup_{x\in\Omega} |\nabla f(x)| < +\infty.$$

Then f is Lipschitz continuous and we have

$$|f|_{C^{0,1}(\Omega)} = M.$$

Proof For every $x, y \in \Omega$, by using the Fundamental Theorem of Calculus we have

$$f(x) - f(y) = \int_0^1 \frac{d}{dt} f(t\,x + (1-t)\,y)\,dt = \int_0^1 \langle \nabla f(t\,x + (1-t)\,y), x - y\rangle\,dt,$$

Observe that $t\,x + (1-t)\,y \in \Omega$ for every $t \in [0,1]$, thanks to the convexity of Ω. By taking the absolute value and using the Cauchy-Schwarz inequality, we get

$$|f(x) - f(y)| \le M\,|x - y|.$$

We also used that $\xi \mapsto |\nabla f(\xi)|$ is bounded, thanks to the assumption. By arbitrariness of $x, y \in \Omega$, the previous inequality proves that $|f|_{C^{0,1}(\Omega)} \le M$.

For the reverse inequality, we first recall that for every $z \in \mathbb{R}^N$, we have

$$|z| = \max_{\omega \in \mathbb{S}^{N-1}} \langle z, \omega \rangle. \tag{5.1.1}$$

We then take a point $x_0 \in \Omega$ and a direction $\omega \in \mathbb{S}^{N-1}$. By using the fact that f is differentiable at x_0, we get

$$f(x_0 + t\,\omega) = f(x_0) + t\,\langle \nabla f(x_0), \omega \rangle + o(|t|), \qquad \text{as } t \to 0.$$

In particular, we get for $t \neq 0$ converging to 0

$$\langle \nabla f(x_0), \omega \rangle = \frac{f(x_0 + t\,\omega) - f(x_0)}{t} + \frac{o(|t|)}{t}.$$

Since Ω is open, we have that $x_0 + t\,\omega \in \Omega$, for t small enough. By using the definition of Lipschitz constant, we get in particular

$$\langle \nabla f(x_0), \omega \rangle \leq |f|_{C^{0,1}(\Omega)} + \frac{o(|t|)}{t}, \qquad \text{as } t \to 0.$$

Thus, by taking the limit as t goes to 0, we obtain

$$\langle \nabla f(x_0), \omega \rangle \leq |f|_{C^{0,1}(\Omega)}.$$

Since this is valid for every $\omega \in \mathbb{S}^{N-1}$, from (5.1.1) we get

$$|\nabla f(x_0)| \leq |f|_{C^{0,1}(\Omega)}.$$

By taking the supremum over $x_0 \in \Omega$, we finally get $M \leq |f|_{C^{0,1}(\Omega)}$, as well. □

Remark 5.1.4 The previous result does not hold for a general open set, see Problem 5.5.13 for a counter-example.

Remark 5.1.5 (Affine Functions) Let $c \in \mathbb{R}$ and $\mathbf{b} \in \mathbb{R}^N$. A function $f : \mathbb{R}^N \to \mathbb{R}$ of the form

$$f(x) = c + \langle \mathbf{b}, x \rangle, \qquad \text{for every } x \in \mathbb{R}^N,$$

is Lipschitz continuous on $\mathbb{R}^N$. This is a particular case of Lemma 5.1.3. The latter also gives that

$$|f|_{C^{0,1}(\mathbb{R}^N)} = |\mathbf{b}|.$$

Let $\Omega \subseteq \mathbb{R}^N$ be an open set. In what follows, we will need the following space

$$C^0_{\mathrm{b}}(\overline{\Omega}) = \Big\{ f : \overline{\Omega} \to \mathbb{R} \,:\, f \text{ is continuous and bounded on } \overline{\Omega} \Big\}.$$

We recall that this is a Banach space, when endowed with the norm

$$\|f\|_{C^0(\overline{\Omega})} := \sup_{x \in \overline{\Omega}} |f(x)|, \qquad \text{for every } f \in C^0_{\mathrm{b}}(\overline{\Omega}),$$

see [35, Theorem 7.9] or Problem 5.5.9 below.

Proposition 5.1.6 *Let $\Omega \subseteq \mathbb{R}^N$ be an open set. Then*

$$C^{0,1}(\overline{\Omega}) = \Big\{ f \in C^0_{\mathrm{b}}(\overline{\Omega}) \ : \ f \text{ is Lipschitz continuous} \Big\},$$

is a vector subspace of $C^0_{\mathrm{b}}(\overline{\Omega})$. Moreover, when endowed with the norm

$$\|f\|_{C^{0,1}(\overline{\Omega})} = \|f\|_{C^0(\overline{\Omega})} + |f|_{C^{0,1}(\overline{\Omega})}, \tag{5.1.2}$$

this is a Banach space.

Proof If $f, g \in C^{0,1}(\overline{\Omega})$ and $\alpha, \beta \in \mathbb{R}$, we clearly have that $\alpha\, f + \beta\, g \in C^0_{\mathrm{b}}(\overline{\Omega})$. Moreover, by triangle inequality and the definition of Lipschitz function, we have for every $x, y \in \overline{\Omega}$

$$\begin{aligned} |\alpha\, f(x) + \beta\, g(x) - \alpha\, f(y) - \beta\, g(y)| &\leq |\alpha|\, |f(x) - f(y)| + |\beta|\, |g(x) - g(y)| \\ &\leq \Big(|\alpha|\, |f|_{C^{0,1}(\overline{\Omega})} + |\beta|\, |g|_{C^{0,1}(\overline{\Omega})} \Big)\, |x - y|, \end{aligned}$$

so that $\alpha\, f + \beta\, g$ is still Lipschitz continuous. This shows that $C^{0,1}(\overline{\Omega})$ is a vector space over $\mathbb{R}$.

The fact that (5.1.2) defines a norm on $C^{0,1}(\overline{\Omega})$ is straighforward. Finally, in order to show that $C^{0,1}(\overline{\Omega})$ is a Banach space, we take $\{f_n\}_{n \in \mathbb{N}} \subseteq C^{0,1}(\overline{\Omega})$ a Cauchy sequence. This means that for every $\varepsilon > 0$, there exists $n_\varepsilon \in \mathbb{N}$ such that

$$\|f_n - f_m\|_{C^{0,1}(\overline{\Omega})} < \varepsilon, \qquad \text{for every } n, m \geq n_\varepsilon. \tag{5.1.3}$$

By definition of the norm (5.1.2), this in particular implies that $\{f_n\}_{n \in \mathbb{N}}$ is a Cauchy sequence in the Banach space $C^0_{\mathrm{b}}(\overline{\Omega})$. Thus, there exists $f \in C^0_{\mathrm{b}}(\overline{\Omega})$ such that

$$\lim_{n \to \infty} \|f_n - f\|_{C^0(\overline{\Omega})} = 0.$$

To conclude, we still need to prove that

$$f \in C^{0,1}(\overline{\Omega}) \qquad \text{and} \qquad \lim_{n \to \infty} |f_n - f|_{C^{0,1}(\overline{\Omega})} = 0.$$

The first fact follows from the uniform convergence of the sequence and (5.1.3). Indeed, take $\varepsilon = 1$, then there exists $n_1 \in \mathbb{N}$ such that

$$|f_n - f_{n_1}|_{C^{0,1}(\overline{\Omega})} < 1, \qquad \text{for every } n \geq n_1.$$

This implies that

$$\begin{aligned}|f_n|_{C^{0,1}(\overline{\Omega})} &\le |f_n - f_{n_1}|_{C^{0,1}(\overline{\Omega})} + |f_{n_1}|_{C^{0,1}(\overline{\Omega})}\\ &< 1 + |f_{n_1}|_{C^{0,1}(\overline{\Omega})} =: L, \qquad \text{for every } n \ge n_1,\end{aligned}$$

and thus

$$|f_n(x) - f_n(y)| \le L\,|x-y|, \qquad \text{for every } x, y \in \overline{\Omega},\ n \ge n_1.$$

By taking the limit as n goes to ∞ and using the uniform convergence, we get

$$|f(x) - f(y)| = \lim_{n\to\infty} |f_n(x) - f_n(y)| \le L\,|x-y|, \qquad \text{for every } x, y \in \overline{\Omega}.$$

This shows that $f \in C^{0,1}(\overline{\Omega})$. We now exploit again (5.1.3) in order to conclude. Indeed, the latter implies that for every $\varepsilon > 0$ we have

$$\frac{|f_n(x) - f_m(x) - (f_n(y) - f_m(y))|}{|x-y|} < \varepsilon,$$
$$\text{for every } x, y \in \overline{\Omega} \text{ with } x \ne y,\ n, m \ge n_\varepsilon.$$

We can in particular pass to the limit in the previous estimate as m goes to ∞ and use again the uniform convergence. This gives

$$\frac{|f_n(x) - f(x) - (f_n(y) - f(y))|}{|x-y|} < \varepsilon,$$
$$\text{for every } x, y \in \overline{\Omega} \text{ with } x \ne y,\ n \ge n_\varepsilon.$$

By taking the supremum over x, y, this is the same as

$$|f_n - f|_{C^{0,1}(\overline{\Omega})} \le \varepsilon, \qquad \text{for every } n \ge n_\varepsilon.$$

This finally shows that

$$\lim_{n\to\infty} |f_n - f|_{C^{0,1}(\overline{\Omega})} = 0,$$

thus the proof is concluded. □

The following compactness result will be useful in the sequel. It is an easy consequence of the classical *Ascoli-Arzelà Theorem* (see [24, Chapter IV, Section 19] or [54, Appendix A, Theorem A5]), but it deserves to be explicitly stated.

Theorem 5.1.7 (Compactness) *Let $\Omega \subseteq \mathbb{R}^N$ be an open bounded set and let $\{f_n\}_{n\in\mathbb{N}} \subseteq C^{0,1}(\overline{\Omega})$ be a family of Lipschitz functions, such that*

$$\|f_n\|_{C^{0,1}(\overline{\Omega})} \le M, \qquad \textit{for every } n \in \mathbb{N}.$$

Then there exists a subsequence $\{f_{n_k}\}_{k\in\mathbb{N}}$ *and a function* $f \in C^{0,1}(\overline{\Omega})$*, such that* f_{n_k} *converges uniformly on* $\overline{\Omega}$ *to* f*. Moreover, we have*

$$|f|_{C^{0,1}(\overline{\Omega})} \leq \liminf_{k\to\infty} |f_{n_k}|_{C^{0,1}(\overline{\Omega})}. \tag{5.1.4}$$

Proof The hypothesis implies that $\{f_n\}_{n\in\mathbb{N}}$ is a family of equi-bounded and equi-continuous functions on the compact set $\overline{\Omega}$. We can thus apply the Ascoli-Arzelà Theorem and infer existence of a subsequence $\{f_{n_k}\}_{k\in\mathbb{N}}$ converging uniformly to a continuous function f.

We need to show that f is Lipschitz and that (5.1.4) holds. For every $x, y \in \overline{\Omega}$, we have

$$|f(x) - f(y)| = \lim_{k\to\infty} |f_{n_k}(x) - f_{n_k}(y)| \leq \liminf_{k\to\infty} |f_{n_k}|_{C^{0,1}(\overline{\Omega})} |x - y|.$$

If $x \neq y$, we can divide by $|x - y|$ and then take the supremum, so to get

$$|f|_{C^{0,1}(\Omega)} \leq \liminf_{k\to\infty} |f_{n_k}|_{C^{0,1}(\overline{\Omega})}.$$

This concludes the proof. □

Proposition 5.1.8 *Let* $\mathcal{X} \subseteq C^{0,1}(E)$ *be a family of Lipschitz functions on* $E \subseteq \mathbb{R}^N$*. Let us suppose that there exist* $x_0 \in E$ *and two constants* $\ell \geq 0$, $m \in \mathbb{R}$ *such that*

$$g(x_0) \geq m \qquad \text{and} \qquad |g|_{C^{0,1}(E)} \leq \ell, \qquad \text{for every } g \in \mathcal{X}.$$

Then the function defined by

$$s(x) = \inf_{g\in\mathcal{X}} g(x), \qquad \text{for every } x \in E,$$

is well-defined and Lipschitz continuous. Moreover, we still have

$$|s|_{C^{0,1}(E)} \leq \ell.$$

Proof We first prove that s is well-defined, i.e. it takes finite values everywhere on E. By using that the functions in $\mathcal{X}$ are equi-Lipschitz, we have

$$|g(x) - g(x_0)| \leq \ell\, |x - x_0|, \quad \text{for every } x \in E, g \in \mathcal{X}.$$

This can be rewritten as

$$g(x_0) - \ell\, |x - x_0| \leq g(x) \leq g(x_0) + \ell\, |x - x_0|, \quad \text{for every } x \in E, g \in \mathcal{X}.$$

From the assumption on $g(x_0)$, we get in particular

$$m - \ell\, |x - x_0| \leq g(x).$$

By taking the infimum over $g \in \mathcal{X}$, we get

$$s(x) \geq m - \ell\,|x - x_0| > -\infty, \qquad \text{for every } x \in E.$$

On the other hand, by definition of s, chosen a function $g \in \mathcal{X}$ we have

$$s(x) \leq g(x) < +\infty, \qquad \text{for every } x \in E.$$

This shows that s is well-defined.

We now prove that s is a Lipschitz function. For every $x, y \in E$ by assumption we have

$$|g(x) - g(y)| \leq \ell\,|x - y|, \qquad \text{for every } g \in \mathcal{X}.$$

Again, this can be rewritten as

$$g(y) - \ell\,|x - y| \leq g(x) \leq g(y) + \ell\,|x - y|, \qquad \text{for every } g \in \mathcal{X}.$$

By taking the infimum over $\mathcal{X}$, we get

$$\inf_{g \in \mathcal{X}} g(y) - \ell\,|x - y| \leq \inf_{g \in \mathcal{X}} g(x) \leq \inf_{g \in \mathcal{X}} g(y) + \ell\,|x - y|.$$

By recalling the definition of s, this is the same as

$$|s(x) - s(y)| \leq \ell\,|x - y|.$$

Since $x, y \in E$ are arbitrary, we get the conclusion. □

Remark 5.1.9 A similar statement holds for the supremum of a family of equi-Lipschitz functions. More precisely, if the family $\mathcal{X} \subseteq C^{0,1}(E)$ is such that

$$g(x_0) \leq M \qquad \text{and} \qquad |g|_{C^{0,1}(\Omega)} \leq \ell, \qquad \text{for every } g \in \mathcal{X},$$

then the function defined by

$$S(x) = \sup_{g \in \mathcal{X}} g(x), \qquad \text{for every } x \in E,$$

is well-defined and Lipschitz continuous, with

$$|S|_{C^{0,1}(E)} \leq \ell.$$

It is sufficient to use the previous result for $-g$ and observe that

$$\sup_{g \in \mathcal{X}} g = -\inf_{g \in \mathcal{X}} (-g).$$

The following *extension result* shows that any Lipschitz function can be extended to the whole space, with the same Lipschitz constant (see also Problem 5.5.11 for a variant of this result).

Theorem 5.1.10 (McShane-Whitney) *Let $\Omega \subseteq \mathbb{R}^N$ be a non-empty set and let $f : \Omega \to \mathbb{R}$ be a Lipschitz function. Then there exists $\widetilde{f} : \mathbb{R}^N \to \mathbb{R}$ such that*

- *$\widetilde{f}$ is still Lipschitz and*

$$|\widetilde{f}|_{C^{0,1}(\mathbb{R}^N)} = |f|_{C^{0,1}(\Omega)};$$

- *$\widetilde{f}(x) = f(x)$ for every $x \in \Omega$.*

Proof We define the function

$$\widetilde{f}(x) = \inf_{y\in\Omega} \Big[f(y) + |f|_{C^{0,1}(\Omega)}\, |x-y| \Big], \qquad \text{for } x \in \mathbb{R}^N.$$

We claim that this function has the desired properties. We observe that the family of functions

$$x \mapsto f_y(x) := f(y) + |f|_{C^{0,1}(\Omega)}\, |x-y|, \qquad \text{for } y \in \Omega,$$

is equi-Lipschitz on $\mathbb{R}^N$, indeed for every $x_0, x_1 \in \mathbb{R}^N$ we have

$$\begin{aligned} |f_y(x_0) - f_y(x_1)| &= \Big| f(y) + |f|_{C^{0,1}(\Omega)}\, |x_0 - y| - f(y) - |f|_{C^{0,1}(\Omega)}\, |x_1 - y| \Big| \\ &= |f|_{C^{0,1}(\Omega)} \Big| |x_0 - y| - |x_1 - y| \Big| \le |f|_{C^{0,1}(\Omega)}\, |x_0 - x_1|, \end{aligned}$$

by the triangle inequality. This in particular shows that

$$|f_y|_{C^{0,1}(\mathbb{R}^N)} \le |f|_{C^{0,1}(\Omega)}, \qquad \text{for every } y \in \Omega.$$

Moreover, by using that f is Lipschitz on Ω and the definition of f_y, by taking $x_0 \in \Omega$ we have

$$f_y(x_0) = f(y) + |f|_{C^{0,1}(\Omega)}\, |x_0 - y| \ge f(x_0), \qquad \text{for every } y \in \Omega. \tag{5.1.5}$$

Thus, we can apply Proposition 5.1.8 to our family $\mathcal{X} = \{f_y\}_{y\in\Omega}$, with

$$m = f(x_0), \qquad \ell = |f|_{C^{0,1}(\Omega)}, \qquad E = \mathbb{R}^N,$$

and obtain that $\widetilde{f}$ is Lipschitz and

$$|\widetilde{f}|_{C^{0,1}(\mathbb{R}^N)} \le |f|_{C^{0,1}(\Omega)}.$$

Let us show that $\widetilde{f}$ extends f. We first observe that by taking the infimum over $y \in \Omega$ in (5.1.5), we get

$$\widetilde{f}(x_0) \geq f(x_0), \qquad \text{for every } x_0 \in \Omega.$$

On the other hand, since $\widetilde{f}$ is defined by an infimum, by choosing $y = x_0$ we obtain

$$\widetilde{f}(x_0) \leq f(x_0) + |f|_{C^{0,1}(\Omega)} |x_0 - x_0| = f(x_0).$$

The last two inequalities, show $\widetilde{f}$ coincides with f on Ω.

Finally, we have already observed that the Lipschitz constant of $\widetilde{f}$ is not larger than that of f. In order to get that they actually coincide, it is sufficient to observe that

$$\begin{aligned} |\widetilde{f}|_{C^{0,1}(\mathbb{R}^N)} &= \sup_{x,y\in\mathbb{R}^N,\, x\neq y} \frac{|\widetilde{f}(x) - \widetilde{f}(y)|}{|x-y|} \\ &\geq \sup_{x,y\in\Omega,\, x\neq y} \frac{|\widetilde{f}(x) - \widetilde{f}(y)|}{|x-y|} \\ &= \sup_{x,y\in\Omega,\, x\neq y} \frac{|f(x) - f(y)|}{|x-y|} = |f|_{C^{0,1}(\Omega)}. \end{aligned} \tag{5.1.6}$$

This concludes the proof. □

In what follows, for a non-empty open set $\Omega \subseteq \mathbb{R}^N$, we consider the space of Lipschitz functions vanishing at the boundary, i.e.

$$C_0^{0,1}(\overline{\Omega}) := \Big\{ u \in C^{0,1}(\overline{\Omega}) \,:\, u = 0 \text{ on } \partial\Omega \Big\}. \tag{5.1.7}$$

For these functions, it is possible to construct a simpler extension operator: we can simply extend them by zero. This is the content of the following

Lemma 5.1.11 (Extension by Zero) *Let $\Omega \subseteq \mathbb{R}^N$ be an open set. We define the "extension by zero" operator*

$$\mathcal{Z} : C_0^{0,1}(\overline{\Omega}) \to C^{0,1}(\mathbb{R}^N),$$

through

$$\mathcal{Z}[u](x) = \begin{cases} u(x), & \text{if } x \in \overline{\Omega}, \\ 0, & \text{otherwise}. \end{cases}$$

Then this is a well-defined linear operator and

$$\Big|\mathcal{Z}[u]\Big|_{C^{0,1}(\mathbb{R}^N)} = |u|_{C^{0,1}(\overline{\Omega})}. \tag{5.1.8}$$

Proof The linearity of $\mathcal{Z}$ is immediate, from its definition. Moreover, for every $u \in C_0^{0,1}(\overline{\Omega})$ it is easy to see that $\mathcal{Z}[u] \in C_b^0(\mathbb{R}^N)$. We need to prove that $\mathcal{Z}[u]$ is Lipschitz on $\mathbb{R}^N$.

The fact that it is Lipschitz separately on $\overline{\Omega}$ and $\mathbb{R}^N \setminus \overline{\Omega}$ follows directly from its definition. Indeed, we have

$$|\mathcal{Z}[u](x) - \mathcal{Z}[u](y)| = |u(x) - u(y)| \le |u|_{C^{0,1}(\overline{\Omega})}\, |x - y|, \qquad \text{for every } x, y \in \overline{\Omega},$$

and

$$|\mathcal{Z}[u](x) - \mathcal{Z}[u](y)| = 0, \qquad \text{for every } x, y \in \mathbb{R}^N \setminus \overline{\Omega}.$$

We now take $x \in \overline{\Omega}$ and $y \in \mathbb{R}^N \setminus \overline{\Omega}$. We consider the segment $\overline{x\,y}$ connecting x to y and observe that this must intersect $\partial\Omega$ in at least a point x_0. Moreover, we clearly have

$$|x - x_0| \le |x - y|,$$

and by assumption $u(x_0) = 0$. We thus obtain

$$\begin{aligned}|\mathcal{Z}[u](x) - \mathcal{Z}[u](y)| &= |u(x)| \\ &= |u(x) - u(x_0)| \\ &\le |u|_{C^{0,1}(\overline{\Omega})}\, |x - x_0| \le |u|_{C^{0,1}(\overline{\Omega})}\, |x - y|,\end{aligned}$$

for every $x \in \overline{\Omega}$ and every $y \in \mathbb{R}^N \setminus \overline{\Omega}$. This discussion entails that $\mathcal{Z}[u] \in C^{0,1}(\mathbb{R}^N)$ and that

$$\Big|\mathcal{Z}[u]\Big|_{C^{0,1}(\mathbb{R}^N)} \le |u|_{C^{0,1}(\overline{\Omega})}.$$

On the other hand, by proceeding as in (5.1.6), we get the reverse estimate, as well. Thus, the identity (5.1.8) follows. □

5.2 Differentiability of Lipschitz Functions

We first show that Lipschitz functions defined on open sets are actually Sobolev functions, in the sense that they have a *weak gradient* in L^∞ (see Definition 3.2.1). More precisely, we have the following

Lemma 5.2.1 *Let $\Omega \subseteq \mathbb{R}^N$ be an open set and let $f : \Omega \to \mathbb{R}$ be a Lipschitz function. Then f has a weak gradient $\nabla f \in L^\infty(\Omega; \mathbb{R}^N)$.*

Proof We first prove the result in the case $\Omega = \mathbb{R}^N$. In the end, we will explain how to get the general case from it.

We fix $i \in \{1, \dots, N\}$ and take $\varphi \in C_0^\infty(\mathbb{R}^N)$. We indicate by K the support of φ, then we observe that for every fixed $h \neq 0$ the function

$$x \mapsto \frac{\varphi(x + h\,\mathbf{e}_i) - \varphi(x)}{h},$$

is still compactly supported. More precisely, if we fix $h_0 > 0$, for every $0 < |h| < h_0$ the support of this function is contained in

$$K_{h_0} := \Big\{x \in \mathbb{R}^N \ : \ \operatorname{dist}(x, K) < h_0\Big\}.$$

Since f is Lipschitz continuous on $\mathbb{R}^N$, then it is bounded on the bounded set K_{h_0} (see Problem 5.5.8 below). Thus, the following integral

$$\int_{\mathbb{R}^N} f(x)\,\frac{\varphi(x + h\,\mathbf{e}_i) - \varphi(x)}{h}\,dx,$$

is well-defined and finite. Moreover, we observe that for every $x \in \mathbb{R}^N$ and every $0 < |h| < h_0$ we have

$$\left|f(x)\,\frac{\varphi(x + h\,\mathbf{e}_i) - \varphi(x)}{h}\right| \le |f(x)|\,|\varphi|_{C^{0,1}(\mathbb{R}^N)} = |f(x)|\,\sup_{\mathbb{R}^N} |\nabla \varphi|,$$

and the last function is in $L^1(K_{h_0})$. Notice that we have used Lemma 5.1.3 for the smooth function φ. Thus, by the Dominated Convergence Theorem, we can infer

$$\int_{\mathbb{R}^N} f(x)\,\frac{\partial \varphi}{\partial x_i}(x)\,dx = \lim_{h \to 0} \int_{\mathbb{R}^N} f(x)\,\frac{\varphi(x + h\,\mathbf{e}_i) - \varphi(x)}{h}\,dx.$$

Thanks to the compactness of the support of φ, we can perform a “discrete” integration by parts in the last integral. Indeed, for $0 < |h| < h_0$ we have

$$\begin{aligned}
&\int_{\mathbb{R}^N} f(x)\,\frac{\varphi(x + h\,\mathbf{e}_i) - \varphi(x)}{h}\,dx \\
&= \int_{K - h\,\mathbf{e}_i} f(x)\,\frac{\varphi(x + h\,\mathbf{e}_i)}{h}\,dx - \int_K f(x)\,\frac{\varphi(x)}{h}\,dx \\
&= \int_K f(y - h\,\mathbf{e}_i)\,\frac{\varphi(y)}{h}\,dy - \int_K f(x)\,\frac{\varphi(x)}{h}\,dx \\
&= \int_{\mathbb{R}^N} \frac{f(x - h\,\mathbf{e}_i) - f(x)}{h}\,\varphi(x)\,dx.
\end{aligned}$$

In conclusion, we obtained

$$\begin{aligned}\int_{\mathbb{R}^N} f(x)\,\frac{\partial\varphi}{\partial x_i}(x)\,dx &= \lim_{h\to 0}\int_{\mathbb{R}^N} f(x)\,\frac{\varphi(x+h\,\mathbf{e}_i)-\varphi(x)}{h}\,dx \\ &= \lim_{h\to 0}\int_{\mathbb{R}^N} \frac{f(x-h\,\mathbf{e}_i)-f(x)}{h}\,\varphi(x)\,dx.\end{aligned} \tag{5.2.1}$$

We observe that for every $x \in \mathbb{R}^N$, by the Lipschitz character of f we can infer

$$\left|\frac{f(x-h\,\mathbf{e}_i)-f(x)}{h}\right| \le |f|_{C^{0,1}(\mathbb{R}^N)}, \qquad \text{for every } 0<|h|<h_0.$$

In conclusion, from (5.2.1) we obtain

$$\left|\int_{\mathbb{R}^N} f\,\frac{\partial\varphi}{\partial x_i}\,dx\right| \le |f|_{C^{0,1}(\mathbb{R}^N)}\,\|\varphi\|_{L^1(\mathbb{R}^N)}, \quad \text{for every } \varphi\in C_0^\infty(\mathbb{R}^N). \tag{5.2.2}$$

By recalling that $C_0^\infty(\mathbb{R}^N)$ is dense[1] in $L^1(\mathbb{R}^N)$ (see [46, Theorem 2.16 & Lemma 2.19]), the estimate (5.2.2) guarantees that we can extend the linear and continuous functional

$$\varphi \mapsto \int_{\mathbb{R}^N} f\,\frac{\partial\varphi}{\partial x_i}\,dx, \tag{5.2.3}$$

to the whole $L^1(\mathbb{R}^N)$, by density. In other words, we define the functional

$$\mathcal{T}_i : L^1(\mathbb{R}^N) \to \mathbb{R},$$

such that for every $\varphi \in L^1(\mathbb{R}^N)$ we have

$$\mathcal{T}_i(\varphi) = \lim_{n\to\infty}\int_{\mathbb{R}^N} f\,\frac{\partial\varphi_n}{\partial x_i}\,dx,$$

where $\{\varphi_n\}_{n\in\mathbb{N}} \subseteq C_0^\infty(\mathbb{R}^N)$ converges to φ in $L^1(\mathbb{R}^N)$. The reader should verify that this definition is well-posed (i.e. such a limit exists and it does not depend on the particular sequence $\{\varphi_n\}_{n\in\mathbb{N}}$ chosen) and that $\mathcal{T}_i$ is still linear and continuous.

In particular, the functional $\mathcal{T}_i$ belongs to the topological dual space of $L^1(\mathbb{R}^N)$ and by construction we have from (5.2.2)

$$|\mathcal{T}_i(\varphi)| \le |f|_{C^{0,1}(\mathbb{R}^N)}\,\|\varphi\|_{L^1(\mathbb{R}^N)}, \qquad \text{for every } \varphi\in L^1(\mathbb{R}^N).$$

[1] That is, for every $\varphi \in L^1(\mathbb{R}^N)$ there exists a sequence $\{\varphi_n\}_{n\in\mathbb{N}} \subseteq C_0^\infty(\mathbb{R}^N)$ such that

$$\lim_{n\to\infty}\|\varphi_n-\varphi\|_{L^1(\mathbb{R}^N)} = 0.$$

By recalling that the topological dual space of $L^1(\mathbb{R}^N)$ can be identified with $L^\infty(\mathbb{R}^N)$ (see for example [46, Theorem 2.14]), we can infer the existence of a (unique!) function $\psi_i \in L^\infty(\mathbb{R}^N)$ such that $\mathcal{T}_i$ can be represented as follows

$$\mathcal{T}_i(\varphi) = \int_{\mathbb{R}^N} \psi_i\, \varphi\, dx, \qquad \text{for every } \varphi \in L^1(\mathbb{R}^N).$$

By setting $g_i = -\psi_i \in L^\infty(\mathbb{R}^N)$ and using that $\mathcal{T}_i$ coincides with (5.2.3) on $C_0^\infty(\mathbb{R}^N)$, we get

$$\int_{\mathbb{R}^N} f\, \frac{\partial \varphi}{\partial x_i}\, dx = \mathcal{T}_i(\varphi) = -\int_{\mathbb{R}^N} g_i\, \varphi\, dx, \qquad \text{for every } \varphi \in C_0^\infty(\mathbb{R}^N).$$

By recalling Definition 3.2.1, this shows that f has a weak gradient in $L^\infty(\mathbb{R}^N; \mathbb{R}^N)$. This concludes the proof in the case $\Omega = \mathbb{R}^N$.

Finally, if f is a Lipschitz function on a general open set $\Omega \subseteq \mathbb{R}^N$, we can consider its extension $\widetilde{f}$ to the whole $\mathbb{R}^N$, provided by Theorem 5.1.10. From the first part of the proof, we know that there exists $\widetilde{g}_1, \dots, \widetilde{g}_N \in L^\infty(\mathbb{R}^N)$ such that

$$\int_{\mathbb{R}^N} \widetilde{f}\, \frac{\partial \varphi}{\partial x_i}\, dx = -\int_{\mathbb{R}^N} \widetilde{g}_i\, \varphi\, dx, \qquad \text{for every } \varphi \in C_0^\infty(\mathbb{R}^N).$$

We now observe that $C_0^\infty(\Omega) \subseteq C_0^\infty(\mathbb{R}^N)$, where we intend that every function $C_0^\infty(\Omega)$ is extended by 0 outside Ω. Then, we may take $\varphi \in C_0^\infty(\Omega)$ in the previous identity. This gives

$$\int_{\Omega} f\, \frac{\partial \varphi}{\partial x_i}\, dx = \int_{\Omega} \widetilde{f}\, \frac{\partial \varphi}{\partial x_i}\, dx = -\int_{\Omega} \widetilde{g}_i\, \varphi\, dx, \qquad \text{for every } \varphi \in C_0^\infty(\Omega).$$

In the first identity, we used that $\widetilde{f} = f$ on Ω. By defining

$$g_i(x) = \widetilde{g}_i(x), \qquad \text{for a. e. } x \in \Omega,$$

we get the desired conclusion $\nabla f \in L^\infty(\Omega; \mathbb{R}^N)$, with

$$\nabla f(x) = \nabla \widetilde{f}(x), \qquad \text{for a. e. } x \in \Omega.$$

This concludes the proof. □

Actually, Lipschitz functions have the following remarkable property: they are differentiable almost everywhere. The proof is taken from [15], up to some additional details. For a different proof, which does not rely on the existence of the weak gradient, the reader is referred for example to [33, Theorem 3.1].

Theorem 5.2.2 (Rademacher) *Let $\Omega \subseteq \mathbb{R}^N$ be an open set and let $f : \Omega \to \mathbb{R}$ be a Lipschitz function. Then f is differentiable almost everywhere in Ω. More precisely, for almost every $x_0 \in \Omega$ the following formula holds*

$$f(x) = f(x_0) + \langle \nabla f(x_0), x - x_0 \rangle + o(|x - x_0|), \qquad \text{for } x \to x_0,$$

where ∇f is the weak gradient of f, given by Lemma 5.2.1.

Proof Thanks to Theorem 5.1.10, we can assume without loss of generality that f is defined on the whole $\mathbb{R}^N$. We indicate by ∇f the weak gradient of f, which belongs to $L^\infty(\mathbb{R}^N; \mathbb{R}^N)$ by Lemma 5.2.1. In particular, for almost every $x_0 \in \mathbb{R}^N$ we have

$$\lim_{r \to 0} \frac{1}{|B_r(x_0)|} \int_{B_r(x_0)} |\nabla f(x) - \nabla f(x_0)| \, dx = 0, \tag{5.2.4}$$

thanks to the Lebesgue Differentiation Theorem, see for example [28, Theorem 1, Section 1.7]. We select one of these points x_0 and define the family of Lipschitz functions $\{f_r\}_{r>0} \subseteq C^{0,1}(\overline{B_1(0)})$ by

$$f_r(x) = \frac{f(x_0 + r\,x) - f(x_0)}{r}, \qquad \text{for } x \in \overline{B_1(0)}.$$

By using the Lipschitz character of f, for every $x, y \in \overline{B_1(0)}$ we have

$$|f_r(x)| = \left| \frac{f(x_0 + r\,x) - f(x_0)}{r} \right| \le |f|_{C^{0,1}(\mathbb{R}^N)} \, |x| \le |f|_{C^{0,1}(\mathbb{R}^N)},$$

and

$$\begin{aligned} |f_r(x) - f_r(y)| &= \left| \frac{f(x_0 + r\,x) - f(x_0)}{r} - \frac{f(x_0 + r\,y) - f(x_0)}{r} \right| \\ &= \left| \frac{f(x_0 + r\,x) - f(x_0 + r\,y)}{r} \right| \\ &\le |f|_{C^{0,1}(\mathbb{R}^N)} \, |x - y|. \end{aligned}$$

This shows that for every $r > 0$ we have

$$\|f_r\|_{C^{0,1}(\overline{B_1(0)})} = \|f_r\|_{C^0(\overline{B_1(0)})} + |f_r|_{C^{0,1}(\overline{B_1(0)})} \le 2\,|f|_{C^{0,1}(\mathbb{R}^N)}.$$

In particular, for every sequence $\{r_n\}_{n \in \mathbb{N}}$ of positive numbers converging to 0, the sequence $\{f_{r_n}\}_{n \in \mathbb{N}}$ is equi-bounded in Lipschitz norm. We can thus apply the compactness criterion of Theorem 5.1.7 and infer that there exists $g \in C^{0,1}(\overline{B_1(0)})$ such that (up to a subsequence) we have

$$\lim_{n \to \infty} \|f_{r_n} - g\|_{C^0(\overline{B_1(0)})} = 0.$$

We can apply Lemma 5.2.1 to g and obtain that this has a weak gradient ∇g. Let us compute it: for every $\varphi \in C_0^\infty(B_1(0))$ and every $i = 1, \dots, N$ we have

$$\begin{aligned}\int_{B_1(0)} g\, \frac{\partial \varphi}{\partial x_i}\, dx &= \lim_{n\to\infty} \int_{B_1(0)} f_{r_n}\, \frac{\partial \varphi}{\partial x_i}\, dx \\ &= \lim_{n\to\infty} \int_{B_1(0)} \frac{f(x_0 + r_n\, x) - f(x_0)}{r_n}\, \frac{\partial \varphi}{\partial x_i}(x)\, dx \\ &= -\lim_{n\to\infty} \int_{B_1(0)} \frac{\partial}{\partial x_i} \frac{f(x_0 + r_n\, x) - f(x_0)}{r_n}\, \varphi(x)\, dx \\ &= -\lim_{n\to\infty} \int_{B_1(0)} \frac{\partial f}{\partial x_i}(x_0 + r_n\, x)\, \varphi(x)\, dx,\end{aligned} \tag{5.2.5}$$

where we used the uniform convergence of f_{r_n}, the definition of weak derivative applied to $x \mapsto (f(x_0 + r_n\, x) - f(x_0))/r$ and the fact that

$$\frac{\partial}{\partial x_i} \frac{f(x_0 + r_n\, x) - f(x_0)}{r_n} = \frac{\partial f}{\partial x_i}(x_0 + r_n\, x).$$

This last fact follows from Theorem 3.2.7. We claim that

$$\lim_{n\to\infty} \int_{B_1(0)} \frac{\partial f}{\partial x_i}(x_0 + r_n\, x)\, \varphi(x)\, dx = \int_{B_1(0)} \frac{\partial f}{\partial x_i}(x_0)\, \varphi(x)\, dx.$$

In order to show it, we perform the change of variable $x_0 + r_n\, x = y$, so to obtain

$$\begin{aligned}&\left| \int_{B_1(0)} \frac{\partial f}{\partial x_i}(x_0 + r_n\, x)\, \varphi(x)\, dx - \int_{B_1(0)} \frac{\partial f}{\partial x_i}(x_0)\, \varphi(x)\, dx \right| \\ &\quad = \frac{1}{r_n^N} \left| \int_{B_{r_n}(x_0)} \left[\frac{\partial f}{\partial x_i}(y) - \frac{\partial f}{\partial x_i}(x_0) \right] \varphi\left(\frac{y - x_0}{r_n} \right) dy \right| \\ &\quad \le \omega_N\, \|\varphi\|_{C^0(\overline{B_1(0)})}\, \frac{1}{|B_{r_n}(x_0)|} \int_{B_{r_n}(x_0)} \left| \frac{\partial f}{\partial x_i}(y) - \frac{\partial f}{\partial x_i}(x_0) \right| dx.\end{aligned}$$

If we now apply (5.2.4), we see that the last integral converges to 0, as n goes to ∞. From (5.2.5), we finally get for every $\varphi \in C_0^\infty(B_1(0))$ and $i = 1, \dots, N$

$$\int_{B_1(0)} g\, \frac{\partial \varphi}{\partial x_i}\, dx = -\int_{B_1(0)} \frac{\partial f}{\partial x_i}(x_0)\, \varphi\, dx.$$

This shows that the weak gradient of g is given by the *constant* vector field

$$\nabla g(x) = \nabla f(x_0), \qquad \text{for a. e. } x \in \overline{B_1(0)}.$$

By observing that the linear function $k(x) = \langle \nabla f(x_0), x\rangle$ is obviously C^1, we can use Proposition 3.2.3 to infer that its weak gradient coincides with the classical one. The latter is given by $\nabla k = \nabla f(x_0)$. Thus, if we set $h = g - k$, this is still a Lipschitz function and its weak gradient is given by

$$\nabla h(x) = \nabla g(x) - \nabla k(x) = \nabla f(x_0) - \nabla f(x_0) = (0, \ldots, 0), \qquad \text{for a. e. } x \in B_1(0).$$

By recalling Proposition 3.2.9, we can infer that h is constant on $\overline{B_1(0)}$, that is g coincides with k up to a constant. Thus, we get

$$g(x) = a + \langle \nabla f(x_0), x\rangle, \qquad \text{for } x \in \overline{B_1(0)},$$

for some constant $a \in \mathbb{R}$. We now recall that the function g is a uniform limit of functions vanishing at $x = 0$, thus it has the same property. This entails that we must have $a = 0$, which means that

$$g(x) = \langle \nabla f(x_0), x\rangle, \qquad \text{for } x \in \overline{B_1(0)}.$$

Up to now, we thus have proven that for every sequence $\{r_n\}_{n\in\mathbb{N}}$ of positive radii converging to 0, we have

$$\lim_{n\to\infty} \|f_{r_n} - g\|_{C^0(\overline{B_1(0)})} = 0,$$

up to a subsequence. Moreover, the limit function g does not depend on the particular sequence $\{r_n\}_{n\in\mathbb{N}}$. This actually implies that

$$\lim_{r\to 0^+} \|f_r - g\|_{C^0(\overline{B_1(0)})} = 0.$$

By recalling the form of g, this means that for every $\varepsilon > 0$ there exists $\delta > 0$ such that for every $0 < r < \delta$ we have

$$\max_{y\in\overline{B_1(0)}} \left| \frac{f(x_0 + r\, y) - f(x_0)}{r} - \langle \nabla f(x_0), y\rangle \right| < \varepsilon.$$

In particular, by taking $x \in B_\delta(x_0) \setminus \{x_0\}$ and using this fact with

$$r = |x - x_0| \qquad \text{and} \qquad y = \frac{x - x_0}{|x - x_0|},$$

we get

$$\left| \frac{f(x) - f(x_0)}{|x - x_0|} - \frac{\langle \nabla f(x_0), x - x_0\rangle}{|x - x_0|} \right| < \varepsilon$$

By appealing to the very definition of limit, this exactly proves that

$$\lim_{x \to x_0} \frac{|f(x) - f(x_0) - \langle \nabla f(x_0), x - x_0 \rangle|}{|x - x_0|} = 0,$$

as desired. □

Remark 5.2.3 From the previous result, we can also obtain

$$\|\nabla f\|_{L^\infty(\Omega;\mathbb{R}^N)} \le |f|_{C^{0,1}(\Omega)}. \tag{5.2.6}$$

In order to prove this, it is sufficient to proceed as in the second part of the proof of Lemma 5.1.3. We leave the details to the reader, as a useful exercise.

5.3 Some Operations on Lipschitz Functions

The following result will be useful.

Proposition 5.3.1 *Let $\Omega \subseteq \mathbb{R}^N$ be an open set and let $u : \overline{\Omega} \to \mathbb{R}$ be a Lipschitz function. If we set*

$$u_+(x) = \max\{u(x), 0\} \qquad \textit{and} \qquad u_-(x) = \min\{u(x), 0\},$$

these are still Lipschitz functions, with

$$|u_+|_{C^{0,1}(\overline{\Omega})} \le |u|_{C^{0,1}(\overline{\Omega})} \qquad \textit{and} \qquad |u_-|_{C^{0,1}(\overline{\Omega})} \le |u|_{C^{0,1}(\overline{\Omega})}.$$

Moreover, for almost every $x \in \Omega$ we have

$$\nabla u_+(x) = \begin{cases} \nabla u(x), & \textit{if } u(x) > 0, \\ 0, & \textit{if } u(x) \le 0, \end{cases} \qquad \textit{and} \quad \nabla u_-(x) = \begin{cases} \nabla u(x), & \textit{if } u(x) < 0, \\ 0, & \textit{if } u(x) \ge 0. \end{cases}$$

Proof We limit ourselves to give the proof for u_+. The proof for u_- is similar and it is left to the reader as an exercise.

Let $x, y \in \overline{\Omega}$, without loss of generality we can suppose that $u(x) \ge u(y)$. This implies that $u_+(x) \ge u_+(y)$, as well. If $u(y) \ge 0$, then we have

$$|u_+(x) - u_+(y)| = u_+(x) - u_+(y) = u(x) - u(y) \le |u|_{C^{0,1}(\overline{\Omega})}\, |x - y|.$$

If $u(x) \le 0$, then we clearly have $u_+(x) = u_+(y) = 0$. Finally, if $u(x) \ge 0 \ge u(y)$, then we get

$$|u_+(x) - u_+(y)| = u_+(x) - u_+(y) = u(x) \le u(x) - u(y) \le |u|_{C^{0,1}(\overline{\Omega})}\, |x - y|.$$

Thus, we get that u_+ is Lipschitz continuous, with a Lipschitz constant not exceeding that of u.

We then compute the gradient of u_+. Let us take $x_0 \in \Omega$ a point of differentiability (in classical sense) for both u and u_+. By Rademacher's Theorem, we know that almost every point of Ω has this property. Observe that, since Ω is open, there exists $t_0 > 0$ such that for every $i \in \{1, \dots, N\}$

$$x_0 + t\,\mathbf{e}_i \in \Omega, \qquad \text{for every } |t| \le t_0.$$

We have two possibilities:

$$\text{either} \quad u(x_0) \neq 0 \quad \text{or} \quad u(x_0) = 0.$$

In the first case, we can suppose that[2] $u(x_0) > 0$. Since u is continuous, we have that

$$u(x_0 + t\,\mathbf{e}_i) > 0, \qquad \text{for every } |t| \text{ small enough.}$$

Accordingly, we get

$$\frac{\partial u_+}{\partial x_i}(x_0) = \lim_{t\to 0} \frac{u_+(x_0 + t\,\mathbf{e}_i) - u_+(x_0)}{t} = \lim_{t\to 0} \frac{u(x_0 + t\,\mathbf{e}_i) - u(x_0)}{t} = \frac{\partial u}{\partial x_i}(x_0).$$

This proves the claimed formula, in the case $u(x_0) \neq 0$.

We still have to take into account the possibility $u(x_0) = 0$. Here as well, we have a dichotomy:

$$\text{either} \quad |\nabla u(x_0)| > 0 \quad \text{or} \quad |\nabla u(x_0)| = 0.$$

If $|\nabla u(x_0)| = 0$, we then obtain

$$u(x) = o(|x - x_0|), \qquad \text{as } x \to x_0,$$

by using that u is differentiable at x_0 and $u(x_0) = 0$. Thus, by applying this formula with $x = x_0 + t\,\mathbf{e}_i$, we can compute

$$\frac{\partial u_+}{\partial x_i}(x_0) = \lim_{t\to 0} \frac{u_+(x_0 + t\,\mathbf{e}_i) - u_+(x_0)}{t} = \lim_{t\to 0} \frac{\max\{o(|t|), 0\}}{t} = 0.$$

This proves again the desired formula for the gradient.

[2] If $u(x_0) < 0$, by continuity we have that $u < 0$ in a neighborhood of x_0. Accordingly, we get that u_+ identically vanishes in this neighborhood. Thus, the claimed formula easily follows.

Finally, let us suppose that $|\nabla u(x_0)| > 0$. Without loss of generality, we can assume that $\langle \nabla u(x_0), \mathbf{e}_1\rangle > 0$. Since u is differentiable at x_0 and $u(x_0) = 0$, we get

$$u(x_0 + t\,\mathbf{e}_1) = \langle \nabla u(x_0), \mathbf{e}_1\rangle\, t + o(|t|), \qquad \text{for } t \to 0.$$

This in turn implies that

$$\begin{aligned}&\lim_{t\to 0^+} \frac{u_+(x_0 + t\,\mathbf{e}_1) - u_+(x_0)}{t}\\ &= \lim_{t\to 0^+} \frac{\max\{\langle \nabla u(x_0), \mathbf{e}_1\rangle\, t + o(|t|), 0\}}{t} = \langle \nabla u(x_0), \mathbf{e}_1\rangle,\end{aligned}$$

while

$$\lim_{t\to 0^-} \frac{u_+(x_0 + t\,\mathbf{e}_1) - u_+(x_0)}{t} = \lim_{t\to 0^-} \frac{\max\{\langle \nabla u(x_0), \mathbf{e}_1\rangle\, t + o(|t|), 0\}}{t} = 0,$$

thus showing that u_+ is not differentiable at x_0. This contradicts the fact that x_0 is a point of differentiability for u_+, thus the possibility $|\nabla u(x_0)| > 0$ and $u(x_0) = 0$ is ruled out. This concludes the proof. □

Remark 5.3.2 In the previous proof, in order to compute the gradients of u_+ and u_- one could also proceed as follows: by Rademacher's Theorem, we know that these functions are both differentiable almost everywhere and their classical gradients coincide with the weak one. Then one could simply apply Proposition 3.4.3 with $p = \infty$. We preferred to give above a more self-contained proof, not appealing to the results of Chap. 3.

Corollary 5.3.3 *Let $\Omega \subseteq \mathbb{R}^N$ be an open set and let $u, v : \overline{\Omega} \to \mathbb{R}$ be two Lipschitz functions. Then*

$$W = \max\{u,\, v\} \qquad \textit{and} \qquad w = \min\{u,\, v\},$$

are still Lipschitz functions, with

$$|W|_{C^{0,1}(\overline{\Omega})} \le \max\left\{|u|_{C^{0,1}(\overline{\Omega})},\ |v|_{C^{0,1}(\overline{\Omega})}\right\},$$

and

$$|w|_{C^{0,1}(\overline{\Omega})} \le \max\left\{|u|_{C^{0,1}(\overline{\Omega})},\ |v|_{C^{0,1}(\overline{\Omega})}\right\}.$$

Moreover, for almost every $x \in \Omega$ we have

$$\nabla W(x) = \begin{cases} \nabla u(x), & \textit{if } u(x) > v(x),\\ \nabla v(x), & \textit{if } v(x) \ge u(x),\end{cases} \quad \textit{and} \quad \nabla w(x) = \begin{cases} \nabla u(x), & \textit{if } u(x) < v(x),\\ \nabla v(x), & \textit{if } u(x) \ge v(x).\end{cases}$$

Proof The fact that W and w are Lipschitz continuous follows from Proposition 5.1.8 and Remark 5.1.9.

In order to compute the gradient of W, it is sufficient to observe that for every $a, b \in \mathbb{R}$ we have

$$\max\{a, b\} = (a - b)_+ + b.$$

This implies that

$$W(x) = (u(x) - v(x))_+ + v(x), \qquad \text{for } x \in \Omega.$$

We can now apply Proposition 5.3.1 and get the desired conclusion. The proof for w is similar and it is left to the reader. □

Remark 5.3.4 As already observed in Remark 3.4.5 in the case of Sobolev functions, if u, v are Lipschitz functions on $\overline{\Omega}$ we still have

$$\nabla u(x) = \nabla v(x), \qquad \text{for a. e. } x \in \{y \in \Omega : u(y) = v(y)\}, \tag{5.3.1}$$

as a consequence of Corollary 5.3.3.

Finally, we come to the absolute value of a Lipschitz function. We first observe that if u is a Lipschitz function, then the function $|u|$ is Lipschitz as well. Indeed, by the triangle inequality we have

$$\Big||u(x)| - |u(y)|\Big| \leq |u(x) - u(y)| \leq |u|_{C^{0,1}(\overline{\Omega})}\, |x - y|.$$

Moreover, we have the following

Corollary 5.3.5 *Let $\Omega \subseteq \mathbb{R}^N$ be an open set and let $u : \Omega \to \mathbb{R}$ be a Lipschitz function. Then*

$$\big|\nabla|u(x)|\big| = \big|\nabla u(x)\big|, \qquad \textit{for a. e. } x \in \Omega.$$

Proof Observe that

$$|u(x)| = \max\{u(x), -u(x)\}.$$

By applying Corollary 5.3.3, we have for almost every $x \in \Omega$

$$\nabla|u(x)| = \begin{cases} \nabla u(x), \text{ if } u(x) > 0, \\ -\nabla u(x), \text{ if } u(x) \leq 0. \end{cases}$$

By taking the modulus on both sides, we get the desired conclusion. □

5.4 Lipschitz Functions VS. Sobolev Functions

In light of Lemma 5.2.1, the next result is certainly not surprising. We recall that the Sobolev space $W^{1,p}(\Omega)$ has been defined in Definition 3.3.1.

Proposition 5.4.1 *Let $\Omega \subseteq \mathbb{R}^N$ be an open set. Then we have*

$$C^{0,1}(\overline{\Omega}) \subseteq W^{1,\infty}(\Omega),$$

with continuous inclusion. More precisely, we have the estimate

$$\|u\|_{W^{1,\infty}(\Omega)} \leq \|u\|_{C^{0,1}(\overline{\Omega})}, \qquad \text{for every } u \in C^{0,1}(\overline{\Omega}).$$

Proof By Lemma 5.2.1, we get that every Lipschitz function has a weak gradient in $L^\infty(\Omega; \mathbb{R}^N)$. Moreover, by (5.2.6) of Remark 5.2.3, we get

$$\|\nabla u\|_{L^\infty(\Omega;\mathbb{R}^N)} \leq |u|_{C^{0,1}(\overline{\Omega})}.$$

Thus, the result immediately follows, by recalling that the elements of $C^{0,1}(\overline{\Omega})$ are bounded functions (see Proposition 5.1.6) and that

$$\|u\|_{C^0(\overline{\Omega})} = \|u\|_{L^\infty(\Omega)},$$

for a continuous function (recall Problem 1.7.31). □

Corollary 5.4.2 *Let $\Omega \subseteq \mathbb{R}^N$ be an open set. For every $1 \leq p < \infty$ we have*

$$C^{0,1}(\overline{\Omega}) \subseteq W^{1,p}_{\mathrm{loc}}(\Omega).$$

Moreover, if Ω has finite measure, then for every $1 \leq p < \infty$ we have

$$C^{0,1}(\overline{\Omega}) \subseteq W^{1,p}(\Omega),$$

with continuous inclusion. More precisely, we have the estimate

$$\|u\|_{W^{1,p}(\Omega)} \leq |\Omega|^{\frac{1}{p}} \|u\|_{C^{0,1}(\overline{\Omega})}, \qquad \text{for every } u \in C^{0,1}(\overline{\Omega}).$$

Proof This follows directly from the previous result, by recalling that

$$W^{1,\infty}(E) \subseteq W^{1,p}(E),$$

for every open set $E \subseteq \mathbb{R}^N$ with finite measure and every $1 \leq p < \infty$, see Problem 3.12.6. □

It is natural to inquire whether the two spaces

$$C^{0,1}(\overline{\Omega}) \qquad \text{and} \qquad W^{1,\infty}(\Omega),$$

coincide or not. This is true, provided the set is "nice". This is the content of the next result.

Proposition 5.4.3 *Let $\Omega \subseteq \mathbb{R}^N$ be an open convex set. Then we have*

$$C^{0,1}(\overline{\Omega}) = W^{1,\infty}(\Omega).$$

Proof In light of Proposition 5.4.1, we only need to prove that

$$C^{0,1}(\overline{\Omega}) \supseteq W^{1,\infty}(\Omega).$$

By recalling that Sobolev spaces are classes of equivalences of functions coinciding almost everywhere, we need to prove that every $\varphi \in W^{1,\infty}(\Omega)$ coincides almost everywhere in Ω with a Lipschitz continuous function. We give the details for the case $\Omega \neq \mathbb{R}^N$. If Ω coincides with the whole space, the proof can be easily adapted (it is actually simpler).

We define

$$r_\Omega = \sup_{x\in\Omega} \operatorname{dist}(x, \partial\Omega),$$

and consider the open set $\operatorname{int}_n(\Omega)$ defined by

$$\operatorname{int}_n(\Omega) = \{x \in \Omega\, :\, \operatorname{dist}(x, \partial\Omega) > 1/n\}, \qquad \text{for } n > \frac{1}{r_\Omega}.$$

We observe that this is a convex set, thanks to the convexity of Ω. We take $\varphi \in W^{1,\infty}(\Omega)$ and consider the following regularized sequence

$$\varphi_n(x) := \varphi * \rho_n(x) = \int_{B_{\frac{1}{n}}(x)} \varphi(y)\, \rho_n(x-y)\, dy, \qquad \text{for } x \in \operatorname{int}_n(\Omega),$$

where

$$\rho_n(x) = n^N\, \rho(n\, x),$$

and ρ is the standard smoothing kernel defined in (1.4.1), as usual. By Proposition 3.2.6, we know that $\varphi_n \in C^\infty(\operatorname{int}_n(\Omega))$ and that

$$\lim_{n\to\infty} \|\varphi_n - \varphi\|_{L^1(\Omega')} = 0, \tag{5.4.1}$$

for every $\Omega' \Subset \Omega$. We fix $n_0 > 1/r_\Omega$ and set[3]

$$\Omega_{n_0} := \mathrm{int}_{n_0}(\Omega) \cap B_{n_0}(0) \Subset \Omega.$$

We can suppose that this intersection is not empty, without loss of generality. This is still an open convex set, bounded by construction. For every $n \geq 2\,n_0$ and every $x \in \Omega_{n_0}$, we have

$$|\varphi_n(x)| \leq \int_{B_{\frac{1}{n}}(x)} |\varphi(y)|\, \rho_n(x-y)\, dy \leq \|\varphi\|_{L^\infty(B_{\frac{1}{n}}(x))} \int_{B_{\frac{1}{n}}(x)} \rho_n(x-y)\, dy.$$

By recalling that the last integral equals 1, this entails

$$\|\varphi_n\|_{C^0(\overline{\Omega_{n_0}})} = \|\varphi_n\|_{L^\infty(\Omega_{n_0})} \leq \|\varphi\|_{L^\infty(\Omega)}, \qquad \text{for every } n \geq 2\,n_0, \tag{5.4.2}$$

thanks to the fact that $B_{1/n}(x) \subseteq \Omega$, for $x \in \Omega_{n_0}$ and $n \geq 2\,n_0$. Observe that we used Problem 1.7.31 for the continuous function φ_n. Moreover, by Proposition 3.2.6 we also know that

$$\nabla \varphi_n = (\nabla \varphi) * \rho_n, \qquad \text{in } \mathrm{int}_n(\Omega).$$

By proceeding as for (5.4.2), we get

$$\|\nabla \varphi_n\|_{L^\infty(\Omega_{n_0};\mathbb{R}^N)} \leq \|\nabla \varphi\|_{L^\infty(\Omega;\mathbb{R}^N)}, \qquad \text{for every } n \geq 2\,n_0, \tag{5.4.3}$$

as well. From Lemma 5.1.3, we know that

$$|\varphi_n|_{C^{0,1}(\Omega_{n_0})} = \sup_{x \in \Omega_{n_0}} |\nabla \varphi_n(x)|,$$

thanks to the convexity of Ω_{n_0}. By combining this identity with (5.4.3) and Problem 5.5.1, we get

$$|\varphi_n|_{C^{0,1}(\overline{\Omega_{n_0}})} \leq \|\nabla \varphi\|_{L^\infty(\Omega;\mathbb{R}^N)}.$$

The last estimate, together with (5.4.2), proves that $\{\varphi_n\}_{n \geq 2\,n_0} \subseteq C^{0,1}(\overline{\Omega_{n_0}})$ is equibounded in Lipschitz norm. Thus, from Theorem 5.1.7 we get that this sequence converges uniformly on $\overline{\Omega_{n_0}}$ to some $\psi \in C^{0,1}(\overline{\Omega_{n_0}})$, up to a subsequence. Moreover, from the semicontinuity property (5.1.4) we have

$$|\psi|_{C^{0,1}(\overline{\Omega_{n_0}})} \leq \liminf_{n\to\infty} |\varphi_n|_{C^{0,1}(\overline{\Omega_{n_0}})} \leq \|\nabla \varphi\|_{L^\infty(\Omega;\mathbb{R}^N)}.$$

[3] If Ω is bounded, we can directly work with $\Omega_{n_0} = \mathrm{int}_{n_0}(\Omega)$.

By using (5.4.1) with $\Omega' = \Omega_{n_0}$, the uniqueness of the limit permits to infer that $\psi = \varphi$ almost everywhere on Ω_{n_0}, with the uniform (in n_0) estimate

$$|\psi|_{C^{0,1}(\overline{\Omega_{n_0}})} \le \|\nabla\varphi\|_{L^\infty(\Omega)}.$$

Observe that the limit function ψ does not depend on the natural index n_0 fixed at the beginning, thanks to the fact that $\Omega_{n_0} \subseteq \Omega_{n_1}$ if $n_0 < n_1$. By arbitrariness of n_0, this eventually proves that ψ is a Lipschitz function on Ω and φ coincides almost everywhere on Ω with it, as desired. □

Remark 5.4.4 The convexity assumption in Proposition 5.4.3 is sufficient, but not necessary. For example, the previous result is still true, provided the open set Ω is *quasiconvex*, in the following sense: there exists a constant $C \ge 1$ such that for every $x_0, x_1 \in \Omega$ there exists a C^1 curve $\gamma : [0, 1] \to \Omega$ with

$$\gamma(0) = x_0, \qquad \gamma(1) = x_1 \qquad \text{and} \qquad \int_0^1 |\gamma'(t)|\, dt \le C\, |x_0 - x_1|.$$

We refer to [33, Theorem 4.1] for the proof of the equivalence, under this geometric assumption.

However, we point out that for a general open set, we may have

$$C^{0,1}(\overline{\Omega}) \ne W^{1,\infty}(\Omega).$$

We refer to Problem 5.5.13 below, for a classical counter-example.

Finally, we have the following result, concerning the space

$$C^{0,1}_0(\overline{\Omega}) := \Big\{u \in C^{0,1}(\overline{\Omega}) \ :\ u = 0 \text{ on } \partial\Omega\Big\},$$

previously encountered.

Proposition 5.4.5 *Let $1 \le p < \infty$ and let $\Omega \subseteq \mathbb{R}^N$ be an open set with finite measure. Then we have*

$$C^{0,1}_0(\overline{\Omega}) \subseteq W^{1,p}_0(\Omega),$$

where $W^{1,p}_0(\Omega)$ has been introduced in Definition 3.7.1.

Proof We already know that $C^{0,1}_0(\overline{\Omega}) \subseteq W^{1,p}(\Omega)$, thanks to Corollary 5.4.2 and the assumption on Ω. In order to prove that $C^{0,1}_0(\overline{\Omega}) \subseteq W^{1,p}_0(\Omega)$, it is sufficient to observe that

$$C^{0,1}_0(\overline{\Omega}) \subseteq \Big\{u \in C^0(\overline{\Omega}) \cap W^{1,p}(\Omega) \ :\ u = 0 \text{ on } \partial\Omega\Big\}.$$

Then the conclusion follows from Problem 3.12.17. □

Remark 5.4.6 In the limit case $p = \infty$ the previous result does not hold: we refer to Problem 5.5.15. We also point out that if $|\Omega| = +\infty$ the previous inclusion does not hold: indeed, functions in $C_0^{0,1}(\Omega)$ are not necessarily p-summable, for $1 \le p < \infty$.

5.5 Problems

Problem 5.5.1 Let $\Omega \subseteq \mathbb{R}^N$ be an open set and let $f : \overline{\Omega} \to \mathbb{R}$ be a Lipschitz continuous function. Prove that

$$|f|_{C^{0,1}(\overline{\Omega})} = |f|_{C^{0,1}(\Omega)}.$$

Problem 5.5.2 Let $\varphi : \mathbb{R}^N \to \mathbb{R}$ be the function defined by

$$\varphi(x) = \max\{1 - |x|^\alpha, 0\}, \qquad \text{for } x \in \mathbb{R}^N,$$

where $\alpha > 0$. Prove that this is a Lipschitz function on $\mathbb{R}^N$ if and only if $\alpha \ge 1$.

Problem 5.5.3 For $\alpha \ge 1$, compute the Lipschitz constant of the function of the previous problem.

Problem 5.5.4 Let $0 \le r < R$ and let $V : [r, R] \to \mathbb{R}$ be a Lipschitz continuous function. Show that the radial function v defined by

$$v(x) = V(|x|), \qquad \text{for } x \in \overline{A_{r,R}} = \{x \in \mathbb{R}^N : r \le |x| \le R\},$$

is Lipschitz continuous. Also prove that

$$|v|_{C^{0,1}(\overline{A_{r,R}})} = |V|_{C^{0,1}([r,R])},$$

and

$$\nabla v(x) = V'(|x|)\,\frac{x}{|x|}, \qquad \text{for a. e. } x \in A_{r,R}.$$

Problem 5.5.5 For a function $g \in L^\infty((a, b))$, we define

$$u(t) = \int_a^t g(\tau)\,d\tau.$$

Show that this is a Lipschitz continuous function on $[a, b]$, with

$$u' = g, \qquad \text{a. e. on } [a, b].$$

Problem 5.5.6 Let $u : [a, b] \to \mathbb{R}$ be a Lipschitz continuous function. Prove that

$$u(b) - u(a) = \int_a^b u'(t)\, dt.$$

Problem 5.5.7 Let $\Omega \subsetneq \mathbb{R}^N$ be an open set. Show that the distance function

$$\mathrm{dist}(x, \partial\Omega) = \inf_{y \in \partial\Omega} |x - y|, \qquad \text{for every } x \in \Omega,$$

is 1-Lipschitz continuous and that

$$|\nabla \mathrm{dist}(x, \partial\Omega)| = 1, \qquad \text{for a.e. } x \in \Omega.$$

Problem 5.5.8 Show that if $\Omega \subseteq \mathbb{R}^N$ is a bounded set, then every Lipschitz function $f : \Omega \to \mathbb{R}$ is bounded and satisfies

$$\sup_{x \in \Omega} |f(x)| \le \inf_{y \in \Omega} |f(y)| + |f|_{C^{0,1}(\Omega)} \,\mathrm{diam}(\Omega).$$

Also prove that if Ω is unbounded, a Lipschitz function on Ω may fail to be bounded.

Problem 5.5.9 Let $\Omega \subseteq \mathbb{R}^N$ be an open set. We define the set

$$C^0_{\mathrm{b}}(\overline{\Omega}) = \Big\{ f : \overline{\Omega} \to \mathbb{R} \,:\, f \text{ is continuous and bounded on } \overline{\Omega} \Big\}.$$

Prove that this is a normed vector space over $\mathbb{R}$, when endowed with the norm

$$\|f\|_{C^0(\overline{\Omega})} = \sup_{x \in \overline{\Omega}} |f(x)|.$$

Show that this is indeed a Banach space.

Problem 5.5.10 Let $\Omega \subseteq \mathbb{R}^N$ be an open set. For an exponent $0 < \alpha < 1$, we denote by $C^{0,\alpha}(\overline{\Omega})$ the space of Hölder continuous functions

$$C^{0,\alpha}(\overline{\Omega}) = \Big\{ f \in C^0_{\mathrm{b}}(\overline{\Omega}) \,:\, |f|_{C^{0,\alpha}(\overline{\Omega})} < +\infty \Big\},$$

where

$$|f|_{C^{0,\alpha}(\overline{\Omega})} = \sup_{x \neq y,\ x, y \in \overline{\Omega}} \frac{|f(x) - f(y)|}{|x - y|^\alpha}.$$

Prove that this is a Banach space, when endowed with the norm

$$\|f\|_{C^{0,\alpha}(\overline{\Omega})} = \|f\|_{C^0(\overline{\Omega})} + |f|_{C^{0,\alpha}(\overline{\Omega})}.$$

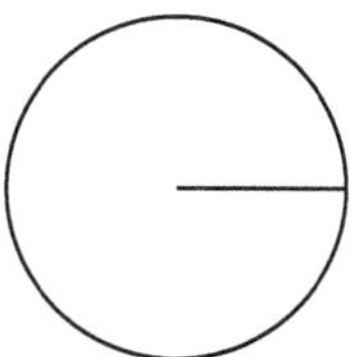

Fig. 5.1 The "slit" disk of Problem 5.5.13

Problem 5.5.11 Prove the following variant of Theorem 5.1.10: let $\Omega \subseteq \mathbb{R}^N$ be a non-empty set and let $f : \Omega \to \mathbb{R}$ be a Lipschitz function. The function defined by

$$f^{\#}(x) = \sup_{y\in\Omega} \Big[f(y) - |f|_{C^{0,1}(\Omega)}\, |x-y| \Big], \qquad \text{for } x \in \mathbb{R}^N,$$

is a extension of f to the whole $\mathbb{R}^N$, having the same Lipschitz constant. Also show that $f^{\#} \le \widetilde{f}$, where $\widetilde{f}$ is the extension defined in Theorem 5.1.10.

Problem 5.5.12 Let $\Omega \subseteq \mathbb{R}^N$ be an open set with finite measure. Let $u \in C^{0,1}_0(\overline{\Omega})$, the latter being defined in (5.1.7). Show that

$$\int_\Omega \frac{\partial u}{\partial x_i}\, dx = 0, \qquad \text{for every } i \in \{1, \dots, N\}.$$

Problem 5.5.13 Let us consider the "slit" disk

$$E = \{(x, y) \in \mathbb{R}^2 \,:\, x^2 + y^2 < 1\} \setminus \{(x, 0) \,:\, x \ge 0\},$$

see Fig. 5.1. Exhibit a function $u \in C^1(E) \cap W^{1,\infty}(E)$ which is not Lipschitz on E, thus showing that

$$C^{0,1}(E) \subsetneq W^{1,\infty}(E).$$

Problem 5.5.14 Let $\Omega \subseteq \mathbb{R}^N$ be an open bounded set and let $\varphi \in C^0(\overline{\Omega})$. Show that there exists a sequence $\{\varphi_k\}_{k\in\mathbb{N}} \subseteq C^{0,1}(\overline{\Omega})$ such that

$$\lim_{k\to\infty} \|\varphi - \varphi_k\|_{C^0(\overline{\Omega})} = 0.$$

Problem 5.5.15 Construct a counterexample to show that $C^{0,1}_0(\overline{\Omega}) \not\subseteq W^{1,\infty}_0(\Omega)$.

Problem 5.5.16 Let $\Omega \subseteq \mathbb{R}^N$ be an open bounded set. Let $u \in W^{1,\infty}(\Omega) \cap C^0(\overline{\Omega})$ be such that its restriction $u_{|\partial\Omega}$ is Lipschitz continuous on $\partial\Omega$. Show that $u \in C^{0,1}(\overline{\Omega})$.

Chapter 6
The Direct Method in Lipschitz Spaces

6.1 Introduction

In this chapter we will apply the idea of the *Direct Method* (see Chap. 4) to variational problems settled on Lipschitz functions. In other words, rather than seeking a minimum in a Sobolev space, we will minimize functionals over a space of Lipschitz functions. Our presentation will essentially follow the contents of [6, Chapter 1], with some adjustments and variants.

The main existence result we will get is not directly comparable to those obtained in Chap. 4: indeed, if on one side we will need to take more restrictive assumptions on the open set $\Omega \subseteq \mathbb{R}^N$ and the boundary datum U, on the other side the existence result will directly give a Lipschitz minimizer, i.e. a more regular function. Moreover, we can allow for variational integrals of the type

$$u \mapsto \int_\Omega F(\nabla u)\,dx,$$

with essentially no requirements on F, apart from convexity and C^1 regularity.[1] This permits for example to consider the *area functional*

$$\int_\Omega \sqrt{1+|\nabla u|^2}\,dx,$$

which is an important functional that *can not* be handled with the methods of Chap. 4. Let us briefly explain why: we observe that the previous functional has *linear growth*, in the sense that

$$\sqrt{1+|z|^2} \sim |z|, \qquad \text{for } |z| \to \infty.$$

[1] This in turn could be dropped: we refer the reader to [6, Chapter 1] for more details.

L. Brasco, *Handbook of Calculus of Variations for Absolute Beginners*,
La Matematica per il 3+2 163, https://doi.org/10.1007/978-3-031-87164-1_6

More precisely, by using the monotonicity and the subadditivity of the square root, we have

$$\int_\Omega |\nabla u|\,dx \le \int_\Omega \sqrt{1+|\nabla u|^2}\,dx \le |\Omega| + \int_\Omega |\nabla u|\,dx.$$

Thus, from the point of view of the Direct Method, the area functional is naturally well-defined on the Sobolev space $W^{1,1}(\Omega)$. However, the latter is a bad space to work with, for a very simple reason: it is no more true that a bounded sequence in $W^{1,1}(\Omega)$ weakly converges, up to a subsequence, to an element of the same space (see Problem 3.12.4 for a simple counter-example). This was an essential ingredient for the Direct Method in Sobolev spaces: recall the discussion of Sect. 3.1 from Chap. 3.

6.2 A Toy Existence Result

By using the results on convex (Chap. 1) and Lipschitz functions (Chap. 5), we can prove a first existence result, for variational problems settled among Lipschitz functions. The result in itself is not very useful, but it will be needed to prove a more general result, i.e. the *Miranda-Stampacchia Theorem* (see Theorem 6.5.1 below).

Theorem 6.2.1 *Let $\Omega \subseteq \mathbb{R}^N$ be an open bounded set and let $F : \mathbb{R}^N \to \mathbb{R}$ be a C^1 convex function. We take U to be a Lipschitz function on $\overline{\Omega}$. Then for every $M > 0$ such that*

$$|U|_{C^{0,1}(\overline{\Omega})} \le M,$$

we have:

(A) the restricted variational problem

$$\inf_{u\in C^{0,1}(\overline{\Omega})} \left\{ \int_\Omega F(\nabla u)\,dx \,:\, u = U \text{ on } \partial\Omega \text{ and } |u|_{C^{0,1}(\overline{\Omega})} \le M \right\}, \tag{6.2.1}$$

admits (at least) a solution v;

(B) moreover, if v is such that

$$|v|_{C^{0,1}(\overline{\Omega})} \le K < M,$$

then v is a solution of

$$\inf_{u\in C^{0,1}(\overline{\Omega})} \left\{ \int_\Omega F(\nabla u)\,dx \,:\, u = U \text{ on } \partial\Omega \right\},$$

as well.

Proof For ease of readability, we divide the proof in two parts.

Proof of (A) We first notice that the class of admissible functions in (6.2.1) is not empty, since the datum U is admissible. By recalling Lemma 5.2.1, we have that every admissible function u has a weak gradient which belongs to $L^\infty(\Omega;\mathbb{R}^N)$. Thus, the integral

$$\int_\Omega F(\nabla u)\,dx,$$

is well-defined and finite. Moreover, by Remark 5.2.3, we know that

$$|\nabla u(x)| \le |u|_{C^{0,1}(\overline{\Omega})} \le M, \qquad \text{for a.e. } x \in \Omega,$$

thus we have

$$\left|\int_\Omega F(\nabla u)\,dx\right| \le \int_\Omega |F(\nabla u)|\,dx \le \left(\max_{z\in\overline{B_M(0)}} |F(z)|\right)|\Omega| < +\infty.$$

If we set for simplicity

$$\mathfrak{m} = \inf_{u\in C^{0,1}(\overline{\Omega})} \left\{\int_\Omega F(\nabla u)\,dx \,:\, u = U \text{ on } \partial\Omega \text{ and } |u|_{C^{0,1}(\overline{\Omega})} \le M\right\},$$

the previous discussion entails that $-\infty < \mathfrak{m} < +\infty$. By definition of infimum, there exists a *minimizing sequence* $\{u_n\}_{n\in\mathbb{N}}$ of admissible functions, such that

$$\lim_{n\to\infty} \int_\Omega F(\nabla u_n)\,dx = \mathfrak{m}. \tag{6.2.2}$$

If we want to apply the Direct Method in order to prove existence, we need to prove that:

1. we can extract a subsequence (not relabeled) such that u_n converges (in a suitable sense) to a function v which is still admissible;
2. our functional is lower semicontinuous with respect to the convergence at point 1, i.e.

$$\liminf_{n\to\infty} \int_\Omega F(\nabla u_n)\,dx \ge \int_\Omega F(\nabla v)\,dx;$$

then these two properties, in conjunction with (6.2.2), prove that

$$\int_\Omega F(\nabla v)\,dx \le \mathfrak{m}.$$

Since the reverse inequality follows by definition of $\mathfrak{m}$, this would establish the existence of a minimizer.

Point 1. above follows from the compactness result of Theorem 5.1.7. Indeed, observe that by construction we have

$$|u_n|_{C^{0,1}(\overline{\Omega})} \leq M, \qquad \text{for every } n \in \mathbb{N}.$$

Moreover, the sequence is bounded in the sup norm, as well. This follows from Problem 5.5.8, which guarantees that (recall that $u_n = U$ on $\partial\Omega$)

$$\begin{aligned}\max_{x\in\overline{\Omega}} |u_n(x)| &\leq \min_{y\in\overline{\Omega}} |u_n(y)| + M \operatorname{diam}(\overline{\Omega})\\ &\leq \min_{y\in\partial\Omega} |U(y)| + M \operatorname{diam}(\overline{\Omega}), \qquad \text{for every } n \in \mathbb{N}.\end{aligned}$$

By Theorem 5.1.7 we obtain that (up to a subsequence) u_n converges uniformly on $\overline{\Omega}$ to a function $v \in C^{0,1}(\overline{\Omega})$. The uniform convergence guarantees that we still have

$$v = U, \qquad \text{on } \partial\Omega.$$

Moreover, by (5.1.4) we also have that

$$|v|_{C^{0,1}(\overline{\Omega})} \leq \liminf_{n\to\infty} |u_n|_{C^{0,1}(\overline{\Omega})} \leq M.$$

This guarantees that v is still admissible for our variational problem.

We now prove point 2., i.e. the lower semicontinuity of our functional. By using the "above tangent" property for convex functions (see Proposition 1.3.4), we have

$$F(\nabla u_n(x)) \geq F(\nabla v(x)) + \langle \nabla F(\nabla v(x)), \nabla u_n(x) - \nabla v(x)\rangle, \quad \text{for a. e. } x \in \overline{\Omega}.$$

By integrating this inequality, we get

$$\int_\Omega F(\nabla u_n)\,dx \geq \int_\Omega F(\nabla v)\,dx + \int_\Omega \langle \nabla F(\nabla v), \nabla u_n - \nabla v\rangle\,dx. \tag{6.2.3}$$

Observe that all the integrals are well-defined, since $\nabla u_n, \nabla v \in L^\infty(\Omega;\mathbb{R}^N)$. In order to conclude, it is sufficient to prove that

$$\lim_{n\to\infty} \int_\Omega \langle \nabla F(\nabla v), \nabla u_n - \nabla v\rangle\,dx = 0. \tag{6.2.4}$$

To this aim, we observe that ∇v is an essentially bounded function (still by Lemma 5.2.1) and $z \mapsto \nabla F(z)$ is continuous, thanks to the fact that F is C^1. This implies that

$$\frac{\partial F}{\partial z_k}(\nabla v) \in L^\infty(\Omega) \subseteq L^2(\Omega), \qquad \text{for } k \in \{1,\dots,N\}. \tag{6.2.5}$$

We also used that Ω is a bounded set. On the other hand, still by Remark 5.2.3 we have

$$\left\| \frac{\partial u_n}{\partial x_k} \right\|_{L^\infty(\Omega)} \le |u_n|_{C^{0,1}(\overline{\Omega})} \le M, \qquad \text{for every } n \in \mathbb{N}.$$

By Theorem B.1.1, we can then extract a further subsequence (still not relabelled) such that

$$\frac{\partial u_n}{\partial x_k} \text{ weakly converges in } L^2(\Omega) \text{ to } \phi_k \in L^2(\Omega).$$

We can prove that ϕ_k actually coincides with the k-th weak partial derivative of v: it is sufficient to repeat the argument in the proof of Theorem 3.3.6. Indeed, let us take $\varphi \in C_0^\infty(\Omega)$, by using the definition of weak derivative, we get

$$\begin{aligned}\int_\Omega \varphi \frac{\partial v}{\partial x_k}\, dx &= -\int_\Omega \frac{\partial \varphi}{\partial x_k}\, v\, dx = -\lim_{n\to\infty} \int_\Omega \frac{\partial \varphi}{\partial x_k}\, u_n\, dx \\ &= \lim_{n\to\infty} \int_\Omega \varphi \frac{\partial u_n}{\partial x_k}\, dx = \int_\Omega \varphi\, \phi_k\, dx.\end{aligned}$$

Since the last identity is true for every $\varphi \in C_0^\infty(\Omega)$, by Lemma 1.4.2 we then get

$$\phi_k = \frac{\partial v}{\partial x_k}, \qquad \text{a. e. in } \Omega,$$

and thus

$$\frac{\partial u_n}{\partial x_k} \text{ weakly converges in } L^2(\Omega) \text{ to } \frac{\partial v}{\partial x_k} \in L^\infty(\Omega) \subseteq L^2(\Omega).$$

By using this fact in connection with (6.2.5), we then get

$$\lim_{n\to\infty} \int_\Omega \frac{\partial F}{\partial z_k}(\nabla v) \left(\frac{\partial u_n}{\partial x_k} - \frac{\partial v}{\partial x_k} \right) dx = 0.$$

This in turn gives (6.2.4) and thus, by (6.2.3), we obtain

$$\liminf_{n\to\infty} \int_\Omega F(\nabla u_n)\, dx \ge \int_\Omega F(\nabla v)\, dx.$$

As explained above, this finally establishes that v is a minimizer.

Proof of (B) Let us assume that

$$|v|_{C^{0,1}(\overline{\Omega})} \le K < M.$$

We take u any Lipschitz function over $\overline{\Omega}$, such that $u = U$ on $\partial\Omega$. Then we build a new admissible function, by taking the convex combination

$$\varphi_t = (1-t)\,v + t\,u, \qquad \text{for } t \in (0,1),$$

which is still Lipschitz and still coincides with U on the boundary of Ω. Moreover, we have

$$\begin{aligned}|\varphi_t(x) - \varphi_t(y)| &\le |(1-t)\,v(x) + t\,u(x) - (1-t)\,v(y) - t\,u(y)| \\ &\le (1-t)\,|v(x) - v(y)| + t\,|u(x) - u(y)| \\ &\le (1-t)\,|v|_{C^{0,1}(\overline{\Omega})}\,|x-y| + t\,|u|_{C^{0,1}(\overline{\Omega})}\,|x-y| \\ &= \Big((1-t)\,|v|_{C^{0,1}(\overline{\Omega})} + t\,|u|_{C^{0,1}(\overline{\Omega})}\Big)\,|x-y|,\end{aligned}$$

so that

$$|\varphi_t|_{C^{0,1}(\overline{\Omega})} \le \Big((1-t)\,|v|_{C^{0,1}(\overline{\Omega})} + t\,|u|_{C^{0,1}(\overline{\Omega})}\Big).$$

It is not difficult to see that we can take $t_0 > 0$ small enough, in such a way that[2]

$$\Big((1-t_0)\,|v|_{C^{0,1}(\overline{\Omega})} + t_0\,|u|_{C^{0,1}(\overline{\Omega})}\Big) \le M.$$

This implies that φ_{t_0} is admissible for the restricted variational problem (6.2.1), which is minimized by v. Thus, we get

$$\int_\Omega F(\nabla v)\,dx \le \int_\Omega F(\nabla \varphi_{t_0})\,dx.$$

By recalling that φ_{t_0} is a convex combination and using that F is convex, we get from the previous inequality

$$\int_\Omega F(\nabla v)\,dx \le (1-t_0)\int_\Omega F(\nabla v)\,dx + t_0\int_\Omega F(\nabla u)\,dx,$$

---

[2] If we have

$$|u|_{C^{0,1}(\overline{\Omega})} \le M,$$

we can take any $t \in (0,1)$. If this is not the case, then it is sufficient to take

$$t_0 = \frac{M - |v|_{C^{0,1}(\overline{\Omega})}}{|u|_{C^{0,1}(\overline{\Omega})} - |v|_{C^{0,1}(\overline{\Omega})}},$$

which is positive and less than 1, thanks to the properties of both u and v.

which is the same as

$$t_0 \int_\Omega F(\nabla v)\,dx \le t_0 \int_\Omega F(\nabla u)\,dx.$$

By recalling that $t_0 > 0$, we can divide by t_0 and obtain that

$$\int_\Omega F(\nabla v)\,dx \le \int_\Omega F(\nabla u)\,dx, \quad \text{for every } u \in C^{0,1}(\overline{\Omega}) \text{ such that } u = U \text{ on } \partial\Omega.$$

This concludes the proof. □

Remark 6.2.2 (Uniqueness of Minimizers) It is not difficult to see that if we further assume F to be strictly convex, then the minimizer v of (6.2.1) is unique. We can repeat the same argument already used several times in the previous chapters: suppose that v_1 and v_2 are two minimizers, we consider the new function

$$w = \frac{v_1 + v_2}{2}.$$

By convexity of F, we get

$$\begin{aligned} \int_\Omega F(\nabla w)\,dx &= \int_\Omega F\left(\frac{\nabla v_1 + \nabla v_2}{2}\right)dx \\ &\le \frac{1}{2}\int_\Omega F(\nabla v_1)\,dx + \frac{1}{2}\int_\Omega F(\nabla v_2)\,dx \\ &= \int_\Omega F(\nabla v_1)\,dx = \int_\Omega F(\nabla v_2)\,dx. \end{aligned} \tag{6.2.6}$$

On the other hand, w is still admissible for the variational problem (6.2.1), thus we have

$$\int_\Omega F(\nabla w)\,dx \ge \int_\Omega F(\nabla v_1)\,dx = \int_\Omega F(\nabla v_2)\,dx.$$

In conclusion, we get that equality must hold in (6.2.6). This fact and the convexity of F imply the following pointwise identity

$$F\left(\frac{\nabla v_1 + \nabla v_2}{2}\right) = \frac{1}{2}F(\nabla v_1) + \frac{1}{2}F(\nabla v_2), \qquad \text{a.e. on } \Omega.$$

By using the strict convexity of F, this in turn implies that

$$\nabla v_1 = \nabla v_2, \qquad \text{a.e. on } \Omega.$$

By Proposition 3.2.9, we get that v_1 and v_2 coincide up to a constant, on every connected component of Ω. By using that

$$v_1 = v_2 = U, \text{ on } \partial\Omega,$$

this finally shows that $v_1 = v_2$.

6.3 The Comparison Principle

Proposition 6.3.1 (Comparison Principle) *Let $F : \mathbb{R}^N \to \mathbb{R}$ be a C^1 strictly convex function. We take U_1, U_2 two Lipschitz functions on $\overline{\Omega}$ such that*

$$|U_i|_{C^{0,1}(\overline{\Omega})} \le M, \qquad \text{for } i = 1, 2,$$

for some $M > 0$. Let v_i be the unique solution of

$$\inf_{u \in C^{0,1}(\overline{\Omega})} \left\{ \int_\Omega F(\nabla u)\, dx \,:\, u = U_i \text{ on } \partial\Omega \text{ and } |u|_{C^{0,1}(\overline{\Omega})} \le M \right\}.$$

Then if $U_1 \le U_2$ on $\partial\Omega$, we also have $v_1 \le v_2$ on $\overline{\Omega}$.

Proof We first notice that the existence of v_1 and v_2 follows from Theorem 6.2.1, while their uniqueness can be inferred by Remark 6.2.2. We consider the function

$$u(x) = \max\{v_1(x), v_2(x)\},$$

which is still Lipschitz and such that

$$|u|_{C^{0,1}(\overline{\Omega})} \le M,$$

thanks to Corollary 5.3.3. We observe that for $x \in \partial\Omega$, we have

$$u(x) = \max\{v_1(x), v_2(x)\} = \max\{U_1(x), U_2(x)\} = U_2(x),$$

thanks to the assumption $U_1 \le U_2$ on $\partial\Omega$. This implies that u is admissible for the minimization problem solved by v_2. Thus, we get

$$\int_\Omega F(\nabla u)\, dx \ge \int_\Omega F(\nabla v_2)\, dx.$$

By recalling the definition of u and using Corollary 5.3.3, this is the same as

$$\int_{\{v_2\geq v_1\}} F(\nabla v_2)\,dx + \int_{\{v_1>v_2\}} F(\nabla v_1)\,dx \geq \int_{\{v_2\geq v_1\}} F(\nabla v_2)\,dx + \int_{\{v_1>v_2\}} F(\nabla v_2)\,dx.$$

By erasing the common factor, this gives

$$\int_{\{v_1>v_2\}} F(\nabla v_1)\,dx \geq \int_{\{v_1>v_2\}} F(\nabla v_2)\,dx.$$

We now add on both sides the term

$$\int_{\{v_2\geq v_1\}} F(\nabla v_1)\,dx,$$

so to get

$$\int_{\{v_2\geq v_1\}} F(\nabla v_1)\,dx + \int_{\{v_1>v_2\}} F(\nabla v_1)\,dx \geq \int_{\{v_2\geq v_1\}} F(\nabla v_1)\,dx + \int_{\{v_1>v_2\}} F(\nabla v_2)\,dx.$$

If we now introduce the function $w(x) = \min\{v_1(x), v_2(x)\}$, use again Corollary 5.3.3 and keep into account (5.3.1), the previous estimate is the same as

$$\int_\Omega F(\nabla v_1) \geq \int_\Omega F(\nabla w)\,dx. \tag{6.3.1}$$

On the other hand, still by Corollary 5.3.3 we know that the function w is Lipschitz continuous, with

$$|w|_{C^{0,1}(\overline{\Omega})} \leq M.$$

Moreover, for every $x \in \partial\Omega$, we have

$$w(x) = \min\{v_1(x), v_2(x)\} = \min\{U_1(x), U_2(x)\} = U_1(x).$$

In other words, w is admissible for the minimization problem uniquely solved by v_1. The estimate (6.3.1) and the uniqueness of the minimizer then imply that

$$v_1 = w = \min\{v_1, v_2\},$$

which is the same as $v_1 \leq v_2$ in $\overline{\Omega}$. □

As an interesting consequence of the previous result, we have the following

Corollary 6.3.2 (Stability with Respect to the Boundary Datum) *Let $F : \mathbb{R}^N \to \mathbb{R}$ be a C^1 strictly convex function. We take U_1, U_2 two Lipschitz functions on $\overline{\Omega}$ such that*

$$|U_i|_{C^{0,1}(\overline{\Omega})} \le M, \qquad \text{for } i = 1, 2,$$

for some $M > 0$. Let v_i be the unique solution of

$$\inf_{u \in C^{0,1}(\overline{\Omega})} \left\{ \int_\Omega F(\nabla u)\, dx \ :\ u = U_i \text{ on } \partial\Omega \text{ and } |u|_{C^{0,1}(\overline{\Omega})} \le M \right\}.$$

Then we have

$$\max_{\overline{\Omega}} |v_2 - v_1| = \max_{\partial\Omega} |v_2 - v_1| = \max_{\partial\Omega} |U_2 - U_1|.$$

Proof For every $x \in \partial\Omega$, we have

$$v_2(x) = v_1(x) + \Big[v_2(x) - v_1(x) \Big] \le v_1(x) + \max_{\partial\Omega}(v_2 - v_1).$$

By construction, the function

$$v(x) = v_1(x) + \max_{\partial\Omega}(v_2 - v_1),$$

is the minimizer of

$$\inf_{u \in C^{0,1}(\overline{\Omega})} \left\{ \int_\Omega F(\nabla u)\, dx \ :\ u = U_1 + \max_{\partial\Omega}(v_2 - v_1) \text{ on } \partial\Omega \text{ and } |u|_{C^{0,1}(\overline{\Omega})} \le M \right\}.$$

Since we have shown that this boundary datum is larger than the one of v_2, by the Comparison Principle of Proposition 6.3.1 we get

$$v_2(x) \le v(x) = v_1(x) + \max_{\partial\Omega}(v_2 - v_1), \qquad \text{for every } x \in \overline{\Omega},$$

that is

$$v_2(x) - v_1(x) \le \max_{\partial\Omega}(v_2 - v_1), \qquad \text{for every } x \in \overline{\Omega}.$$

By taking the maximum over x, we get

$$\max_{\overline{\Omega}}(v_2 - v_1) \le \max_{\partial\Omega}(v_2 - v_1),$$

and thus equality holds, since the opposite inequality is trivially true (thanks to the fact that $\partial\Omega \subseteq \overline{\Omega}$). It is now sufficient to exchange the role of v_2 and v_1, in order to get the claimed result. □

We conclude this section with a technical lemma, which will be quite useful for the main result of this chapter.

Lemma 6.3.3 (Radó) *Let $F : \mathbb{R}^N \to \mathbb{R}$ be a C^1 strictly convex function. We take U to be a Lipschitz function on $\overline{\Omega}$ and M a positive number such that*

$$|U|_{C^{0,1}(\overline{\Omega})} \leq M.$$

Let v be the unique solution of

$$\inf_{u\in C^{0,1}(\overline{\Omega})} \left\{ \int_\Omega F(\nabla u)\, dx \ :\ u = U \text{ on } \partial\Omega \text{ and } |u|_{C^{0,1}(\overline{\Omega})} \leq M \right\}.$$

Then

$$|v|_{C^{0,1}(\overline{\Omega})} = \sup_{x\in\overline{\Omega},\, y\in\partial\Omega} \left\{ \frac{|v(x)-v(y)|}{|x-y|} \ :\ x \neq y \right\}.$$

Proof We first observe that we trivially have

$$|v|_{C^{0,1}(\overline{\Omega})} \geq \sup_{x\in\overline{\Omega},\, y\in\partial\Omega} \left\{ \frac{|v(x)-v(y)|}{|x-y|} \ :\ x \neq y \right\},$$

since the supremum on the right-hand side is performed on a smaller set. In order to conclude, we need to prove that

$$|v|_{C^{0,1}(\overline{\Omega})} \leq \sup_{x\in\overline{\Omega},\, y\in\partial\Omega} \left\{ \frac{|v(x)-v(y)|}{|x-y|} \ :\ x \neq y \right\}.$$

We recall that

$$|v|_{C^{0,1}(\overline{\Omega})} = |v|_{C^{0,1}(\Omega)},$$

see Problem 5.5.1. Thus, we just have to show that for every $x_0, x_1 \in \Omega$, there exist two distinct points $y_0 \in \overline{\Omega}$ and $y_1 \in \partial\Omega$ such that

$$|x_0 - x_1| = |y_0 - y_1| \qquad \text{and} \qquad |v(x_0) - v(x_1)| \leq |v(y_0) - v(y_1)|. \tag{6.3.2}$$

Let us fix some notation. For every $\tau \in \mathbb{R}^N$, we set

$$\Omega_\tau := \Omega + \tau = \left\{ x \in \mathbb{R}^N \ :\ \exists y \in \Omega \text{ such that } x = y + \tau \right\}.$$

We now fix two distinct points $x_0, x_1 \in \Omega$ and choose $\tau = x_0 - x_1$. We observe that by definition

$$x_0 = \tau + x_1 \in \Omega_\tau \qquad \text{and also} \qquad x_0 \in \Omega.$$

Thus, the set $\Omega \cap \Omega_\tau$ is open and non-empty, since it contains x_0. We define the new function

$$v_\tau(x) := v(x - \tau), \qquad \text{for } x \in \Omega_\tau.$$

By construction and minimality of v, we see that v_τ is the unique solution of

$$\inf_{u \in C^{0,1}(\overline{\Omega_\tau})} \left\{ \int_{\Omega_\tau} F(\nabla u)\, dx \,:\, u = U_\tau \text{ on } \partial\Omega_\tau \text{ and } |u|_{C^{0,1}(\overline{\Omega_\tau})} \le M \right\},$$

where we have set $U_\tau(x) = U(x - \tau)$. Observe that both v and v_τ minimize the functional

$$u \mapsto \int_{\Omega \cap \Omega_\tau} F(\nabla u)\, dx,$$

in the class of M-Lipschitz functions, each one with its own boundary datum. By Corollary 6.3.2, we then obtain

$$\max_{\overline{\Omega \cap \Omega_\tau}} |v - v_\tau| = \max_{\partial(\Omega \cap \Omega_\tau)} |v - v_\tau|.$$

This implies in particular that

$$\begin{aligned} |v(x_0) - v(x_1)| &= |v(x_0) - v(x_0 + (x_1 - x_0))| \\ &= |v(x_0) - v_\tau(x_0)| \le \max_{\partial(\Omega \cap \Omega_\tau)} |v - v_\tau|. \end{aligned}$$

Then there exists $z_0 \in \partial(\Omega \cap \Omega_\tau)$ such that

$$|v(x_0) - v(x_1)| \le |v(z_0) - v_\tau(z_0)| = |v(z_0) - v(z_0 - \tau)|.$$

By construction, we have that at least one of the points z_0 and $z_0 - \tau$ belongs to $\partial\Omega$ (see Problem 6.8.6 below). Thus, the pair of points $z_0, z_0 - \tau$ satisfies (6.3.2). The proof is concluded. □

6.4 The Bounded Slope Condition

The existence result of Theorem 6.2.1 was not very useful, because of the unnatural restriction given by the uniform bound on the Lipschitz constant of admissible functions. In this section, we introduce a "geometric" condition on the boundary datum U, which will permit to drop such a restriction and prove an existence result in the space of Lipschitz functions taking the boundary datum, without any further requirement.

Such a condition is introduced in the following

Definition 6.4.1 Let $\Omega \subseteq \mathbb{R}^N$ be an open bounded set. We say that $U : \partial\Omega \to \mathbb{R}$ satisfies the *bounded slope condition of rank K* if the following condition is satisfied:

- for every $y \in \partial\Omega$, there exist $\mathbf{a}_y, \mathbf{b}_y \in \mathbb{R}^N$ such that

$$|\mathbf{a}_y| \leq K, \qquad |\mathbf{b}_y| \leq K,$$

 and

$$U(y) + \langle \mathbf{a}_y, x - y\rangle \leq U(x) \leq U(y) + \langle \mathbf{b}_y, x - y\rangle, \qquad \text{for every } x \in \partial\Omega.$$

Remark 6.4.2 It is easy to see that if $U : \partial\Omega \to \mathbb{R}$ coincides with the restriction to $\partial\Omega$ of an affine function, then U satisfies the BSC. Indeed, if

$$U(x) = c + \langle \mathbf{v}, x\rangle, \qquad x \in \partial\Omega,$$

then Definition 6.4.1 is verified by taking $\mathbf{a}_y = \mathbf{b}_y = \mathbf{v}$, for every $y \in \partial\Omega$. In this case, the rank of the bounded slope condition is given by $K = |\mathbf{v}|$.

The bounded slope condition (*BSC* for short) is not always easy to verify. On the other hand, as announced above, it will play a crucial role in the main existence result of this chapter. For this reason, we want to explore some necessary or sufficient conditions for the BSC to hold.

Lemma 6.4.3 (A Necessary Condition for BSC) *Let $\Omega \subseteq \mathbb{R}^N$ be an open bounded set. Let $U : \partial\Omega \to \mathbb{R}$ be a function which satisfies the BSC of rank K. If U is not the restriction to $\partial\Omega$ of an affine function, then Ω must be convex.*

Proof Let us fix $y \in \partial\Omega$, by definition of BSC, there exists two vectors $\mathbf{a}, \mathbf{b}$ such that

$$U(y) + \langle \mathbf{a}, x - y\rangle \leq U(x) \leq U(y) + \langle \mathbf{b}, x - y\rangle, \qquad \text{for every } x \in \partial\Omega.$$

For simplicity, we wrote $\mathbf{a}$ and $\mathbf{b}$, in place of $\mathbf{a}_y$ and $\mathbf{b}_y$. This in particular implies that

$$\langle \mathbf{b} - \mathbf{a}, x - y\rangle \geq 0, \qquad \text{for every } x \in \partial\Omega. \tag{6.4.1}$$

Since U is not the restriction of an affine function, then $\mathbf{a} \neq \mathbf{b}$. We now take $z \in \Omega$ and observe that there exists at least one point $x \in \partial\Omega \setminus \{y\}$ such that z belongs to the segment connecting y to x, i.e. there exists $\lambda \in (0, 1)$ such that

$$z = (1 - \lambda)\, y + \lambda\, x.$$

We then have

$$\langle \mathbf{b} - \mathbf{a}, z - y\rangle = \langle \mathbf{b} - \mathbf{a}, \lambda\, (x - y)\rangle = \lambda\, \langle \mathbf{b} - \mathbf{a}, x - y\rangle \geq 0,$$

thanks to (6.4.1). By arbitrariness of $z \in \Omega$, this shows that

$$\Omega \subseteq \{x \in \mathbb{R}^N \,:\, \langle \mathbf{b} - \mathbf{a}, x - y\rangle \geq 0\} := \Pi_y^+.$$

This proves that for every $y \in \partial\Omega$, the set $\partial\Pi_y^+$ is a *supporting hyperplane* for Ω passing from y: in light of Problem 6.8.7, we conclude that Ω is convex. □

We now present a couple of *sufficient* conditions for the BSC to hold.

Proposition 6.4.4 (A Sufficient Condition for BSC—part I) *Let $\Omega \subseteq \mathbb{R}^N$ be an open convex set. Let $f : \mathbb{R}^N \to \mathbb{R}$ be a C^1 convex function and let $g : \mathbb{R}^N \to \mathbb{R}$ be a C^1 concave function. Let us suppose that*

$$f = g \qquad \text{on } \partial\Omega.$$

Then the function $U = f = g : \partial\Omega \to \mathbb{R}$ satisfies the BSC, with rank

$$K = \max\left\{\max_{x\in\overline{\Omega}} |\nabla f(x)|,\ \max_{x\in\overline{\Omega}} |\nabla g(x)|\right\}.$$

Proof By using the "above tangent" property for convex functions (Proposition 1.3.4), we have for every $x, y \in \partial\Omega$

$$U(x) = f(x) \geq f(y) + \langle\nabla f(y), x - y\rangle = U(y) + \langle\nabla f(y), x - y\rangle. \qquad (6.4.2)$$

On the other hand, by using the "below tangent" property for concave functions,[3] we have for every $x, y \in \partial\Omega$

$$U(x) = g(x) \leq g(y) + \langle\nabla g(y), x - y\rangle = U(y) + \langle\nabla g(y), x - y\rangle. \qquad (6.4.3)$$

We fix $y \in \partial\Omega$ and introduce the two affine functions

$$w_-(x) = U(y) + \langle\nabla f(y), x - y\rangle \qquad \text{and} \qquad w_+(x) = U(y) + \langle\nabla g(y), x - y\rangle.$$

In light of (6.4.2) and (6.4.3), we get

$$w_-(x) \leq U(x) \leq w_+(x) \text{ for } x \in \partial\Omega \qquad \text{and} \qquad w_-(y) = U(y) = w_+(y).$$

Moreover, by using Remark 5.1.5 for the two affine functions w_+ and w_-, we have

$$|w_-|_{C^{0,1}(\overline{\Omega})} = |\nabla f(y)| \leq \max_{\overline{\Omega}} |\nabla f|,$$

[3] It is sufficient to observe that if g is concave, then $-g$ is convex: then one applies the "above tangent" property to $-g$.

and

$$|w_+|_{C^{0,1}(\overline{\Omega})} = |\nabla g(y)| \leq \max_{\overline{\Omega}} |\nabla g|.$$

If we now set

$$K = \max\left\{ \max_{x\in\overline{\Omega}} |\nabla f(x)|,\ \max_{x\in\overline{\Omega}} |\nabla g(x)| \right\},$$

we just have shown that U satisfies the BSC of rank K. □

Proposition 6.4.5 (A Sufficient Condition for BSC—part II) *Let $\Omega \subseteq \mathbb{R}^N$ be an open bounded convex set, with boundary of class C^2. Let us suppose that Ω has the following geometric property:*

- *there exists $R > 0$ such that for every $x \in \partial\Omega$, there exists a ball $B \supseteq \Omega$ of radius R such that*

$$\partial B \cap \partial\Omega = \{x\}.$$

Then, for every $\varphi \in C^2(\overline{\Omega})$ its restriction to the boundary $\varphi_{|\partial\Omega}$ satisfies the BSC.

Proof We recall that an open convex set has the following property: through each boundary point $y \in \partial\Omega$, there is a *supporting hyperplane* for Ω (see Problem 6.8.7 below for the definition). This is a classical fact, we refer for example to [55, Theorem 1.3.2].

Let us fix $y \in \partial\Omega$, without loss of generality we can suppose that y coincides with the origin 0. Up to rotating Ω, we can assume that the hyperplane

$$\{x \in \mathbb{R}^N\ :\ x_N = 0\},$$

is a supporting hyperplane for Ω at the point 0 and that

$$\Omega \subseteq B_R(R\,\mathbf{e}_N) = \{x \in \mathbb{R}^N\ :\ |x - R\,\mathbf{e}_N| < R\}, \tag{6.4.4}$$

with $\partial B_R(R\,\mathbf{e}_N)$ and $\partial\Omega$ touching only at 0 (see Fig. 6.1).

We have to show that there exists a constant $K > 0$ depending only on the dimension N, the datum φ and the geometric properties of Ω, such that the graph of $\varphi_{|\partial\Omega}$ is “trapped” in between the graph of two affine functions passing though the point $(0, \varphi(0)) \in \mathbb{R}^{N+1}$ and having slopes not exceeding K. In other words, we need to show that there exists two vectors $\mathbf{a}, \mathbf{b} \in \mathbb{R}^N$ such that

$$\varphi(0) + \langle \mathbf{a}, x\rangle \leq \varphi(x) \leq \varphi(0) + \langle \mathbf{b}, x\rangle, \qquad \text{for every } x \in \partial\Omega,$$

with

$$|\mathbf{a}| \leq K \qquad \text{and} \qquad |\mathbf{b}| \leq K,$$

and $K > 0$ a uniform constant.

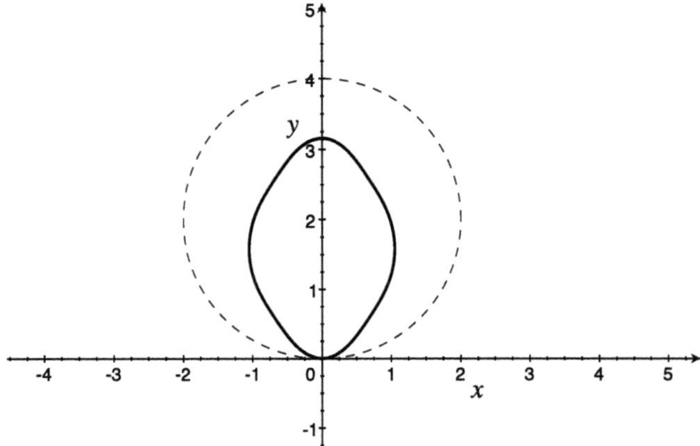

Fig. 6.1 The geometric construction for the proof of Proposition 6.4.5. The bold line stands for the boundary of Ω

We observe that[4] for every $x \in \partial\Omega$ we have

$$\varphi(x) = \varphi(0) + \langle \nabla\varphi(0), x \rangle + \int_0^1 \langle D^2\varphi(t\,x) \cdot x, x \rangle\,(1-t)\,dt.$$

By using the Cauchy-Schwarz inequality, we have

$$\begin{aligned}
|\langle D^2\varphi(t\,x) \cdot x, x \rangle| &\le |D^2\varphi(t\,x) \cdot x|\,|x| \\
&= |x| \sqrt{\sum_{i=1}^N \left|\sum_{j=1}^N \frac{\partial^2\varphi}{\partial x_i\,\partial x_j}(t\,x)\,x_j\right|^2} \\
&\le |x| \sqrt{\sum_{i=1}^N \left(\sum_{j=1}^N |x_j|^2\right)\left(\sum_{j=1}^N \left|\frac{\partial^2\varphi}{\partial x_i\,\partial x_j}(t\,x)\right|^2\right)} \\
&= |x|^2 \sqrt{\sum_{i,j=1}^N \left|\frac{\partial^2\varphi}{\partial x_i\,\partial x_j}(t\,x)\right|^2} \\
&\le |x|^2 \max_{y\in\overline{\Omega}} \sqrt{\sum_{i,j=1}^N \left|\frac{\partial^2\varphi}{\partial x_i\,\partial x_j}(y)\right|^2} =: C\,|x|^2,
\end{aligned}$$

where C is a constant depending on the C^2 norm of φ. This implies that

[4] We use the following Taylor-type identity $g(1) = g(0) + g'(0) + \int_0^1 g''(t)\,(1-t)\,dt$, with the function $g(t) = \varphi(t\,x)$.

$$\varphi(x) \leq \varphi(0) + \langle \nabla\varphi(0), x \rangle + C\,|x|^2, \tag{6.4.5}$$

and

$$\varphi(x) \geq \varphi(0) + \langle \nabla\varphi(0), x \rangle - C\,|x|^2.$$

We now use our geometric assumption on the set Ω: by (6.4.4), we have that for every $x = (x', x_N) \in \partial\Omega$

$$R - \sqrt{R^2 - |x'|^2} \leq x_N \leq R + \sqrt{R^2 - |x'|^2}.$$

By using the concavity inequality[5]

$$\sqrt{1-t} \leq 1 - \frac{t}{2}, \qquad \text{for every } t \in [0, 1],$$

we can infer that

$$R - \sqrt{R^2 - |x'|^2} = R\left(1 - \sqrt{1 - \frac{|x'|^2}{R^2}}\right) \geq \frac{1}{2\,R}\,|x'|^2.$$

In particular, we obtain that for every $x \in \partial\Omega$, we have

$$\frac{1}{2\,R}\,|x'|^2 \leq x_N \qquad \text{that is} \qquad |x'|^2 \leq 2\,R\,x_N.$$

By summing x_N^2 on both sides, this implies

$$|x|^2 \leq (2\,R + x_N)\,x_N, \qquad \text{for every } x \in \partial\Omega.$$

We finally recall that $\Omega \subseteq B_R$, thus if $x \in \partial\Omega$ then $x \in \overline{B_R}$ which implies that

$$x_N \leq 2\,R.$$

By using this information in the previous estimate, we finally end up with

$$|x|^2 \leq 4\,R\,x_N, \qquad \text{for every } x \in \partial\Omega.$$

[5] If we set $h(t) = \sqrt{1-t}$, this is a concave function for $t \leq 1$. By using the "below tangent" property of concave functions, we have

$$h(t) \leq h(0) + h'(0)\,t, \qquad \text{for } t \leq 1,$$

which is exactly the claimed inequality.

We insert this estimate in (6.4.5) to obtain

$$\varphi(x) \leq \varphi(0) + \langle \nabla\varphi(0), x\rangle + 4\,C\,R\,x_N, \qquad \text{for every } x \in \partial\Omega. \tag{6.4.6}$$

We now take the vector

$$\mathbf{b} = \nabla\varphi(0) + 4\,C\,R\,\mathbf{e}_N,$$

and observe that

$$|\mathbf{b}| \leq \max_{x\in\overline{\Omega}} |\nabla\varphi(x)| + 4\,C\,R =: K,$$

with the last quantity K depending only on R and the C^2 norm of φ. Moreover, by construction we have for every $x \in \partial\Omega$

$$\begin{aligned}\varphi(0) + \langle \mathbf{b}, x\rangle &= \varphi(0) + \langle \nabla\varphi(0), x\rangle + 4\,C\,R\,\langle \mathbf{e}_N, x\rangle \\ &= \varphi(0) + \langle \nabla\varphi(0), x\rangle + 4\,C\,R\,x_N \geq \varphi(x),\end{aligned}$$

where in the last estimate we used (6.4.6). To construct the vector **a**, we proceed similarly: the proof is left to the reader, as a useful exercise. □

Remark 6.4.6 The assumption $\varphi \in C^2(\overline{\Omega})$ in the previous result is not optimal, but it can not be lowered too much. For example, Proposition 6.4.5 already fails in dimension $N = 2$ if Ω is a disk and φ is only Lipschitz continuous (see Problem 6.8.11 below). We refer the interested reader to the classical paper [77] by Hartman, for a more detailed study on the BSC.

6.5 The Miranda-Stampacchia Existence Theorem

We recall the notation

$$C^{0,1}_0(\overline{\Omega}) = \left\{u \in C^{0,1}(\overline{\Omega}) \,:\, u = 0 \text{ on } \partial\Omega\right\}.$$

The following existence result has been proved by Stampacchia in [89], then generalized by Miranda in [83] (we also refer to [78] for more general results).

Theorem 6.5.1 (Miranda-Stampacchia) *Let $\Omega \subseteq \mathbb{R}^N$ be an open bounded set and let $F : \mathbb{R}^N \to \mathbb{R}$ be a C^1 strictly convex function. Let us suppose that $U : \partial\Omega \to \mathbb{R}$ satisfies the BSC of rank K. Then the variational problem*

$$\inf_{u\in C^{0,1}(\overline{\Omega})} \left\{\int_\Omega F(\nabla u)\,dx \,:\, u = U \text{ on } \partial\Omega\right\},$$

admits a unique solution v. *Moreover, we have that*

$$|v|_{C^{0,1}(\overline{\Omega})} \leq K. \tag{6.5.1}$$

Finally, v *satisfies the following* Euler-Lagrange equation *in weak form*

$$\int_\Omega \langle \nabla F(\nabla v), \nabla \varphi \rangle \, dx = 0, \qquad \text{for every } \varphi \in C_0^{0,1}(\overline{\Omega}). \tag{6.5.2}$$

Proof We divide the proof in two parts: we first prove existence of a solution, together with the estimate on its Lipschitz constant; then we show that it satisfies the claimed Euler-Lagrange equation.

Existence By recalling the definition of BSC, for every $y \in \partial\Omega$, there exist $\mathbf{a}_y, \mathbf{b}_y \in \mathbb{R}^N$ such that

$$|\mathbf{a}_y| \leq K, \qquad |\mathbf{b}_y| \leq K, \tag{6.5.3}$$

and

$$U(y) + \langle \mathbf{a}_y, x - y \rangle \leq U(x) \leq U(y) + \langle \mathbf{b}_y, x - y \rangle, \qquad \text{for every } x \in \partial\Omega. \tag{6.5.4}$$

If we set

$$\psi(x) = \inf_{y \in \partial\Omega} \Big[U(y) + \langle \mathbf{b}_y, x - y \rangle \Big],$$

then this is a K-Lipschitz function, thanks to Proposition 5.1.8. We observe that for every $x \in \partial\Omega$, we have by (6.5.4)

$$\psi(x) = \inf_{y \in \partial\Omega} \Big[U(y) + \langle \mathbf{b}_y, x - y \rangle \Big] \geq U(x).$$

On the other hand, for every $x \in \partial\Omega$ by taking $y = x$ we have

$$\psi(x) = \inf_{y \in \partial\Omega} \Big[U(y) + \langle \mathbf{b}_y, x - y \rangle \Big] \leq \Big[U(x) + \langle \mathbf{b}_y, x - x \rangle \Big] = U(x).$$

We thus obtain that ψ is a K-Lipschitz function, coinciding with U on $\partial\Omega$. We now fix $M > K$ and consider the restricted problem

$$\inf_{u \in C^{0,1}(\overline{\Omega})} \left\{ \int_\Omega F(\nabla u) \, dx \ : \ u = U \text{ on } \partial\Omega \text{ and } |u|_{C^{0,1}(\overline{\Omega})} \leq M \right\}.$$

Observe that this problem is well-posed, since there exists at least an admissible function, given by ψ. By point (A) of Theorem 6.2.1, we can infer existence of a minimizer v. In order to conclude, it is sufficient to prove that v satisfies (6.5.1): in this case, by point (B) of Theorem 6.2.1, we would get that v is a minimizer for

$$\inf_{u\in C^{0,1}(\overline{\Omega})} \left\{ \int_\Omega F(\nabla u)\,dx \,:\, u = U \text{ on } \partial\Omega \right\},$$

as well, thanks to the choice of M. Uniqueness of v would then follow from Remark 6.2.2.

In order to prove (6.5.1), we first observe that by Problem 6.8.1 the two affine functions

$$x \mapsto U(y) + \langle \mathbf{a}_y, x - y\rangle \qquad \text{and} \qquad x \mapsto U(y) + \langle \mathbf{b}_y, x - y\rangle,$$

are minimizers of $\int_\Omega F(\nabla u)\,dx$, with respect to their own boundary data. In light of (6.5.4), we can thus apply the Comparison Principle (see Proposition 6.3.1) and obtain that for every $y \in \partial\Omega$ we have

$$U(y) + \langle \mathbf{a}_y, x - y\rangle \le v(x) \le U(y) + \langle \mathbf{b}_y, x - y\rangle, \qquad \text{for every } x \in \overline{\Omega}.$$

This entails that for every $y \in \partial\Omega$, we have

$$\langle \mathbf{a}_y, x - y\rangle \le v(x) - v(y) \le \langle \mathbf{b}_y, x - y\rangle, \qquad \text{for every } x \in \overline{\Omega},$$

since $U = v$ on the boundary $\partial\Omega$. We divide everything by $|x-y|$ for $x \neq y$ and use the Cauchy-Schwarz inequality, so to get

$$-|\mathbf{a}_y| \le \frac{v(x) - v(y)}{|x-y|} \le |\mathbf{b}_y|, \qquad \text{for every } x \in \overline{\Omega},\ y \in \partial\Omega \text{ with } x \neq y.$$

By further using (6.5.3), we then get

$$\frac{|v(x) - v(y)|}{|x-y|} \le K, \qquad \text{for every } x \in \overline{\Omega},\ y \in \partial\Omega \text{ with } x \neq y.$$

We can now apply Rado's Lemma (i.e. Lemma 6.3.3), so to finally get

$$|v|_{C^{0,1}(\overline{\Omega})} = \sup_{x\in\overline{\Omega},\, y\in\partial\Omega} \left\{ \frac{|v(x) - v(y)|}{|x-y|} \,:\, x \neq y \right\} \le K,$$

i.e. the desired estimate (6.5.1).

Euler-Lagrange Equation We take the minimizer v and for every function $\varphi \in C^{0,1}_0(\overline{\Omega})$ we observe that

$$\varphi_t(x) = v(x) + t\,\varphi(x),$$

is still admissible in the variational problem. Indeed, it is still Lipschitz continuous and it satisfies the boundary condition $\varphi_t = U$ on $\partial\Omega$. By minimality of v, we have

$$\int_\Omega F(\nabla \varphi_t)\,dx \geq \int_\Omega F(\nabla v)\,dx,$$

i.e. the function of one variable

$$g(t) = \int_\Omega F(\nabla \varphi_t)\,dx,$$

is minimal at $t = 0$. We now show that g is differentiable at $t = 0$ and compute its derivative: for every $t \neq 0$, we have

$$\frac{g(t) - g(0)}{t} = \int_\Omega \frac{F(\nabla \varphi_t) - F(\nabla v)}{t}\,dx = \int_\Omega \frac{F(\nabla v + t\,\nabla \varphi) - F(\nabla v)}{t}\,dx.$$

By the Fundamental Theorem of Calculus, for every $t \neq 0$ and for almost every $x \in \Omega$, we have

$$\begin{aligned} F(\nabla v(x) + t\,\nabla \varphi(x)) - F(\nabla v(x)) &= \int_0^t \frac{d}{d\tau} F(\nabla v(x) + \tau\,\nabla \varphi(x))\,d\tau \\ &= \int_0^t \langle \nabla F(\nabla v(x) + \tau\,\nabla \varphi(x)), \nabla \varphi(x)\rangle\,d\tau. \end{aligned}$$

By setting

$$\Theta = \|\nabla v\|_{L^\infty(\Omega)} + \|\nabla \varphi\|_{L^\infty(\Omega)},$$

and observing that for $|\tau| \leq 1$ we have

$$|\nabla v(x) + \tau\,\nabla \varphi(x)| \leq \Theta, \qquad \text{for a. e. } x \in \Omega,$$

we get in particular that

$$|\nabla F(\nabla v(x) + \tau\,\nabla \varphi(x))| \leq \max_{|\xi| \leq \Theta} |\nabla F(\xi)| = C < +\infty.$$

Indeed, recall that F is a C^1 function. We thus obtain

$$\begin{aligned} &\left| \frac{F(\nabla v(x) + t\,\nabla \varphi(x)) - F(\nabla v(x))}{t} \right| \\ &\quad = \left| \frac{1}{t} \int_0^t \langle \nabla F(\nabla v(x) + \tau\,\nabla \varphi(x)), \nabla \varphi(x)\rangle\,d\tau \right| \\ &\quad \leq \frac{|\nabla \varphi(x)|}{|t|} \int_0^{|t|} |\nabla F(\nabla v(x) + \tau\,\nabla \varphi(x))|\,d\tau \\ &\quad \leq C\,|\nabla \varphi(x)|. \end{aligned}$$

The last function is summable on Ω and independent of the parameter $t \neq 0$. Thus, we can apply the Dominated Convergence Theorem and obtain

$$g'(0) = \lim_{t\to 0} \frac{g(t) - g(0)}{t} = \int_\Omega \lim_{t\to 0} \frac{F(\nabla v + t\,\nabla\varphi) - F(\nabla v)}{t}\,dx$$
$$= \int_\Omega \langle \nabla F(\nabla v), \nabla\varphi\rangle\,dx.$$

On the other hand, since g is minimal at $t = 0$, we must have $g'(0) = 0$. The previous computation then proves the validity of (6.5.2). □

Remark 6.5.2 (Elliptic Equations) By recalling Definition 1.5.5, we can say that (6.5.2) is the *weak formulation* of the equation

$$-\operatorname{div}(\nabla F(\nabla v)) = 0, \qquad \text{in } \Omega.$$

According to Proposition 1.6.2, this equation is *elliptic*. Theorem 6.5.1 guarantees the existence of a (unique) weak solution of the boundary value problem

$$\begin{cases} -\operatorname{div}(\nabla F(\nabla v)) = 0, & \text{in } \Omega, \\ v = U, & \text{on } \partial\Omega, \end{cases}$$

in the class of Lipschitz functions and should be compared with Theorem 4.3.1, where existence is obtained in the framework of Sobolev spaces.

Exactly as for the results of Chap. 4, we should remark that the solution v found in this way is only a weak solution and *not* a classical solution. In other words, in general we do not know whether $v \in C^2(\Omega) \cap C^0(\overline{\Omega})$ or not. This is the goal of the *Regularity Theory*: proving that a weak solution v, which in principle is only Lipschitz, is actually more regular. We will briefly tackle this subject in Chap. 7.

6.6 Two Special Cases

There are a couple of special cases of Theorem 6.5.1 which deserve to be singled out, since they correspond to some problems already presented in the Preface. The first case is linked to the existence of harmonic functions with prescribed boundary data, while the second one is connected with Plateau's problem in non-parametric form. The first case corresponds to choose

$$F(z) = \frac{1}{2}\,|z|^2, \qquad \text{for } z \in \mathbb{R}^N,$$

in Theorem 6.5.1. We thus obtain the following

Corollary 6.6.1 (Harmonic Functions) *Let $\Omega \subseteq \mathbb{R}^N$ be an open bounded set. Let us suppose that $U : \partial\Omega \to \mathbb{R}$ satisfies the BSC of rank K. Then the minimization problem*

$$\inf_{u \in C^{0,1}(\overline{\Omega})} \left\{ \frac{1}{2} \int_\Omega |\nabla u|^2 \, dx \, : \, u = U \text{ on } \partial\Omega \right\},$$

admits a unique solution v. Moreover, v satisfies the Euler-Lagrange equation

$$\int_\Omega \langle \nabla v, \nabla \varphi \rangle \, dx = 0, \qquad \text{for every } \varphi \in C_0^{0,1}(\overline{\Omega}),$$

i.e. v is a weak solution of the boundary value problem

$$\begin{cases} -\Delta v = 0, & \text{in } \Omega, \\ v = U, & \text{on } \partial\Omega. \end{cases}$$

Suppose that we are interested in the following problem, already encountered in the Preface: given a set Ω and a boundary datum U on $\partial\Omega$, find a function which coincides with U on $\partial\Omega$ and whose graph has the least possible area. The previous problem can be rephrased as

$$\inf \left\{ \int_\Omega \sqrt{1 + |\nabla u|^2} \, dx \, : \, u = U \text{ on } \partial\Omega \right\}.$$

By taking

$$F(z) = \sqrt{1 + |z|^2}, \qquad \text{for } z \in \mathbb{R}^N,$$

in Theorem 6.5.1, we get the following

Corollary 6.6.2 (Cartesian Minimal Surfaces) *Let $\Omega \subseteq \mathbb{R}^N$ be an open bounded set. Let us suppose that $U : \partial\Omega \to \mathbb{R}$ satisfies the BSC of rank K. Then the minimization problem*

$$\inf_{u \in C^{0,1}(\overline{\Omega})} \left\{ \int_\Omega \sqrt{1 + |\nabla u|^2} \, dx \, : \, u = U \text{ on } \partial\Omega \right\},$$

admits a unique solution v. Moreover, v satisfies the Euler-Lagrange equation

$$\int_\Omega \left\langle \frac{\nabla v}{\sqrt{1 + |\nabla v|^2}}, \nabla \varphi \right\rangle dx = 0, \qquad \text{for every } \varphi \in C_0^{0,1}(\overline{\Omega}),$$

i.e. v is a weak solution of the boundary value problem

$$\begin{cases} -\operatorname{div}\left(\dfrac{\nabla v}{\sqrt{1+|\nabla v|^2}}\right) = 0, & in\ \Omega, \\ \qquad\qquad\qquad\qquad v = U, & on\ \partial\Omega. \end{cases}$$

Remark 6.6.3 From Remark 1.6.3, we recall that the operator

$$-\operatorname{div}\left(\frac{\nabla v}{\sqrt{1+|\nabla v|^2}}\right),$$

is elliptic, but not strictly elliptic, and it is called *mean curvature operator*.

6.7 Two Counter-examples to Existence

In this section we show that when the BSC fails, in general we can not expect to have existence of Lipschitz minimizers.

6.7.1 Harmonic Functions

We consider the two-dimensional "Pac-Man" set[6]

$$P = \left\{(x, y) \in \mathbb{R}^2 \,:\, x^2 + y^2 < 1\right\} \setminus \left\{(x, y) \in \mathbb{R}^2 \,:\, x \geq 0 \text{ and } y \leq 0\right\}.$$

This set is better described by using polar coordinates $x = \varrho\, \cos\vartheta$ and $y = \varrho\, \sin\vartheta$, in fact we have

$$P = \left\{(\varrho, \vartheta) \,:\, 0 < \varrho < 1,\ \vartheta \in \left(0, \frac{3}{2}\pi\right)\right\},$$

see Fig. 6.2. We take the boundary datum

$$U(\varrho, \vartheta) = \begin{cases} 0, & \text{if } \vartheta = 0 \text{ or } \vartheta = \dfrac{3}{2}\pi, \\ \sin\left(\dfrac{2}{3}\vartheta\right), & \text{if} \varrho = 1 \text{ and } 0 < \vartheta < \dfrac{3}{2}\pi, \end{cases}$$

[6] For readers born after the 80s of the twentieth century, it could be difficult to understand the reason for this weird name.

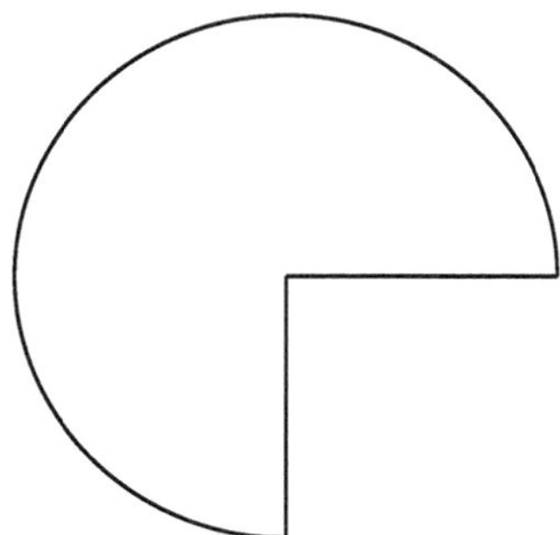

Fig. 6.2 The "Pac-Man" set P

and consider the minimization problem

$$\inf_{\varphi \in C^{0,1}(\overline{P})} \left\{ \frac{1}{2} \int_P |\nabla \varphi|^2 \, dx \, : \, \varphi = U \text{ on } \partial P \right\}. \tag{6.7.1}$$

Observe that we can not apply Corollary 6.6.1, since U *does not* satisfy the BSC. Indeed, U is not the restriction of an affine function and Ω is not convex. Thus by Lemma 6.4.3, U can not satisfy the BSC.

We will show in this section that problem (6.7.1) does not admit a solution. We will need to use the definition of Sobolev spaces $W^{1,2}(\Omega)$ and $W_0^{1,2}(\Omega)$, from Chap. 3. At first, we present a useful preliminary result.

Lemma 6.7.1 *The function given in polar coordinates by*

$$v(\varrho, \vartheta) = \varrho^{\frac{2}{3}} \sin\left(\frac{2}{3}\vartheta\right),$$

is such that $v = U$ on ∂P and is harmonic in classical sense on P. Moreover, it is the unique solution of the following problem

$$\inf_{\varphi \in W^{1,2}(P)} \left\{ \frac{1}{2} \int_P |\nabla \varphi|^2 \, dx \, : \, \varphi - U \in W_0^{1,2}(P) \right\}.$$

Proof The fact that $v = U$ on ∂P is immediate. In order to show that v is harmonic on P, it is sufficient to compute the Laplacian of v, in polar coordinates. By recalling that (see Appendix A)

$$\Delta v = \frac{\partial^2 v}{\partial \varrho^2} + \frac{1}{\varrho}\frac{\partial v}{\partial \varrho} + \frac{1}{\varrho^2}\frac{\partial^2 v}{\partial \vartheta^2},$$

we have for $\varrho > 0$

$$\Delta\left(\varrho^{\frac{2}{3}}\sin\left(\frac{2}{3}\vartheta\right)\right)=-\frac{2}{9}\varrho^{-\frac{4}{3}}\sin\left(\frac{2}{3}\vartheta\right)$$
$$+\frac{2}{3}\varrho^{-\frac{4}{3}}\sin\left(\frac{2}{3}\vartheta\right)-\frac{4}{9}\varrho^{-\frac{4}{3}}\sin\left(\frac{2}{3}\vartheta\right)=0.$$

The function v is bounded on P, thus in particular we have $v \in L^2(P)$, since P is a bounded set. Moreover, it is not difficult to see that the classical gradient ∇v belongs to $L^2(P;\mathbb{R}^N)$: indeed, in polar coordinates we have (see formula (A.1.5))

$$|\nabla v|^2=\left(\frac{\partial v}{\partial\varrho}\right)^2+\frac{1}{\varrho^2}\left(\frac{\partial v}{\partial\vartheta}\right)^2$$
$$=\frac{4}{9}\varrho^{-\frac{2}{3}}\sin^2\left(\frac{2}{3}\vartheta\right)+\frac{4}{9}\varrho^{-\frac{2}{3}}\cos^2\left(\frac{2}{3}\vartheta\right)=\frac{4}{9}\varrho^{-\frac{2}{3}}.$$

Thus, we get

$$\int_P|\nabla v|^2\,dx=\int_0^{\frac{3}{2}\pi}\int_0^1\frac{4}{9}\varrho^{-\frac{2}{3}}\,\varrho\,d\varrho\,d\vartheta=\frac{\pi}{2}<+\infty.$$

Thus, we obtain that $v \in W^{1,2}(P)$, by Proposition 3.2.3. From Proposition 5.4.1, we also have $U \in C^{0,1}(\overline{P}) \subseteq W^{1,\infty}(P) \subseteq W^{1,2}(P)$. Then, the difference $v-U$ belongs to $W^{1,2}(P)$, is continuous on $\overline{P}$ and it vanishes on ∂P. In light of Problem 3.12.16, we get that $v-U \in W^{1,2}_0(P)$, as claimed.

We still have to show that v is a solution of the minimization problem. This follows from Proposition 4.3.5, by observing that $v-U \in W^{1,2}_0(P)$ and v is a weak solution of the relevant Euler-Lagrange equation, i.e.

$$\int_P\langle\nabla v,\nabla\varphi\rangle\,dx=0,\qquad\text{for every }\varphi\in C^\infty_0(P).$$

The latter follows from the fact that v is harmonic in classical sense in P. Indeed, for every $\varphi \in C^\infty_0(P)$ by the Divergence Theorem we have

$$0=\int_P\Delta v\,\varphi\,dx=\int_P\operatorname{div}(\nabla v\,\varphi)\,dx-\int_P\langle\nabla v,\nabla\varphi\rangle\,dx=\int_P\langle\nabla v,\nabla\varphi\rangle\,dx,$$

thanks to the fact that φ is compactly supported. The uniqueness of the minimizer follows from the strict convexity of the function $z\mapsto|z|^2$, as in Remark 6.2.2. □

Proposition 6.7.2 *With the notation above, problem* (6.7.1) *does not admit a solution.*

Proof We argue by contradiction and assume that a solution $w \in C^{0,1}(\overline{P})$ exists. By Problem 6.8.10 below, we know that

$$\inf_{\varphi\in W^{1,2}(\Omega)}\left\{\int_P |\nabla\varphi|^2\,dx\ :\ \varphi-U\in W_0^{1,2}(P)\right\}$$
$$=\inf_{\varphi\in C^{0,1}(\overline{P})}\left\{\int_P |\nabla\varphi|^2\,dx\ :\ \varphi=U \text{ on } \partial P\right\},$$

and that w must coincide with the unique solution of

$$\inf_{\varphi\in W^{1,2}(\Omega)}\left\{\int_P |\nabla\varphi|^2\,dx\ :\ \varphi-U\in W_0^{1,2}(P)\right\}.$$

By Lemma 6.7.1, we thus get $w = v$. We now show that $v \notin C^{0,1}(\overline{P})$: this will give the desired contradiction. To this aim, we take for example a point of the form $(x,-x)$ with $x<0$ then we have (recall that v vanishes at the origin)

$$|v(x,-x)-v(0,0)| = |v(x,-x)| = \left(\sqrt{x^2+x^2}\right)^{\frac{2}{3}}\left|\sin\left(\frac{2}{3}\cdot\frac{3}{4}\pi\right)\right| = \sqrt[3]{2}\,|x|^{\frac{2}{3}}.$$

We then get

$$|v|_{C^{0,1}(\overline{P})} \geq \sup_{0<x\leq 1}\frac{|v(x,-x)-v(0,0)|}{|(x,-x)-(0,0)|} = \sup_{0<x\leq 1}\frac{\sqrt[3]{2}\,|x|^{\frac{2}{3}}}{\sqrt{2}\,|x|} = +\infty,$$

as desired (see Fig. 6.3). □

6.7.2 Cartesian Minimal Surfaces

The example contained in this section is taken from [6, Example 1.1]. We take $0 < r < R$ and consider the two-dimensional spherical shell

$$A_{r,R} = \{x\in\mathbb{R}^2\ :\ r<|x|<R\}.$$

We fix a constant $M>0$ and the boundary datum $U:\partial A_{r,R}\to\mathbb{R}$

$$U(x) = M \text{ if } |x| = r, \qquad U(x) = 0 \text{ if } |x| = R,$$

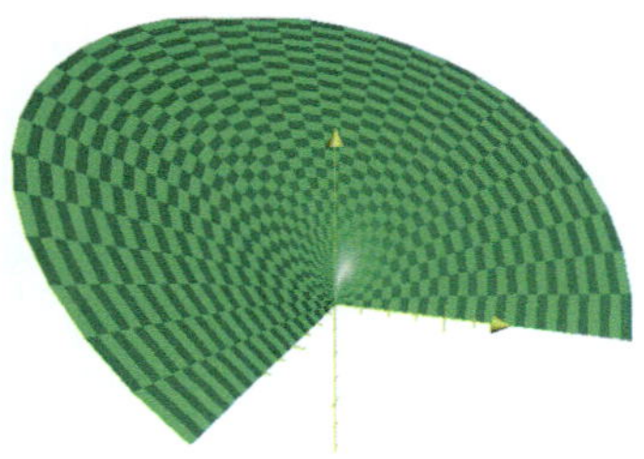

Fig. 6.3 The graph of the function v of Lemma 6.7.1. We have that $|\nabla v(x)|\sim|x|^{-1/3}$ for $|x|\to 0$, thus v is not Lipschitz on $\overline{P}$

and look for the function coinciding with U on the boundary and whose graph has the least possible area. Thus, we consider the minimization problem

$$\inf_{\varphi\in C^{0,1}(\overline{A_{r,R}})} \left\{ \int_{A_{r,R}} \sqrt{1+|\nabla\varphi|^2}\,dx \ :\ \varphi = U \text{ on } \partial A_{r,R} \right\}. \tag{6.7.2}$$

In this case as well, we can not apply Corollary 6.6.2, since U does not satisfy the BSC. Indeed, U is not the restriction of an affine function and Ω is not convex. As before, U can not satisfy the BSC by appealing to Lemma 6.4.3.

We will show in this section that (6.7.2) does not admit a solution. In the following result, we collect some necessary optimality conditions for this minimization problem.

Lemma 6.7.3 *Let us assume that problem* (6.7.2) *admits a solution* v. *Then such a solution is unique and radially symmetric, i.e. there exists a Lipschitz function* $V : [r, R] \to [0, M]$ *such that*

$$v(x) = V(|x|), \qquad \text{for } r \le |x| \le R.$$

Moreover, there exists a constant $C > 0$ *such that the function* V *verifies*

$$\frac{V'(t)}{\sqrt{1+|V'(t)|^2}} = -\frac{C}{t}, \qquad \text{for a. e. } t \in [r, R]. \tag{6.7.3}$$

Proof The uniqueness and the radial symmetry follow from Problem 6.8.2 below. Moreover, the latter also gives that V is non-increasing. In order to prove the last point, we observe that by radial symmetry of v, we obtain that

$$\begin{aligned}\inf_{u\in C^{0,1}(\overline{A_{r,R}})} &\left\{ \int_{A_{r,R}} \sqrt{1+|\nabla u|^2}\,dx \ :\ u = U \text{ on } \partial A_{r,R} \right\} \\ &= \inf_{u\in C^{0,1}_{\text{rad}}(\overline{A_{r,R}})} \left\{ \int_{A_{r,R}} \sqrt{1+|\nabla u|^2}\,dx \ :\ u = U \text{ on } \partial A_{r,R} \right\},\end{aligned}$$

where

$$C^{0,1}_{\text{rad}}(\overline{A_{r,R}}) = \left\{ u \in C^{0,1}(\overline{A_{r,R}}) \ :\ u \text{ is radially symmetric} \right\}.$$

Indeed, since

$$C^{0,1}_{\text{rad}}(\overline{A_{r,R}}) \subseteq C^{0,1}(\overline{A_{r,R}}),$$

one inequality between the two infima is obvious. The fact that they coincide is a consequence of the radial symmetry of the solution v.

By using polar coordinates, the minimization problem on radial functions can be rewritten as the one-dimensional problem

$$\inf_{\psi\in C^{0,1}([r,R])}\left\{\int_r^R \sqrt{1+|\psi'(\varrho)|^2}\,\varrho\,d\varrho\ :\ \psi(r)=M,\ \psi(R)=0\right\}.$$

From the first part of the proof, we know that this is uniquely solved by V. By proceeding as in the proof of the optimality condition of Theorem 6.5.1, we get that V is a weak solution of

$$-\left(\frac{V'(\varrho)}{\sqrt{1+|V'(\varrho)|^2}}\,\varrho\right)'=0,\qquad \text{in } [r,\,R].$$

In other words, we have

$$\int_r^R \frac{V'(\varrho)\,\varphi'(\varrho)}{\sqrt{1+|V'(\varrho)|^2}}\,\varrho\,d\varrho=0,\qquad \text{for every } \varphi\in C_0^{0,1}([r,\,R]).$$

We take $\eta\in C_0^\infty((r,\,R))$ and use the previous identity with the choice

$$\varphi(\varrho)=\int_r^\varrho \eta(t)\,dt-\left(⨍_r^R \eta(t)\,dt\right)(\varrho-r),$$

whose derivative is given by

$$\varphi'(\varrho)=\eta(\varrho)-\left(⨍_r^R \eta(t)\,dt\right).$$

Observe that $\varphi(r)=\varphi(R)=0$, by construction. We then obtain

$$\int_r^R \frac{V'(\varrho)\,\eta(\varrho)}{\sqrt{1+|V'(\varrho)|^2}}\,\varrho\,d\varrho=\left(⨍_r^R \eta(t)\,dt\right)\int_r^R \frac{V'(\varrho)}{\sqrt{1+|V'(\varrho)|^2}}\,\varrho\,d\varrho.$$

If we set

$$C:=-⨍_r^R \frac{V'(\varrho)}{\sqrt{1+|V'(\varrho)|^2}}\,\varrho\,d\varrho,$$

then the previous equation can be rewritten as

$$\int_r^R \frac{V'(\varrho)\,\eta(\varrho)}{\sqrt{1+|V'(\varrho)|^2}}\,\varrho\,d\varrho=-C\int_r^R \eta(t)\,dt.$$

Finally, we observe that this is the same as

$$\int_r^R \left[\frac{V'(\varrho)}{\sqrt{1+|V'(\varrho)|^2}}\, \varrho + C \right] \eta(\varrho)\, d\varrho = 0.$$

This is valid for every $\eta \in C_0^\infty((r, R))$, thus by the Du Bois-Reymond Lemma (see Lemma 1.4.2), we get

$$\frac{V'(\varrho)}{\sqrt{1+|V'(\varrho)|^2}}\, \varrho = -C, \qquad \text{for a. e. } \varrho \in [r, R].$$

This gives the optimality condition (6.7.3).

Finally, we observe that the previous identity implies that V' has constant sign on $[r, R]$. Moreover, from the boundary conditions $V(r) = M > 0 = V(R)$ and Problem 5.5.6, we get

$$\int_r^R V'(\varrho)\, d\varrho = -M < 0.$$

This shows that $V' < 0$ on a set of positive measure. In particular, it must result $C > 0$. □

Proposition 6.7.4 *With the previous notation, let us suppose that r, R and M are linked through the relation*

$$M = r \log \left(\frac{R + \sqrt{R^2 - r^2}}{r} \right). \tag{6.7.4}$$

Then problem (6.7.2) *does not admit a solution.*

Proof We crucially exploit the condition (6.7.3). We observe that the function

$$g(\tau) = \frac{\tau}{\sqrt{1+\tau^2}}, \qquad \tau \in \mathbb{R},$$

is strictly increasing and takes values in $(-1, 1)$. This implies that if a solution v of (6.7.2) exists, from (6.7.3) we must have

$$-1 < -\frac{C}{t} < 0, \text{ for every } t \in [r, R] \qquad \text{that is} \qquad r > C > 0, \tag{6.7.5}$$

otherwise (6.7.3) could not hold. By observing that g is invertible, with inverse function given by

$$g^{-1}(s) = \frac{s}{\sqrt{1-s^2}}, \qquad \text{for } s \in (-1, 1),$$

we would get from (6.7.3)

$$V'(t) = g^{-1}\left(-\frac{C}{t}\right) = -\frac{C}{\sqrt{t^2 - C^2}}, \qquad \text{for a. e. } t \in [r, R].$$

By integrating this identity, thanks to Problem 5.5.6 we would get

$$V(\varrho) - V(r) = -\int_r^{\varrho} \frac{C}{\sqrt{t^2 - C^2}}\, dt,$$

that is[7]

$$V(\varrho) - M = -C\, \operatorname{arg\,cosh}\left(\frac{\varrho}{C}\right) + C\, \operatorname{arg\,cosh}\left(\frac{r}{C}\right).$$

We thus obtained

$$V(\varrho) = M - C\, \log\left(\frac{\varrho + \sqrt{\varrho^2 - C^2}}{r + \sqrt{r^2 - C^2}}\right), \qquad \text{for } \varrho \in [r, R].$$

We can now determine the crucial constant C, by imposing the other boundary condition $V(R) = 0$. This gives

$$M = C\, \log\left(\frac{R + \sqrt{R^2 - C^2}}{r + \sqrt{r^2 - C^2}}\right).$$

By recalling (6.7.4), we thus get

$$r\, \log\left(\frac{R + \sqrt{R^2 - r^2}}{r}\right) = C\, \log\left(\frac{R + \sqrt{R^2 - C^2}}{r + \sqrt{r^2 - C^2}}\right). \tag{6.7.6}$$

We now observe that the function

$$t \mapsto t\, \log\left(\frac{R + \sqrt{R^2 - t^2}}{r + \sqrt{r^2 - t^2}}\right),$$

is strictly increasing on its domain of definition $[-r, r]$, see Lemma 6.7.5 below. Thus (6.7.6) would imply that $C = r$, which gives a contradiction with (6.7.5). We refer to Fig. 6.4 for the graph of the solution v: we have obtained that its gradient must blow-up when approaching the internal radius r. □

[7] We recall that

$$\operatorname{arg\,cosh}(t) = \log\left(t + \sqrt{t^2 - 1}\right), \qquad \text{for } t \geq 1.$$

The next technical result was instrumental to the previous one.

Lemma 6.7.5 *Let* $0 < r < R$*, then the function*

$$h(t) = t \log\left(\frac{R + \sqrt{R^2 - t^2}}{r + \sqrt{r^2 - t^2}}\right), \qquad \textit{for } t \in [-r, r],$$

is strictly increasing.

Proof We first observe that h is an odd function, thus we can confine ourselves to study it on the interval $[0, r]$. We first take into account the function

$$t \mapsto \log\left(\frac{R + \sqrt{R^2 - t^2}}{r + \sqrt{r^2 - t^2}}\right), \qquad \text{for } t \in [0, r]. \tag{6.7.7}$$

It is sufficient to prove that the latter is strictly increasing: indeed, if this is true, then for every $0 \le t < s \le r$ we get

$$t \log\left(\frac{R + \sqrt{R^2 - t^2}}{r + \sqrt{r^2 - t^2}}\right) \le t \log\left(\frac{R + \sqrt{R^2 - s^2}}{r + \sqrt{r^2 - s^2}}\right) < s \log\left(\frac{R + \sqrt{R^2 - s^2}}{r + \sqrt{r^2 - s^2}}\right),$$

which shows that our function is strictly increasing on $[0, r]$, as well.

We rewrite the function (6.7.7) as follows

$$t \mapsto \log\left(R + \sqrt{R^2 - t^2}\right) - \log\left(r + \sqrt{r^2 - t^2}\right).$$

For $0 \le t < r$, the derivative of this function is thus given by

$$-\frac{t}{\sqrt{R^2 - t^2}} \frac{1}{R + \sqrt{R^2 - t^2}} + \frac{t}{\sqrt{r^2 - t^2}} \frac{1}{r + \sqrt{r^2 - t^2}},$$

which is non-negative if and only if

$$t\left(r + \sqrt{r^2 - t^2}\right)\sqrt{r^2 - t^2} \le t\left(R + \sqrt{R^2 - t^2}\right)\sqrt{R^2 - t^2}.$$

Fig. 6.4 The graph of the solution to problem (6.7.2) corresponding to the choice (6.7.4) for M. Observe that the graph has a vertical tangent on the internal circle $|x| = r$, thus the solution is not Lipschitz

We now observe that $r < R$, thus the last inequality is trivially true for every $t \in [0, r)$. Moreover, the inequality sign is strict for every $t \in (0, r)$. This discussion gives that the function in (6.7.7) is strictly increasing, as required. □

6.8 Problems

Problem 6.8.1 Let $\Omega \subseteq \mathbb{R}^N$ be an open bounded set and let $F : \mathbb{R}^N \to \mathbb{R}$ be a convex function. For $a \in \mathbb{R}$ and $\mathbf{b} \in \mathbb{R}^N$, we consider the affine function

$$U(x) = a + \langle \mathbf{b}, x \rangle, \qquad \text{for } x \in \mathbb{R}^N.$$

Prove that U is a solution of

$$\inf_{u \in C^{0,1}(\overline{\Omega})} \left\{ \int_\Omega F(\nabla u)\, dx \, : \, u = U \text{ on } \partial\Omega \right\}.$$

If F is additionally supposed to be strictly convex, prove that U is the unique solution.

Problem 6.8.2 Let $f : [0, +\infty) \to \mathbb{R}$ be a strictly convex increasing function. Given $0 < r_1 < r_2$, we consider the spherical shell

$$\overline{A_{r_1,r_2}} = \{x \in \mathbb{R}^N \, : \, r_1 \le |x| \le r_2\}.$$

For a pair of constants $C_1 \neq C_2$, let us suppose that the following problem

$$\inf_{u \in C^{0,1}(\overline{A_{r_1,r_2}})} \left\{ \int_{A_{r_1,r_2}} f(|\nabla u|)\, dx \, : \, u = C_i \text{ on } \partial B_{r_i}(0) \text{ for } i = 1, 2 \right\},$$

admits a solution v. Prove that such a solution must be unique and radially symmetric, i.e. there exists a Lipschitz function V of one variable such that

$$v(x) = V(|x|), \qquad \text{for every } x \in \overline{A_{r_1,r_2}}.$$

Also prove that V is:

- non-decreasing if $C_1 < C_2$;
- non-increasing if $C_1 > C_2$.

Problem 6.8.3 Show that in Theorem 6.5.1 we can remove the strict convexity assumption on F and still get existence of a minimizer for

$$\inf_{u \in C^{0,1}(\overline{\Omega})} \left\{ \int_\Omega F(\nabla u)\, dx \, : \, u = U \text{ on } \partial\Omega \right\}.$$

Problem 6.8.4 Show that, under the assumptions of Problem 6.8.3, in general minimizers are not unique. Prove that if v_1 and v_2 are two minimizers, then

$$w = \min\{v_1, v_2\} \qquad \text{and} \qquad W = \max\{v_1, v_2\},$$

are minimizers, as well.

Problem 6.8.5 Under the assumptions of Problem 6.8.3, let us further assume that F is strictly convex "at infinity", i.e. there exists $\delta > 0$ such that F is strictly convex outside the ball $B_\delta(0)$. This means that

$$F(t\,z + (1-t)\,w) < t\,F(z) + (1-t)\,F(w),$$

for every $t \in (0,1)$ and every pair $z \neq w$ such that at least one of these belong to $\mathbb{R}^N \setminus B_\delta(0)$.

Prove that every minimizer v satisfies the estimate

$$\|\nabla v\|_{L^\infty(\Omega)} \le K + 2\,\delta,$$

where K is the rank of the BSC for the boundary datum.

Problem 6.8.6 Let $\Omega \subseteq \mathbb{R}^N$ be an open bounded set. For $\tau \in \mathbb{R}^N \setminus \{0\}$, we set

$$\Omega_\tau := \Omega + \tau.$$

Show that if $z_0 \in \partial(\Omega \cap \Omega_\tau)$, then at least one of the two points z_0 and $z_0 - \tau$ belongs to $\partial\Omega$.

Problem 6.8.7 Let $\Omega \subseteq \mathbb{R}^N$ be an open set with the following property: for every $y \in \partial\Omega$, there exists a direction $\omega_y \in \mathbb{S}^{N-1}$ such that if we set

$$\Pi_y^+ = \Big\{x \in \mathbb{R}^N \,:\, \langle x - y, \omega_y\rangle \ge 0\Big\},$$

then

$$y \in \partial\Pi_y^+ \qquad \text{and} \qquad \Omega \subseteq \Pi_y^+.$$

Show that Ω is convex. The set $\partial\Pi_y^+$ is called *supporting hyperplane for Ω at the point y*.

Problem 6.8.8 Let $\Omega \subseteq \mathbb{R}^N$ be an open bounded convex set, with the following property: for every $x_0 \in \partial\Omega$ there exists a direction $\omega_{x_0} \in \mathbb{S}^{N-1}$ such that if we set

$$\Pi_{x_0}^+ = \Big\{x \in \mathbb{R}^N \,:\, \langle x - x_0, \omega_{x_0}\rangle \ge 0\Big\},$$

then

$$|x - x_0|^2 \le C_\Omega \operatorname{dist}(x, \partial \Pi^+_{x_0}), \qquad \text{for every } x \in \partial\Omega,$$

for a uniform constant $C_\Omega > 0$. Show that the restriction on $\partial\Omega$ of a function $U \in C^2(\mathbb{R}^N)$ satisfies the BSC.

Problem 6.8.9 Let $\Omega \subseteq \mathbb{R}^N$ be an open bounded set and let $U : \partial\Omega \to \mathbb{R}$ satisfies the BSC of rank K. Prove that U is Lipschitz on $\partial\Omega$, with a Lipschitz constant not exceeding K.

Problem 6.8.10 Let $\Omega \subseteq \mathbb{R}^N$ be an open bounded set and let $g \in C^{0,1}(\overline{\Omega})$. Show that for every $1 < p < \infty$ we have[8]

$$\inf_{\varphi \in W^{1,p}(\Omega)} \left\{ \int_\Omega |\nabla\varphi|^p \, dx \, : \, \varphi - g \in W^{1,p}_0(\Omega) \right\}$$
$$= \inf_{\varphi \in C^{0,1}(\overline{\Omega})} \left\{ \int_\Omega |\nabla\varphi|^p \, dx \, : \, \varphi = g \text{ on } \partial\Omega \right\},$$

where the space $W^{1,p}_0(\Omega)$ has been defined in Definition 3.7.1. Also show that if the problem on the right-hand side admits a solution, then this must coincide with the minimizer of the problem on the left-hand side.

Problem 6.8.11 Let $B_1(0) = \{(x, y) \in \mathbb{R}^2 \, : \, x^2 + y^2 < 1\}$ and let $U : \overline{B_1(0)} \to \mathbb{R}$ be the function defined by

$$U(x, y) = |y|, \qquad \text{for every } (x, y) \in \overline{B_1(0)}.$$

Then:

- show that the restriction of U to $\partial B_1(0)$ is Lipschitz continuous on $\partial B_1(0)$;
- prove that the problem

$$\inf_{u \in C^{0,1}(\overline{B_1(0)})} \left\{ \frac{1}{2} \int_{B_1(0)} |\nabla u|^2 \, dx\, dy \, : \, u = U \text{ on } \partial B_1(0) \right\},$$

  does not admit a solution;
- finally, deduce that the restriction of U to $\partial B_1(0)$ does not satisfy the BSC.

[8] In the language of Calculus of Variations, this implies that the so-called *Lavrentiev phenomenon* does not occur for the problem of minimizing the functional $\varphi \mapsto \int_\Omega |\nabla\varphi|^p \, dx$ among functions with a given boundary datum. We refer the reader to [66] for a general overview on this subject.

Chapter 7
Excerpts from Regularity Theory

7.1 Introduction

In Chaps. 4 and 6, we obtained existence of weak solutions to quite general elliptic equations, through some minimization problems settled in the space of Sobolev or Lipschitz functions. We now have to inquire to which extent such minimizers are actually solutions of the relevant equation in a classical sense. This is the goal of the *Regularity Theory*: proving that a function, thanks to its minimality properties, is actually more regular than just merely Sobolev or Lipschitz.

It is not the scope of this book to introduce the reader to the deepest realm of the theory, for which we rather refer to [6]. For this reason, we will avoid any general discussion and rather stick to the very simple, yet meaningful, example that accompanied us throughout this journey. In other words, we still consider our model problem

$$\min_{u\in W^{1,2}(\Omega)} \left\{\frac{1}{2}\int_\Omega |\nabla u|^2\,dx \,:\, u-g\in W^{1,2}_0(\Omega)\right\}, \tag{7.1.1}$$

where $\Omega\subseteq\mathbb{R}^N$ is an open bounded set and $g\in W^{1,2}(\Omega)$ is a given function. We have seen that the unique minimizer v of this problem is a *weak solution* of the Laplace equation, i.e. it satisfies

$$\int_\Omega \langle\nabla v,\nabla\varphi\rangle\,dx=0,\qquad \text{for every } \varphi\in C^\infty_0(\Omega).$$

We now would like to know whether such a weak solution is a harmonic function in the usual classical sense. By Remark 1.5.6, we thus need to show that v is actually of class C^2.

There are many ways to prove this: we will choose to use the celebrated *De Giorgi-Moser's techniques*, which were introduced and exploited by Ennio De

L. Brasco, *Handbook of Calculus of Variations for Absolute Beginners*,
La Matematica per il 3+2 163, https://doi.org/10.1007/978-3-031-87164-1_7

Giorgi in order to tackle the XIXth Hilbert's problem (see [72]). More precisely, we will follow the approach by Jürgen Moser, introduced in [84]. Even if this may seem the most difficult way to prove regularity in the case of harmonic functions, this technique is purely variational in nature and can be widely generalized to cover much more complicated and general cases, even nonlinear, like those studied in Theorems 4.3.1 and 6.5.1. This is the reason why we decided to take this path.[1] Thus, we will essentially use the case of harmonic function as a training session to learn the basics of these techniques. We will then try to gently guide the reader through some generalizations of these techniques, by means of a small (yet meaningful) list of problems in the last section of the chapter.

The reader who will be willing to deepen his knowledge in the modern Regularity Theory, could consult [17], [29, Chapter 8], [31, 47, Chapters 3 & 4] or the survey paper [82], among others.

We start by recalling a couple of definitions that are crucial for the sequel.

Definition 7.1.1 Let $\Omega \subseteq \mathbb{R}^N$ be an open set, we say that $u \in W^{1,2}_{\rm loc}(\Omega)$ is *weakly harmonic in* Ω if

$$\int_\Omega \langle \nabla u, \nabla \varphi \rangle \, dx = 0, \qquad \text{for every } \varphi \in C_0^\infty(\Omega). \tag{7.1.2}$$

By a standard density argument, the previous identity continues to hold for every $\varphi \in W_0^{1,2}(\Omega')$ and every $\Omega' \Subset \Omega$.

We will also need the following definition

Definition 7.1.2 Let $\Omega \subseteq \mathbb{R}^N$ be an open set, we say that $u \in W^{1,2}_{\rm loc}(\Omega)$ is *weakly subharmonic in* Ω if

$$\int_\Omega \langle \nabla u, \nabla \varphi \rangle \, dx \le 0, \qquad \text{for every } \varphi \in C_0^\infty(\Omega) \text{ such that } \varphi \ge 0.$$

As above, we can use a density argument (and Problem 3.12.26) to assure that the previous inequality continues to hold for every $\varphi \in W_0^{1,2}(\Omega')$ such that $\varphi \ge 0$ and every $\Omega' \Subset \Omega$.

7.2 The Caccioppoli Inequality

The next result will be extremely important: we refer the reader to Remark 7.2.2 below, for a comment.

[1] For a simpler proof, which is however quite specific of the Laplace equation and can not be generalized to nonlinear equations, the interested reader could consult [56, Chapter 2, Section 3]. This is the so-called *Weyl Lemma*. However, despite the usual attribution to Hermann Weyl, it should be noticed that this result can be traced back for the first time in the literature in a paper by Sergej L'vovič Sobolev.

Theorem 7.2.1 *Let $u \in W^{1,2}_{loc}(\Omega)$ be a weakly harmonic function. Let $F : \mathbb{R} \to \mathbb{R}$ be a C^1 convex function, with bounded first derivative. Then $F \circ u \in W^{1,2}_{loc}(\Omega)$ is a weakly subharmonic function on Ω. Moreover, if F is non-negative the following inequality holds*

$$\int_\Omega |\nabla(F \circ u)|^2 \varphi^2 \, dx \le 4 \int_\Omega |\nabla \varphi|^2 \, |F(u)|^2 \, dx, \qquad \textit{for every } \varphi \in C_0^\infty(\Omega). \tag{7.2.1}$$

Proof The fact that $F \circ u \in W^{1,2}_{loc}(\Omega)$ follows from Proposition 3.4.1 (also recall Remark 3.4.2). In order to prove that $F \circ u$ is weakly subharmonic, we further assume for the moment that

$$F \in C^2(\mathbb{R}) \qquad \text{and} \qquad |F''(t)| \le C, \qquad \text{for every } t \in \mathbb{R}. \tag{7.2.2}$$

We take $\eta \in C_0^\infty(\Omega)$, then there exists $\Omega' \Subset \Omega$ such that η has compact support in Ω', as well. Observe that $F' : \mathbb{R} \to \mathbb{R}$ is a C^1 function with bounded first derivative, thanks to (7.2.2). Thus, we have $F' \circ u \in W^{1,2}(\Omega')$, again by Proposition 3.4.1. Moreover, by Problem 3.12.18, we know that

$$\varphi = \eta \, (F' \circ u) \in W_0^{1,2}(\Omega').$$

We can then use this test function in (7.1.2), so to get

$$0 = \int_\Omega \langle \nabla u, \nabla(\eta \, (F' \circ u)) \rangle \, dx = \int_\Omega |\nabla u|^2 \, F''(u) \, \eta \, dx + \int_\Omega \langle \nabla u, \nabla \eta \rangle \, F'(u) \, dx.$$

If we further suppose that $\eta \ge 0$ and use that F is convex, we have $F''(u) \, \eta \ge 0$. Thus, from the previous identity we get

$$\int_\Omega \langle \nabla u, \nabla \eta \rangle \, F'(u) \, dx \le 0, \qquad \text{for every } \eta \in C_0^\infty(\Omega') \text{ such that } \eta \ge 0. \tag{7.2.3}$$

By using the chain rule, we have $F'(u) \, \nabla u = \nabla(F \circ u)$ and thus Eq. (7.2.3) becomes

$$\int_\Omega \langle \nabla(F \circ u), \nabla \eta \rangle \, dx \le 0, \qquad \text{for every } \eta \in C_0^\infty(\Omega) \text{ such that } \eta \ge 0. \tag{7.2.4}$$

This shows that $F \circ u$ is weakly subharmonic in Ω, under the further assumptions (7.2.2). We will show below how to remove this assumption.

Let us now suppose that $F \circ u$ is weakly subharmonic, with F non-negative as in the statement: we are going to prove the validity of (7.2.1). As previously observed, we have that (7.2.4) still holds for functions $\eta \in W_0^{1,2}(\Omega')$, for every $\Omega' \Subset \Omega$. We now take $\psi \in C_0^\infty(\Omega)$ and use in (7.2.4) the test function

$$\eta = \psi^2 \, F(u),$$

which is admissible, by the same arguments as above. We then obtain

$$\int_\Omega |\nabla(F\circ u)|^2\,\psi^2\,dx + 2\int_\Omega \langle \nabla(F\circ u), \nabla\psi\rangle\,\psi\,F(u)\,dx \le 0.$$

From this, we can obtain

$$\begin{aligned}\int_\Omega |\nabla(F\circ u)|^2\,\psi^2\,dx &\le -2\int_\Omega \langle \nabla(F\circ u), \nabla\psi\rangle\,\psi\,F(u)\,dx\\ &\le 2\int_\Omega |\nabla(F\circ u)|\,|\nabla\psi|\,|F(u)|\,|\psi|\,dx\\ &\le 2\left(\int_\Omega |\nabla(F\circ u)|^2\,\psi^2\,dx\right)^{\frac{1}{2}}\left(\int_\Omega |\nabla\psi|^2\,|F(u)|^2\,dx\right)^{\frac{1}{2}},\end{aligned}$$

thanks to Hölder's inequality. Observe that we can simplify on both sides the term containing $\nabla(F\circ u)$, so to obtain

$$\int_\Omega |\nabla(F\circ u)|^2\,\psi^2\,dx \le 4\int_\Omega |\nabla\psi|^2\,|F(u)|^2\,dx,$$

which is the desired estimate.

In order to complete the proof, we are only left with removing the additional assumptions (7.2.2) that we used to get (7.2.4). To this aim, if F is a C^1 convex function with bounded first derivative, we need to show that we can suitably approximate it by a family of C^2 convex functions, with bounded second order derivative. We define at first the convolution

$$H_\varepsilon(t) = F * \rho_\varepsilon(t) = \int_{\mathbb{R}} F(t-\tau)\,\rho_\varepsilon(\tau)\,d\tau, \qquad \text{for every } t\in\mathbb{R},$$

where

$$\rho_\varepsilon(\tau) = \frac{1}{\varepsilon}\,\rho\left(\frac{\tau}{\varepsilon}\right),$$

and ρ is the standard smoothing kernel in dimension 1, given by (1.4.1). We have that $H_\varepsilon\in C^\infty(\mathbb{R})$ and it is still convex, with bounded first derivative H'_ε. More precisely, we get

$$|H'_\varepsilon(t)| = |F' * \rho_\varepsilon(t)| \le \int_{\mathbb{R}} |F'(t-\tau)|\,\rho_\varepsilon(\tau)\,d\tau \le \|F'\|_{L^\infty(\mathbb{R})}. \tag{7.2.5}$$

We also have that

$$\lim_{\varepsilon\to 0}\max_{t\in[-M,M]}\Big[|H_\varepsilon(t)-F(t)| + |H'_\varepsilon(t)-F'(t)|\Big] = 0, \qquad \text{for every } M>0, \tag{7.2.6}$$

(*verify this assertion as an exercise*). We then set

$$F_\varepsilon(t) = F(0) + F'(0)\,t + \int_0^t \int_0^s H_\varepsilon''(\tau)\,\chi(\varepsilon\,\tau)\,d\tau\,ds,$$

where $\chi \in C_0^\infty(\mathbb{R})$ is such that $0 \le \chi \le 1$, $|\chi'| \le C$ and

$$\chi(t) = 1, \quad \text{if } |t| \le 1, \qquad \chi(t) = 0, \quad \text{if } |t| \ge 2.$$

Observe that

$$F_\varepsilon''(t) = H_\varepsilon''(t)\,\chi(\varepsilon\,t) \ge 0,$$

thus F_ε is a C^2 convex function, with bounded second order derivative, since

$$|F_\varepsilon''(t)| \le \max_{t\in\left[-\frac{2}{\varepsilon},\frac{2}{\varepsilon}\right]} |H_\varepsilon''(t)| < +\infty.$$

We notice that the last bound depends on $\varepsilon > 0$. On the other hand, we have

$$\begin{aligned}
|F_\varepsilon'(t)| &= \left| F'(0) + \int_0^t H_\varepsilon''(\tau)\,\chi(\varepsilon\,\tau)\,d\tau \right| \\
&\le |F'(0)| + \int_0^t H_\varepsilon''(\tau)\,d\tau \\
&= |F'(0)| + H_\varepsilon'(t) - H_\varepsilon'(0), \qquad \text{for } t \ge 0,
\end{aligned}$$

and

$$\begin{aligned}
|F_\varepsilon'(t)| &= \left| F'(0) + \int_0^t H_\varepsilon''(\tau)\,\chi(\varepsilon\,\tau)\,d\tau \right| \\
&\le |F'(0)| + \int_t^0 H_\varepsilon''(\tau)\,d\tau \\
&= |F'(0)| + H_\varepsilon'(0) - H_\varepsilon'(t), \qquad \text{for } t < 0,
\end{aligned}$$

Thanks to (7.2.5), we get that F_ε' is bounded, uniformly in $\varepsilon > 0$.

We now claim that $(F_\varepsilon, F_\varepsilon')$ converges to (F, F') as ε goes to 0, uniformly on compact sets. We start from the derivatives: by using an integration by parts, we have

$$\begin{aligned}
F_\varepsilon'(t) &= F'(0) + \int_0^t H_\varepsilon''(\tau)\,\chi(\varepsilon\,\tau)\,d\tau \\
&= F'(0) + H_\varepsilon'(t)\,\chi(\varepsilon\,t) - H_\varepsilon'(0)\,\chi(0) - \varepsilon \int_0^t H_\varepsilon'(\tau)\,\chi'(\varepsilon\,\tau)\,d\tau.
\end{aligned}$$

Thus, by using (7.2.5), the fact that χ' is bounded on $\mathbb{R}$ and $\chi(0) = 1$, we get

$$|F_\varepsilon'(t) - F'(t)| \le |F'(0) - H_\varepsilon'(0)| + |H_\varepsilon'(t)\,\chi(\varepsilon\, t) - F'(t)| + C\,\varepsilon\, t,$$

for some $C > 0$ independent of $\varepsilon > 0$. The claimed uniform convergence on intervals of the type $[-M, M]$ now follows from this estimate, by using (7.2.6). Finally, by using that $F_\varepsilon(0) = F(0)$ and the Fundamental Theorem of Calculus, for every $M > 0$ and every $t \in [-M, M]$ we have

$$\begin{aligned}|F_\varepsilon(t) - F(t)| = \left|\int_0^t (F_\varepsilon'(\tau) - F'(\tau))\,d\tau\right| &\le \int_{-M}^{M} |F_\varepsilon'(\tau) - F'(\tau)|\,d\tau \\ &\le 2\,M \max_{\tau\in[-M,M]} |F_\varepsilon'(\tau) - F'(\tau)|.\end{aligned}$$

By using the uniform convergence of F_ε', we get the uniform convergence of F_ε, as well. In particular, we finally get that

$$\lim_{\varepsilon\to 0} F_\varepsilon'(t) = F'(t) \qquad \text{and} \qquad \lim_{\varepsilon\to 0} F_\varepsilon(t) = F(t), \qquad \text{for every } t \in \mathbb{R}.$$

We can now complete the proof of the main statement. With the previous notation, by the first part of the proof, we know that

$$\int_\Omega \langle \nabla(F_\varepsilon \circ u), \nabla\varphi\rangle\,dx \le 0, \qquad \text{for every } \varphi \in C_0^\infty(\Omega) \text{ such that } \varphi \ge 0. \tag{7.2.7}$$

We need to pass to the limit as ε goes to 0. We have previously observed that

$$|F_\varepsilon'(t)| \le C, \qquad \text{for every } t \in \mathbb{R},$$

with $C > 0$ independent of ε. This entails that

$$\Big|\langle \nabla(F_\varepsilon \circ u), \nabla\varphi\rangle\Big| \le C\,|\nabla u|\,|\nabla\varphi| \in L^1(\Omega),$$

for every $\varphi \in C_0^\infty(\Omega)$. We can thus apply the Dominated Convergence Theorem to pass to the limit in (7.2.7). We obtain for every $\varphi \in C_0^\infty(\Omega)$ such that $\varphi \ge 0$

$$\begin{aligned}0 \ge \lim_{\varepsilon\to 0} \int_\Omega \langle \nabla(F_\varepsilon \circ u), \nabla\varphi\rangle\,dx &= \lim_{\varepsilon\to 0} \int_\Omega F_\varepsilon'(u)\,\langle \nabla u, \nabla\varphi\rangle\,dx \\ &= \int_\Omega F'(u)\,\langle \nabla u, \nabla\varphi\rangle\,dx = \int_\Omega \langle \nabla(F \circ u), \nabla\varphi\rangle\,dx.\end{aligned}$$

This shows that $F \circ u$ is weakly subharmonic. □

Remark 7.2.2 The inequality (7.2.1) is called *Caccioppoli inequality*, named after the italian mathematician Renato Caccioppoli. Observe that from the point of view of the subharmonic function $F \circ u$, such an inequality is quite unnatural: indeed, apart for the presence of the (arbitrary!) weights φ^2 and $|\nabla \varphi|^2$, this inequality is telling us that the values of $F \circ u$ control *from above* those of $\nabla(F \circ u)$, in the L^2 sense. In other words, this may be seen as a *reverse Poincaré inequality*.

This is going to be a crucial ingredient of the regularity estimates proved in the next sections.

By slightly restricting the class of convex functions F, the previous result still holds for functions which are only weakly subharmonic. The proof is a very minor modification of that of Theorem 7.2.1: the reader is invited to write down the details.

Proposition 7.2.3 *Let $u \in W^{1,2}_{\rm loc}(\Omega)$ be a weakly subharmonic function. Let $F : \mathbb{R} \to \mathbb{R}$ be a C^1 convex non-decreasing function, with bounded first derivative. Then $F \circ u \in W^{1,2}_{\rm loc}(\Omega)$ is still weakly subharmonic in Ω. Moreover, if F is non-negative the Caccioppoli inequality* (7.2.1) *still holds.*

Remark 7.2.4 If the convex function F is decreasing, in general the previous result can not hold. For example, take the function

$$u(x) = |x|^2, \qquad \text{for every } x \in \mathbb{R}^N.$$

This is a C^2 function such that

$$-\Delta u(x) = -2\,N, \qquad \text{for every } x \in \mathbb{R}^N,$$

thus it is a subharmonic function. On the other hand, if we take the convex decreasing function $F(t) = e^{-t}$, we have $F \circ u(x) = e^{-|x|^2}$ and its Laplacian is given by

$$-\Delta(F \circ u)(x) = 2\,(N - 2\,|x|^2)\,e^{-|x|^2}.$$

In particular, the function $F \circ u$ is not subharmonic in the ball $\{x : |x| < \sqrt{N/2}\}$.

7.3 Boundedness

The first non-trivial regularity result concerns the *local boundedness* of weakly harmonic functions: this comes with an explicit estimate. The proof presented below is based on the so-called *Moser iteration technique*, introduced in [84].

The desired result will follow from the following one, valid for subharmonic functions (see Corollary 7.3.3).

Theorem 7.3.1 *Let $u \in W^{1,2}_{\mathrm{loc}}(\Omega)$ be a non-negative weakly subharmonic function. Then $u \in L^\infty_{\mathrm{loc}}(\Omega)$ and for every pair of concentric balls $B_{r_0}(x_0) \Subset B_{R_0}(x_0) \Subset \Omega$ we have the estimate*

$$\|u\|_{L^\infty(B_{r_0}(x_0))} \le C \left(\frac{R_0}{R_0 - r_0}\right)^{\frac{N}{2}} \left(\fint_{B_{R_0}(x_0)} u^2\,dx\right)^{\frac{1}{2}}, \tag{7.3.1}$$

for a constant $C > 0$ depending only on N.

Proof We divide the proof in two parts: we first prove that $u \in L^q_{\mathrm{loc}}(\Omega)$ for every $2 \le q < \infty$. Then we will show that we actually have $u \in L^\infty_{\mathrm{loc}}(\Omega)$, together with the a priori estimate. For simplicity, we will consider the case $N \ge 3$, only. The two-dimensional case $N = 2$ can be treated with minor modifications, the details are left as an exercise (see Problem 7.8.1).

Part 1: Higher Integrability For every $\beta \ge 1$ and $\varepsilon, M > 0$, we define the continuous increasing function

$$f^\varepsilon_{\beta,M}(t) = \begin{cases} 0, & \text{if } t < 0,\\ \beta\,(\varepsilon + t^2)^{\frac{\beta-2}{2}}\, t, & \text{if } 0 \le t \le M,\\ \beta\,(\varepsilon + M^2)^{\frac{\beta-2}{2}}\, M, & \text{if } t > M, \end{cases}$$

and take the primitive[2]

$$F^\varepsilon_{\beta,M}(t) = \int_0^t f^\varepsilon_{\beta,M}(\tau)\,d\tau, \qquad t \in \mathbb{R},$$

see Fig. 7.1. Observe that $F^\varepsilon_{\beta,M}$ is non-negative and identically vanishes in $(-\infty, 0]$. Also, by noticing that

$$0 \le f^\varepsilon_{\beta,M}(t) \le \beta\,(\varepsilon + t^2)^{\frac{\beta-2}{2}}\, t, \qquad \text{for every } t \ge 0,$$

we get

$$\begin{aligned} 0 \le F^\varepsilon_{\beta,M}(t) &\le \int_0^t \beta\,(\varepsilon + \tau^2)^{\frac{\beta-2}{2}}\,\tau\,d\tau \\ &= (\varepsilon + t^2)^{\frac{\beta}{2}} - \varepsilon^{\frac{\beta}{2}} \le (\varepsilon + t^2)^{\frac{\beta}{2}}, \qquad \text{for every } t \ge 0. \end{aligned} \tag{7.3.2}$$

[2] The reader should think of the function $F^\varepsilon_{\beta,M}$ as an approximation of the power function $t \mapsto t^\beta$, for $t \ge 0$. The role of the parameter M is that of making $F^\varepsilon_{\beta,M}$ linear "at infinity", so that its first derivative is bounded. The parameter $\varepsilon > 0$ is only needed to regularize the power around the origin, so that $F^\varepsilon_{\beta,M}$ is C^1 also in the case $\beta = 1$.

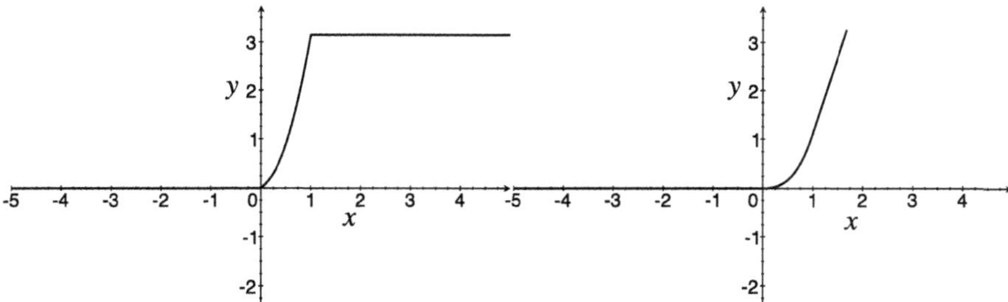

Fig. 7.1 The functions $f^{\varepsilon}_{\beta,M}$ (left) and $F^{\varepsilon}_{\beta,M}$ (right) of Theorem 7.3.1. Here $\beta = 3$, $\varepsilon = 0.1$ and $M = 1$

Moreover, by construction, this is a C^1 convex non-decreasing function, with bounded first derivative. By Proposition 7.2.3, we then know that $F^{\varepsilon}_{\beta,M} \circ u \in W^{1,2}_{\mathrm{loc}}(\Omega)$ is weakly subharmonic in Ω and there holds

$$\int_{\Omega} |\nabla(F^{\varepsilon}_{\beta,M} \circ u)|^2\, \varphi^2\, dx \le 4 \int_{\Omega} |\nabla\varphi|^2\, |F^{\varepsilon}_{\beta,M}(u)|^2\, dx, \qquad \text{for every } \varphi \in C^{\infty}_0(\Omega).$$

In the previous estimate, we sum on both sides the term

$$\int_{\Omega} |\nabla\varphi|^2\, |F^{\varepsilon}_{\beta,M}(u)|^2\, dx,$$

so to get

$$\int_{\Omega} |\nabla(F^{\varepsilon}_{\beta,M} \circ u)|^2\, \varphi^2\, dx + \int_{\Omega} |\nabla\varphi|^2\, |F^{\varepsilon}_{\beta,M}(u)|^2\, dx \le 5 \int_{\Omega} |\nabla\varphi|^2\, |F^{\varepsilon}_{\beta,M}(u)|^2\, dx. \tag{7.3.3}$$

Then we observe that

$$\begin{aligned} \Big|\nabla\big((F^{\varepsilon}_{\beta,M} \circ u)\,\varphi\big)\Big|^2 &= \Big|\varphi\, \nabla(F^{\varepsilon}_{\beta,M} \circ u) + F^{\varepsilon}_{\beta,M}(u)\, \nabla\varphi\Big|^2 \\ &\le 2\, |\nabla(F^{\varepsilon}_{\beta,M} \circ u)|^2\, |\varphi|^2 + 2\, |F^{\varepsilon}_{\beta,M}(u)|^2\, |\nabla\varphi|^2. \end{aligned}$$

In the last inequality we used Problem 1.7.2. Thus, from (7.3.3) we get

$$\int_{\Omega} \Big|\nabla\big((F^{\varepsilon}_{\beta,M} \circ u)\,\varphi\big)\Big|^2 dx \le 10 \int_{\Omega} |\nabla\varphi|^2\, |F^{\varepsilon}_{\beta,M}(u)|^2\, dx. \tag{7.3.4}$$

This holds for every $\varphi \in C^{\infty}_0(\Omega)$. We now observe that $(F^{\varepsilon}_{\beta,M} \circ u)\,\varphi \in W^{1,2}_0(\Omega)$, by Problem 3.12.18. We can thus apply the Sobolev embedding of Theorem 3.8.1 (with $p = 2 < N$ and $q = 2^*$) and obtain from (7.3.4)

$$\frac{1}{\mathcal{S}^2} \left(\int_{\Omega} |F^{\varepsilon}_{\beta,M}(u)\, \varphi|^{2^*}\, dx\right)^{\frac{2}{2^*}} \le 10 \int_{\Omega} |\nabla\varphi|^2\, |F^{\varepsilon}_{\beta,M}(u)|^2\, dx, \tag{7.3.5}$$

where[3]

$$2^* = \frac{2N}{N-2}.$$

The previous estimate is valid for every $\varphi \in C_0^\infty(\Omega)$. We now take a pair of concentric balls $B_r(x_0) \Subset B_R(x_0) \Subset \Omega$ and consider a cut-off function $\varphi \in C_0^\infty(B_R(x_0))$, i.e. a function such that

$$0 \le \varphi \le 1, \qquad \varphi \equiv 1 \text{ on } B_r(x_0), \qquad |\nabla\varphi| \le \frac{C}{R-r},$$

for some constant $C > 0$. By inserting such a function in (7.3.5) and using its properties, we get

$$\left(\int_{B_r(x_0)} |F^\varepsilon_{\beta,M}(u)|^{2^*}\,dx\right)^{\frac{2}{2^*}} \le \frac{C}{(R-r)^2}\int_{B_R(x_0)} |F^\varepsilon_{\beta,M}(u)|^2\,dx, \tag{7.3.6}$$

possibly for a different constant $C > 0$, which only depends on N. Before proceeding further, we observe that (7.3.6) is an extremely powerful estimate! Indeed, by recalling the definition of $F^\varepsilon_{\beta,M}$, heuristically we have that

$$F^\varepsilon_{\beta,M}(u) \simeq u^\beta,$$

thus (7.3.6), roughly speaking, is saying that the $L^{2\beta}$ norm of u is controlling *from above* the $L^{2^*\beta}$ norm of u. In other words, inequality (7.3.6) is a *reverse Hölder inequality*, which at each step increases the integrability of u. Observe that there is a small price to be paid in this process: there is a gain in integrability, but the support of the integral is shrinked (i.e. we pass from $B_R(x_0)$ to the smaller ball $B_r(x_0)$).

These were the heuristics, let us now try to implement this idea, in order to get that $u \in L^q_{\rm loc}(\Omega)$, for every $2 \le q < \infty$. We define the sequence of exponents $\{\beta_i\}_{i\in\mathbb{N}}$ as follows

$$\beta_0 = 1, \qquad \beta_{i+1} = \frac{2^*}{2}\,\beta_i = \left(\frac{2^*}{2}\right)^{i+1}, \text{ for } i \in \mathbb{N}.$$

We then prove the following implication

$$u \in L^{2\beta_i}_{\rm loc}(\Omega) \qquad \Longrightarrow \qquad u \in L^{2\beta_{i+1}}_{\rm loc}(\Omega). \tag{7.3.7}$$

Let us suppose that $u \in L^{2\beta_i}_{\rm loc}(\Omega)$, we fix an open set $\Omega' \Subset \Omega$ and set

$$\delta = \mathrm{dist}(\overline{\Omega'}, \partial\Omega) > 0.$$

[3] It is here that the choice $N \ge 3$ enters.

Then we can cover $\overline{\Omega'}$ by a finite number of balls with radius $r = \delta/4$, centered at some points belonging to $\overline{\Omega'}$. We take one of these balls $B_r(x_0)$ and observe that by construction, we have

$$B_r(x_0) \Subset B_{2r}(x_0) \Subset \Omega.$$

We use (7.3.6) with $\beta = \beta_i$ and this pair of balls, so to obtain

$$\begin{aligned}\left(\int_{B_r(x_0)} |F^{\varepsilon}_{\beta_i,M}(u)|^{2^*}\, dx\right)^{\frac{2}{2^*}} &\leq \frac{C}{r^2}\int_{B_{2r}(x_0)} |F^{\varepsilon}_{\beta_i,M}(u)|^2\, dx \\ &\leq \frac{C}{r^2}\int_{B_{2r}(x_0)} (\varepsilon + u^2)^{\beta_i}\, dx.\end{aligned}$$

We also used (7.3.2), in the second inequality. For every fixed $\varepsilon > 0$, we can now pass to the limit as M goes to $+\infty$ on the left-hand side: by observing that for every $t \geq 0$ we have

$$\lim_{M\to+\infty} F^{\varepsilon}_{\beta_i,M}(t) = (\varepsilon + t^2)^{\frac{\beta_i}{2}} - \varepsilon^{\frac{\beta_i}{2}} \geq t^{\beta_i},$$

and recalling that $u \geq 0$, an application of Fatou's Lemma leads to

$$\left(\int_{B_r(x_0)} u^{2^*\beta_i}\, dx\right)^{\frac{2}{2^*}} \leq \frac{C}{r^2}\int_{B_{2r}(x_0)} (\varepsilon + u^2)^{\beta_i}\, dx,$$

for every $\varepsilon > 0$. This estimate and the fact that $u \in L^{2\beta_i}_{\mathrm{loc}}(\Omega)$ show that $u \in L^{2^*\beta_i}(B_r(x_0))$. Moreover, since these balls cover the set Ω', we get $u \in L^{2^*\beta_i}(\Omega')$. By arbitrariness of $\Omega' \Subset \Omega$ and observing that $2^*\beta_i = 2\,\beta_{i+1}$ by construction, we finally obtain

$$u \in L^{2\beta_{i+1}}_{\mathrm{loc}}(\Omega),$$

as desired.

By observing that β_i is an increasing sequence diverging to $+\infty$ and that $u \in L^{2\beta_0}_{\mathrm{loc}}(\Omega)$ by assumption (indeed we have that $2\,\beta_0 = 2$), we can iterate (7.3.7) as many times as we wish, so to get

$$u \in L^q_{\mathrm{loc}}(\Omega), \qquad \text{for every } 2 \leq q < \infty.$$

Part 2: Boundedness It is sufficient to prove the estimate (7.3.1), then the fact that $u \in L^\infty_{\mathrm{loc}}(\Omega)$ easily follows by a covering argument, like in the previous part of the proof.

In order to prove (7.3.1), we fix two balls $B_{r_0}(x_0)$ and $B_{R_0}(x_0)$ as in the statement and we still indicate

$$\beta_i = \left(\frac{2^*}{2}\right)^i, \qquad \text{for } i \in \mathbb{N}.$$

We define the following sequence

$$r_i = r_0 + \frac{R_0 - r_0}{2^i}, \qquad \text{for } i \in \mathbb{N},$$

and apply (7.3.6) with $\beta = \beta_i$, $r = r_{i+1}$ and $R = r_i$. We get

$$\begin{aligned}\left(\int_{B_{r_{i+1}}(x_0)} |F^{\varepsilon}_{\beta_i,M}(u)|^{2^*}\,dx\right)^{\frac{2}{2^*}} &\le \frac{C\,4^i}{(R_0 - r_0)^2}\int_{B_{r_i}(x_0)} |F^{\varepsilon}_{\beta_i,M}(u)|^2\,dx\\ &\le \frac{C\,4^i}{(R_0 - r_0)^2}\int_{B_{r_i}(x_0)} (\varepsilon + u^2)^{\beta_i}\,dx.\end{aligned}$$

By taking the limit as M goes to $+\infty$ as above, we get

$$\left(\int_{B_{r_{i+1}}(x_0)} u^{2^*\beta_i}\,dx\right)^{\frac{2}{2^*}} \le \frac{C\,4^i}{(R_0 - r_0)^2}\int_{B_{r_i}(x_0)} (\varepsilon + u^2)^{\beta_i}\,dx.$$

This holds for every $\varepsilon > 0$: by taking the limit as ε goes to 0 and then raising both sides to the power $1/(2\,\beta_i)$, we obtain

$$\left(\int_{B_{r_{i+1}}(x_0)} u^{2^*\beta_i}\,dx\right)^{\frac{1}{2^*\beta_i}} \le \left(\frac{C\,4^i}{(R_0 - r_0)^2}\right)^{\frac{1}{2\beta_i}}\left(\int_{B_{r_i}(x_0)} u^{2\beta_i}\,dx\right)^{\frac{1}{2\beta_i}}.$$

Finally, by recalling that $2^*\beta_i = 2\beta_{i+1}$, the previous estimate can be rewritten as

$$\|u\|_{L^{2\beta_{i+1}}(B_{r_{i+1}}(x_0))} \le \left(\frac{C\,4^i}{(R_0 - r_0)^2}\right)^{\frac{1}{2\beta_i}} \|u\|_{L^{2\beta_i}(B_{r_i}(x_0))}, \qquad \text{for } i \in \mathbb{N}. \tag{7.3.8}$$

We now iterate n times (7.3.8), by starting from $i = 0$: this gives

$$\begin{aligned}\|u\|_{L^{2\beta_{n+1}}(B_{r_{n+1}}(x_0))} &\le \prod_{i=0}^{n}\left(\frac{C\,4^i}{(R_0 - r_0)^2}\right)^{\frac{1}{2\beta_i}} \|u\|_{L^2(B_{R_0}(x_0))}\\ &= \left(\frac{C}{(R_0 - r_0)^2}\right)^{\sum\limits_{i=0}^{n}\frac{1}{2\beta_i}}\left(\prod_{i=0}^{n}(4^i)^{\frac{1}{2\beta_i}}\right)\|u\|_{L^2(B_{R_0}(x_0))}.\end{aligned} \tag{7.3.9}$$

We wish to let the number of iterations n go to infinity, but we have to pay attention to the constants appearing in the right-hand side of (7.3.9). We need to ensure that they remain bounded. We observe that

$$\lim_{n\to\infty}\sum_{i=0}^{n}\frac{1}{2\beta_i}=\sum_{i=0}^{\infty}\frac{1}{2}\left(\frac{2}{2^*}\right)^i=\frac{1}{2}\,\frac{1}{1-\dfrac{2}{2^*}}=\frac{1}{2}\,\frac{2^*}{2^*-2}=\frac{N}{4},$$

and

$$\begin{aligned}\lim_{n\to\infty}\prod_{i=0}^{n}(4^i)^{\frac{1}{2\beta_i}}&=\lim_{n\to\infty}\exp\left(\log\prod_{i=0}^{n}(4^i)^{\frac{1}{2\beta_i}}\right)\\&=\lim_{n\to\infty}\exp\left(\sum_{i=0}^{n}\frac{i}{2\beta_i}\log 4\right)\\&=\lim_{n\to\infty}\exp\left(\frac{\log 4}{2}\sum_{i=0}^{n}i\left(\frac{2}{2^*}\right)^i\right)=\exp\left(\frac{(N-2)\,N}{4}\log 2\right).\end{aligned}$$

This shows that

$$\limsup_{n\to\infty}\|u\|_{L^{2\beta_{n+1}}(B_{r_{n+1}}(x_0))}\le\frac{C}{(R_0-r_0)^{\frac{N}{2}}}\,\|u\|_{L^2(B_{R_0}(x_0))},$$

possibly for a different constant $C=C(N)>0$. By further observing that

$$\limsup_{n\to\infty}\|u\|_{L^{2\beta_{n+1}}(B_{r_{n+1}}(x_0))}\ge\limsup_{n\to\infty}\|u\|_{L^{2\beta_{n+1}}(B_{r_0}(x_0))},$$

Problem 1.7.30 gives that $u\in L^\infty(B_{r_0}(x_0))$ and we have the estimate

$$\|u\|_{L^\infty(B_{r_0}(x_0))}\le\frac{C}{(R_0-r_0)^{\frac{N}{2}}}\,\|u\|_{L^2(B_{R_0}(x_0))}.$$

Multiplying and dividing by $R_0^{N/2}$, this is the same as (7.3.1), up to renaming the constant C. This concludes the proof. □

Remark 7.3.2 Apart for certain technical facts that could obscure the central idea in the previous proof, the reader should notice that the cornerstone is the combination of the following two facts:

- a *natural* “size” estimate on functions in terms of their gradients (i.e. the Sobolev inequality), which holds for any Sobolev function;
- an *unnatural* reverse Poincaré inequality (the Caccioppoli inequality), which only holds for solutions or subsolutions.

The main outcome of this combination is the recursive scheme of *reverse Hölder’s inequalities* (7.3.8), which eventually leads to the desired result.

We can now give the claimed result for weakly harmonic functions.

Corollary 7.3.3 *Let $u \in W^{1,2}_{\mathrm{loc}}(\Omega)$ be a weakly harmonic function. Then $u \in L^\infty_{\mathrm{loc}}(\Omega)$ and for every pair of concentric balls $B_{r_0}(x_0) \Subset B_{R_0}(x_0) \Subset \Omega$ we still have the estimate* (7.3.1).

Proof For every $\varepsilon > 0$, we introduce the function

$$F(t) = \sqrt{\varepsilon + t^2}, \qquad \text{for every } t \in \mathbb{R}.$$

Observe that this a C^1 convex function with bounded first derivative. According to Theorem 7.2.1, we have that $u_\varepsilon = F_\varepsilon \circ u$ is a non-negative weakly subharmonic function. Thus, from Theorem 7.3.1 we get $u_\varepsilon \in L^\infty_{\mathrm{loc}}(\Omega)$. By observing that

$$|u| \le \sqrt{\varepsilon + |u|^2} = u_\varepsilon,$$

we get that $u \in L^\infty_{\mathrm{loc}}(\Omega)$, as well. Finally, by using (7.3.1) for u_ε we get

$$\begin{aligned}\|u\|_{L^\infty(B_{r_0}(x_0))} \le \|u_\varepsilon\|_{L^\infty(B_{r_0}(x_0))} &\le C\left(\frac{R_0}{R_0 - r_0}\right)^{\frac{N}{2}}\left(\fint_{B_{R_0}(x_0)} u_\varepsilon^2\,dx\right)^{\frac{1}{2}}\\ &= C\left(\frac{R_0}{R_0 - r_0}\right)^{\frac{N}{2}}\left(\fint_{B_{R_0}(x_0)} \left(\varepsilon + |u|^2\right)dx\right)^{\frac{1}{2}}.\end{aligned}$$

By taking the limit as ε goes to 0, we get that u satisfies (7.3.1), as claimed. □

7.4 A Harnack-type Estimate

We first need a technical result for solutions with a definite sign. This is similar to the logarithmic estimate we obtained in the proof of the minimum principle in Lemma 4.4.5.

Lemma 7.4.1 (Moser's Truncated Logarithm) *Let $u \in W^{1,2}_{\mathrm{loc}}(\Omega)$ be a weakly harmonic function, such that $u \ge 0$ almost everywhere in an open subset $\Omega' \subseteq \Omega$. For every $0 < \varepsilon < 1$, we set*

$$G_\varepsilon(t) = \max\{-\log(t + \varepsilon),\, 0\}, \qquad \text{for } t \ge 0.$$

Then $G_\varepsilon \circ u \in W^{1,2}_{\mathrm{loc}}(\Omega')$ is weakly subharmonic in Ω'. Moreover, for every $B_R(x_0) \Subset \Omega'$ and every $0 < \sigma < 1$ there holds

$$\fint_{B_{\sigma R}(x_0)} |\nabla(G_\varepsilon \circ u)|^2\,dx \le \left(\frac{1}{\sigma^N} - 1\right)\frac{1}{(1-\sigma)^2}\,\frac{C}{R^2}, \tag{7.4.1}$$

for a constant $C > 0$ depending only on the dimension N.

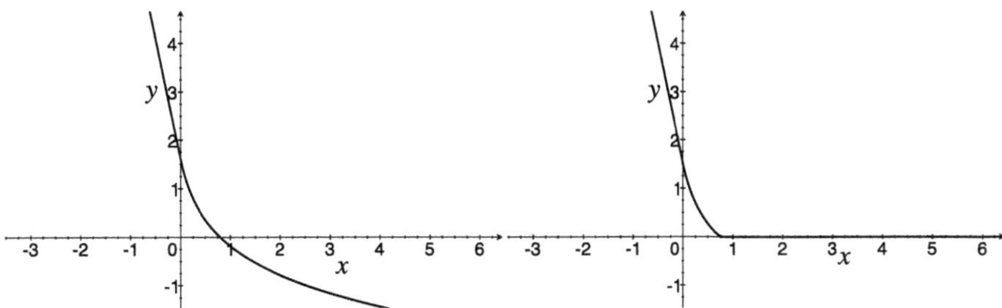

Fig. 7.2 The functions g_ε (on the left) and $G_{\varepsilon,\delta}$ (on the right)

Proof We divide the proof in two parts, for ease of readability.

Part 1: Subharmonicity We observe that

$$g_\varepsilon(t) = \begin{cases} -\dfrac{t}{\varepsilon} - \log \varepsilon, \text{ if } t < 0, \\ \\ -\log(t+\varepsilon), \text{ if } t \geq 0, \end{cases}$$

is a C^1 convex function, with bounded first derivative (see Fig. 7.2). Thus, we have $g_\varepsilon \circ u \in W^{1,2}_{\text{loc}}(\Omega')$ by Proposition 3.4.1 and Remark 3.4.2. By using that $u \geq 0$ almost everywhere in Ω', by definition we have

$$G_\varepsilon \circ u = \max\{-\log(u+\varepsilon),\, 0\} = \max\{g_\varepsilon \circ u,\, 0\}, \qquad \text{a. e. in } \Omega'.$$

According to Propositions 3.4.3 and 3.4.1, we get that $G_\varepsilon \circ u \in W^{1,2}_{\text{loc}}(\Omega')$ and for almost every $x \in \Omega'$ we have

$$\begin{aligned} \nabla(G_\varepsilon \circ u)(x) &= \begin{cases} g'_\varepsilon(u(x))\, \nabla u(x), \text{ if } g_\varepsilon(u(x)) > 0, \\ \qquad\qquad\quad 0, \text{ if } g_\varepsilon(u(x)) \leq 0, \end{cases} \\ &= \begin{cases} -\dfrac{\nabla u(x)}{\varepsilon + u(x)}, \text{ if } u(x) < 1-\varepsilon, \\ \\ \qquad\qquad 0, \text{ if } u(x) \geq 1-\varepsilon. \end{cases} \end{aligned} \tag{7.4.2}$$

In order to show that $G_\varepsilon \circ u$ is weakly subharmonic in Ω', we observe that G_ε is a convex function. Unfortunately, it is not C^1 and this prevents us from applying directly Theorem 7.2.1. However, this is a minor issue.[4] Indeed, we consider the following regularization of G_ε: for every $\delta > 0$, we set

[4] The reader is invited to skip this technical point, in a first reading, and jump directly to *Part 2*.

$$G_{\varepsilon,\delta}(t) = \begin{cases} \dfrac{\log \varepsilon}{\sqrt{\delta^2 + (\log \varepsilon)^2}} \dfrac{t}{\varepsilon} + \sqrt{\delta^2 + (\log \varepsilon)^2} - \delta, & \text{if } t < 0, \\ \sqrt{\delta^2 + (\log(t+\varepsilon))^2} - \delta, & \text{if } 0 \le t \le 1-\varepsilon, \\ 0, & \text{if } t > 1-\varepsilon, \end{cases}$$

see Fig. 7.2. It is not difficult to see that $G_{\varepsilon,\delta}$ is a non-negative C^1 convex function, with bounded first derivative. Thus, by Theorem 7.2.1 we know that

$$\int_{\Omega'} \langle \nabla(G_{\varepsilon,\delta} \circ u), \nabla\varphi \rangle \, dx \le 0, \qquad \text{for every } \varphi \in C_0^\infty(\Omega') \text{ such that } \varphi \ge 0. \tag{7.4.3}$$

We observe that for almost every $x \in \Omega'$, we have by Proposition 3.4.1

$$\begin{aligned} \nabla(G_{\varepsilon,\delta} \circ u)(x) &= G'_{\varepsilon,\delta}(u(x)) \, \nabla u(x) \\ &= \begin{cases} \dfrac{\log(u(x)+\varepsilon)}{\sqrt{\delta^2 + (\log(u+\varepsilon))^2}} \dfrac{\nabla u(x)}{\varepsilon + u(x)}, & \text{if } u(x) < 1-\varepsilon, \\ 0, & \text{if } u(x) \ge 1-\varepsilon, \end{cases} \end{aligned}$$

Thus, by recalling (7.4.2), we get

$$\lim_{\delta \to 0^+} \nabla(G_{\varepsilon,\delta} \circ u) = \nabla(G_\varepsilon \circ u), \qquad \text{a. e. in } \Omega'.$$

Moreover, for every $\delta > 0$ and every $\varphi \in C_0^\infty(\Omega')$ we have

$$|\langle \nabla(G_{\varepsilon,\delta} \circ u), \nabla\varphi \rangle| \le |\nabla(G_{\varepsilon,\delta} \circ u)| \, |\nabla\varphi| \le \frac{|\nabla u|}{u+\varepsilon} |\nabla\varphi| \le \frac{|\nabla u|}{\varepsilon} |\nabla\varphi| \in L^1(\Omega').$$

We can thus apply the Dominated Convergence Theorem and pass to the limit as δ goes to 0 in (7.4.3), so to obtain that

$$\int_{\Omega'} \langle \nabla(G_\varepsilon \circ u), \nabla\varphi \rangle \, dx \le 0, \qquad \text{for every } \varphi \in C_0^\infty(\Omega') \text{ such that } \varphi \ge 0,$$

i.e. $G_\varepsilon \circ u$ is weakly subhamornic in Ω'.

Part 2: Energy Estimate In order to prove (7.4.1), we now fix a ball $B_R(x_0) \Subset \Omega'$ as in the statement and take $\eta \in C_0^\infty(B_R(x_0))$. We use

$$\int_{\Omega'} \langle \nabla u, \nabla\varphi \rangle \, dx = 0 \qquad \text{with } \varphi = \frac{\eta^2}{u+\varepsilon}.$$

Observe that such a choice is feasible, i.e. $\varphi \in W_0^{1,2}(B_R(x_0))$ thanks to Problem 3.12.25. We then get

$$-\int_{B_R(x_0)} \left|\frac{\nabla u}{u+\varepsilon}\right|^2 \eta^2\,dx + 2\int_{B_R(x_0)} \left\langle \frac{\nabla u}{u+\varepsilon}, \nabla\eta\right\rangle \eta\,dx = 0.$$

This in turn implies

$$\int_{B_R(x_0)} \left|\frac{\nabla u}{u+\varepsilon}\right|^2 \eta^2\,dx \le 2\int_{B_R(x_0)} \left|\frac{\nabla u}{u+\varepsilon}\right| |\nabla\eta|\,|\eta|\,dx.$$

By using the generalized Young's inequality (1.2.6), we have

$$\left|\frac{\nabla u}{u+\varepsilon}\right| |\nabla\eta|\,|\eta| \le \frac{\delta}{2}\left|\frac{\nabla u}{u+\varepsilon}\right|^2 \eta^2 + \frac{1}{2\,\delta}\,|\nabla\eta|^2,$$

thus we obtain for every $\delta > 0$

$$\int_{B_R(x_0)} \left|\frac{\nabla u}{u+\varepsilon}\right|^2 \eta^2\,dx \le \delta\int_{B_R(x_0)} \left|\frac{\nabla u}{u+\varepsilon}\right|^2 \eta^2\,dx + \frac{1}{\delta}\int_{B_R(x_0)} |\nabla\eta|^2\,dx.$$

We choose $\delta = 1/2$ and simplify a term on the right-hand side, so to obtain

$$\frac{1}{2}\int_{B_R(x_0)} \left|\frac{\nabla u}{u+\varepsilon}\right|^2 \eta^2\,dx \le 2\int_{B_R(x_0)} |\nabla\eta|^2\,dx. \tag{7.4.4}$$

This is valid for every $\eta \in C_0^\infty(B_R(x_0))$. We fix $0 < \sigma < 1$ and take $\eta \in C_0^\infty(B_R(x_0))$ to be a cut-off function, such that $0 \le \eta \le 1$ and

$$\eta \equiv 1 \text{ on } B_{\sigma R}(x_0), \qquad |\nabla\eta| \le \frac{C}{R\,(1-\sigma)},$$

for some universal constant $C > 0$. Inserted in (7.4.4), this gives

$$\begin{aligned}\int_{B_{\sigma R}(x_0)} \left|\frac{\nabla u}{u+\varepsilon}\right|^2 dx \le 4\int_{B_R(x_0)} |\nabla\eta|^2\,dx &= 4\int_{B_R(x_0)\setminus B_{\sigma R}(x_0)} |\nabla\eta|^2\,dx\\ &\le |B_R(x_0)\setminus B_{\sigma R}(x_0)|\,\frac{4\,C^2}{R^2\,(1-\sigma)^2}\\ &= \frac{4\,C^2}{R^2\,(1-\sigma)^2}\,\omega_N\,R^N\,(1-\sigma^N)\\ &= \frac{4\,C^2\,(1-\sigma^N)}{(1-\sigma)^2\,\sigma^N}\,\frac{1}{R^2}\,|B_{\sigma R}(x_0)|.\end{aligned}$$

In order to conclude, we can observe that by (7.4.2) we have

$$\int_{B_{\sigma R}(x_0)} \left| \frac{\nabla u}{u + \varepsilon} \right|^2 dx \geq \int_{B_{\sigma R}(x_0)} |\nabla (G_\varepsilon \circ u)|^2 \, dx.$$

This gives the desired estimate. □

The previous result is expedient to the next one, which is a sort of *Harnack-type inequality* (we refer to [29, Theorem 2.5] for the classical Harnack inequality). This is due to Moser, see [84, Theorem 2]. It asserts that if a non-negative solution is "large" in a "big" portion of a ball, then it can not suddenly jump to 0 in a smaller ball. As we will see, by suitably iterating this fact, we will eventually get some continuity properties of the solution.

Proposition 7.4.2 *Let $u \in W^{1,2}_{\mathrm{loc}}(\Omega)$ be a weakly harmonic function, such that $u \geq 0$ almost everywhere in a ball $B_{R_0}(x_0) \subseteq \Omega$. Let us suppose that*

$$|\{x \in B_{R_0/2}(x_0) \, : \, u(x) \geq 1\}| \geq \frac{|B_{R_0/2}(x_0)|}{2}. \tag{7.4.5}$$

Then there exists a constant $0 < c_0 < 1$ depending on N only, such that

$$u(x) \geq c_0, \qquad \text{for a. e. } x \in B_{R_0/4}(x_0).$$

Proof We still consider the function $G_\varepsilon \circ u$ of Lemma 7.4.1 and set

$$v = G_\varepsilon \circ u,$$

for simplicity. From Lemma 7.4.1 applied with $\Omega' = B_{R_0}(x_0)$, we know that v is a weakly subharmonic function in $B_{R_0}(x_0)$. Moreover, by construction it is non-negative. Thus, from Theorem 7.3.1 we get

$$v \in L^\infty_{\mathrm{loc}}(B_{R_0}(x_0)),$$

with the estimate

$$\|v\|_{L^\infty(B_r(y_0))} \leq C \left(\frac{R}{R-r} \right)^N \left(\fint_{B_R(y_0)} |v|^2 \, dx \right)^{\frac{1}{2}},$$

for every $B_r(y_0) \subseteq B_R(y_0) \Subset B_{R_0}(x_0)$. We now use this estimate with

$$y_0 = x_0, \qquad R = \frac{R_0}{2} \qquad \text{and} \qquad r = \frac{R_0}{4},$$

where R_0 still denotes the radius in the statement. By recalling that v is non-negative by construction, this gives

$$\sup_{B_{R_0/4}(x_0)} v \leq C \left(\fint_{B_{R_0/2}(x_0)} v^2 \, dx \right)^{\frac{1}{2}}. \tag{7.4.6}$$

We now observe that the hypothesis (7.4.5) implies that

$$\begin{aligned} |\{x \in B_{R_0/2}(x_0) : v(x) = 0\}| &= |\{x \in B_{R_0/2}(x_0) : u(x) \geq 1 - \varepsilon\}| \\ &\geq |\{x \in B_{R_0/2}(x_0) : u(x) \geq 1\}| \geq \frac{|B_{R_0/2}(x_0)|}{2}. \end{aligned}$$

We can thus use the Poincaré inequality of Corollary 3.10.5 for v and get

$$\begin{aligned} \int_{B_{R_0/2}(x_0)} |v|^2 \, dx &\leq \frac{R_0^2}{\mu_{N,2}} \left(\frac{|B_{R_0/2}(x_0)|}{|\{x \in B_{R_0/2}(x_0) : v(x) = 0\}|} \right) \int_{B_{R_0/2}(x_0)} |\nabla v|^2 \, dx \\ &\leq \frac{2}{\mu_{N,2}} \frac{\cancel{|B_{R_0/2}(x_0)|}}{\cancel{|B_{R_0/2}(x_0)|}} R_0^2 \int_{B_{R_0/2}(x_0)} |\nabla v|^2 \, dx \\ &= \frac{2}{\mu_{N,2}} R_0^2 \int_{B_{R_0/2}(x_0)} |\nabla v|^2 \, dx. \end{aligned}$$

By inserting this estimate in (7.4.6), we get

$$\sup_{B_{R_0/4}(x_0)} v \leq C \left(R_0^2 \fint_{B_{R_0/2}(x_0)} |\nabla v|^2 \, dx \right)^{\frac{1}{2}},$$

possibly with a different constant $C > 0$, which still depends on N only. By recalling that $v = G_\varepsilon \circ u$, we can use the estimate (7.4.1) of Lemma 7.4.1 in order to estimate the last integral. This gives[5]

$$\sup_{B_{R_0/4}(x_0)} v \leq C \left(R_0^2 \fint_{B_{R_0/2}(x_0)} |\nabla v|^2 \, dx \right)^{\frac{1}{2}} \leq \widetilde{C},$$

[5] We can observe that if we apply Lemma 7.4.1 with $\Omega' = B_{R_0}(x_0)$, then $B_{(3R_0)/4}(x_0) \Subset \Omega'$ and we may use (7.4.1) with $R = (3\,R_0)/4$ and $\sigma = 2/3$, so that in this case

$$B_{\sigma R}(x_0) = B_{\frac{R_0}{2}}(x_0).$$

for a constant $\widetilde{C} > 0$ depending on N only. By recalling the definition of G_ε, the previous estimate implies that

$$0 \le \max\{-\log(u(x)+\varepsilon),\, 0\} \le \widetilde{C}, \qquad \text{for a. e. } x \in B_{R_0/4}(x_0).$$

This in turn yields

$$u(x) + \varepsilon \ge e^{-\widetilde{C}}, \qquad \text{for a. e. } x \in B_{R_0/4}(x_0).$$

Since this is valid for every $1 > \varepsilon > 0$, we get the desired conclusion by choosing $c_0 = e^{-\widetilde{C}} > 0$, which only depends on the dimension N. □

7.5 Hölder Continuity

We are ready for the main result of this section: weakly harmonic functions are actually *continuous* functions. More precisely, they are Hölder continuous (recall Problem 5.5.10).

Theorem 7.5.1 *Let $u \in W^{1,2}_{\rm loc}(\Omega)$ be a weakly harmonic function. Then there exists $0 < \alpha < 1$ such that for every open set $\mathcal{O} \Subset \Omega$, we have $u \in C^{0,\alpha}(\overline{\mathcal{O}})$.*

Proof We follow the proof by De Giorgi. Let $\mathcal{O} \Subset \Omega$ be an open set, we define $\tau = \mathrm{dist}(\overline{\mathcal{O}}, \partial\Omega) > 0$. For every $0 < t < \tau$, we also set

$$\mathcal{O}_t = \Big\{x \in \Omega \,:\, \mathrm{dist}(x, \overline{\mathcal{O}}) < t\Big\}.$$

Thanks to Corollary 7.3.3, we have in particular that

$$\|u\|_{L^\infty(\mathcal{O}_t)} < +\infty, \qquad \text{for every } 0 < t < \tau.$$

We set

$$R = \frac{\tau}{8}, \tag{7.5.1}$$

we fix a point $x_0 \in \mathcal{O}_{\tau/2}$ and consider the ball $B_\varrho(x_0)$, with $\varrho \le R$. We observe in particular that, by construction, we have $B_{2\varrho}(x_0) \Subset \Omega$. Then we set

$$\omega(\varrho) = M(\varrho) - m(\varrho) := \sup_{B_\varrho(x_0)} u - \inf_{B_\varrho(x_0)} u, \qquad \text{for } \varrho \le R.$$

Observe that the quantity $\omega(\varrho)$ is finite, since $u \in L^\infty_{\rm loc}(\Omega)$. More precisely, we have

$$\omega(\varrho) \le 2\,\|u\|_{L^\infty(\mathcal{O}_{\frac{5}{8}\tau})}, \qquad \text{for } \varrho \le R. \tag{7.5.2}$$

Moreover, by definition it is monotone non-decreasing.

We claim that there exists a constant $0 < \theta < 1$, depending only on the dimension N, such that

$$\omega\left(\frac{\varrho}{4}\right) \leq \theta\,\omega(\varrho), \qquad \text{for } \varrho \leq R. \tag{7.5.3}$$

Let us take $0 < \varrho \leq R$, we can clearly suppose that $\omega(\varrho) > 0$, otherwise such an inequality trivially holds. We introduce the new functions

$$v = 1 + \frac{u - \dfrac{M(\varrho) + m(\varrho)}{2}}{\dfrac{M(\varrho) - m(\varrho)}{2}} \qquad \text{and} \qquad w = 1 - \frac{u - \dfrac{M(\varrho) + m(\varrho)}{2}}{\dfrac{M(\varrho) - m(\varrho)}{2}},$$

and observe that both functions are still weakly harmonic in Ω, since they are obtained from u by addition and multiplication by constants. Moreover, by construction, it is not difficult to see that

$$v \geq 0 \quad \text{and} \quad w \geq 0, \qquad \text{a. e. in } B_\varrho(x_0).$$

We now observe that we have two possibilities: either

$$\left|\left\{x \in B_{\varrho/2}(x_0) \,:\, u(x) \geq \frac{M(\varrho) + m(\varrho)}{2}\right\}\right| \geq \frac{|B_{\varrho/2}(x_0)|}{2}, \tag{7.5.4}$$

or

$$\left|\left\{x \in B_{\varrho/2}(x_0) \,:\, u(x) \leq \frac{M(\varrho) + m(\varrho)}{2}\right\}\right| \geq \frac{|B_{\varrho/2}(x_0)|}{2}. \tag{7.5.5}$$

If (7.5.4) occurs, by definition of v we have that this is a weakly harmonic function, non-negative in $B_\varrho(x_0)$ and such that

$$|\{x \in B_{\varrho/2}(x_0) \,:\, v(x) \geq 1\}| \geq \frac{|B_{\varrho/2}(x_0)|}{2}.$$

By applying Proposition 7.4.2, we have that

$$v(x) \geq c_0, \qquad \text{for a. e. } x \in B_{\varrho/4}(x_0),$$

with $0 < c_0 < 1$ constant depending on N, only. By recalling the definition of v, this is the same as

$$u - \frac{M(\varrho) + m(\varrho)}{2} \geq (c_0 - 1)\,\frac{M(\varrho) - m(\varrho)}{2}, \qquad \text{in } B_{\varrho/4}(x_0).$$

By taking the infimum over $B_{\varrho/4}(x_0)$ and multiplying both sides by -1, we get

$$\begin{aligned}-m(\varrho/4) &\leq -\frac{M(\varrho)+m(\varrho)}{2} + (1-c_0)\,\frac{M(\varrho)-m(\varrho)}{2}\\ &= -\frac{M(\varrho)+m(\varrho)}{2} + \frac{1-c_0}{2}\,\omega(\varrho).\end{aligned}$$

By recalling the definition of oscillation $\omega(\varrho)$, the previous estimate implies

$$\begin{aligned}\omega\left(\frac{\varrho}{4}\right) &= M(\varrho/4) - m(\varrho/4)\\ &\leq M(\varrho) - m(\varrho/4)\\ &\leq M(\varrho) - \frac{M(\varrho)+m(\varrho)}{2} + \frac{1-c_0}{2}\,\omega(\varrho) = \frac{M(\varrho)-m(\varrho)}{2} + \frac{1-c_0}{2}\,\omega(\varrho),\end{aligned}$$

that is

$$\omega\left(\frac{\varrho}{4}\right) \leq \left(1-\frac{c_0}{2}\right)\,\omega(\varrho).$$

This is exactly (7.5.3), with

$$\theta = \left(1-\frac{c_0}{2}\right).$$

On the other hand, if the alternative (7.5.5) occurs, by definition of w we have that this is a weakly harmonic function, non-negative in $B_\varrho(x_0)$ and such that

$$|\{x \in B_{\varrho/2}(x_0)\,:\, w(x) \geq 1\}| \geq \frac{|B_{\varrho/2}(x_0)|}{2}.$$

By applying again Proposition 7.4.2, this time to w, we have that

$$w(x) \geq c_0, \qquad \text{for a. e. } x \in B_{\varrho/4}(x_0),$$

still with the same universal constant $0 < c_0 < 1$. By recalling the definition of w, this is the same as

$$u - \frac{M(\varrho)+m(\varrho)}{2} \leq (1-c_0)\,\frac{M(\varrho)-m(\varrho)}{2}, \qquad \text{in } B_{\varrho/4}(x_0).$$

By taking the supremum over $B_{\varrho/4}(x_0)$, we get

$$\begin{aligned}M(\varrho/4) &\leq \frac{M(\varrho)+m(\varrho)}{2} + (1-c_0)\,\frac{M(\varrho)-m(\varrho)}{2}\\ &\leq \frac{M(\varrho)+m(\varrho)}{2} + \frac{1-c_0}{2}\,\omega(\varrho).\end{aligned}$$

As above, by recalling the definition of oscillation ω, the previous estimate implies

$$\begin{aligned}\omega\left(\frac{\varrho}{4}\right) &= M(\varrho/4) - m(\varrho/4) \\ &\leq M(\varrho/4) - m(\varrho) \\ &\leq \frac{M(\varrho) + m(\varrho)}{2} + \frac{1 - c_0}{2}\,\omega(\varrho) - m(\varrho) \\ &= \frac{M(\varrho) - m(\varrho)}{2} + \frac{1 - c_0}{2}\,\omega(\varrho) = \left(1 - \frac{c_0}{2}\right)\omega(\varrho).\end{aligned}$$

Thus, the estimate (7.5.3) holds in this case, as well, again with constant

$$\theta = \left(1 - \frac{c_0}{2}\right).$$

We now fix $0 < \varrho \leq R$ and take another radius $0 < r \leq \varrho$. Then there exists $k \in \mathbb{N}$ such that

$$\frac{\varrho}{4^{k+1}} < r \leq \frac{\varrho}{4^k}.$$

By using that $r \mapsto \omega(r)$ is monotone non-decreasing and iterating k times the estimate (7.5.3), we obtain

$$\omega(r) \leq \omega\left(\frac{\varrho}{4^k}\right) \leq \theta^k\,\omega(\varrho).$$

We observe that if we set

$$\alpha = -\log_4 \theta, \tag{7.5.6}$$

we have[6]

$$\theta^k = 4^{k \log_4 \theta} = \left(\frac{1}{4^k}\right)^{\alpha} \leq 4^{\alpha}\left(\frac{r}{\varrho}\right)^{\alpha}.$$

This finally implies the following crucial decay estimate for the oscillation of the solution u on balls

$$\omega(r) \leq 4^{\alpha}\left(\frac{r}{\varrho}\right)^{\alpha}\omega(\varrho), \qquad \text{for every } 0 < r \leq \varrho \leq R. \tag{7.5.7}$$

This estimate is the key ingredient in order to reach the claimed regularity property of u.

---

[6] In the last estimate, we use that $\varrho/4^{k+1} < r$, which entails in particular that

$$\frac{1}{4^k} < 4\,\frac{r}{\varrho}.$$

To this aim, a topological fact at first: since $\overline{\mathcal{O}}$ is a compact set, we can cover it with a finite number of balls having radius R defined in (7.5.1), centered at points of $\overline{\mathcal{O}}$. Let us indicate by $B_R(y_1), \dots, B_R(y_\ell)$ these balls, with $y_1, \dots, y_\ell \in \overline{\mathcal{O}}$. Observe in particular that we have

$$B_R(y_j) \Subset \mathcal{O}_{\tau/2}, \qquad \text{for every } j \in \{1, \dots, \ell\},$$

thanks to the choice of R. In order to conclude the proof of the statement, it is sufficient to prove that there exists $C > 0$ such that

$$|u|_{C^{0,\alpha}(\overline{B_R(y_j)})} \le C, \qquad \text{for every } j \in \{1, \dots, \ell\}, \tag{7.5.8}$$

where α is the same exponent as in (7.5.6).

We take two distinct points $x, y \in \overline{B_R(y_j)}$ and distinguish two cases:

- either $|x - y| \le R/2$;
- or $|x - y| > R/2$.

In the first case, we consider the middle point

$$x_0 = \frac{x + y}{2}.$$

Then we observe that $x, y \in B_\varrho(x_0)$, for every radius $\varrho > |x - y|/2$. Moreover, since a ball is a convex set, we have that

$$x_0 \in B_R(y_j) \Subset \mathcal{O}_{\tau/2}.$$

In particular, the decay estimate (7.5.7) holds for balls centered at x_0, with radii $0 < r \le \varrho \le R$. By using this estimate with $r = |x - y|$ and $\varrho = R$, we obtain

$$|u(x) - u(y)| \le \omega(|x - y|) \le 4^\alpha \left(\frac{|x - y|}{R}\right)^\alpha \omega(R) = \left[\left(\frac{4}{R}\right)^\alpha \omega(R)\right] |x - y|^\alpha.$$

By recalling (7.5.2), we get

$$|u(x) - u(y)| \le \left[2 \left(\frac{4}{R}\right)^\alpha \|u\|_{L^\infty(\mathcal{O}_{\frac{5}{8}\tau})}\right] |x - y|^\alpha.$$

The second case $|x - y| > R/2$ is simpler: we simply have

$$\begin{aligned} |u(x) - u(y)| \le |u(x)| + |u(y)| &\le 2\, \|u\|_{L^\infty(B_R(y_j))} \\ &\le 2\, \|u\|_{L^\infty(\mathcal{O}_{\tau/2})} \\ &\le 2 \left(\frac{2}{R}\right)^\alpha \|u\|_{L^\infty(\mathcal{O}_{\tau/2})}\, |x - y|^\alpha. \end{aligned}$$

In conclusion, we obtain (7.5.8), with

$$C = 2\left(\frac{4}{R}\right)^{\alpha} \|u\|_{L^\infty(O_{\frac{5}{8}\tau})},$$

which conclude the proof. □

7.6 Higher Differentiability

We are going to show that a weakly harmonic function enjoys higher differentiability properties, at least in weak sense. This will be the occasion for the reader to get acquainted with the so-called *Nirenberg's difference quotient technique* (see [85]), in a very simple model case. It consists in differentiating the equation in a discrete sense.

In the proof, the contents of Sect. 3.11 from Chap. 3 will play a crucial role.

Theorem 7.6.1 *Let $u \in W^{1,2}_{\rm loc}(\Omega)$ be a weakly harmonic function. Then $\nabla u \in W^{1,2}_{\rm loc}(\Omega; \mathbb{R}^N)$ and every weak partial derivative*

$$\frac{\partial u}{\partial x_k}, \qquad \textit{for } k \in \{1, \dots, N\},$$

is weakly harmonic in Ω.

Proof We divide the proof in various parts.

Part 1: Discrete Differentiation Let $B_R(x_0) \Subset \Omega$ and set

$$h_0 = \operatorname{dist}(B_R(x_0), \partial\Omega)/4 > 0.$$

For every $\varphi \in C^\infty_0(B_R(x_0))$, we observe that the translated function

$$\mathrm{T}_{-h}\varphi(x) = \varphi(x - h),$$

is still in $C^\infty_0(\Omega)$, for every $h \in \mathbb{R}^N$ such that $|h| \le h_0$. By inserting φ and $\mathrm{T}_{-h}\varphi$ into (7.1.2), we have

$$\int_\Omega \langle \nabla u, \nabla \varphi\rangle \, dx = 0, \tag{7.6.1}$$

and

$$\int_\Omega \langle \nabla u, \nabla(\mathrm{T}_{-h}\varphi)\rangle \, dx = 0.$$

We can make the change of variable $x - h = y$ in the second integral. By using that φ has compact support we thus get

$$\int_{\Omega} \langle \nabla(\mathrm{T}_h u), \nabla \varphi \rangle \, dx = 0. \tag{7.6.2}$$

By subtracting (7.6.1) from (7.6.2), we get

$$\int_{\Omega} \langle \nabla(\mathrm{T}_h u) - \nabla u, \nabla \varphi \rangle \, dx = 0, \qquad \text{for every } \varphi \in C_0^\infty(B_R(x_0)),\ |h| \le h_0. \tag{7.6.3}$$

As usual, by density, we can even admit test functions $\varphi \in W_0^{1,2}(B_R(x_0))$ in (7.6.3). We now take $\eta \in C_0^\infty(B_R(x_0))$ and use the previous identity with

$$\varphi = \eta^2\,(\mathrm{T}_h u - u),$$

which belongs to $W_0^{1,2}(B_R(x_0))$, thanks to Problem 3.12.18. This gives

$$\int_{\Omega} |\nabla(\mathrm{T}_h u) - \nabla u|^2\, \eta^2 \, dx + 2 \int_{\Omega} \langle \nabla(\mathrm{T}_h u) - \nabla u, \nabla \eta \rangle\, \eta\, (\mathrm{T}_h u - u) \, dx = 0.$$

By observing that $\nabla(\mathrm{T}_h u) = \mathrm{T}_h(\nabla u)$, with simple manipulations we get

$$\int_{\Omega} |\mathrm{T}_h(\nabla u) - \nabla u|^2\, \eta^2 \, dx \le 2 \int_{\Omega} |\mathrm{T}_h(\nabla u) - \nabla u|\, |\nabla \eta|\, |\eta|\, |\mathrm{T}_h u - u| \, dx. \tag{7.6.4}$$

By using the generalized Young inequality (1.2.6), we have for every $\delta > 0$

$$|\mathrm{T}_h(\nabla u) - \nabla u|\, |\nabla \eta|\, |\eta|\, |\mathrm{T}_h u - u| \le \frac{\delta}{2}\, |\mathrm{T}_h(\nabla u) - \nabla u|^2\, \eta^2 + \frac{1}{2\,\delta}\, |\mathrm{T}_h u - u|^2\, |\nabla \eta|^2.$$

Thus, from (7.6.4) we get

$$\int_{\Omega} |\mathrm{T}_h(\nabla u) - \nabla u|^2\, \eta^2 \, dx \le \delta \int_{\Omega} |\mathrm{T}_h(\nabla u) - \nabla u|^2\, \eta^2 + \frac{1}{\delta} \int_{\Omega} |\nabla \eta|^2\, |\mathrm{T}_h u - u|^2 \, dx.$$

If we choose $\delta = 1/2$, we can obtain (after a simplification)

$$\int_{\Omega} |\mathrm{T}_h(\nabla u) - \nabla u|^2\, \eta^2 \, dx \le 4 \int_{\Omega} |\nabla \eta|^2\, |\mathrm{T}_h u - u|^2 \, dx.$$

This is valid for every $\eta \in C_0^\infty(B_R(x_0))$ and $0 < |h| \le h_0$. We divide by $|h|^2$ the previous estimate, this yields

$$\int_{\Omega} \left| \frac{\mathrm{T}_h(\nabla u) - \nabla u}{|h|} \right|^2 \eta^2 \, dx \le 4 \int_{\Omega} |\nabla \eta|^2 \left| \frac{\mathrm{T}_h u - u}{|h|} \right|^2 dx. \tag{7.6.5}$$

We now take $\eta \in C_0^\infty(B_R(x_0))$ to be a cut-off function such that

$$0 \le \eta \le 1, \qquad \eta \equiv 1 \text{ on } B_{R/2}(x_0), \qquad |\nabla \eta| \le \frac{C}{R}.$$

From (7.6.5), we get for every $0 < |h| \le h_0$

$$\int_{B_{R/2}(x_0)} \left| \frac{\mathrm{T}_h(\nabla u) - \nabla u}{|h|} \right|^2 dx \le \frac{4\, C^2}{R^2} \int_{B_R(x_0)} \left| \frac{\mathrm{T}_h u - u}{|h|} \right|^2 dx.$$

By applying Proposition 3.11.1 on the right-hand side, we get

$$\int_{B_{R/2}(x_0)} \left| \frac{\mathrm{T}_h(\nabla u) - \nabla u}{|h|} \right|^2 dx$$
$$\le \frac{C}{R^2} \left(\int_{B_{R+h_0}(x_0)} |\nabla u|^2\, dx + \frac{1}{h_0^2} \int_{B_{R+3\,h_0}(x_0)} |u|^2\, dx \right),$$

possibly with a different constant $C > 0$, still depending on N only. We now choose $h = t\, \mathbf{e}_i$ for some $i \in \{1, \dots, N\}$, thus we obtain for every $0 < |t| \le h_0$

$$\int_{B_{R/2}(x_0)} \left| \frac{\mathrm{T}_{t\,\mathbf{e}_i}(\nabla u) - \nabla u}{|t|} \right|^2 dx$$
$$\le \frac{C}{R^2} \left(\int_{B_{R+h_0}(x_0)} |\nabla u|^2\, dx + \frac{1}{h_0^2} \int_{B_{R+3\,h_0}(x_0)} |u|^2\, dx \right).$$

Part 2: The Gradient Is in a Sobolev Space By Proposition 3.11.2, we can thus conclude that

$$\frac{\partial \nabla u}{\partial x_i} \in L^2(B_{R/2}(x_0); \mathbb{R}^N).$$

By arbitrariness of i, this finally gives

$$\nabla u \in W^{1,2}(B_{R/2}(x_0); \mathbb{R}^N).$$

If we now take $\Omega' \Subset \Omega$ and set $\delta_0 = \operatorname{dist}(\Omega'; \partial\Omega) > 0$, we have that Ω' admits a finite covering made by balls

$$B_{\frac{\delta_0}{8}}(x_1), \dots, B_{\frac{\delta_0}{8}}(x_\ell),$$

with centers $x_1, \dots, x_\ell \in \Omega'$, for a suitable $\ell \in \mathbb{N} \setminus \{0\}$. By construction we have

$$B_{\frac{\delta_0}{2}}(x_j) \Subset \Omega, \qquad \text{for } j = 1, \dots, \ell,$$

thus we can use the result above with $R = \delta_0/2$, to infer that

$$\nabla u \in W^{1,2}(B_{\frac{\delta_0}{4}}(x_j); \mathbb{R}^N), \qquad \text{for } j = 1, \dots, \ell. \tag{7.6.6}$$

Since these balls still cover Ω', we are going to show that (7.6.6) permits to infer that $\nabla u \in W^{1,2}(\Omega'; \mathbb{R}^N)$. Indeed, for every $\varphi \in C_0^\infty(\Omega')$ and every $i, k \in \{1, \dots, N\}$, we consider

$$\int_{\Omega'} \frac{\partial \varphi}{\partial x_k} \frac{\partial u}{\partial x_i} \, dx.$$

Then from Lemma 7.6.2 below,[7] we know that there exists a family of functions $\{\varphi_j\}_{j=1}^{\ell}$ such that

- $\varphi_j \in C_0^\infty(B_{\frac{\delta_0}{4}}(x_j))$, for every $j \in \{1, \dots, \ell\}$;
- $0 \le \varphi_j \le 1$ for every $j \in \{1, \dots, \ell\}$;
- they satisfy

$$\sum_{j=1}^{\ell} \varphi_j(x) = 1, \qquad \text{for every } x \in \Omega'.$$

We can then rewrite

$$\int_{\Omega'} \frac{\partial \varphi}{\partial x_k} \frac{\partial u}{\partial x_i} \, dx = \int_{\Omega'} \frac{\partial}{\partial x_k} \left(\varphi \sum_{j=1}^{\ell} \varphi_j \right) \frac{\partial u}{\partial x_i} \, dx = \sum_{j=1}^{\ell} \int_{\Omega'} \frac{\partial}{\partial x_k} \left(\varphi \, \varphi_j \right) \frac{\partial u}{\partial x_i} \, dx.$$

We observe that $\varphi \, \varphi_j \in C_0^\infty(B_{\delta_0/4}(x_j))$ by construction. Thus, by recalling (7.6.6) and using the definition of weak partial derivative in each integral, we get

$$\begin{aligned}
\int_{\Omega'} \frac{\partial \varphi}{\partial x_k} \frac{\partial u}{\partial x_i} \, dx &= \sum_{j=1}^{\ell} \int_{B_{\frac{\delta_0}{4}}(x_j)} \frac{\partial}{\partial x_k} \left(\varphi \, \varphi_j \right) \frac{\partial u}{\partial x_i} \, dx \\
&= -\sum_{j=1}^{\ell} \int_{B_{\frac{\delta_0}{4}}(x_j)} \varphi \, \varphi_j \, \frac{\partial}{\partial x_k} \frac{\partial u}{\partial x_i} \, dx \\
&= -\int_{\Omega'} \varphi \left(\sum_{j=1}^{\ell} \varphi_j \right) \frac{\partial}{\partial x_k} \frac{\partial u}{\partial x_i} \, dx = -\int_{\Omega'} \varphi \, \frac{\partial}{\partial x_k} \frac{\partial u}{\partial x_i} \, dx.
\end{aligned}$$

[7] We use it with $E_0 = \emptyset$, $r = \delta_0/8$ and $R = 2r = \delta_0/4$.

This shows that $\nabla u \in W^{1,2}(\Omega';\mathbb{R}^N)$. By arbitrariness of $\Omega' \Subset \Omega$, we get that $\nabla u \in W^{1,2}_{\rm loc}(\Omega;\mathbb{R}^N)$, as claimed.

Part 3: The Gradient Is Harmonic In order to conclude, we need to show that each weak partial derivative is weakly harmonic. We take $h = t\,\mathbf{e}_i$ in (7.6.3), with $0 < |t| \le \delta_0/4$ and divide it by t. Then we get

$$\int_{B_R(x_0)} \left\langle \frac{\mathrm{T}_{t\,\mathbf{e}_i}(\nabla u) - \nabla u}{t}, \nabla\varphi \right\rangle dx = 0, \qquad \text{for every } \varphi \in C^\infty_0(B_R(x_0)).$$

Still by Proposition 3.11.2, we know that

$$\frac{\mathrm{T}_{t\,\mathbf{e}_i}(\nabla u) - \nabla u}{t} \quad \text{weakly converges in } L^2(B_R(x_0)) \text{ to } \quad \frac{\partial \nabla u}{\partial x_k} \quad \text{as } t \to 0,$$

up to taking a sequence. By taking the limit in the above identity, we then obtain

$$\int_{B_R(x_0)} \left\langle \frac{\partial \nabla u}{\partial x_i}, \nabla\varphi \right\rangle dx = 0, \qquad \text{for every } \varphi \in C^\infty_0(B_R(x_0)).$$

We observe that (*prove it as an exercise*)

$$\frac{\partial \nabla u}{\partial x_i} = \nabla \frac{\partial u}{\partial x_i}, \qquad \text{for a. e. } x \in \Omega.$$

Thus, we finally obtain

$$\int_{B_R(x_0)} \left\langle \nabla\frac{\partial u}{\partial x_i}, \nabla\varphi \right\rangle dx = 0, \qquad \text{for every } \varphi \in C^\infty_0(B_R(x_0)).$$

This shows that each weak partial derivative is a weakly harmonic function in $B_R(x_0)$. By using the arbitrariness of the ball and a covering argument, coupled with Lemma 7.6.2 below, we get the desired assertion. The reader should try to complete the missing details as a useful exercise. □

Lemma 7.6.2 (Partition of Unity) *Let $E \subseteq \mathbb{R}^N$ be an open bounded set. Let $B_r(x_1), \dots, B_r(x_\ell)$ be a family of balls and $E_0 \Subset E$ an open set compactly contained in E, such that*[8]

$$E \Subset E_0 \cup \bigcup_{j=1}^{\ell} B_r(x_j).$$

[8] It is admitted that the set E_0 could be empty. In this case, there is no function φ_0.

Then for every $0 < t < \mathrm{dist}(E_0, \partial E)$ *and* $R > r$*, there exists a family of functions* $\{\varphi_j\}_{j=0}^{\ell}$ *such that*

- $\varphi_0 \in C_0^\infty(E_0^t)$*, where* E_0^t *is the following open set*

$$E_0^t := \{x \in E \,:\, \mathrm{dist}(x, E_0) < t\};$$

- $\varphi_j \in C_0^\infty(B_R(x_j))$*, for every* $j \in \{1, \ldots, \ell\}$*;*
- $0 \le \varphi_j \le 1$ *for every* $j \in \{0, \ldots, \ell\}$*;*
- *they form a* partition of unity *on* E*, i.e.*

$$\sum_{j=0}^{\ell} \varphi_j(x) = 1, \qquad \textit{for every } x \in E.$$

Proof We fix $0 < t < \mathrm{dist}(E_0, \partial E)$ and $R > r$, we observe at first that the balls $B_R(x_1), \ldots, B_R(x_\ell)$ and the set E_0^t still cover E (see Fig. 7.3). For every $j \in \{1, \ldots, \ell\}$, we choose a cut-off function $\eta_j \in C_0^\infty(B_R(x_j))$ such that

$$0 \le \eta_j \le 1, \qquad \eta_j \equiv 1 \text{ on } B_r(x_j).$$

We also choose a cut-off function $\eta_0 \in C_0^\infty(E_0^t)$ such that

$$0 \le \eta_0 \le 1, \qquad \eta_0 \equiv 1 \text{ on } E_0.$$

We then define recursively the functions

$$\begin{cases} \varphi_0 = \eta_0, \\ \\ \varphi_j = \eta_j \displaystyle\prod_{i=0}^{j-1} (1 - \eta_i), & \text{for } j \ge 1. \end{cases}$$

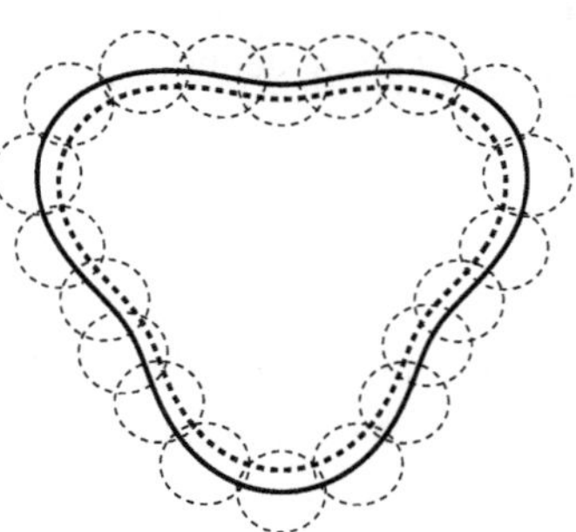

Fig. 7.3 The bold line entoures the set E, while the dotted lines indicate the covering made by $E_0^t, B_R(x_1), \ldots, B_R(x_\ell)$

We observe that each φ_j is a C^∞ function, with compact support either in $B_R(x_j)$ or in E_0^t and such that $0 \le \varphi_j \le 1$. Moreover, by construction, we have

$$\sum_{j=0}^{\ell} \varphi_j = \eta_0 + \sum_{j=1}^{\ell}\left(\eta_j \prod_{i=0}^{j-1}(1-\eta_i)\right) = 1 - \prod_{j=0}^{\ell}(1-\eta_j). \tag{7.6.7}$$

This can be easily proved by using induction on ℓ. Indeed, for $\ell = 0$, this is trivially true. If we assume the formula to hold for ℓ, then for $\ell + 1$ we would get

$$\begin{aligned}\sum_{j=0}^{\ell+1} \varphi_j &= \sum_{j=0}^{\ell} \varphi_j + \varphi_{\ell+1} \\ &= 1 - \prod_{j=0}^{\ell}(1-\eta_j) + \eta_{\ell+1} \prod_{j=0}^{\ell}(1-\eta_j) \\ &= 1 - (1-\eta_{\ell+1}) \prod_{j=0}^{\ell}(1-\eta_j) = 1 - \prod_{j=0}^{\ell+1}(1-\eta_j),\end{aligned}$$

as desired.

We now show that

$$\sum_{j=0}^{\ell} \varphi_j(x) = 1, \qquad \text{for every } x \in E.$$

Take $x \in E$, then either $x \in E_0$ or there exists a ball $B_r(x_j)$ such that $x \in B_r(x_j)$. This implies that

$$\text{either} \quad \eta_0(x) = 1 \quad \text{or} \quad \eta_j(x) = 1,$$

and by (7.6.7), we easily get the desired conclusion, since the product on the right-hand side in (7.6.7) vanishes. □

7.7 Classical Regularity

Finally, we can prove that weakly harmonic functions are harmonic in a classical sense. Namely, we have the following

Theorem 7.7.1 *Let $u \in W^{1,2}_{\mathrm{loc}}(\Omega)$ be a weakly harmonic function. Then there exists $0 < \alpha < 1$ such that for every $\mathcal{O} \Subset \Omega$ open set, we have $u \in C^{k,\alpha}(\overline{\mathcal{O}})$, for every $k \in \mathbb{N}$. In particular, u is a C^∞ function in Ω and thus it is harmonic in classical sense.*

Proof It is sufficient to use iteratively Theorems 7.5.1 and 7.6.1. □

The claimed regularity for our model minimization problem (7.1.1) is now a direct consequence of the previous result. We recall that the *bounded slope condition* has been introduced in Definition 6.4.1.

Corollary 7.7.2 *Let $\Omega \subseteq \mathbb{R}^N$ be an open bounded set and let $g \in W^{1,2}(\Omega)$. The unique solution v of the minimization problem*

$$\min_{u\in W^{1,2}(\Omega)} \left\{ \frac{1}{2}\int_\Omega |\nabla u|^2\,dx \,:\, u-g \in W_0^{1,2}(\Omega) \right\},$$

is an harmonic function in Ω in classical sense.

Moreover, if $g \in C^{0,1}(\overline{\Omega})$ and its restriction to $\partial\Omega$ is additionally supposed to satisfy the bounded slope condition of rank K, then v is Lipschitz continuous on $\overline{\Omega}$ and

$$|v|_{C^{0,1}(\overline{\Omega})} \le K.$$

Proof The first part of the statement follows directly from Theorem 7.7.1. In order to prove the second statement, i.e. the global Lipschitz character of v, by Theorem 6.5.1 we know that

$$\min_{u\in C^{0,1}(\overline{\Omega})} \left\{ \frac{1}{2}\int_\Omega |\nabla u|^2\,dx \,:\, u = g \text{ on } \partial\Omega \right\},$$

admits a unique solution, that we call $w \in C^{0,1}(\overline{\Omega})$. By Problem 6.8.10, we also know that this minimum coincides with

$$\min_{u\in W^{1,2}(\Omega)} \left\{ \frac{1}{2}\int_\Omega |\nabla u|^2\,dx \,:\, u-g \in W_0^{1,2}(\Omega) \right\},$$

and $v = w$. This is enough to conclude. □

Remark 7.7.3 We notice that if the BSC assumption on g is dropped, the solution v could fail to be Lipschitz continuous on $\overline{\Omega}$. In particular, the sole assumption $g \in C^{0,1}(\overline{\Omega})$ *is not enough* in general to entail that $v \in C^{0,1}(\overline{\Omega})$. Indeed, the reader should to keep in mind the counter-examples of Proposition 6.7.2 and Problem 6.8.11.

7.8 Problems

Problem 7.8.1 Write down the details of the proof of Theorem 7.3.1 for the case $N = 2$.

Problem 7.8.2 Let $\Omega \subseteq \mathbb{R}^N$ be an open set and let $f \in W^{1,2}_{\rm loc}(\Omega)$. If $u \in W^{1,2}_{\rm loc}(\Omega)$ is a weak solution of

$$-\Delta u = f, \qquad \text{in } \Omega,$$

prove that $\nabla u \in W^{1,2}_{\rm loc}(\Omega; \mathbb{R}^N)$. Also prove that for every $i \in \{1, \dots, N\}$ it holds

$$\int_\Omega \left\langle \nabla \left(\frac{\partial u}{\partial x_i}\right), \nabla \varphi \right\rangle dx = \int_\Omega \varphi \, \frac{\partial f}{\partial x_i}\, dx, \qquad \text{for every } \varphi \in C^\infty_0(\Omega),$$

i.e. for every $i \in \{1, \dots, N\}$ the i-th component of ∇u is a weak solution of

$$-\Delta \frac{\partial u}{\partial x_i} = \frac{\partial f}{\partial x_i}, \qquad \text{in } \Omega.$$

Problem 7.8.3 Let $\Omega \subseteq \mathbb{R}^N$ be an open bounded connected set and let $v \in W^{1,2}_0(\Omega)$ be the torsion function of Ω, i.e. the unique weak solution of

$$\begin{cases} -\Delta u = 1, & \text{in } \Omega, \\ \quad\; u = 0, & \text{on } \partial\Omega, \end{cases}$$

see Proposition 4.5.1. Prove that v is a C^∞ function in Ω.

Problem 7.8.4 Let $\Omega \subseteq \mathbb{R}^N$ be an open bounded connected set and let $v \in W^{1,2}_0(\Omega)$ be an eigenfunction of the Laplacian-Dirichlet on Ω (see Definition 4.7.1). Prove that $v \in C^\infty(\Omega)$.

Problem 7.8.5 Let $\Omega \subseteq \mathbb{R}^N$ be an open set and let $\mathrm{Sym}_N(\mathbb{R})$ be the space of $N \times N$ symmetric matrices, with real entries. Let $A : \Omega \to \mathrm{Sym}_N(\mathbb{R})$ be a matrix-valued measurable function, such that

$$\lambda\, |\xi|^2 \le \langle A(x) \cdot \xi, \xi\rangle \le \Lambda\, |\xi|^2, \qquad \text{for a.e. } x \in \Omega \text{ and every } \xi \in \mathbb{R}^N,$$

for some $0 < \lambda \le \Lambda$. We consider the equation

$$-\operatorname{div}(A \cdot \nabla u) = 0, \qquad \text{in } \Omega, \tag{7.8.1}$$

we say that $u \in W^{1,2}_{\rm loc}(\Omega)$ is a:

- *weak solution* of (7.8.1) if it satisfies

$$\int_\Omega \langle A \cdot \nabla u, \nabla \varphi\rangle \, dx = 0, \qquad \text{for every } \varphi \in C^\infty_0(\Omega);$$

- *weak subsolution* of (7.8.1) if it satisfies

$$\int_\Omega \langle A \cdot \nabla u, \nabla \varphi \rangle \, dx \leq 0, \qquad \text{for every } \varphi \in C_0^\infty(\Omega) \text{ such that } \varphi \geq 0.$$

Prove the following generalization of Theorem 7.3.1: if $u \in W^{1,2}_{\text{loc}}(\Omega)$ is a non-negative weak subsolution, then $u \in L^\infty_{\text{loc}}(\Omega)$. Deduce the same result for a weak solution, without sign assumption.

Problem 7.8.6 (De Giorgi-Nash-Moser Theorem) Generalize Theorem 7.5.1 to a weak solution $u \in W^{1,2}_{\text{loc}}(\Omega)$ of the equation

$$-\operatorname{div}(A \cdot \nabla u) = 0, \qquad \text{in } \Omega,$$

with A as in Problem 7.8.5.

Problem 7.8.7 Let $1 < p < \infty$ and let $\Omega \subseteq \mathbb{R}^N$ be an open set, we say that $u \in W^{1,p}_{\text{loc}}(\Omega)$ is:

- *weakly p-harmonic in* Ω if

$$\int_\Omega \langle |\nabla u|^{p-2} \nabla u, \nabla \varphi \rangle \, dx = 0, \qquad \text{for every } \varphi \in C_0^\infty(\Omega);$$

- *weakly p-subharmonic in* Ω if

$$\int_\Omega \langle |\nabla u|^{p-2} \nabla u, \nabla \varphi \rangle \, dx \leq 0, \qquad \text{for every } \varphi \in C_0^\infty(\Omega) \text{ such that } \varphi \geq 0.$$

Prove the analogue of Theorem 7.2.1 for a weakly p-harmonic function u, i.e. if $F : \mathbb{R} \to \mathbb{R}$ is a C^1 convex function with bounded first derivative, then $F \circ u \in W^{1,p}_{\text{loc}}(\Omega)$ is a weakly p-subharmonic function on Ω. Moreover, if F is non-negative the following inequality holds

$$\int_\Omega |\nabla F(u)|^p \, \varphi^p \, dx \leq p^p \int_\Omega |\nabla \varphi|^p \, |F(u)|^p \, dx, \qquad \text{for every } \varphi \in C_0^\infty(\Omega). \tag{7.8.2}$$

Problem 7.8.8 Prove the analogue of Proposition 7.2.3 for a non-negative weakly p-subharmonic function.

Problem 7.8.9 Let $1 < p < \infty$ and let $\Omega \subseteq \mathbb{R}^N$ be an open set. Prove the analogue of Theorem 7.3.1 for a non-negative weakly p-subharmonic function u, i.e. $u \in L^\infty_{\text{loc}}(\Omega)$, with the a priori estimate (valid for every $B_{R_0}(x_0) \Subset \Omega$)

$$\|u\|_{L^\infty(B_{r_0}(x_0))} \le C \left(\frac{R_0}{R_0 - r_0}\right)^N \left(\fint_{B_{R_0}(x_0)} |u|^p \, dx\right)^{\frac{1}{p}},$$

for a constant $C > 0$, depending on N and p, only.

Problem 7.8.10 Let $1 < p < \infty$ and let $\Omega \subseteq \mathbb{R}^N$ be an open set. Let $u \in W^{1,p}_{\mathrm{loc}}(\Omega)$ be a weakly p-harmonic function. Prove that there exists $0 < \alpha < 1$ such that for every $\mathcal{O} \Subset \Omega$ open set, we have $u \in C^{0,\alpha}(\overline{\mathcal{O}})$.

Problem 7.8.11 Let $1 < p \le N$ and let $\Omega \subseteq \mathbb{R}^N$ be an open bounded set. Prove that for every $q > N/p$ and every $f \in L^q(\Omega)$ there exists a unique weak solution $v \in W^{1,p}_0(\Omega)$ of the problem

$$\begin{cases} -\Delta_p u = f, & \text{in } \Omega, \\ u = 0, & \text{on } \partial\Omega. \end{cases}$$

Show that we have $v \in L^\infty(\Omega)$.

Problem 7.8.12 For $p \ge 2$ and $\alpha > 0$, we consider the convex function

$$F_\alpha(z) = \frac{1}{p} \left(\alpha + |z|^2\right)^{\frac{p}{2}}, \qquad \text{for } z \in \mathbb{R}^N.$$

Let $\Omega \subseteq \mathbb{R}^N$ be an open set and let $u \in W^{1,p}_{\mathrm{loc}}(\Omega)$ be a weak solution of the equation

$$-\operatorname{div}(\nabla F_\alpha(\nabla u)) = 0, \qquad \text{in } \Omega,$$

i.e. u satisfies

$$\int_\Omega \langle \nabla F_\alpha(\nabla u), \nabla \varphi \rangle \, dx = 0, \qquad \text{for every } \varphi \in C^\infty_0(\Omega).$$

Prove that $\nabla u \in W^{1,2}_{\mathrm{loc}}(\Omega; \mathbb{R}^N)$.

Chapter 8
Solutions to Problems

8.1 Problems of Chap. 1

1.7.1 We observe that the function $f(t) = t^\alpha$ is concave on $[0, +\infty)$, indeed we have

$$f''(t) = \alpha\,(\alpha - 1)\,t^{\alpha-2} < 0, \qquad \text{for every } t > 0.$$

By concavity, we then get

$$\left(\frac{t+s}{2}\right)^\alpha \geq \frac{1}{2}\,t^\alpha + \frac{1}{2}\,s^\alpha, \qquad \text{for every } t, s \geq 0,$$

which is the same as

$$2^{\alpha-1}\,(t^\alpha + s^\alpha) \leq (t+s)^\alpha, \qquad \text{for every } t, s \geq 0.$$

This proves the first inequality in (1.7.1).

In order to prove the second one, we first observe that this is trivial if $t = 0$ or $s = 0$. We then suppose that $t > 0$ and $s > 0$. We now observe that

$$(t+s)^\alpha \leq t^\alpha + s^\alpha \qquad \Longleftrightarrow \qquad \left(1 + \frac{s}{t}\right)^\alpha \leq 1 + \left(\frac{s}{t}\right)^\alpha.$$

If we call $\tau = s/t$, then the inequality we want to prove is equivalent to prove that

$$(1+\tau)^\alpha \leq 1 + \tau^\alpha, \qquad \text{for every } \tau > 0,$$

L. Brasco, *Handbook of Calculus of Variations for Absolute Beginners*,
La Matematica per il 3+2 163, https://doi.org/10.1007/978-3-031-87164-1_8

that is

$$\frac{1+\tau^\alpha}{(1+\tau)^\alpha} \ge 1, \qquad \text{for every } \tau > 0.$$

We then consider the function

$$h(\tau) = \frac{1+\tau^\alpha}{(1+\tau)^\alpha},$$

and compute its infimum. We have

$$h'(\tau) = \frac{\alpha\,\tau^{\alpha-1}\,(1+\tau)^\alpha - \alpha\,(1+\tau^\alpha)\,(1+\tau)^{\alpha-1}}{(1+\tau)^{2\alpha}} = \alpha\,\frac{\tau^{\alpha-1}-1}{(1+\tau)^{\alpha+1}},$$

which shows that h is increasing on $(0, 1)$ and decreasing on $(1, +\infty)$. By observing that

$$h(0) = 1 \qquad \text{and} \qquad \lim_{\tau\to+\infty} h(\tau) = 1,$$

we thus get

$$h(\tau) \ge h(0) = 1, \qquad \text{for every } \tau > 0,$$

as desired.

1.7.2 This time, the function $f(t) = t^\alpha$ is convex on $[0, +\infty)$, indeed we have

$$f''(t) = \alpha\,(\alpha-1)\,t^{\alpha-2} > 0, \qquad \text{for every } t > 0.$$

By convexity, we then get

$$\left(\frac{t+s}{2}\right)^\alpha \le \frac{1}{2}\,t^\alpha + \frac{1}{2}\,s^\alpha, \qquad \text{for every } t, s \ge 0,$$

which is the same as

$$(t+s)^\alpha \le 2^{\alpha-1}\,(t^\alpha + s^\alpha), \qquad \text{for every } t, s \ge 0.$$

This proves the second inequality in (1.7.2).

In order to prove the first one, we can proceed as in the previous exercise. It is then sufficient to show that

$$\frac{1+\tau^\alpha}{(1+\tau)^\alpha} \le 1, \qquad \text{for every } \tau > 0.$$

The details are left to the reader.

1.7.3 Without loss of generality, we can suppose that $t \geq s$. We distinguish three cases: $s \geq 0$, $t \leq 0$ and $s < 0 \leq t$. In the case $s \geq 0$, from the rightmost inequality of Problem 1.7.1 we get

$$t^\alpha = (t - s + s)^\alpha \leq (t - s)^\alpha + s^\alpha,$$

that is

$$t^\alpha - s^\alpha \leq (t - s)^\alpha,$$

thus proving the claimed inequality. The case $t \leq 0$ can be treated similarly: again by Problem 1.7.1 we have

$$|s|^\alpha = (|s| - |t| + |t|)^\alpha \leq (|s| - |t|)^\alpha + |t|^\alpha,$$

which is the same as

$$|s|^\alpha - |t|^\alpha \leq (|s| - |t|)^\alpha.$$

In view of the assumptions on t and s, this is again the claimed inequality.

Finally, if $s < 0 \leq t$ we have

$$\Big| |t|^{\alpha-1}\, t - |s|^{\alpha-1}\, s \Big| = t^\alpha + |s|^\alpha \leq 2^{1-\alpha}\, (t + |s|)^\alpha = 2^{1-\alpha}\, |t - s|^\alpha,$$

by using this time the leftmost inequality of Problem 1.7.1.

1.7.4 The case $\alpha = 1$ is straightforward. For the case $\alpha > 1$, we observe that the function $\tau \mapsto |\tau|^{\alpha-1}\, \tau$ is C^1. We can suppose without loss of generality that $t \geq s$, then by the Fundamental Theorem of Calculus we have that

$$|t|^{\alpha-1}\, t - |s|^{\alpha-1}\, s = \alpha \int_s^t |\tau|^{\alpha-1}\, d\tau.$$

By recalling that $t > s$, this is the same as

$$\Big| |t|^{\alpha-1}\, t - |s|^{\alpha-1}\, s \Big| = \alpha \int_s^t |\tau|^{\alpha-1}\, d\tau.$$

We now observe that $\alpha - 1 > 0$, thus we get

$$\alpha \int_s^t |\tau|^{\alpha-1}\, d\tau \leq \alpha\, \max\{|t|, |s|\}^{\alpha-1}\, |t - s| \leq \alpha\, (|t| + |s|)^{\alpha-1}\, |t - s|.$$

This concludes the proof.

1.7.5 We can suppose without loss of generality that $t \geq s$. Thanks to the assumption on G, we have

$$|G(t) - G(s)| \leq \int_s^t |G'(\tau)|\,d\tau \leq A \int_s^t (|\tau|^{\alpha-1} + 1)\,d\tau$$
$$\leq A \left(\max\{|t|, |s|\}^{\alpha-1} + 1\right)(t - s)$$
$$\leq A\,(|t| + |s|)^{\alpha-1}\,(t - s) + A\,(t - s).$$

This shows the first estimate. In order to prove the second one, we observe that by taking $s = 0$ in the previous estimate, we get

$$|G(t) - G(0)| \leq A\,|t|^\alpha + A\,|t|.$$

Thus, we obtain

$$|G(t)| \leq |G(0)| + |G(t) - G(0)| \leq |G(0)| + A\,|t|^\alpha + A\,|t|.$$

For $\alpha = 1$, this gives the desired result with

$$\widetilde{A} = \max\{2\,A, |G(0)|\}.$$

For $\alpha > 1$, by using Young's inequality with exponents α and α' on the term $A\,|t|$, we get

$$|G(t)| \leq |G(0)| + |G(t) - G(0)| \leq |G(0)| + \left(A + \frac{A}{\alpha}\right)|t|^\alpha + \frac{A}{\alpha'}.$$

We then get the desired conclusion, by setting

$$\widetilde{A} = \max\left\{A + \frac{A}{\alpha}, \frac{A}{\alpha'} + |G(0)|\right\}.$$

1.7.6 Let $t \in (a, b)$, then t can be written as a convex combination of a and b. More precisely, we can write

$$t = \frac{b-t}{b-a}\,a + \frac{t-a}{b-a}\,b.$$

By convexity of the function f, we thus get

$$f(t) = f\left(\frac{b-t}{b-a}\,a + \frac{t-a}{b-a}\,b\right) \leq \frac{b-t}{b-a}\,f(a) + \frac{t-a}{b-a}\,f(b)$$

In particular, we get for every $t \in (a, b)$

$$f(t) \leq \left(\frac{b-t}{b-a} + \frac{t-a}{b-a}\right) \max\{f(a), f(b)\} = \max\{f(a), f(b)\}.$$

This gives that f is bounded from above on $[a, b]$. In order to show that f is also bounded from below, we take $t \in (a, b)$. We set

$$t_0 = \frac{a+b}{2},$$

and first suppose that $t \geq (a+b)/2$. Thus, we can write the middle point t_0 as a convex combination of a and t, i.e. we have

$$t_0 = \lambda\, t + (1-\lambda)\, a, \qquad \text{with } \lambda = \frac{t_0 - a}{t-a}.$$

By using the convexity of f, we get

$$f(t_0) \leq \lambda\, f(t) + (1-\lambda)\, f(a) \leq \lambda\, f(t) + (1-\lambda)\, |f(a)| \leq \lambda\, f(t) + |f(a)|.$$

This shows that

$$\lambda\, f(t) \geq f(t_0) - |f(a)|, \qquad \text{for every } t \in \left[\frac{a+b}{2}, b\right].$$

We now observe that by construction we have $\lambda \leq 1$ and

$$\lambda = \frac{t_0 - a}{t-a} = \frac{b-a}{2\,(t-a)} \geq \frac{1}{2}.$$

Thus, from the previous estimate we get[1]

$$f(t) \geq \min\big\{0, 2\,(f(t_0) - |f(a)|)\big\}, \qquad \text{for every } t \in \left[\frac{a+b}{2}, b\right].$$

If $t \leq (a+b)/2$, we can repeat the same argument, by writing the middle point t_0 as a convex combination of t and b. This gives

$$f(t) \geq \min\big\{0, 2\,(f(t_0) - |f(b)|)\big\}, \qquad \text{for every } t \in \left[a, \frac{a+b}{2}\right].$$

[1] We use that either $f(t) \geq 0$; or $f(t) < 0$ and thus

$$f(t_0) - |f(a)| \leq \lambda\, f(t) \leq \frac{1}{2}\, f(t) \qquad \text{that is} \qquad 2\,(f(t_0) - |f(a)|) \leq f(t).$$

The two previous estimates finally shows that f is bounded from below on $[a, b]$.

Finally, a convex function on the open interval (a, b) is not necessarily bounded: it is sufficient to take

$$f(t) = \frac{1}{t-a}, \qquad \text{for } t \in (a, b),$$

as a counterexample.

1.7.7 Let $z \in \mathbb{R}^N$, for every $t \in \mathbb{R}$ we define the function of one real variable

$$h(t) = H(t\, z).$$

By using an integration by parts, we have the second order Taylor-type formula

$$h(1) = h(0) + h'(0) + \int_0^1 h''(\tau)\,(1-\tau)\,d\tau.$$

In terms of H, this rewrites as follows

$$H(z) = H(0) + \langle \nabla H(0), z\rangle + \int_0^1 \langle D^2 H(\tau\, z)\cdot z, z\rangle\,(1-\tau)\,d\tau.$$

In particular, by using the assumption on H, we get

$$\begin{aligned} H(z) &\geq H(0) + \langle \nabla H(0), z\rangle + \lambda\,|z|^2 \int_0^1 (1-\tau)\,d\tau \\ &= H(0) + \langle \nabla H(0), z\rangle + \frac{\lambda}{2}\,|z|^2. \end{aligned}$$

By using the Cauchy-Schwarz inequality and the generalized Young inequality (1.2.6), for every $z \in \mathbb{R}^N$ and every $\varepsilon > 0$ we have

$$\langle \nabla H(0), z\rangle \geq -|\nabla H(0)|\,|z| \geq -\frac{1}{2\,\varepsilon}\,|\nabla H(0)|^2 - \frac{\varepsilon}{2}\,|z|^2.$$

By choosing for example $\varepsilon = \lambda/2$, we finally get

$$H(z) \geq \frac{\lambda}{4}\,|z|^2 + \left(H(0) - \frac{1}{\lambda}\,|\nabla H(0)|^2\right). \tag{$*$}$$

This estimate shows at first that H is bounded from below. Moreover, H has bounded sublevel sets, thus we can prove existence of a minimizer by reproducing the proof of Weierstrass' Theorem. We set

$$\mathfrak{m} = \inf_{z\in\mathbb{R}^N} H(z),$$

then for every $n \in \mathbb{N}$, there exists $z_n \in \mathbb{R}^N$ such that

$$H(z_n) \leq \mathfrak{m} + \frac{1}{n+1}, \qquad \text{for every } n \in \mathbb{N}.$$

This fact and $(*)$ show that $\{z_n\}_{n\in\mathbb{N}}$ is a bounded sequence. By Bolzano-Weierstrass Theorem, we get that this converges (up to a subsequence) to some $\overline{z} \in \mathbb{R}^N$. By construction and using the continuity of H, we finally get

$$\mathfrak{m} = \lim_{n\to\infty} \left(\mathfrak{m} + \frac{1}{n+1}\right) \geq \lim_{n\to\infty} H(z_n) = H(\overline{z}).$$

Thus, we get that $\overline{z}$ is a minimum point for H over $\mathbb{R}^N$. The uniqueness of the minimum point follows from Remark 1.3.9.

1.7.8 Let us take two real numbers $0 \leq t_0 < t_1$. By using the monotonicity and the strict convexity, we get

$$f(t_0) \leq f\left(\frac{t_0+t_1}{2}\right) < \frac{1}{2}\, f(t_0) + \frac{1}{2}\, f(t_1) \leq \left(\frac{1}{2} + \frac{1}{2}\right) f(t_1) = f(t_1),$$

which proves that f is actually strictly monotone.

We now prove that F is convex. Indeed, for every $z_0, z_1 \in \mathbb{R}^N$ and $\lambda \in [0,1]$, we have

$$F((1-\lambda)\, z_0 + \lambda\, z_1) = f(|(1-\lambda)\, z_0 + \lambda\, z_1|) \leq f((1-\lambda)\, |z_0| + \lambda\, |z_1|),$$

thanks to the triangle inequality for the Euclidean norm and the monotonicity assumption on f. We also observe that, from the first part of the exercise, we have that the inequality is strict, whenever

$$|(1-\lambda)\, z_0 + \lambda\, z_1| < (1-\lambda)\, |z_0| + \lambda\, |z_1|.$$

If we further use the convexity of f, we then obtain

$$F((1-\lambda)\, z_0+\lambda\, z_1) \leq (1-\lambda)\, f(|z_0|)+\lambda\, f(|z_1|) = (1-\lambda)\, F(z_0)+\lambda\, F(z_1). \qquad (*)$$

Again, we notice that the inequality is strict if $\lambda \in (0,1)$ and

$$|z_0| \neq |z_1|.$$

From $(*)$ we get that F is convex on $\mathbb{R}^N$. Moreover, the previous analysis of the equality cases entails that, if for some $\lambda \in (0,1)$ and $z_0 \neq z_1$ equality holds in $(*)$, we must have

$$|(1-\lambda)\, z_0 + \lambda\, z_1| = (1-\lambda)\, |z_0| + \lambda\, |z_1| \qquad \text{and} \qquad |z_0| = |z_1|.$$

These two facts imply that

$$|(1-\lambda)\, z_0 + \lambda\, z_1| = |z_0| = |z_1|.$$

By raising to the power 2 and using that $|z_0| = |z_1|$, this is the same as

$$(1-\lambda)^2\, |z_0|^2 + \lambda^2\, |z_0|^2 + 2\,\lambda\,(1-\lambda)\,\langle z_0, z_1\rangle = |z_0|^2,$$

that is

$$2\,\lambda\,(1-\lambda)\,\langle z_0, z_1\rangle = 2\,\lambda\,(1-\lambda)\,|z_0|^2 = \lambda\,(1-\lambda)\,(|z_0|^2 + |z_1|^2).$$

By observing that $\lambda\,(1-\lambda) \neq 0$, the latter can be recast into

$$|z_0|^2 + |z_1|^2 - 2\,\langle z_0, z_1\rangle = 0.$$

By observing that the left-hand side coincides with $|z_0-z_1|^2$, we would immediately get $z_0 = z_1$, i.e. a contradiction. Thus, the function F is actually strictly convex on $\mathbb{R}^N$.

We are left with the last point. Let us suppose that there exists $a > 0$ such that f is increasing on $[0, a]$ and decreasing on $[a, +\infty)$. Then by convexity we would get

$$f(a) = f\left(\frac{1}{2}\,0 + \frac{1}{2}\,2\,a\right) < \frac{1}{2}\,f(0) + \frac{1}{2}\,f(2\,a),$$

while by monotonicity we would have

$$f(a) > f(0) \qquad \text{and} \qquad f(a) > f(2\,a).$$

The last two equations in display give a contradiction. Thus, if f has two intervals of strict monotonicity, the only possibility is that f is decreasing on $[0, a]$ and increasing on $[a, +\infty)$. Accordingly, we take a point $z \in \mathbb{R}^N \setminus \{0\}$ such that $|z| = a$ and observe that

$$F(z) = f(a) < f(0) = F(0) = F\left(\frac{1}{2}\,z + \frac{1}{2}\,(-z)\right),$$

while

$$\frac{1}{2}\,F(z) + \frac{1}{2}\,F(-z) = \frac{1}{2}\,f(a) + \frac{1}{2}\,f(a) = f(a) = F(z).$$

This shows that we actually have

$$F\left(\frac{1}{2}\,z + \frac{1}{2}\,(-z)\right) > \frac{1}{2}\,F(z) + \frac{1}{2}\,F(-z),$$

and thus F is not convex.

1.7.9 We first observe that $z \mapsto |\nabla H(z)|$ is a continuous function, thus it attains a maximum on the compact set $\overline{B_1(0)}$. By indicating with M such a maximum, we trivially get

$$|\nabla H(z)| \leq M \leq M\,(|z|^{p-1} + 1), \qquad \text{for every } |z| \leq 1.$$

Thus, the desired property holds for $|z| \leq 1$. Let us now take $z \in \mathbb{R}^N$ such that $|z| > 1$. By using the "above tangent" property (Proposition 1.3.4), we have for every $\omega \in \mathbb{S}^{N-1}$

$$H(z + |z|\,\omega) \geq H(z) + \langle \nabla H(z), \omega\rangle\,|z|.$$

In particular, by using the assumption on H, we get

$$C_2\left(\left|z + |z|\,\omega\right|^p + 1\right) \geq C_1\,(|z|^p - 1) + \langle \nabla H(z), \omega\rangle\,|z|.$$

We now observe that by the triangle inequality and using that ω has unit norm, we have

$$\left|z + |z|\,\omega\right|^p \leq (|z| + |z|)^p = 2^p\,|z|^p.$$

The last two equations imply that

$$\langle \nabla H(z), \omega\rangle\,|z| \leq (2^p\,C_2 - C_1)\,|z|^p + (C_1 + C_2), \qquad \text{for every } |z| > 1,\ \omega \in \mathbb{S}^{N-1}.$$

In particular, by dividing both sides by $|z| > 1$ we get

$$\langle \nabla H(z), \omega\rangle \leq (2^p\,C_2 - C_1)\,|z|^{p-1} + \frac{C_1 + C_2}{|z|} \leq C_3\,(|z|^{p-1} + 1),$$

where we set $C_3 = \max\{2^p\,C_2 - C_1,\ C_1 + C_2\} > 0$. This is now enough to conclude, thanks to the fact that

$$\max_{\omega \in \mathbb{S}^{N-1}} \langle \nabla H(z), \omega\rangle = |\nabla H(z)|.$$

1.7.10 We proceed by induction over k. We observe that for $k = 2$ the inequality (1.7.3) is a consequence of the usual Young's inequality (see Lemma 1.2.9) with $p = p' = 2$. We now suppose that (1.7.3) holds for a $k \geq 2$ and show that this implies that (1.7.3) holds true for $k + 1$, as well. We start by writing

$$\prod_{i=1}^{k+1} a_i = \prod_{i=1}^{k} a_i \cdot a_{k+1},$$

then we apply Young's inequality with conjugate exponents

$$\frac{k+1}{k} \qquad \text{and} \qquad k+1.$$

We thus obtain

$$\prod_{i=1}^{k+1} a_i \le \frac{k}{k+1} \prod_{i=1}^{k} a_i^{\frac{k+1}{k}} + \frac{1}{k+1}\, a_{k+1}^{k+1}. \tag{$*$}$$

On the other hand, by using the inductive assumption, we can infer

$$\prod_{i=1}^{k} a_i^{\frac{k+1}{k}} \le \frac{1}{k}\left(\sum_{i=1}^{k}\left(a_i^{\frac{k+1}{k}}\right)^k\right) = \frac{1}{k}\left(\sum_{i=1}^{k} a_i^{k+1}\right).$$

By inserting this estimate into ($*$), we get

$$\prod_{i=1}^{k+1} a_i \le \frac{1}{k+1}\left(\sum_{i=1}^{k} a_i^{k+1}\right) + \frac{1}{k+1}\, a_{k+1}^{k+1},$$

which shows the validity of (1.7.3) for $k+1$, as desired.

1.7.11 For every $\varphi \in C_0^\infty((a,b))$, consider the new function

$$\psi(t) = \int_a^t \varphi(\tau)\, d\tau - \frac{t-a}{b-a}\int_a^b \varphi(\tau)\, d\tau.$$

Observe that $\psi \in C^1([a,b])$ by construction and moreover it verifies

$$\psi(a) = 0 = \psi(b).$$

By assumption, we thus obtain

$$\int_a^b f(t)\, \psi'(t)\, dt = 0.$$

We now observe that

$$\psi'(t) = \varphi(t) - \frac{1}{b-a}\int_a^b \varphi(\tau)\, d\tau.$$

By using this fact in the previous identity, we get

$$0 = \int_a^b f(t)\,\varphi(t)\,dt - \frac{1}{b-a}\left(\int_a^b \varphi(\tau)\,d\tau\right)\int_a^b f(t)\,dt.$$

If we now set

$$c = \frac{1}{b-a}\int_a^b f(t)\,dt,$$

the previous equation can be rewritten as

$$0 = \int_a^b f(t)\,\varphi(t)\,dt - \int_a^b c\,\varphi(t)\,dt \qquad \text{that is} \qquad \int_a^b \Big[f(t) - c\Big]\,\varphi(t)\,dt = 0.$$

Since this is true for every $\varphi \in C_0^\infty((a,b))$, by Lemma 1.4.1 we conclude that

$$f(t) - c = 0, \qquad \text{for every } t \in [a,b].$$

This gives the desired conclusion.

1.7.12 We fix $\eta \in C_0^\infty((a,b))$ such that

$$\int_a^b \eta(t)\,dt = 1.$$

Then, similarly to the previous proof, for every $\varphi \in C_0^\infty((a,b))$ we define

$$\psi(t) = \int_a^t \varphi(\tau)\,d\tau - \left(\int_a^b \varphi(\tau)\,d\tau\right)\int_a^t \eta(\tau)\,d\tau.$$

Observe that $\psi \in C_0^\infty((a,b))$ by construction. We can now reproduce the computations used in the solution of Problem 1.7.11. Indeed, by hypothesis, we have

$$\int_a^b f(t)\,\psi'(t)\,dt = 0.$$

By definition of ψ, it holds

$$\psi'(t) = \varphi(t) - \eta(t)\int_a^b \varphi(\tau)\,d\tau.$$

By using this fact into the previous identity, we get

$$0 = \int_a^b f(t)\,\varphi(t)\,dt - \int_a^b \varphi(\tau)\,d\tau \int_a^b \eta(t)\,f(t)\,dt.$$

This time we set

$$c = \int_a^b \eta(t)\, f(t)\, dt,$$

then the previous equation can be rewritten as

$$0 = \int_a^b f(t)\,\varphi(t)\, dt - \int_a^b c\,\varphi(t)\, dt \qquad \text{that is} \qquad \int_a^b \Big[f(t) - c\Big]\,\varphi(t)\, dt = 0.$$

Since this is true for every $\varphi \in C_0^\infty((a,b))$, we conclude by Lemma 1.4.1 again that

$$f(t) - c = 0, \qquad \text{for every } t \in [a,b].$$

This concludes the exercise.

1.7.13 For every $\psi \in C_0^\infty((a,b))$, we have

$$\int_a^b \psi'(t)\, dt = 0.$$

Since $\psi' \in C_0^\infty((a,b))$, we can use the claimed property of f with $\varphi = \psi'$. This gives

$$\int_a^b f(t)\,\psi'(t)\, dt = 0, \qquad \text{for every } \psi \in C_0^\infty((a,b)).$$

Then we get the conclusion from Problem 1.7.12.

1.7.14 We prove at first the special case $a = 0$ and $b = 1$. For every $\varphi \in C_0^\infty((0,1))$, we define

$$\psi(t) = \int_0^t \left(\int_0^\tau \varphi(s)\, ds\right) d\tau + A\, t^3 + B\, t^2,$$

where A, B are given by

$$A = -\int_0^1 \varphi(s)\, ds + 2\int_0^1 \left(\int_0^\tau \varphi(s)\, ds\right) d\tau,$$

and

$$B = \int_0^1 \varphi(s)\, ds - 3\int_0^1 \left(\int_0^\tau \varphi(s)\, ds\right) d\tau.$$

Observe that by construction we have $\psi \in C^2([0,1])$ and

$$\psi(0) = \psi'(0) = 0, \qquad \psi(1) = \psi'(1) = 0.$$

Thus, by assumption we get

$$0 = \int_0^1 f(t)\,\psi''(t)\,dt.$$

On the other hand, by computing the second derivative of ψ we get

$$\psi''(t) = \varphi(t) + 6\,A\,t + 2\,B, \qquad \text{for every } t \in [0,1].$$

By comparing the last two equations in display, we obtain

$$0 = \int_0^1 f(t)\,[\varphi(t) + 6\,A\,t + 2\,B]\,dt, \qquad \text{for every } \varphi \in C_0^\infty((0,1)).$$

This can be rewritten as

$$\int_0^1 f(t)\,\varphi(t)\,dt = -6\,A \int_0^1 f(t)\,t\,dt - 2\,B \int_0^1 f(t)\,dt. \tag{$*$}$$

We now observe that, by using an integration by parts, we can rewrite

$$\begin{aligned}\int_0^1 \left(\int_0^\tau \varphi(s)\,ds\right) d\tau &= \left[\tau \int_0^\tau \varphi(s)\,ds\right]_0^1 - \int_0^1 \tau\,\varphi(\tau)\,d\tau \\ &= \int_0^1 (1-\tau)\,\varphi(\tau)\,d\tau.\end{aligned}$$

By using this fact in the definitions of both A and B, we can write

$$A = \int_0^1 (1 - 2\,\tau)\,\varphi(\tau)\,d\tau \qquad \text{and} \qquad B = \int_0^1 (3\,\tau - 2)\,\varphi(\tau)\,d\tau.$$

If we now introduce the two constants

$$\alpha = -6 \int_0^1 f(t)\,t\,dt \qquad \text{and} \qquad \beta = -2 \int_0^1 f(t)\,dt,$$

and use the expressions for A and B obtained above, from $(*)$ we get

$$\int_0^1 f(t)\,\varphi(t)\,dt = \alpha \int_0^1 (1-2\,t)\,\varphi(t)\,dt + \beta \int_0^1 (3\,t - 2)\,\varphi(t)\,dt.$$

In other words, by rearranging the terms, we obtain

$$\int_0^1 \Big[f(t) - (3\,\beta - 2\,\alpha)\,t - \alpha + 2\,\beta\Big]\,\varphi(t)\,dt = 0, \qquad \text{for every } \varphi \in C_0^\infty((0,1)).$$

By Lemma 1.4.1, we can finally infer that f must coincide with the affine function $t \mapsto (2\,\alpha - 3\,\beta)\,t + \alpha - 2\,\beta$.

For a general interval $[a, b]$, it is sufficient to use a change of variable: let us suppose that

$$\int_a^b f(t)\,\varphi''(t)\,dt = 0,$$

for every $\varphi \in C^2([a, b])$ as in the statement. We now take any function $\psi \in C^2([0, 1])$ with $\psi(0) = \psi(1) = \psi'(0) = \psi'(1) = 0$ and use the previous identity for the function

$$\varphi(t) = \psi\left(\frac{t-a}{b-a}\right), \qquad \text{for } t \in [a, b].$$

We then get

$$0 = \int_a^b f(t)\,\frac{1}{(b-a)^2}\,\psi''\left(\frac{t-a}{b-a}\right)\,dt = \frac{1}{b-a}\int_0^1 f((b-a)\,s + a)\,\psi''(s)\,ds,$$

where we used the change of variable $t = (b - a)\,s + a$. By the first part of the proof, we get that

$$\widetilde{f}(s) := f((b-a)\,s + a), \qquad \text{for } s \in [0, 1],$$

is an affine function of the variable s, on the interval $[0, 1]$. By going back to the original variable t, we get the desired conclusion.

1.7.15 We can proceed exactly as in the proof of Lemma 1.4.1. We argue by contradiction and assume that there exists a point $\xi \in \Omega$ such that $f(\xi) < 0$. By continuity of f, there exists a ball $B_r(\xi) \subseteq \Omega$ such that

$$f(x) < 0, \qquad \text{for every } x \in B_r(\xi).$$

We now take again the function

$$\rho_r(x) = \rho\left(\frac{x-\xi}{r}\right),$$

where ρ is the standard mollifier. By assumption, we have

$$0 \leq \int_{\Omega} f(x)\, \rho_r(x)\, dx = \int_{B_r(\xi)} f(x)\, \rho_r(x)\, dx.$$

Since f is strictly negative in $B_r(\xi)$ and ρ_r is strictly positive there, we get the desired contradiction.

1.7.16 We take $\varphi \in C_0^\infty((-1, 1))$ and consider for $\varepsilon \in \mathbb{R}$

$$\mathcal{F}(u + \varepsilon\, \varphi) = \frac{1}{4} \int_{-1}^{1} |u' + \varepsilon\, \varphi'|^4\, dt - \int_{-1}^{1} f\, u\, dt - \varepsilon \int_{-1}^{1} f\, \varphi\, dt.$$

At first, we need to show that we can differentiate this quantity with respect to ε, at $\varepsilon = 0$. Then, by imposing that this derivative vanishes, we would get the weak formulation of the Euler-Lagrange equation.

Observe that

$$\varepsilon \mapsto \int_{-1}^{1} f\, u\, dt + \varepsilon \int_{-1}^{1} f\, \varphi\, dt,$$

thus this part of the functional is obviously differentiable with respect to ε, with (constant) derivative given by

$$\int_{-1}^{1} f\, \varphi\, dt.$$

For the integral containing the first derivative of u, we observe that by the Fundamental Theorem of Calculus,[2] for every $t \in [-1, 1]$ and every ε, we have

$$|u'(t) + \varepsilon\, \varphi'(t)|^4 = |u'(t)|^4 + 4\, \varepsilon\, \varphi'(t) \int_0^1 |u'(t) + s\, \varepsilon\, \varphi'(t)|^2\, (u'(t) + s\, \varepsilon\, \varphi'(t))\, ds.$$

We thus get for $t \in [-1, 1]$ and $|\varepsilon| < 1$

[2] We use that

$$g(1) = g(0) + \int_0^1 g'(s)\, ds,$$

where, for every fixed $t \in [-1, 1]$ and $\varepsilon \in \mathbb{R}$, the function g is given by

$$g(s) = |u'(t) + s\, \varepsilon\, \varphi'(t)|^4.$$

$$\begin{aligned}\left|\frac{|u'(t)+\varepsilon\,\varphi'(t)|^4-|u'(t)|^4}{\varepsilon}\right| &\le 4\,|\varphi'(t)|\int_0^1 |u'(t)+s\,\varepsilon\,\varphi'(t)|^3\,ds\\ &\le 4\,|\varphi'(t)|\int_0^1 \Big(|u'(t)|+s\,|\varepsilon|\,|\varphi'(t)|\Big)^3\,ds\\ &\le 16\,|\varphi'(t)|\int_0^1 \Big(|u'(t)|^3+s^3\,|\varphi'(t)|^3\Big)\,ds\\ &= 16\,|\varphi'(t)|\left[|u'(t)|^3+\frac{1}{4}\,|\varphi'(t)|^3\right].\end{aligned}$$

In the third inequality we used Problem 1.7.2 with $\alpha = 3$. Observe that the last term is in $L^1([-1,1])$, since $u \in C^1([-1,1])$ and $\varphi \in C_0^\infty((-1,1))$. Moreover, it is independent of ε. We can thus apply the Dominated Convergence Theorem, to infer that

$$\lim_{\varepsilon\to 0}\frac{1}{\varepsilon}\left[\frac{1}{4}\int_{-1}^1 |u'+\varepsilon\,\varphi'|^4\,dt-\frac{1}{4}\int_{-1}^1 |u'|^4\,dt\right]=\int_{-1}^1 |u'|^2\,u'\,\varphi'\,dt.$$

This shows that $\mathcal{F}(u+\varepsilon\,\varphi)$ is differentiable at $\varepsilon = 0$, with derivative given by

$$\int_{-1}^1 |u'|^2\,u'\,\varphi'\,dt-\int_{-1}^1 f\,\varphi\,dt.$$

We thus have computed the first variation of our functional at a point u, which is given by

$$\delta\mathcal{F}(u)[\varphi]=\int_{-1}^1 |u'|^2\,u'\,\varphi'\,dt-\int_{-1}^1 f\,\varphi\,dt.$$

The weak formulation of the Euler-Lagrange equation is thus

$$\delta\mathcal{F}(u)[\varphi]=0,\qquad \text{for every } \varphi\in C_0^\infty((-1,1)),$$

that is

$$\int_{-1}^1 |u'|^2\,u'\,\varphi'\,dt-\int_{-1}^1 f\,\varphi\,dt=0,\qquad \text{for every } \varphi\in C_0^\infty((-1,1)).$$

Finally, in order to find the classical formulation, it is sufficient to use an integration by parts in the first integral, to get

$$-\int_{-1}^1 (|u'|^2\,u')'\,\varphi\,dt-\int_{-1}^1 f\,\varphi\,dt=0,\qquad \text{for every } \varphi\in C_0^\infty((-1,1)).$$

By the Du Bois-Reymond Lemma (i.e. Lemma 1.4.1), we thus get

$$-\left(|u'|^2\,u'\right)' = f, \qquad \text{in } (-1,1).$$

Of course, the last integration by parts can be performed only under the additional regularity property $|u'|^2\,u' \in C^1([-1,1])$.

1.7.17 We first try to guess a solution, by using some formal manipulations. Then we will rigorously verify that our guess is actually a weak solution. Our equation now reads

$$-\left(|u'|^2\,u'\right)' = 1, \qquad \text{in } (-1,1).$$

Taking a primitive, we look for a solution u such that

$$|u'|^2\,u' = -C - t, \qquad \text{in } (-1,1), \tag{$*$}$$

for a suitable constant C. We now observe that the function

$$J(t) = |t|^2\,t = t^3,$$

is strictly monotone increasing. Its inverse function is given by

$$J^{-1}(t) = t^{\frac{1}{3}}.$$

From $(*)$, we then get

$$u'(t) = J^{-1}(-C - t) = -(C + t)^{\frac{1}{3}}, \qquad \text{for } t \in (-1,1).$$

We integrate this identity on $(-1, x)$ and take into account that we have to impose the condition $u(-1) = 0$. We obtain

$$u(x) = -\int_{-1}^{x} (C + t)^{\frac{1}{3}}\,dt = -\frac{3}{4}\left[(C + t)^{\frac{4}{3}}\right]_{-1}^{x},$$

that is

$$u(x) = \frac{3}{4}\left((C - 1)^{\frac{4}{3}} - (C + x)^{\frac{4}{3}}\right).$$

In order to complete our guess, we still have to determine the constant C. By imposing the further condition $u(1) = 0$, that is

$$\frac{3}{4}\left((C - 1)^{\frac{4}{3}} - (C + 1)^{\frac{4}{3}}\right) = 0,$$

we get that $C = 0$. We thus obtain

$$u(x) = \frac{3}{4}\left(1 - x^{\frac{4}{3}}\right), \qquad \text{for } x \in [-1, 1].$$

We now verify that this is a true weak solution of our Euler-Lagrange equation. We have

$$u'(x) = -x^{\frac{1}{3}} \qquad \text{so that} \qquad |u'(x)|^2\, u'(x) = -x.$$

Then, for every $\varphi \in C_0^\infty((-1, 1))$, we have

$$\begin{aligned}\int_{-1}^{1} |u'|^2\, u'\, \varphi'\, dx - \int_{-1}^{1} \varphi\, dx &= \int_{-1}^{1} -x\, \varphi'\, dx - \int_{-1}^{1} \varphi\, dx \\ &= \Big[-x\,\varphi\Big]_{-1}^{1} + \int_{-1}^{1} \varphi\, dx - \int_{-1}^{1} \varphi\, dx = 0.\end{aligned}$$

This shows that u is a weak solution. Also observe that $u \notin C^2((-1, 1))$, since u' is not differentiable at $x = 0$.

As for uniqueness, let us suppose that $v \in C^1([-1, 1])$ is another weak solution of our equation, with $v(-1) = v(1) = 0$. We then have

$$\int_{-1}^{1} |u'|^2\, u'\, \varphi'\, dx - \int_{-1}^{1} \varphi\, dx = 0, \qquad \text{for every } \varphi \in C_0^\infty((-1, 1)),$$

and

$$\int_{-1}^{1} |v'|^2\, v'\, \varphi'\, dx - \int_{-1}^{1} \varphi\, dx = 0, \qquad \text{for every } \varphi \in C_0^\infty((-1, 1)).$$

By subtracting these two equations, we obtain

$$\int_{-1}^{1} \left(|u'|^2\, u' - |v'|^2\, v'\right) \varphi'\, dx = 0, \qquad \text{for every } \varphi \in C_0^\infty((-1, 1)).$$

By using Problem 1.7.12, we then get that $|u'|^2\, u' - |v'|^2\, v'$ is constant on $[-1, 1]$. By recalling that $|u'(x)|^2\, u'(x) = -x$, we then get

$$|v'(x)|^2\, v'(x) = -C - x, \qquad \text{for } x \in [-1, 1],$$

for a constant C. By starting from this identity and proceeding as above, we then end up with $u = v$.

1.7.18 We have to show that for every $u \in C^1([-1,1])$ and $\varphi \in C_0^\infty((-1,1))$, the function

$$\varepsilon \mapsto \mathcal{F}(u+\varepsilon\,\varphi) = \frac{1}{4}\int_{-1}^{1} |u' + \varepsilon\,\varphi'|^4\,dt - \frac{1}{2}\int_{-1}^{1} |u+\varepsilon\,\varphi|^2\,dt,$$

is differentiable at $\varepsilon = 0$ and compute the derivative. For the first term, we can proceed as in Problem 1.7.16. For the second term, we can simply expand the square

$$\frac{1}{2}\int_{-1}^{1} |u+\varepsilon\,\varphi|^2\,dt = \frac{1}{2}\int_{-1}^{1} |u|^2\,dt + \varepsilon\int_{-1}^{1} u\,\varphi\,dt + \frac{\varepsilon}{2}\int_{-1}^{1} |\varphi|^2\,dt,$$

which is just a second order polynomial in ε. As such, it is clearly differentiable and its derivative computed at $\varepsilon = 0$ is given by

$$\int_{-1}^{1} u\,\varphi\,dt.$$

We thus get

$$\delta\mathcal{F}(u)[\varphi] = \int_{-1}^{1} |u'|^2\,u'\,\varphi'\,dt - \int_{-1}^{1} u\,\varphi\,dt, \qquad \text{for every } \varphi \in C_0^\infty((-1,1)).$$

The Euler-Lagrange equation in weak form is then given by

$$\int_{-1}^{1} |u'|^2\,u'\,\varphi'\,dt - \int_{-1}^{1} u\,\varphi\,dt = 0, \qquad \text{for every } \varphi \in C_0^\infty((-1,1)).$$

An integration by parts and the Du Bois-Reymond lemma give the classical formulation

$$-\left(|u'|^2\,u'\right)' = u, \qquad \text{in } (-1,1).$$

This concludes the exercise.

1.7.19 We start by writing the weak formulation of our equation: it is sufficient to multiply it by $\varphi(t)$, for $\varphi \in C_0^\infty((0,1))$, then integrate over $(0,1)$ and perform an integration by parts. Namely, we have

$$\begin{aligned}\int_0^1 |u|^5\,u\,\varphi\,dt &= -\int_0^1 \left(t\,u'' + u'\right)\varphi\,dt\\ &= -\int_0^1 \left(t\,u'\right)'\varphi\,dt = \int_0^1 t\,u'\,\varphi'\,dt.\end{aligned}$$

The weak formulation is thus given by

$$\int_0^1 t\,u'\,\varphi'\,dt - \int_0^1 |u|^5\,u\,\varphi\,dt = 0, \qquad \text{for every } \varphi \in C_0^\infty((0,1)).$$

We now have to find a functional $\mathcal{F}: C^1([0,1]) \to \mathbb{R}$ such that its first variation is given by

$$\delta\mathcal{F}(u)[\varphi] = \int_0^1 t\,u'\,\varphi'\,dt - \int_0^1 |u|^5\,u\,\varphi\,dt, \qquad \text{for every } \varphi \in C_0^\infty((0,1)).$$

One can verify that by taking

$$\mathcal{F}(u) = \frac{1}{2}\int_0^1 t\,|u'|^2\,dt - \frac{1}{7}\int_0^1 |u|^7\,dt,$$

we obtain the desired conclusion.

1.7.20 It is not difficult to see that the functional

$$\mathcal{F}(u) = \frac{1}{p}\int_{-1}^1 |u'|^p\,dt - \int_{-1}^1 u\,dt,$$

is convex on $C^1([-1,1])$, in the sense that

$$\mathcal{F}(\lambda\,u + (1-\lambda)\,v) \le \lambda\,\mathcal{F}(u) + (1-\lambda)\,\mathcal{F}(v),$$

for every $\lambda \in [0,1]$ and every $u, v \in C^1([-1,1])$. This follows from the convexity of the function $\tau \mapsto |\tau|^p$ for $1 < p < \infty$ and the linearity of the term

$$u \mapsto -\int_{-1}^1 u\,dt.$$

The heuristic idea to show existence of a solution is then quite easy: try to find a critical point of $\mathcal{F}$, in the sense of Definition 1.5.8. Then we can hope to exploit the idea that "*for a convex function, critical points are actually minimum points*", as in the Dirichlet principle (i.e. Theorem 1.5.3). The only difference with the Dirichlet principle of Theorem 1.5.3 is that we enlarged the class of admissible functions to $C^1([-1,1])$, rather than $C^2([-1,1])$.

These were the heuristics, let us try to formalize them. We start by computing the first variation of $\mathcal{F}$: we need to show that, for every $u \in C^1([-1,1])$ and every $\varphi \in C_0^\infty((-1,1))$, the function of one real variable

$$\varepsilon \mapsto \mathcal{F}(u + \varepsilon\,\varphi),$$

is differentiable at $\varepsilon = 0$ and compute such a derivative. By repeating the same arguments of Problem 1.7.16 (which was concerned with the particular case $p = 4$), we thus find

$$\delta\mathcal{F}(u)[\varphi] = \int_{-1}^{1} |u'|^{p-2}\, u'\, \varphi'\, dt - \int_{-1}^{1} \varphi\, dt.$$

A critical point of $\mathcal{F}$ is a function $u \in C^1([-1, 1])$ such that

$$\int_{-1}^{1} |u'|^{p-2}\, u'\, \varphi'\, dt - \int_{-1}^{1} \varphi\, dt = 0, \qquad \text{for every } \varphi \in C_0^\infty((-1, 1)).$$

This is the Euler-Lagrange equation in weak form, which has to be coupled with the boundary conditions $u(-1) = u(1) = 0$. Let us now compute a weak solution u_0 of the Euler-Lagrange equation, with $u_0(-1) = u_0(1) = 0$. By proceeding exactly as in Problem 1.7.17 (which deals with the particular case $p = 4$), we find

$$u_0(t) = \frac{1}{p'}\left(1 - |t|^{p'}\right), \qquad \text{for } t \in [-1, 1].$$

We claim that u_0 is the desired minimizer. We take a function u admissible for our minimization problem and use the "above tangent" property (i.e. Proposition 1.2.3) for the function

$$f(\tau) = \frac{1}{p}\, |\tau|^p, \qquad \tau \in \mathbb{R},$$

as follows

$$\frac{1}{p}\, |u'(t)|^p \geq \frac{1}{p}\, |u_0'(t)|^p + |u_0'(t)|^{p-2}\, u_0'(t)\, (u'(t) - u_0'(t)), \qquad \text{for } t \in [-1, 1].$$

By integrating over $[-1, 1]$, we get

$$\frac{1}{p}\int_{-1}^{1} |u'|^p\, dt \geq \frac{1}{p}\int_{-1}^{1} |u_0'|^p\, dt + \int_{-1}^{1} |u_0'|^{p-2}\, u_0'\, (u' - u_0')\, dt. \qquad (*)$$

From the explicit expression of u_0, we get

$$|u_0'|^{p-2}\, u_0' = -t, \qquad \text{for } t \in [-1, 1],$$

thus by integrating by parts and using that $u(-1) = u_0(-1)$, $u(1) = u_0(1)$, we obtain

$$\int_{-1}^{1} |u_0'|^{p-2}\, u_0'\, (u' - u_0')\, dt = \int_{-1}^{1} -t\, (u' - u_0')\, dt = \int_{-1}^{1} (u - u_0)\, dt.$$

If we now insert this identity into (∗), we get

$$\frac{1}{p} \int_{-1}^{1} |u'|^p\, dt \geq \frac{1}{p} \int_{-1}^{1} |u_0'|^p\, dt + \int_{-1}^{1} (u - u_0)\, dt,$$

which can be rewritten as

$$\mathcal{F}(u) \geq \mathcal{F}(u_0).$$

In other words, the critical point u_0 is a minimizer. In order to show that u_0 is actually the *unique* minimizer, one can proceed by using the same argument as in Remark 1.5.2: observe that the function

$$F(u, z) = \frac{1}{p}\, |z|^p - u,$$

is convex and strictly convex in the variable z. Let us suppose that $v_0 \in C^1([-1, 1])$ is another minimizer, the function

$$w = \frac{u_0 + v_0}{2},$$

is still admissible for the minimization problem. By using the convexity, we obtain

$$\frac{1}{p}\, |w'|^p - w \leq \frac{1}{2} \left(\frac{1}{p}\, |u_0'|^p - u_0 \right) + \frac{1}{2} \left(\frac{1}{p}\, |v_0'|^p - v_0 \right). \qquad (**)$$

When integrated over $[-1, 1]$, we get

$$\frac{1}{p} \int_{-1}^{1} |w'|^p\, dt - \int_{-1}^{1} w\, dt \leq \frac{1}{2} \left[\frac{1}{p} \int_{-1}^{1} |u_0'|^p\, dt - \int_{-1}^{1} u_0\, dt \right] + \frac{1}{2} \left[\frac{1}{p} \int_{-1}^{1} |v_0'|^p\, dt - \int_{-1}^{1} v_0\, dt \right]$$

By using that both u_0 and v_0 are minimizers, we obtain that w is a minimizer, as well. Thus, we must have equality everywhere, in particular in the pointwise convexity inequality (∗∗). By strict convexity, this gives in particular that

$$u_0'(t) = v_0'(t), \qquad \text{for every } t \in [-1, 1],$$

Thus, $u_0 - v_0$ is a constant function on $[-1, 1]$. Since u_0 and v_0 coincide at the endpoints $t = -1$ and $t = 1$, we must have $u_0 = v_0$.

Finally, observe that u_0 is of class C^2 if and only if $p' \geq 2$ that is if and only if $1 < p \leq 2$.

1.7.21 We observe that the boundary datum $U(x, y) = x^2 - y^2$ is actually a harmonic function. Moreover, observe that

$$\nabla U(x, y) = (2\,x, -2\,y),$$

so that

$$\int_{B_1(0)} |\nabla U|^2\, dx\, dy = 4 \int_{B_1(0)} (x^2 + y^2)\, dx\, dy = 4\,\mathcal{I}.$$

In order to conclude, we thus have to prove that

$$\int_{B_1(0)} |\nabla u|^2\, dx\, dy \geq \int_{B_1(0)} |\nabla U|^2\, dx\, dy,$$
$$\text{for } u \in C^1(\overline{B_1(0)}) \text{ s. t. } u = U \text{ on } \partial B_1(0).$$

In other words, we need to prove that the harmonic function U solves the following minimization problem

$$\inf_{u \in C^1(\overline{B_1(0)})} \left\{ \int_{B_1(0)} |\nabla u|^2\, dx\, dy \ :\ u = U \text{ on } \partial B_1(0) \right\}.$$

As in the previous problem, the only difference with the Dirichlet principle of Theorem 1.5.1 is that we enlarged the class of admissible functions to $C^1(\overline{B_1(0)})$. By coping verbatim the proof of Theorem 1.5.1, we observe that for every admissible $u \in C^1(\overline{B_1(0)})$ we have

$$-\,\Delta U(x, y)\,(u(x, y) - U(x, y)) = 0, \qquad \text{for } (x, y) \in B_1(0).$$

We integrate over $B_1(0)$ and use the Divergence Theorem (in conjunction with the fact that $u = U$ on $\partial B_1(0)$), so to get

$$\int_{B_1(0)} \langle \nabla U, \nabla u - \nabla U \rangle\, dx\, dy = 0.$$

By using this identity and the "above tangent" property (i.e. Proposition 1.3.4) for the convex function $z \mapsto |z|^2$, we get

$$\begin{aligned}\int_{B_1(0)} |\nabla u|^2\, dx\, dy &\geq \int_{B_1(0)} |\nabla U|^2\, dx\, dy + 2\int_{B_1(0)} \langle \nabla U, \nabla u - \nabla U\rangle\, dx\, dy\\ &= \int_{B_1(0)} |\nabla U|^2\, dx\, dy.\end{aligned}$$

This proves the desired minimality of U.

1.7.22 We first consider the case $p = \infty$, which is simpler. In this case we have

$$\int_E |f|^r\, dx = \int_E |f|^{r-q}\, |f|^q\, dx \leq \|f\|_{L^\infty(E)}^{r-q} \int_E |f|^q\, dx,$$

which implies

$$\|f\|_{L^r(E)} = \left(\int_E |f|^r\, dx\right)^{\frac{1}{r}} \leq \|f\|_{L^\infty(E)}^{\frac{r-q}{r}}\, \|f\|_{L^q(E)}^{\frac{q}{r}}.$$

This proves (1.7.4) for $p = \infty$.
We now suppose $p < \infty$. We observe that if $q < r < p$ then there exists $\alpha \in (0, 1)$ such that

$$r = (1-\alpha)\, q + \alpha\, p.$$

With a simple computation, we find

$$\alpha = \frac{r-q}{p-q}.$$

We now write

$$\int_E |f|^r\, dx = \int_E |f|^{\alpha\, p}\, |f|^{q\,(1-\alpha)}\, dx,$$

then we use Hölder's inequality with conjugate exponents $1/\alpha$ and $1/(1-\alpha)$. This gives

$$\int_E |f|^r\, dx \leq \left(\int_E |f|^p\, dx\right)^\alpha \left(\int_E |f|^q\, dx\right)^{1-\alpha}.$$

By taking the power $1/r$ on both sides and recalling the definition of α, we get the desired conclusion (1.7.4) for $p < \infty$, as well.

1.7.23 We first observe that we can suppose

$$\int_E |f_i|^k\, dx \neq 0, \qquad \text{for every } i \in \{1, \dots, k\},$$

otherwise there is nothing to prove. We then introduce the functions

$$g_i(x) = \frac{f_i(x)}{\|f_i\|_{L^k(E)}}, \qquad \text{for every } i \in \{1, \dots, k\}.$$

We apply Problem 1.7.10 with the choice

$$a_i = |g_i(x)|,$$

so to get

$$\prod_{i=1}^{k} |g_i(x)| \le \frac{1}{k} \sum_{i=1}^{k} |g_i(x)|^k, \qquad \text{for a. e. } x \in E.$$

We integrate this inequality over E: this gives

$$\int_E \prod_{i=1}^{k} |g_i|\, dx \le \frac{1}{k} \sum_{i=1}^{k} \int_E |g_i|^k\, dx.$$

By recalling the definition of g_i, we have

$$\int_E |g_i|^k\, dx = \frac{\|f_i\|_{L^k(E)}^k}{\|f_i\|_{L^k(E)}^k} = 1, \qquad \text{for every } i \in \{1, \dots, k\}.$$

Thus, we obtain

$$\int_E \prod_{i=1}^{k} |g_i|\, dx \le \frac{1}{k} \sum_{i=1}^{k} \int_E |g_i|^k\, dx = 1.$$

Finally, by using again the definition of g_i, the latter is the same as

$$\int_E \frac{\prod_{i=1}^{k} |f_i|}{\prod_{i=1}^{k} \|f_i\|_{L^k(E)}}\, dx \le 1 \quad \text{that is} \quad \int_E \prod_{i=1}^{k} |f_i|\, dx \le \prod_{i=1}^{k} \|f_i\|_{L^k(E)},$$

as desired.

1.7.24 This technical result is due to Emilio Gagliardo (see [75]) and it will be expedient in order to prove one of the main results of Chap. 3: the Sobolev inequality.

We prove (1.7.5) by induction over N. For $N = 2$, inequality (1.7.5) turns into an equality: indeed, it simply follows from the Fubini-Tonelli Theorem that

$$\int_{\mathbb{R}^2} \Phi_1\, \Phi_2\, dx = \int_{\mathbb{R}} \left(\int_{\mathbb{R}} \Phi_1\, \Phi_2\, dx_1 \right) dx_2$$
$$= \int_{\mathbb{R}} \Phi_1 \left(\int_{\mathbb{R}} \Phi_2\, dx_1 \right) dx_2 = \left(\int_{\mathbb{R}} \Phi_1\, dx_2 \right) \left(\int_{\mathbb{R}} \Phi_2\, dx_1 \right).$$

We used that Φ_1 only depends on x_2, while Φ_2 only depends on x_1.

Let us now assume that (1.7.5) is true for a given $N \geq 2$ and consider the case $N + 1$. Again by the Fubini-Tonelli Theorem, we have

$$\int_{\mathbb{R}^{N+1}} \prod_{k=1}^{N+1} \Phi_k\, dx = \int_{\mathbb{R}^N} \Phi_{N+1} \left(\int_{\mathbb{R}} \prod_{k=1}^{N} \Phi_k\, dx_{N+1} \right) d\widehat{x}_{N+1}$$
$$\leq \left(\int_{\mathbb{R}^N} \Phi_{N+1}^N\, d\widehat{x}_{N+1} \right)^{\frac{1}{N}}$$
$$\times \left(\int_{\mathbb{R}^N} \left(\int_{\mathbb{R}} \prod_{k=1}^{N} \Phi_k\, dx_{N+1} \right)^{\frac{N}{N-1}} d\widehat{x}_{N+1} \right)^{\frac{N-1}{N}},$$

where we used Hölder's inequality, with exponents N and $N/(N-1)$. In the second integral, we apply the generalized Hölder inequality (see Problem 1.7.23)

$$\left(\int_{\mathbb{R}} \prod_{k=1}^{N} \Phi_k\, dx_{N+1} \right)^{\frac{N}{N-1}} \leq \prod_{k=1}^{N} \left(\int_{\mathbb{R}} \Phi_k^N\, dx_{N+1} \right)^{\frac{1}{N-1}}.$$

We thus have obtained

$$\int_{\mathbb{R}^{N+1}} \prod_{k=1}^{N+1} \Phi_k\, dx \leq \left(\int_{\mathbb{R}^N} \Phi_{N+1}^N\, d\widehat{x}_{N+1} \right)^{\frac{1}{N}}$$
$$\times \left(\int_{\mathbb{R}^N} \prod_{k=1}^{N} \left(\int_{\mathbb{R}} \Phi_k^N\, dx_{N+1} \right)^{\frac{1}{N-1}} d\widehat{x}_{N+1} \right)^{\frac{N-1}{N}}.$$

We can now apply the inductive assumption on the second integral, since it contains the product of the N functions

$$g_k = \left(\int_{\mathbb{R}} \Phi_k^N\, dx_{N+1} \right)^{\frac{1}{N-1}}, \qquad \text{for } k \in \{1, \dots, N\},$$

each one not depending on the x_k variable. This yields[3]

$$\int_{\mathbb{R}^{N+1}} \prod_{k=1}^{N+1} \Phi_k \, dx \le \left(\int_{\mathbb{R}^N} \Phi_{N+1}^N \, d\widehat{x}_{N+1} \right)^{\frac{1}{N}} \prod_{k=1}^{N} \left(\int_{\mathbb{R}^{N-1}} \left(\int_{\mathbb{R}} \Phi_k^N \, dx_{N+1} \right) d\widehat{x}_k \right)^{\frac{1}{N}}$$

$$= \left(\int_{\mathbb{R}^N} \Phi_{N+1}^N \, d\widehat{x}_{N+1} \right)^{\frac{1}{N}} \prod_{k=1}^{N} \left(\int_{\mathbb{R}^N} \Phi_k^N \, d\widehat{x}_k \right)^{\frac{1}{N}}$$

$$= \prod_{k=1}^{N+1} \left(\int_{\mathbb{R}^N} \Phi_k^N \, d\widehat{x}_k \right)^{\frac{1}{N}},$$

which is (1.7.5) for $N + 1$. This finally establishes the statement for every $N \ge 2$.

1.7.25 We deal with the case $q < \infty$, the extremal case $q = \infty$ being left to the reader. Let us show that $N_{p,q}$ verifies the properties of a norm. It is clear that $N_{p,q}$ is non-negative, moreover if $N_{p,q}(u, \phi) = 0$, we get

$$\|u\|_{L^p(E;\mathbb{R}^k)} = \|\phi\|_{L^p(E;\mathbb{R}^n)} = 0,$$

thanks to the fact that $\| \cdot \|_{\ell^q}$ is a norm on $\mathbb{R}^2$. In turn, the previous identity gives that both u and ϕ vanish almost everywhere in E.

For every $\lambda \in \mathbb{R}$ and $(u, \phi) \in L^p(E; \mathbb{R}^k) \times L^p(E; \mathbb{R}^n)$, we have

$$\begin{aligned} N_{p,q}(\lambda\, u, \lambda\, \phi) &= \left\| \left(\|\lambda\, u\|_{L^p(E;\mathbb{R}^k)}, \|\lambda\, \phi\|_{L^p(E;\mathbb{R}^n)} \right) \right\|_{\ell^q} \\ &= \left\| \left(|\lambda|\, \|u\|_{L^p(E;\mathbb{R}^k)}, |\lambda|\, \|\phi\|_{L^p(E;\mathbb{R}^n)} \right) \right\|_{\ell^q} \\ &= |\lambda| \left\| \left(\|u\|_{L^p(E;\mathbb{R}^k)}, \|\phi\|_{L^p(E;\mathbb{R}^n)} \right) \right\|_{\ell^q} = |\lambda|\, N_{p,q}(u, \phi). \end{aligned}$$

Finally, we show that $N_{p,q}$ satisfies the triangle inequality. To this aim, we first need to observe that the ℓ^q norm has the following property

$$\|(x_0, y_0)\|_{\ell^q} \le \|(x_1, y_1)\|_{\ell^q}, \qquad \text{for every } x_1 \ge x_0 \ge 0,\ y_1 \ge y_0 \ge 0, \qquad (*)$$

which follows immediately from its definition.

For every $(u, \phi), (v, \psi) \in L^p(E; \mathbb{R}^k) \times L^p(E; \mathbb{R}^n)$, we have

$$\begin{aligned} & N_{p,q}(u + v, \phi + \psi) \\ & \quad = \left\| \left(\|u + v\|_{L^p(E;\mathbb{R}^k)}, \|\phi + \psi\|_{L^p(E;\mathbb{R}^n)} \right) \right\|_{\ell^q} \\ & \quad \le \left\| \left(\|u\|_{L^p(E;\mathbb{R}^k)} + \|v\|_{L^p(E;\mathbb{R}^k)}, \|\phi\|_{L^p(E;\mathbb{R}^n)} + \|\psi\|_{L^p(E;\mathbb{R}^n)} \right) \right\|_{\ell^q}, \end{aligned}$$

[3] Observe that in $\mathbb{R}^{N+1}$, we have $d\widehat{x}_{N+1} = dx_1 \dots dx_N$. Accordingly, here by the symbol $d\widehat{x}_k$ we mean the $(N-1)$-dimensional element volume, obtained by "deleting" dx_k from $d\widehat{x}_{N+1}$.

where we used the triangle inequality for the L^p norms and the property $(*)$. If we now use the triangle inequality for the ℓ^q norm, we finally obtain

$$\begin{aligned}
&N_{p,q}(u+v,\phi+\psi)\\
&\le \left\|\left(\|u\|_{L^p(E;\mathbb{R}^k)},\|\phi\|_{L^p(E;\mathbb{R}^n)}\right)\right\|_{\ell^q}\\
&\quad+\left\|\left(\|v\|_{L^p(E;\mathbb{R}^k)},\|\psi\|_{L^p(E;\mathbb{R}^n)}\right)\right\|_{\ell^q}=N_{p,q}(u,\phi)+N_{p,q}(v,\psi),
\end{aligned}$$

as desired.

In order to prove the last point, we observe that for $1<q<\infty$ we have by Problem 1.7.1

$$N_{p,q}(u,\phi)=\left(\|u\|^q_{L^p(E;\mathbb{R}^k)}+\|\phi\|^q_{L^p(E;\mathbb{R}^n)}\right)^{\frac{1}{q}}\le\|u\|_{L^p(E;\mathbb{R}^k)}+\|\phi\|_{L^p(E;\mathbb{R}^n)},$$

and

$$\begin{aligned}
N_{p,q}(u,\phi)&=\left(\|u\|^q_{L^p(E;\mathbb{R}^k)}+\|\phi\|^q_{L^p(E;\mathbb{R}^n)}\right)^{\frac{1}{q}}\\
&\ge 2^{\frac{1}{q}-1}\left(\|u\|_{L^p(E;\mathbb{R}^k)}+\|\phi\|_{L^p(E;\mathbb{R}^n)}\right).
\end{aligned}$$

For the extremal case $q=\infty$, by observing that

$$N_{p,\infty}(u,\phi)=\max\left\{\|u\|_{L^p(E;\mathbb{R}^k)},\|\phi\|_{L^p(E;\mathbb{R}^n)}\right\},$$

and using that

$$\frac{a+b}{2}\le\max\{a,b\}\le a+b,\qquad\text{for every } a,b\ge 0,$$

we get the conclusion.

1.7.26 We observe that $\partial\Omega$ is a closed subset of $\mathbb{R}^N$ and it is not empty, by assumption. Moreover, for every fixed $x\in\Omega$, we have by the triangle inequality

$$|x-y|\ge|y|-|x|.$$

Thus, if for every $n\in\mathbb{N}$ we take $y_n\in\partial\Omega$ such that

$$|y_n-x|\le\operatorname{dist}(x,\partial\Omega)+\frac{1}{n+1},$$

the previous estimate entails that $\{y_n\}_{n\in\mathbb{N}}\subseteq\partial\Omega$ is a bounded sequence. Thus, up to a subsequence, it converges to some $\overline{y}$, which still belongs to $\partial\Omega$, since the latter is closed. This finally gives

$$|\overline{y} - x| = \lim_{n\to\infty} |y_n - x| \leq \lim_{n\to\infty} \left(\operatorname{dist}(x, \partial\Omega) + \frac{1}{n+1}\right) = \operatorname{dist}(x, \partial\Omega),$$

which proves that $\overline{y}$ attains the infimum in the definition of $\operatorname{dist}(x, \partial\Omega)$. By arbitrariness of $x \in \Omega$, we get the first statement.

In order to deal with the second statement, we first notice that $x \mapsto \operatorname{dist}(x, \partial\Omega)$ is a continuous function over Ω. Indeed, let us fix $x \in \Omega$ and take a sequence $\{x_n\}_{n\in\mathbb{N}} \subseteq \Omega$ such that

$$\lim_{n\to\infty} |x_n - x| = 0.$$

Correspondingly, we take $\{y_n\}_{n\in\mathbb{N}} \subseteq \partial\Omega$ such that

$$\operatorname{dist}(x_n, \partial\Omega) = |x_n - y_n|.$$

By using the definition and the triangle inequality, we get for every $n \in \mathbb{N}$

$$\operatorname{dist}(x, \partial\Omega) \leq |x - y_n| \leq |x - x_n| + |y_n - x_n| = |x - x_n| + \operatorname{dist}(x_n, \partial\Omega).$$

By using that $\{x_n\}_{n\in\mathbb{N}}$ converges to x, we thus obtain

$$\operatorname{dist}(x, \partial\Omega) \leq \liminf_{n\to\infty} \operatorname{dist}(x_n, \partial\Omega).$$

Similarly, for every $z \in \partial\Omega$ we also have

$$\limsup_{n\to\infty} \operatorname{dist}(x_n, \partial\Omega) \leq \lim_{n\to\infty} |x_n - z| = |x - z|,$$

and by taking the minimum over $z \in \partial\Omega$, we get

$$\limsup_{n\to\infty} \operatorname{dist}(x_n, \partial\Omega) \leq \operatorname{dist}(x, \partial\Omega),$$

as well. This finally shows the asserted continuity[4] of the distance function.

We now come to the proof of the claimed limit. We observe that

$$\operatorname{int}_n(\Omega) \subseteq \operatorname{int}_{n+1}(\Omega), \qquad \text{for every } n \in \mathbb{N},$$

[4] This could also be deduced by observing that, by its very definition, the distance function is an infimum of non-negative *Lipschitz continuous functions*, see Proposition 5.1.8 and Problem 5.5.7. We prefer here to give an elementary proof.

and each $\mathrm{int}_n(\Omega)$ is an open set, thanks to the continuity of the distance function. Thus, each $\mathrm{int}_n(\Omega)$ is measurable, which entails that

$$\bigcup_{n=1}^{\infty} \mathrm{int}_n(\Omega),$$

is measurable, as well. By the basic properties of the Lebesgue measure, we have

$$\lim_{n\to\infty} |\mathrm{int}_n(\Omega)| = \left|\bigcup_{n=1}^{\infty} \mathrm{int}_n(\Omega)\right|.$$

We now observe that

$$\bigcup_{n=1}^{\infty} \mathrm{int}_n(\Omega) = \Omega,$$

thus, the last two equations show that

$$\lim_{n\to\infty} |\mathrm{int}_n(\Omega)| = |\Omega|.$$

Moreover, we have that for every $n \in \mathbb{N}$ the set Ω can be written as the disjoint union

$$\Omega = \Big(\Omega \setminus \mathrm{int}_n(\Omega)\Big) \cup \mathrm{int}_n(\Omega),$$

and by the additivity of the Lebesgue measure, this gives

$$|\Omega \setminus \mathrm{int}_n(\Omega)| = |\Omega| - |\mathrm{int}_n(\Omega)|.$$

The desired conclusion now follows.

1.7.27 We first observe that if $\Omega = \mathbb{R}^N$, it is sufficient to take

$$\Omega_k := B_k(0) = \{x \in \mathbb{R}^N\,:\, |x| < k\}, \qquad \text{for every } k \in \mathbb{N}.$$

Thus, we can assume that $\Omega \neq \mathbb{R}^N$. We start by fixing $x_0 \in \Omega$ and setting $\delta_0 = \mathrm{dist}(x_0, \partial\Omega)$. Observe that since Ω is an open set, we have $\delta_0 > 0$. We then define for every $k \in \mathbb{N}$ the set

$$\mathcal{O}_k = \left\{x \in \Omega\,:\, |x - x_0| < k + 2 \quad \text{and} \quad \mathrm{dist}(x, \partial\Omega) > \frac{\delta_0}{k+2}\right\}.$$

This is an open bounded set which is not empty, thanks to the fact that $x_0 \in \mathcal{O}_k$, by construction. Moreover, we have

$$O_k \subseteq O_{k+1} \qquad \text{and} \qquad O_k \Subset \Omega, \qquad \text{for every } k \in \mathbb{N},$$

and also

$$\Omega = \bigcup_{k \in \mathbb{N}} O_k. \tag{$*$}$$

In order to show the last property, we take $y \in \Omega$, then we have $\mathrm{dist}(y, \partial\Omega) > 0$. In particular, by choosing $k_0 \in \mathbb{N}$ such that

$$k_0 > \max\left\{\frac{\delta_0}{\mathrm{dist}(y, \partial\Omega)}, |x_0 - y|\right\} - 2,$$

we have

$$|x_0 - y| < k + 2 \qquad \text{and} \qquad \mathrm{dist}(y, \partial\Omega) > \frac{\delta_0}{k+2}, \qquad \text{for every } k \geq k_0.$$

This means that $y \in O_k$ for every $k \geq k_0$ and thus in particular

$$y \in \bigcup_{k \in \mathbb{N}} O_k.$$

By arbitrariness of $y \in \Omega$, we thus get

$$\Omega \subseteq \bigcup_{k \in \mathbb{N}} O_k.$$

The reverse inclusion is trivial by the definition of each O_k and thus we get $(*)$.

However, in general the sets O_k are not connected. To fix this issue and produce the required exhaustion of the connected set Ω, we take

$$\Omega_k = \text{"the connected component of } O_k \text{ containing } x_0\text{"}.$$

We recall that for an open subset of the Euclidean space, such a connected component is still open (see for example [60, Corollary 27.10]). Moreover, by construction we still have

$$\Omega_k \subseteq \Omega_{k+1} \qquad \text{and} \qquad \Omega_k \Subset \Omega, \qquad \text{for every } k \in \mathbb{N}.$$

We are left with proving that we still have

$$\Omega = \bigcup_{k \in \mathbb{N}} \Omega_k. \tag{$**$}$$

Let us take $y \in \Omega$, by connectedness of Ω we know that there exists a continuous curve[5] $\gamma : [0, 1] \to \Omega$ such that

$$\gamma(0) = x_0 \qquad \text{and} \qquad \gamma(1) = y.$$

Since $\gamma([0, 1])$ is a compact set contained in the open set Ω, we know that

$$\max_{t\in[0,1]} |x_0 - \gamma(t)| < +\infty \qquad \text{and} \qquad \min_{t\in[0,1]} \operatorname{dist}(\gamma(t), \partial\Omega) > 0.$$

Thus, by $(*)$ there exists $k_0 \in \mathbb{N}$ such that

$$\gamma([0, 1]) \subseteq \mathcal{O}_k, \qquad \text{for every } k \geq k_0.$$

However, $\gamma([0, 1])$ is a closed connected set containing x_0, thus by definition of connected component and by the previous property, we must have

$$\gamma([0, 1]) \subseteq \Omega_k, \qquad \text{for every } k \geq k_0,$$

as well. This is enough to get the conclusion $(**)$.

1.7.28 We first observe that $L^1(E) \cap L^\infty(E) \subseteq L^p(E)$, thanks to Problem 1.7.22. Let us now take $1 \leq p < \infty$, at first we prove the statement for $u \in L^p(E)$ *non-negative* almost everywhere in E. In this case, we define for every $n \in \mathbb{N}$ the following sequence of functions

$$u_n = \min\{u, n\}\, 1_{B_n(0)\cap E}.$$

It is easily seen that

$$\|u_n\|_{L^\infty(E)} \leq n \qquad \text{and} \qquad \|u_n\|_{L^1(E)} \leq n\, |B_n(0)|,$$

thus $u_n \in L^1(E) \cap L^\infty(E)$. Moreover, by construction we have

$$u_n(x) \leq u_{n+1}(x) \leq u(x), \qquad \text{for a. e. } x \in E,$$

and

$$\lim_{n\to\infty} u_n(x) = u(x), \qquad \text{for a. e. } x \in E.$$

Thus, by using the Monotone Convergence Theorem, we have

[5] We are using the following well-known fact: *an open set $\Omega \subseteq \mathbb{R}^N$ is connected if and only if it is arcwise connected*, see for example [60, Corollary 27.6].

$$\lim_{n\to\infty} \int_E |u_n - u|^p \, dx = \lim_{n\to\infty} \int_E (u - u_n)^p \, dx = 0,$$

as desired.

Let us now assume that $u \in L^p(E)$ is such that both sets

$$\Big\{x \in E \, : \, u(x) > 0\Big\} \qquad \text{and} \qquad \Big\{x \in E \, : \, u(x) < 0\Big\},$$

have positive measure. We then set

$$u_+ = \max\{u, 0\} \qquad \text{and} \qquad u_- = \max\{-u, 0\},$$

so that we can write $u = u_+ - u_-$ almost everywhere in E. Both functions u_+, u_- are non-negative and belongs to $L^p(E)$. Thus, by the first part above, we know that there exists $\{\varphi_n\}_{n\in\mathbb{N}}, \{\psi_n\}_{n\in\mathbb{N}} \subseteq L^1(E) \cap L^\infty(E)$ such that

$$\lim_{n\to\infty} \int_E |\varphi_n - u_+|^p \, dx = \lim_{n\to\infty} \int_E |\psi_n - u_-|^p \, dx = 0.$$

Thus, by setting $u_n = \varphi_n - \psi_n$, we get by Minkowski's inequality

$$\begin{aligned}\lim_{n\to\infty} \left(\int_E |u_n - u|^p \, dx\right)^{\frac{1}{p}} &\le \lim_{n\to\infty} \left(\int_E |\varphi_n - u_+|^p \, dx\right)^{\frac{1}{p}} \\ &\quad + \lim_{n\to\infty} \left(\int_E |\psi_n - u_-|^p \, dx\right)^{\frac{1}{p}} = 0,\end{aligned}$$

as desired. This proves the claim.

Finally, we give a counterexample for the case L^∞: we take $E = \mathbb{R}$ and $u \equiv 1$. Let us suppose that there exists $\{u_n\}_{n\in\mathbb{N}} \subseteq L^1(\mathbb{R}) \cap L^\infty(\mathbb{R})$ such that

$$\lim_{n\to\infty} \|1 - u_n\|_{L^\infty(\mathbb{R})} = 0.$$

This in particular implies that there exists $n_0 \in \mathbb{N}$ such that

$$\|1 - u_n\|_{L^\infty(\mathbb{R})} < \frac{1}{2} \qquad \text{for every } n \ge n_0,$$

that is

$$\frac{1}{2} < u_n(x) < \frac{3}{2}, \qquad \text{for a. e. } x \in \mathbb{R} \text{ and for every } n \ge n_0.$$

The lower bound clearly contradicts the fact that $u_n \in L^1(\mathbb{R})$.

1.7.29 The main point is showing that the assertion is true, *without taking a subsequence*. To this aim, we will use that $L^1(E) \cap L^\infty(E)$ is dense in $L^{q'}(E)$, thanks to Problem 1.7.28. That is, for every $\varphi \in L^{q'}(E)$, there exists $\{\varphi_n\}_{n\in\mathbb{N}} \subseteq L^1(E) \cap L^\infty(E)$ such that

$$\lim_{n\to\infty} \|\varphi_n - \varphi\|_{L^{q'}(E)} = 0.$$

For every $n, k \in \mathbb{N}$, we have

$$\begin{aligned}\left|\int_E f_n\,\varphi\,dx - \int_E f\,\varphi\,dx\right| &\le \left|\int_E f_n\,(\varphi - \varphi_k)\,dx\right| \\ &\quad + \left|\int_E (f_n - f)\,\varphi_k\,dx\right| + \left|\int_E f\,(\varphi_k - \varphi)\,dx\right| \\ &\le \left(M + \|f\|_{L^q(E)}\right)\,\|\varphi_k - \varphi\|_{L^{q'}(E)} \\ &\quad + \left|\int_E (f_n - f)\,\varphi_k\,dx\right|.\end{aligned}$$

By observing that $\varphi_k \in L^1(E) \cap L^\infty(E) \subseteq L^{p'}(E)$ (thanks to Problem 1.7.22), by using the assumption on the weak convergence we get

$$\lim_{n\to\infty}\left|\int_E (f_n - f)\,\varphi_k\,dx\right| = 0.$$

Thus, the estimate above gives

$$\limsup_{n\to\infty}\left|\int_E f_n\,\varphi\,dx - \int_E f\,\varphi\,dx\right| \le \left(M + \|f\|_{L^q(E)}\right)\,\|\varphi_k - \varphi\|_{L^{q'}(E)},$$

which holds for every $k \in \mathbb{N}$. By taking the limit as k goes to ∞ and using that the last term converges to 0, we conclude.

1.7.30 As in Problem 1.7.28, for every $n \in \mathbb{N}$, we set

$$u_n = \min\{|u|,\, n\}\, 1_{B_n(0)\cap E}.$$

By construction, we have $u_n \in L^\infty(E)$ and $0 \le u_n \le |u|$ almost everywhere in E. Thus, we obtain

$$M \ge \limsup_{p\nearrow\infty} \|u\|_{L^p(E)} \ge \limsup_{p\nearrow\infty} \|u_n\|_{L^p(E)}.$$

On the other hand, we have

$$\|u_n\|_{L^p(E)} = \|u_n\|_{L^p(E\cap B_n(0))},$$

and since the set $E \cap B_n(0)$ is bounded, we know that (see for example [24, Chapter V, Proposition 4.2])

$$\lim_{p \nearrow \infty} \|u_n\|_{L^p(E\cap B_n(0))} = \|u_n\|_{L^\infty(E\cap B_n(0))} = \|u_n\|_{L^\infty(E)}.$$

Thus, by joining the two previous facts, we have obtained

$$\|u_n\|_{L^\infty(E)} \le M, \qquad \text{for every } n \in \mathbb{N}.$$

By recalling the definition of u_n, this means that for every $n \in \mathbb{N}$

$$\min\{|u(x)|,\, n\}\, 1_{B_n(0)} \le M, \qquad \text{for a. e. } x \in E.$$

By letting n go to ∞, we obtain that $u \in L^\infty(E)$, with $\|u\|_{L^\infty(E)} \le M$.

1.7.31 Let $u : \overline{\Omega} \to \mathbb{R}$ be a continuous function. We point out that $\overline{\Omega}$ is not necessarily compact, thus the supremum of u on $\overline{\Omega}$ is not necessarily attained. We first observe that $\Omega \subseteq \overline{\Omega}$, thus

$$|u(x)| \le \sup_{\Omega} |u| \le \sup_{\overline{\Omega}} |u|, \qquad \text{for every } x \in \Omega.$$

This implies in particular that

$$\|u\|_{L^\infty(\Omega)} \le \sup_{\Omega} |u| \le \sup_{\overline{\Omega}} |u| := M.$$

Let us first suppose that $M < +\infty$. Then, for every $\varepsilon > 0$ there exists a point $x_\varepsilon \in \overline{\Omega}$ such that

$$|u(x_\varepsilon)| > M - 2\,\varepsilon.$$

Since $x_\varepsilon \in \overline{\Omega}$ and Ω is an open set, there exists a sequence of points $\{x_{n,\varepsilon}\}_{n\in\mathbb{N}} \subseteq \Omega$, which converges to x_ε. Thus, by continuity of u, we can claim the existence of an index $n_0 \in \mathbb{N}$ such that

$$|u(x_{n_0,\varepsilon})| > M - \varepsilon.$$

This shows that the set $\{x \in \Omega : |u(x)| > M - \varepsilon\}$ is not empty. Moreover, it is open, thanks to the continuity of u. Thus, it has positive Lebesgue measure. By definition of essential supremum, we get that

$$\|u\|_{L^\infty(\Omega)} > M - \varepsilon.$$

By arbitrariness of $\varepsilon > 0$, this finally gives $\|u\|_{L^\infty(\Omega)} \ge M$, as well.

The case $M = +\infty$ is simpler. In this case, for every $n \in \mathbb{N}$ there exists a point $x_n \in \overline{\Omega}$ such that $|u(x_n)| > n$. With the same argument as before, applied to any point x_n, we can find a point $y_n \in \Omega$ such that

$$|u(y_n)| > n - 1.$$

Then, the set $\{x \in \Omega \, : \, |u(x)| > n - 1\}$ is not empty and open. Thus, as above we get

$$\|u\|_{L^\infty(\Omega)} > n - 1, \qquad \text{for every } n \in \mathbb{N}.$$

By arbitrariness of $n \in \mathbb{N}$, we get that the L^∞ norm is $+\infty$, as well. Finally, we have obtained

$$\sup_{\overline{\Omega}} |u| \le \|u\|_{L^\infty(\Omega)} \le \sup_{\Omega} |u| \le \sup_{\overline{\Omega}} |u|,$$

which permits to conclude.

8.2 Problems of Chap. 2

2.8.1 We take the functional

$$\mathcal{F}(\gamma) = \int_0^T |\gamma'(t)|^2 \, dt.$$

For a given curve $\gamma \in C^1([0, T]; \mathbb{R}^N)$, we take $\varepsilon \in \mathbb{R}$ and $\varphi \in C^\infty_0((0, T); \mathbb{R}^N)$. We need to compute the derivative at $\varepsilon = 0$ for the function

$$\varepsilon \mapsto \int_0^T |\gamma'(t) + \varepsilon \, \varphi'(t)|^2 \, dt.$$

It is sufficient to observe that

$$\int_0^T |\gamma'(t) + \varepsilon \, \varphi'(t)|^2 \, dt = \int_0^T |\gamma'(t)|^2 \, dt + 2 \, \varepsilon \int_0^T \langle \gamma'(t), \varphi'(t) \rangle \, dt$$
$$+ \varepsilon^2 \int_0^T |\varphi'(t)|^2,$$

which is a polynomial of degree two in ε. The derivative, evaluated at $\varepsilon = 0$, is thus given by

$$\int_0^T \langle \gamma'(t), \varphi'(t)\rangle\, dt = 0, \qquad \text{for every } \varphi \in C_0^\infty((0,T);\mathbb{R}^N).$$

This is the Euler-Lagrange equation in weak form. In order to write down the classic form of this equation, we further suppose that $\gamma \in C^2([0,T];\mathbb{R}^N)$ and observe that

$$0 = \int_0^T \langle \gamma'(t), \varphi'(t)\rangle\, dt = \sum_{i=1}^N \int_0^T \gamma_i'(t)\,\varphi_i'(t)\, dt = -\sum_{i=1}^N \int_0^T \gamma_i''(t)\,\varphi_i(t)\, dt,$$

where we used the definition of scalar product and an integration by parts. If we now apply component-wise the Du Bois-Reymond lemma (Lemma 1.4.1), we find

$$\gamma_1''(t) = \cdots = \gamma_N''(t) = 0, \qquad \text{for } t \in (0,T).$$

This concludes the exercise.

2.8.2 The proof is exactly the same of Theorem 2.2.2. It is sufficient to use Jensen's inequality (Proposition 1.3.10)

$$\begin{aligned}\int_0^T F(\gamma'(t))\, dt &= T \fint_0^T F(\gamma'(t))\, dt \\ &\geq T\, F\left(\fint_0^T \gamma'(t)\, dt\right) = T\, F\left(\frac{\gamma(T)-\gamma(0)}{T}\right) \\ &= T\, F\left(\frac{x_1 - x_0}{T}\right).\end{aligned}$$

Then observe that the previous lower bound is attained by γ_{opt}, thanks to the fact that its speed is the constant vector $(x_1 - x_0)/T$. The uniqueness follows from the strict convexity of F.

2.8.3 We proceed point by point.

1. We fix $\gamma \in C^1([0,1];\mathbb{R}^2)$ and $\varphi \in C_0^\infty((0,1);\mathbb{R}^2)$, then for every $\varepsilon \in \mathbb{R}$ we consider the function of one real variable

$$\varepsilon \mapsto \mathcal{F}(\gamma + \varepsilon\,\varphi).$$

By using the bilinearity of the scalar product, we have

$$\begin{aligned}\mathcal{F}(\gamma+\varepsilon\,\varphi) &= \frac{1}{2}\int_0^1 \langle A(t)\cdot(\gamma'(t)+\varepsilon\,\varphi'(t)),\gamma'(t)+\varepsilon\,\varphi'(t)\rangle\,dt\\ &= \frac{1}{2}\int_0^1 \langle A(t)\cdot\gamma'(t),\gamma'(t)\rangle\,dt\\ &+\frac{\varepsilon}{2}\int_0^1 \Big[\langle A(t)\cdot\gamma'(t),\varphi'(t)\rangle+\langle A(t)\cdot\varphi'(t),\gamma'(t)\rangle\Big]dt\\ &+\frac{\varepsilon^2}{2}\int_0^1 \langle A(t)\cdot\varphi'(t),\varphi'(t)\rangle\,dt.\end{aligned}$$

Thus, this is just a polynomial of degree two in the variable ε and as such, it is differentiable. By computing its derivative at $\varepsilon = 0$, we thus get the expression for the first variation

$$\delta\mathcal{F}(\gamma)[\varphi] = \frac{1}{2}\int_0^1 \Big[\langle A(t)\cdot\gamma'(t),\varphi'(t)\rangle\,dt+\langle A(t)\cdot\varphi'(t),\gamma'(t)\rangle\Big]\,dt.$$

Before going further, we observe that this expression can be simplified: by recalling the properties of the transposed matrix with respect to the scalar product[6] and using that A is symmetric, we have

$$\langle A(t)\cdot\varphi'(t),\gamma'(t)\rangle = \langle\varphi'(t),A(t)\cdot\gamma'(t)\rangle = \langle A(t)\cdot\gamma'(t),\varphi'(t)\rangle.$$

Thus, the first variation can be rewritten as

$$\delta\mathcal{F}(\gamma)[\varphi] = \int_0^1 \langle A(t)\cdot\gamma'(t),\varphi'(t)\rangle\,dt,\qquad \text{for every } \varphi\in C^\infty_0((0,1);\mathbb{R}^2).$$

The Euler-Lagrange equation in weak form is then given by

$$\int_0^1 \langle A(t)\cdot\gamma'(t),\varphi'(t)\rangle\,dt = 0,\qquad \text{for every } \varphi\in C^\infty_0((0,1);\mathbb{R}^2).\qquad (*)$$

In order to find the classical form of this equation, we suppose that $\gamma \in C^2([0,1];\mathbb{R}^2)$ and then use an integration by parts in the previous identity. In other words, by using the definition of $A(t)$, for every $\varphi \in C^\infty_0((0,1);\mathbb{R}^2)$ we have

[6] Recall that for every matrix $A \in \mathcal{M}_{N\times N}(\mathbb{R})$ and every $\mathbf{v},\mathbf{w}\in\mathbb{R}^N$, we have

$$\langle A\cdot\mathbf{v},\mathbf{w}\rangle = \langle\mathbf{v},A^{\mathrm{T}}\cdot\mathbf{w}\rangle,$$

where A^{T} is the transposed matrix.

$$0 = \int_0^1 \langle A(t) \cdot \gamma'(t), \varphi'(t) \rangle \, dt = \int_0^1 \left\langle \begin{bmatrix} t\, \gamma_1'(t) + \gamma_2'(t) \\ \gamma_1'(t) + 2\, \gamma_2'(t) \end{bmatrix}, \begin{bmatrix} \varphi_1'(t) \\ \varphi_2'(t) \end{bmatrix} \right\rangle dt$$

$$= \int_0^1 \Big(t\, \gamma_1'(t) + \gamma_2'(t) \Big)\, \varphi_1'(t) \, dt$$

$$+ \int_0^1 \Big(\gamma_1'(t) + 2\, \gamma_2'(t) \Big)\, \varphi_2'(t) \, dt$$

$$= - \int_0^1 \Big(t\, \gamma_1'(t) + \gamma_2'(t) \Big)'\, \varphi_1(t) \, dt$$

$$- \int_0^1 \Big(\gamma_1'(t) + 2\, \gamma_2'(t) \Big)'\, \varphi_2(t) \, dt.$$

By using the Du Bois-Reymond Lemma (i.e. Lemma 1.4.1), we finally get the classical form of the Euler-Lagrange equation

$$\begin{cases} (t\, \gamma_1'(t))' + \gamma_2''(t) = 0 \\ \gamma_1''(t) + 2\, \gamma_2''(t) = 0. \end{cases} \tag{$**$}$$

Observe that since γ is vector-valued, the Euler-Lagrange equation is actually given by a system of ordinary differential equations.

2. We separately find the general solution of $(*)$ and $(**)$. In order to find the general solution of $(**)$, we use the second equation to infer that

$$\gamma_2''(t) = -\frac{1}{2}\, \gamma_1''(t).$$

By replacing this into the first equation, we get

$$(t\, \gamma_1'(t))' - \frac{1}{2}\, \gamma_1''(t) = 0.$$

This is the same as

$$\gamma_1'(t) + \left(t - \frac{1}{2} \right) \gamma_1''(t) = 0.$$

The latter can also be rewritten as

$$\left(\left(t - \frac{1}{2} \right) \gamma_1'(t) \right)' = 0.$$

This implies that there exists a constant $C \in \mathbb{R}$ such that

$$\left(t - \frac{1}{2}\right) \gamma_1'(t) = C, \qquad \text{for } t \in (0, 1).$$

By computing the left-hand side for $t = 1/2$, we then get that such a constant must be $C = 0$, i.e. we must have

$$\left(t - \frac{1}{2}\right) \gamma_1'(t) = 0, \qquad \text{for } t \in (0, 1).$$

Since the first term on the left-hand side is different from zero for $t \neq 1/2$, we obtain

$$\gamma_1'(t) = 0, \qquad \text{for } t \in (0, 1) \setminus \left\{\frac{1}{2}\right\}.$$

Finally, since we are assuming that γ_1' is a continuous function, we can finally infer that

$$\gamma_1'(t) = 0, \qquad \text{for } t \in (0, 1),$$

so that

$$\gamma_1(t) = C_1, \qquad \text{for } t \in (0, 1),$$

for a constant $C_1 \in \mathbb{R}$. In order to find γ_2, it is sufficient to use this information into the second equation of $(**)$, so to obtain

$$\gamma_2''(t) = 0, \qquad \text{for } t \in (0, 1),$$

i.e. γ_2 has to be an affine function. In conclusion, the general solution of the Euler-Lagrange equation in classical form is given by

$$\gamma(t) = (C_1, C_2\, t + C_3), \qquad \text{for } t \in (0, 1),$$

where C_1, C_2, C_3 are real constants.

We now analyze the solutions of the equation in weak form[7] $(*)$. We have previously seen that this can be rewritten as

$$\int_0^1 \left(t\, \gamma_1'(t) + \gamma_2'(t)\right) \varphi_1'(t)\, dt + \int_0^1 \left(\gamma_1'(t) + 2\, \gamma_2'(t)\right) \varphi_2'(t)\, dt = 0.$$

[7] Recall that the equation in weak form may have more solutions than that in classical form, as seen for example in Problem 1.7.20.

By using the result *à la* Du Bois-Reymond of Problem 1.7.12, we then find that

$$t\,\gamma_1'(t) + \gamma_2'(t) = c_1 \qquad \text{and} \qquad \gamma_1'(t) + 2\,\gamma_2'(t) = c_2,$$

i.e. both quantities are constant on $(0, 1)$. From the second equation, we get

$$\gamma_2'(t) = \frac{c_2}{2} - \frac{\gamma_1'(t)}{2},$$

which inserted in the first one gives

$$\left(t - \frac{1}{2}\right)\gamma_1'(t) = c_1 - \frac{c_2}{2}.$$

This implies that the left-hand side must be constant. By proceeding as in the case of classical solutions, we obtain

$$\gamma_1'(t) = 0, \qquad \text{for } t \in (0, 1).$$

By using the relation above between γ_1' and γ_2', we obtain that γ_2' must be constant and thus again

$$\gamma(t) = (C_1, C_2\,t + C_3), \qquad \text{for } t \in (0, 1),$$

for $C_1, C_2, C_3 \in \mathbb{R}$.

3. We argue by contraction and assume that, for $\alpha \neq 0$, there exists $\gamma \in C^1([0, 1]; \mathbb{R}^2)$ which solves (2.8.1). Then γ must be a critical point of $\mathcal{F}$: indeed, by minimality we get

$$\mathcal{F}(\gamma + \varepsilon\,\varphi) \geq \mathcal{F}(\gamma), \qquad \text{for every } \varepsilon \in \mathbb{R},\ \varphi \in C_0^\infty((0, 1); \mathbb{R}^2).$$

This implies that the differentiable function

$$\varepsilon \mapsto \mathcal{F}(\gamma + \varepsilon\,\varphi),$$

is minimal for $\varepsilon = 0$. By definition of first variation, this exactly gives that

$$\delta\mathcal{F}(\gamma)[\varphi] = 0, \qquad \text{for every } \varphi \in C_0^\infty((0, 1); \mathbb{R}^2).$$

Since γ is a weak solution of the Euler-Lagrange, we get from the previous point that it must result

$$\gamma(t) = (C_1, C_2\,t + C_3), \qquad \text{for } t \in (0, 1).$$

We now impose the boundary conditions: we must have

$$(0,0)=\gamma(0)=(C_1,C_3),$$

and

$$(\alpha,1)=\gamma(1)=(C_1,C_2+C_3).$$

It is easily seen that if $\alpha\neq 0$, then the resulting system does not have any solution. This gives the desired contradiction.

The exercise is over.

2.8.4 We observe that F is a C^2 function on its domain. Thus, we can appeal to Proposition 1.3.8 and Remark 1.3.9: it is sufficient to prove that the Hessian matrix D^2F of F is definite positive.

We first compute the gradient of F: this is given by

$$\frac{\partial F}{\partial x_1}(x_1,x_2)=-\frac{1}{x_1^3}\frac{1}{\sqrt{\dfrac{1}{x_1^2}+4\,x_2^2}}=-\frac{1}{x_1^3}\frac{1}{F(x_1,x_2)},$$

and

$$\frac{\partial F}{\partial x_2}(x_1,x_2)=\frac{4\,x_2}{\sqrt{\dfrac{1}{x_1^2}+4\,x_2^2}}=\frac{4\,x_2}{F(x_1,x_2)}.$$

Accordingly, the second order partial derivatives are given by

$$\begin{aligned}\frac{\partial^2 F}{\partial x_1^2}(x_1,x_2)&=\frac{3}{x_1^4}\frac{1}{F(x_1,x_2)}+\frac{1}{x_1^3}\frac{1}{F(x_1,x_2)^2}\frac{\partial F}{\partial x_1}(x_1,x_2)\\&=\frac{3}{x_1^4}\frac{1}{F(x_1,x_2)}-\frac{1}{x_1^6}\frac{1}{F(x_1,x_2)^3},\end{aligned}$$

$$\frac{\partial^2 F}{\partial x_1\,\partial x_2}=-\frac{4\,x_2}{F(x_1,x_2)^2}\frac{\partial F}{\partial x_1}(x_1,x_2)=\frac{4\,x_2}{x_1^3}\frac{1}{F(x_1,x_2)^3},$$

and

$$\frac{\partial^2 F}{\partial x_2^2}=\frac{4}{F(x_1,x_2)}-\frac{4\,x_2}{F(x_1,x_2)^2}\frac{\partial F}{\partial x_2}(x_1,x_2)=\frac{4}{F(x_1,x_2)}-\frac{16\,x_2^2}{F(x_1,x_2)^3}.$$

The trace of the Hessian matrix is then given by

$$\begin{aligned}
\operatorname{tr}(D^2F(x_1,x_2)) &= \left(\frac{3}{x_1^4}\,\frac{1}{F(x_1,x_2)} - \frac{1}{x_1^6}\,\frac{1}{F(x_1,x_2)^3}\right) \\
&\quad + \left(\frac{4}{F(x_1,x_2)} - \frac{16\,x_2^2}{F(x_1,x_2)^3}\right) \\
&= \left(\frac{3}{x_1^4}+4\right)\frac{1}{F(x_1,x_2)} - \left(\frac{1}{x_1^6}+16\,x_2^2\right)\frac{1}{F(x_1,x_2)^3} \\
&= \frac{1}{F(x_1,x_2)^3}\left[\left(\frac{3}{x_1^4}+4\right)F(x_1,x_2)^2 - \left(\frac{1}{x_1^6}+16\,x_2^2\right)\right].
\end{aligned}$$

By using that

$$F(x_1,x_2)^2 = \frac{1}{x_1^2} + 4\,x_2^2,$$

we easily get from the above computation that the trace is always positive.

The determinant of the Hessian matrix is given by

$$\begin{aligned}
\det D^2F(x_1,x_2) &= \left(\frac{3}{x_1^4}\,\frac{1}{F(x_1,x_2)} - \frac{1}{x_1^6}\,\frac{1}{F(x_1,x_2)^3}\right)\left(\frac{4}{F(x_1,x_2)} - \frac{16\,x_2^2}{F(x_1,x_2)^3}\right) \\
&\quad - \frac{16\,x_2^2}{x_1^6}\,\frac{1}{F(x_1,x_2)^6} \\
&= \frac{12}{x_1^4}\,\frac{1}{F(x_1,x_2)^2} - \frac{48\,x_2^2}{x_1^4}\,\frac{1}{F(x_1,x_2)^4} - \frac{4}{x_1^6}\,\frac{1}{F(x_1,x_2)^4} \\
&= \frac{4}{x_1^4}\,\frac{1}{F(x_1,x_2)^4}\left[3\,F(x_1,x_2)^2 - \left(\frac{1}{x_1^2}+12\,x_2^2\right)\right].
\end{aligned}$$

By using again the definition of F^2, we get that the determinant is positive, as well. By recalling that a 2×2 symmetric matrix is positive definite if and only its trace and determinant are positive, this finally shows that $D^2F(x_1,x_2)$ is positive definite.

2.8.5 We need to compute (at least formally) the derivative of the function

$$g(\varepsilon) = \mathcal{F}(u + \varepsilon\,\varphi),$$

at $\varepsilon = 0$. Here u is an admissible function and $\varphi \in C_0^\infty((0,1))$. We introduce the notation

$$G(x_1, x_2) = \frac{\sqrt{1 + x_2^2}}{\sqrt{x_1}},$$

then we can rewrite

$$g(\varepsilon) = \int_0^1 G\big(u(x) + \varepsilon\, \varphi(x), u'(x) + \varepsilon\, \varphi'(x)\big)\, dx.$$

By proceeding informally, we get from the Chain Rule formula

$$\begin{aligned} g'(\varepsilon) = \int_0^1 \Bigg[& \frac{\partial G}{\partial x_1}(u(x) + \varepsilon\, \varphi(x), u'(x) + \varepsilon\, \varphi'(x))\, \varphi(x) \\ & + \frac{\partial G}{\partial x_2}(u(x) + \varepsilon\, \varphi(x), u'(x) + \varepsilon\, \varphi'(x))\, \varphi'(x) \Bigg]\, dx. \end{aligned}$$

By taking $\varepsilon = 0$, we get the expression

$$\delta\mathcal{F}(u)[\varphi] = g'(0) = \int_0^1 \left[\frac{\partial G}{\partial x_1}(u(x), u'(x))\, \varphi(x) + \frac{\partial G}{\partial x_2}(u(x), u'(x))\, \varphi'(x) \right] dx,$$

for the first variation of the brachistocrone functional. Observe that

$$\frac{\partial G}{\partial x_1}(x_1, x_2) = -\frac{1}{2}\, \frac{\sqrt{1 + x_2^2}}{x_1\, \sqrt{x_1}} \qquad \text{and} \qquad \frac{\partial G}{\partial x_2}(x_1, x_2) = \frac{x_2}{\sqrt{x_1}\, \sqrt{1 + x_2^2}},$$

thus, the first variation can be rewritten as

$$\delta\mathcal{F}(u)[\varphi] = \int_0^1 \left[-\frac{1}{2}\, \frac{\sqrt{1 + |u'|^2}}{u\, \sqrt{u}}\, \varphi + \frac{u'}{\sqrt{u}\, \sqrt{1 + |u'|^2}}\, \varphi' \right] dx.$$

For notational simplicity, we omit to indicate the argument x. The Euler-Lagrange equation in weak form is then given by

$$\int_0^1 \left[-\frac{1}{2}\, \frac{\sqrt{1 + |u'|^2}}{u\, \sqrt{u}}\, \varphi + \frac{u'}{\sqrt{u}\, \sqrt{1 + |u'|^2}}\, \varphi' \right] dx = 0,$$

for every $\varphi \in C_0^\infty((0, 1))$. In order to find the equation in classical form, we need to integrate by parts the second term in the integral above, i.e. the one containing φ'. Thanks to the fact that φ has compact support, we then obtain

$$
\begin{aligned}
0 &= \int_0^1 \left[-\frac{1}{2} \frac{\sqrt{1+|u'|^2}}{u\,\sqrt{u}}\,\varphi + \frac{u'}{\sqrt{u}\,\sqrt{1+|u'|^2}}\,\varphi' \right] dx \\
&= \int_0^1 \left[-\frac{1}{2} \frac{\sqrt{1+|u'|^2}}{u\,\sqrt{u}}\,\varphi - \left(\frac{u'}{\sqrt{u}\,\sqrt{1+|u'|^2}} \right)' \varphi \right] dx.
\end{aligned}
$$

This integral identity holds for every $\varphi \in C_0^\infty((0,1))$, thus by the Du Bois-Reymond lemma (i.e. Lemma 1.4.1), we get

$$
-\left(\frac{u'(t)}{\sqrt{u(t)}\,\sqrt{1+|u'(t)|^2}} \right)' = \frac{1}{2} \frac{\sqrt{1+|u'(t)|^2}}{u(t)\,\sqrt{u(t)}}, \qquad \text{for every } t \in (0,1). \qquad (*)
$$

This is the Euler-Lagrange equation in classical form. However, at first sight it looks quite different from what we expected! In order to arrive at the claimed equation, let us use some manipulations. By computing the derivative on the left-hand side, we have

$$
\begin{aligned}
\left(\frac{u'(t)}{\sqrt{u(t)}\,\sqrt{1+|u'(t)|^2}} \right)' &= \frac{u''(t)}{\sqrt{u(t)}\,\sqrt{1+|u'(t)|^2}} - \frac{1}{2} \frac{|u'(t)|^2}{u(t)\,\sqrt{u(t)}\,\sqrt{1+|u'(t)|^2}} \\
&\quad - \frac{|u'(t)|^2\,u''(t)}{\sqrt{u(t)}\,(1+|u'(t)|^2)^{\frac{3}{2}}} \\
&= \frac{2\,u''(t)\,u(t)\,(1+|u'(t)|^2) - |u'(t)|^2\,(1+|u'(t)|^2)}{2\,u(t)\,\sqrt{u(t)}\,(1+|u'(t)|^2)^{\frac{3}{2}}} \\
&\quad - \frac{2\,u''(t)\,u(t)\,|u'(t)|^2}{2\,u(t)\,\sqrt{u(t)}\,(1+|u'(t)|^2)^{\frac{3}{2}}} \\
&= \frac{2\,u''(t)\,u(t) - |u'(t)|^2\,(1+|u'(t)|^2)}{2\,u(t)\,\sqrt{u(t)}\,(1+|u'(t)|^2)^{\frac{3}{2}}}.
\end{aligned}
$$

Thus from $(*)$, we get

$$
-\frac{2\,u''(t)\,u(t) - |u'(t)|^2\,(1+|u'(t)|^2)}{2\,u(t)\,\sqrt{u(t)}\,(1+|u'(t)|^2)^{\frac{3}{2}}} = \frac{1}{2} \frac{\sqrt{1+|u'(t)|^2}}{u(t)\,\sqrt{u(t)}}.
$$

We now multiply both sides by the strictly positive quantity $2\,u(t)\,\sqrt{u(t)}\,(1+|u'(t)|^2)^{3/2}$, the previous equation is equivalent to

$$
-\,2\,u''(t)\,u(t) + |u'(t)|^2\,(1+|u'(t)|^2) = (1+|u'(t)|^2)^2.
$$

By computing the square on the right-hand side and erasing the common factors, we finally get

$$-2\,u''(t)\,u(t) = 1 + |u'(t)|^2, \qquad \text{for every } t \in (0,1),$$

as claimed.

2.8.6 We first perform the change of variable

$$\sqrt{\frac{\tau}{C-\tau}} = s \qquad \text{that is} \qquad \tau = C\,\frac{s^2}{s^2+1}.$$

Thus we get

$$\int_0^T \sqrt{\frac{\tau}{C-\tau}}\,d\tau = C\int_0^{\sqrt{\frac{T}{C-T}}} \frac{2\,s^2}{(1+s^2)^2}\,ds.$$

By using an integration by parts in the last integral, we get

$$\begin{aligned}\int_0^{\sqrt{\frac{T}{C-T}}} \frac{2\,s^2}{(1+s^2)^2}\,ds &= \left[-\frac{1}{1+s^2}\,s\right]_0^{\sqrt{\frac{T}{C-T}}} + \int_0^{\sqrt{\frac{T}{C-T}}} \frac{1}{1+s^2}\,ds \\ &= -\frac{\sqrt{T\,(C-T)}}{C} + \arctan\sqrt{\frac{T}{C-T}}.\end{aligned}$$

By putting everything together, we finally get

$$\int_0^T \sqrt{\frac{\tau}{C-\tau}}\,d\tau = -\sqrt{T\,(C-T)} + C\,\arctan\sqrt{\frac{T}{C-T}},$$

see Fig. 8.1. This concludes the exercise.

2.8.7 This is very easy, it is sufficient to observe at first that

$$\frac{\displaystyle\int_0^1 |\varphi'(t)|^2\,dt}{\displaystyle\int_0^1 |\varphi(t)|^2\,dt} \geq 0,$$

for every admissible φ, thus the infimum is non-negative. On the other hand, since $\alpha \neq 0$, the constant function $\varphi_0(t) = \alpha$ for every $t \in [0,1]$ is admissible for the minimization problem. Thus we have

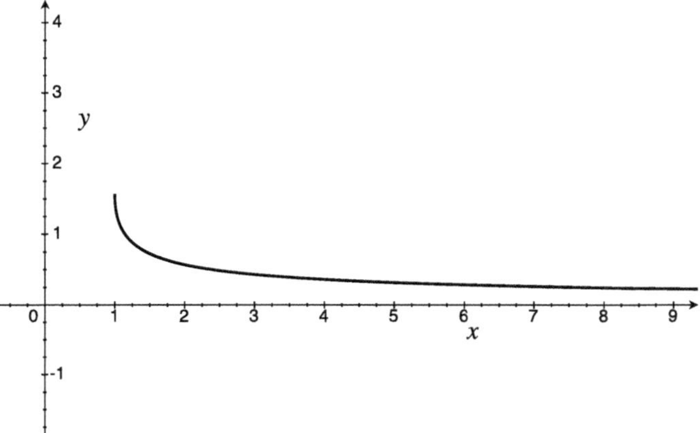

Fig. 8.1 The graph of the function $C \mapsto \int_0^T \sqrt{\frac{\tau}{C-\tau}}\,d\tau$ of Problem 2.8.6, defined on $[T, +\infty)$. Here we have chosen $T = 1$

$$\inf_{\varphi \in C^1([0,1])\setminus\{0\}} \left\{ \frac{\displaystyle\int_0^1 |\varphi'(t)|^2\,dt}{\displaystyle\int_0^1 |\varphi(t)|^2\,dt} \,:\, \varphi(0) = \alpha \right\} \leq \frac{\displaystyle\int_0^1 |\varphi_0'(t)|^2\,dt}{\displaystyle\int_0^1 |\varphi_0(t)|^2\,dt} = \frac{0}{\alpha^2} = 0.$$

This gives the desired conclusion.

2.8.8 We first show that the infimum is strictly positive. It is sufficient to reproduce the proof of Lemma 2.4.1. Thus, for every admissible φ, we can write

$$\varphi(t) = \varphi(t) - \varphi(0) = \int_0^t \varphi'(\tau)\,d\tau, \qquad \text{for every } t \in [0, 1].$$

By taking the absolute value, we thus get

$$|\varphi(t)| = \left|\int_0^t \varphi'(\tau)\,d\tau\right| \leq \int_0^t |\varphi'(\tau)|\,d\tau \leq \int_0^1 |\varphi'(\tau)|\,d\tau \leq \|\varphi'\|_{L^\infty([0,1])}.$$

Since this is true for every $t \in [0, 1]$, we can pass to the supremum of $|\varphi(t)|$ for $t \in [0, 1]$ and get

$$\|\varphi\|_{L^\infty([0,1])} \leq \|\varphi'\|_{L^\infty([0,1])}.$$

In other words, we get

$$\frac{\|\varphi'\|_{L^\infty([0,1])}}{\|\varphi\|_{L^\infty([0,1])}} \geq 1, \qquad \text{for every } \varphi \in C^1([0, 1]) \setminus \{0\} \text{ with } \varphi(0) = 0.$$

This shows that

$$\inf_{\varphi\in C^1([0,1])\setminus\{0\}}\left\{\frac{\|\varphi'\|_{L^\infty([0,1])}}{\|\varphi\|_{L^\infty([0,1])}}\,:\,\varphi(0)=0\right\}\geq 1.$$

In order to prove that this infimum is attained and actually coincides with 1, let us try to insert the most simple admissible function: an affine one! We use $\varphi_0(t)=t$ as a test, thus we get

$$\inf_{\varphi\in C^1([0,1])\setminus\{0\}}\left\{\frac{\|\varphi'\|_{L^\infty([0,1])}}{\|\varphi\|_{L^\infty([0,1])}}\,:\,\varphi(0)=0\right\}\leq \frac{\|\varphi_0'\|_{L^\infty([0,1])}}{\|\varphi_0\|_{L^\infty([0,1])}}=1.$$

This concludes the exercise.

2.8.9 As in the previous problem, we first show that the infimum is strictly positive, by reproducing the proof of Lemma 2.4.1. We will be slightly more precise in the estimate, in order to prove a finer result. For every admissible φ, we start by writing

$$\varphi(t)=\varphi(t)-\varphi(0)=\int_0^t\varphi'(\tau)\,d\tau,\qquad \text{for every } t\in[0,1].$$

By taking the absolute value, we thus get

$$|\varphi(t)|=\left|\int_0^t\varphi'(\tau)\,d\tau\right|\leq\int_0^t|\varphi'(\tau)|\,d\tau.$$

If we now integrate over $[0,1]$, we get from the previous estimate

$$\int_0^1|\varphi(t)|\,dt\leq\int_0^1\left(\int_0^t|\varphi'(\tau)|\,d\tau\right)dt.$$

We now exchange the order of integration in the last integral: we have

$$\int_0^1\left(\int_0^t|\varphi'(\tau)|\,d\tau\right)dt=\int_0^1\left(\int_\tau^1|\varphi'(\tau)|\,dt\right)d\tau=\int_0^1(1-\tau)\,|\varphi'(\tau)|\,d\tau.$$

Thus, we obtain

$$\int_0^1|\varphi(t)|\,dt\leq\int_0^1(1-\tau)\,|\varphi'(\tau)|\,d\tau\leq\int_0^1|\varphi'(\tau)|\,d\tau. \tag{$*$}$$

In other words, we get

$$\frac{\displaystyle\int_0^1|\varphi'(t)|\,dt}{\displaystyle\int_0^1|\varphi(t)|\,dt}\geq 1,\qquad \text{for every } \varphi\in C^1([0,1])\setminus\{0\} \text{ with } \varphi(0)=0.$$

This shows that

$$\inf_{\varphi\in C^1([0,1])\setminus\{0\}}\left\{\frac{\displaystyle\int_0^1|\varphi'(t)|\,dt}{\displaystyle\int_0^1|\varphi(t)|\,dt}\ :\ \varphi(0)=0\right\}\geq 1.$$

In order to prove that this infimum actually coincides with 1, we take for every $0<\varepsilon<1$ the function

$$\psi_\varepsilon(t)=\min\left\{\frac{t}{\varepsilon},\,1\right\},\qquad \text{for } t\in[0,1].$$

Observe that for ε converging to 0, this is converging pointwise on $(0,1]$ to the constant function 1. This function is not really admissible for our problem, since this is not in $C^1([0,1])$, but only piecewise C^1. However, admitting for a moment that we could take it as an admissible function, we would get

$$\lim_{\varepsilon\to 0}\frac{\displaystyle\int_0^1|\psi_\varepsilon'(t)|\,dt}{\displaystyle\int_0^1|\psi_\varepsilon(t)|\,dt}=\lim_{\varepsilon\to 0}\frac{\displaystyle\int_0^\varepsilon\frac{1}{\varepsilon}\,dt}{\displaystyle\int_\varepsilon^1 dt+\int_0^\varepsilon\frac{t}{\varepsilon}\,dt}=1.$$

In order to fix the regularity issue, we can use a smoothing argument: observe that we can write[8]

$$\psi_\varepsilon(t)=\min\left\{\frac{t}{\varepsilon},\,1\right\}=1+\min\left\{\frac{t}{\varepsilon}-1,\,0\right\}=1+\frac{1}{2}\left(\frac{t}{\varepsilon}-1-\left|\frac{t}{\varepsilon}-1\right|\right).$$

We need to regularize the absolute value: the standard trick is to approximate

$$\tau\mapsto|\tau|\qquad \text{by}\qquad \tau\mapsto\sqrt{\delta^2+\tau^2},$$

where $\delta>0$ is a small parameter. We choose $\delta=\sqrt{\varepsilon}$, for simplicity. We then consider the function

$$\widetilde{\psi}_\varepsilon(t)=1+\frac{1}{2}\left(\frac{t}{\varepsilon}-1-\sqrt{\varepsilon+\left(\frac{t}{\varepsilon}-1\right)^2}\right)+\frac{\sqrt{\varepsilon+1}-1}{2},$$

[8] We use that for a real number $\alpha\in\mathbb{R}$ we have

$$\min\{\alpha,0\}=\frac{\alpha-|\alpha|}{2}.$$

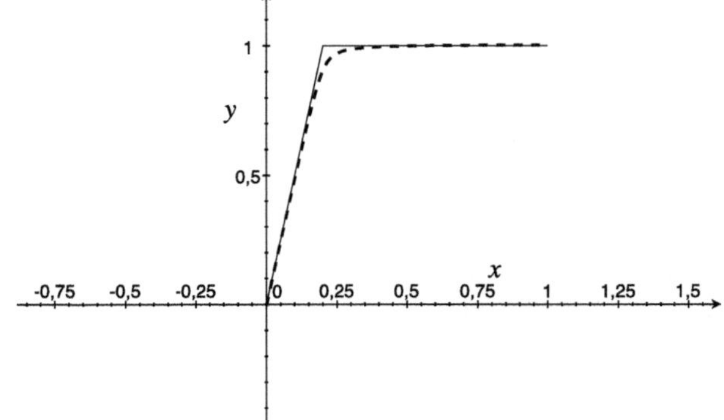

Fig. 8.2 The function ψ_ε and (in dashed line) its smooth approximation $\widetilde{\psi}_\varepsilon$

see Fig. 8.2. Observe that $\widetilde{\psi}_\varepsilon \in C^1([0, 1])$ and

$$\widetilde{\psi}_\varepsilon(0) = 1 + \frac{1}{2}\left(-1 - \sqrt{\varepsilon + 1}\right) + \frac{\sqrt{\varepsilon + 1} - 1}{2} = 0,$$

thus it is admissible. We now claim that as ε goes to 0, we still have

$$\lim_{\varepsilon \to 0} \frac{\displaystyle\int_0^1 |\widetilde{\psi}_\varepsilon'(t)|\, dt}{\displaystyle\int_0^1 |\widetilde{\psi}_\varepsilon(t)|\, dt} = 1. \tag{$**$}$$

We first observe that the derivative is given by

$$\widetilde{\psi}_\varepsilon'(t) = \frac{1}{2\varepsilon}\left(1 - \frac{\left(\dfrac{t}{\varepsilon} - 1\right)}{\sqrt{\varepsilon + \left(\dfrac{t}{\varepsilon} - 1\right)^2}}\right) = \frac{1}{2\varepsilon}\left(1 - \frac{t - \varepsilon}{\sqrt{\varepsilon^3 + (t - \varepsilon)^2}}\right).$$

Thus, we have

$$\begin{aligned}\int_0^1 |\widetilde{\psi}_\varepsilon'(t)|\, dt = \int_0^1 \widetilde{\psi}_\varepsilon'(t)\, dt &= \frac{1}{2\varepsilon} - \frac{1}{2\varepsilon}\int_0^1 \frac{t - \varepsilon}{\sqrt{\varepsilon^3 + (t - \varepsilon)^2}}\, dt \\ &= \frac{1}{2\varepsilon} - \frac{1}{2\varepsilon}\left[\sqrt{\varepsilon^3 + (t - \varepsilon)^2}\right]_0^1 \\ &= \frac{1 - \sqrt{\varepsilon^3 + (1 - \varepsilon)^2} + \varepsilon\sqrt{\varepsilon + 1}}{2\varepsilon}.\end{aligned}$$

In particular, we get

$$\lim_{\varepsilon\to 0}\int_0^1 |\widetilde{\psi}_\varepsilon'(t)|\,dt = \lim_{\varepsilon\to 0}\frac{1-(1-\varepsilon)+\varepsilon+o(\varepsilon)}{2\,\varepsilon} = 1.$$

In order to conclude, we are going to prove that

$$\lim_{\varepsilon\to 0}\int_0^1 |\widetilde{\psi}_\varepsilon(t)-1|\,dt = 0 \qquad \text{so that} \qquad \lim_{\varepsilon\to 0}\int_0^1 |\widetilde{\psi}_\varepsilon(t)|\,dt = 1,$$

as well. This would give $(**)$, as desired. We observe that

$$\begin{aligned}|\widetilde{\psi}_\varepsilon(t)-1| &\le \frac{1}{2}\left|\left(\frac{t}{\varepsilon}-1\right)-\sqrt{\varepsilon+\left(\frac{t}{\varepsilon}-1\right)^2}\right| + \frac{\sqrt{\varepsilon+1}-1}{2}\\ &= \frac{1}{2}\left(\sqrt{\varepsilon+\left(\frac{t}{\varepsilon}-1\right)^2}-\left(\frac{t}{\varepsilon}-1\right)\right)+\frac{\sqrt{\varepsilon+1}-1}{2}\\ &\le \frac{1}{2}\left(\sqrt{\varepsilon}+\left|\frac{t}{\varepsilon}-1\right|-\left(\frac{t}{\varepsilon}-1\right)\right)+\frac{\sqrt{\varepsilon}}{2},\end{aligned}$$

where we used the subadditivity of the square root (i.e. Problem 1.7.1 with $\alpha = 1/2$). By integrating over $[0, 1]$, we obtain

$$\begin{aligned}\int_0^1 |\widetilde{\psi}_\varepsilon(t)-1|\,dt &\le \sqrt{\varepsilon}+\frac{1}{2}\int_0^1\left|\frac{t}{\varepsilon}-1\right|\,dt-\frac{1}{2}\int_0^1\left(\frac{t}{\varepsilon}-1\right)\,dt\\ &= \sqrt{\varepsilon}+\int_0^\varepsilon\left(1-\frac{t}{\varepsilon}\right)\,dt\\ &= \sqrt{\varepsilon}+\varepsilon\int_0^1(1-s)\,ds.\end{aligned}$$

The last quantity clearly converges to 0, as ε goes to 0. We thus obtain $(**)$, as previously explained.

We are left with proving that the sharp Poincaré constant is not attained in $C^1([0, 1])$. Assume by contradiction that there exists $v \in C^1([0, 1]) \setminus \{0\}$ such that $v(0) = 0$ and

$$\frac{\displaystyle\int_0^1 |v'(t)|\,dt}{\displaystyle\int_0^1 |v(t)|\,dt} = 1.$$

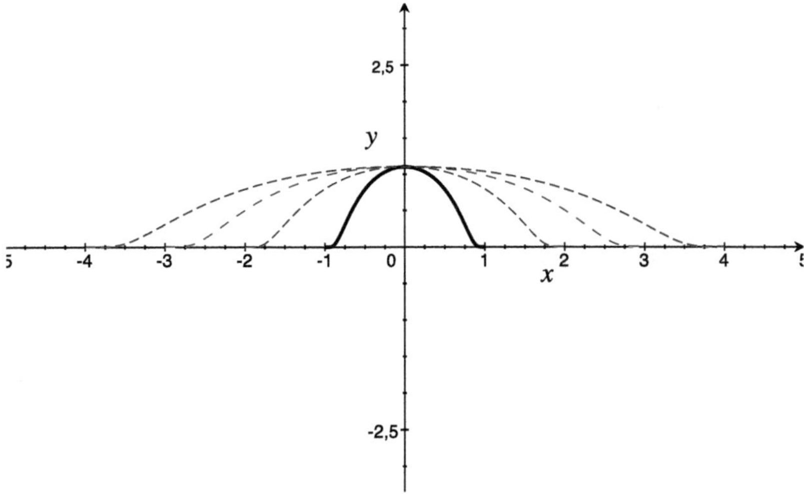

Fig. 8.3 A smooth compactly supported function φ (in bold black) and some of its scalings φ_ℓ

This entails that for v we have equality in $(*)$. In particular, we get that

$$\int_0^1 (1-\tau)\,|v'(\tau)|\,d\tau = \int_0^1 |v'(\tau)|\,d\tau \quad \text{that is} \quad \int_0^1 \tau\,|v'(\tau)|\,d\tau = 0.$$

Since the function τ is strictly positive on $(0, 1]$, the previous identity and the continuity of v' imply that we must have

$$v'(t) = 0, \qquad \text{for every } t \in [0, 1].$$

Thus v is constant, but since it vanishes at $t = 0$, we get that v identically vanishes. This yields the desired contradiction.

2.8.10 Let $\varphi \in C_0^\infty(\mathbb{R}) \setminus \{0\}$, for every $\ell > 0$ we define the rescaled function

$$\varphi_\ell(t) = \varphi(\ell\, t), \qquad \text{for } t \in \mathbb{R},$$

see Fig. 8.3. We thus obtain

$$\frac{\displaystyle\int_{\mathbb{R}} |\varphi_\ell'(t)|^p\,dt}{\displaystyle\int_{\mathbb{R}} |\varphi_\ell(t)|^p\,dt} = \frac{\ell^p \displaystyle\int_{\mathbb{R}} |\varphi'(\ell\, t)|^p\,dt}{\displaystyle\int_{\mathbb{R}} |\varphi(\ell\, t)|^p\,dt}$$

$$= \frac{\ell^{p-1} \displaystyle\int_{\mathbb{R}} |\varphi'(s)|^p\,ds}{\ell^{-1} \displaystyle\int_{\mathbb{R}} |\varphi(s)|^p\,ds} = \ell^p\, \frac{\displaystyle\int_{\mathbb{R}} |\varphi'(s)|^p\,ds}{\displaystyle\int_{\mathbb{R}} |\varphi(s)|^p\,ds}.$$

By taking the limit as ℓ goes to 0, we thus get

$$\lim_{\ell\to 0^+} \frac{\displaystyle\int_{\mathbb{R}} |\varphi_\ell'(t)|^p\,dt}{\displaystyle\int_{\mathbb{R}} |\varphi_\ell(t)|^p\,dt} = 0.$$

This is enough to get the desired conclusion.

2.8.11 For the necessary prerequisites on Fourier series, we refer the reader to [57, Chapters 2 & 3]. Let $\varphi \in C^1([0,1])$ be such that

$$\varphi(0) = \varphi(1) \qquad \text{and} \qquad \int_0^1 \varphi(t)\,dt = 0.$$

We can extend it periodically to the whole $\mathbb{R}$. Observe that thanks to the condition $\varphi(0) = \varphi(1)$, its periodic extension is continuous in $\mathbb{R}$. Thus, if we define

$$\widehat{\varphi}(n) = \int_0^1 \varphi(t)\,e^{-2\pi\,i\,n\,t}\,dt, \qquad \text{for every } n \in \mathbb{Z},$$

we have

$$\varphi(t) = \sum_{n\in\mathbb{Z}} \widehat{\varphi}(n)\,e^{2\pi\,i\,n\,t}, \qquad \text{for } t \in [0,1], \tag{$*$}$$

and the convergence of the Fourier series is total on $[0,1]$, since $\varphi \in C^1([0,1]) \cap C^0(\mathbb{R})$. On the other hand, for the derivative of φ we have

$$\varphi'(t) = \sum_{n\in\mathbb{Z}} \widehat{\varphi'}(n)\,e^{2\pi\,i\,n\,t},$$

where now the convergence of the series has to be intended in the $L^2([0,1])$ sense. By Plancherel's identity, we know that

$$\int_0^1 |\varphi(t)|^2\,dt = \sum_{n\in\mathbb{Z}} |\widehat{\varphi}(n)|^2 \qquad \text{and} \qquad \int_0^1 |\varphi'(t)|^2\,dt = \sum_{n\in\mathbb{Z}} |\widehat{\varphi'}(n)|^2.$$

We now observe that by definition of Fourier coefficients and using an integration by parts, we get for every $n \neq 0$

$$\begin{aligned}\widehat{\varphi'}(n) &= \int_0^1 \varphi'(t)\, e^{-2\pi\, \mathrm{i}\, n t}\, dt \\ &= \Big[\varphi(t)\, e^{-2\pi\, \mathrm{i}\, n t}\Big]_0^1 + 2\pi\, \mathrm{i}\, n \int_0^1 \varphi(t)\, e^{-2\pi\, \mathrm{i}\, n t}\, dt \\ &= 2\pi\, \mathrm{i}\, n \int_0^1 \varphi(t)\, e^{-2\pi\, \mathrm{i}\, n t}\, dt = 2\pi\, \mathrm{i}\, n\, \widehat{\varphi}(n),\end{aligned}$$

where we used that

$$\varphi(1) = \varphi(0) \qquad \text{and} \qquad e^{-2\pi\, \mathrm{i}\, n} = e^0 = 1.$$

As for the coefficients corresponding to $n = 0$, we have

$$\widehat{\varphi'}(0) = \int_0^1 \varphi'(t)\, dt = \varphi(1) - \varphi(0) = 0,$$

and also

$$\widehat{\varphi}(0) = \int_0^1 \varphi(t)\, dt = 0,$$

thanks to the assumptions on φ. In conclusion, we can infer that

$$\begin{aligned}\int_0^1 |\varphi'(t)|^2\, dt &= \sum_{n\neq 0} |\widehat{\varphi'}(n)|^2 = 4\,\pi^2 \sum_{n\neq 0} n^2\, |\widehat{\varphi}(n)|^2 \\ &\geq 4\,\pi^2 \sum_{n\neq 0} |\widehat{\varphi}(n)|^2 = 4\,\pi^2 \int_0^1 |\varphi(t)|^2\, dt.\end{aligned} \tag{$**$}$$

This shows that for every $\varphi \in C^1([0, 1])$ such that $\varphi(0) = \varphi(1)$ and $\int_0^1 \varphi(t)\, dt = 0$, we have

$$\frac{\displaystyle\int_0^1 |\varphi'(t)|^2\, dt}{\displaystyle\int_0^1 |\varphi(t)|^2\, dt} \geq 4\,\pi^2.$$

Moreover, from $(**)$ we get that equality can hold if and only if φ is such that

$$\widehat{\varphi}(n) = 0 \qquad \text{for every } |n| \geq 2.$$

By going back to $(*)$, this means that the lower bound $4\,\pi^2$ is attained if and only if

$$\begin{aligned}\varphi(t) &= \alpha\, e^{2\pi\, \mathrm{i}\, t} + \beta\, e^{-2\pi\, \mathrm{i}\, t}\\ &= \alpha\, \Big(\cos(2\,\pi\, t) + \mathrm{i}\, \sin(2\,\pi\, t)\Big) + \beta\, \Big(\cos(2\,\pi\, t) - \mathrm{i}\, \sin(2\,\pi\, t)\Big)\\ &= (\alpha+\beta)\, \cos(2\,\pi\, t) + \mathrm{i}\, (\alpha-\beta)\, \sin(2\,\pi\, t),\end{aligned}$$

for $\alpha, \beta \in \mathbb{C}$.

2.8.12 We first show that, for every $\varphi \in C^1([0,1])$, the infimum

$$\inf_{c\in\mathbb{R}} \int_0^1 |\varphi(t) - c|^2\, dt.$$

is attained. Indeed, we set

$$g(c) = \int_0^1 |\varphi(t) - c|^2\, dt,$$

by expanding the square we get

$$g(c) = \int_0^1 |\varphi(t)|^2\, dt - 2\, c \int_0^1 \varphi(t)\, dt + c^2.$$

Thus, the quantity g to be minimized is nothing but a second order convex polynomial in c. As such, g admits a global minimum, which coincides with its unique critical point, by Lemma 1.2.6. This is found by imposing that

$$g'(c) = 0 \qquad \text{i.e.} \qquad -2\int_0^1 \varphi(t)\, dt + 2\, c = 0.$$

We thus get that

$$c_\varphi = \int_0^1 \varphi(t)\, dt,$$

is the unique minimizer, so that

$$\inf_{c\in\mathbb{R}} \int_0^1 |\varphi(t) - c|^2\, dt = \int_0^1 \left|\varphi(t) - \int_0^1 \varphi(\tau)\, d\tau\right|^2\, dt. \qquad (*)$$

It is now sufficient to observe that for every $\varphi \in C^1([0,1])$, the function

$$\widetilde{\varphi}(t) = \varphi(t) - \int_0^1 \varphi(\tau)\, d\tau,$$

is such that

$$\int_0^1 \widetilde{\varphi}(t)\,dt = 0 \qquad \text{and} \qquad \widetilde{\varphi}'(t) = \varphi'(t).$$

From Theorem 2.4.9 applied to $\widetilde{\varphi}$, we thus obtain

$$\begin{aligned}\inf_{c\in\mathbb{R}} \int_0^1 |\varphi(t) - c|^2\,dt &= \int_0^1 \left|\varphi(t) - \int_0^1 \varphi(\tau)\,d\tau\right|^2 dt \\ &= \int_0^1 |\widetilde{\varphi}(t)|^2\,dt \\ &\le \frac{1}{\pi^2}\int_0^1 |\widetilde{\varphi}'(t)|^2\,dt = \frac{1}{\pi^2}\int_0^1 |\varphi'(t)|^2\,dt.\end{aligned}$$

This proves the claimed inequality. In order to prove that the constant is sharp, it is sufficient to take a function which is optimal for Theorem 2.4.9, i.e. for example

$$\phi(t) = \cos(\pi\, t).$$

Indeed, we already know that

$$\int_0^1 |\phi'(t)|^2\,dt = \pi^2 \int_0^1 |\phi(t)|^2\,dt,$$

then it is left to observe that

$$\int_0^1 \phi(t)\,dt = 0,$$

thus from $(*)$, we get

$$\inf_{c\in\mathbb{R}} \int_0^1 |\phi(t) - c|^2\,dt = \int_0^1 |\phi(t)|^2\,dt.$$

This concludes the exercise.

2.8.13 Let $\alpha \neq 2$ be a positive exponent. We take $v \in C^1([-1,1]) \setminus \{0\}$ such that $v(-1) = v(1) = 0$. Then we consider the function $v_C(t) = C\,v(t)$, for every $C > 0$. Observe that this function is still admissible and we have

$$\frac{\left(\displaystyle\int_{-1}^1 v_C\,dt\right)^\alpha}{\displaystyle\int_{-1}^1 |v_C'|^2\,dt} = C^{\alpha-2}\,\frac{\left(\displaystyle\int_{-1}^1 v\,dt\right)^\alpha}{\displaystyle\int_{-1}^1 |v'|^2\,dt}.$$

We then obtain

$$\sup_{\varphi\in C^1([-1,1])\setminus\{0\}} \left\{ \frac{\left(\displaystyle\int_{-1}^{1} \varphi\, dt\right)^{\alpha}}{\displaystyle\int_{-1}^{1} |\varphi'|^2\, dt} : \varphi(-1) = \varphi(1) = 0 \right\} \geq C^{\alpha-2}\, \frac{\left(\displaystyle\int_{-1}^{1} v\, dt\right)^{\alpha}}{\displaystyle\int_{-1}^{1} |v'|^2\, dt}.$$

Thus, if $0 < \alpha < 2$, we obtain that the previous supremum is $+\infty$, by taking the limit as $C \to 0^+$. If $\alpha > 2$, we take instead the limit as C goes to $+\infty$, so to obtain the same conclusion.
We now take $\alpha = 2$ and proceed point by point.

1. Let φ be an admissible function, by using Jensen's inequality (Proposition 1.2.11) we have

$$\left(\int_{-1}^{1} \varphi\, dt\right)^2 = 4\left(⨍_{-1}^{1} \varphi\, dt\right)^2 \leq 4 ⨍_{-1}^{1} |\varphi|^2\, dt = 2\int_{-1}^{1} |\varphi|^2\, dt.$$

We then use the Poincaré inequality for C^1 functions on $[-1, 1]$ vanishing at the boundary of this interval. By Corollary 2.4.7, we have

$$\int_{-1}^{1} |\varphi|^2\, dt \leq \frac{1}{\lambda([-1, 1])} \int_{-1}^{1} |\varphi'|^2\, dt.$$

The last two displays shows that

$$\frac{\left(\displaystyle\int_{-1}^{1} \varphi\, dt\right)^2}{\displaystyle\int_{-1}^{1} |\varphi'(t)|^2\, dt} \leq \frac{2}{\lambda([-1, 1])},$$

and by taking the supremum on the left-hand side, we get the desired estimate on $\mathcal{T}([-1, 1])$. For completeness, we observe that by Corollary 2.4.7 we also know that

$$\lambda([-1, 1]) = \frac{\pi^2}{4}.$$

Thus, the previous upper bound reads

$$\mathcal{T}([-1, 1]) \leq \frac{8}{\pi^2}.$$

2. For this point, we define

$$\mathcal{F}(\varphi) = 2\int_{-1}^{1} \varphi\, dt - \int_{-1}^{1} |\varphi'|^2\, dt,$$

for $\varphi \in C^1([-1, 1])$ such that $\varphi(-1) = \varphi(1) = 0$. We observe at first that

$$\mathcal{F}(0) = 0.$$

On the other hand, if φ is admissible, not identically vanishing and such that $\int_{-1}^{1} \varphi\, dt > 0$, then $\alpha\,\varphi$ is still admissible for every $\alpha > 0$ and

$$\begin{aligned}\mathcal{F}(\alpha\,\varphi) &= 2\,\alpha \int_{-1}^{1} \varphi\, dt - \alpha^2 \int_{-1}^{1} |\varphi'|^2\, dt \\ &= 2\,\alpha \int_{-1}^{1} \varphi\, dt + o(\alpha) > 0, \qquad \text{for } \alpha \to 0^+.\end{aligned}$$

The last two facts prove that

$$\begin{aligned}&\sup_{\varphi\in C^1([-1,1])} \Big\{\mathcal{F}(\varphi) \,:\, \varphi(-1) = \varphi(1) = 0\Big\} \\ &\qquad = \sup_{\varphi\in C^1([-1,1])\setminus\{0\}} \Big\{\mathcal{F}(\varphi) \,:\, \varphi(-1) = \varphi(1) = 0\Big\}.\end{aligned}$$

In order to get the claimed equality, we still exploit this idea of "vertically" scale the admissible functions. Indeed, since for every $\alpha \in \mathbb{R}$ the function $\alpha\,\varphi$ is still admissible for the maximization problem, we must have

$$\begin{aligned}&\sup_{\varphi\in C^1([-1,1])} \Big\{\mathcal{F}(\varphi) \,:\, \varphi(-1) = \varphi(1) = 0\Big\} \\ &\qquad = \sup_{\varphi\in C^1([-1,1])\setminus\{0\}} \Big\{\sup_{\alpha\in\mathbb{R}} \mathcal{F}(\alpha\,\varphi) \,:\, \varphi(-1) = \varphi(1) = 0\Big\}.\end{aligned} \tag{$*$}$$

Observe that we also used that the maximization problem on the left-hand side is equivalently settled on $C^1([-1, 1]) \setminus \{0\}$, as shown above. Given an admissible φ, let us now compute

$$\sup_{\alpha\in\mathbb{R}} \mathcal{F}(\alpha\,\varphi) = \sup_{\alpha\in\mathbb{R}} \left[2\,\alpha \int_{-1}^{1} \varphi\, dt - \alpha^2 \int_{-1}^{1} |\varphi'|^2\, dt\right].$$

Observe that the function of one real variable

$$g(\alpha) = 2\,\alpha \int_{-1}^{1} \varphi\, dt - \alpha^2 \int_{-1}^{1} |\varphi'|^2\, dt,$$

is concave and its unique critical point coincides with its global maximum point. Such a point is given by

$$g'(\alpha) = 0 \qquad \text{that is} \qquad 2\int_{-1}^{1} \varphi\, dt - 2\alpha \int_{-1}^{1} |\varphi'|^2\, dt = 0.$$

In other words, by setting[9]

$$\alpha_0 = \frac{\displaystyle\int_{-1}^{1} \varphi\, dt}{\displaystyle\int_{-1}^{1} |\varphi'|^2\, dt},$$

we have

$$\begin{aligned}
\sup_{\alpha\in\mathbb{R}} \mathcal{F}(\alpha\,\varphi) &= \mathcal{F}(\alpha_0\,\varphi) \\
&= 2\,\frac{\displaystyle\int_{-1}^{1} \varphi\, dt}{\displaystyle\int_{-1}^{1} |\varphi'|^2\, dt} \int_{-1}^{1} \varphi\, dt - \left(\frac{\displaystyle\int_{-1}^{1} \varphi\, dt}{\displaystyle\int_{-1}^{1} |\varphi'|^2\, dt}\right)^2 \int_{-1}^{1} |\varphi'|^2\, dt \\
&= \frac{\left(\displaystyle\int_{-1}^{1} \varphi\, dt\right)^2}{\displaystyle\int_{-1}^{1} |\varphi'|^2\, dt}.
\end{aligned}$$

By recalling $(*)$ and the definition of $\mathcal{T}([-1,1])$, we get

$$\begin{aligned}
&\sup_{\varphi\in C^1([-1,1])} \Big\{\mathcal{F}(\varphi)\,:\,\varphi(-1)=\varphi(1)=0\Big\} \\
&= \sup_{\varphi\in C^1([-1,1])\setminus\{0\}} \left\{ \frac{\left(\displaystyle\int_{-1}^{1} \varphi\, dt\right)^2}{\displaystyle\int_{-1}^{1} |\varphi'(t)|^2\, dt} \,:\, \varphi(-1)=\varphi(1)=0 \right\} = \mathcal{T}([-1,1]),
\end{aligned}$$

[9] Observe that this is well-defined, since for $\varphi \in C^1([-1,1]) \setminus \{0\}$ with $\varphi(-1) = \varphi(1) = 0$, we have

$$\int_{-1}^{1} |\varphi'|^2\, dt \neq 0.$$

as desired.

3. Here, we can take advantage of the previous point. Indeed, we can write

$$\begin{aligned}\mathcal{T}([-1,1]) &= \sup_{\varphi\in C^1([-1,1])} \left\{ 2\int_{-1}^{1} \varphi\,dt - \int_{-1}^{1} |\varphi'|^2\,dt \,:\, \varphi(-1)=\varphi(1)=0 \right\} \\ &= -2 \inf_{\varphi\in C^1([-1,1])} \left\{ \frac{1}{2}\int_{-1}^{1} |\varphi'|^2\,dt - \int_{-1}^{1} \varphi\,dt \,:\, \varphi(-1)=\varphi(1)=0 \right\}.\end{aligned}$$

We now observe that the last problem has been considered in Problem 1.7.20. There we showed that (for $p = 2$)

$$u_0(t) = \frac{1}{2}\left(1 - t^2\right),$$

is the unique solution. We are only left with computing

$$\begin{aligned}\frac{1}{2}\int_{-1}^{1} |u_0'(t)|^2\,dt - \int_{-1}^{1} u_0(t)\,dt &= \frac{1}{2}\int_{-1}^{1} t^2\,dt - \frac{1}{2}\int_{-1}^{1} (1 - t^2)\,dt \\ &= \int_{-1}^{1} |t|^2\,dt - \frac{1}{2}\int_{-1}^{1} dt = -\frac{1}{3}.\end{aligned}$$

We finally obtain

$$\mathcal{T}([-1,1]) = -2\left(\frac{1}{2}\int_{-1}^{1} |u_0'(t)|^2\,dt - \int_{-1}^{1} u_0(t)\,dt\right) = \frac{2}{3}.$$

2.8.14 We first observe that $\mathcal{T}([a,b])$ is invariant by translations of the interval $[a,b]$, thus it is sufficient to determine

$$\mathcal{T}([-L,L]), \qquad \text{where } L = \frac{b-a}{2}.$$

To this aim, we use a scaling argument. We now observe that every $\varphi \in C^1([-L,L])$ such that $\varphi(-L) = \varphi(L) = 0$, can be written in the form

$$\varphi(t)=\psi_L(t) := \psi(t/L), \qquad \text{where } \psi \in C^1([-1,1]) \text{ such that } \psi(-1)=\psi(1)=0.$$

Thus, we easily get

$$\begin{aligned}\mathcal{T}([-L,L]) &= \sup_{\psi\in C^1([-1,1])\setminus\{0\}} \left\{ \frac{\left(\displaystyle\int_{-L}^{L} \psi_L\,dt\right)^2}{\displaystyle\int_{-L}^{L} |\psi_L'|^2\,dt} \,:\, \psi(-L)=\psi(L)=0 \right\} \\ &= \sup_{\psi\in C^1([-1,1])\setminus\{0\}} \left\{ \frac{L^2\left(\displaystyle\int_{-1}^{1} \psi\,dt\right)^2}{L^{-1}\displaystyle\int_{-1}^{1} |\psi'|^2\,dt} \,:\, \psi(-1)=\psi(1)=0 \right\} \\ &= L^3\,\mathcal{T}([-1,1]).\end{aligned}$$

By using the previous problem and the definition of L, we then get

$$\mathcal{T}([a,b]) = \mathcal{T}([-L,L]) = L^3\,\mathcal{T}([-1,1]) = L^3\,\frac{2}{3} = \frac{(b-a)^3}{12}.$$

2.8.15 Let $u \in C^1(\overline{B_1(0)})$ be such that $u = 0$ on $\partial B_1(0)$. By using polar coordinates

$$x = \varrho\,\cos\vartheta, \qquad y = \varrho\,\sin\vartheta,$$

we have

$$\iint_{B_1(0)} \frac{1}{\sqrt{x^2+y^2}}\,|\nabla u|^2\,dx\,dy = \int_0^{2\pi}\int_0^1 \frac{1}{\varrho}\left[\left(\frac{\partial u}{\partial \varrho}\right)^2 + \frac{1}{\varrho^2}\left(\frac{\partial u}{\partial \vartheta}\right)^2\right]\varrho\,d\varrho\,d\vartheta,$$

where we used (A.1.3) and (A.1.4) from Appendix A, to write the modulus of the gradient in polar coordinates. We can bound from below the last integral by simply dropping the term with the derivative with respect to ϑ. In other word, we have

$$\begin{aligned}\iint_{B_1(0)} \frac{1}{\sqrt{x^2+y^2}}\,|\nabla u|^2\,dx\,dy &\ge \int_0^{2\pi}\int_0^1 \left(\frac{\partial u}{\partial \varrho}\right)^2 d\varrho\,d\vartheta \\ &= \int_0^{2\pi}\left(\int_0^1 \left(\frac{\partial u}{\partial \varrho}\right)^2 d\varrho\right) d\vartheta.\end{aligned} \tag{$*$}$$

We now observe that for every fixed $\vartheta \in [0, 2\pi]$, the function of one variable

$$\varrho \mapsto u(\varrho, \vartheta),$$

is defined on the interval $[0, 1]$ and by construction vanishes at the endpoint $\varrho = 1$. We can thus apply the Poincaré inequality of Theorem 2.4.2 (recall Remark 2.4.5) and infer that

$$\int_0^1 \left(\frac{\partial u}{\partial \varrho}\right)^2 d\varrho \geq \left(\frac{\pi}{2}\right)^2 \int_0^1 |u|^2 \, d\varrho. \qquad (**)$$

By integrating with respect to ϑ and using $(*)$, we then get

$$\iint_{B_1(0)} \frac{1}{\sqrt{x^2+y^2}} |\nabla u|^2 \, dx \, dy \geq \left(\frac{\pi}{2}\right)^2 \int_0^{2\pi} \int_0^1 |u|^2 \, d\varrho \, d\vartheta.$$

We are only left to observe that, by going back to cartesian variables, we have

$$\int_0^{2\pi} \int_0^1 |u|^2 \, d\varrho \, d\vartheta = \int_0^{2\pi} \int_0^1 \frac{1}{\varrho} |u|^2 \, \varrho \, d\varrho \, d\vartheta = \iint_{B_1(0)} \frac{1}{\sqrt{x^2+y^2}} |u|^2 \, dx \, dy,$$

thus proving the claimed weighted Poincaré inequality.

In order to identify the equality cases, we observe that if u satisfies the equality in the previous inequality, then in particular we must have equality in $(*)$. This implies that we must have

$$\frac{\partial u}{\partial \vartheta} = 0, \qquad \text{in } B_1(0).$$

In other words, u must be a radial function. Moreover, we must have equality in the Poincaré inequality $(**)$, as well. By recalling the equality cases of this last inequality (see Remark 2.4.5), we get the conclusion.

2.8.16 Let us argue by contradiction and suppose that there exists a universal constant $C > 0$ such that

$$|\Omega| \leq C \, (L(\Omega))^\alpha.$$

For every simply connected open bounded set $\Omega \subseteq \mathbb{R}^2$, we consider its scaled copy

$$\lambda \, \Omega = \left\{ x \in \mathbb{R}^2 \, : \, x = \lambda \, y, \text{ for some } y \in \Omega \right\},$$

where $\lambda > 0$. By the scaling properties of both area and perimeter, we have

$$|\lambda \, \Omega| = \lambda^2 \, |\Omega| \qquad \text{and} \qquad L(\lambda \, \Omega) = \lambda \, L(\Omega).$$

We then obtain

$$\lambda^2 \, |\Omega| = |\lambda \, \Omega| \leq C \, (L(\lambda \, \Omega))^\alpha = C \, \lambda^\alpha \, (L(\Omega))^\alpha.$$

By the arbitrariness of $\lambda > 0$, such an inequality is possible only if $\alpha = 2$. This gives the desired contradiction.

2.8.17 We proceed to prove each point separately.

1. Here the argument is similar to the one we used in Problem 2.8.12. We consider the function of two variables

$$g(x_1, x_2) = \mathcal{I}_{\partial\Omega}(x_1, x_2) = \int_{\partial\Omega} \Big((x - x_1)^2 + (y - x_2)^2\Big)\, d\ell(x, y).$$

By expanding the square, this can be written as

$$\begin{aligned} g(x_1, x_2) &= \int_{\partial\Omega} (x^2 + y^2)\, d\ell(x, y) \\ &\quad - 2\,x_1 \left(\int_{\partial\Omega} x\, d\ell(x, y)\right) - 2\,x_2 \left(\int_{\partial\Omega} y\, d\ell(x, y)\right) + (x_1^2 + x_2^2)\, L(\Omega). \end{aligned}$$

It is not difficult to see that g is a strictly convex function, indeed its Hessian matrix is given by the constant diagonal matrix

$$D^2 g(x_1, x_2) = \begin{bmatrix} 2\,L(\Omega) & 0 \\ 0 & 2\,L(\Omega) \end{bmatrix},$$

which is positive definite. The unique critical point of g is found by imposing

$$\nabla g(x_1, x_2) = (0, 0) \qquad \text{that is} \qquad \begin{cases} \displaystyle\int_{\partial\Omega} x\, d\ell(x, y) = x_1\, L(\Omega), \\ \displaystyle\int_{\partial\Omega} y\, d\ell(x, y) = x_2\, L(\Omega), \end{cases}$$

and this must be the unique minimizer of g, by Lemma 1.3.7. We call (x_Ω, y_Ω) this point and observe that this is given by

$$x_\Omega = \fint_{\partial\Omega} x\, d\ell(x, y), \qquad y_\Omega = \fint_{\partial\Omega} y\, d\ell(x, y).$$

In other words, (x_Ω, y_Ω) is the *center of mass* of $\partial\Omega$. In conclusion, we get

$$\inf_{(x_1, x_2) \in \mathbb{R}^2} \mathcal{I}_{\partial\Omega}(x_1, x_2) = \int_{\partial\Omega} \Big((x - x_\Omega)^2 + (y - y_\Omega)^2\Big)\, d\ell(x, y),$$

where (x_Ω, y_Ω) is the center of mass of $\partial\Omega$.

2. The proof of the claimed isoperimetric-type inequality is now already contained in the proof of Theorem 2.6.1. Indeed, take a parametrization $\gamma : [0, 1] \to \mathbb{R}^2$ of the boundary $\partial\Omega$. As in that proof, we can suppose that

$$|\gamma'(t)| = L, \qquad \text{for } t \in [0, 1],$$

where we used the shortcut notation $L = L(\Omega)$. Then the polar moment of inertia rewrites as

$$\begin{aligned} \mathcal{I}(\partial\Omega) &= \int_{\partial\Omega} ((x - x_\Omega)^2 + (y - y_\Omega)^2)\, d\ell(x, y) \\ &= L \int_0^1 |\gamma_1(t) - x_\Omega|^2\, dt + L \int_0^1 |\gamma_2(t) - y_\Omega|^2\, dt. \end{aligned} \tag{$*$}$$

We now observe that the barycenter of $\partial\Omega$ can be written in terms of γ as

$$x_\Omega = \fint_{\partial\Omega} x\, d\ell(x, y) = \int_0^1 \gamma_1(t)\, dt,$$

and

$$y_\Omega = \fint_{\partial\Omega} y\, d\ell(x, y) = \int_0^1 \gamma_2(t)\, dt.$$

We can thus apply Theorem 2.4.11 to the functions

$$\gamma_i - \int_0^1 \gamma_i\, dt, \qquad \text{for } i = 1, 2,$$

and infer that

$$\int_0^1 |\gamma_1(t) - x_\Omega|^2\, dt \le \frac{1}{4\pi^2} \int_0^1 |\gamma_1'(t)|^2\, dt,$$

and

$$\int_0^1 |\gamma_2(t) - y_\Omega|^2\, dt \le \frac{1}{4\pi^2} \int_0^1 |\gamma_2'(t)|^2\, dt.$$

In light of ($*$), we thus obtain

$$\mathcal{I}(\partial\Omega) \le \frac{L}{4\pi^2} \int_0^1 \left[|\gamma_1'(t)|^2 + |\gamma_2'(t)|^2 \right] dt = \frac{L^3}{4\pi^2},$$

that is

$$\frac{\mathcal{I}(\partial\Omega)}{(L(\Omega))^3} \le \frac{1}{4\pi^2}. \tag{$**$}$$

We now compute the quantity on the left-hand side for a disk D centered at a point (x_0, y_0) and having radius $R > 0$. We have

$$L(D) = 2\pi\, R.$$

Moreover, by using the parametrization $\eta(\vartheta) = (x_0 + R\cos\vartheta, y_0 + R\sin\vartheta)$, we can compute

$$\begin{aligned}\mathcal{I}(\partial D) = \min_{(x_1,x_2)\in\mathbb{R}^2} \mathcal{I}_{\partial D}(x_1, x_2) &= \int_{\partial D} \Big((x - x_0)^2 + (y - y_0)^2\Big)\, d\ell(x, y)\\ &= \int_0^{2\pi} R^2 \cdot R\, d\vartheta = 2\pi\, R^3.\end{aligned}$$

Observe that we used that the minimum of $(x_1, x_2) \mapsto \mathcal{I}_{\partial D}(x_1, x_2)$ is (uniquely) attained at the barycenter of ∂D, which coincides with the center of the disk. This entails that

$$\frac{\mathcal{I}(\partial D)}{(L(D))^3} = \frac{2\pi\, R^3}{8\pi^3\, R^3} = \frac{1}{4\pi^2}.$$

By recalling the inequality (∗∗), this permits to obtain the desired estimate.

This eventually concludes the exercise.

2.8.18 We consider the new function

$$\underline{v}(t) = \int_0^t |v'(\tau)|\, d\tau, \qquad \text{for } t \in [0, 1].$$

Observe that by construction we still have $\underline{v} \in C^1([0, 1])$ and $\underline{v}(0) = 0$. Moreover, by definition, we have

$$\underline{v}(t) = \int_0^t |v'(\tau)|\, d\tau \geq \left|\int_0^t v'(\tau)\, d\tau\right| = |v(t) - v(0)| = |v(t)|,$$

and thus

$$\underline{v}(t) \geq v(t), \qquad \text{for every } t \in [0, 1].$$

By also observing that

$$|\underline{v}'(t)| = |v'(t)|, \qquad \text{for every } t \in [0, 1],$$

the assumption on F implies that

$$F(t, \underline{v}(t), |\underline{v}'(t)|) \leq F(t, v(t), |v'(t)|), \qquad \text{for every } t \in [0, 1].$$

Thus, $\underline{v}$ is still a minimizer and by uniqueness we get $v = \underline{v}$. This concludes the exercise.

8.3 Problems of Chap. 3

3.12.1 We indicate by

$$\psi'(x) = \alpha\,|x|^{\alpha-2}\,x, \qquad \text{for } x \in (-1,1)\setminus\{0\},$$

the derivative in classical sense. We wish to prove that this coincides with the weak derivative, i.e. that it holds

$$\int_{-1}^{1} \psi(x)\,\varphi'(x)\,dx = -\int_{-1}^{1} \psi'(x)\,\varphi(x)\,dx, \qquad \text{for every } \varphi \in C_0^\infty((-1,1)). \tag{$*$}$$

Then the fact that this weak derivative is in L^2 would simply follows from a direct computation: indeed

$$\int_{-1}^{1} |\psi'(x)|^2\,dx = \alpha^2 \int_{-1}^{1} |x|^{2\alpha-2}\,dx$$
$$= 2\,\alpha^2 \int_0^1 x^{2\alpha-2}\,dx = 2\,\alpha^2 \lim_{\varepsilon\to 0^+} \left[\frac{x^{2\alpha-1}}{2\,\alpha-1}\right]_\varepsilon^1 = \frac{2\,\alpha^2}{2\,\alpha-1} < +\infty,$$

thanks to the fact that $\alpha > 1/2$.

Let us now prove $(*)$: we first observe that since ψ is continuous, for every $\varphi \in C_0^\infty((-1,1))$ we have

$$\int_{-1}^{1} \varphi'(x)\,\psi(x)\,dx = \lim_{\varepsilon\to 0^+} \int_{-1}^{-\varepsilon} \varphi'(x)\,\psi(x)\,dx + \lim_{\varepsilon\to 0^+} \int_{\varepsilon}^{1} \varphi'(x)\,\psi(x)\,dx. \tag{$**$}$$

On the intervals $[-1,-\varepsilon]$ and $[\varepsilon, 1]$ the function ψ is C^1, thus we can apply an integration by parts. This gives

$$\int_{-1}^{-\varepsilon} \varphi'(x)\,\psi(x)\,dx = \Big[\psi(x)\,\varphi(x)\Big]_{-1}^{-\varepsilon} - \int_{-1}^{-\varepsilon} \varphi(x)\,\psi'(x)\,dx$$
$$= \varphi(-\varepsilon)\,\varepsilon^\alpha - \int_{-1}^{-\varepsilon} \varphi(x)\,\psi'(x)\,dx,$$

and

$$\int_{\varepsilon}^{1} \varphi'(x)\,\psi(x)\,dx = \Big[\psi(x)\,\varphi(x)\Big]_{\varepsilon}^{1} - \int_{\varepsilon}^{1} \varphi(x)\,\psi'(x)\,dx$$
$$= -\varphi(\varepsilon)\,\varepsilon^{\alpha} - \int_{\varepsilon}^{1} \varphi(x)\,\psi'(x)\,dx.$$

By inserting these informations in $(**)$, we get

$$\begin{aligned}\int_{-1}^{1} \varphi'(x)\,\psi(x)\,dx &= \lim_{\varepsilon\to 0^+}\left[\varphi(-\varepsilon)\,\varepsilon^{\alpha} - \int_{-1}^{-\varepsilon} \varphi(x)\,\psi'(x)\,dx\right]\\ &\quad + \lim_{\varepsilon\to 0^+}\left[-\varphi(\varepsilon)\,\varepsilon^{\alpha} - \int_{\varepsilon}^{1} \varphi(x)\,\psi'(x)\,dx\right]\\ &= -\lim_{\varepsilon\to 0^+}\int_{-1}^{-\varepsilon} \varphi(x)\,\psi'(x)\,dx - \lim_{\varepsilon\to 0^+}\int_{\varepsilon}^{1} \varphi(x)\,\psi'(x)\,dx.\end{aligned}$$

By observing that $\psi'\,\varphi \in L^1((-1,1))$, we have that the last two integrals are converging to

$$-\int_{-1}^{0} \varphi(x)\,\psi'(x)\,dx - \int_{0}^{1} \varphi(x)\,\psi'(x)\,dx = -\int_{-1}^{1} \varphi(x)\,\psi'(x)\,dx,$$

thus proving $(**)$.

Finally, let us show that the function $\psi(x) = \sqrt{|x|}$ does not have a weak derivative in $L^2((-1,1))$. We argue by contradiction and assume that there exists $g \in L^2((-1,1))$ such that

$$-\int_{-1}^{1} \psi(x)\,\varphi'(x)\,dx = \int_{-1}^{1} g(x)\,\varphi(x)\,dx, \qquad \text{for every } \varphi \in C_0^{\infty}((-1,1)).$$

In particular, we can use this identity for functions $\varphi \in C_0^{\infty}((-1,-\delta)\cup(\delta,1))$ and every $0<\delta<1$: by using an integration by parts, we would get

$$\begin{aligned}\int_{-1}^{1} g(x)\,\varphi(x)\,dx &= -\int_{-1}^{1} \psi(x)\,\varphi'(x)\,dx\\ &= -\int_{-1}^{-\delta} \psi(x)\,\varphi'(x)\,dx - \int_{\delta}^{1} \psi(x)\,\varphi'(x)\,dx\\ &= \int_{-1}^{-\delta} \psi'(x)\,\varphi(x)\,dx + \int_{\delta}^{1} \psi'(x)\,\varphi(x)\,dx,\end{aligned}$$

thanks to the fact that ψ is C^1 on $[-1,-\delta]\cup[\delta,1]$. By arbitrariness of φ, from Lemma 1.4.2 we get that

$$g(x) = \psi'(x), \qquad \text{for a. e. } x \in (-1, -\delta) \cup (\delta, 1),$$

and every $\delta > 0$. This in turn gives that g must coincide almost everywhere in $(-1, 1)$ with the classical derivative ψ'. However, we have

$$\psi'(x) = \frac{1}{2\sqrt{|x|}} \frac{x}{|x|}, \qquad \text{for a. e. } x \in (-1, 1),$$

and this function does not belong to $L^2((-1, 1))$, thus getting a contradiction.

3.12.2 This is quite similar to the previous exercise, it is sufficient to carefully repeat the above reasoning, with the due adaptations to the higher dimensional case. We indicate by

$$\nabla\psi(x, y) = \alpha\,(x^2 + y^2)^{\frac{\alpha-2}{2}}\,(x, y), \qquad \text{for } (x, y) \in B_1(0) \setminus \{0\},$$

the gradient in classical sense. We wish to prove that this coincides with the weak gradient, i.e. that for every $\varphi \in C_0^\infty(B_1(0))$ it holds

$$\int_{B_1(0)} \frac{\partial\varphi}{\partial x}(x, y)\,\psi(x, y)\,dx\,dy = -\int_{B_1(0)} \varphi(x, y)\,\frac{\partial\psi}{\partial x}(x, y)\,dx\,dy, \tag{$*$}$$

and

$$\int_{B_1(0)} \frac{\partial\varphi}{\partial y}(x, y)\,\psi(x, y)\,dx\,dy = -\int_{B_1(0)} \varphi(x, y)\,\frac{\partial\psi}{\partial y}(x, y)\,dx\,dy. \tag{$**$}$$

Then the fact that this weak gradient belongs to $L^2(B_1(0))$ would simply follows from a direct computation, whose details are left to the reader. It is sufficient to observe that

$$|\nabla\psi(x, y)|^2 = \alpha^2\,(x^2 + y^2)^{\alpha-1},$$

and then use polar coordinates to compute its integral over $B_1(0)$.

We focus on proving $(*)$, the proof of $(**)$ being exactly the same. As in the previous exercise, we observe that since ψ is continuous, for every $\varphi \in C_0^\infty(B_1(0))$ we have

$$\int_{B_1(0)} \frac{\partial\varphi}{\partial x}(x, y)\,\psi(x, y)\,dx\,dy = \lim_{\varepsilon\to 0^+} \int_{B_1(0)\setminus B_\varepsilon(0)} \frac{\partial\varphi}{\partial x}(x, y)\,\psi(x, y)\,dx\,dy.$$

Observe that for every fixed $\varepsilon > 0$, the function ψ is C^1 on the annular set $B_1(0) \setminus B_\varepsilon(0)$. Thus we can apply the Divergence Theorem as follows

$$
\begin{aligned}
&\int_{B_1(0)\setminus B_\varepsilon(0)} \frac{\partial \varphi}{\partial x}(x,y)\,\psi(x,y)\,dx\,dy \\
&= \int_{B_1(0)\setminus B_\varepsilon(0)} \operatorname{div}\Big(\varphi(x,y)\,\psi(x,y)\,\mathbf{e}_1\Big)\,dx\,dy \\
&- \int_{B_1(0)\setminus B_\varepsilon(0)} \varphi(x,y)\,\frac{\partial \psi}{\partial x}(x,y)\,dx\,dy \\
&= \int_{\partial B_1(0)} \langle \varphi(x,y)\,\psi(x,y)\,\mathbf{e}_1, (x,y)\rangle\,d\ell(x,y) \\
&- \frac{1}{\varepsilon}\int_{\partial B_\varepsilon(0)} \langle \varphi(x,y)\,\psi(x,y)\,\mathbf{e}_1, (x,y)\rangle\,d\ell(x,y) \\
&- \int_{B_1(0)\setminus B_\varepsilon(0)} \varphi(x,y)\,\frac{\partial \psi}{\partial x}(x,y)\,dx\,dy.
\end{aligned}
$$

In the second identity, we used that the unit outer normal vector ν to the boundary of $B_1(0) \setminus B_\varepsilon(0)$ is given by

$$
\nu(x,y) = (x,y), \qquad \text{for } (x,y) \in \partial B_1(0),
$$

and

$$
\nu(x,y) = \left(-\frac{x}{\varepsilon}, -\frac{y}{\varepsilon}\right), \qquad \text{for } (x,y) \in \partial B_\varepsilon(0).
$$

By recalling that $\varphi \in C_0^\infty(B_1(0))$, we get that

$$
\int_{\partial B_1(0)} \langle \varphi(x,y)\,\psi(x,y)\,\mathbf{e}_1, (x,y)\rangle\,d\ell(x,y) = 0.
$$

Thus, up to now we have obtained

$$
\begin{aligned}
\int_{B_1(0)} \frac{\partial \varphi}{\partial x}(x,y)\,\psi(x,y)\,dx\,dy &= \lim_{\varepsilon\to 0^+} \int_{B_1(0)\setminus B_\varepsilon(0)} \frac{\partial \varphi}{\partial x}(x,y)\,\psi(x,y)\,dx\,dy \\
&= -\lim_{\varepsilon\to 0^+} \frac{1}{\varepsilon}\int_{\partial B_\varepsilon(0)} \varphi(x,y)\,\psi(x,y)\,x\,d\ell(x,y) \\
&- \lim_{\varepsilon\to 0^+} \int_{B_1(0)\setminus B_\varepsilon(0)} \varphi(x,y)\,\frac{\partial \psi}{\partial x}(x,y)\,dx\,dy.
\end{aligned}
$$

We now observe that ψ is a radial function, this entails that

$$
\psi(x,y) = \varepsilon^\alpha, \qquad \text{for } (x,y) \in \partial B_\varepsilon(0).
$$

Moreover, we also have

$$|x| \leq \sqrt{x^2 + y^2} = \varepsilon, \qquad \text{for } (x, y) \in \partial B_\varepsilon(0).$$

Thus, we can infer

$$\left| \frac{1}{\varepsilon} \int_{\partial B_\varepsilon(0)} \varphi(x, y)\, \psi(x, y)\, x\, d\ell(x, y) \right| \leq \varepsilon^\alpha \int_{\partial B_\varepsilon(0)} |\varphi(x, y)|\, d\ell(x, y).$$

The last quantity converges to 0, as ε goes to 0 (recall that $\alpha > 0$). In conclusion, we get

$$\int_{B_1(0)} \frac{\partial \varphi}{\partial x}(x, y)\, \psi(x, y)\, dx\, dy = - \lim_{\varepsilon \to 0+} \int_{B_1(0) \setminus B_\varepsilon(0)} \varphi(x, y)\, \frac{\partial \psi}{\partial x}(x, y)\, dx\, dy.$$

It is only left to observe that

$$\varphi(x, y)\, \frac{\partial \psi}{\partial x}(x, y) \in L^1(B_1(0)),$$

which can be obtained by using polar coordinates. Then from the previous identity we get $(*)$, as desired.

3.12.3 We take $\varphi \in C_0^\infty((a, b))$, we need to prove that

$$\int_a^b \varphi'(x)\, \psi(x)\, dx = - \int_a^b \varphi(x)\, \psi'(x)\, dx,$$

where ψ' denotes the derivative in classical sense, which is well defined for $x \in [a, b] \setminus \{t_1, \dots, t_{n-1}\}$ (see Fig. 8.4). We start by decomposing the first integral as follows

$$\int_a^b \varphi'(x)\, \psi(x)\, dx = \sum_{i=0}^{n-1} \int_{t_i}^{t_{i+1}} \varphi'(x)\, \psi(x)\, dx.$$

By assumption, we have $\psi \in C^1([t_i, t_{i+1})$, thus we can use an integration by parts on each interval. In other words, we have

$$\begin{aligned} \int_{t_i}^{t_{i+1}} \varphi'(x)\, \psi(x)\, dx &= \Big[\varphi(x)\, \psi(x)\Big]_{t_i}^{t_{i+1}} - \int_{t_i}^{t_{+1}} \varphi(x)\, \psi'(x)\, dx \\ &= \Big[\varphi(t_{i+1})\, \psi(t_{i+1}) - \varphi(t_i)\, \psi(t_i)\Big] - \int_{t_i}^{t_{+1}} \varphi(x)\, \psi'(x)\, dx. \end{aligned}$$

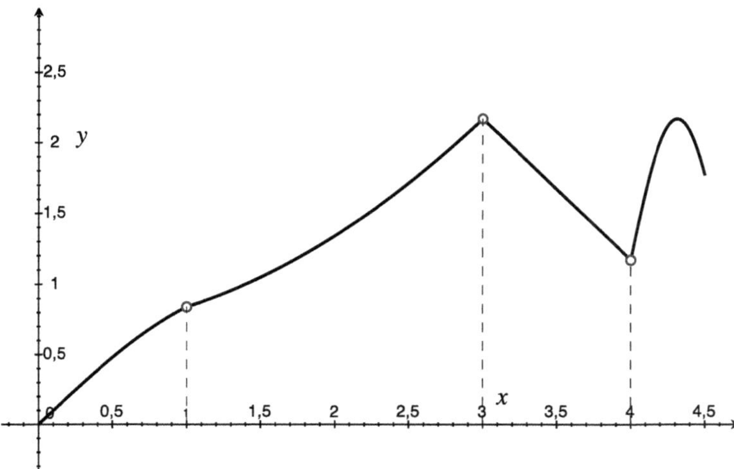

Fig. 8.4 An example of function which satisfies the assumptions of Problem 3.12.3. We highlight the "junction" points where the function is not differentiable in classical sense

By summing up, we get

$$
\begin{aligned}
\int_a^b \varphi'(x)\,\psi(x)\,dx &= \sum_{i=0}^{n-1}\Big[\varphi(t_{i+1})\,\psi(t_{i+1}) - \varphi(t_i)\,\psi(t_i)\Big] \\
&\quad - \sum_{i=0}^{n-1}\int_{t_i}^{t_{+1}} \varphi(x)\,\psi'(x)\,dx \\
&= \sum_{i=0}^{n-1}\Big[\varphi(t_{i+1})\,\psi(t_{i+1}) - \varphi(t_i)\,\psi(t_i)\Big] - \int_a^b \varphi(x)\,\psi'(x)\,dx.
\end{aligned}
$$

In order to conclude, we only need to prove that the last sum vanishes. For this, it is sufficient to observe that

$$
\varphi(t_n) = \varphi(b) = 0 \qquad \text{and} \qquad \varphi(t_0) = \varphi(a) = 0,
$$

because φ is compactly supported and the terms which are left form a telescopic sum. This concludes the exercise.

3.12.4 We verify at first that each function ψ_n belongs to $W^{1,1}((-1, 1))$. By observing that each of these functions is a piecewise C^1 function on $[0, 1]$, we get the desired conclusion from Problem 3.12.3. From the latter, we also know that its weak derivative is given by the function

$$\psi_n'(t) = \begin{cases} n, & \text{if } t < 1/n, \\ 0, & \text{if } 1/n < t < 1. \end{cases}$$

Thus, we get

$$\|\psi_n'\|_{L^1((0,1))} = \int_0^{\frac{1}{n}} n\,dt = 1, \qquad \text{for every } n \in \mathbb{N},$$

and

$$\|\psi_n\|_{L^1((0,1))} = \int_0^{\frac{1}{n}} n\,t\,dt + \int_{\frac{1}{n}}^1 dt = \frac{1}{2n} + 1 - \frac{1}{n} \leq 1, \qquad \text{for every } n \in \mathbb{N}.$$

This shows that

$$\|\psi_n\|_{W^{1,1}((0,1))} \leq 2, \qquad \text{for every } n \in \mathbb{N},$$

and thus the first part of the statement follows. Let us now argue by contradiction and suppose that there exists a subsequence (not relabeled) and a function $\psi \in W^{1,1}((0,1))$ such that

$$\lim_{n\to\infty} \int_0^1 \psi_n\,\varphi\,dt = \int_0^1 \psi\,\varphi\,dt, \qquad \text{for every } \varphi \in L^\infty((0,1)),$$

and

$$\lim_{n\to\infty} \int_0^1 \psi_n'\,\varphi\,dt = \int_0^1 \psi'\,\varphi\,dt, \qquad \text{for every } \varphi \in L^\infty((0,1)).$$

Observe that

$$\int_0^1 |\psi_n - 1|\,dt = \int_0^{\frac{1}{n}} (1 - n\,t)\,dt = \frac{1}{n} - \frac{1}{2n},$$

thus we get that ψ_n converges strongly in $L^1((0,1))$ to the constant function 1. This in turn implies that the weak limit ψ must coincide with the constant function 1, whose weak derivative is the null function on the interval $(0,1)$. From this discussion, we thus have obtained

$$\lim_{n\to\infty} \int_0^1 \psi_n'\,\varphi\,dt = 0, \qquad \text{for every } \varphi \in L^\infty((0,1)).$$

By recalling the expression for the weak derivative ψ_n', this is the same as

$$\lim_{n\to\infty} n \int_0^{\frac{1}{n}} \varphi\, dt = 0, \qquad \text{for every } \varphi \in L^\infty((0,1)).$$

If we now test this identity with $\varphi \equiv 1$, we get

$$\lim_{n\to\infty} n \int_0^{\frac{1}{n}} dt = 0,$$

which is a plain contradiction.

3.12.5 It is easily seen that $\widetilde{u} \in L^1((-1,1))$. We first observe that[10]

$$\widetilde{u} \in W^{1,1}((-1,0) \cup (0,1)).$$

Indeed, on $(0,1)$ this function coincides with u and thus $\widetilde{u}' = u' \in L^1((0,1))$, by assumption. On the interval $(-1,0)$, the function $\widetilde{u}$ is the composition of u with an affine change of variable, thus it follows from Theorem 3.2.7 that $\widetilde{u}' \in L^1((-1,0))$ and

$$\widetilde{u}'(x) = -u'(-x), \qquad \text{for a. e. } x \in (-1,0). \tag{$*$}$$

We will show that $\widetilde{u}$ has a weak derivative in the whole $(-1,1)$, given by the L^1 function

$$\begin{cases} u'(x), & \text{for a. e. } x \in (0,1), \\ -u'(-x), & \text{for a. e. } x \in (-1,0). \end{cases}$$

To this aim, for every $\varepsilon > 0$ small enough, we take a cut-off function $\eta_\varepsilon \in C_0^\infty((-\varepsilon,\varepsilon))$ such that

$$0 \le \eta_\varepsilon \le 1, \qquad \eta_\varepsilon \equiv 1 \text{ in } \left[-\frac{\varepsilon}{2}, \frac{\varepsilon}{2}\right], \qquad \|\eta_\varepsilon'\|_{L^\infty} \le \frac{C}{\varepsilon}.$$

We further suppose that η_ε is even, i.e.

$$\eta_\varepsilon(x) = \eta_\varepsilon(-x), \qquad \text{for } x \in (-\varepsilon,\varepsilon).$$

Let $\varphi \in C_0^\infty((-1,1))$, then by the Dominated Convergence Theorem

$$\int_{-1}^1 \widetilde{u}\,\varphi'\, dx = \lim_{\varepsilon\to 0^+} \int_{-1}^1 \widetilde{u}\,(1-\eta_\varepsilon)\,\varphi'\, dx.$$

[10] The unexperienced reader should observe that this is *not* sufficient in order to conclude that $\widetilde{u} \in W^{1,1}((-1,1))$. Indeed, recall the example of Remark 3.2.5.

In particular, we get

$$\begin{aligned}\int_{-1}^{1} \widetilde{u}\,\varphi'\,dx &= \lim_{\varepsilon\to 0^+} \int_{-1}^{1} \widetilde{u}\,(1-\eta_\varepsilon)\,\varphi'\,dx \\ &= \lim_{\varepsilon\to 0^+} \int_{-1}^{1} \widetilde{u}\,\Big((1-\eta_\varepsilon)\,\varphi\Big)'\,dx + \lim_{\varepsilon\to 0^+} \int_{-1}^{1} \widetilde{u}\,\eta_\varepsilon'\,\varphi\,dx.\end{aligned} \tag{$**$}$$

We prove at first that the last limit is zero. Indeed, by using the definition of $\widetilde{u}$ and the properties of η_ε, we have

$$\begin{aligned}\int_{-1}^{1} \widetilde{u}\,\eta_\varepsilon'\,\varphi\,dx = \int_{-\varepsilon}^{\varepsilon} \widetilde{u}\,\eta_\varepsilon'\,\varphi\,dx &= \int_{0}^{\varepsilon} \widetilde{u}\,\eta_\varepsilon'\,\varphi\,dx + \int_{-\varepsilon}^{0} \widetilde{u}\,\eta_\varepsilon'\,\varphi\,dx \\ &= \int_0^\varepsilon u(x)\,\eta_\varepsilon'(x)\,\varphi(x)\,dx \\ &\quad + \int_0^\varepsilon u(x)\,\eta_\varepsilon'(-x)\,\varphi(-x)\,dx.\end{aligned}$$

In the last identity, we used a simple change of variable. We then observe that η_ε' is odd by construction, thus $\eta_\varepsilon'(-x) = -\eta_\varepsilon'(x)$ and we can write

$$\int_{-1}^{1} \widetilde{u}\,\eta_\varepsilon'\,\varphi\,dx = \int_0^\varepsilon u(x)\,\eta_\varepsilon'(x)\,\Big(\varphi(x) - \varphi(-x)\Big)\,dx.$$

By using that for $x \in [0, \varepsilon]$ we have

$$|\varphi(x) - \varphi(-x)| = \left|\int_{-x}^{x} \varphi'(t)\,dt\right| \le \int_{-x}^{x} |\varphi'(t)|\,dt \le 2\,\|\varphi'\|_{L^\infty((-1,1))}\,x,$$

we can thus infer

$$\begin{aligned}\left|\int_{-1}^{1} \widetilde{u}\,\eta_\varepsilon'\,\varphi\,dx\right| &\le \frac{2\,C\,\|\varphi'\|_{L^\infty((-1,1))}}{\varepsilon} \int_0^\varepsilon |u(x)|\,x\,dx \\ &\le 2\,C\,\|\varphi'\|_{L^\infty((-1,1))} \int_0^\varepsilon |u(x)|\,dx.\end{aligned}$$

This finally implies that

$$\lim_{\varepsilon\to 0^+} \int_{-1}^{1} \widetilde{u}\,\eta_\varepsilon'\,\varphi\,dx = 0,$$

and thus by $(**)$ we get

$$\int_{-1}^{1} \widetilde{u}\,\varphi'\,dx = \lim_{\varepsilon\to 0^+} \int_{-1}^{1} \widetilde{u}\,\Big((1-\eta_\varepsilon)\,\varphi\Big)'\,dx.$$

We now observe that $(1-\eta_\varepsilon)\,\varphi \in C_0^\infty((-1,0)\cup(0,1))$, by construction. Thus, by using the definition of weak derivative on both intervals $(-1,0)$ and $(0,1)$, we get

$$\begin{aligned}\int_{-1}^{1} \widetilde{u}\left((1-\eta_\varepsilon)\,\varphi\right)' dx &= \int_{-1}^{0} \widetilde{u}\left((1-\eta_\varepsilon)\,\varphi\right)' dx + \int_{0}^{1} u\left((1-\eta_\varepsilon)\,\varphi\right)' dx \\ &= -\int_{-1}^{0} \widetilde{u}'\,(1-\eta_\varepsilon)\,\varphi\,dx - \int_{0}^{1} u'\,(1-\eta_\varepsilon)\,\varphi\,dx.\end{aligned}$$

In turn, by the Dominated Convergence Theorem, we get

$$\begin{aligned}\int_{-1}^{1} \widetilde{u}\,\varphi'\,dx &= \lim_{\varepsilon\to 0^+} \int_{-1}^{1} \widetilde{u}\left((1-\eta_\varepsilon)\,\varphi\right)' dx \\ &= \lim_{\varepsilon\to 0^+} \left[-\int_{-1}^{0} \widetilde{u}'\,(1-\eta_\varepsilon)\,\varphi\,dx - \int_{0}^{1} u'\,(1-\eta_\varepsilon)\,\varphi\,dx\right] \\ &= -\int_{-1}^{0} \widetilde{u}'\,\varphi\,dx - \int_{0}^{1} u'\,\varphi\,dx.\end{aligned}$$

By recalling $(*)$, we obtain in particular

$$\begin{aligned}\int_{-1}^{1} \widetilde{u}\,\varphi'\,dx &= -\int_{-1}^{0} \widetilde{u}'\,\varphi\,dx - \int_{0}^{1} u'\,\varphi\,dx \\ &= -\int_{-1}^{0} (-u'(-x))\,\varphi(x)\,dx - \int_{0}^{1} u'\,\varphi\,dx,\end{aligned}$$

as desired.

3.12.6 This is quite easy, it is sufficient to use the inclusion properties of L^p spaces, i.e. the fact that for $p > q$ we have

$$L^p(\Omega) \subseteq L^q(\Omega),$$

whenever Ω has finite Lebesgue measure. We recall that, for $p < \infty$, this simply follows by using Jensen's inequality (i.e. Proposition 1.2.11) with the convex function $\tau \mapsto \tau^{p/q}$, that is

$$\begin{aligned}\left(\int_\Omega |f|^q\,dx\right)^{\frac{p}{q}} &= |\Omega|^{\frac{p}{q}} \left(\fint |f|^q\,dx\right)^{\frac{p}{q}} \\ &\le |\Omega|^{\frac{p}{q}} \fint |f|^p\,dx, \qquad \text{for every } f \in L^p(\Omega).\end{aligned}$$

The case $p = \infty$ is even simpler, it is sufficient to observe that

$$\int_\Omega |f|^q\,dx \le \|f\|^q_{L^\infty(\Omega)}\,|\Omega|, \qquad \text{for every } f \in L^\infty(\Omega).$$

Thus, for every $u \in W^{1,p}(\Omega)$, we obtain (again we consider the case $p < \infty$ for simplicity)

$$\begin{aligned}&\left(\int_\Omega |u|^q\,dx\right)^{\frac{p}{q}} + \left(\int_\Omega |\nabla u|^q\,dx\right)^{\frac{p}{q}}\\ &\le |\Omega|^{\frac{p-q}{q}}\left(\int_\Omega |u|^p\,dx + \int_\Omega |\nabla u|^p\,dx\right) < +\infty,\end{aligned}$$

which implies that $u \in W^{1,q}(\Omega)$, as well.

Let us now take $\Omega = \mathbb{R}$ and consider the function

$$u(x) = \frac{1}{\sqrt{1+x^2}}, \qquad \text{for } x \in \mathbb{R}.$$

This is a C^1 function such that

$$u \in L^p(\mathbb{R}), \quad \text{for every } p > 1 \qquad \text{and} \qquad u \notin L^1(\mathbb{R}).$$

Moreover, its classical derivative is given by

$$u'(x) = -\frac{x}{(1+x^2)^{\frac{3}{2}}},$$

and this belongs to every $L^p(\mathbb{R})$, for $1 \le p \le \infty$. We thus obtain for example that

$$u \in W^{1,2}(\mathbb{R}) \qquad \text{but} \qquad u \notin W^{1,1}(\mathbb{R}).$$

3.12.7 We first observe that since $u \in W^{1,p}(\Omega)$, in particular we get $u \in L^p(\Omega)$, i.e.

$$\int_\Omega |u|^p\,dx < +\infty.$$

This can also be rephrased as $|u|^p \in L^1(\Omega)$. In order to conclude, we thus need to show that $\nabla |u|^p \in L^1(\Omega;\mathbb{R}^N)$ and that this weak gradient is given by $p\,|u|^{p-2}\,u\,\nabla u$. The proof is similar to that of Proposition 3.4.1.

We take $\varphi \in C^\infty_0(\Omega)$ and consider the sequence $\{u_n\}_{n\in\mathbb{N}}$ of Proposition 3.2.6. Thanks to the fact that φ is compactly supported in Ω, there exists $\Omega' \Subset \Omega$ such that Ω' contains the support of φ. We clearly have that $\Omega' \Subset \mathrm{int}_n(\Omega)$, for n large enough.

By Proposition 3.2.6 we have $|u_n|^p \in C^1(\mathrm{int}_n(\Omega))$, with the classical gradient given by

$$\nabla |u_n|^p = p\,|u_n|^{p-2}\,u_n\,\nabla u_n.$$

By Proposition 3.2.3, we get that this is the weak gradient, as well. Thus, we obtain

$$\int_\Omega \frac{\partial \varphi}{\partial x_k}\,|u_n|^p\,dx = -\int_\Omega \varphi\,\frac{\partial}{\partial x_k}|u_n|^p\,dx = -p\int_\Omega \varphi\,|u_n|^{p-2}\,u_n\,\frac{\partial u_n}{\partial x_k}\,dx, \tag{$*$}$$

for n large enough. We now recall that from Proposition 3.2.6, we have

$$\lim_{n\to\infty} \|u_n - u\|_{L^p(\Omega')} = 0 \qquad \text{and} \qquad \lim_{n\to\infty} \|\nabla u_n - \nabla u\|_{L^p(\Omega';\mathbb{R}^N)} = 0.$$

From the first information, by using Problem 1.7.4 and the fact that the support of φ is contained in Ω', we get

$$\begin{aligned}
\lim_{n\to\infty} &\left|\int_\Omega \frac{\partial \varphi}{\partial x_k}\,|u_n|^p\,dx - \int_\Omega \frac{\partial \varphi}{\partial x_k}\,|u|^p\,dx\right| \\
&\le \lim_{n\to\infty} \|\nabla\varphi\|_{L^\infty(\Omega';\mathbb{R}^N)} \int_{\Omega'} \big||u_n|^p - |u_n|^p\big|\,dx \\
&\le p\,\|\nabla\varphi\|_{L^\infty(\Omega';\mathbb{R}^N)} \lim_{n\to\infty} \int_{\Omega'} \big||u_n| - |u|\big|\,(|u_n| + |u|)^{p-1}\,dx \\
&\le p\,\|\nabla\varphi\|_{L^\infty(\Omega';\mathbb{R}^N)} \lim_{n\to\infty} \int_{\Omega'} |u_n - u|\,(|u_n| + |u|)^{p-1}\,dx \\
&\le \|\nabla\varphi\|_{L^\infty(\Omega';\mathbb{R}^N)} \lim_{n\to\infty} \|u_n - u\|_{L^p(\Omega')}\,\big\||u_n| + |u|\big\|_{L^p(\Omega')}^{p-1} \\
&\le C \lim_{n\to\infty} \|u_n - u\|_{L^p(\Omega')} = 0.
\end{aligned}$$

As for the right-hand side of $(*)$, we have

$$\begin{aligned}
&\left|\int_\Omega \varphi\,|u_n|^{p-2}\,u_n\,\frac{\partial u_n}{\partial x_k}\,dx - \int_\Omega \varphi\,|u|^{p-2}\,u\,\frac{\partial u}{\partial x_k}\,dx\right| \\
&\qquad \le \|\varphi\|_{L^\infty(\Omega')} \int_{\Omega'} \left||u_n|^{p-2}\,u_n\,\frac{\partial u_n}{\partial x_k} - |u|^{p-2}\,u\,\frac{\partial u}{\partial x_k}\right|\,dx \\
&\qquad \le \|\varphi\|_{L^\infty(\Omega')} \int_{\Omega'} |u_n|^{p-1}\left|\frac{\partial u_n}{\partial x_k} - \frac{\partial u}{\partial x_k}\right|\,dx \\
&\qquad + \|\varphi\|_{L^\infty(\Omega')} \int_{\Omega'} \big||u_n|^{p-2}\,u_n - |u|^{p-2}\,u\big|\,\left|\frac{\partial u}{\partial x_k}\right|\,dx.
\end{aligned}$$

By using that $\{u_n\}_{n\in\mathbb{N}}$ is bounded in $L^p(\Omega')$ and the strong convergence of the gradients, we get by Hölder's inequality that

$$\lim_{n\to\infty}\int_{\Omega'}|u_n|^{p-1}\left|\frac{\partial u_n}{\partial x_k}-\frac{\partial u}{\partial x_k}\right|dx\le C\lim_{n\to\infty}\left\|\frac{\partial u_n}{\partial x_k}-\frac{\partial u}{\partial x_k}\right\|_{L^p(\Omega')}=0.$$

For the second integral, we get from Hölder's inequality

$$\int_{\Omega'}\left||u_n|^{p-2}u_n-|u|^{p-2}u\right|\left|\frac{\partial u}{\partial x_k}\right|dx\le\left\|\frac{\partial u}{\partial x_k}\right\|_{L^p(\Omega')}\left\||u_n|^{p-2}u_n-|u|^{p-2}u\right\|_{L^{p'}(\Omega')}.$$

In order to show that the last $L^{p'}$ norm converges to 0, we need to distinguish two cases:

- if $1<p\le 2$, we have from Problem 1.7.3 with $\alpha=p-1$

  $$\left||u_n|^{p-2}u-|u|^{p-2}u\right|\le 2^{2-p}\,|u_n-u|^{p-1}.$$

  This shows that

  $$\lim_{n\to\infty}\left\||u_n|^{p-2}u_n-|u|^{p-2}u\right\|_{L^{p'}(\Omega')}\le 2^{2-p}\lim_{n\to\infty}\|u_n-u\|^{p-1}_{L^p(\Omega')}=0;$$

- if $p>2$, we use Problem 1.7.4 with $\alpha=p-1$, so to get

  $$\left||u_n|^{p-2}u-|u|^{p-2}u\right|\le(p-1)\,(|u_n|+|u|)^{p-2}\,|u_n-u|.$$

  This implies that

  $$\begin{aligned}&\left\||u_n|^{p-2}u_n-|u|^{p-2}u\right\|^{p'}_{L^{p'}(\Omega')}\\&\le(p-1)^{\frac{p}{p-1}}\int_{\Omega'}(|u_n|+|u|)^{\frac{p(p-2)}{p-1}}\,|u_n-u|^{\frac{p}{p-1}}\,dx\\&\le(p-1)^{\frac{p}{p-1}}\left(\int_{\Omega'}(|u_n|+|u|)^p\,dx\right)^{\frac{p-2}{p-1}}\\&\quad\times\left(\int_{\Omega'}|u_n-u|^p\,dx\right)^{\frac{1}{p-1}},\end{aligned}$$

  thanks to Hölder's inequality with exponents $(p-1)$ and $(p-1)/(p-2)$. This shows again that

  $$\lim_{n\to\infty}\left\||u_n|^{p-2}u_n-|u|^{p-2}u\right\|_{L^{p'}(\Omega')}=0.$$

By taking the limit on both sides of (∗), we then get

$$\int_\Omega \frac{\partial \varphi}{\partial x_k}\,|u|^p\,dx = -p\int_\Omega \varphi\,|u|^{p-2}\,u\,\frac{\partial u}{\partial x_k}\,dx.$$

This is valid for an arbitrary $\varphi \in C_0^\infty(\Omega)$. Moreover, we have $|u|^{p-2}\,u\,\nabla u \in L^1(\Omega;\mathbb{R}^N)$, since

$$\int_\Omega \big||u|^{p-2}\,u\,\nabla u\big|\,dx = \int_\Omega |u|^{p-1}\,|\nabla u|\,dx \le \left(\int_\Omega |u|^p\,dx\right)^{\frac{p-1}{p}}\left(\int_\Omega |\nabla u|^p\,dx\right)^{\frac{1}{p}}.$$

This shows that $|u|^p$ has a weak gradient in $L^1(\Omega;\mathbb{R}^N)$ and this is given by

$$\nabla|u|^p = p\,|u|^{p-2}\,u\,\nabla u.$$

This is enough to conclude.

3.12.8 We try to mimick the proof of Theorem 3.5.6. We first establish the validity of the inequality for $\varphi \in C_0^\infty(\mathbb{H}_+^N)$. For $\alpha > 0$, we set $h(t) = t^\alpha$, for every $t > 0$. Then the function $H(x) = h(x_N)$ is such that

$$\nabla H(x) = (0,\dots,0,h'(x_N)),\qquad \text{for every } x \in \mathbb{H}_+^N,$$

and thus its p-Laplacian is given for every $x_N > 0$ by

$$\begin{aligned}-\Delta_p H(x) = -(|h'(x_N)|^{p-2}\,h'(x_N))' &= \alpha^{p-1}\,(1-\alpha)\,(p-1)\,x_N^{\alpha\,(p-1)-p}\\ &= \alpha^{p-1}\,(1-\alpha)\,(p-1)\,\frac{H(x)^{p-1}}{x_N^p}.\end{aligned}$$

We now take a function $\varphi \in C_0^\infty(\mathbb{H}_+^N)$, multiply the previous equation by $|\varphi|^p/H^{p-1}$ and integrate over $\mathbb{H}_+^N$. By using the Divergence Theorem as in the proof of Theorem 3.5.6 and the fact that φ is compactly supported, we get

$$\alpha^{p-1}\,(1-\alpha)\,(p-1)\int_{\mathbb{H}_+^N}\frac{|\varphi|^p}{x_N^p}\,dx = \int_{\mathbb{H}_+^N}\left\langle |\nabla H|^{p-2}\,\nabla H, \nabla\left(\frac{|\varphi|^p}{H^{p-1}}\right)\right\rangle dx.$$

We can apply the Picone inequality of Lemma 1.3.5 in the right-hand side. This gives

$$\alpha^{p-1}\,(1-\alpha)\,(p-1)\int_{\mathbb{H}_+^N}\frac{|\varphi|^p}{x_N^p}\,dx \le \int_{\mathbb{H}_+^N}|\nabla\varphi|^p\,dx.$$

Observe that the right-hand side above is independent of α. We can then optimize with respect to α: by observing that the function

$$\alpha \mapsto \alpha^{p-1}\,(1-\alpha),$$

is maximal for $\alpha = (p-1)/p$, we get the claimed inequality for a function in $C_0^\infty(\mathbb{H}_+^N)$.

We now show that the same inequality still holds for a function $\varphi \in W_0^{1,p}(\mathbb{H}_+^N)$. Indeed, by definition there exists a sequence $\{\varphi_n\}_{n\in\mathbb{N}} \subseteq C_0^\infty(\mathbb{H}_+^N)$ such that

$$\lim_{n\to\infty} \|\varphi_n - \varphi\|_{L^p(\mathbb{H}_+^N)} = \lim_{n\to\infty} \|\nabla\varphi_n - \nabla\varphi\|_{L^p(\mathbb{H}_+^N;\mathbb{R}^N)} = 0.$$

In particular, up to a subsequence, we can also suppose that

$$\lim_{n\to\infty} \varphi_n(x) = \varphi(x), \qquad \text{for a. e. } x \in \mathbb{H}_+^N.$$

By using these facts and Hardy's inequality for each φ_n, a further application of Fatou's Lemma gives

$$\begin{aligned}\left(\frac{p-1}{p}\right)^p \int_{\mathbb{H}_+^N} \frac{|\varphi|^p}{x_N^p}\,dx &\le \liminf_{n\to\infty} \left(\frac{p-1}{p}\right)^p \int_{\mathbb{H}_+^N} \frac{|\varphi_n|^p}{x_N^p}\,dx \\ &\le \lim_{n\to\infty} \int_{\mathbb{H}_+^N} |\nabla\varphi_n|^p\,dx = \int_{\mathbb{H}_+^N} |\nabla\varphi|^p\,dx.\end{aligned}$$

This concludes the proof.

3.12.9 The idea is that, whenever such an inequality holds, it must remain true for "vertical" scalings of the functions, i.e. by replacing φ with

$$\lambda\,\varphi(x), \qquad \text{for } \lambda > 0$$

and for "horizontal" scalings of the functions, i.e. by replacing φ with

$$\varphi(\lambda\,x), \qquad \text{for } \lambda > 0.$$

The first operation imposes that the two terms

$$\left(\int_{\mathbb{R}^N} |\varphi|^q\,dx\right)^{\frac{\alpha}{q}} \qquad \text{and} \qquad \int_{\mathbb{R}^N} |\nabla\varphi|^2\,dx,$$

must have the same *homogeneity*, while the second one imposes that these two terms must have the same *scaling*, i.e. the same physical dimensions.

These were the heuristics, now let us try to apply them: let $\varphi \in C_0^\infty(\mathbb{R}^N)$ be a nontrivial function. We then use the claimed inequality

$$\left(\int_{\mathbb{R}^N} |\varphi|^q \, dx\right)^{\frac{\alpha}{q}} \leq C \int_{\mathbb{R}^N} |\nabla \varphi|^2 \, dx,$$

with the function $\lambda\,\varphi$ for $\lambda > 0$, which is still in $C_0^\infty(\mathbb{R}^N)$. We get

$$\lambda^\alpha \left(\int_{\mathbb{R}^N} |\varphi|^q \, dx\right)^{\frac{\alpha}{q}} \leq C\,\lambda^2 \int_{\mathbb{R}^N} |\nabla \varphi|^2 \, dx,$$

that is

$$\lambda^{\alpha-2} \left(\int_{\mathbb{R}^N} |\varphi|^q \, dx\right)^{\frac{\alpha}{q}} \leq C \int_{\mathbb{R}^N} |\nabla \varphi|^2 \, dx.$$

If we assume that $\alpha > 2$, then by taking the limit as λ goes to $+\infty$ we would get a contradiction, from the last display. Similarly, if $\alpha < 2$ we would get again a contradiction, by taking this time the limit as λ goes to 0. This shows that we must have

$$\alpha = 2.$$

Thus, our assumption amounts to the validity of the inequality

$$\left(\int_{\mathbb{R}^N} |\varphi|^q \, dx\right)^{\frac{2}{q}} \leq C \int_{\mathbb{R}^N} |\nabla \varphi|^2 \, dx, \qquad \text{for every } \varphi \in C_0^\infty(\mathbb{R}^N),$$

with $C > 0$ universal constant, independent of φ. We now take again a nontrivial function $\varphi \in C_0^\infty(\mathbb{R}^N)$ and apply the previous inequality to the function $\varphi_\lambda(x) := \varphi(\lambda\,x)$, for $\lambda > 0$. We get

$$\left(\int_{\mathbb{R}^N} |\varphi(\lambda\,x)|^q \, dx\right)^{\frac{2}{q}} \leq C\,\lambda^2 \int_{\mathbb{R}^N} |\nabla \varphi(\lambda\,x)|^2 \, dx.$$

We make the change of variable $\lambda\,x = y$, thus from the previous inequality we get

$$\lambda^{-N\,\frac{2}{q}} \left(\int_{\mathbb{R}^N} |\varphi(y)|^q \, dy\right)^{\frac{2}{q}} \leq C\,\lambda^{2-N} \int_{\mathbb{R}^N} |\nabla \varphi(y)|^2 \, dy,$$

that is

$$\lambda^{N-2-N\frac{2}{q}} \left(\int_{\mathbb{R}^N} |\varphi|^q \, dy \right)^{\frac{2}{q}} \le C \int_{\mathbb{R}^N} |\nabla \varphi|^2 \, dy.$$

We now observe that

$$\text{if } N - 2 - N\frac{2}{q} < 0 \qquad \text{then} \qquad \lim_{\lambda \to 0^+} \lambda^{N-2-N\frac{2}{q}} = +\infty,$$

$$\text{if } N - 2 - N\frac{2}{q} > 0 \qquad \text{then} \qquad \lim_{\lambda \to +\infty} \lambda^{N-2-N\frac{2}{q}} = +\infty.$$

In both cases, we would get a contradiction. Thus, it must result

$$N - 2 - N\frac{2}{q} = 0 \qquad \text{i.e.} \qquad q = \frac{2N}{N-2}.$$

This concludes the exercise.

3.12.10 This is similar to the previous problem. It is sufficient to play with the "horizontal" scalings. We take a nontrivial function $\varphi \in C_0^\infty(\mathbb{R}^N)$ and suppose that the claimed inequality holds true: for simplicity we take $q < \infty$. The case $q = \infty$ requires minor modifications and is left to the reader. By applying such inequality to the function $\varphi_\lambda(x) := \varphi(\lambda\, x)$, for $\lambda > 0$, we get

$$\left(\int_{\mathbb{R}^N} |\varphi(\lambda\, x)|^q \, dx \right)^{\frac{p}{q}} \le C\, \lambda^p \int_{\mathbb{R}^N} |\nabla \varphi(\lambda\, x)|^p \, dx.$$

By making the change of variable $\lambda\, x = y$, we get

$$\lambda^{-N\frac{p}{q}} \left(\int_{\mathbb{R}^N} |\varphi(y)|^q \, dy \right)^{\frac{p}{q}} \le C\, \lambda^{p-N} \int_{\mathbb{R}^N} |\nabla \varphi(y)|^p \, dy,$$

that is

$$\lambda^{N-p-N\frac{p}{q}} \left(\int_{\mathbb{R}^N} |\varphi|^q \, dy \right)^{\frac{p}{q}} \le C \int_{\mathbb{R}^N} |\nabla \varphi|^p \, dy.$$

By observing that the exponent of λ is negative, if we now take the limit as λ goes to 0, we get a contradiction.

3.12.11 Thanks to the definition of $W_0^{1,p}(\Omega)$, it is enough to prove the inequality for functions in $C_0^\infty(\Omega)$. We first observe that we can not appeal to Proposition 3.5.2, since an open set with finite measure might be unbounded in every direction $\omega \in \mathbb{S}^{N-1}$.

We distinguish four possibilities:

- if $1 \le p < N$, by Sobolev's inequality (Theorem 3.6.1) we know that

$$\|\varphi\|^p_{L^{p^*}(\Omega)} \le \mathcal{S}^p \, \|\nabla\varphi\|^p_{L^p(\Omega;\mathbb{R}^N)}, \qquad \text{for every } \varphi \in C_0^\infty(\Omega).$$

  We observe that $p < p^*$, thus by Hölder's inequality we have

$$\|\varphi\|^p_{L^p(\Omega)} \le |\Omega|^{1-\frac{p}{p^*}} \, \|\varphi\|^p_{L^{p^*}(\Omega)}.$$

  By joining the last inequalities and observing that

$$1 - \frac{p}{p^*} = 1 - \frac{N-p}{N} = \frac{p}{N},$$

  we get the claimed inequality with constant $C = \mathcal{S}$;
- if $p = N$, by choosing $q = 2N$ in the Ladyzhenskaya inequality (Theorems 3.6.4 and 3.6.6), we know that

$$\|\varphi\|_{L^{2N}(\Omega)} \le \mathcal{L} \, \|\varphi\|^{\frac{1}{2}}_{L^N(\Omega)} \, \|\nabla\varphi\|^{\frac{1}{2}}_{L^N(\Omega;\mathbb{R}^N)},$$

  for every $\varphi \in C_0^\infty(\Omega)$. As before, we can use Hölder's inequality to infer that

$$\|\varphi\|_{L^N(\Omega)} \le |\Omega|^{\frac{1}{2N}} \, \|\varphi\|_{L^{2N}(\Omega)}.$$

  By joining the last two estimates, we thus get

$$\|\varphi\|_{L^N(\Omega)} \le \mathcal{L} \, |\Omega|^{\frac{1}{2N}} \, \|\varphi\|^{\frac{1}{2}}_{L^N(\Omega)} \, \|\nabla\varphi\|^{\frac{1}{2}}_{L^N(\Omega;\mathbb{R}^N)}.$$

  We can simplify the term containing the L^N norm of φ on the right-hand side, which yields

$$\|\varphi\|^{\frac{1}{2}}_{L^N(\Omega)} \le \mathcal{L} \, |\Omega|^{\frac{1}{2N}} \, \|\nabla\varphi\|^{\frac{1}{2}}_{L^N(\Omega;\mathbb{R}^N)}.$$

  By raising both sides to the power 2, this entails the claimed inequality with constant $C = \mathcal{L}^2$;
- if $N < p < \infty$, by the second Morrey inequality (Theorem 3.6.8) we have

$$\|\varphi\|_{L^\infty(\Omega)} \le \mathcal{M}_2 \, \|\nabla\varphi\|^{\frac{N}{p}}_{L^p(\Omega;\mathbb{R}^N)} \, \|\varphi\|^{\frac{p-N}{p}}_{L^p(\Omega)},$$

  for every $\varphi \in C_0^\infty(\Omega)$. On the other hand, we have

$$\|\varphi\|_{L^p(\Omega)} \le |\Omega|^{\frac{1}{p}} \, \|\varphi\|_{L^\infty(\Omega)}.$$

By joining the last two estimates, we get

$$\|\varphi\|_{L^p(\Omega)} \leq \mathcal{M}_2\, |\Omega|^{\frac{1}{p}}\, \|\nabla\varphi\|^{\frac{N}{p}}_{L^p(\Omega;\mathbb{R}^N)}\, \|\varphi\|^{\frac{p-N}{p}}_{L^p(\Omega)}.$$

By simplifying the L^p norm of φ on both sides and then raising to the power p/N, we get the conclusion with constant $C = \mathcal{M}_2^{p/N}$;

- if $p = \infty$, we consider the quantity

$$r_\Omega = \sup_{x\in\Omega} \operatorname{dist}(x, \partial\Omega),$$

already introduced in (3.2.3). Since $|\Omega| < +\infty$, it is not difficult to see that $r_\Omega < +\infty$: indeed, for every $x \in \Omega$ we have that $B_\delta(x) \subseteq \Omega$ for $\delta = \operatorname{dist}(x, \partial\Omega)$. Thus, we obtain

$$\omega_N\, \delta^N = |B_\delta(x)| \leq |\Omega|.$$

By recalling the definition of δ, this is the same as

$$\operatorname{dist}(x, \partial\Omega) \leq \left(\frac{|\Omega|}{\omega_N}\right)^{\frac{1}{N}}. \qquad (*)$$

We now take $x \in \Omega$ and $y \in \partial\Omega$. In particular, if $\varphi \in C_0^\infty(\Omega)$, we have $\varphi(y) = 0$. By Remark 3.6.10, we have

$$|\varphi(x)| = |\varphi(x) - \varphi(y)| \leq |x - y|\, \|\nabla\varphi\|_{L^\infty(\Omega;\mathbb{R}^N)}.$$

This holds for every $y \in \partial\Omega$: by taking the infimum over $y \in \partial\Omega$, we thus obtain

$$|\varphi(x)| \leq \operatorname{dist}(x, \partial\Omega)\, \|\nabla\varphi\|_{L^\infty(\Omega;\mathbb{R}^N)}.$$

If we now apply $(*)$, we get the desired estimate with constant $C = (1/\omega_N)^{1/N}$.

This concludes the exercise.

3.12.12 We start by observing that $u \in W^{1,p}((-1, 1))$: it is sufficient to apply the result of Problem 3.12.3. From the latter, we also know that the weak derivative is given by the piecewise constant function

$$u'(x) = \begin{cases} -1, & \text{if } 0 < x < 1, \\ 1, & \text{if } -1 < x < 0. \end{cases}$$

By definition of $W_0^{1,p}((-1, 1))$, we have to construct a sequence $\{u_n\}_{n\in\mathbb{N}} \subseteq C_0^\infty((-1, 1))$ such that

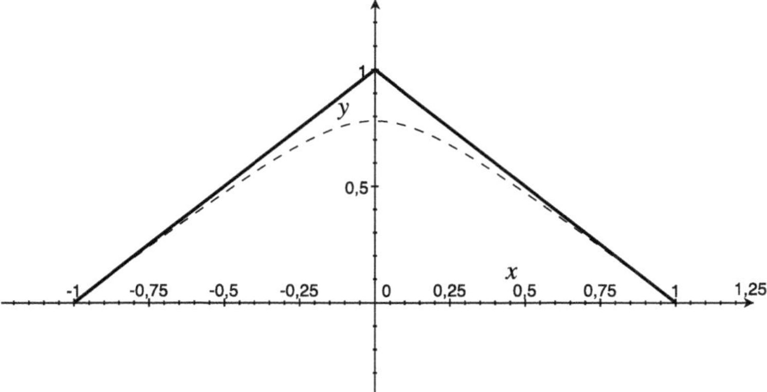

Fig. 8.5 The function $u(x) = 1 - |x|$ (*bold line*) and its smooth approximation v_n (*dashed line*). Observe that v_n vanishes at $x = -1$ and $x = 1$, but it does not have compact support in $(-1, 1)$

$$\lim_{n\to\infty} \|u_n - u\|_{W^{1,p}((-1,1))} = 0. \qquad (*)$$

We construct the sequence by using the usual trick to approximate the absolute value: we first set

$$v_n(x) = \sqrt{\frac{1}{n^2} + 1} - \sqrt{\frac{1}{n^2} + |x|^2},$$

which is a non-negative C^∞ function on the interval $[-1, 1]$, vanishing at the two endpoints (see Fig. 8.5). Observe that

$$|v_n'(x)| = \frac{|x|}{\sqrt{\dfrac{1}{n^2} + |x|^2}} \le 1, \qquad \text{for } x \in [-1, 1]. \qquad (**)$$

We also claim that

$$v_n(x) \le u(x) = 1 - |x|, \qquad \text{for } x \in [-1, 1]. \qquad (***)$$

Indeed, if $x \in [-1, 0]$ from the Fundamental Theorem of Calculus and the fact that $v_n(-1) = 0$, we get

$$v_n(x) = \int_{-1}^{x} v_n'(t)\, dt \le \int_{-1}^{x} dt = (1 + x) = u(x),$$

thanks to $(**)$. Since both v_n and u are even functions, the previous inequality holds on $[0, 1]$, as well.

Then we take a cut-off function $\varphi_n \in C_0^\infty((-1, 1))$ such that

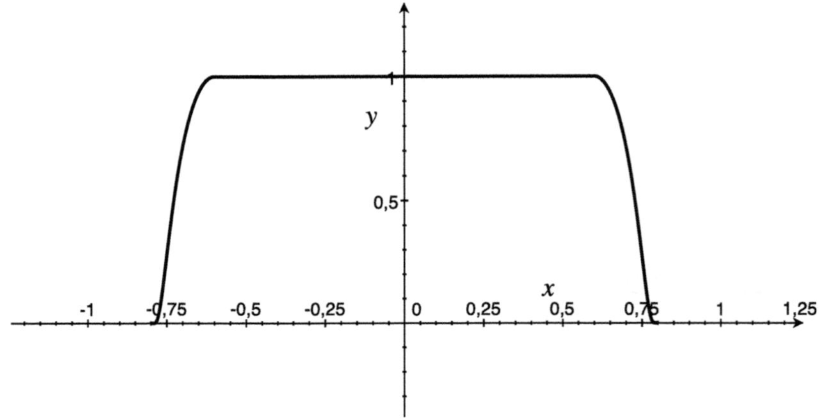

Fig. 8.6 A cut-off function φ_n, i.e. a smooth approximation of the unity over the interval $[-1, 1]$

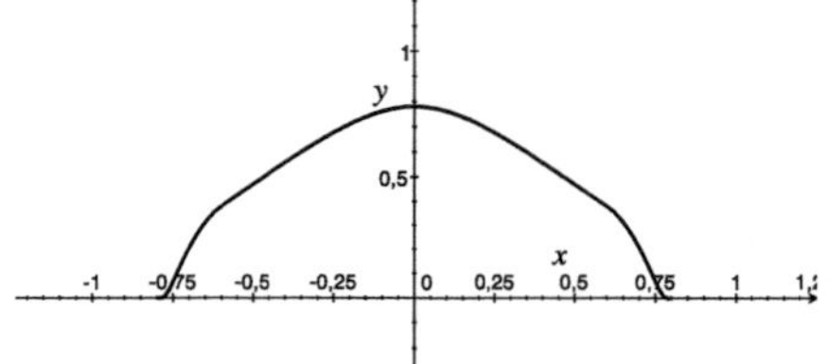

Fig. 8.7 The multiplication of the smooth approximation v_n by the cut-off φ_n: the resulting function is a smooth compactly supported approximation of $u(x) = 1 - |x|$

$$0 \le \varphi_n(x) \le 1, \qquad \text{for } x \in [-1, 1],$$

$$\varphi_n(x) \equiv 1, \qquad \text{for } x \in \left[-1 + \frac{2}{n}, 1 - \frac{2}{n}\right],$$

$$\varphi_n(x) \equiv 0, \qquad \text{for } x \in \left[-1, -1 + \frac{1}{n}\right] \cup \left[1 - \frac{1}{n}, 1\right]$$

and

$$|\varphi_n'(x)| \le C\, n,$$

for a universal constant $C > 0$ (see Fig. 8.6).

We finally set (see Fig. 8.7)

$$u_n = v_n\, \varphi_n,$$

and claim that for such a sequence we have $(*)$. We take $1 \le p < \infty$ and estimate the L^p norm as follows

$$\lim_{n\to\infty}\int_{-1}^{1}|u-u_n|^p\,dx$$
$$\leq 2^{p-1}\lim_{n\to\infty}\int_{-1}^{1}|u-v_n|^p\,dx+2^{p-1}\lim_{n\to\infty}\int_{-1}^{1}|v_n-v_n\,\varphi_n|^p\,dx.$$

We used the definition of u_n, the triangle inequality and Problem 1.7.2. By using the properties of the cut-off function φ_n, we have

$$\int_{-1}^{1}|v_n-v_n\,\varphi_n|^p\,dx=\int_{-1}^{1}|v_n|^p\,(1-\varphi_n)^p\,dx\leq\int_{\left\{1-\frac{2}{n}\leq|x|\leq 1\right\}}|v_n|^p\,dx\leq\frac{4}{n},$$

thanks to the fact that $|v_n(x)|\leq 1$ for every $x\in[-1,1]$. Thus we get

$$\lim_{n\to\infty}\int_{-1}^{1}|v_n-v_n\,\varphi_n|^p\,dx=0.$$

As for the other integral, it is sufficient to observe that v_n converges uniformly to u on $[-1,1]$: indeed, by recalling $(***)$, for every $x\in[-1,1]$ we have

$$\begin{aligned}|u(x)-v_n(x)|=u(x)-v_n(x)&=\left(\sqrt{\frac{1}{n^2}+|x|^2}-|x|\right)+\left(1-\sqrt{\frac{1}{n^2}+1}\right)\\&\leq\left(\sqrt{\frac{1}{n^2}+|x|^2}-|x|\right)\\&\leq\left(\frac{1}{n}+|x|-|x|\right)=\frac{1}{n}.\end{aligned}$$

In the second inequality we used that $\sqrt{a+b}\leq\sqrt{a}+\sqrt{b}$, for every $a,b\geq 0$ (this is a particular case of Problem 1.7.1). We thus have obtained that

$$\lim_{n\to\infty}\int_{-1}^{1}|u-u_n|^p\,dx=0.$$

In order to complete the proof of $(*)$, we still have to prove that

$$\lim_{n\to\infty}\int_{-1}^{1}|u'-u_n'|^p\,dx=0.$$

By observing that

$$u_n'=v_n'\,\varphi_n+v_n\,\varphi_n',$$

we can estimate the required integral as follows

$$
\begin{aligned}
\int_{-1}^{1} |u' - u_n'|^p \, dx &\le 2^{p-1} \int_{-1}^{1} |u' - v_n' \, \varphi_n|^p \, dx + 2^{p-1} \int_{-1}^{1} |v_n|^p \, |\varphi_n'|^p \, dx \\
&\le 4^{p-1} \int_{-1}^{1} |u' - v_n'|^p \, dx + 4^{p-1} \int_{-1}^{1} |v_n'|^p \, (1 - \varphi_n)^p \, dx \\
&+ 2^{p-1} \int_{-1}^{1} |v_n|^p \, |\varphi_n'|^p \, dx =: \mathcal{I}_1 + \mathcal{I}_2 + \mathcal{I}_3.
\end{aligned}
$$

We start from the last integral: by using the properties of φ_n, we get

$$
\mathcal{I}_3 \le C^p \, n^p \int_{\left\{1 - \frac{2}{n} \le |x| \le 1\right\}} |v_n|^p \, dx.
$$

By recalling $(***)$, we have

$$
\mathcal{I}_3 \le C^p \, n^p \int_{\left\{1 - \frac{2}{n} \le |x| \le 1\right\}} \Big|1 - |x|\Big|^p \, dx \le C^p \, \cancel{n^p} \, \frac{2^p}{\cancel{n^p}} \, \frac{4}{n},
$$

which converges to 0, as n goes to ∞.

The estimate for $\mathcal{I}_2$ is simpler: it is sufficient to recall $(**)$ and the fact that φ_n coincides with 1 on the interval $[-1 + 2/n, 1 - 2/n]$. These yield

$$
\mathcal{I}_2 \le 4^{p-1} \int_{\left\{1 - \frac{2}{n} \le |x| \le 1\right\}} dx = \frac{4^p}{n}.
$$

Finally, for the integral $\mathcal{I}_1$, we can use the Dominated Convergence Theorem. Indeed, we have

$$
|u'(x) - v_n'(x)|^p \le 2^{p-1} \Big(|u'(x)| + |v_n'(x)|\Big) \le 2^p, \qquad \text{for a. e. } x \in [-1, 1],
$$

and

$$
\lim_{n \to \infty} v_n'(x) = u'(x), \qquad \text{for a. e. } x \in [-1, 1].
$$

Thus, we get that

$$
\mathcal{I}_1 = 4^{p-1} \int_{-1}^{1} |u' - v_n'|^p \, dx,
$$

converges to 0, as well. This finally establishes $(*)$ and the proof is over.

3.12.13 From Problem 3.12.3, we know that $u \in W^{1,\infty}((-1,1))$, with weak derivative given by

$$u'(x) = \begin{cases} -1, & \text{if } 0 < x < 1, \\ 1, & \text{if } -1 < x < 0. \end{cases}$$

We argue by contradiction and assume that $u \in W_0^{1,\infty}((-1,1))$, as well. By definition, this means that there exists a sequence $\{u_n\}_{n\in\mathbb{N}} \subseteq C_0^\infty((-1,1))$ such that

$$\lim_{n\to\infty} \Big[\|u_n - u\|_{L^\infty((-1,1))} + \|u_n' - u'\|_{L^\infty((-1,1))} \Big] = 0.$$

In particular, from the second term and the expression of u', we get that

$$\lim_{n\to\infty} \|u_n' - 1\|_{L^\infty((0,1))} = 0.$$

On the other hand, since each u_n is compactly supported in $(-1,1)$, we get that for every $n \in \mathbb{N}$ there exists $0 < \delta_n < 1$ such that u_n' identically vanishes on the interval $(1-\delta_n, 1)$. This implies that for every $n \in \mathbb{N}$ we have

$$\|u_n' - 1\|_{L^\infty((0,1))} \geq \|u_n' - 1\|_{L^\infty((1-\delta_n,1))} = 1.$$

This gives the desired contradiction.

3.12.14 By proceeding as in Problem 3.12.2, one can prove that the weak gradient of u is given by the L^∞ vector field

$$\nabla u(x) = \frac{x}{|x|}, \qquad \text{for a. e. } x \in B_1(0).$$

Then it is sufficient to proceed as in the solutions of the two previous problems. In particular, the approximating sequence $\{u_n\}_{n\in\mathbb{N}} \subseteq C_0^\infty(B_1(0))$ can be constructed as

$$u_n = v_n\, \varphi_n,$$

where

$$v_n(x) = \sqrt{\frac{1}{n^2} + 1} - \sqrt{\frac{1}{n^2} + |x|^2}, \qquad \text{for } x \in B_1(0),$$

and the cut-off function $\varphi_n \in C_0^\infty(B_1(0))$ is such that

$$0 \leq \varphi_n \leq 1, \qquad \varphi_n(x) \equiv 1, \quad \text{in } B_{1-\frac{2}{n}}(0),$$

$$\varphi_n \equiv 0, \quad \text{in } B_1(0) \setminus B_{1-\frac{1}{n}}(0), \qquad |\nabla \varphi_n(x)| \le C\, n,$$

for a universal constant $C > 0$. The reader is invited to complete the details, by proceeding similarly as in Problem 3.12.12.

3.12.15 We take $u \in C^0(\overline{\Omega}) \cap C^1(\Omega)$ such that $u = 0$ on $\partial\Omega$ and its classical gradient is bounded on Ω, i.e.

$$\sup_{x\in\Omega} |\nabla u(x)| = M < +\infty. \tag{$*$}$$

In light of Remark 3.7.3, it is sufficient to construct a sequence $\{u_n\}_{n\in\mathbb{N}} \subseteq C^1_0(\Omega)$, such that

$$\lim_{n\to\infty} \|u_n - u\|_{W^{1,p}(\Omega)} = 0. \tag{$**$}$$

In order to construct such a sequence, for every $k \in \mathbb{N} \setminus \{0\}$ large enough, we set

$$\mathrm{int}_k(\Omega) = \{x \in \Omega \,:\, \mathrm{dist}(x, \partial\Omega) > 1/k\}.$$

We recall that $\mathrm{int}_k(\Omega)$ is an open set, thanks to the fact that the function $x \mapsto \mathrm{dist}(x, \partial\Omega)$ is continuous (see Problem 1.7.26). Moreover, we have $\mathrm{int}_k(\Omega) \Subset \Omega$. We now take a function $\eta_n \in C^\infty_0(\mathrm{int}_{4n}(\Omega))$ such that

$$\eta_n \equiv 1 \text{ on } \mathrm{int}_n(\Omega), \qquad 0 \le \eta_n \le 1, \qquad |\nabla \eta_n| \le C\, n.$$

Such a function can be constructed just by taking the convolution

$$\eta_n = 1_{\mathrm{int}_{2n}(\Omega)} * \rho_{6n},$$

where as usual

$$\rho_k(x) = k^N\, \rho(k\, x),$$

and ρ is the standard smoothing kernel defined in (1.4.1). Indeed, we already know that this produces a C^∞ function. Moreover, its support is such that

$$\mathrm{spt}(1_{\mathrm{int}_{2n}(\Omega)} * \rho_{6n}) \subseteq \overline{\mathrm{int}_{2n}(\Omega)} + \overline{B_{\frac{1}{6n}}(0)},$$

and the latter is a compact set, contained in $\mathrm{int}_{4n}(\Omega)$. The other properties of η_n are easily verified.

We now set $u_n = u\, \eta_n$ and claim that this sequence has the property $(**)$. We first observe that by construction $u_n \in C^1_0(\Omega)$. We compute the L^p norm of the difference: we have

$$\int_\Omega |u_n - u|^p\,dx = \int_\Omega |u\,\eta_n - u|^p\,dx = \int_\Omega |u|^p\,|1-\eta_n|^p\,dx$$
$$= \int_{\Omega\setminus \mathrm{int}_n(\Omega)} |u|^p\,|1-\eta_n|^p\,dx$$
$$\le \int_{\Omega\setminus \mathrm{int}_n(\Omega)} |u|^p\,dx.$$

By using that (see Problem 1.7.26)

$$\lim_{n\to\infty} |\Omega \setminus \mathrm{int}_n(\Omega)| = 0,$$

we obtain that

$$\lim_{n\to\infty} \int_{\Omega\setminus \mathrm{int}_n(\Omega)} |u|^p\,dx = 0,$$

thanks to the Dominated Convergence Theorem. This in turn shows that

$$\lim_{n\to\infty} \|u_n - u\|_{L^p(\Omega)} = 0.$$

To complete the proof of (∗∗), we need to compute the L^p norm of the difference of the gradients: we have

$$\int_\Omega |\nabla u_n - \nabla u|^p\,dx = \int_\Omega |u\,\nabla\eta_n + \eta_n\,\nabla u - \nabla u|^p\,dx$$
$$\le 2^{p-1}\int_\Omega |\nabla u|^p\,|1-\eta_n|^p\,dx + 2^{p-1}\int_\Omega |\nabla \eta_n|^p\,|u|^p\,dx$$
$$\le 2^{p-1}\int_{\Omega\setminus \mathrm{int}_n(\Omega)} |\nabla u|^p\,dx + 2^{p-1}\,C^p\,n^p\int_{\Omega\setminus \mathrm{int}_n(\Omega)} |u|^p\,dx.$$

We have

$$\lim_{n\to\infty} \int_{\Omega\setminus \mathrm{int}_n(\Omega)} |\nabla u|^p\,dx = 0,$$

thanks to the same argument as above. The second integral is more delicate, due to the presence of the diverging factor n^p in front of the integral. Here the hypothesis that $u = 0$ on $\partial\Omega$ will be crucial. We take $x \in \Omega \setminus \mathrm{int}_n(\Omega)$ and pick a point $y \in \partial\Omega$ such that

$$|x - y| = \mathrm{dist}(x, \partial\Omega).$$

Observe that such a point y exists, see Problem 1.7.26. By using the Mean Value Theorem and the fact that $u \in C^0(\overline{\Omega}) \cap C^1(\Omega)$, we know that there exists a point ξ which belongs to the interior of the segment connecting x and y, such that

$$u(x) = u(y) + \langle \nabla u(\xi), x - y \rangle.$$

We observe that the interior of the segment entirely lies in Ω, otherwise the distance of x from the boundary would be strictly smaller that $|x - y|$. Accordingly, we have $\xi \in \Omega$. By using that $u(y) = 0$, the previous identity implies that for every $x \in \Omega \setminus \mathrm{int}_n(\Omega)$ we have

$$|u(x)| \leq |\nabla u(\xi)|\,|x - y| \leq M\,\mathrm{dist}(x, \partial\Omega) \leq M\,\frac{1}{n},$$

thanks to (*). This finally permits to infer that

$$\lim_{n\to\infty} n^p \int_{\Omega\setminus\mathrm{int}_n(\Omega)} |u|^p\,dx \leq \lim_{n\to\infty} \left(n^p\,M^p\,\frac{1}{n^p}\,|\Omega \setminus \mathrm{int}_n(\Omega)| \right) = 0.$$

We have thus proved (**).

3.12.16 Let $u \in X_0^{1,p}(\Omega)$, we first observe that if we set

$$u_+(x) = \max\{u(x), 0\} \qquad \text{and} \qquad u_-(x) = \min\{u(x), 0\},$$

these are two continuous functions on $\overline{\Omega}$, which still belong to $W^{1,p}(\Omega)$ (thanks to Proposition 3.4.3). Moreover, they still vanish on the boundary $\partial\Omega$, that is

$$u_+, u_- \in X_0^{1,p}(\Omega) \qquad \text{and} \qquad u = u_+ + u_-.$$

These preliminary observations show that we can suppose without loss of generality that $u \geq 0$ in Ω, not identically vanishing. We now define for every $n \in \mathbb{N}$

$$u_n(x) = \max\left\{u(x) - \frac{1}{n+1}, 0\right\},$$

which belongs to $W^{1,p}(\Omega)$, again by Proposition 3.4.3. In addition, this is still a continuous function on $\overline{\Omega}$. We set

$$\Omega_n := \{x \in \Omega \,:\, u_n(x) > 0\} = \left\{x \in \Omega \,:\, u(x) > \frac{1}{n+1}\right\},$$

which is a non-empty open bounded set, provided that n is large enough. Since $u \in C^0(\overline{\Omega})$ and it vanishes on $\partial\Omega$, we get that

$$\delta_n := \mathrm{dist}(\Omega_n, \partial\Omega) > 0.$$

We then consider the usual family $\{\rho_n\}_{n\geq 1}$ of smoothing kernels: by recalling that each ρ_n is supported in the ball centered at the origin and with radius $1/n$, for every $n \in \mathbb{N}$ we choose $k_n \in \mathbb{N}$ such that

$$\frac{1}{k_n} \leq \frac{\delta_n}{2}.$$

This guarantees that

$$\widetilde{u}_{n,m} = u_n * \rho_m \in C_0^\infty(\Omega), \qquad \text{for every } m \geq k_n.$$

We then observe that

$$\lim_{m\to\infty} \|\widetilde{u}_{n,m} - u_n\|_{W^{1,p}(\Omega)} = 0.$$

It is sufficient to use the properties of convolutions and proceed as in the proof of Proposition 3.2.6. This in particular shows that $u_n \in W_0^{1,p}(\Omega)$, as a limit of functions in $C_0^\infty(\Omega)$.

We then observe that (recall that both u_n and u are non-negative and that $u_n \leq u$)

$$|u_n - u|^p \leq (u_n + u)^p \leq 2^p\, u^p \in L^1(\Omega),$$

and that by construction

$$\lim_{n\to\infty} u_n(x) = u(x), \qquad \text{for every } x \in \Omega.$$

Thus, we get by the Dominated Convergence Theorem that

$$\lim_{n\to\infty} \left(\int_\Omega |u_n - u|^p\, dx \right)^{\frac{1}{p}} = 0.$$

Moreover, from Proposition 3.4.3 we also have that

$$\nabla u_n(x) = \begin{cases} \nabla u(x), & \text{if } x \in \Omega_n, \\ 0, & \text{if } x \in \Omega \setminus \Omega_n. \end{cases}$$

This yields

$$|\nabla u_n - \nabla u|^p \leq |\nabla u|^p \in L^1(\Omega),$$

and

$$\lim_{n\to\infty} \nabla u_n(x) = \nabla u(x), \qquad \text{for a. e. } x \in \Omega.$$

Again the Dominated Convergence Theorem yields

$$\lim_{n\to\infty}\left(\int_\Omega |\nabla u_n - \nabla u|^p\,dx\right)^{\frac{1}{p}} = 0.$$

This shows that $u \in W_0^{1,p}(\Omega)$, as a strong limit of functions in $W_0^{1,p}(\Omega)$.

3.12.17 We show that the conclusion of Problem 3.12.16 still holds by removing the boundedness assumption on Ω. The case of Problem 3.12.15 is similar and left to the reader.

Let $u \in C^0(\overline{\Omega}) \cap W^{1,p}(\Omega)$ such that $u = 0$ on $\partial\Omega$. We set

$$n_0 = \min\{n \in \mathbb{N} : \Omega \cap B_n(0) \neq \emptyset\}.$$

For every $n \geq n_0$ we take a cut-off function $\eta_n \in C_0^\infty(B_{n+1}(0))$ such that

$$0 \leq \eta_n \leq 1, \qquad \eta_n \equiv 1 \text{ on } B_n(0), \qquad |\nabla \eta_n| \leq C,$$

for a universal constant $C > 0$. We then set

$$\Omega_n = \Omega \cap B_{n+1}(0), \qquad u_n = u\,\eta_n, \qquad \text{for } n \geq n_0.$$

By Proposition 3.4.7, we have that $u_n \in W^{1,p}(\Omega_n)$ and its weak gradient is given by

$$\nabla u_n = \eta_n\,\nabla u + u\,\nabla \eta_n.$$

Moreover, by construction we have that $u_n \in C^0(\overline{\Omega_n})$, as a product of two continuous functions. Finally, we also have that

$$u_n = 0, \qquad \text{on } \partial\Omega_n,$$

since u vanishes on $\partial\Omega$ and η_n vanishes on $\partial B_{n+1}(0)$. We can apply the conclusion of Problem 3.12.16 to the open bounded set Ω_n and get that $\{u_n\}_{n\geq n_0} \subseteq W_0^{1,p}(\Omega_n)$. Since $\Omega_n \subseteq \Omega$, thanks to Lemma 3.7.9 we also have that $\{u_n\}_{n\geq n_0} \subseteq W_0^{1,p}(\Omega)$, by considering the functions to be extended by 0 outside Ω_n. In order to conclude that $u \in W_0^{1,p}(\Omega)$, it is now sufficient to show that

$$\lim_{n\to\infty} \|u_n - u\|_{L^p(\Omega)} = \lim_{n\to\infty} \|\nabla u_n - \nabla u\|_{L^p(\Omega;\mathbb{R}^N)} = 0.$$

We have

$$\int_\Omega |u_n - u|^p\,dx = \int_\Omega |u|^p\,|1-\eta_n|^p\,dx \leq \int_{\Omega\setminus B_n(0)} |u|^p\,dx.$$

The last integral converges to 0 as n goes to ∞, by using that $u \in L^p(\Omega)$ and the Dominated Convergence Theorem. Similarly, for the gradients we have

$$\begin{aligned}\int_\Omega |\nabla u_n - \nabla u|^p\,dx &\le 2^{p-1}\int_\Omega |\nabla u|^p\,|1-\eta_n|^p\,dx + 2^{p-1}\int_\Omega |\nabla \eta_n|^p\,|u|^p\,dx\\ &\le 2^{p-1}\int_{\Omega\setminus B_n(0)} |\nabla u|^p\,dx\\ &+ 2^{p-1}\,C^p\int_{\Omega\cap(B_{n+1}(0)\setminus B_n(0))} |u|^p\,dx.\end{aligned}$$

By using the integrability of both u and ∇u, we get the desired conclusion.

3.12.18 By observing that $C^1_0(\Omega) \subseteq C^1(\Omega)\cap W^{1,\infty}(\Omega)$, we can apply Proposition 3.4.7 and infer that $\eta\,u \in W^{1,p}_{\mathrm{loc}}(\Omega)$. Moreover, thanks to the compact support of η, we have

$$\eta\,u \in L^p(\Omega) \qquad \text{and} \qquad \nabla(\eta\,u) = u\,\nabla\eta + \eta\,\nabla u \in L^p(\Omega;\mathbb{R}^N),$$

thus we actually get $\eta\,u \in W^{1,p}(\Omega)$. We have to show that $\eta\,u \in W^{1,p}_0(\Omega)$: by Remark 3.7.3, it is enough to find a sequence $\{\varphi_n\}_{n\in\mathbb{N}} \subseteq C^1_0(\Omega)$ such that

$$\lim_{n\to\infty} \|\varphi_n - \eta\,u\|_{W^{1,p}(\Omega)} = 0.$$

To this aim, we pick the sequence $\{u_n\}_{n\in\mathbb{N}}$ of Proposition 3.2.6 and define

$$\varphi_n = \eta\,u_n \in C^1_0(\Omega).$$

By Proposition 3.2.6, we know that for every $\Omega' \Subset \Omega$

$$\lim_{n\to\infty}\left(\|u_n - u\|_{L^p(\Omega')} + \|\nabla u_n - \nabla u\|_{L^p(\Omega')}\right) = 0. \tag{$*$}$$

If we indicate by K the compact support of η, we then have that there exists $\Omega' \Subset \Omega$ such that $K \subseteq \Omega'$. By using this fact, the definition of φ_n, the Leibniz rule and the Minkowski inequality, we get

$$\begin{aligned}\|\varphi_n - \eta\,u\|_{W^{1,p}(\Omega)} &= \|\eta\,u_n - \eta\,u\|_{L^p(\Omega')} + \|\nabla(\eta\,u_n) - \nabla(\eta\,u)\|_{L^p(\Omega';\mathbb{R}^N)}\\ &\le \|\eta\|_{L^\infty(\Omega')}\,\|u_n - u\|_{L^p(\Omega')}\\ &+ \|\nabla\eta\|_{L^\infty(\Omega';\mathbb{R}^N)}\,\|u_n - u\|_{L^p(\Omega')}\\ &+ \|\eta\|_{L^\infty(\Omega')}\,\|\nabla u_n - \nabla u\|_{L^p(\Omega';\mathbb{R}^N)}.\end{aligned}$$

The desired conclusion now follows from ($*$).

3.12.19 Again by Proposition 3.4.7, we have that $\eta\, u \in W^{1,p}(\Omega)$, with weak gradient given by

$$\nabla(\eta\, u) = u\, \nabla\eta + \eta\, \nabla u.$$

As in the previous exercise, we only need to find a sequence $\{\varphi_n\}_{n\in\mathbb{N}} \subseteq C_0^1(\Omega)$ such that

$$\lim_{n\to\infty} \|\varphi_n - \eta\, u\|_{W^{1,p}(\Omega)} = 0.$$

Since $u \in W_0^{1,p}(\Omega)$, by definition of this space and Remark 3.7.3, we know that there exists $\{u_n\}_{n\in\mathbb{N}} \subseteq C_0^1(\Omega)$ such that

$$\lim_{n\to\infty} \|u_n - u\|_{W^{1,p}(\Omega)} = 0. \tag{$*$}$$

We claim that $\varphi_n = \eta\, u_n$ has the required property. The fact that it belongs to $C_0^1(\Omega)$ is immediate. Moreover, we have

$$\|\eta\, u_n - \eta\, u\|_{L^p(\Omega)} \le \|\eta\|_{L^\infty(\Omega)}\, \|u_n - u\|_{L^p(\Omega)},$$

and

$$\begin{aligned}\|\nabla(\eta\, u_n) - \nabla\,(\eta\, u)\|_{L^p(\Omega;\mathbb{R}^N)} &= \|\eta\,(\nabla u_n - \nabla u) + (u_n - u)\,\nabla\eta\|_{L^p(\Omega;\mathbb{R}^N)}\\ &\le \|\eta\,(\nabla u_n - \nabla u)\|_{L^p(\Omega;\mathbb{R}^N)}\\ &+ \|(u_n - u)\,\nabla\eta\|_{L^p(\Omega;\mathbb{R}^N)}\\ &\le \|\eta\|_{L^\infty(\Omega)}\|\nabla u_n - \nabla u\|_{L^p(\Omega;\mathbb{R}^N)}\\ &+ \|\nabla\eta\|_{L^\infty(\Omega;\mathbb{R}^N)}\, \|u_n - u\|_{L^p(\Omega)}.\end{aligned}$$

By using ($*$), we get

$$\lim_{n\to\infty} \|\eta\, u_n - \eta\, u\|_{W^{1,p}(\Omega)} = 0,$$

as desired.

3.12.20 In order to extend the inequality to the space $W_0^{1,p}(B_R(x_0))$, we can use the same density argument as in the solution of Problem 3.12.8. We now prove (3.12.1). The implication "$\Longrightarrow$" follows from Hardy's inequality (3.5.13), which is valid for $W_0^{1,p}(B_R(x_0))$, as shown. For the converse implication we proceed similarly as in Problem 3.12.15. We will show that for every $u \in W^{1,p}$ such that

$$\int_{B_R(x_0)} \frac{|u|^p}{(R-|x-x_0|)^p}\,dx < +\infty,$$

there exists $\{u_n\}_{n\in\mathbb{N}} \subseteq W^{1,p}_0(B_R(x_0))$ converging to u in the norm of $W^{1,p}(B_R(x_0))$. Thanks to the fact that $W^{1,p}_0(B_R(x_0))$ is closed, this will be enough to conclude.

In order to construct such a sequence, for every $k \in \mathbb{N}\setminus\{0\}$, we set $R_k = R-1/k$. Then, we take a function $\eta_n \in C^\infty_0(B_{R_{2n}}(x_0))$ such that

$$\eta_n \equiv 1 \text{ on } B_{R_n}(x_0), \qquad 0 \le \eta_n \le 1, \qquad |\nabla \eta_n| \le C\,n.$$

Finally, we set $u_n = u\,\eta_n$: observe that $u_n \in W^{1,p}_0(B_R(x_0))$, thanks to Problem 3.12.18. For the L^p norm of the difference, we have

$$\int_{B_R(x_0)} |u_n-u|^p\,dx = \int_{B_R(x_0)\setminus B_{R_n}(x_0)} |u|^p\,|1-\eta_n|^p\,dx \le \int_{B_R(x_0)\setminus B_{R_n}(x_0)} |u|^p\,dx.$$

By using that the measure of the annular region $B_R(x_0)\setminus B_{R_n}(x_0)$ goes to zero, we get

$$\lim_{n\to\infty} \|u_n-u\|_{L^p(B_R)} = 0,$$

thanks to the Dominated Convergence Theorem. We compute the L^p norm of the gradients: we have

$$\begin{aligned}
\int_{B_R(x_0)} |\nabla u_n - \nabla u|^p\,dx &= \int_{B_R(x_0)} |u\,\nabla\eta_n + \eta_n\,\nabla u - \nabla u|^p\,dx \\
&\le 2^{p-1}\int_{B_R(x_0)} |\nabla u|^p\,|1-\eta_n|^p\,dx \\
&\quad + 2^{p-1}\int_{B_R(x_0)} |\nabla\eta_n|^p\,|u|^p\,dx \\
&\le 2^{p-1}\int_{B_R(x_0)\setminus B_{R_n}(x_0)} |\nabla u|^p\,dx \\
&\quad + 2^{p-1}\,C\,n^p\int_{B_R(x_0)\setminus B_{R_n}(x_0)} |u|^p\,dx.
\end{aligned}$$

We now observe that

$$R-|x-x_0| \le \frac{1}{n}, \qquad \text{for every } x \in B_R(x_0)\setminus B_{R_n}(x_0).$$

This entails that

$$n^p \int_{B_R(x_0)\setminus B_{R_n}(x_0)} |u|^p\,dx \le \int_{B_R(x_0)\setminus B_{R_n}(x_0)} \frac{|u|^p}{(R-|x-x_0|)^p}\,dx.$$

Thanks to the integrability assumption on u, we have by the Dominated Convergence Theorem

$$\lim_{n\to\infty} \int_{B_R(x_0)\setminus B_{R_n}(x_0)} \frac{|u|^p}{(R-|x-x_0|)^p}\,dx = 0.$$

This is enough to conclude. We remark that the equivalence (3.12.1) is true for more general sets than just balls, see for example [40, Example 9.12].

3.12.21 If Ω is bounded this has already been proved in Theorem 3.8.3. Thus, we assume that Ω is unbounded and $|\Omega| < +\infty$. We try to reproduce step by step the proof of Theorem 3.8.3: again the fact that we have

$$W_0^{1,p}(\Omega) \hookrightarrow L^p(\Omega),$$

is straightforward. In order to prove the compactness of the embedding, we want to appeal again to the Riesz-Fréchet-Kolmogorov Theorem, by possibly taking into account Remark B.2.2. We take a sequence $\{u_n\}_{n\in\mathbb{N}} \subseteq W_0^{1,p}(\Omega)$ such that

$$\|u_n\|_{W^{1,p}(\Omega)} \le M, \qquad \text{for every } n \in \mathbb{N}.$$

Thanks to Lemma 3.7.9 we can extend these functions by zero outside Ω and consider them as elements of $W_0^{1,p}(\mathbb{R}^N) = W^{1,p}(\mathbb{R}^N)$. Hypothesis (H1) of Theorem B.2.1 is obviously satisfied. As for hypothesis (H3), by Lemma 3.7.8 we still have

$$\int_{\mathbb{R}^N} |u_n(x+h)-u_n(x)|^p\,dx \le |h|^p \int_{\mathbb{R}^N} |\nabla u_n|^p\,dx \le M^p\,|h|^p, \qquad \text{for every } n \in \mathbb{N}.$$

This shows that

$$\lim_{|h|\to 0} \sup_{n\in\mathbb{N}} \int_{\mathbb{R}^N} |u_n(x+h)-u_n(x)|^p\,dx \le \lim_{|h|\to 0} M^p\,|h|^p = 0,$$

and thus here the assumption that Ω is unbounded has no bearing.

On the contrary, hypothesis (H2) of the same theorem now *is not* verified. By Remark B.2.2, we notice that Theorem B.2.1 still applies, provided $\{u_n\}_{n\in\mathbb{N}}$ verifies (H2bis). This reads as follows: for every $\varepsilon > 0$, there exists $R_\varepsilon > 0$ such that

$$\int_{\mathbb{R}^N\setminus B_{R_\varepsilon}(0)} |u_n|^p\,dx < \varepsilon, \qquad \text{for every } n \in \mathbb{N}.$$

In order to get this property, we first observe that by Theorem 3.8.1, we have that there exists an exponent $q > p$ and a constant $C = C(N, p, q) > 0$ such that

$$\|u\|_{L^q(\Omega)} \leq C\, \|u\|_{W^{1,p}(\Omega)}, \qquad \text{for every } u \in W_0^{1,p}(\Omega). \tag{$*$}$$

Then, for every $R > 0$ and every $n \in \mathbb{N}$, by Hölder's inequality and $(*)$ we get

$$\begin{aligned}\int_{\mathbb{R}^N\setminus B_R(0)} |u_n|^p\, dx = \int_{\Omega\setminus B_R(0)} |u_n|^p\, dx &\leq |\Omega \setminus B_R(0)|^{1-\frac{p}{q}} \left(\int_{\Omega\setminus B_R(0)} |u_n|^q\, dx\right)^{\frac{p}{q}}\\ &\leq |\Omega \setminus B_R(0)|^{1-\frac{p}{q}} \left(\int_{\Omega} |u_n|^q\, dx\right)^{\frac{p}{q}}\\ &\leq C\, |\Omega \setminus B_R(0)|^{1-\frac{p}{q}}\, \|u_n\|^p_{W^{1,p}(\Omega)},\end{aligned}$$

for some constant $C > 0$ independent of both R and n. If we now use that the last norms are uniformly bounded, we can infer the following uniform estimate for every $n \in \mathbb{N}$ and $R > 0$

$$\int_{\mathbb{R}^N\setminus B_R(0)} |u_n|^p\, dx \leq C\, |\Omega \setminus B_R(0)|^{1-\frac{p}{q}}\, M^p,$$

with a constant $C = C(N, p, q) > 0$. Since Ω has finite measure, we must have

$$\lim_{R\to+\infty} |\Omega \setminus B_R(0)| = 0.$$

The last two equations finally show that the sequence $\{u_n\}_{n\in\mathbb{N}}$ verifies the required hypothesis (H2bis). We can thus apply Theorem B.2.1 and get that there exists $u \in L^p(\mathbb{R}^N)$ such that, up to subsequences, we have

$$\lim_{n\to\infty} \|u_n - u\|_{L^p(\mathbb{R}^N)} = 0.$$

The conclusion now follows as in the proof of Theorem 3.8.3.

3.12.22 We construct a sequence $\{\varphi_n\}_{n\in\mathbb{N}} \subseteq W_0^{1,N}(B_1(0))$ such that

$$\lim_{n\to\infty} \|\varphi_n\|_{W^{1,N}(B_1(0))} = 0,$$

while

$$\lim_{n\to\infty} \|\varphi_n\|_{L^\infty(B_1(0))} = 1, \qquad \text{for every } n \in \mathbb{N}.$$

This would show that $W_0^{1,N}(B_1(0))$ is not continuously embedded in $L^\infty(B_1(0))$.

We start by taking the radially symmetric function for $n \in \mathbb{N} \setminus \{0\}$

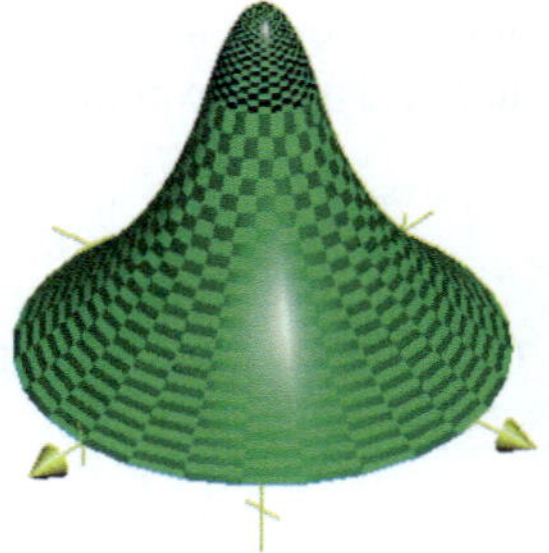

Fig. 8.8 The function φ_n of Problem 3.12.22

$$u_n(x) = -\frac{\log|x|}{\log n}, \qquad \text{for } \frac{1}{n} \le |x| \le 1,$$

and observe that this is a C^1 function defined on the annulus

$$\overline{B_1(0)} \setminus B_{\frac{1}{n}}(0) = \left\{x \in \mathbb{R}^N \, : \, \frac{1}{n} \le |x| \le 1\right\},$$

such that

$$u_n(x) = 1, \quad \text{for } |x| = \frac{1}{n}, \qquad\qquad u_n(x) = 0, \quad \text{for } |x| = 1.$$

We “fill the hole” by considering the radially symmetric function

$$\psi_n(x) = 1 + \frac{n^2}{2\,\log n}\left(\frac{1}{n^2} - |x|^2\right), \qquad \text{for } |x| \le \frac{1}{n},$$

see Fig. 8.8. There is no mystery in the choice of this function, this has been constructed so that

$$\varphi_n(x) = \begin{cases} \psi_n(x), & \text{if } |x| \le 1/n, \\ u_n(x), & \text{if } 1/n \le |x| \le 1, \end{cases}$$

is C^1 on $\overline{B_1(0)}$. Moreover, by construction this function vanishes on $\partial B_1(0)$. Thus, by Problem 3.12.15, we know that $\varphi_n \in W_0^{1,N}(B_1(0))$. Observe that φ_n is a non-negative radially symmetric decreasing function, thus

$$\|\varphi_n\|_{L^\infty(B_1(0))} = \varphi_n(0) = 1 + \frac{1}{2\,\log n}, \qquad \text{for every } n \in \mathbb{N} \setminus \{0\},$$

which obviously implies

$$\lim_{n\to\infty} \|\varphi_n\|_{L^\infty(B_1(0))} = 1.$$

As for the Sobolev norm, we first observe that by Theorem 3.7.6, we have

$$\|\varphi_n\|_{W^{1,N}(B_1(0))} \leq C\,\|\nabla\varphi_n\|_{L^N(B_1(0);\mathbb{R}^N)}.$$

Thus, in order to conclude, it is sufficient to show that

$$\lim_{n\to\infty}\int_{B_1(0)} |\nabla\varphi_n|^N\,dx = 0.$$

By computing this integral in spherical coordinates, we get

$$\begin{aligned}
\int_{B_1(0)} |\nabla\varphi_n|^N\,dx &= \int_{B_{\frac{1}{n}}(0)} |\nabla\psi_n|^N\,dx + \int_{B_1(0)\setminus B_{\frac{1}{n}}(0)} |\nabla u_n|^N\,dx \\
&= \left(\frac{n^2}{\log n}\right)^N \int_{B_{\frac{1}{n}}(0)} |x|^N\,dx + \frac{1}{(\log n)^N}\int_{B_1(0)\setminus B_{\frac{1}{n}}(0)} \frac{1}{|x|^N}\,dx \\
&= N\,\omega_N \left(\frac{n^2}{\log n}\right)^N \int_0^{\frac{1}{n}} \varrho^{2N-1}\,d\varrho + \frac{N\,\omega_N}{(\log n)^N}\int_{\frac{1}{n}}^1 \frac{1}{\varrho}\,d\varrho \\
&= \frac{\omega_N}{2}\,\frac{1}{(\log n)^N} + \frac{N\,\omega_N}{(\log n)^{N-1}}.
\end{aligned}$$

It is now easily seen that the last quantities goes to 0, as n goes to ∞, thanks to the fact that $N \geq 2$.

3.12.23 We divide the solution in two parts.

Continuous Embedding At first, we recall that from Theorem 3.6.6 we have

$$\|\varphi\|_{C^0_b(\mathbb{R})} = \|\varphi\|_{L^\infty(\mathbb{R})} \leq \frac{1}{2}\int_{\mathbb{R}} |\varphi'|\,dx, \qquad \text{for every } \varphi \in C_0^\infty(\mathbb{R}). \tag{$*$}$$

Notice that we used that the sup norm and the L^∞ norm coincide for continuous functions (recall Problem 1.7.31). We now take $u \in W_0^{1,1}((a,b))$, by definition there exists a sequence $\{u_n\}_{n\in\mathbb{N}} \subseteq C_0^\infty((a,b))$ such that

$$\lim_{n\to\infty} \|u_n - u\|_{W^{1,1}((a,b))} = 0.$$

In particular, $\{u_n'\}_{n\in\mathbb{N}}$ is a Cauchy sequence in $L^1((a,b))$. By using this fact and inequality $(*)$ for $u_n - u_m$, we get that

$$\|u_n - u_m\|_{C^0_b(\mathbb{R})} \leq \frac{1}{2}\int_{\mathbb{R}} |u_n' - u_m'|\,dx,$$

so that

$$\{u_n\}_{n\in\mathbb{N}} \subseteq C_0^\infty((a,b)) \subseteq C_b^0([a,b]),$$

is a Cauchy sequence, with respect to the sup-norm. By using that $C_b^0([a,b])$ endowed with the sup-norm is a Banach space (see for example [35, Theorem 7.9] or Problem 5.5.1), we get that u_n converges uniformly on $[a,b]$. Such a limit function is u itself. This shows that[11]

$$W_0^{1,1}((a,b)) \subseteq C_b^0([a,b]).$$

Moreover, such an embedding is continuous: it is sufficient to use $(*)$ for u_n and pass to the limit as n goes to ∞.

Compact Embeddings We assume that the interval (a,b) is *bounded*. From Corollary 3.8.5, we already know that for every $1 \le q < \infty$ the embedding

$$W_0^{1,1}((a,b)) \hookrightarrow L^q((a,b)),$$

is compact. On the contrary, the embedding

$$W_0^{1,1}((a,b)) \hookrightarrow L^\infty((a,b)),$$

is continuous (in light of the first part above), but not compact. As a counterexample, we take for simplicity $(a,b) = (-1,1)$ and consider the sequence

$$u_n(x) = \max\{1 - n\,|x|, 0\}, \qquad \text{for } x \in (-1,1),\ n \ge 1.$$

By Problem 3.12.3 we have that $u_n \in W^{1,1}((-1,1))$, with

$$u_n'(x) = \begin{cases} -n, \text{ if } 0 < x < 1/n, \\ n, \text{ if } -1/n < x < 0, \\ 0, \text{ if } 1/n < |x| < 1. \end{cases}$$

Moreover, each u_n is continuous on $[-1,1]$ and vanishes at the endpoints. Thus, by Problem 3.12.16 we have $\{u_n\}_{n\ge1} \subseteq W_0^{1,1}((-1,1))$. We observe that

$$\int_{-1}^{1} |u_n'|\,dx = \int_{-\frac{1}{n}}^{\frac{1}{n}} n\,dx = n\,\frac{2}{n} = 2,$$

and for every $1 \le q < \infty$

[11] We have already observed in Problem 3.12.22 that for $N \ge 2$ and $p = N$ *it is not true* that $W_0^{1,N}(\Omega)$ is embedded in a space of continuous functions. We see here that the one-dimensional case is exceptional.

$$\int_{-1}^{1} |u_n|^q \, dx = \int_{-\frac{1}{n}}^{\frac{1}{n}} (1 - n\,|x|)^q \, dx \le \int_{-\frac{1}{n}}^{\frac{1}{n}} dx = \frac{2}{n},$$

Thus, the sequence $\{u_n\}_{n\ge 1}$ is bounded in $W_0^{1,1}((-1,1))$ and converges strongly to 0 in $L^q((-1,1))$, for every $1 \le q < \infty$. However, since

$$\|u_n\|_{L^\infty((-1,1))} = 1, \qquad \text{for every } n \in \mathbb{N},$$

we do not have strong convergence in $L^\infty((-1,1))$ to 0. This shows that the embedding $W_0^{1,1}((-1,1)) \hookrightarrow L^\infty((-1,1))$ is not compact.

3.12.24 We already know by Theorem 3.2.7 and Remark 3.2.8 that $u_\Phi \in W^{1,p}(\Omega)$, thanks to the fact that $\mathcal{O} := \Phi^{-1}(\Omega) = \Omega$. We need to show that u_Φ can be approximated in Sobolev norm by a sequence of smooth compactly supported functions.

By definition of $W_0^{1,p}(\Omega)$, there exists a sequence $\{u_n\}_{n\in\mathbb{N}} \subseteq C_0^\infty(\Omega)$ such that

$$\lim_{n\to\infty} \|u_n - u\|_{L^p(\Omega)} = \lim_{n\to\infty} \|\nabla u_n - \nabla u\|_{L^p(\Omega;\mathbb{R}^N)} = 0.$$

We define the new sequence $\{u_{n,\Phi}\}_{n\in\mathbb{N}} \subseteq C_0^\infty(\Omega)$ by

$$u_{n,\Phi}(x) := u_n(\Phi(x)) = u_n(x \cdot A + \mathbf{b}), \qquad \text{for } x \in \Omega,\ n \in \mathbb{N}.$$

By using the change of variable $y = x \cdot A + \mathbf{b}$, we get

$$\begin{aligned}\int_\Omega |u_{n,\Phi} - u_\Phi|^p \, dx &= \int_\Omega |u_n(x \cdot A + \mathbf{b}) - u(x \cdot A + \mathbf{b})|^p \, dx \\ &= \frac{1}{|\det A|} \int_\Omega |u_n(y) - u(y)|^p \, dy.\end{aligned}$$

Similarly, by using the Chain Rule formula for the smooth function $u_{n,\Phi}$ and Theorem 3.2.7, we have

$$\begin{aligned}\int_\Omega |\nabla u_{n,\Phi} - \nabla u_\Phi|^p \, dx &= \int_\Omega |\nabla u_n(x \cdot A + \mathbf{b}) \cdot A - \nabla u(x \cdot A + \mathbf{b}) \cdot A|^p \, dx \\ &= \frac{1}{|\det A|} \int_\Omega |(\nabla u_n(y) - \nabla u(y) \cdot A)|^p \, dy.\end{aligned}$$

As in the proof of Theorem 3.2.7 we have

$$|(\nabla u_n(y) - \nabla u(y)) \cdot A| \le |\nabla u_n(y) - \nabla u(y)| \sqrt{\sum_{i,j=1}^{N} |a_{ji}|^2}.$$

This observation permits to infer that

$$\int_\Omega |\nabla u_{n,\Phi} - \nabla u_\Phi|^p\, dx = \frac{1}{|\det A|} \int_\Omega |(\nabla u_n - \nabla u)\cdot A|^p\, dy$$

$$\leq \frac{\left(\sum\limits_{i,j=1}^{N} |a_{ji}|^2\right)^{\frac{p}{2}}}{|\det A|} \int_\Omega |\nabla u_n - \nabla u|^p\, dy.$$

This finally shows that the sequence $\{u_{n,\Phi}\}_{n\in\mathbb{N}} \subseteq C_0^\infty(\Omega)$ has the desired properties.

3.12.25 For every $\varepsilon > 0$, we set

$$f_\varepsilon(t) = \begin{cases} \dfrac{1}{(t+\varepsilon)^{p-1}}, & \text{if } t \geq 0, \\[2ex] \dfrac{1}{\varepsilon^{p-1}} - (p-1)\,\dfrac{t}{\varepsilon^p}, & \text{if } t < 0, \end{cases}$$

then we have

$$\frac{1}{(\varepsilon+u)^{p-1}} = f_\varepsilon \circ u.$$

Since f_ε is a C^1 function with bounded derivative, we get that $1/(\varepsilon+u)^{p-1} \in W^{1,p}_{\mathrm{loc}}(\Omega)$ by Proposition 3.4.1 and Remark 3.4.2. Moreover, its weak gradient is given by

$$\nabla\left(\frac{1}{(\varepsilon+u)^{p-1}}\right) = -(p-1)\,\frac{\nabla u}{(\varepsilon+u)^p}.$$

Observe that $|\eta|^p \in C_0^1(\Omega)$, then by Problem 3.12.18 we get

$$\frac{|\eta|^p}{(\varepsilon+u)^{p-1}} \in W_0^{1,p}(\Omega),$$

and its weak gradient is given by

$$\nabla\left(\frac{|\eta|^p}{(\varepsilon+u)^{p-1}}\right) = p\,\frac{|\eta|^{p-2}\,\eta\,\nabla\eta}{(\varepsilon+u)^{p-1}} - (p-1)\,|\eta|^p\,\frac{\nabla u}{(\varepsilon+u)^p}.$$

By observing that

$$\left\langle |\nabla u|^{p-2}\,\nabla u, \nabla\left(\frac{|\eta|^p}{(\varepsilon+u)^{p-1}}\right)\right\rangle = \left\langle |\nabla(\varepsilon+u)|^{p-2}\,\nabla(\varepsilon+u), \nabla\left(\frac{|\eta|^p}{(\varepsilon+u)^{p-1}}\right)\right\rangle,$$

we can now simply repeat the computations in the proof of the classical Picone inequality, i.e. Lemma 1.3.5.

3.12.26 Let $u \in W_0^{1,p}(\Omega)$ be such that $u \geq 0$ almost everywhere in Ω. We give the details for the case $1 \leq p < \infty$: the reader is kindly asked to take care of the case $p = \infty$. By definition there exists a sequence $\{\varphi_n\}_{n\in\mathbb{N}} \subseteq C_0^\infty(\Omega)$ such that

$$\lim_{n\to\infty} \|\varphi_n - u\|_{W^{1,p}(\Omega)} = 0.$$

Up to pass to a subsequence, we can further assume that

$$\lim_{n\to\infty} \varphi_n(x) = u(x), \qquad \text{for a. e. } x \in \Omega.$$

For every $n \in \mathbb{N}\setminus\{0\}$, we introduce the function

$$f_n(t) = \sqrt{\frac{1}{n^2} + t^2} - \frac{1}{n},$$

which is C^∞ and non-negative, vanishing only at the origin. This is a smooth approximation of the absolute value, already encountered in other places. Observe that

$$|f_n'(t)| = \frac{|t|}{\sqrt{\dfrac{1}{n^2} + t^2}} \leq 1, \qquad \text{for every } t \in \mathbb{R},$$

thus in particular we get

$$|f_n(t) - f_n(s)| = \left|\int_s^t f_n'(\tau)\,d\tau\right| \leq |t-s|, \qquad \text{for every } t, s \in \mathbb{R}. \tag{$*$}$$

Moreover, we have

$$\sqrt{\frac{1}{n^2} + t^2} \leq \frac{1}{n} + |t|, \qquad \text{for every } t \in \mathbb{R},$$

thanks to Problem 1.7.1 with $\alpha = 1/2$. In particular, this gives

$$\Big|f_n(t) - |t|\Big| = |t| + \frac{1}{n} - \sqrt{\frac{1}{n^2} + t^2}, \qquad \text{for every } t \in \mathbb{R}. \tag{$**$}$$

We define the sequence

$$u_n = f_n \circ \varphi_n,$$

and claim that this has the desired properties. By construction, we still have $u_n \in C_0^\infty(\Omega)$. Moreover, since f_n is non-negative, we clearly have $u_n \geq 0$ in Ω. We need to prove that

$$\lim_{n\to\infty} \|u_n - u\|_{L^p(\Omega)} = \lim_{n\to\infty} \|\nabla u_n - \nabla u\|_{L^p(\Omega;\mathbb{R}^N)} = 0.$$

By using the triangle inequality, Problem 1.7.2, $(*)$ and $(**)$, we get

$$\begin{aligned}\int_\Omega |u_n - u|^p\,dx &= \int_\Omega |f_n(\varphi_n) - u|^p\,dx \\ &\leq 2^{p-1}\int_\Omega |f_n(\varphi_n) - f_n(u)|^p\,dx + 2^{p-1}\int_\Omega |f_n(u) - u|^p\,dx \\ &\leq 2^{p-1}\int_\Omega |\varphi_n - u|^p\,dx + 2^{p-1}\int_\Omega \left(u + \frac{1}{n} - \sqrt{\frac{1}{n^2} + u^2}\right)^p\,dx.\end{aligned}$$

Observe that we also used that $|u| = u$, thanks to the assumption on u. The first integral on the right-hand side converges to 0 as n goes to 0, by virtue of the properties of the sequence $\{\varphi_n\}_{n\in\mathbb{N}}$. For the second one, it is sufficient to observe that

$$\lim_{n\to\infty}\left(u(x) + \frac{1}{n} - \sqrt{\frac{1}{n^2} + u(x)^2}\right)^p = 0, \qquad \text{for a. e. } x \in \Omega,$$

and

$$\left(u + \frac{1}{n} - \sqrt{\frac{1}{n^2} + u^2}\right)^p \leq u^p \in L^1(\Omega), \qquad \text{a. e. in } \Omega.$$

Thus, by the Dominated Convergence Theorem, we get that this integral converges to 0, as well. In conclusion, we get

$$\lim_{n\to\infty} \|u_n - u\|_{L^p(\Omega)} = 0.$$

We now consider the gradients: by the triangle inequality and the definition of u_n, we get

$$
\begin{aligned}
\int_\Omega |\nabla u_n - \nabla u|^p \, dx &= \int_\Omega |f_n'(\varphi_n)\, \nabla \varphi_n - \nabla u|^p \, dx \\
&\le 2^{p-1} \int_\Omega |f_n'(\varphi_n)|^p \, |\nabla \varphi_n - \nabla u|^p \, dx \\
&+ 2^{p-1} \int_\Omega |f_n'(\varphi_n) - 1|^p \, |\nabla u|^p \, dx \\
&\le 2^{p-1} \int_\Omega |\nabla \varphi_n - \nabla u|^p \, dx \\
&+ 2^{p-1} \int_\Omega |f_n'(\varphi_n) - 1|^p \, |\nabla u|^p \, dx.
\end{aligned}
$$

In the last estimate, we used that $|f_n'| \le 1$. Thus, the first integral on the right-hand side converges to 0, as n goes to ∞. For the second integral, we can use the Dominated Convergence Theorem, but some care is needed: we recall that by (3.4.7), we have

$$
\nabla u(x) = 0, \qquad \text{for a. e. } x \in \{y \in \Omega \, : \, u(y) = 0\}.
$$

Thus, if we set $\Omega_+ = \{y \in \Omega \, : \, u(y) > 0\}$, we get

$$
\int_\Omega |f_n'(\varphi_n) - 1|^p \, |\nabla u|^p \, dx = \int_{\Omega_+} |f_n'(\varphi_n) - 1|^p \, |\nabla u|^p \, dx.
$$

We then observe that, since $|f_n'| \le 1$, we have

$$
|f_n'(\varphi_n) - 1|^p \, |\nabla u|^p \le 2^p \, |\nabla u|^p \in L^1(\Omega),
$$

while

$$
\lim_{n \to \infty} f_n'(\varphi_n(x)) = \lim_{n \to \infty} \frac{\varphi_n(x)}{\sqrt{\dfrac{1}{n^2} + \varphi_n(x)^2}} = \frac{u(x)}{|u(x)|} = 1, \qquad \text{for a. e. } x \in \Omega_+.
$$

By the Dominated Convergence Theorem, we then obtain

$$
\lim_{n \to \infty} \int_\Omega |f_n'(\varphi_n) - 1|^p \, |\nabla u|^p \, dx = \lim_{n \to \infty} \int_{\Omega_+} |f_n'(\varphi_n) - 1|^p \, |\nabla u|^p \, dx = 0.
$$

This in turn implies that

$$
\lim_{n \to \infty} \|\nabla u_n - \nabla u\|_{L^p(\Omega;\mathbb{R}^N)} = 0,
$$

as well.

3.12.27 We just need to prove that

$$u \in L^q(B),$$

under the standing assumptions on u. To this aim, we will make a suitable use of Sobolev embeddings. Let us introduce the sequence of exponents

$$\gamma_i = \frac{N}{N-i}, \qquad \text{for } i \in \{1, \ldots, N-1\}.$$

We observe that by construction we have[12]

$$\gamma_{i+1} = \gamma_i^* = \frac{N\,\gamma_i}{N-\gamma_i}, \qquad \text{for } i \in \{1, \ldots, N-2\},$$

i.e. γ_{i+1} coincides with the critical Sobolev exponent of γ_i. The assumptions on u imply in particular that $u \in W^{1,1}(B)$. By using Theorem 3.9.3 with $\Omega = B$ and $p = 1$, we then get that

$$u \in L^{1^*}(B) = L^{\gamma_1}(B).$$

If $1 < q \leq \gamma_1$, then this is sufficient to get in particular that $u \in L^q(B)$, by Hölder's inequality. If on the contrary $q > \gamma_1$, we can apply once more Theorem 3.9.3, this time with $\Omega = B$ and $p = \gamma_1$. This gives that

$$u \in L^{\gamma_1^*}(B) = L^{\gamma_2}(B).$$

We proceed iteratively in this way, by distinguishing two possibilities:

- if $1 < q \leq N$, we stop whenever we reach γ_i such that $q \leq \gamma_i$. Observe that this is possible, thanks to the fact that $\gamma_{N-1} = N$. In this case, we conclude as above;
- if $q > N$, we iterate $N-2$ times the above reasoning and obtain

$$u \in L^{\gamma_{N-2}^*}(B) = L^{\gamma_{N-1}}(B) = L^N(B).$$

This implies that $u \in W^{1,N}(B)$. A further application of Theorem 3.9.3, with $\Omega = B$, $p = N$ and q as in the statement, would permit to conclude.

Remark 8.3.1 The conclusion of Problem 3.12.27 still holds if we replace the ball B with an open bounded set $\Omega \subseteq \mathbb{R}^N$ having C^1 boundary. On the contrary, this might fail to be true for an open set having an "irregular" boundary (see for example [50, Example 1, page 7]).

[12] This makes sense for $N \geq 3$. In dimension $N = 2$, observe that we directly have $\gamma_1 = 2 = N$.

8.4 Problems of Chap. 4

4.10.1 For the uniqueness part it is sufficient to reproduce the proof of Proposition 4.5.1. Let us suppose that $v, w \in W^{1,p}(\Omega)$ are two minimizers. We aim at proving that $w = v$: observe that for every $x \in \Omega$ the function

$$F(u, z) = H(z) - f(x)\, u,$$

is convex in the variables (u, z) and strictly convex in the variable z. We consider the convex combination

$$\psi = \frac{v + w}{2} \in W^{1,p}(\Omega),$$

which still verifies $\psi - U \in W_0^{1,p}(\Omega)$. By convexity, we get that

$$F(\psi, \nabla\psi) \le \frac{1}{2}\, F(v, \nabla v)\, dx + \frac{1}{2}\, F(w, \nabla w), \qquad \text{a.e. on } \Omega. \tag{$*$}$$

If we integrate over Ω, we thus obtain

$$\int_\Omega F(\psi, \nabla\psi)\, dx \le \frac{1}{2} \int_\Omega F(v, \nabla v)\, dx + \frac{1}{2} \int_\Omega F(w, \nabla w)\, dx.$$

By minimality of both v and w, we get that

$$\int_\Omega F(\psi, \nabla\psi)\, dx = \frac{1}{2} \int_\Omega F(v, \nabla v)\, dx + \frac{1}{2} \int_\Omega F(w, \nabla w)\, dx.$$

The latter implies that we must have equality almost everywhere in $(*)$. Hence the strict convexity of F in the gradient variable implies that

$$\nabla v = \nabla w, \qquad \text{a.e. on } \Omega,$$

that is

$$v - w = (v - U) - (w - U) \in W_0^{1,p}(\Omega),$$

is such that $\nabla(v - w) = (0, \dots, 0)$ almost everywhere on Ω. By Proposition 3.7.11 we get that v and w coincide almost everywhere on Ω.

Once uniqueness is established, the radial symmetry of v is readily obtained. Indeed, assume by contradiction that v is not radially symmetric. This implies that there exists an $N \times N$ orthogonal matrix O such that, if we define the new function

$$v_{\mathrm{O}}(x) := v(x \cdot \mathrm{O}),$$

then $v_{\mathrm{O}} \not\equiv v$. By Theorem 3.2.7, we have

$$\nabla v_{\mathrm{O}}(x) = \nabla v(x \cdot \mathrm{O}) \cdot \mathrm{O}.$$

This yields

$$\begin{aligned}|\nabla v_{\mathrm{O}}(x)| &= \sqrt{\langle \nabla v_{\mathrm{O}}(x), \nabla v_{\mathrm{O}}(x)\rangle} \\ &= \sqrt{\langle \nabla v(x \cdot \mathrm{O}) \cdot \mathrm{O}, \nabla v(x \cdot \mathrm{O}) \cdot \mathrm{O}\rangle} \\ &= \sqrt{\langle \nabla v(x \cdot \mathrm{O}) \cdot \mathrm{O}\,\mathrm{O}^T, \nabla v(x \cdot \mathrm{O})\rangle} = |\nabla v(x \cdot \mathrm{O})|.\end{aligned}$$

Here we used that for an orthogonal matrix we have $\mathrm{O}\,\mathrm{O}^{\mathrm{T}} = \mathrm{Id}_N$. In particular, since $H(z)$ depends only on $|z|$, we have

$$H(\nabla v_{\mathrm{O}}(x)) = H(\nabla v(x \cdot \mathrm{O})), \qquad \text{for a. e. } x \in \Omega. \tag{$*$}$$

By Problem 3.12.24, we still have $v_{\mathrm{O}} - U \in W^{1,p}_0(\Omega)$, thanks to the radial symmetry of U. By using that

$$|\det \mathrm{O}| = 1,$$

formula ($*$) and the rotational symmetry of f, we would get

$$\begin{aligned}&\int_\Omega H(\nabla v_{\mathrm{O}}(x))\,dx - \int_\Omega f(x)\,v_{\mathrm{O}}(x)\,dx \\ &\qquad= \int_\Omega H(\nabla v(x \cdot \mathrm{O}))\,dx - \int_\Omega f(x)\,v(x \cdot \mathrm{O})\,dx \\ &\qquad= \int_\Omega H(\nabla v(y))\,dy - \int_\Omega f(y)\,v(y)\,dy,\end{aligned}$$

thanks to the change of variable $y = x \cdot \mathrm{O}$. This shows that v_O is another minimizer, different from v. By uniqueness, we get a contradiction.

4.10.2 The existence of a weak solution can be inferred by reproducing almost verbatim the proof of Sect. 4.1, i.e. we can use the Direct Method to show existence of a solution to

$$\min_{u \in W^{1,2}(\Omega)} \left\{ \frac{1}{2} \int_\Omega |\nabla u|^2\,dx \,:\, u - g \in W^{1,2}_0(\Omega) \right\}. \tag{$*$}$$

We only have to pay attention to a detail: the set Ω is not necessarily bounded, not even in one direction $\omega \in \mathbb{S}^{N-1}$, thus we can not use the Poincaré inequality of Theorem 3.7.6 to infer that a minimizing sequence is bounded in $W^{1,2}(\Omega)$. However, in place of this, we can use the Poincaré inequality of Problem 3.12.11

$$\int_\Omega |u|^2\,dx \le C\,|\Omega|^{\frac{2}{N}} \int_\Omega |\nabla u|^2\,dx, \qquad \text{for every } u \in W^{1,2}_0(\Omega).$$

Then the proof carries over without any further modifications and show the existence of a weak solution v.

Let us now show uniqueness of the weak solution. Here we can proceed by adapting to the weak setting the corresponding observation of Remark 1.5.2. In other words, we first prove that any weak solution is necessarily a minimizer of $(*)$ and then use the strict convexity of the functional to infer uniqueness. Thus, let us suppose that $w \in W^{1,2}(\Omega)$ such that $w - g \in W^{1,2}_0(\Omega)$ is another weak solution of our problem. By Proposition 4.3.5, we know that w is a minimizer, as well. In order to prove uniqueness, it is now sufficient to use the strict convexity of the functional, by proceeding as in the proof of Proposition 4.5.1.

4.10.3 We already know from the previous problem that the solution is unique. Moreover, the solution of the previous problem shows that finding a minimizer for our problem is equivalent to find a weak solution of

$$\begin{cases} -\Delta v = 0, \text{ in } A_{r,R}, \\ \quad v = g, \text{ on } \partial A_{r,R}. \end{cases}$$

Observe that the boundary datum g is radially symmetric and such that

$$g(x) = 0, \quad \text{if } |x| = R, \qquad g(x) = 1, \quad \text{if } |x| = r.$$

We seek an harmonic function, taking the value 1 on $\partial B_r(0)$ and 0 on $\partial B_R(0)$. Due to the rotational symmetry of both the set $A_{r,R}$ and the boundary datum, by Problem 4.10.1 we know that such a solution must be radially symmetric. By using polar coordinates, we are thus lead to find a solution of the form

$$v(x) = \psi(|x|),$$

with ψ solving the following ordinary differential equation (see Appendix A for the expression of the Laplacian in polar coordinates)

$$\psi''(\varrho) + \frac{1}{\varrho}\,\psi'(\varrho) = 0, \qquad \text{for } \varrho \in (r, R). \tag{$*$}$$

This has to be coupled with the boundary conditions

$$\psi(r) = 1, \qquad \psi(R) = 0.$$

In order to find a solution, we observe that Eq. $(*)$ can be equivalently rewritten as

$$\varrho\,\psi''(\varrho) + \psi'(\varrho) = 0, \qquad \text{for } \varrho \in (r, R),$$

it is sufficient to multiply both sides by ϱ. In turn, the left-hand side can be rewritten as the derivative of a product, i.e.

$$(\varrho\,\psi'(\varrho))' = 0, \qquad \text{for } \varrho \in (r, R).$$

This implies that there exists a constant A such that

$$\varrho\,\psi'(\varrho) = A, \qquad \text{for } \varrho \in (r, R).$$

We divide both sides by ϱ, so to get

$$\psi'(\varrho) = \frac{A}{\varrho}, \qquad \text{for } \varrho \in (r, R).$$

This implies that solutions of $(*)$ take the form

$$\psi(\varrho) = B + A\,\log\varrho, \qquad \text{for } \varrho \in (r, R).$$

Here A, B are two constants. We now impose the boundary conditions in order to determine A, B: we have

$$1 = \psi(r) = B + A\,\log r \qquad \text{and} \qquad 0 = B + A\,\log R.$$

From the second equation, we obtain

$$B = -A\,\log R,$$

and by substituting this relation into the first equation, we get

$$1 = -A\,\log R + A\,\log r \qquad \text{that is} \qquad A = \frac{1}{\log r - \log R} = \frac{1}{\log \dfrac{r}{R}}.$$

Finally, we get that

$$\psi(\varrho) = -\frac{1}{\log \dfrac{r}{R}}\,\log R + \frac{1}{\log \dfrac{r}{R}}\,\log\varrho = \frac{1}{\log \dfrac{r}{R}}\,\log\frac{\varrho}{R}, \qquad \text{for } \varrho \in (r, R),$$

and thus, by going back to our original problem

$$v(x) = \psi(|x|) = \frac{1}{\log \dfrac{r}{R}}\,\log\frac{|x|}{R}, \qquad \text{for } x \in A_{r,R},$$

see Fig. 8.9. We only need to verify that

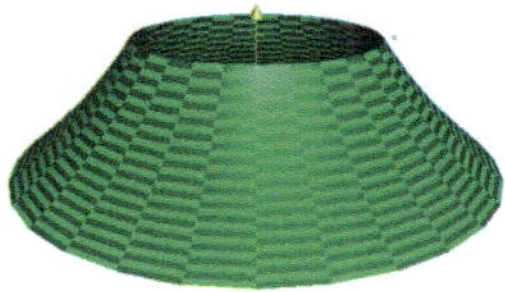

Fig. 8.9 The solution of Problem 4.10.2, with $r = 1$ and $R = 2$

$$v \in W^{1,2}(A_{r,R}) \qquad \text{and} \qquad v - g \in W_0^{1,2}(A_{r,R}). \tag{$**$}$$

By observing that $v \in C^1(\overline{A_{r,R}})$ and

$$v - g \in \left\{ u \in C^0(\overline{A_{r,R}}) \cap C^1(A_{r,R}) \, : \, u = 0 \text{ on } \partial A_{r,R}, \ \nabla u \in L^\infty(A_{r,R}; \mathbb{R}^2) \right\},$$

we get $(**)$ by using Proposition 3.2.3 and Problem 3.12.15, respectively.

Remark 8.4.1 In Electromagnetism, the minimum value

$$\min_{u \in W^{1,2}(A_{r,R})} \left\{ \frac{1}{2} \int_{A_{r,R}} |\nabla u|^2 \, dx \, : \, u - g \in W_0^{1,2}(A_{r,R}) \right\}.$$

appearing in the previous problem is called *electrostatic capacity of the condenser* made by the two circular plates $\partial B_r(0)$ and $\partial B_R(0)$. This is the electrical energy stored by the condenser, once an electrical potential difference 1 is established between the two plates. The relevant optimal function

$$x \mapsto \frac{1}{\log \dfrac{r}{R}} \log \frac{|x|}{R},$$

is called *capacitary potential.* We have already presented this problem in the Preface of the book.

4.10.4 Let us denote by $B_R \subseteq \mathbb{R}^2$ a disk of radius R. Without loss of generality, we can suppose that the disk is centered at the origin. We already know from Proposition 4.5.1 that such a function exists, is unique and can be obtained by solving

$$\min_{u \in W_0^{1,2}(B_R)} \left\{ \frac{1}{2} \int_{B_R} |\nabla u|^2 \, dx - \int_{B_R} u \, dx \right\}.$$

We need to compute it, i.e. we need to compute the weak solution of

$$\begin{cases} -\Delta v = 1, \text{ in } B_R, \\ \quad v = 0, \text{ on } \partial B_R. \end{cases}$$

Moreover, by Problem 4.10.1 we know that such a solution is radially symmetric. By using polar coordinates, we are thus led to find a solution of the form

$$v(x) = \psi(|x|),$$

with ψ solving the ordinary differential equation

$$-\psi''(\varrho) - \frac{1}{\varrho}\,\psi'(\varrho) = 1, \qquad \text{in } (0, R). \tag{$*$}$$

This has to be coupled with the boundary condition $\psi(R) = 0$. In order to solve $(*)$, we proceed as in Problem 4.10.3. By multiplying both sides by ϱ, the equation rewrites as

$$-\left(\varrho\,\psi'(\varrho)\right)' = \varrho, \qquad \text{in } (0, R).$$

This implies that there exists a constant A such that

$$\varrho\,\psi'(\varrho) = -\frac{\varrho^2}{2} + A, \qquad \text{in } (0, R).$$

We divide both sides by ϱ and take the primitives, so to get

$$\psi(\varrho) = -\frac{\varrho^2}{4} + A\,\log\varrho + B, \qquad \text{for } \varrho \in (0, R).$$

Here A, B are two constants. Observe that there is something strange: we have two constants at our disposal, but only one boundary condition. However, since we want our function

$$u(x) = \psi(|x|),$$

to belong to the Sobolev space $W_0^{1,2}(B_R)$, we also need to impose that

$$A\,\log\varrho = 0 \qquad \text{that is} \qquad A = 0.$$

Indeed, it is easy to see that $\log|x| \not\in W^{1,2}(B_R)$ (*verify this fact as an exercise*). We are left with imposing the boundary condition in order to determine B: we have

$$0 = \psi(R) = -\frac{R^2}{4} + B \qquad \text{that is} \qquad B = \frac{R^2}{4}.$$

Finally, we get that

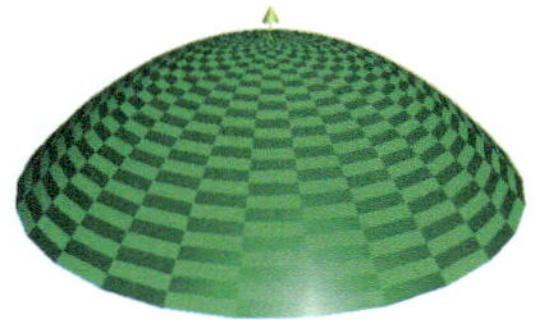

Fig. 8.10 The graph of the torsion function v for a disk centered at the origin

$$v(x) = \psi(|x|) = \frac{R^2 - |x|^2}{4}, \qquad \text{for } x \in B_R,$$

see Fig. 8.10. By construction, this is a classical solution of $-\Delta v = 1$ and thus a weak solution, as well. The fact that $v \in W_0^{1,2}(B_R)$ follows by observing that

$$v \in \left\{ u \in C^0(\overline{B_R}) \cap C^1(B_R) \,:\, u = 0 \text{ on } \partial B_R,\ \nabla u \in L^\infty(B_R) \right\},$$

and using Problem 3.12.15.

4.10.5 We proceed as in the previous problem and look for a radial solution

$$v(x) = \psi(|x|).$$

From the solution of the previous problem, we already know that

$$\psi(\varrho) = -\frac{\varrho^2}{4} + A \log \varrho + B, \qquad \text{for } \varrho \in (r, R).$$

By imposing the boundary conditions

$$\psi(r) = \psi(R) = 0,$$

we get

$$\psi(\varrho) = \frac{r^2 - \varrho^2}{4} + \frac{R^2 - r^2}{4 \log \dfrac{R}{r}} \log \frac{\varrho}{r}, \qquad \text{for } \varrho \in (r, R),$$

so that

$$v(x) = \psi(|x|) = \frac{r^2 - |x|^2}{4} + \frac{R^2 - r^2}{4 \log \dfrac{R}{r}} \log \frac{|x|}{r}, \qquad \text{for } x \in A_{r,R},$$

is the desired torsion function (see Fig. 8.11).

4.10.6 We first observe that by setting

Fig. 8.11 The graph of the torsion function v for the annulus $A_{r,R}$, with $r = 1/2$ and $R = 2$

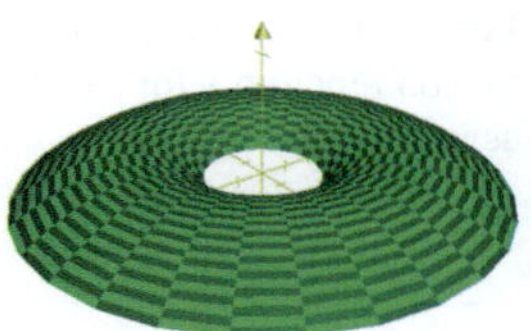

$$f(x, y) = \frac{1}{\sqrt{x^2 + y^2}}, \qquad \text{for a. e. } (x, y) \in B_1(0),$$

the problem under consideration coincides with the Euler-Lagrange equation of the functional

$$\frac{1}{2} \int_{B_1(0)} |\nabla u|^2 \, dx\, dy - \int_{B_1(0)} f\, u \, dx\, dy, \qquad \text{for } u \in W_0^{1,2}(B_1(0)).$$

Thus, we can get existence of a weak solution by minimizing this functional on $W_0^{1,2}(B_1(0))$. By using polar coordinates, we see that for every $1 \leq \gamma < 2$ we have

$$\int_{B_1(0)} |f|^\gamma \, dx\, dy = 2\pi \int_0^1 \frac{1}{\varrho^\gamma} \, \varrho \, d\varrho = 2\pi \int_0^1 \varrho^{1-\gamma} \, d\varrho = \frac{2\pi}{2-\gamma} < +\infty.$$

Thus, we can apply Theorem 4.3.1, with

$$H(z) = \frac{1}{2} |z|^2, \qquad 1 < \gamma < 2, \qquad G(t) = t,$$

and get existence of a weak solution. By Proposition 4.3.4, we know that any other weak solution would be a minimizer of the previous functional. On the other hand, by Problem 4.10.1, we know that such a minimizer is unique and radially symmetric. This shows the uniqueness and radial symmetry of the weak solution to our equation, as well.

By using polar coordinates, we are thus lead to find a solution of the form

$$v(x) = \psi(|x|),$$

with ψ solving the ordinary differential equation

$$-\psi''(\varrho) - \frac{1}{\varrho} \psi'(\varrho) = \frac{1}{\varrho}, \qquad \text{in } (0, 1).$$

This has to be coupled with the boundary condition $\psi(1) = 0$. We multiply both sides by ϱ, so to get

$$-(\varrho\, \psi'(\varrho))' = 1, \qquad \text{in } (0, 1).$$

By integrating, we then get

$$\psi'(\varrho) = -1 + \frac{A}{\varrho}, \qquad \text{in } (0,1).$$

A further integration then leads to

$$\psi(\varrho) = -\varrho + A\,\log\varrho + B, \qquad \text{in } (0,1).$$

As in Problem 4.10.4, we must have $A = 0$, in order to have $v \in W^{1,2}(B_1(0))$. Finally, by imposing the condition $\psi(1) = 0$, we obtain

$$B = 1.$$

Thus, the function $v(x) = 1 - |x|$ is the desired solution. Observe that $v \in W^{1,2}_0(B_1(0))$ thanks to Problem 3.12.14.

4.10.7 As in Problem 4.10.6, we can observe that if we set

$$f(x,y) = (x^2+y^2)^{\frac{\alpha}{2}}, \qquad \text{for a. e. } (x,y) \in B_1(0),$$

a weak solution can be found by minimizing on $W^{1,2}_0(B_1(0))$ the following functional

$$\frac{1}{2}\int_{B_1(0)} |\nabla u|^2\,dx - \int_{B_1(0)} f\,u\,dx, \qquad \text{for } u \in W^{1,2}_0(B_1(0)).$$

By using polar coordinates as before, it is easily seen that $f \in L^\gamma(B_1(0))$ for every

$$\begin{cases} 1 \le \gamma, & \text{if } \alpha \ge 0,\\ 1 \le \gamma < 2/(-\alpha), & \text{if } -2 < \alpha < 0. \end{cases}$$

Thanks to the assumption on α, we have $2/(-\alpha) > 1$ and thus we can choose in particular $\gamma > 1$. Thus, we can apply Theorem 4.3.1, with

$$H(z) = \frac{1}{2}|z|^2, \qquad 1 < \gamma, \qquad G(t) = t,$$

and get existence of a weak solution. Uniqueness and radial symmetry of the solution follows as in Problem 4.10.6. In order to determine the explicit solution for $\alpha = 2$, we proceed as in the previous problem. We only need to solve

$$-\psi''(\varrho) - \frac{1}{\varrho}\,\psi'(\varrho) = \varrho^2, \qquad \text{in } (0,1),$$

with the boundary condition $\psi(1) = 0$. By proceeding as before, we find this time

$$\psi(\varrho) = -\frac{\varrho^4}{16} - A\,\log\varrho + B, \qquad \text{in } (0,1).$$

As in Problem 4.10.4, we must have $A = 0$, in order to have $v \in W^{1,2}(B_1(0))$. Finally, by imposing the condition $\psi(1) = 0$, we obtain

$$B = \frac{1}{16}.$$

Thus, the function $v(x) = (1 - |x|^4)/16$ is the desired solution. Observe that $v \in W^{1,2}_0(B_1(0))$ thanks to Problem 3.12.15.

4.10.8 We first observe that $\lambda_1(B_1(0)) > 0$, thanks to Remark 4.7.2. The weak formulation of the equation under consideration is given by

$$\int_{B_1(0)} \langle \nabla v, \nabla\varphi\rangle\,dx = \int_{B_1(0)} a\,v\,\varphi\,dx + \int_{B_1(0)} f\,\varphi\,dx, \quad \text{for every } \varphi \in C^\infty_0(B_1(0)).$$

This is nothing but the Euler-Lagrange equation for the functional

$$\begin{aligned}\mathcal{F}(u) &= \frac{1}{2}\int_{B_1(0)} |\nabla u|^2\,dx \\ &\quad - \frac{1}{2}\int_{B_1(0)} a\,|u|^2\,dx - \int_{B_1(0)} f\,u\,dx, \quad \text{for every } u \in W^{1,2}(B_1(0)).\end{aligned}$$

In order to prove existence of a weak solution for the claimed equation, it would be sufficient to show that the minimization problem

$$\inf_{u \in W^{1,2}_0(B_1(0))} \mathcal{F}(u),$$

admits a solution.

We can not directly appeal to the existence result of Theorem 4.3.1, because of the term containing $|u|^2$. However, we can easily adapt the proof of this result, thanks to the assumption on a. Indeed, by Hölder's inequality we have

$$\mathcal{F}(u) \ge \frac{1}{2}\int_{B_1(0)} |\nabla u|^2\,dx - \frac{\|a\|_{L^\infty(B_1(0))}}{2}\int_{B_1(0)} |u|^2\,dx - \|f\|_{L^q(B_1(0))}\,\|u\|_{L^{q'}(B_1(0))}.$$

Observe that $2 < q' < \infty$, since $1 < q < 2$. The last negative term can be controlled as in the proof of Theorem 4.3.1, by using the generalized Young inequality (1.2.6)

$$-\|f\|_{L^q(B_1(0))}\,\|u\|_{L^{q'}(B_1(0))} \ge -\frac{1}{q}\,\varepsilon^{1-q}\,\|f\|^q_{L^q(B_1(0))} - \frac{\varepsilon}{q'}\,\|u\|^{q'}_{L^{q'}(B_1(0))}.$$

In turn, by using Sobolev embeddings (i.e. Theorem 3.8.1 with $p = N = 2$) we have

$$\|u\|^{q'}_{L^{q'}(B_1(0))} \le C_q\, \|u\|^2_{L^2(B_1(0))}\, \|\nabla u\|^{q'-2}_{L^2(B_1(0);\mathbb{R}^2)},$$

for a constant depending on q, only. We can further estimate the L^2 norm on the right-hand side above by using the definition of λ_1, that is

$$\|u\|^2_{L^2(B_1(0))} \le \frac{1}{\lambda_1(B_1(0))}\, \|\nabla u\|^2_{L^2(B_1(0);\mathbb{R}^2)}.$$

By putting all the estimates together, we thus obtain

$$\begin{aligned} & - \|f\|_{L^q(B_1(0))}\, \|u\|_{L^{q'}(B_1(0))} \\ & \ge -\frac{1}{q}\, \varepsilon^{1-q}\, \|f\|^q_{L^q(B_1(0))} - \frac{\varepsilon}{q'}\, \frac{C_q}{\lambda_1(B_1(0))}\, \|\nabla u\|^2_{L^2(B_1(0);\mathbb{R}^2)}. \end{aligned}$$

On the other hand, again by definition of λ_1 we have

$$-\frac{\|a\|_{L^\infty(B_1(0))}}{2} \int_{B_1(0)} |u|^2\, dx \ge -\frac{1}{2}\, \frac{\|a\|_{L^\infty(B_1(0))}}{\lambda_1(B_1(0))} \int_{B_1(0)} |\nabla u|^2\, dx.$$

In conclusion, for every $u \in W^{1,2}_0(B_1(0))$ we have

$$\begin{aligned} \mathcal{F}(u) \ge & \frac{1}{2} \left(1 - \frac{\|a\|_{L^\infty(B_1(0))}}{\lambda_1(B_1(0))} - \frac{2\,\varepsilon}{q'}\, \frac{C_q}{\lambda_1(B_1(0))}\right) \\ & \times \int_{B_1(0)} |\nabla u|^2\, dx - \frac{1}{q}\, \varepsilon^{1-q}\, \|f\|^q_{L^q(B_1(0))}. \end{aligned}$$

It is now crucial to observe that the assumption on a entails that

$$1 - \frac{\|a\|_{L^\infty(B_1(0))}}{\lambda_1(B_1(0))} > 0.$$

Thus, it is possible to choose $\varepsilon > 0$ small enough in such a way that the coefficient in front of the L^2 norm of ∇u is still positive. With such a choice, we get

$$\mathcal{F}(u) \ge c \int_{B_1(0)} |\nabla u|^2\, dx - C, \qquad \text{for every } u \in W^{1,2}_0(B_1(0)),$$

with $c, C > 0$ depending only on $\lambda_1(B_1(0))$, q, $\|f\|_{L^q(B_1(0))}$ and $\|a\|_{L^\infty(B_1(0))}$, but not on the particular function u.

This shows that the infimum of $\mathcal{F}$ on $W_0^{1,2}(B_1(0))$ is finite and that the functional $\mathcal{F}$ is weakly coercive. We can use the Direct Method, by reproducing the proof of Theorem 4.3.1 in order to infer existence of a minimizer.

The uniqueness part needs some care. Indeed, observe that the functional $\mathcal{F}$ is not convex, again because of the term containing $|u|^2$. We suppose that there exists two solutions $v_1, v_2 \in W_0^{1,2}(B_1(0))$. Thus, we have

$$\int_{B_1(0)} \langle \nabla v_1, \nabla \varphi \rangle \, dx = \int_{B_1(0)} a\, v_1\, \varphi \, dx + \int_{B_1(0)} f\, \varphi \, dx, \qquad \text{for every } \varphi \in C_0^\infty(B_1(0)).$$

and

$$\int_{B_1(0)} \langle \nabla v_2, \nabla \varphi \rangle \, dx = \int_{B_1(0)} a\, v_2\, \varphi \, dx + \int_{B_1(0)} f\, \varphi \, dx, \qquad \text{for every } \varphi \in C_0^\infty(B_1(0)).$$

In particular, by subtracting the two equations we get

$$\int_{B_1(0)} \langle \nabla (v_1 - v_2), \nabla \varphi \rangle \, dx = \int_{B_1(0)} a\, (v_1 - v_2)\, \varphi \, dx, \quad \text{for every } \varphi \in C_0^\infty(B_1(0)).$$

As usual, by a density argument we can admit test functions $\varphi \in W_0^{1,2}(B_1(0))$. In particular, we can take $\varphi = v_1 - v_2$ and get

$$\int_{B_1(0)} |\nabla (v_1 - v_2)|^2 \, dx = \int_{B_1(0)} a\, |v_1 - v_2|^2 \, dx \le \|a\|_{L^\infty(B_1(0))} \int_D |v_1 - v_2|^2 \, dx.$$

By using the definition of $\lambda_1(B_1(0))$ on the left-hand side, we get

$$\lambda_1(B_1(0)) \int_{B_1(0)} |v_1 - v_2|^2 \, dx \le \|a\|_{L^\infty(B_1(0))} \int_{B_1(0)} |v_1 - v_2|^2 \, dx,$$

that is

$$\Big(\lambda_1(B_1(0)) - \|a\|_{L^\infty(B_1(0))}\Big) \int_{B_1(0)} |v_1 - v_2|^2 \, dx \le 0.$$

By using again the assumption on a, this gives that v_1 and v_2 must coincide almost everywhere.

4.10.9 The existence of a solution follows directly from Theorem 4.3.1 with

$$H(z) = \frac{1}{2}|z|^2, \qquad G(t) = t, \qquad \text{and} \qquad \gamma = 2.$$

In other words, a weak solution can be obtained by solving the minimization problem

$$\min_{u \in W_0^{1,2}(\Omega)} \left\{ \frac{1}{2} \int_\Omega |\nabla u|^2 \, dx - \int_\Omega f \, u \, dx \right\}. \tag{$*$}$$

In order to prove uniqueness, we proceed as in the case of Proposition 4.5.1: if $w \in W_0^{1,2}(\Omega)$ is another weak solution, by Proposition 4.3.4 it must be another minimizer of $(*)$. By using the strict convexity of the functional, we obtain that $u = w$. The details are exactly as in the uniqueness part of Proposition 4.5.1.

We only need to prove the claimed estimate on the solution v. Since v is the minimizer of $(*)$ and observing that $t\,v$ is still admissible for the same problem (for every $t \in \mathbb{R}$), we have that

$$\frac{1}{2} \int_\Omega |\nabla(t\,v)|^2 \, dx - \int_\Omega f\,(t\,v) \, dx \ge \frac{1}{2} \int_\Omega |\nabla v|^2 \, dx - \int_\Omega f \, v \, dx$$

that is

$$(1-t) \int_\Omega f \, v \, dx \ge \frac{1-t^2}{2} \int_\Omega |\nabla v|^2 \, dx.$$

We take $0 < t < 1$, simplify $1 - t$ on both sides and then take the limit as t goes to 1. This gives

$$\int_\Omega f \, v \, dx \ge \int_\Omega |\nabla v|^2 \, dx.$$

By using Hölder's inequality in the right-hand side and the definition of $\lambda_1(\Omega)$, we get

$$\begin{aligned} \int_\Omega |\nabla v|^2 \, dx \le \int_\Omega f \, v \, dx &\le \|f\|_{L^2(\Omega)} \, \|v\|_{L^2(\Omega)} \\ &\le \|f\|_{L^2(\Omega)} \frac{1}{\sqrt{\lambda_1(\Omega)}} \left(\int_\Omega |\nabla v|^2 \, dx \right)^{\frac{1}{2}}. \end{aligned}$$

By simplifying an L^2 norm of ∇v on both sides, we obtain

$$\left(\int_\Omega |\nabla v|^2 \, dx \right)^{\frac{1}{2}} \le \|f\|_{L^2(\Omega)} \frac{1}{\sqrt{\lambda_1(\Omega)}}.$$

Finally, by recalling Remark 4.7.2, we conclude.

4.10.10 This is a direct consequence of the Poincaré inequality of Problem 3.12.11. Indeed, the latter gives that for every $\varphi \in C_0^\infty(\Omega)$ we have

$$\int_\Omega |\varphi|^2\,dx \leq C\,|\Omega|^{\frac{2}{N}} \int_\Omega |\nabla\varphi|^2\,dx,$$

for some $C = C(N) > 0$. By recalling the definition of $\lambda_1(\Omega)$, we get the desired result.

We remark that the constant found in this way is not the optimal one. The sharpest constant corresponds to

$$C = \lambda_1(B)\,|B|^{\frac{2}{N}},$$

where $B \subseteq \mathbb{R}^N$ in any N-dimensional open ball. In other words, we have the so-called *Faber-Krahn inequality*

$$\lambda_1(\Omega) \geq \left(\lambda_1(B)\,|B|^{\frac{2}{N}}\right)\,|\Omega|^{-\frac{2}{N}},$$

for every open set $\Omega \subseteq \mathbb{R}^N$ with finite measure. We refer the reader to [34, Chapter 3] for more details.

4.10.11 A weak solution v can be found as a minimizer of the following problem

$$\inf_{u\in W_0^{1,2}(B_1(0))} \left\{\frac{1}{2}\int_{B_1(0)} |\nabla u|^2\,dx - \int_{B_1(0)} f\,u\,dx\right\}.$$

Existence can be inferred directly from Theorem 4.3.1, with the choices

$$H(z) = \frac{1}{2}\,|z|^2, \qquad \gamma = \frac{6}{5}, \qquad q = 1, \qquad G(t) = t.$$

Indeed, observe that in dimension $N = 3$ we have $2^* = 6$ and thus

$$\frac{6}{5} = \left(\frac{6}{q}\right)'.$$

Uniqueness can be inferred by the usual convexity argument as in the proof of Proposition 4.5.1, we leave the details to the reader.

We now assume that $f \geq 0$ almost everywhere in B. In particular, we have

$$-\,f\,|v| = -|f\,v| \leq -f\,v, \qquad \text{a. e. in } B_1(0),$$

and by integrating

$$-\int_{B_1(0)} f\,|v|\,dx \le -\int_{B_1(0)} f\,v\,dx.$$

We observe that $|v| \in W_0^{1,2}(B_1(0))$ by Proposition 3.7.13 and

$$|\nabla |v|| = |\nabla v|, \qquad \text{a. e. in } B_1(0),$$

thanks to Corollary 3.4.6. These facts imply that

$$\frac{1}{2}\int_{B_1(0)} |\nabla |v||^2\,dx - \int_{B_1(0)} f\,|v|\,dx \le \frac{1}{2}\int_{B_1(0)} |\nabla v|^2\,dx - \int_{B_1(0)} f\,v\,dx.$$

This shows that $|v|$ is a minimizer, as well. By uniqueness, we get $v = |v|$ almost everywhere and thus $v \ge 0$ almost everywhere. Finally, from the equation and the fact that $f \ge 0$, we get that v is weakly superharmonic. Thus, the conclusion follows from Lemma 4.4.5.

4.10.12 The weak formulation of the equation is given by

$$\int_{\mathbb{R}^2} \langle \nabla v, \nabla \varphi\rangle\,dx + \int_{\mathbb{R}^2} V\,v\,\varphi\,dx = \int_{\mathbb{R}^2} f\,\varphi\,dx, \qquad \text{for every } \varphi \in C_0^\infty(\mathbb{R}^2).$$

This is the Euler-Lagrange equation of the functional

$$\mathcal{F}(u) = \frac{1}{2}\int_{\mathbb{R}^2} |\nabla u|^2\,dx + \frac{1}{2}\int_{\mathbb{R}^2} V\,|u|^2\,dx - \int_{\mathbb{R}^2} f\,u\,dx,$$

thus it will be sufficient to show that

$$\inf_{u \in W^{1,2}(\mathbb{R}^2)} \mathcal{F}(u),$$

admits a minimizer. Such a minimizer will be the desired weak solution. In order to show existence of a minimizer, we use the Direct Method. We first show that $\mathcal{F}$ is well-defined on $W^{1,2}(\mathbb{R}^2)$. By recalling that (see Proposition 3.7.10)

$$W_0^{1,2}(\mathbb{R}^2) = W^{1,2}(\mathbb{R}^2),$$

from Theorem 3.8.1 (with $p = N = 2$) we get

$$W^{1,2}(\mathbb{R}^2) \subseteq L^\gamma(\mathbb{R}^2), \qquad \text{for every } 2 < \gamma < \infty.$$

Moreover, for every $u \in W^{1,2}(\mathbb{R}^2)$ we have

$$\|u\|_{L^\gamma(\mathbb{R}^2)} \le C_\gamma\,\|u\|_{L^2(\mathbb{R}^2)}^{\frac{2}{\gamma}}\,\|\nabla u\|_{L^2(\mathbb{R}^2;\mathbb{R}^2)}^{1-\frac{2}{\gamma}}, \qquad (*)$$

for a constant $C_\gamma > 0$, depending on γ only. Thus, the two terms

$$\int_{\mathbb{R}^2} V\,|u|^2\,dx \qquad \text{and} \qquad \int_{\mathbb{R}^2} f\,u\,dx,$$

are well-defined and finite, under the standing assumptions on both V and f.

Let us show that $\mathcal{F}$ is bounded from below on $W^{1,2}(\mathbb{R}^2)$. It is sufficient to control the term containing f: if $f \in L^2(\mathbb{R}^2)$, then we simply have for every $\varepsilon > 0$

$$-\int_{\mathbb{R}^2} f\,u\,dx \geq -\|f\|_{L^2(\mathbb{R}^2)}\,\|u\|_{L^2(\mathbb{R}^2)} \geq -\frac{1}{2\,\varepsilon}\,\|f\|^2_{L^2(\mathbb{R}^2)} - \frac{\varepsilon}{2}\,\|u\|^2_{L^2(\mathbb{R}^2)},$$

thanks to Hölder's and Young's inequalities. For the case $f \in L^q(\mathbb{R}^2)$ with $1 < q < 2$, we have to be more careful: we have

$$-\int_{\mathbb{R}^2} f\,u\,dx \geq -\|f\|_{L^q(\mathbb{R}^2)}\,\|u\|_{L^{q'}(\mathbb{R}^2)} \geq -C_{q'}\,\|f\|_{L^q(\mathbb{R}^2)}\,\|u\|^{\frac{2}{q'}}_{L^2(\mathbb{R}^2)}\,\|\nabla u\|^{1-\frac{2}{q'}}_{L^2(\mathbb{R}^2;\mathbb{R}^2)},$$

thanks to $(*)$. On the last two terms, we use Young's inequality with exponents

$$\frac{q'}{2} \qquad \text{and} \qquad \frac{q'}{q'-2},$$

so to get

$$-\int_{\mathbb{R}^2} f\,u\,dx \geq -C_{q'}\,\|f\|_{L^q(\mathbb{R}^2)}\left[\frac{q'}{2}\,\|u\|_{L^2(\mathbb{R}^2)} + \frac{q'-2}{q'}\,\|\nabla u\|_{L^2(\mathbb{R}^2;\mathbb{R}^2)}\right].$$

Finally, on the last term, we can use the generalized Young's inequality (1.2.6) with exponent 2, thus yielding

$$-\int_{\mathbb{R}^2} f\,u\,dx \geq -\frac{C_{q'}^2}{2\,\varepsilon}\,\|f\|^2_{L^q(\mathbb{R}^2)} - \frac{\varepsilon}{2}\left[\frac{q'}{2}\,\|u\|_{L^2(\mathbb{R}^2)} + \frac{q'-2}{q'}\,\|\nabla u\|_{L^2(\mathbb{R}^2;\mathbb{R}^2)}\right]^2.$$

By using this estimate and the properties of V, we get in any case

$$\begin{aligned}
\mathcal{F}(u) &= \frac{1}{2}\int_{\mathbb{R}^2} |\nabla u|^2\,dx + \frac{1}{2}\int_{\mathbb{R}^2} V\,|u|^2\,dx - \int_{\mathbb{R}^2} f\,u\,dx \\
&\geq \frac{1}{2}\int_{\mathbb{R}^2} |\nabla u|^2\,dx + \frac{c}{2}\int_{\mathbb{R}^2} |u|^2\,dx \\
&\quad - \frac{C_{q'}^2}{2\,\varepsilon}\,\|f\|^2_{L^q(\mathbb{R}^2)} - \frac{\varepsilon}{2}\left[\frac{q'}{2}\,\|u\|_{L^2(\mathbb{R}^2)} + \frac{q'-2}{q'}\,\|\nabla u\|_{L^2(\mathbb{R}^2;\mathbb{R}^2)}\right]^2
\end{aligned}$$

With simple algebraic manipulations, we get that we can choose $\varepsilon > 0$ small enough, such that

$$\mathcal{F}(u) \geq \widetilde{c}\left(\int_{\mathbb{R}^2} |\nabla u|^2\, dx + \int_{\mathbb{R}^2} |u|^2\, dx\right) - \widetilde{C},$$

for two constants $\widetilde{c} > 0$ and $\widetilde{C} > 0$, depending only on q, V and f. This shows that $\mathcal{F}$ is bounded from below and thus its infimum over $W^{1,2}(\mathbb{R}^2)$ is finite.

Let us call $\mathfrak{m}$ such an infimum. We take a minimizing sequence $\{u_n\}_{n\in\mathbb{N}} \subseteq W^{1,2}(\mathbb{R}^2)$ such that

$$\mathcal{F}(u_n) < \mathfrak{m} + \frac{1}{n+1}, \qquad \text{for every } n \in \mathbb{N}.$$

Thanks to the lower bound proved above, we get in particular that

$$\int_{\mathbb{R}^2} |\nabla u_n|^2\, dx + \int_{\mathbb{R}^2} |u_n|^2\, dx \leq C, \qquad \text{for every } n \in \mathbb{N},$$

for a constant $C > 0$ independent of n. By Theorem 3.3.6, we have that this sequence weakly converges in $W^{1,2}(\mathbb{R}^2)$ to a function $v \in W^{1,2}(\mathbb{R}^2)$, up to a subsequence. We observe that $\{u_n\}_{n\in\mathbb{N}}$ is uniformly bounded in $L^\gamma(\mathbb{R}^2)$ for every $2 \leq \gamma < \infty$, by owing to the uniform bound on the $W^{1,2}$ norms and $(*)$. Thus, by Problem 1.7.29 we have that actually this sequence weakly converges to v in $L^\gamma(\mathbb{R}^2)$, *for every* $2 \leq \gamma < \infty$ (without further extractions of subsequences).

We have to prove that v is the desired minimizer: we need to show that $\mathcal{F}$ is weakly lower semicontinuous. The fact that

$$\liminf_{n\to\infty} \frac{1}{2}\int_{\mathbb{R}^2} |\nabla u_n|^2\, dx \geq \frac{1}{2}\int_{\mathbb{R}^2} |\nabla v|^2\, dx,$$

easily follows from the weak convergence in $L^2(\mathbb{R}^2)$ of the gradients and Proposition 1.3.11. The term containing f is actually continuous, i.e.

$$\lim_{n\to\infty} \int_{\mathbb{R}^2} f\, u_n\, dx = \int_{\mathbb{R}^2} f\, v\, dx,$$

by using that $\{u_n\}_{n\in\mathbb{N}}$ weakly converges in $L^{q'}(\mathbb{R}^2)$, as we said above. For the term containing V, we could proceed in various ways, we act as follows: we observe that

$$V\, v \in L^{\frac{3}{2}}(\mathbb{R}^2).$$

Indeed, we have by Hölder's inequality with exponents $4/3$ and 4

$$\int_{\mathbb{R}^2} |V\,v|^{\frac{3}{2}}\,dx \leq \left(\int_{\mathbb{R}^2} |V|^2\,dx\right)^{\frac{3}{4}} \left(\int_{\mathbb{R}^2} |v|^6\,dx\right)^{\frac{1}{4}},$$

and the last integral is finite, again thanks to $(*)$. By using that $\{u_n\}_{n\in\mathbb{N}}$ weakly converges to v in $L^3(\mathbb{R}^2)$, we get

$$\lim_{n\to\infty} \int_{\mathbb{R}^2} V\,v\,(u_n - v)\,dx = 0,$$

by recalling that $L^{3/2}(\mathbb{R}^2)$ can be identified with the topological dual space of $L^3(\mathbb{R}^2)$ (see for example [46, Theorem 2.14]). By using this fact and the "above tangent" property (i.e. Proposition 1.2.3)

$$|u_n|^2 \geq |v|^2 + 2\,v\,(u_n - v),$$

we then obtain

$$\begin{aligned}\liminf_{n\to\infty} \int_{\mathbb{R}^2} V\,|u_n|^2\,dx &\geq \int_{\mathbb{R}^2} V\,|v|^2\,dx + 2\,\liminf_{n\to\infty} \int_{\mathbb{R}^2} V\,v\,(u_n - v)\,dx \\ &= \int_{\mathbb{R}^2} V\,|v|^2\,dx.\end{aligned}$$

In conclusion, we obtain that $\mathcal{F}$ is lower semicontinuous with respect to the convergences inferred above. Consequently, we get

$$\mathfrak{m} \leq \mathcal{F}(v) \leq \liminf_{n\to\infty} \mathcal{F}(u_n) \leq \lim_{n\to\infty} \left(\mathfrak{m} + \frac{1}{n+1}\right) = \mathfrak{m},$$

and v is the desired minimizer.

The uniqueness of the weak solution can be obtained by using the convexity of the functional $\mathcal{F}$, as in the proof of Proposition 4.5.1.

4.10.13 To prove the first point, it is sufficient to reproduce the first part of the proof of Theorem 4.2.1. Let $\{u_n\}_{n\in\mathbb{N}} \subseteq W_0^{1,2}(\Omega)$ be a sequence weakly converging to $u \in W_0^{1,2}(\Omega)$. From the "above tangent" property (i.e. Proposition 1.3.4) we have

$$\frac{1}{2}\,|\nabla u_n(x)|^2 \geq \frac{1}{2}\,|\nabla u(x)|^2 + \langle \nabla u(x), \nabla u_n(x) - \nabla u(x)\rangle, \qquad \text{for a. e. } x \in \Omega.$$

By multiplying the previous inequality by $a(x)$ and the integrating over Ω, we obtain

$$\frac{1}{2}\int_{\Omega} a\,|\nabla u_n|^2\,dx \geq \frac{1}{2}\int_{\Omega} a\,|\nabla u|^2\,dx + \int_{\Omega} a\,\langle \nabla u, \nabla u_n - \nabla u\rangle\,dx,$$

thanks to the fact that a is positive. By observing that $a\,\nabla u \in L^2(\Omega;\mathbb{R}^N)$ and using the weak convergence, we get

$$\liminf_{n\to\infty} \frac{1}{2}\int_\Omega a\,|\nabla u_n|^2\,dx \geq \frac{1}{2}\int_\Omega a\,|\nabla u|^2\,dx,$$

thus establishing the desired semicontinuity property. Observe that for this part, the weaker assumption

$$a(x) \geq 0, \qquad \text{for a. e. } x \in \Omega,$$

would still be sufficient.

In order to show existence of a weak solution to the claimed equation, we observe that the latter coincides with the Euler-Lagrange equation of the functional

$$\mathcal{F}(u) = \frac{1}{2}\int_\Omega a\,|\nabla u|^2\,dx - \int_\Omega f\,u\,dx, \qquad \text{for every } u \in W^{1,2}_0(\Omega).$$

It is thus sufficient to prove existence of a minimizer of $\mathcal{F}$ over $W^{1,2}_0(\Omega)$. This can be done by applying the Direct Method and reproducing verbatim the proof of Theorem 4.3.1. We do not give all the details, we just show how to infer that

$$\inf_{u\in W^{1,2}_0(\Omega)} \mathcal{F}(u) > -\infty.$$

Thanks to the assumption on a, for every $u \in W^{1,2}_0(\Omega)$, we have

$$\mathcal{F}(u) = \frac{1}{2}\int_\Omega a\,|\nabla u|^2\,dx - \int_\Omega f\,u\,dx \geq \frac{\overline{a}}{2}\int_\Omega |\nabla u|^2\,dx - \|f\|_{L^2(\Omega)}\,\|u\|_{L^2(\Omega)}.$$

On the last term, we can apply Poincaré inequality (see Remark 3.7.7)

$$\|u\|_{L^2(\Omega)} \leq \operatorname{diam}(\Omega)\,\|\nabla u\|_{L^2(\Omega;\mathbb{R}^N)},$$

thanks to the boundedness of Ω. Thus, we get

$$\mathcal{F}(u) \geq \frac{\overline{a}}{2}\int_\Omega |\nabla u|^2\,dx - \operatorname{diam}(\Omega)\,\|f\|_{L^2(\Omega)}\,\|\nabla u\|_{L^2(\Omega;\mathbb{R}^N)}.$$

We now apply the generalized Young inequality (1.2.6), which gives

$$\mathcal{F}(u) \geq \frac{\overline{a}}{2}\int_\Omega |\nabla u|^2\,dx - \frac{1}{2\,\varepsilon}\,\|f\|^2_{L^2(\Omega)} - \frac{\varepsilon(\operatorname{diam}(\Omega))^2}{2}\,\|\nabla u\|^2_{L^2(\Omega;\mathbb{R}^N)},$$

for every $\varepsilon > 0$. By choosing

$$\varepsilon = \frac{\overline{a}}{2\,(\mathrm{diam}(\Omega))^2},$$

we finally get

$$\mathcal{F}(u) \geq \frac{\overline{a}}{4} \int_\Omega |\nabla u|^2\, dx - C,$$

where $C > 0$ is a constant that only depends on $\mathrm{diam}(\Omega)$, f and $\overline{a}$.

Finally, for the uniqueness we can adapt the usual argument as in Proposition 4.5.1, by observing that for almost every $x \in \Omega$ the function

$$(u, z) \mapsto \frac{1}{2}\, a(x)\, |z|^2 - f(x)\, u,$$

is convex and strictly convex in the z variable.

4.10.14 For the first part, we just need to prove that there exists a constant $C > 0$ such that

$$\int_\Omega |\varphi|^2\, dx\, dy \leq C \int_\Omega |\nabla\varphi|^2\, dx\, dy, \qquad \text{for every } \varphi \in C_0^\infty(\Omega),$$

i.e. Ω supports Poincaré inequality. Observe that the set

$$\Omega = \Big(\mathbb{R} \times (-1, 1)\Big) \cup \Big((-1, 1) \times \mathbb{R}\Big),$$

is unbounded in every direction, see Fig. 8.12. Moreover, it is such that $|\Omega| = +\infty$. Thus, we can not appeal neither to Problem 3.12.11 nor to Proposition 3.5.2. However, we can suitably adapt the proof of the latter. To this aim, we decompose Ω in 5 subsets:

$$\Omega_0 = (-1, 1) \times (-1, 1), \quad \Omega_1 = [1, +\infty) \times (-1, 1), \quad \Omega_2 = (-\infty, -1] \times (-1, 1)$$

$$\Omega_3 = (-1, 1) \times [1, +\infty), \quad \Omega_4 = (-1, 1) \times (-\infty, -1].$$

We then take $\varphi \in C_0^\infty(\Omega)$ and observe that for every $(x, y) \in \Omega_1$ we have

$$|\varphi(x, y)| = |\varphi(x, y) - \varphi(x, 1)| = \left| \int_x^1 \frac{\partial \varphi}{\partial y}(x, t)\, dt \right| \leq \int_{-1}^1 \left| \frac{\partial \varphi}{\partial y}(x, t) \right|\, dt.$$

By raising to the power 2 and using Jensen's inequality, we get

$$|\varphi(x, y)| \leq 2 \int_{-1}^1 \left| \frac{\partial \varphi}{\partial y}(x, t) \right|^2\, dt.$$

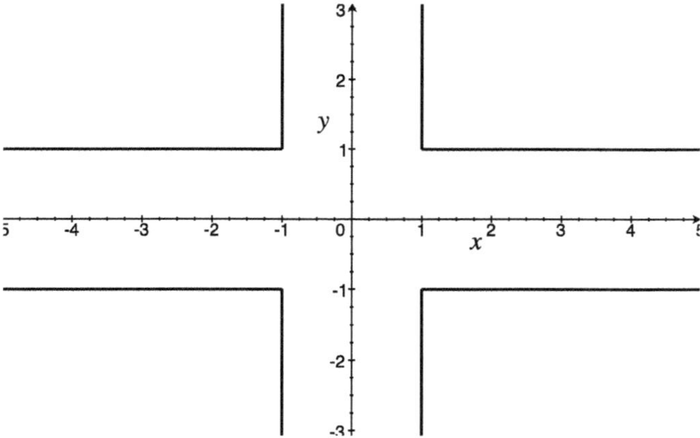

Fig. 8.12 The set Ω of Problem 4.10.14

By integrating over Ω_1, we obtain

$$\int_{\Omega_1} |\varphi(x, y)|^2 \, dx \, dy \leq 4 \int_{\Omega_1} \left| \frac{\partial \varphi}{\partial y}(x, y) \right|^2 \, dx \, dy \leq 4 \int_{\Omega_1} |\nabla \varphi|^2 \, dx \, dy.$$

By proceeding similarly for Ω_2, Ω_3 and Ω_4, we can finally obtain

$$\int_{\Omega_i} |\varphi|^2 \, dx \, dy \leq 4 \int_{\Omega_i} |\nabla \varphi|^2 \, dx \, dy, \qquad \text{for } i = 1, 2, 3, 4. \tag{$*$}$$

We are only left with estimating the L^2 norm on Ω_0: for every $(x, y) \in \Omega_0$, we write

$$\begin{aligned} |\varphi(x, y)| &\leq |\varphi(x, y) - \varphi(x + 2, y)| + |\varphi(x + 2, y)| \\ &= \left| \int_x^{x+2} \frac{\partial \varphi}{\partial x}(t, y) \, dt \right| + |\varphi(x + 2, y)| \\ &\leq \int_{-1}^{3} \left| \frac{\partial \varphi}{\partial x}(t, y) \right| \, dt + |\varphi(x + 2, y)|. \end{aligned}$$

We used that if $x \in (-1, 1)$, then $x \geq -1$ and $x + 2 \leq 3$. We raise to the power 2 and use Problem 1.7.2. This yields

$$\begin{aligned} |\varphi(x, y)|^2 &\leq 2 \left(\int_{-1}^{3} \left| \frac{\partial \varphi}{\partial x}(t, y) \right| \, dt \right)^2 + 2 \, |\varphi(x + 2, y)|^2 \\ &\leq 8 \int_{-1}^{3} \left| \frac{\partial \varphi}{\partial x}(t, y) \right|^2 \, dt + 2 \, |\varphi(x + 2, y)|^2. \end{aligned}$$

In the second inequality, we used Jensen's inequality. We integrate this inequality over $\Omega_0 = (-1, 1) \times (-1, 1)$. We get

$$\begin{aligned}
\int_{\Omega_0} |\varphi(x, y)|^2 \, dx \, dy &\le 8 \int_{\Omega_0} \left(\int_{-1}^{3} \left| \frac{\partial \varphi}{\partial x}(t, y) \right|^2 dt \right) dx \, dy \\
&\quad + 2 \int_{\Omega_0} |\varphi(x + 2, y)|^2 \, dx \, dy \\
&= 16 \int_{(1,3)\times(-1,1)} \left| \frac{\partial \varphi}{\partial x}(t, y) \right|^2 dt \, dy \\
&\quad + 2 \int_{(1,3)\times(-1,1)} |\varphi(\tau, y)|^2 \, d\tau \, dy \\
&\le 16 \int_{\Omega_1} |\nabla \varphi(t, y)|^2 \, dt \, dy \\
&\quad + 2 \int_{(1,3)\times(-1,1)} |\varphi(\tau, y)|^2 \, d\tau \, dy.
\end{aligned}$$

In the last integral we used the change of variable $\tau = x + 2$. We can now apply $(*)$ with $i = 1$ to the rightmost integral, so to get

$$\int_{\Omega_0} |\varphi|^2 \, dx \, dy \le 24 \int_{\Omega_1} |\nabla \varphi|^2 \, dx \, dy,$$

By combining this estimate with $(*)$, we now get

$$\int_{\Omega} |\varphi|^2 \, dx \, dy \le C \int_{\Omega} |\nabla \varphi|^2 \, dx \, dy,$$

for an explicit constant $C > 0$, not depending on anything.

We now take $f \in L^2(\Omega)$. In order to prove existence of a weak solution, it is sufficient to show that the minimization problem

$$\inf_{u \in W_0^{1,2}(\Omega)} \left\{ \frac{1}{2} \int_{\Omega} |\nabla u|^2 \, dx - \int_{\Omega} f \, u \, dx \right\},$$

admits a solution: a minimizer would be the desired weak solution. Observe that we can not directly appeal to the existence result of Theorem 4.3.1, because the set Ω is not bounded. However, we can still proceed by using the Direct Method: we set

$$\mathfrak{m} := \inf_{u \in W_0^{1,2}(\Omega)} \left\{ \frac{1}{2} \int_{\Omega} |\nabla u|^2 \, dx - \int_{\Omega} f \, u \, dx \right\},$$

and first prove that $\mathfrak{m} > -\infty$. Indeed, for every $u \in W_0^{1,2}(\Omega)$

$$\begin{aligned}\frac{1}{2}\int_\Omega |\nabla u|^2\,dx - \int_\Omega f\,u\,dx &\ge \frac{1}{2}\int_\Omega |\nabla u|^2\,dx - \|f\|_{L^2(\Omega)}\,\|u\|_{L^2(\Omega)}\\ &\ge \frac{1}{2}\int_\Omega |\nabla u|^2\,dx - \|f\|_{L^2(\Omega)}\,\frac{\|\nabla u\|_{L^2(\Omega;\mathbb{R}^2)}}{\sqrt{\lambda_1(\Omega)}}.\end{aligned}$$

We crucially exploited that $\lambda_1(\Omega) > 0$, thanks to the first part. We use now the generalized Young inequality, to so get

$$\begin{aligned}\frac{1}{2}\int_\Omega |\nabla u|^2\,dx - \int_\Omega f\,u\,dx &\ge \frac{1}{2}\int_\Omega |\nabla u|^2\,dx - \|f\|_{L^2(\Omega)}\,\frac{\|\nabla u\|_{L^2(\Omega;\mathbb{R}^2)}}{\sqrt{\lambda_1(\Omega)}}\\ &\ge \left(\frac{1}{2} - \frac{\varepsilon}{2\,\lambda_1(\Omega)}\right)\int_\Omega |\nabla u|^2\,dx\\ &\quad - \frac{1}{2\,\varepsilon}\,\|f\|^2_{L^2(\Omega)},\end{aligned}$$

which holds for every $\varepsilon > 0$. We choose

$$\varepsilon = \frac{\lambda_1(\Omega)}{2},$$

and conclude that

$$\frac{1}{2}\int_\Omega |\nabla u|^2\,dx - \int_\Omega f\,u\,dx \ge \frac{1}{4}\int_\Omega |\nabla u|^2\,dx - \frac{1}{\lambda_1(\Omega)}\,\|f\|^2_{L^2(\Omega)}.$$

This is enough to get that $\mathfrak{m} > -\infty$.

We take a minimizing sequence $\{u_n\}_{n\in\mathbb{N}} \subseteq W_0^{1,2}(\Omega)$ such that

$$\frac{1}{2}\int_\Omega |\nabla u_n|^2\,dx - \int_\Omega f\,u_n\,dx < \mathfrak{m} + \frac{1}{n+1}, \qquad \text{for every } n \in \mathbb{N}.$$

In light of the lower bound proved above, we can infer

$$\int_\Omega |\nabla u_n|^2\,dx \le C, \qquad \text{for every } n \in \mathbb{N},$$

for a constant $C > 0$ independent of n. Thanks to the first part, we also have a uniform bound on the L^2 norms of $\{u_n\}_{n\in\mathbb{N}}$. This shows that $\{u_n\}_{n\in\mathbb{N}}$ is a bounded sequence in $W_0^{1,2}(\Omega)$. We can apply the Banach-Alaoglu Theorem 3.3.6 and infer weak convergence (up to a subsequence) to a function $v \in W^{1,2}(\Omega)$. Moreover, we have $v \in W_0^{1,2}(\Omega)$, since this is a weakly closed space. By using that $\{u_n\}_{n\in\mathbb{N}}$ is a minimizing sequence, recalling the lower semicontinuity of the L^2 norm (Proposition 1.3.11) and using that by weak convergence we have

$$\lim_{n\to\infty}\int_\Omega f\, u_n\, dx = \int_\Omega f\, v\, dx,$$

we can infer that

$$\mathfrak{m} = \liminf_{n\to\infty}\left[\frac{1}{2}\int_\Omega |\nabla u_n|^2\, dx - \int_\Omega f\, u_n\, dx\right] \geq \frac{1}{2}\int_\Omega |\nabla v|^2\, dx - \int_\Omega f\, v\, dx.$$

Thus, v is the desired minimizer. Finally, for the uniqueness part it is sufficient to proceed as in the proof of Proposition 4.5.1.

4.10.15 We observe that the weak formulation of the equation is given by

$$\int_\Omega \langle |\nabla w|^{p-2}\,\nabla w, \nabla\varphi\rangle\, dx = \int_\Omega \varphi\, dx, \qquad \text{for every } \varphi \in C_0^\infty(\Omega).$$

This corresponds to the Euler-Lagrange equation of the functional

$$\frac{1}{p}\int_\Omega |\nabla u|^p\, dx - \int_\Omega u\, dx, \qquad \text{for every } u \in W_0^{1,p}(\Omega).$$

Thus, a weak solution w can be obtained by solving

$$\inf_{u\in W_0^{1,p}(\Omega)}\left\{\frac{1}{p}\int_\Omega |\nabla u|^p\, dx - \int_\Omega u\, dx\right\}. \tag{$*$}$$

For such a functional we can directly apply Theorem 4.3.1, this time with

$$H(z) = \frac{1}{p}\,|z|^p, \qquad G(t) = t, \qquad f \equiv 1.$$

To show uniqueness, we can copy almost verbatim the proof of Proposition 4.5.1. Let $v \in W_0^{1,p}(\Omega)$ be another weak solution of the equation, by Proposition 4.3.4 we get that v is another solution of the minimization problem $(*)$. In order to prove that $w = v$, we now exploit that

$$F(u, z) = \frac{1}{p}\,|z|^p - u,$$

is convex in the variables (u, z) and strictly convex in the variable z. As usual, we consider the convex combination

$$U = \frac{v+w}{2} \in W_0^{1,p}(\Omega).$$

By convexity, we get that

$$F(U, \nabla U) \le \frac{1}{2} F(v, \nabla v)\,dx + \frac{1}{2} F(w, \nabla w), \qquad \text{a. e. in } \Omega. \tag{$**$}$$

If we integrate over Ω, we thus obtain

$$\int_\Omega F(U, \nabla U)\,dx \le \frac{1}{2} \int_\Omega F(v, \nabla v)\,dx + \frac{1}{2} \int_\Omega F(w, \nabla w)\,dx.$$

By using that v, w are both minimizers and that U is admissible, it must result

$$\int_\Omega F(U, \nabla U)\,dx = \frac{1}{2} \int_\Omega F(v, \nabla v)\,dx + \frac{1}{2} \int_\Omega F(w, \nabla w)\,dx.$$

The latter implies that we must have equality almost everywhere in ($**$). Hence the strict convexity of F in the gradient variable implies that

$$\nabla v = \nabla w, \qquad \text{a. e. in } \Omega,$$

that is $v - w \in W_0^{1,p}(\Omega)$ is such that $\nabla(v-w) = (0, \dots, 0)$ almost everywhere on Ω. We can apply again Proposition 3.7.11 to infer that $v = w$.

In order to show that $w > 0$ almost everywhere, we observe that from the equation we get

$$\int_\Omega \langle |\nabla w|^{p-2}\, \nabla w, \nabla \varphi \rangle\,dx \ge 0, \qquad \text{for every } \varphi \in C_0^\infty(\Omega) \text{ such that } \varphi \ge 0,$$

so that $w \in W_0^{1,p}(\Omega)$ is weakly p-superharmonic. Moreover, we have $w_- \in W_0^{1,p}(\Omega)$, as well, thanks to Proposition 3.7.13. Thus, by Lemma 4.4.3 we can infer that $w \ge 0$ almost everywhere in Ω. The desired conclusion now follows from Lemma 4.4.5, as in the case $p = 2$.

4.10.16 Without loss of generality, we can suppose that the ball is centered at the origin. According to Problem 4.10.15, we seek the minimizer w of

$$\min_{u \in W_0^{1,p}(B_R)} \left\{ \frac{1}{p} \int_{B_R} |\nabla u|^p\,dx - \int_{B_R} u\,dx \right\}.$$

In light of Problem 4.10.1, we look for a radial minimizer, i.e.

$$w(x) = \psi(|x|), \qquad \text{for } |x| < R.$$

The idea is to proceed as in the solution of Problem 4.10.4, thus we would need the expression of the p-Laplacian in spherical coordinates. Since this is a bit complicated, we will avoid it by using a variational argument.

In any case, it is useful to observe that, thanks to Problem 4.10.15, we already know that such a solution exists and is *unique*. By uniqueness, it is then sufficient to find a method to guess its expression, even by proceeding a bit informally: the *a posteriori* verification that this solves the claimed equation will be sufficient to get a rigorous justification.

We observe at first that for radial functions $u(x) = \varphi(|x|)$, the functional can be rewritten as

$$\begin{aligned}\frac{1}{p}\int_{B_R} |\nabla u|^p\,dx - \int_{B_R} u\,dx \\ &= N\,\omega_N\left[\frac{1}{p}\int_0^R |\varphi'|^p\,\varrho^{N-1}\,d\varrho - \int_0^R \varphi\,\varrho^{N-1}\,d\varrho\right] \\ &=: N\,\omega_N\,\mathfrak{F}(\varphi),\end{aligned}$$

where we used polar coordinates. The original condition on $\partial B_R(0)$ implies that in the last one-dimensional problem we restrict the minimization to the functions φ such that $\varphi(R) = 0$. Since $w(x) = \psi(|x|)$ is a radial minimizer, we get that ψ must minimize the functional $\mathfrak{F}$, defined on functions of one variable. Thus, the first variation of such a functional computed at ψ must vanish. In particular, this implies that

$$\frac{d}{dt}\mathfrak{F}(\psi + t\,\varphi)_{|t=0} = 0, \qquad \text{for every } \varphi \in C^1([0,R]) \text{ such that } \varphi(R) = 0.$$

By computing this derivative, we get

$$\int_0^R |\psi'|^{p-2}\,\psi'\,\varphi'\,\varrho^{N-1}\,d\varrho - \int_0^R \varphi\,\varrho^{N-1}\,d\varrho = 0,$$

which holds for every $\varphi \in C^1([0,R])$ such that $\varphi(R) = 0$. It is not difficult to see that this is the weak form of the following nonlinear ordinary differential equation

$$-\left(\varrho^{N-1}\,|\psi'(\varrho)|^{p-2}\,\psi'(\varrho)\right)' = \varrho^{N-1}, \qquad \text{for } \varrho \in (0,R).$$

This has to be coupled with the following conditions:

$$\psi(R) = 0 \qquad \text{and} \qquad \lim_{\varrho\to 0^+} \varrho^{N-1}\,|\psi'(\varrho)|^{p-2}\,\psi'(\varrho) = 0.$$

By integrating with respect to ϱ, we get

$$-\varrho^{N-1}\,|\psi'(\varrho)|^{p-2}\,\psi'(\varrho) = \frac{\varrho^N}{N} + C, \qquad \text{for } \varrho \in (0, R).$$

Taking into account the condition at $\varrho = 0$, we must have $C = 0$. That is

$$-\,|\psi'(\varrho)|^{p-2}\,\psi'(\varrho) = \frac{\varrho}{N}, \qquad \text{for } \varrho \in (0, R).$$

Observe that since the right-hand side is positive, we get that $\psi' \le 0$. Thus, the previous identity can also be written as

$$(-\psi'(\varrho))^{p-1} = \frac{\varrho}{N}, \qquad \text{for } \varrho \in (0, R).$$

By further integrating and using that $\psi(R) = 0$, we get

$$\psi(\varrho) = \int_\varrho^R (-\psi'(\tau))\,d\tau = \int_\varrho^R \left(\frac{\tau}{N}\right)^{\frac{1}{p-1}} d\tau = \frac{p-1}{p}\left(\frac{1}{N}\right)^{\frac{1}{p-1}} \left(R^{\frac{p}{p-1}} - \varrho^{\frac{p}{p-1}}\right).$$

Finally, we get that

$$w(x) = \frac{p-1}{p}\left(\frac{1}{N}\right)^{\frac{1}{p-1}} \left(R^{\frac{p}{p-1}} - |x|^{\frac{p}{p-1}}\right).$$

We now verify that w is a weak solution. Observe that

$$w \in \left\{u \in C^0(\overline{B_R}) \cap C^1(B_R) \,:\, u = 0 \text{ on } \partial B_R,\ \nabla u \in L^\infty(B_R;\mathbb{R}^N)\right\} \subseteq W_0^{1,p}(B_R),$$

where the last inclusion follows from Problem 3.12.15. Moreover, we have

$$|\nabla w|^{p-2}\,\nabla w = \left|-\left(\frac{1}{N}\right)^{\frac{1}{p-1}} |x|^{\frac{p}{p-1}-2}\,x\right|^{p-2} \left(-\left(\frac{1}{N}\right)^{\frac{1}{p-1}} |x|^{\frac{p}{p-1}-2}\,x\right) = -\frac{1}{N}\,x.$$

In particular, for every $\varphi \in C_0^\infty(B_R)$, we have

$$\begin{aligned}
\int_{B_R} \langle |\nabla w|^{p-2}\,\nabla w, \nabla\varphi\rangle\,dx &= -\frac{1}{N}\int_{B_R} \langle x, \nabla\varphi\rangle\,dx \\
&= -\frac{1}{N}\int_{B_R} \operatorname{div}(x\,\varphi)\,dx + \frac{1}{N}\int_{B_R} \varphi\,\operatorname{div} x\,dx \\
&= -\frac{1}{N}\int_{\partial B_R} \varphi\,\langle x, \nu_{B_R}\rangle\,d\sigma + \int_{B_R} \varphi\,dx = \int_{B_R} \varphi\,dx,
\end{aligned}$$

thanks to the Divergence Theorem. This shows that v is the desired weak solution. Observe that for $p = N = 2$, we get back the function of Problem 4.10.4.

4.10.17 It is sufficient to proceed exactly as in the proof of Theorem 4.7.3. The problem (4.10.1) is equivalent to the following one

$$\inf_{u\in W_0^{1,p}(\Omega)} \left\{ \int_\Omega |\nabla u|^p\,dx \;:\; \int_\Omega |u|^p\,dx = 1 \right\}. \tag{$*$}$$

We then prove existence of a minimizer by using the Direct Method. We take a minimizing sequence $\{u_n\}_{n\in\mathbb{N}} \subseteq W_0^{1,p}(\Omega)$ for problem $(*)$, i.e.

$$\lim_{n\to\infty} \int_\Omega |\nabla u_n|^p\,dx = \lambda_{1,p}(\Omega) \qquad \text{and} \qquad \int_\Omega |u_n|^p\,dx = 1, \qquad \text{for every } n \in \mathbb{N}.$$

In particular, such a sequence is bounded in $W_0^{1,p}(\Omega)$. By Theorem 3.3.6 and Lemma 3.8.7, we know that $\{u_n\}_{n\in\mathbb{N}}$ converges weakly in $W^{1,p}(\Omega)$ and strongly in $L^p(\Omega)$ to a limit function $v \in W_0^{1,p}(\Omega)$, up to a subsequence. Moreover, $v \in W_0^{1,p}(\Omega)$ since the latter is weakly closed (see Theorem 3.7.4).

The strong L^p convergence entails that v is still admissible in $(*)$. Moreover, by the lower semicontinuity of the L^p norm with respect to weak convergence (see Proposition 1.3.11), we have

$$\int_\Omega |\nabla v|^p\,dx \le \lim_{n\to\infty} \int_\Omega |\nabla u_n|^p\,dx = \lambda_{1,p}(\Omega).$$

Thus, v is a minimizer of both $(*)$ and (4.10.1).

We now have to prove that any minimizer $v \in W_0^{1,p}(\Omega) \setminus \{0\}$ verifies the optimality condition

$$\int_\Omega \langle |\nabla v|^{p-2}\,\nabla v, \nabla\varphi\rangle\,dx = \lambda_{1,p}(\Omega) \int_\Omega |v|^{p-2}\,v\,\varphi\,dx, \qquad \text{for every } \varphi \in W_0^{1,p}(\Omega).$$

By using the same homogeneity trick as in the proof of Theorem 4.7.3, we get that v is a minimizer of the functional

$$\mathcal{F}_p(u) := \int_\Omega |\nabla u|^p\,dx - \lambda_{1,p}(\Omega) \int_\Omega |u|^p\,dx \ge 0, \qquad \text{for every } u \in W_0^{1,p}(\Omega),$$

as well. It is now sufficient to compute the first variation of this functional, in order to conclude. Indeed, by minimality of v, for every $t \in \mathbb{R}$ and $\varphi \in C_0^\infty(\Omega)$, we have

$$\mathcal{F}_p(v + t\,\varphi) \ge \mathcal{F}_p(v).$$

Thus, if we set $g(t) = \mathcal{F}(v + t\,\varphi)$, we must have $g'(0) = 0$. In order to compute this derivative, we need now to use the Dominated Convergence Theorem. It is sufficient to repeat the argument from the final part of the proof of Theorem 4.3.1.

4.10.18 By assumption, we have that

$$\int_\Omega \langle |\nabla u_1|^{p-2}\,\nabla u_1, \nabla\varphi\rangle\,dx = \int_\Omega f_1\,\varphi\,dx, \qquad \text{for every } \varphi \in C_0^\infty(\Omega),$$

and

$$\int_\Omega \langle |\nabla u_2|^{p-2}\,\nabla u_2, \nabla\varphi\rangle\,dx = \int_\Omega f_2\,\varphi\,dx, \qquad \text{for every } \varphi \in C_0^\infty(\Omega).$$

Thanks to the fact that $|\nabla u_i|^{p-2}\,\nabla u_i \in L^{p'}(\Omega;\mathbb{R}^N)$, by a density argument we can even admit test functions $\varphi \in W_0^{1,p}(\Omega)$. We subtract the two equations above and test with $\varphi = (u_1 - u_2)_+$, which is feasible by the initial hypothesis. We obtain in this way

$$\int_\Omega \langle |\nabla u_1|^{p-2}\,\nabla u_1 - |\nabla u_2|^{p-2}\,\nabla u_2, \nabla(u_1 - u_2)_+\rangle\,dx = \int_\Omega (f_1 - f_2)\,(u_1 - u_2)_+\,dx.$$

Thanks to the assumption on f_1 and f_2, the last integral is non-positive. Thus, we obtain

$$\int_{\{x\in\Omega\,:\,u_1(x)>u_2(x)\}} \langle |\nabla u_1|^{p-2}\,\nabla u_1 - |\nabla u_2|^{p-2}\,\nabla u_2, \nabla u_1 - \nabla u_2\rangle\,dx \le 0,$$

where we also used Proposition 3.4.3 to compute the weak gradient of $(u_1 - u_2)_+$. By recalling that the p-Laplacian is an elliptic operator (see Proposition 1.6.2), the inequality above implies that we must have

$$\nabla u_1 = \nabla u_2, \qquad \text{a. e. in } \{x \in \Omega \,:\, u_1(x) > u_2(x)\}.$$

By using again Proposition 3.4.3, this implies that

$$\nabla(u_1 - u_2)_+ = (0,\ldots,0), \qquad \text{a. e. in } \Omega.$$

Finally, by recalling that $(u_1 - u_2)_+ \in W_0^{1,p}(\Omega)$, we get by Proposition 3.7.11 that $(u_1 - u_2)_+$ vanishes almost everywhere in Ω. This is the same as saying that $u_1 \le u_2$ almost everywhere in Ω, as desired.

4.10.19 Existence of a weak solution v can be inferred by solving the minimization problem

$$\min_{u \in W_0^{1,p}(B_R)} \left\{ \frac{1}{p} \int_{B_R} |\nabla u|^p - \int_{B_R} f\, u\, dx \right\}.$$

It is thus sufficient to apply Theorem 4.3.1 with the choices

$$H(z) = \frac{|z|^p}{p}, \qquad G(t) = t, \quad q = 1, \quad \gamma = p'.$$

Once we get existence of a weak solution, its uniqueness follows by the convexity of the function

$$(u, z) \mapsto \frac{1}{p}\, |z|^p - f(x)\, u,$$

see for example the solution of Problem 4.10.15.

We now assume that $f \in L^\infty(B_R)$ and come to the second statement: we need to show that $v \in L^\infty(B_R)$. For $p > N$, this is a plain consequence of Sobolev embeddings, i.e. Theorem 3.8.1. On the contrary, for $1 < p \leq N$ this is a *regularity result*: by using the equation, we are going to prove that v is actually more regular than a generic function of $W_0^{1,p}(B_R)$. We can employ the *comparison principle* of Problem 4.10.18. As a comparison function, we take

$$\mathrm{W} = \|f\|_{L^\infty(B_R)}^{\frac{1}{p-1}}\, w \in W_0^{1,p}(B_R),$$

where w is the p-torsion function of B_R, determined in Problem 4.10.16. Observe that W is a weak solution of

$$-\Delta_p \mathrm{W} = \|f\|_{L^\infty(B_R)} \geq f, \qquad \text{in } B_R.$$

Moreover, we have $v - \mathrm{W} \in W_0^{1,p}(B_R)$ and thus in particular

$$(v - \mathrm{W})_+ \in W_0^{1,p}(B_R),$$

by Proposition 3.7.13. We can apply the Comparison Principle (see Problem 4.10.18) and conclude that

$$v \leq \mathrm{W}, \qquad \text{a.e. in } B_R.$$

We can repeat the argument by using this time $-\mathrm{W}$ and get

$$v \geq -\mathrm{W}, \qquad \text{a.e. in } B_R,$$

as well. In conclusion, we can infer that $|v| \leq \mathrm{W}$ almost everywhere in B_R and thus

$$\|v\|_{L^\infty(B_R)} \le \|\mathrm{W}\|_{L^\infty(B_R)} = \|f\|_{L^\infty(B_R)}^{\frac{1}{p-1}} \|w\|_{L^\infty(B_R)}.$$

By recalling the explicit expression of w (see the solution of Problem 4.10.16), it is easily seen that

$$\|w\|_{L^\infty(B_R)} = \frac{p-1}{p}\left(\frac{1}{N}\right)^{\frac{1}{p-1}} R^{\frac{p}{p-1}},$$

and thus we conclude.

4.10.20 For $2 \le p < \infty$, the existence of a weak solution $v \in W_0^{1,p}(B_1(0))$ can be inferred directly from Theorem 4.3.1, with the following choices

$$N = 3, \quad H(z) = \frac{|z|^p}{p}, \quad G(t) = t, \quad q = 1, \quad \gamma = 2.$$

Indeed, observe that with these choices we have for $2 \le p < 3$

$$\left(\frac{p^*}{q}\right)' = \left(\frac{3\,p}{3-p}\right)' = \frac{3\,p}{4\,p-3},$$

which is smaller than or equal to $2 = \gamma$. On the other hand, for $p \ge 3$ Theorem 4.3.1 simply requires that $f \in L^\gamma(B_1(0))$ with $\gamma > 1$. In other words, the solution v is the minimizer over $W_0^{1,p}(B_1(0))$ of the following functional

$$\frac{1}{p}\int_{B_1(0)} |\nabla u|^p\,dx - \int_{B_1(0)} f\,u\,dx, \qquad \text{for every } u \in W_0^{1,p}(B_1(0)).$$

Once we get existence of a weak solution, its uniqueness follows with the usual convexity argument, see for example the solution of Problem 4.10.15.

Observe that all that we said above is still true for $p < 2$, provided that

$$\left(\frac{p^*}{q}\right)' = \frac{3\,p}{4\,p-3} \le 2,$$

which is true if and only if $p \ge 6/5$. Thus, we still have existence and uniqueness of the weak solution for

$$\frac{6}{5} \le p < 2.$$

On the contrary, for $1 < p < 6/5$ this existence proof does not work anymore and in fact the equation may fail to have a weak solution in $W_0^{1,p}(B_1(0))$, if $f \in L^2(B_1(0))$.

4.10.21 We observe that the weak formulation of the equation is given by

$$\int_\Omega \left[\left|\frac{\partial v}{\partial x}\right|^{p-2} \frac{\partial v}{\partial x}\frac{\partial \varphi}{\partial x} + \left|\frac{\partial v}{\partial y}\right|^{p-2} \frac{\partial v}{\partial y}\frac{\partial \varphi}{\partial y}\right] dx\, dy = \int_\Omega f\, \varphi\, dx\, dy,$$

for every $\varphi \in C_0^\infty(\Omega)$. This is the Euler-Lagrange equation of the functional

$$\mathcal{F}(u) = \frac{1}{p}\int_\Omega \left[\left|\frac{\partial u}{\partial x}\right|^p + \left|\frac{\partial u}{\partial y}\right|^p\right] dx\, dy - \int_\Omega f\, u\, dx\, dy.$$

Existence of a weak solution can then be inferred by proving existence of a minimizer for the following problem

$$\min_{u \in W_0^{1,p}(\Omega)} \mathcal{F}(u).$$

To this aim, we can simply apply Theorem 4.3.1, with the choices

$$N = 2, \quad H(z) = \frac{1}{p}|z_1|^p + \frac{1}{p}|z_2|^p, \quad G(t) = t, \quad q = 1, \quad \gamma = 1.$$

Observe that since $p > 2 = N$, then the assumption $f \in L^1(\Omega)$ is enough to infer existence. We also observe that, by Problem 1.7.2 with $\alpha = p/2 > 1$, we have

$$H(z) = \frac{1}{p}|z_1|^p + \frac{1}{p}|z_2|^p \le \frac{1}{p}\left(|z_1|^2 + |z_2|^2\right)^{\frac{p}{2}} = \frac{1}{p}|z|^p,$$

and

$$H(z) \ge \frac{1}{p\, 2^{\frac{p-2}{2}}}|z|^p. \qquad (*)$$

Thus, H satisfies the assumptions of Theorem 4.3.1. Once the existence of a solution $v \in W_0^{1,p}(\Omega)$ is obtained, it is sufficient to apply Theorem 3.8.1 (with $p > 2 = N$ and $q = \infty$) to get that $v \in L^\infty(\Omega)$. In particular, we get

$$\|v\|_{L^\infty(\Omega)} \le C\, \|v\|_{L^p(\Omega)}^{1-\frac{2}{p}}\, \|\nabla v\|_{L^p(\Omega;\mathbb{R}^2)}^{\frac{2}{p}},$$

for a constant $C = C(p) > 0$. Observe that, since Ω is bounded, we have the Poincaré inequality at our disposal and by Remark 3.7.7 we know that

$$\|v\|_{L^p(\Omega)} \le \mathrm{diam}(\Omega)\, \|\nabla v\|_{L^p(\Omega;\mathbb{R}^2)}.$$

Thus, we have

$$\|v\|_{L^\infty(\Omega)} \le C\ (\mathrm{diam}(\Omega))^{\frac{p-2}{p}} \left(\int_\Omega |\nabla v|^p\, dx\, dy\right)^{\frac{1}{p}}.$$

Moreover, by appealing to (∗), we also get

$$\|v\|_{L^\infty(\Omega)} \le C\ (\mathrm{diam}(\Omega))^{\frac{p-2}{p}}\ 2^{\frac{p-2}{2p}} \left(\int_\Omega \left[\left|\frac{\partial v}{\partial x}\right|^p + \left|\frac{\partial v}{\partial y}\right|^p\right] dx\, dy\right)^{\frac{1}{p}}.$$

On the other hand, if we test the equation with $\varphi = v$ itself, we obtain

$$\int_\Omega \left[\left|\frac{\partial v}{\partial x}\right|^p + \left|\frac{\partial v}{\partial y}\right|^p\right] dx\, dy = \int_\Omega f\, v\, dx\, dy \le \|f\|_{L^1(\Omega)}\, \|v\|_{L^\infty(\Omega)}.$$

By joining the last two inequalities, we get the claimed L^∞ estimate in terms of f.

4.10.22 Let us suppose that $u \in C^2(\Omega)$ is superharmonic in classical sense, i.e.

$$-\Delta u(x) \ge 0, \qquad \text{for every } x \in \Omega.$$

We multiply the previous inequality by $\varphi \in C_0^\infty(\Omega)$ such that $\varphi \ge 0$ and integrate over Ω. This gives

$$\begin{aligned} 0 \le \int_\Omega (-\Delta u)\, \varphi\, dx &= \sum_{i=1}^N \int_{\mathbb{R}^N} -\frac{\partial}{\partial x_i}\left(\frac{\partial u}{\partial x_i}\right) \varphi\, dx \\ &= \sum_{i=1}^N \int_{\mathbb{R}^N} \frac{\partial u}{\partial x_i}\, \frac{\partial \varphi}{\partial x_i}\, dx = \int_\Omega \langle \nabla u, \nabla \varphi\rangle\, dx, \end{aligned}$$

where we used an integration by parts and the compact support of φ. This shows that u is weakly superharmonic, as well.

On the contrary, let us suppose that $u \in C^2(\Omega)$ is such that

$$\int_\Omega \langle \nabla u, \nabla \varphi\rangle\, dx \ge 0, \qquad \text{for every } \varphi \in C_0^\infty(\Omega) \text{ such that } \varphi \ge 0.$$

By using the previous chain of identities, this is the same as

$$\int_\Omega (-\Delta u)\, \varphi\, dx \ge 0, \qquad \text{for every } \varphi \in C_0^\infty(\Omega) \text{ such that } \varphi \ge 0.$$

This implies that the continuous function $x \mapsto -\Delta u(x)$ is non-negative on Ω, thanks to Problem 1.7.15.

4.10.23 For every $n \in \mathbb{N} \setminus \{0\}$, we consider the partial sum

$$S_n = \sum_{k=1}^{n} \alpha_k\, v_k \in W_0^{1,2}(\Omega).$$

For every $m > n \geq 1$, we have in particular

$$S_m - S_n = \sum_{k=n+1}^{m} \alpha_k\, v_k,$$

and thus

$$\begin{aligned}\|\nabla S_m - \nabla S_n\|^2_{L^2(\Omega;\mathbb{R}^N)} = \int_\Omega \left|\sum_{k=n+1}^{m} \alpha_k\, \nabla v_k\right|^2 dx &= \sum_{k=n+1}^{m} |\alpha_k|^2 \int_\Omega |\nabla v_k|^2\, dx \\ &= \sum_{k=n+1}^{m} \lambda_k(\Omega)\, |\alpha_k|^2,\end{aligned}$$

thanks to the orthogonality relations of the family $\{v_k\}_{k\in\mathbb{N}\setminus\{0\}}$. By using the hypothesis on the convergence of the series with coefficients $\lambda_k(\Omega)\, |\alpha_k|^2$, the previous identity shows that $\{\nabla S_n\}_{n\in\mathbb{N}\setminus\{0\}}$ is a Cauchy sequence in $L^2(\Omega; \mathbb{R}^N)$. Moreover, under the assumptions of Theorem 4.8.1, we have that

$$u \mapsto \|\nabla u\|_{L^2(\Omega;\mathbb{R}^N)},$$

is an equivalent norm on $W_0^{1,2}(\Omega)$ (see Theorem 3.7.6). Thus, $\{S_n\}_{n\in\mathbb{N}\setminus\{0\}}$ is a Cauchy sequence in $W_0^{1,2}(\Omega)$. Since the latter is a Banach space (recall Theorem 3.7.4), we obtain that $\{S_n\}_{n\in\mathbb{N}\setminus\{0\}}$ converges strongly in $W_0^{1,2}(\Omega)$ to a function S belonging to this space. This is enough to conclude the proof of the first claim.

For the second part, by using the previous notation and the strong convergence of the partial sum, for every $j \in \mathbb{N} \setminus \{0\}$ we have

$$\widehat{S}(j) = \int_\Omega S\, v_j\, dx = \lim_{n\to\infty} \int_\Omega S_n\, v_j\, dx = \lim_{n\to\infty} \sum_{k=1}^{n} \alpha_k \int_\Omega v_k\, v_j\, dx = \lim_{n\to\infty} \alpha_j = \alpha_j,$$

as desired.

4.10.24 Let us suppose that

$$u(x, t) = X(x)\, T(t),$$

is a standing heat wave. We first observe that, in order to satisfy the boundary condition, we need to impose

$$X(x) = 0, \qquad \text{for } x \in \partial\Omega.$$

By inserting $u(x,t) = X(x)\,T(t)$ in the heat equation, it must result

$$X(x)\,T'(t) = T(t)\,\Delta X(x),$$

that is, by supposing that T and X do not vanish, we get

$$\frac{T'(t)}{T(t)} = \frac{\Delta X(x)}{X(x)}. \tag{$*$}$$

Observe that the function on the left-hand side only depends on t, whereas the function on the right-hand side only depends on x. Thus, Eq. $(*)$ can only be true provided that there exists a constant $c \in \mathbb{R}$ such that

$$\frac{T'(t)}{T(t)} = c \qquad \text{and} \qquad c = \frac{\Delta X(x)}{X(x)}.$$

Or, in other words, that there exists a constant $c \in \mathbb{R}$ such that the two equations

$$T'(t) = c\,T(t) \qquad \text{and} \qquad -\Delta X(x) = -c\,X(x),$$

admit nontrivial solutions. Observe that the second equation has to be complemented with the boundary condition $X = 0$ on $\partial\Omega$.

It is here that Theorem 4.8.1 comes into play: indeed, we know that the second equation admits non-trivial (weak) solutions in $W^{1,2}_0(\Omega)$ if and only if $-c = \lambda_k(\Omega)$, for some $k \in \mathbb{N} \setminus \{0\}$. With such a choice, X must be a k-th eigenfunction of the Dirichlet-Laplacian on Ω. Then, we use the choice $c = -\lambda_k(\Omega)$ in the equation for the time variable, i.e.

$$-T'(t) = \lambda_k(\Omega)\,T(t), \qquad \text{for } t > 0.$$

Every solution of this ordinary differential equation takes the form

$$T(t) = A\,e^{-\lambda_k(\Omega)\,t}, \qquad \text{for some constant } A \in \mathbb{R}.$$

In conclusion, we end up with the standing heat waves

$$u(x,t) = A\,e^{-\lambda_k(\Omega)\,t}\,u_k(x),$$

with u_k a k-th eigenfunction and $A \in \mathbb{R}$.

We observe that the temperature u decays to 0 exponentially fast as time goes to $+\infty$, with rate of convergence dictated by the eigenvalue $\lambda_k(\Omega)$. By recalling that $\lambda_k(\Omega)$ is non-decreasing with respect to k, we get that the slowest rate of heat dissipation corresponds to the first eigenvalue $\lambda_1(\Omega)$.

4.10.25 This is similar to the previous exercise. We seek all non-trivial solutions to the wave equation taking the form

$$u(x,t) = X(x)\,T(t).$$

In order to satisfy the requirement that the membrane stays fixed along the boundary, we need to impose

$$X(x) = 0, \qquad \text{for } x \in \partial\Omega.$$

We insert $u(x,t) = X(x)\,T(t)$ in the wave equation, so to get

$$X(x)\,T''(t) = T(t)\,\Delta X(x),$$

that is, by supposing again that T and X do not vanish, we get

$$\frac{T''(t)}{T(t)} = \frac{\Delta X(x)}{X(x)}. \tag{$*$}$$

As before, the only possibility for $(*)$ to be true is that there exists a constant $c \in \mathbb{R}$ such that

$$\frac{T''(t)}{T(t)} = c \qquad \text{and} \qquad c = \frac{\Delta X(x)}{X(x)}.$$

These are the same as saying that

$$T''(t) = c\,T(t) \qquad \text{and} \qquad \Delta X(x) = c\,X(x),$$

admit nontrivial solutions. The second equation has to be complemented with the boundary condition $X = 0$ on $\partial\Omega$.

Again, we can use Theorem 4.8.1: the second equation admits non-trivial (weak) solutions in $W^{1,2}_0(\Omega)$ if and only if $-c = \lambda_k(\Omega)$, for some $k \in \mathbb{N} \setminus \{0\}$. With such a choice, X must be a k-th eigenfunction of the Dirichlet-Laplacian on Ω. By making the choice $c = -\lambda_k(\Omega)$ in the equation for the time variable, i.e.

$$-\,T''(t) = \lambda_k(\Omega)\,T(t), \qquad \text{for } t > 0,$$

we get that T must have the form

$$T(t) = A\,\cos\left(\sqrt{\lambda_k(\Omega)}\,t\right) + B\,\sin\left(\sqrt{\lambda_k(\Omega)}\,t\right), \qquad \text{for some constants } A, B \in \mathbb{R}.$$

In conclusion, we end up with the stationary solutions

$$u(x,t) = \left(A\,\cos\left(\sqrt{\lambda_k(\Omega)}\,t\right) + B\,\sin\left(\sqrt{\lambda_k(\Omega)}\,t\right)\right) u_k(x),$$

with u_k a k-th eigenfunction and $A, B \in \mathbb{R}$.

As for the frequency of vibration of these solutions: by observing that

$$t \mapsto A \cos\left(\sqrt{\lambda_k(\Omega)}\, t\right) + B \sin\left(\sqrt{\lambda_k(\Omega)}\, t\right),$$

is $(2\pi/\sqrt{\lambda_k(\Omega)})$-periodic, the frequency is given by

$$\frac{\sqrt{\lambda_k(\Omega)}}{2\pi}.$$

Thus, in this physical model, the first eigenvalue $\lambda_1(\Omega)$ has to be interpreted as the lowest frequency admitting a standing vibration.

4.10.26 This is quite similar to the case of the heat equation (which corresponds to $m = 1$). We only have to pay attention to the fact that now the equation is *nonlinear*, due to the presence of the power $m > 1$. We first use a trick: we introduce the change of variable

$$w = u^m \qquad \text{that is} \qquad u = w^{\frac{1}{m}},$$

and perform some formal manipulations. In terms of w the porous medium equation rewrites as

$$\frac{1}{m}\, w^{\frac{1}{m}-1}\, \frac{\partial w}{\partial t} = \Delta w.$$

We now look for a positive nontrivial solution of this equation having the form $w(x,t) = X(x)\, T(t)$. By inserting this expression in the equation above, we get

$$\frac{1}{m}\, X(x)^{\frac{1}{m}-1}\, T(t)^{\frac{1}{m}-1}\, X(x)\, T'(t) = T(t)\, \Delta X(x).$$

We separate variables, so to get

$$\frac{1}{m}\, T(t)^{\frac{1}{m}-2}\, T'(t) = \frac{\Delta X(x)}{X(x)^{\frac{1}{m}}}.$$

As above, this is possible if and only if there exists a constant $c \in \mathbb{R}$ such that

$$\frac{1}{m}\, T(t)^{\frac{1}{m}-2}\, T'(t) = c \qquad \text{and} \qquad c = \frac{\Delta X(x)}{X(x)^{\frac{1}{m}}}.$$

These are the same as saying that

$$\frac{1}{m}\, T(t)^{\frac{1}{m}-2}\, T'(t) = c \qquad \text{and} \qquad \Delta X(x) = c\, X(x)^{\frac{1}{m}}.$$

The second equation has to be coupled with the boundary condition $X = 0$ on $\partial\Omega$.

We may now invoke Proposition 4.6.1 in order to find a nontrivial positive (weak) solution of the second equation: indeed, by taking $c = -1$, this is nothing but the sublinear Lane-Emden equation

$$-\Delta X(x) = X(x)^{\frac{1}{m}},$$

with power nonlinearity

$$q = \frac{1}{m} + 1.$$

Observe that this power is bigger than 1 and smaller than 2, thanks to the fact that $m > 1$. Let us call v the positive solution provided by Proposition 4.6.1. By taking $c = -1$ in the equation for T, we get

$$\frac{1}{m}\, T(t)^{\frac{1}{m}-2}\, T'(t) = -1,$$

which is the same as

$$\frac{m}{1-m}\,\frac{d}{dt}\left(T(t)^{\frac{1}{m}-1}\right) = -m \qquad \text{that is} \qquad \frac{d}{dt}\left(T(t)^{\frac{1}{m}-1}\right) = m-1.$$

Then we must have

$$T(t)^{\frac{1}{m}-1} = (m-1)\,t + A,$$

that is

$$T(t) = \Big((m-1)\,t + A\Big)^{\frac{m}{1-m}},$$

for some constant A. We should pay attention to the fact that, since the exponent $1 - m < 0$, the constant A has to be taken strictly positive, if we want a solution well-defined up to time $t = 0$. In conclusion, we obtain the positive solution

$$w(x,t) = \Big((m-1)\,t + A\Big)^{\frac{m}{1-m}}\, v(x),$$

with separated variables. Finally, by recalling that $w = u^m$, we get

$$u(x,t) = w(x,t)^{\frac{1}{m}} = \Big((m-1)\,t + A\Big)^{\frac{1}{1-m}}\, v(x)^{\frac{1}{m}}, \qquad \text{with } A > 0,$$

for the original porous medium equation. We recall that v is the (unique!) non-trivial weak solution of

$$-\Delta v = |v|^{\frac{1}{m}}, \qquad \text{in } \Omega,$$

vanishing at the boundary $\partial\Omega$.

As a final remark, we could notice that u decays to 0 polinomially fast as time goes to $+\infty$, with rate of convergence dictated by the (negative) exponent $1/(1-m)$. Observe that this rate gets slower and slower as the parameter m gets larger and larger.

8.5 Problems of Chap. 5

5.5.1 By using that $\Omega \subseteq \overline{\Omega}$, we clearly have

$$|f|_{C^{0,1}(\overline{\Omega})} \geq |f|_{C^{0,1}(\Omega)}.$$

On the other hand, for every $x, y \in \overline{\Omega}$ we can find two sequences $\{x_n\}_{n\in\mathbb{N}}, \{y_n\}_{n\in\mathbb{N}} \subseteq \Omega$ such that

$$\lim_{n\to\infty} x_n = x \qquad \text{and} \qquad \lim_{n\to\infty} y_n = y.$$

By continuity of f, we thus have

$$|f(x)-f(y)| = \lim_{n\to\infty} |f(x_n)-f(y_n)| \leq |f|_{C^{0,1}(\Omega)} \lim_{n\to\infty} |x_n-y_n| = |f|_{C^{0,1}(\Omega)}\, |x-y|.$$

The arbitrariness of $x, y \in \overline{\Omega}$ entails that $|f|_{C^{0,1}(\overline{\Omega})} \leq |f|_{C^{0,1}(\Omega)}$, as well.

5.5.2 Let us take $0 < \alpha < 1$, by choosing

$$x_n = \left(\frac{1}{n}, 0, \dots, 0\right), \qquad n \in \mathbb{N} \setminus \{0\},$$

we have

$$|\varphi|_{C^{0,1}(\mathbb{R}^N)} \geq \frac{|\varphi(0) - \varphi(x_n)|}{|x_n|} = \left(1 - \left(1 - \left(\frac{1}{n}\right)^{\alpha}\right)\right) n = n^{1-\alpha}.$$

For $0 < \alpha < 1$, the last quantity diverges to $+\infty$ for n going to ∞. This shows that φ is not Lipschitz in this case.

We now fix $\alpha \geq 1$. For every $x, y \in \overline{B_1(0)}$, we have

$$|\varphi(x) - \varphi(y)| = \Big||x|^{\alpha} - |y|^{\alpha}\Big| \leq \alpha\, (|x| + |y|)^{\alpha-1} \Big||x| - |y|\Big|,$$

thanks to Problem 1.7.4. By using the triangle inequality and the fact that $x, y \in \overline{B_1(0)}$, the last estimate in turn gives

$$|\varphi(x) - \varphi(y)| \le 2^{\alpha-1}\,\alpha\,|x-y|, \qquad \text{for } x, y \in \overline{B_1(0)}.$$

This shows that φ is Lipschitz continuous on $\overline{B_1(0)}$. This is true in $\mathbb{R}^N \setminus \overline{B_1(0)}$, as well, since φ identically vanishes there. Finally, if $x \in \overline{B_1(0)}$ and $y \in \mathbb{R}^N \setminus \overline{B_1(0)}$, we have again by Problem 1.7.4

$$\begin{aligned}|\varphi(x) - \varphi(y)| = |\varphi(x)| &= 1 - |x|^\alpha \\ &\le \alpha\,(1+|x|)^{\alpha-1}\,(1-|x|) \\ &\le 2^{\alpha-1}\,\alpha\,(|y|-|x|) \le 2^{\alpha-1}\,\alpha\,|x-y|.\end{aligned}$$

In conclusion, we proved that for $\alpha \ge 1$, we have

$$|\varphi(x) - \varphi(y)| \le 2^{\alpha-1}\,\alpha\,|x-y|, \qquad \text{for every } x, y \in \mathbb{R}^N.$$

5.5.3 From the discussion above, we already know that

$$|\varphi|_{C^{0,1}(\mathbb{R}^N)} \le 2^{\alpha-1}\,\alpha.$$

However, we will show that this estimate is not optimal. We observe that $\varphi \in C^{0,1}_0(\overline{B_1}(0))$, i.e. it vanishes on the boundary $\partial B_1(0)$. By Lemma 5.1.11 we have

$$|\varphi|_{C^{0,1}(\mathbb{R}^N)} = |\varphi|_{C^{0,1}(\overline{B_1(0)})}.$$

Since φ is a C^1 function on the open convex set $B_1(0)$, by Lemma 5.1.3 we have

$$|\varphi|_{C^{0,1}(\overline{B_1(0)})} = |\varphi|_{C^{0,1}(B_1(0))} = \sup_{x\in B_1(0)} |\nabla\varphi(x)| = \alpha \sup_{x\in B_1(0)} |x|^{\alpha-1} = \alpha.$$

In the first equality, we used Problem 5.5.1. This finally proves that $|\varphi|_{C^{0,1}(\mathbb{R}^N)} = \alpha$.

5.5.4 Let $x, y \in \overline{A_{r,R}}$, by using the definition of v and the triangle inequality, we immediately get

$$|v(x)-v(y)| = |V(|x|)-V(|y|)| \le |V|_{C^{0,1}([r,R])}\,\Big||x|-|y|\Big| \le |V|_{C^{0,1}([r,R])}\,|x-y|.$$

This proves that v is Lipschitz continuous and

$$|v|_{C^{0,1}(\overline{A_{r,R}})} \le |V|_{C^{0,1}([r,R])}.$$

On the other hand, for every $t, s \in [r, R]$, we take the two points $x = t\,\mathbf{e}_1$ and $y = s\,\mathbf{e}_1$. We have $|x| = t$, $|y| = s$ and $|x - y| = |s - t|$. Thus, we get

$$\begin{aligned}|v|_{C^{0,1}(\overline{A_{r,R}})}\,|s-t| &= |v|_{C^{0,1}(\overline{A_{r,R}})}\,|x-y| \\ &\geq |v(x)-v(y)| = |V(|x|)-V(|y|)| = |V(t)-V(s)|.\end{aligned}$$

By taking the supremum over $t, s \in [r, R]$, we get the reverse inequality

$$|v|_{C^{0,1}(\overline{A_{r,R}})} \geq |V|_{C^{0,1}([r,R])},$$

and thus the conclusion.

We now compute the gradient of v. By Rademacher's Theorem, we know that V is differentiable for almost every $r_0 \in (r, R)$. Let us take $x_0 \in A_{r,R}$ such that $r_0 = |x_0|$ is a differentiability point for V. We then obtain

$$\begin{aligned}v(x) = V(|x|) &= V(|x_0|) + V'(|x_0|)\,(|x|-|x_0|) + o(|x|-|x_0|) \\ &= v(x_0) + V'(|x_0|)\,(|x|-|x_0|) + o(|x|-|x_0|), \qquad \text{for } x \to x_0,\end{aligned}$$

from the differentiability of V. We observe that $|x_0| \neq 0$, thus the function $x \mapsto |x|$ is differentiable at x_0. This implies that

$$|x| - |x_0| = \left\langle \frac{x_0}{|x_0|}, x - x_0 \right\rangle + o(|x-x_0|), \qquad \text{for } x \to x_0.$$

By inserting this information into the identity above, we get

$$\begin{aligned}v(x) = v(x_0) &+ \left\langle V'(|x_0|)\,\frac{x_0}{|x_0|}, x - x_0 \right\rangle \\ &+ V'(|x_0|)\,o(|x-x_0|) + o(|x|-|x_0|), \qquad \text{for } x \to x_0.\end{aligned}$$

We now observe that

$$V'(|x_0|)\,o(|x-x_0|) = o(|x-x_0|), \qquad o(|x|-|x_0|) = o(|x-x_0|), \qquad \text{for } x \to x_0.$$

This is enough to conclude.

5.5.5 The integral defining u has to be understood in the sense of Lebesgue integration, of course. Thus, the function u is well-defined. For every $t, s \in [a, b]$ with $s < t$, we have

$$
\begin{aligned}
|u(t)-u(s)| &= \left|\int_a^t g(\tau)\,d\tau - \int_a^s g(\tau)\,d\tau\right| = \left|\int_s^t g(\tau)\,d\tau\right| \\
&\le \int_s^t |g(\tau)|\,d\tau \le |t-s|\,\|g\|_{L^\infty((a,b))}.
\end{aligned}
$$

This shows that u is Lipschitz continuous. In order to prove the second statement, we take $\varphi \in C_0^\infty((a,b))$. By using the definition of u and the Fubini-Tonelli Theorem, we then have

$$
\begin{aligned}
\int_a^b u(t)\,\varphi'(t)\,dt = \int_a^b \left(\int_a^t g(\tau)\,d\tau\right)\varphi'(t)\,dt &= \int_a^b \left(\int_\tau^b \varphi'(t)\,dt\right) g(\tau)\,d\tau \\
&= -\int_a^b \varphi(\tau)\,g(\tau)\,d\tau.
\end{aligned}
$$

This is valid for every $\varphi \in C_0^\infty((a,b))$, thus by appealing to the definition of weak derivative we get that $u' = g$ almost everywhere in (a,b).

5.5.6 By using Lemma 5.2.1 and the definition of weak derivative, we have

$$
\int_a^b u'(t)\,\varphi(t)\,dt = -\int_a^b u(t)\,\varphi'(t)\,dt, \qquad \text{for every } \varphi \in C_0^\infty((a,b)).
$$

We first observe that the same identity is still true if $\varphi \in W_0^{1,1}((a,b))$ (recall Definition 3.7.1 for this space). Indeed, by definition for every $\varphi \in W_0^{1,1}((a,b))$ there exists a sequence $\{\varphi_n\}_{n\in\mathbb{N}} \subseteq C_0^\infty((a,b))$ such that

$$
\lim_{n\to\infty} \|\varphi' - \varphi_n'\|_{L^1((a,b))} = \lim_{n\to\infty} \|\varphi - \varphi_n\|_{L^1((a,b))} = 0.
$$

By using that both u and u' are essentially bounded, we get

$$
\lim_{n\to\infty} \left|\int_a^b u'(t)\,\big(\varphi(t) - \varphi_n(t)\big)\,dt\right| \le \|u'\|_{L^\infty((a,b))} \lim_{n\to\infty} \|\varphi - \varphi_n\|_{L^1((a,b))} = 0,
$$

and similarly

$$
\lim_{n\to\infty} \left|\int_a^b u(t)\,\big(\varphi'(t) - \varphi_n'(t)\big)\,dt\right| \le \|u\|_{L^\infty((a,b))} \lim_{n\to\infty} \|\varphi' - \varphi_n'\|_{L^1((a,b))} = 0.
$$

We thus obtain

$$
\int_a^b u'(t)\,\varphi(t)\,dt = -\int_a^b u(t)\,\varphi'(t)\,dt, \qquad \text{for every } \varphi \in W_0^{1,1}((a,b)).
$$

We now take $n \in \mathbb{N} \setminus \{0\}$ and make the following choice for the test function: we define

$$\varphi_n(t) = \begin{cases} n\,(t-a), & \text{if } a \le t < a + \dfrac{1}{n}, \\[2ex] 1, & \text{if } a + \dfrac{1}{n} \le t < b - \dfrac{1}{n}, \\[2ex] n\,(b-t), & \text{if } b - \dfrac{1}{n} \le t \le b. \end{cases}$$

Observe that this is nothing but a piecewise affine approximation of the function identically equal to 1 on $[a, b]$, with functions vanishing at the endpoints. By using Problems 3.12.3 and 3.12.16, we have that $\varphi_n \in W_0^{1,1}((a, b))$ and that its weak derivative is given by

$$\varphi_n'(t) = \begin{cases} n\,, & \text{if } a \le t < a + \dfrac{1}{n}, \\[2ex] 0, & \text{if } a + \dfrac{1}{n} \le t < b - \dfrac{1}{n}, \\[2ex] -n, & \text{if } b - \dfrac{1}{n} \le t \le b. \end{cases}$$

Thus we obtain

$$\int_a^b u'(t)\,\varphi_n(t)\,dt = -\int_a^b u(t)\,\varphi_n'(t)\,dt = -n\int_a^{a+\frac{1}{n}} u(t)\,dt + n\int_{b-\frac{1}{n}}^b u(t)\,dt. \tag{$*$}$$

By using the continuity of u, we have

$$\lim_{n\to\infty} n\int_a^{a+\frac{1}{n}} u(t)\,dt = u(a) \qquad \text{and} \qquad \lim_{n\to\infty} n\int_{b-\frac{1}{n}}^b u(t)\,dt = u(b).$$

On the other hand, by using that

$$\lim_{n\to\infty} u'(t)\,\varphi_n(t) = u'(t), \qquad \text{for a.e. } t \in [a, b],$$

and

$$|u'(t)\,\varphi_n(t)| \le |u'(t)| \in L^1((a, b)), \qquad \text{for a.e. } t \in [a, b],$$

we can apply the Lebesgue Dominated Convergence Theorem and obtain

$$\lim_{n\to\infty}\int_a^b u'(t)\,\varphi_n(t) = \int_a^b u'(t)\,dt.$$

Thus, by passing to the limit as n goes to ∞ in $(*)$, we can conclude.

5.5.7 The first fact simply follows from Proposition 5.1.8: it is sufficient to observe that for every $y \in \partial\Omega$, the function $x \mapsto |x-y|$ is bounded from below (since it is non-negative) and 1-Lipschitz.

For the second fact: by Rademacher's Theorem (see Theorem 5.2.2) we know that d_Ω is differentiable almost everywhere. Moreover, by using that d_Ω is 1-Lipschitz and (5.2.6), we have

$$|\nabla d_\Omega(x)| \le 1, \qquad \text{for a. e. } x \in \Omega.$$

In order to prove that equality holds in the previous estimate, we take $x_0 \in \Omega$ a point of differentiability for d_Ω. This implies in particular that for every $\omega \in \mathbb{S}^{N-1}$ we have

$$d_\Omega(x_0 + t\,\omega) = d_\Omega(x_0) + \langle \nabla d_\Omega(x_0), \omega\rangle\, t + o(t), \qquad \text{for } t \to 0. \tag{$*$}$$

We take $y_0 \in \partial\Omega$ such that $d_\Omega(x_0) = |x_0 - y_0|$ and set

$$\omega_0 = \frac{y_0 - x_0}{|y_0 - x_0|}.$$

Observe that such a point y_0 exists by Problem 1.7.26. By construction, it is not difficult to see that

$$d_\Omega(x_0 + t\,\omega_0) = d_\Omega(x_0) - t, \qquad \text{for } t \in (0, |x_0 - y_0|). \tag{$**$}$$

Indeed, by definition of d_Ω and ω_0 we have

$$d_\Omega(x_0 + t\,\omega_0) \le |x_0 + t\,\omega_0 - y_0| = d_\Omega(x_0) - t,$$

for every t as above. On the other hand, by the triangle inequality, we also have for every $y \in \partial\Omega$

$$|x_0 + t\,\omega_0 - y| \ge |x_0 - y| - |t\,\omega_0| = |x_0 - y| - t \ge d_\Omega(x_0) - t.$$

Thus, by taking the minimum over $y \in \partial\Omega$, we get

$$d_\Omega(x_0 + t\,\omega_0) \ge d_\Omega(x_0) - t,$$

as well. By using $(**)$ in $(*)$, we get

$$d_\Omega(x_0) - t = d_\Omega(x_0 + t\,\omega_0) = d_\Omega(x_0) + \langle \nabla d_\Omega(x_0), \omega_0\rangle\, t + o(t), \qquad \text{for } t \to 0.$$

This implies

$$\langle \nabla d_\Omega(x_0), \omega_0\rangle = -1 + \frac{o(t)}{t}, \qquad \text{for } t \to 0,$$

which in turn gives

$$\langle \nabla d_\Omega(x_0), \omega_0\rangle = -1.$$

Thanks to the Cauchy-Schwarz inequality, this proves that $|\nabla d_\Omega(x_0)| \ge 1$, as well. This concludes the proof.

5.5.8 For every $x, y \in \Omega$ we have

$$|f(x)| \le |f(y)| + |f(x) - f(y)| \le |f(y)| + |f|_{C^{0,1}(\Omega)}\, |x - y|.$$

By recalling that the diameter of Ω is defined by

$$\operatorname{diam}(\Omega) = \sup_{x,y\in\Omega} |x - y|,$$

and that this is finite by assumption, we get

$$|f(x)| \le |f(y)| + |f|_{C^{0,1}(\Omega)} \operatorname{diam}(\Omega), \qquad \text{for every } x, y \in \Omega.$$

By taking the infimum over y and the supremum over x, we get the conclusion.

Let us now suppose that Ω is unbounded. Thus, it contains a sequence $\{x_n\}_{n\in\mathbb{N}}$ such that

$$\lim_{n\to\infty} |x_n| = +\infty.$$

The function $f(x) = |x|$ is 1-Lipschitz continuous and

$$\lim_{n\to\infty} f(x_n) = \lim_{n\to\infty} |x_n| = +\infty.$$

This shows that f is unbounded.

5.5.9 The fact that $C^0_{\mathrm{b}}(\overline{\Omega})$ is a vector space is straightforward. Also the fact that $\|\cdot\|_{C^0(\overline{\Omega})}$ defines a norm is immediate, it just follows from the properties of the absolute value.

In order to prove that $C^0_{\mathrm{b}}(\overline{\Omega})$ is a Banach space, we take a Cauchy sequence $\{f_n\}_{n\in\mathbb{N}} \subseteq C^0_{\mathrm{b}}(\overline{\Omega})$. Thus, for every $\varepsilon > 0$ there exists $n_\varepsilon \in \mathbb{N}$ such that

$$\|f_n - f_m\|_{C^0(\overline{\Omega})} < \varepsilon, \qquad \text{for every } n, m \ge n_\varepsilon.$$

We first observe that this implies that there exists $M > 0$ such that

$$\|f_n\|_{C^0(\overline{\Omega})} \leq M, \qquad \text{for every } n \in \mathbb{N}. \tag{$*$}$$

Indeed, by taking $\varepsilon = 1$, we get that there exists $n_1 \in \mathbb{N}$ such that

$$\|f_n\|_{C^0(\overline{\Omega})} \leq \|f_n - f_{n_1}\|_{C^0(\overline{\Omega})} + \|f_{n_1}\|_{C^0(\overline{\Omega})} < 1 + \|f_{n_1}\|_{C^0(\overline{\Omega})}, \qquad \text{for every } n \geq n_1.$$

By defining

$$M = 1 + \max\Big\{\|f_k\|_{C^0(\overline{\Omega})} \,:\, k = 1, \dots, n_1\Big\},$$

we get the claim. We now use that for every $x \in \overline{\Omega}$, the sequence of real numbers $\{f_n(x)\}_{n\in\mathbb{N}}$ is actually a Cauchy sequence, by assumption and by definition of the norm. Thus, by completeness of $\mathbb{R}$, we get that this sequence converges to a real number, that we shall indicate by $f(x)$. We need to prove that $f \in C^0_{\mathrm{b}}(\overline{\Omega})$ and that

$$\lim_{n\to\infty} \|f_n - f\|_{C^0(\overline{\Omega})} = 0. \tag{$**$}$$

By using the pointwise convergence and the uniform bound $(*)$, we get that

$$|f(x)| = \lim_{n\to\infty} |f_n(x)| \leq M.$$

This shows that f is bounded. Again from the pointwise convergence, for every $x \in \overline{\Omega}$ and every $\varepsilon > 0$, there exists $m_{\varepsilon,x} \in \mathbb{N}$ such that

$$|f_n(x) - f(x)| < \varepsilon, \qquad \text{for every } n \geq m_{\varepsilon,x}.$$

We set $k_\varepsilon = \max\{n_\varepsilon, m_{\varepsilon,x}\}$. Thus, for every $n \geq n_\varepsilon$ we have

$$|f_n(x) - f(x)| \leq |f_n(x) - f_{k_\varepsilon}(x)| + |f_{k_\varepsilon}(x) - f(x)| < 2\,\varepsilon.$$

This proves $(**)$. We are only left with proving that f is continuous: from $(**)$, for every $\varepsilon > 0$, there exists $\ell_\varepsilon \in \mathbb{N}$ such that

$$\|f_n - f\|_{C^0(\overline{\Omega})} < \varepsilon, \qquad \text{for every } n \geq \ell_\varepsilon.$$

Thus, for every $x, y \in \overline{\Omega}$ and $n \geq \ell_\varepsilon$, we have

$$\begin{aligned} |f(x) - f(y)| &\leq |f(x) - f_n(x)| + |f_n(x) - f_n(y)| + |f_n(y) - f(y)| \\ &\leq \|f_n - f\|_{C^0(\overline{\Omega})} + |f_n(x) - f_n(y)| + \|f_n - f\|_{C^0(\overline{\Omega})} \\ &< 2\,\varepsilon + |f_n(x) - f_n(y)|. \end{aligned}$$

On the other hand, by using the continuity of f_n at x, we get that there exists $\delta > 0$ such that if $|x - y| < \delta$, then

$$|f_n(x) - f_n(y)| < \varepsilon.$$

This finally shows that

$$|f(x) - f(y)| < 3\,\varepsilon, \qquad \text{for every } y \in \overline{\Omega} \text{ such that } |x - y| < \delta,$$

which gives the desired continuity of f at the generic point $x \in \overline{\Omega}$.

5.5.10 It is sufficient to reproduce verbatim the proof of Proposition 5.1.6.

5.5.11 We can proceed as in the proof of Theorem 5.1.10, up to some small modifications. We start by observing that the family of functions

$$x \mapsto f_y(x) := f(y) - |f|_{C^{0,1}(\Omega)}\,|x - y|,$$

is equi-Lipschitz on $\mathbb{R}^N$. More precisely, we have

$$|f_y|_{C^{0,1}(\mathbb{R}^N)} \le |f|_{C^{0,1}(\Omega)}, \qquad \text{for every } y \in \Omega.$$

By using that f is Lipschitz on Ω and the definition of f_y, we get

$$f_y(x_0) = f(y) - |f|_{C^{0,1}(\Omega)}\,|x_0 - y| \le f(x_0), \qquad \text{for every } y \in \Omega. \tag{$*$}$$

Thus, we can apply Remark 5.1.9 to the family $\mathcal{X} = \{f_y\}_{y\in\Omega}$, with

$$M = f(x_0), \qquad \ell = |f|_{C^{0,1}(\Omega)}, \qquad E = \mathbb{R}^N.$$

This shows that $f^{\#}$ is Lipschitz and

$$|f^{\#}|_{C^{0,1}(\mathbb{R}^N)} \le |f|_{C^{0,1}(\Omega)}.$$

We now show that $f^{\#} = f$ on Ω. By taking the supremum over $y \in \Omega$ in $(*)$, we get

$$f^{\#}(x_0) \le f(x_0), \qquad \text{for every } x_0 \in \Omega.$$

On the other hand, since $f^{\#}$ is defined by a supremum, we obtain for $x \in \Omega$

$$f^{\#}(x) \ge f(x) - |f|_{C^{0,1}(\Omega)}\,|x - x| = f(x).$$

The last two inequalities show that $f^{\#}$ coincides with f on Ω. The fact that the Lipschitz constant of $f^{\#}$ coincides with that of f follows as in Theorem 5.1.10.

We now prove that $f^\# \leq \widetilde{f}$. To this aim, it is sufficient to observe that the Lipschitz continuity of $\widetilde{f}$ implies in particular that for $x \in \mathbb{R}^N$ we have

$$\widetilde{f}(y) - |f|_{C^{0,1}(\Omega)}\, |x-y| \leq \widetilde{f}(x), \qquad \text{for every } y \in \mathbb{R}^N.$$

In particular, by using that $f^\# = \widetilde{f} = f$ on Ω, we get

$$f(y) - |f|_{C^{0,1}(\Omega)}\, |x-y| \leq \widetilde{f}(x), \qquad \text{for every } y \in \Omega.$$

By taking the supremum over $y \in \Omega$ and recalling the definition of $f^\#$, we get

$$f^\#(x) \leq \widetilde{f}(x),$$

as desired.

5.5.12 Since $u \in C^{0,1}_0(\overline{\Omega})$, by Proposition 5.4.5 we have in particular that $u \in W^{1,1}_0(\Omega)$. Thanks to the definition of this space (see Definition 3.7.1), there exists a sequence $\{u_n\}_{n\in\mathbb{N}} \subseteq C^\infty_0(\Omega)$ such that

$$\lim_{n\to\infty} \|u - u_n\|_{W^{1,1}(\Omega)} = 0.$$

This in particular implies that

$$\lim_{n\to\infty} \int_\Omega \frac{\partial u_n}{\partial x_i}\, dx = \int_\Omega \frac{\partial u}{\partial x_i}\, dx, \qquad \text{for every } i \in \{1,\dots,N\}.$$

On the other hand, since each u_n belongs to $C^\infty_0(\Omega)$, by considering it extended by zero outside Ω, we have

$$\int_\Omega \frac{\partial u_n}{\partial x_i}\, dx = \int_{\mathbb{R}^N} \frac{\partial u_n}{\partial x_i}\, dx = \int_{\mathbb{R}^{N-1}} \left(\int_{-\infty}^{+\infty} \frac{\partial u_n}{\partial x_i}\, dx_i\right) d\widehat{x_i} = 0.$$

We used the notation $d\widehat{x_i} := dx_1 \dots dx_{i-1}\, dx_{i+1} \dots dx_N$. By joining the last two informations, we get the desired conclusion.

5.5.13 We divide E into three subsets

$$E_0 = \{(x,y) \in E \,:\, x < 0\},$$

$$E_+ = \{(x,y) \in E \,:\, x \geq 0 \text{ and } y > 0\} \quad \text{and}$$
$$E_- = \{(x,y) \in E \,:\, x \geq 0 \text{ and } y < 0\}.$$

We then take the function

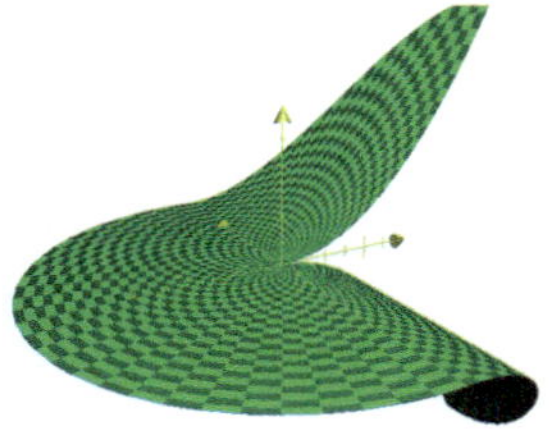

Fig. 8.13 The function of Problem 5.5.13 defined on the slit disk

$$u(x,y) = \begin{cases} 0, & \text{if } (x,y) \in E_0, \\ x^2, & \text{if } (x,y) \in E_+, \\ -x^2, & \text{if } (x,y) \in E_-, \end{cases}$$

which belongs to $C^1(E)$, see Fig. 8.13. By construction, the classical gradient of u on E is given by

$$\nabla u(x,y) = \begin{cases} (0,0), & \text{if } (x,y) \in E_0, \\ 2\,x\,\mathbf{e}_1, & \text{if } (x,y) \in E_+, \\ -2\,x\,\mathbf{e}_1, & \text{if } (x,y) \in E_-. \end{cases}$$

This is a bounded vector field on E, thus by Proposition 3.2.3 we have $u \in W^{1,\infty}(E)$. On the other hand, we have $u \notin C^{0,1}(E)$. Indeed, by taking the pair of points[13] $(1/2, \varepsilon)$ and $(1/2, -\varepsilon)$ for $\varepsilon > 0$, we have

$$\left| u\left(\frac{1}{2}, \varepsilon\right) - u\left(\frac{1}{2}, -\varepsilon\right) \right| = \left| \left(\frac{1}{2}\right)^2 + \left(\frac{1}{2}\right)^2 \right| = \frac{1}{2},$$

Dividing by the distance between the points

$$\left| \left(\frac{1}{2}, \varepsilon\right) - \left(\frac{1}{2}, -\varepsilon\right) \right| = 2\,\varepsilon,$$

and taking the supremum over $\varepsilon > 0$, we get

$$|u|_{C^{0,1}(E)} \geq \sup_{\varepsilon>0} \frac{\left| u\left(\frac{1}{2}, \varepsilon\right) - u\left(\frac{1}{2}, -\varepsilon\right) \right|}{\left| \left(\frac{1}{2}, \varepsilon\right) - \left(\frac{1}{2}, -\varepsilon\right) \right|} = \sup_{\varepsilon>0} \frac{1}{4\,\varepsilon} = +\infty.$$

Thus, u is not a Lipschitz function on E.

[13] Observe that the two points lie on different sides of the "slit".

5.5.14 For every $k \in \mathbb{N}$, we define the function

$$\varphi_k(x) = \min_{y \in \overline{\Omega}} \Big[\varphi(y) + k\,|x - y| \Big], \qquad \text{for every } x \in \overline{\Omega}.$$

We observe that, for every $y \in \overline{\Omega}$, the function

$$x \mapsto \varphi(y) + k\,|x - y|,$$

is Lipschitz continuous, with Lipschitz constant equal to k. Moreover, we also have

$$\varphi(y) + k\,|x - y| \geq \varphi(y) \geq \min_{\overline{\Omega}} \varphi =: m.$$

Accordingly, φ_k is Lipschitz continuous, as an infimum of a family of equi-Lipschitz functions, uniformly bounded from below (see Proposition 5.1.8). We also observe that by choosing $y = x$, we have

$$\varphi_k(x) = \min_{y \in \overline{\Omega}} \Big[\varphi(y) + k\,|x - y| \Big] \leq \varphi(x). \qquad (*)$$

It is easily seen that the sequence $\{\varphi_k\}_{k \in \mathbb{N}}$ is non-decreasing, i.e.

$$\varphi_k(x) \leq \varphi_{k+1}(x), \qquad \text{for every } x \in \overline{\Omega},\ k \in \mathbb{N}.$$

This implies that

$$\lim_{k \to \infty} \varphi_k(x),$$

exists for every $x \in \overline{\Omega}$. Moreover, we have from $(*)$

$$\lim_{k \to \infty} \varphi_k(x) \leq \varphi(x).$$

We first prove that the limit actually coincide with $\varphi(x)$. Indeed, for every $x \in \overline{\Omega}$ and $k \in \mathbb{N}$, we indicate by $y_{k,x} \in \overline{\Omega}$ a point which attains the minimum in the definition of $\varphi_k(x)$. We then have from $(*)$

$$\varphi(y_{k,x}) + k\,|x - y_{k,x}| \leq \varphi(x). \qquad (**)$$

This in turn implies that

$$\min_{\overline{\Omega}} \varphi + k\,|x - y_{k,x}| \leq \max_{\overline{\Omega}} \varphi,$$

that is

$$|x - y_{k,x}| \le \frac{1}{k}\left(\max_{\overline{\Omega}} \varphi - \min_{\overline{\Omega}} \varphi\right).$$

The last estimate implies that

$$\lim_{k\to\infty} y_{k,x} = x, \qquad \text{for every } x \in \overline{\Omega}.$$

By spending this information into $(**)$ and using the continuity of φ, we get

$$\varphi(x) = \lim_{k\to\infty} \varphi(y_{k,x}) \le \lim_{k\to\infty} [\varphi(y_{k,x}) + k\,|x - y_{k,x}|] = \lim_{k\to\infty} \varphi_k(x) \le \varphi(x).$$

This finally proves that $\{\varphi_k\}_{k\in\mathbb{N}}$ converges pointwise to φ. In order to upgrade the convergence to the uniform one, we can appeal to the monotonicity of the sequence and conclude thanks to the so-called *Dini Theorem* (see for example [38, Theorem 12.1]).

5.5.15 Let $x_0 \in \Omega$, then there exists a ball $B_r(x_0) \subseteq \Omega$. The function

$$u(x) = \max\{r - |x - x_0|, 0\}, \qquad \text{for } x \in \overline{\Omega},$$

is Lipschitz continuous on $\overline{\Omega}$, thanks to Proposition 5.3.1. Moreover, by construction we have $u \equiv 0$ on $\overline{\Omega} \setminus B_r(x_0)$. Thus, in particular we have $u = 0$ on the boundary $\partial\Omega$, i.e. $u \in C_0^{0,1}(\overline{\Omega})$. On the other hand, we have $u \notin W_0^{1,\infty}(\Omega)$: we could proceed as in the solutions of Problems 3.12.13 and 3.12.14, but we prefer to give here a slightly more sophisticated argument.

We argue by contradiction and assume that there exists a sequence $\{u_n\}_{n\in\mathbb{N}} \subseteq C_0^\infty(\Omega)$ such that

$$\lim_{n\to\infty} \|u_n - u\|_{W^{1,\infty}(\Omega)} = 0.$$

In particular, by recalling the definition of the norm of $W^{1,\infty}(\Omega)$, we get that $\{\nabla u_n\}_{n\in\mathbb{N}}$ is a Cauchy sequence in $L^\infty(\Omega; \mathbb{R}^N)$. Observe that the components of ∇u_n are continuous and bounded functions on $\overline{\Omega}$: by appealing to Problem 1.7.31, we obtain in particular that

$$\left\{\frac{\partial u_n}{\partial x_k}\right\}_{n\in\mathbb{N}}, \qquad \text{for } k \in \{1, \ldots, N\},$$

are Cauchy sequences in the Banach space $C_b^0(\overline{\Omega})$. Thus, there exists $v_1, \ldots, v_N \in C_b^0(\overline{\Omega})$ such that

$$\lim_{n\to\infty} \left\| \frac{\partial u_n}{\partial x_k} - v_k \right\|_{C^0(\overline{\Omega})} = 0, \qquad \text{for } k \in \{1, \ldots, N\}.$$

We appeal again to Problem 1.7.31, so that the previous fact can be rewritten as

$$\lim_{n\to\infty}\left\|\frac{\partial u_n}{\partial x_k}-v_k\right\|_{L^\infty(\Omega)}=0,\qquad \text{for } k\in\{1,\ldots,N\}.$$

By the triangle inequality, if we set $v=(v_1,\ldots,v_N)$ we then obtain

$$\lim_{n\to\infty}\|\nabla u_n-v\|_{L^\infty(\Omega;\mathbb{R}^N)}=0.$$

By uniqueness of the limit, we finally obtain that

$$\nabla u=v,\qquad \text{a. e. in } \Omega,$$

that is ∇u coincides almost everywhere with a continuous vector field. We are going to show that this is not possible, thus giving the desired contradiction.

Indeed, observe that in $B_r(x_0)$ we have

$$\nabla u(x)=-\frac{x-x_0}{|x-x_0|},\qquad \text{for a. e. } x\in B_r(x_0).$$

This implies that

$$|v(x)|=|\nabla u(x)|=1,\qquad \text{for a. e. } x\in B_r(x_0),$$

and since $|v|$ is a continuous functions, we actually have $|v(x)|=1$ *for every* $x\in B_r(x_0)$. In particular, we get $|v(x_0)|=1$. From the previous discussion we get

$$\frac{1}{|B_\varepsilon(x_0)|}\int_{B_\varepsilon(x_0)}\langle\nabla u(x),v(x_0)\rangle\,dx=\frac{1}{|B_\varepsilon(x_0)|}\int_{B_\varepsilon(x_0)}\langle v(x),v(x_0)\rangle\,dx,$$

for every $0<\varepsilon<r$. By using the continuity of v, we know that the limit as ε goes to 0 of the integral on the right-hand side exists and it is given by $|v(x_0)|^2=1$. Thus, we obtain

$$\lim_{\varepsilon\to 0}\frac{1}{|B_\varepsilon(x_0)|}\int_{B_\varepsilon(x_0)}\langle\nabla u(x),v(x_0)\rangle\,dx=1, \tag{**}$$

as well. We now use the expression of ∇u and write this integral in spherical coordinates centered at x_0, i.e. $x=x_0+\varrho\,\omega$, with $0\le\varrho\le\varepsilon$ and $\omega\in\mathbb{S}^{N-1}$. This gives for $0<\varepsilon<r$

$$
\begin{aligned}
\frac{1}{|B_\varepsilon(x_0)|} \int_{B_\varepsilon(x_0)} & \langle \nabla u(x), \nu(x_0)\rangle \, dx \\
&= \frac{1}{\varepsilon^N \, \omega_N} \int_{\mathbb{S}^{N-1}} \left(\int_0^\varepsilon \langle -\omega, \nu(x_0)\rangle \, \varrho^{N-1} \, d\varrho \right) d\sigma(\omega) \\
&= -\frac{1}{N \, \omega_N} \int_{\mathbb{S}^{N-1}} \langle \omega, \nu(x_0)\rangle \, d\sigma(\omega) = 0.
\end{aligned}
$$

This contradicts $(**)$.

5.5.16 We consider the restriction $u_{|\partial\Omega}$ and take its Lipschitz extension to the whole space $\mathbb{R}^N$, given by Theorem 5.1.10. We call U this function and observe that

$$
u - U \equiv 0, \qquad \text{on } \partial\Omega,
$$

just by construction. We notice that $U \in C^{0,1}(\overline{\Omega})$: indeed, we already know that it is Lipschitz continuous; moreover, it is bounded on $\overline{\Omega}$ by Problem 5.5.8. Thus, according to Proposition 5.4.1 we have $U \in W^{1,\infty}(\Omega)$, as well. Up to now, we can assure that

$$
u - U \in \left\{ \varphi \in C^0(\overline{\Omega}) \cap W^{1,\infty}(\Omega) \ : \ \varphi = 0 \text{ on } \partial\Omega \right\}.
$$

By using that $W^{1,\infty}(\Omega) \subseteq W^{1,1}(\Omega)$, we can infer from Problem 3.12.16 that

$$
u - U \in W^{1,1}_0(\Omega).
$$

We now consider the extension by zero outside Ω of $u - U$: we indicate it by $\mathcal{Z}[u - U]$. Thanks to Lemma 3.7.9 with $E = \mathbb{R}^N$, we know that this is a Sobolev function on the whole $\mathbb{R}^N$, whose weak gradient is given by

$$
\nabla \mathcal{Z}[u - U](x) = \begin{cases} \nabla u(x) - \nabla U(x), \text{ if} x \in \Omega, \\ \qquad (0, \dots, 0), \text{ otherwise.} \end{cases}
$$

In particular, we get that

$$
\|\mathcal{Z}[u - U]\|_{W^{1,\infty}(\mathbb{R}^N)} = \|u - U\|_{W^{1,\infty}(\Omega)},
$$

i.e. $\mathcal{Z}[u - U]$ belongs to $W^{1,\infty}(\mathbb{R}^N)$. Since $\mathbb{R}^N$ is an open convex set, in light of Proposition 5.4.3 we get $\mathcal{Z}[u-U] \in C^{0,1}(\mathbb{R}^N)$. By recalling that $\mathcal{Z}[u-U]$ coincide with $u - U$ on $\overline{\Omega}$, we then obtain in particular

$$
u - U \in C^{0,1}(\overline{\Omega}).
$$

Finally, by recalling that $U \in C^{0,1}(\overline{\Omega})$, this is enough to conclude.

8.6 Problems of Chap. 6

6.8.1 By using Jensen's inequality (see Proposition 1.3.10), for every admissible function u we have

$$\int_\Omega F(\nabla u)\,dx = |\Omega| \fint_\Omega F(\nabla u)\,dx \ge |\Omega|\, F\left(\fint_\Omega \nabla u\,dx\right). \qquad (*)$$

By using that $u-U$ is a Lipschitz function identically vanishing on $\partial\Omega$, we get from Problem 5.5.12

$$\begin{aligned}\int_\Omega \frac{\partial u}{\partial x_i}\,dx &= \int_\Omega \frac{\partial(u-U)}{\partial x_i}\,dx + \int_\Omega \frac{\partial U}{\partial x_i}\,dx\\ &= \int_\Omega \frac{\partial U}{\partial x_i}\,dx, \qquad \text{for every } i = 1,\dots,N.\end{aligned}$$

We use this identity in $(*)$, so to get

$$\int_\Omega F(\nabla u)\,dx \ge |\Omega|\, F\left(\fint_\Omega \nabla U\,dx\right).$$

On the other hand, since U is affine we have $\nabla U = \mathbf{b}$, i.e. the gradient is a constant vector. This implies that

$$\begin{aligned}|\Omega|\, F\left(\fint_\Omega \nabla U\,dx\right) &= |\Omega|\, F\left(\fint_\Omega \mathbf{b}\,dx\right) = |\Omega|\, F(\mathbf{b}) = \int_\Omega F(\mathbf{b})\,dx\\ &= \int_\Omega F(\nabla U)\,dx.\end{aligned}$$

By keeping everything together, for every admissible function u we obtained

$$\int_\Omega F(\nabla u)\,dx \ge \int_\Omega F(\nabla U)\,dx.$$

This proves the minimality of U. If F is strictly convex, uniqueness follows as in Remark 6.2.2.

6.8.2 We first observe that by Problem 1.7.8 the function $F : \mathbb{R}^N \to \mathbb{R}$ defined by

$$F(z) = f(|z|), \qquad \text{for every } z \in \mathbb{R}^N,$$

is strictly convex. Then uniqueness of the solution v follows from Remark 6.2.2.

Once uniqueness is established, the radial symmetry of v is readily obtained. It is sufficient to proceed as in the solution of Problem 4.10.1.

We show that V is non-decreasing if $C_1 < C_2$, the other case being left to the reader. We define the new function

$$\widetilde{V}(t) = C_1 + \int_{r_1}^{t} |V'(\tau)|\,d\tau, \qquad \text{for } t \in [r_1, r_2].$$

This is Lipschitz continuous, thanks to Problem 5.5.5. By construction, $\widetilde{V}$ is non-decreasing. Also, in light of Problem 5.5.5 we have

$$\widetilde{V}'(t) = |V'(t)|, \qquad \text{for a. e. } t \in [r_1, r_2].$$

Moreover, we have

$$\widetilde{V}(r_1) = C_1 \qquad \text{and} \qquad \widetilde{V}(r_2) \geq C_1 + \left| \int_{r_1}^{r_2} V'(\tau)\,d\tau \right| = C_1 + |C_2 - C_1| = C_2,$$

where we used that $C_2 - C_1 > 0$. We now set

$$W(t) = \min\left\{\widetilde{V}(t), C_2\right\},$$

which is still a non-decreasing function. Moreover, it is still Lipschitz continuous, thanks to Corollary 5.3.3. Thus, by Problem 5.5.4 the function $\widetilde{v}(x) = W(|x|)$ is admissible for the initial minimization problem. Moreover, we have

$$W'(t) \leq \widetilde{V}'(t) = |V'(t)|, \qquad \text{for a. e. } t \in [r_1, r_2],$$

still thanks to Corollary 5.3.3. Thus, from the monotone behaviour of f we would get

$$\int_{A_{r_1,r_2}} f(|\nabla v|)\,dx \geq \int_{A_{r_1,r_2}} f(|\nabla \widetilde{v}|)\,dx.$$

Observe that we used that

$$|\nabla v(x)| = |V'(|x|)|, \qquad |\nabla \widetilde{v}(x)| = W'(|x|), \qquad \text{for a. e. } x \in A_{r_1,r_2},$$

which follows from Problem 5.5.4. By uniqueness of the minimizer v, we get $v = \widetilde{v}$ and thus $V = W$ is monotone non-decreasing.

6.8.3 We take the same assumptions as in Theorem 6.5.1, except for the fact that $F : \mathbb{R}^N \to \mathbb{R}$ is C^1 and convex, but not necessarily strictly convex. For every $0 < \varepsilon < 1$ we introduce

$$F_\varepsilon(z) = F(z) + \frac{\varepsilon}{2}|z|^2, \qquad \text{for every } z \in \mathbb{R}^N.$$

By virtue of the additional term containing $|z|^2$, we get that this new function is C^1 and strictly convex. Accordingly, we consider the new problem

$$\inf_{\varphi\in C^{0,1}(\overline{\Omega})}\left\{\int_\Omega F_\varepsilon(\nabla\varphi)\,dx\ :\ \varphi=U \text{ on } \partial\Omega\right\}.$$

By Theorem 6.5.1, this admits a unique minimizer v_ε. By using the minimality of v_ε and the definition of F_ε, we get

$$\int_\Omega F_\varepsilon(\nabla\varphi)\,dx \ge \int_\Omega F_\varepsilon(\nabla v_\varepsilon)\,dx \ge \int_\Omega F(\nabla v_\varepsilon)\,dx, \tag{$*$}$$

for every admissible function φ. We observe that thanks to (6.5.1), we have

$$|v_\varepsilon|_{C^{0,1}(\overline{\Omega})} \le K, \qquad \text{for every } 0<\varepsilon<1.$$

Moreover, if we fix a point $x_0\in\partial\Omega$, by Problem 5.5.8 we know that

$$\begin{aligned}\max_{\overline{\Omega}}|v_\varepsilon| &\le |v_\varepsilon(x_0)| + |v_\varepsilon|_{C^{0,1}(\overline{\Omega})}\operatorname{diam}(\overline{\Omega})\\ &\le |U(x_0)| + K\operatorname{diam}(\overline{\Omega}), \qquad \text{for every } 0<\varepsilon<1.\end{aligned}$$

By virtue of Theorem 5.1.7, there exists an infinitesimal decreasing sequence $\{\varepsilon_k\}_{k\in\mathbb{N}}\subseteq(0,1)$ such that $\{v_{\varepsilon_k}\}_{k\in\mathbb{N}}$ converges uniformly in $\overline{\Omega}$ to a Lipschitz function v. More precisely, we have

$$|v|_{C^{0,1}(\overline{\Omega})} \le \liminf_{k\to\infty}|v_{\varepsilon_k}|_{C^{0,1}(\overline{\Omega})} \le K.$$

Thanks to the uniform convergence we still have $v=U$ on $\partial\Omega$. Moreover, by proceeding as in the proof of Theorem 6.2.1, we can further suppose that $\{\nabla v_{\varepsilon_k}\}_{k\in\mathbb{N}}$ weakly converges in $L^2(\Omega;\mathbb{R}^N)$ to ∇v, possibly passing to a subsequence. We claim that v is the desired minimizer.

Indeed, observe that for every $\varphi\in C^{0,1}(\overline{\Omega})$ we have

$$\lim_{k\to\infty}\int_\Omega F_{\varepsilon_k}(\nabla\varphi)\,dx) = \int_\Omega F(\nabla\varphi)\,dx + \lim_{k\to\infty}\frac{\varepsilon_k}{2}\int_\Omega|\nabla\varphi|^2\,dx = \int_\Omega F(\nabla\varphi)\,dx.$$

From $(*)$, we thus get that for every $\varphi\in C^{0,1}(\overline{\Omega})$ such that $\varphi=U$ on $\partial\Omega$, we have

$$\int_\Omega F(\nabla\varphi)\,dx \ge \liminf_{k\to\infty}\int_\Omega F(\nabla v_{\varepsilon_k})\,dx.$$

In particular, by using the "above tangent" property for the convex function F and the weak convergence of the gradients, we have

$$\int_\Omega F(\nabla\varphi)\,dx \geq \liminf_{k\to\infty}\left[\int_\Omega F(\nabla v)\,dx + \int_\Omega \langle \nabla F(\nabla v), \nabla v_{\varepsilon_k} - \nabla v\rangle\,dx\right]$$
$$= \int_\Omega F(\nabla v)\,dx.$$

By arbitrariness of φ, this shows that v is a minimizer.

6.8.4 Let us consider for example the C^1 convex function

$$F(z) = (|z|-1)_+^2, \qquad \text{for every } z \in \mathbb{R}^N.$$

We indicate by $(\cdot)_+$ the positive part, as usual. Observe that this is not strictly convex, because it identically vanishes in the ball $\overline{B_1(0)}$. For an open bounded set $\Omega \subseteq \mathbb{R}^N$, we consider the minimization problem

$$\min_{u\in C^{0,1}(\overline{\Omega})}\left\{\int_\Omega (|\nabla u|-1)_+^2\,dx \,:\, u = 0 \text{ on } \partial\Omega\right\}.$$

We immediately notice that $u \equiv 0$ is a solution of the previous problem. At the same time, by taking any other 1-Lipschitz continuous function u which vanish on the boundary of Ω, by (5.2.6) we get

$$|\nabla u(x)| \leq 1, \qquad \text{for a. e. } x \in \Omega,$$

and thus

$$\int_\Omega (|\nabla u|-1)_+^2\,dx = 0.$$

This proves that each of these functions is a minimizer, as well.

We now consider the minimization problem

$$\mathfrak{m} := \inf_{\varphi\in C^{0,1}(\overline{\Omega})}\left\{\int_\Omega F(\nabla\varphi)\,dx \,:\, \varphi = U \text{ on } \partial\Omega\right\},$$

under the assumptions of Problem 6.8.3. We suppose that v_1 and v_2 are two distinct solutions. By Corollary 5.3.3, we have that both functions

$$w = \min\{v_1, v_2\} \qquad \text{and} \qquad W = \max\{v_1, v_2\},$$

are admissible for the minimization problem above. We observe that

$$\int_\Omega F(\nabla v_1)\,dx = \mathfrak{m} \leq \int_\Omega F(\nabla w)\,dx = \int_{\{v_1\leq v_2\}} F(\nabla v_1)\,dx + \int_{\{v_1>v_2\}} F(\nabla v_2)\,dx, \tag{$*$}$$

thanks to Corollary 5.3.3 and the property (5.3.1). Thus, by writing

$$\int_{\Omega} F(\nabla v_1)\,dx = \int_{\{v_1 \le v_2\}} F(\nabla v_1)\,dx + \int_{\{v_1 > v_2\}} F(\nabla v_1)\,dx,$$

and erasing the common factor $\int_{\{v_1 \le v_2\}} F(\nabla v_1)\,dx$, we obtain from $(*)$

$$\int_{\{v_1 > v_2\}} F(\nabla v_1)\,dx \le \int_{\{v_1 > v_2\}} F(\nabla v_2)\,dx.$$

By adding on both sides the term $\int_{\{v_1 \le v_2\}} F(\nabla v_2)\,dx$ and appealing again to Corollary 5.3.3 and (5.3.1), we obtain

$$\mathfrak{m} \le \int_{\Omega} F(\nabla W)\,dx \le \int_{\Omega} F(\nabla v_2)\,dx = \mathfrak{m}.$$

This shows that W is a minimizer, as well. Observe that this entails that all the inequalities above must hold as equalities. In particular, we must have equality in $(*)$, i.e.

$$\int_{\Omega} F(\nabla v_1)\,dx = \mathfrak{m} = \int_{\Omega} F(\nabla w)\,dx,$$

thus proving that also w is a minimizer,

6.8.5 By Problem 6.8.3, we already know that there exists at least a solution v. Moreover, we have seen that

$$\|\nabla v\|_{L^\infty(\Omega;\mathbb{R}^N)} \le |v|_{C^{0,1}(\overline{\Omega})} \le K.$$

The first inequality follows from (5.2.6), as usual. Let us now suppose that there exists another minimizer w. By using the usual trick of considering

$$\frac{v+w}{2},$$

we see that this is still admissible and by convexity

$$\int_{\Omega} F\left(\frac{\nabla v + \nabla w}{2}\right)\,dx \le \frac{1}{2}\int_{\Omega} F(\nabla v) + \frac{1}{2}\int_{\Omega} F(\nabla w)\,dx.$$

The minimality of both v and w implies that $(v+w)/2$ must be a minimizer, as well. Thus, equality must hold in the inequality above. By convexity of F, this means that we must have

$$F\left(\frac{\nabla v(x)+\nabla w(x)}{2}\right)=\frac{1}{2}\,F(\nabla v(x))+\frac{1}{2}\,F(\nabla w(x)),\qquad \text{for a. e. } x\in\Omega.$$

The assumption of strict convexity "at infinity" implies that for almost every $x \in \Omega$ we must have

$$\text{either}\qquad \nabla v(x)=\nabla w(x)\qquad \text{or}\qquad |\nabla v(x)|<\delta \text{ and } |\nabla w(x)|<\delta.$$

In both cases, we get that

$$|\nabla w(x)|\le|\nabla v(x)|+|\nabla w(x)-\nabla v(x)|\le|\nabla v(x)|+2\,\delta,$$

as desired.

6.8.6 If $z_0 \in \partial\Omega$, then there is nothing to prove. Let us suppose that $z_0 \not\in \partial\Omega$. By basic topological facts, we know that

$$\partial(\Omega\cap\Omega_\tau)\subseteq(\partial\Omega\cap\overline{\Omega_\tau})\cup(\overline{\Omega}\cap\partial\Omega_\tau),$$

and thus

$$z_0\in\overline{\Omega}\cap\partial\Omega_\tau.$$

In particular, we have $z_0 \in \partial\Omega_\tau$. By construction, the latter is given by $\partial\Omega_\tau = \partial\Omega + \tau$. In conclusion, we get $z_0 \in \partial\Omega + \tau$, which is the same as $z_0 - \tau \in \partial\Omega$, as desired.

6.8.7 We need to prove that for every distinct points $x_0, x_1 \in \Omega$ we have

$$(1-t)\,x_0+t\,x_1\in\Omega,\qquad \text{for every } t\in(0,1).$$

We argue by contradiction and suppose that there exists $x_0 \neq x_1$ both belonging to Ω such that the segment connecting them is not entirely contained in Ω. In particular, there exists $\bar{t} \in (0, 1)$ such that $(1-\bar{t})\,x_0 + \bar{t}\,x_1 \in \partial\Omega$. By assumption, there exists a closed half-space Π^+ such that

$$(1-\bar{t})\,x_0+\bar{t}\,x_1\in\partial\Pi^+\qquad \text{and}\qquad \Omega\subseteq\Pi^+.$$

The closed half-space Π^+ can be written as

$$\Pi^+=\{x\in\mathbb{R}^N\,:\,\langle x-(1-\bar{t})\,x_0-\bar{t}\,x_1,\omega\rangle\ge 0\},$$

for a suitable $\omega \in \mathbb{S}^{N-1}$. Since $\Omega \subseteq \Pi^+$ and Ω is an open set, we thus must have

$$\langle x-(1-\bar{t})\,x_0-\bar{t}\,x_1,\omega\rangle>0,\qquad \text{for every } x\in\Omega.$$

In particular, by taking $x = x_0$ and $x = x_1$ in the previous inequality, we have both

$$\bar{t}\,\langle x_0 - x_1, \omega\rangle > 0 \qquad \text{that is} \qquad \langle x_0 - x_1, \omega\rangle > 0,$$

and

$$(1-\bar{t})\,\langle x_1 - x_0, \omega\rangle > 0 \qquad \text{that is} \qquad \langle x_0 - x_1, \omega\rangle < 0.$$

We thus get a contradiction.

6.8.8 This result is taken from [83, Proposizione 6.2]. We will proceed similarly as in the proof of Proposition 6.4.5. We first need some geometric properties of Ω, which follow from our assumption. The latter implies that

$$\partial\Pi^+_{x_0} \cap \partial\Omega = \{x_0\} \quad \text{and} \quad \partial\Pi^+_{x_0} \cap \Omega = \emptyset, \qquad \text{for every } x_0 \in \partial\Omega. \tag{\#}$$

Indeed, if $x \neq x_0$ belongs to $\partial\Pi^+_{x_0} \cap \partial\Omega$, we would get

$$0 < |x - x_0|^2 \le C_\Omega \operatorname{dist}(x, \partial\Pi^+_{x_0}) = 0,$$

i.e. a contradiction. On the other hand, if there would exist $x \in \partial\Pi^+_{x_0} \cap \Omega$, let us consider the half-line

$$\ell = \{(1-t)\,x_0 + t\,x \,:\, t \ge 0\} \subseteq \partial\Pi^+_{x_0}.$$

Since Ω is bounded, this would intersect $\partial\Omega$ in another point, different from x_0, thus contradicting the first property in (#).

Finally, it is not difficult to see that Ω must entirely lie inside one of the two open half-spaces

$$\mathring{\Pi}^+_{x_0} = \Big\{x \in \mathbb{R}^N \,:\, \langle x - x_0, \omega_{x_0}\rangle > 0\Big\}, \qquad \mathring{\Pi}^-_{x_0} = \Big\{x \in \mathbb{R}^N \,:\, \langle x - x_0, \omega_{x_0}\rangle < 0\Big\},$$

separated by $\partial\Pi^+_{x_0}$. In other words, $\partial\Pi^+_{x_0}$ is a supporting hyperplane for Ω at the point x_0 (recall Problem 6.8.7). Indeed, we argue again by contradiction: let us suppose that there exist

$$z \in \mathring{\Pi}^+_{x_0} \cap \Omega \qquad \text{and} \qquad y \in \mathring{\Pi}^-_{x_0} \cap \Omega.$$

By convexity, Ω must entirely contain the segment connecting these two points. On the other hand, such a segment must intersect $\partial\Pi^+_{x_0}$. We thus get a contradiction with the second property in (#).

We now take $U \in C^2(\mathbb{R}^N)$ and come to the proof of the claim. We fix $x_0 \in \partial\Omega$: without loss of generality we can suppose that $x_0 = 0$. Up to rotating Ω, we can assume that

$$\Pi^+_{x_0} = \{x \in \mathbb{R}^N \,:\, x_N \geq 0\},$$

and that, in light of the previous discussion, we have

$$\Omega \subseteq \Pi^+_{x_0}. \tag{$*$}$$

We need to find two vectors $\mathbf{a}, \mathbf{b} \in \mathbb{R}^N$ such that

$$U(0) + \langle \mathbf{a}, x\rangle \leq U(x) \leq U(0) + \langle \mathbf{b}, x\rangle, \qquad \text{for every } x \in \partial\Omega,$$

with

$$|\mathbf{a}| \leq K \qquad \text{and} \qquad |\mathbf{b}| \leq K,$$

for some uniform constant $K > 0$.

By proceeding exactly as in the proof of Proposition 6.4.5, we obtain

$$U(x) \leq U(0) + \langle \nabla U(0), x\rangle + C\,|x|^2, \tag{$**$}$$

and

$$U(x) \geq U(0) + \langle \nabla U(0), x\rangle - C\,|x|^2,$$

for a constant C depending only on the C^2 norm of U. We now use our assumption on Ω: we have that for every $x = (x', x_N) \in \partial\Omega$

$$|x|^2 = |x - x_0|^2 \leq C_\Omega \operatorname{dist}(x, \partial\Pi^+_{x_0}) = C_\Omega\, x_N.$$

Observe that in the last identity we used ($*$), so that $x_N = |x_N|$ for every $(x', x_N) \in \partial\Omega$. By using this estimate in ($**$), we obtain

$$U(x) \leq U(0) + \langle \nabla U(0), x\rangle + C\,C_\Omega\, x_N, \qquad \text{for every } x \in \partial\Omega. \tag{$***$}$$

We can take the vector

$$\mathbf{b} = \nabla U(0) + C\,C_\Omega\,\mathbf{e}_N,$$

and observe that

$$|\mathbf{b}| \leq \max_{\overline{\Omega}} |\nabla U| + C\,C_\Omega =: K,$$

with the last quantity K depending only on Ω and the C^2 norm of U. By construction, we have for every $x \in \partial\Omega$

$$
\begin{aligned}
U(0) + \langle \mathbf{b}, x \rangle &= U(0) + \langle \nabla U(0), x \rangle + C\, C_\Omega \langle \mathbf{e}_N, x \rangle \\
&= U(0) + \langle \nabla U(0), x \rangle + C\, C_\Omega\, x_N \geq U(x),
\end{aligned}
$$

where in the last estimate we used $(***)$. The vector $\mathbf{a}$ can be constructed in a similar way.

6.8.9 From the definition of BSC, we know that for every $y \in \partial\Omega$, there exists $\mathbf{a}_y, \mathbf{b}_y \in \mathbb{R}^N$ such that

$$
U(y) + \langle \mathbf{a}_x, y - x \rangle \leq U(x) \leq U(y) + \langle \mathbf{b}_x, y - x \rangle, \qquad \text{for every } x \in \partial\Omega.
$$

Moreover, we have $|\mathbf{a}_y| \leq K$ and $|\mathbf{b}_y| \leq K$. In particular, we get

$$
U(x) - U(y) \leq |\mathbf{b}_y|\, |x - y| \leq K\, |x - y|,
$$

and

$$
U(x) - U(y) \geq -|\mathbf{a}_y|\, |x - y| \geq -K\, |x - y|.
$$

Thus, we have obtained

$$
|U(x) - U(y)| \leq K\, |x - y|, \qquad \text{for every } x, y \in \partial\Omega.
$$

This gives the desired conclusion.

6.8.10 We recall that $W_0^{1,p}(\Omega)$ is the closure of $C_0^\infty(\Omega)$ in the norm of $W^{1,p}(\Omega)$. For every $\varphi \in C^{0,1}(\overline{\Omega})$ such that $\varphi = g$ on $\partial\Omega$, we have

$$
\varphi - g \in C_0^{0,1}(\overline{\Omega}) \subseteq W_0^{1,p}(\Omega),
$$

thanks to Proposition 5.4.5. Moreover, we know that $C^{0,1}(\overline{\Omega}) \subseteq W^{1,p}(\Omega)$, thanks to Corollary 5.4.2. This shows that

$$
\left\{\varphi \in C^{0,1}(\overline{\Omega}) \,:\, \varphi = g \text{ on } \partial\Omega\right\} \subseteq \left\{\varphi \in W^{1,p}(\Omega) \,:\, \varphi - g \in W_0^{1,p}(\Omega)\right\}. \qquad (*)
$$

This immediately gives that

$$
\begin{aligned}
&\inf_{\varphi \in W^{1,p}(\Omega)} \left\{ \int_\Omega |\nabla \varphi|^p\, dx \,:\, \varphi - g \in W_0^{1,p}(\Omega) \right\} \\
&\qquad \leq \inf_{\varphi \in C^{0,1}(\overline{\Omega})} \left\{ \int_\Omega |\nabla \varphi|^p\, dx \,:\, \varphi = g \text{ on } \partial\Omega \right\}.
\end{aligned}
$$

In order to prove the reverse inequality, we take $\varphi \in W^{1,p}(\Omega)$ such that $\varphi - g \in W_0^{1,p}(\Omega)$. By definition of $W_0^{1,p}(\Omega)$, there exists $\{\psi_n\}_{n\in\mathbb{N}} \subseteq C_0^\infty(\Omega)$ such that

$$\lim_{n\to\infty} \left(\|\psi_n - (\varphi - g)\|_{L^p(\Omega)} + \|\nabla\psi_n - \nabla(\varphi - g)\|_{L^p(\Omega;\mathbb{R}^N)} \right) = 0. \qquad (**)$$

We observe in particular that $\psi_n + g \in C^{0,1}(\overline{\Omega})$ and this function coincides with g on $\partial\Omega$. Thus, for every $n \in \mathbb{N}$ we have

$$\inf_{\varphi\in C^{0,1}(\overline{\Omega})} \left\{ \int_\Omega |\nabla\varphi|^p \, dx \; : \; \varphi = g \text{ on } \partial\Omega \right\} \le \int_\Omega |\nabla\psi_n + \nabla g|^p \, dx.$$

Moreover, from $(**)$ we have

$$\lim_{n\to\infty} \int_\Omega |\nabla\psi_n + \nabla g|^p \, dx = \int_\Omega |\nabla\varphi|^p \, dx.$$

By taking the limit as n goes to ∞, we then obtain that

$$\inf_{\varphi\in C^{0,1}(\overline{\Omega})} \left\{ \int_\Omega |\nabla\varphi|^p \, dx \; : \; \varphi = g \text{ on } \partial\Omega \right\} \le \int_\Omega |\nabla\varphi|^p \, dx.$$

This holds for every $\varphi \in W^{1,p}(\Omega)$ such that $\varphi - g \in W_0^{1,p}(\Omega)$. Thus, we get

$$\begin{aligned} &\inf_{\varphi\in W^{1,p}(\Omega)} \left\{ \int_\Omega |\nabla\varphi|^p \, dx \; : \; \varphi - g \in W_0^{1,p}(\Omega) \right\} \\ &\qquad \ge \inf_{\varphi\in C^{0,1}(\overline{\Omega})} \left\{ \int_\Omega |\nabla\varphi|^p \, dx \; : \; \varphi = g \text{ on } \partial\Omega \right\}, \end{aligned}$$

as well.

Finally, let us call v the unique minimizer of the problem settled on the Sobolev space $W^{1,p}(\Omega)$. This exists by Theorem 4.3.3, while uniqueness follows from the usual convexity argument (see Remark 6.2.2). We suppose that the problem settled on Lipschitz functions admits a minimizer, as well: we call it w. From the equality of the infima proved in the first part, we get that

$$\int_\Omega |\nabla v|^p \, dx = \int_\Omega |\nabla w|^p \, dx.$$

Moreover, by recalling $(*)$, we get that w is a minimizer for the problem settled over $W^{1,p}(\Omega)$, as well. By uniqueness, we conclude that $v = w$.

6.8.11 We start by verifying that U is Lipschitz continuous on $\partial B_1(0)$. This is quite easy: by the triangle inequality, for every $(x, y), (t, s) \in \partial B_1(0)$ we have

$$\big|U(x,y) - U(t,s)\big| = \big||y| - |s|\big| \le |y-s| \le \sqrt{(x-t)^2 + (y-s)^2}.$$

In order to prove that the problem

$$\inf_{u\in C^{0,1}(\overline{B_1(0)})} \left\{ \frac{1}{2}\int_{B_1(0)} |\nabla u|^2\, dx\, dy \,:\, u = U \text{ on } \partial B_1(0) \right\},$$

does not have a solution, we will proceed as in Proposition 6.7.2: we will exhibit a solution of

$$\begin{cases} -\Delta u = 0, & \text{in } B_1(0), \\ \quad\ u = U, & \text{on } \partial B_1(0), \end{cases} \tag{$*$}$$

which is not Lipschitz continuous on $\overline{B_1(0)}$. To solve this boundary value problem, we will use polar coordinates

$$x = \varrho\, \cos\vartheta, \qquad y = \varrho\, \sin\vartheta.$$

In these coordinates, the boundary datum U can be written as

$$U(1,\vartheta) = |\sin\vartheta|, \qquad \text{for } \vartheta \in [0, 2\pi).$$

By observing that this function is π-periodic, it can be expanded in Fourier series

$$U(1,\vartheta) = \sum_{n\in\mathbb{Z}} \widehat{U}(n)\, e^{2\,n\,\mathrm{i}\,\vartheta},$$

where the Fourier coefficients $\{\widehat{U}(n)\}_{n\in\mathbb{Z}}$ are given by

$$\widehat{U}(n) = \frac{1}{\pi}\int_0^{\pi} \sin\vartheta\, e^{-2\,n\,\mathrm{i}\,\vartheta}\, d\vartheta = \frac{2}{\pi}\,\frac{1}{1-4\,n^2}, \qquad \text{for every } n \in \mathbb{Z}.$$

For the necessary facts on Fourier series, we refer the reader for example to [57, Chapters 2 & 3]. Since $\widehat{U}(n) = \widehat{U}(-n)$, the Fourier series can be rearranged into a series only involving cosine functions, i.e.

$$U(1,\vartheta) = \frac{2}{\pi} + \frac{4}{\pi}\sum_{n=1}^{\infty} \frac{1}{1-4\,n^2}\,\cos(2\,n\,\vartheta).$$

Observe that since $U(1,\cdot) \in C^1([0,\pi]) \cap C^0(\mathbb{R})$, by the general theory of Fourier series we know that the series converges totally on every compact interval of $\mathbb{R}$. We now define the following function on $B_1(0)$

$$v(\varrho, \vartheta) = \frac{2}{\pi} + \frac{4}{\pi} \sum_{n=1}^{\infty} \frac{1}{1-4n^2} \varrho^{2n} \cos(2n\vartheta), \qquad \text{for } (\varrho, \vartheta) \in [0, 1) \times [0, 2\pi).$$

We notice that, in light of the previous discussion, this series converges uniformly even for $\varrho = 1$ and thus

$$v = U, \qquad \text{on } \partial B_1(0).$$

By Lemma A.3.2 in Appendix A, we know that v solves $(*)$ in classical sense. Moreover, by the same result we also have that $v \in W^{1,2}(B_1(0))$ and

$$\int_{B_1(0)} |\nabla v|^2 \, dx \, dy = 4 \sum_{n=1}^{\infty} \frac{2n}{(4n^2-1)^2} < +\infty.$$

We now claim that v is the unique minimizer of

$$\min_{u \in W^{1,2}(B_1(0))} \left\{ \frac{1}{2} \int_{B_1(0)} |\nabla u|^2 \, dx \, dy \, : \, u - U \in W_0^{1,2}(B_1(0)) \right\}.$$

Indeed, a minimizer w of such a problem exists (by Theorem 4.3.3) and is unique, by strict convexity of the functional. Moreover, it coincides with the (unique, see Problem 4.10.2) weak solution of $(*)$, i.e. it satisfies

$$w - U \in W_0^{1,2}(B_1(0)) \quad \text{and} \quad \int_{B_1(0)} \langle \nabla w, \nabla \varphi \rangle \, dx = 0,$$
$$\text{for every } \varphi \in C_0^\infty(B_1(0)).$$

On the other hand, since v is a harmonic function in classical sense, we have that it satisfies the previous weak formulation, as well (recall Remark 1.5.6). We wish to use the uniqueness of the solution to the boundary value problem in order to conclude that $w = v$: for this, we still need to know that $v - U \in W_0^{1,2}(B_1(0))$. By observing that

$$v - U \in \left\{ u \in C^0(\overline{B_1(0)}) \cap W^{1,2}(B_1(0)) \, : \, u = 0 \text{ on } \partial B_1(0) \right\},$$

the conclusion follows from Problem 3.12.16. Thus, the function v is the unique minimizer of the problem above.

Let us now argue by contradiction and suppose that the restriction of U on $\partial B_1(0)$ satisfies the BSC. By Problem 6.8.10, we know that

$$\min_{u\in W^{1,2}(B_1(0))}\left\{\frac{1}{2}\int_{B_1(0)}|\nabla u|^2\,dx\,dy\,:\,u-U\in W_0^{1,2}(B_1(0))\right\}$$
$$=\min_{u\in C^{0,1}(\overline{B}_1(0))}\left\{\frac{1}{2}\int_{B_1(0)}|\nabla u|^2\,dx\,dy\,:\,u=U\text{ on }B_1(0)\right\},$$

and that the minimizers of the two problems coincide. Observe that the problem on the right-hand side admits a solution by Theorem 6.5.1, because we are supposing that U satisfies the BSC.

Since v is a minimizer of the first problem, we get in particular that $v \in C^{0,1}(\overline{B_1(0)})$. Thus, in order to get a contradiction and conclude, it is "only" left to prove that v *is not* Lipschitz continuous on $\overline{B_1(0)}$. This would give the desired contradiction and conclude this part of the exercise (at last!).

To this aim, still by using polar coordinates, it is sufficient to show that

$$\lim_{\varrho\to 1^-}\frac{|v\,(\varrho,0)-v\,(1,0)|}{1-\varrho}=+\infty. \tag{**}$$

For $0<\varrho<1$, by using the definition of v we have

$$\begin{aligned}\frac{|v\,(\varrho,0)-v\,(1,0)|}{1-\varrho}&=\frac{1}{1-\varrho}\left|\frac{2}{\pi}+\frac{4}{\pi}\sum_{n=1}^{\infty}\frac{1}{1-4n^2}\varrho^{2n}-\frac{2}{\pi}-\frac{4}{\pi}\sum_{n=1}^{\infty}\frac{1}{1-4n^2}\right|\\&=\frac{4}{\pi}\frac{1}{1-\varrho}\left|\sum_{n=1}^{\infty}\frac{1}{1-4n^2}(\varrho^{2n}-1)\right|\\&=\frac{4}{\pi}\frac{1}{1-\varrho}\left|\sum_{n=1}^{\infty}\frac{1}{4n^2-1}(1-\varrho^{2n})\right|=\frac{4}{\pi}\sum_{n=1}^{\infty}\frac{1}{4n^2-1}\frac{1-\varrho^{2n}}{1-\varrho}.\end{aligned}$$

By the Fundamental Theorem of Calculus and Jensen's inequality, we have (notice that $2n-1\geq 1$ for every $n\geq 1$)

$$\begin{aligned}\frac{1-\varrho^{2n}}{1-\varrho}=(2n-1)\fint_{\varrho}^{1}t^{2n-1}\,dt&\geq(2n-1)\left(\fint_{\varrho}^{1}t\,dt\right)^{2n-1}\\&=(2n-1)\left(\frac{1-\varrho^2}{2}\frac{1}{1-\varrho}\right)^{2n-1}\\&=(2n-1)\left(\frac{1+\varrho}{2}\right)^{2n-1}.\end{aligned}$$

In light of this estimate, we thus get

$$
\begin{aligned}
\frac{|v(\varrho,0)-v(1,0)|}{1-\varrho} &\geq \frac{4}{\pi}\sum_{n=1}^{\infty}\frac{1}{2n+1}\left(\frac{1+\varrho}{2}\right)^{2n-1} \\
&= \frac{4}{\pi}\sum_{n=1}^{\infty}\frac{1}{2n+1}\left(\frac{1+\varrho}{2}\right)^{2n+2}\left(\frac{1+\varrho}{2}\right)^{-3} \\
&\geq \frac{2}{\pi}\left(\frac{2}{1+\varrho}\right)^{3}\sum_{n=1}^{\infty}\frac{1}{n+1}\left(\left(\frac{1+\varrho}{2}\right)^{2}\right)^{n+1}.
\end{aligned}
$$

The last series can be computed, by integrating a geometric series: indeed, for every $0 < t < 1$ we have

$$
\sum_{n=1}^{\infty}\frac{t^{n+1}}{n+1}=\sum_{n=0}^{\infty}\frac{t^{n+1}}{n+1}-t=\int_0^t\sum_{n=0}^{\infty}\tau^n\,d\tau-t=\int_0^t\frac{1}{1-\tau}\,d\tau-t=-\log(1-t)-t.
$$

By using this identity with $t=(1+\varrho)^2/4$, we finally get[14]

$$
\begin{aligned}
&\lim_{\varrho\to 1^-}\sum_{n=1}^{\infty}\frac{1}{n+1}\left(\left(\frac{1+\varrho}{2}\right)^{2}\right)^{n+1} \\
&= \lim_{\varrho\to 1^-}\left[-\log\left(1-\left(\frac{1+\varrho}{2}\right)^{2}\right)-\left(\frac{1+\varrho}{2}\right)^{2}\right]=+\infty.
\end{aligned}
$$

This proves the claimed property (∗∗). We point out that quite a similar example is contained in [71, Example 1.5].

Finally, we observe that in light of Theorem 6.5.1 and of the previous point, the restriction of U to the boundary can not satisfy the BSC. The exercise is over.

8.7 Problems of Chap. 7

7.8.1 *Hints:* start as in the case $N \geq 3$ and, once arrived at (7.3.4), multiply both sides of this estimate by the term

$$
\int_\Omega |F_{\beta,M}(u)\,\varphi|^2\,dx.
$$

[14] A subtle detail: the computation shows that the solution v fails to have a bounded gradient, just by a logarithmic factor. This is a general fact for harmonic functions on smooth sets, having a Lipschitz continuous boundary datum: the interested reader could have a look at [79, Theorem 2], for this fact.

Then, in place of Sobolev's inequality, use Ladyzhenskaya's inequality (i.e. Theorem 3.8.1 for $p = N = 2$) with q any exponent larger than 2.

7.8.2 *Hints:* try to mimick the proof of Theorem 7.6.1. In place of (7.6.3), one should now have

$$\int_\Omega \langle \nabla(\mathrm{T}_h u) - \nabla u, \nabla\varphi\rangle\, dx = \int_\Omega (\mathrm{T}_h f - f)\,\varphi\, dx,$$
$$\text{for every } \varphi \in C_0^\infty(B_R(x_0)),\quad |h| \le h_0/4.$$

By using the same test function as for the case $f \equiv 0$, one obtains

$$\int_\Omega |\nabla(\mathrm{T}_h u) - \nabla u|^2\,\eta^2\, dx \le 2\int_\Omega |\nabla(\mathrm{T}_h u) - \nabla u|\,|\nabla\eta|\,|\eta|\,|\mathrm{T}_h u - u|\, dx + \int_\Omega |\mathrm{T}_h f - f|\,|\mathrm{T}_h u - u|\,\eta^2\, dx,$$

in place of (7.6.4). Use Hölder's inequality and Proposition 3.11.1 to estimate the new term.

7.8.3 *Hints:* first observe that $\nabla v \in W^{1,2}_{\mathrm{loc}}(\Omega;\mathbb{R}^N)$, by using Problem 7.8.2 with $f \equiv 1$. Then observe that each component of ∇v is locally weakly harmonic, still by Problem 7.8.2. Apply Theorem 7.7.1 to each component.

7.8.4 *Hints:* observe that one can apply the result of Problem 7.8.2 with $f = \lambda_1(\Omega)\,u$. Thus, $\nabla u \in W^{1,2}_{\mathrm{loc}}(\Omega;\mathbb{R}^N)$ and each component of the gradient is a local weak solution of

$$-\Delta\frac{\partial u}{\partial x_i} = \lambda_1(\Omega)\,\frac{\partial u}{\partial x_i},\qquad \text{in } \Omega,$$

i.e. it still solves the initial equation. Iterate the argument to prove that u is weakly differentiable infinitely many times. Use recursively Sobolev embeddings and Problem 3.12.27 to deduce that u is smooth.

7.8.5 *Hints:* first obtain a Caccioppoli inequality, by reproducing the proof of Theorem 7.2.1 and Proposition 7.2.3. This time, start from the equation

$$\int_\Omega \langle A\cdot\nabla u, \nabla\varphi\rangle\, dx \le 0,\qquad \text{for every } \varphi \in C_0^\infty(\Omega),\ \varphi \ge 0.$$

Then try to repeat the proof of Theorem 7.3.1. The following two facts should be useful:

$$\lambda\,|\xi|^2 \le \langle A\cdot\xi, \xi\rangle \le \Lambda\,|\xi|^2,\qquad \text{for every } \xi \in \mathbb{R}^N,$$

and

$$|\langle A\cdot\xi,\zeta\rangle| \le \sqrt{\langle A\cdot\xi,\xi\rangle}\,\sqrt{\langle A\cdot\zeta,\zeta\rangle}, \qquad \text{for every } \xi,\zeta\in\mathbb{R}^N.$$

The second one is just a generalization of the Cauchy-Schwarz inequality. One should obtain estimates like (7.2.1) and (7.3.1), with constants C now depending on the ratio Λ/λ, as well.

Finally, the result for a weak solution can be obtained as in the proof of Corollary 7.3.3.

7.8.6 *Hints:* in order to generalize the proof of Theorem 7.5.1, it is sufficient to use Problem 7.8.5 and then try to reproduce the proof of the Harnack-type estimate of Proposition 7.4.2.

7.8.7 *Hints:* try to mimick the proof of Theorem 7.2.1. In order to prove the first fact, it could be useful to use the test function

$$\varphi=\eta\,|F'(u)|^{p-2}\,F'(u),$$

thanks to the fact that

$$\langle|\nabla u|^{p-2}\,\nabla u,\nabla\eta\rangle\,|F'(u)|^{p-2}\,F'(u)=\langle|\nabla(F\circ u)|^{p-2}\,\nabla(F\circ u),\nabla\eta\rangle.$$

However, carefully discuss the admissibility of this test function, by paying particular attention to the case $1<p<2$ (in this case the above choice should be a bit regularized).

In order to prove the second fact, it would be sufficient to use the test function

$$\varphi=|\eta|^p\,F'(u),$$

in the equation for $F(u)$.

7.8.8 *Hints:* proceed as in the solution of the previous problem, by paying attention to the fact that now u only satisfies

$$\int_\Omega \langle|\nabla u|^{p-2}\,\nabla u,\nabla\varphi\rangle\,dx\le 0,$$

and that test functions must be non-negative (here the restriction $F'\ge 0$ will be important).

7.8.9 *Hints:* distinguish three cases, $1<p<N$, $p=N$ and $p>N$. For $1<p<N$ and $p=N$, proceed as for the case $p=2$, by making the necessary adjustments.

For $p>N$, we can rely on the second Morrey inequality (3.6.12). Indeed, it is sufficient to observe that $u\,\eta\in W^{1,p}_0(\Omega)$, for every $\eta\in C^\infty_0(\Omega)$ (see Problem 3.12.18). In particular, by taking

$$R_1=r_0+\frac{R_0-r_0}{2}=\frac{R_0+r_0}{2},$$

and $\eta \in C_0^\infty(B_{R_1}(x_0))$ such that

$$\eta \equiv 1 \text{ in } B_{r_0}(x_0), \qquad 0 \le \eta \le 1, \qquad |\nabla \eta| \le \frac{C}{R_1 - r_0} = \frac{2\,C}{R_0 - r_0},$$

and using (3.6.12) of Theorem 3.6.8, one would get

$$\begin{aligned}\|u\|_{L^\infty(B_{r_0}(x_0))} &\le \|u\,\eta\|_{L^\infty(B_{R_1}(x_0))}\\ &\le C\left(\int_{B_{R_1}(x_0)} |\nabla(u\,\eta)|^p\,dx\right)^{\frac{N}{p^2}} \left(\int_{B_{R_1}(x_0)} |u\,\eta|^p\,dx\right)^{\frac{p-N}{p^2}},\end{aligned}$$

for a constant $C = C(N,p) > 0$. By using the properties of η, one obtains

$$\int_{B_{R_1}(x_0)} |u\,\eta|^p\,dx \le \int_{B_{R_0}(x_0)} |u|^p\,dx,$$

and

$$\begin{aligned}\left(\int_{B_{R_1}(x_0)} |\nabla(u\,\eta)|^p\,dx\right)^{\frac{1}{p}} &\le \frac{2\,C}{R_0 - r_0}\left(\int_{B_{R_0}(x_0)} |u|^p\,dx\right)^{\frac{1}{p}}\\ &\quad + \left(\int_{B_{R_1}(x_0)} |\nabla u|^p\,|\eta|^p\,dx\right)^{\frac{1}{p}}.\end{aligned}$$

In order to estimate the last term, one can now rely on the Caccioppoli inequality (7.8.2), with $F(u) = u$ and $\varphi = \eta$. By keeping everything together, one obtains the claimed estimate.

7.8.10 *Hints:* as before, distinguish three cases. The case $p > N$ is a direct consequence of the first Morrey inequality, i.e. (3.6.11). For the cases $1 < p < N$ and $p = N$, proceed as in Sect. 7.5. We also point out that for $p = N$ a shorter proof is possible, see for example [47, Chapter 3, Section 2].

7.8.11 *Hints:* we observe at first that this result considerably improves that of Problem 4.10.19. The existence and uniqueness of the solution can be inferred by using the results of Chap. 4.

For proving that $v \in L^\infty(\Omega)$, the idea is to perform a Moser iteration, as in the proof of Theorem 7.3.1. Observe that we now want a *global* L^∞ estimate, thus it is not necessary to use cut-off functions to localize the estimates on balls. Thus, the idea would be to use the weak formulation

$$\int_\Omega \langle |\nabla v|^{p-2}\,\nabla v, \nabla\varphi\rangle\,dx = \int_\Omega f\,\varphi\,dx,$$

with $\varphi = |v|^{\beta-1}\, v$, for $\beta \geq 1$. After some simple algebraic manipulations, this would give

$$\beta\left(\frac{p}{\beta+p-1}\right)^p \int_\Omega \left|\nabla\left(|v|^{\frac{\beta+p-1}{p}}\right)\right|^p dx = \int_\Omega f\, |v|^{\beta-1}\, v\, dx.$$

By using Hölder's inequality on the right-hand side and Sobolev's inequality on the left-hand side (we consider for simplicity the case $1 < p < N$), one would get

$$\left(\int_\Omega \left(|v|^{\frac{\beta+p-1}{p}}\right)^{p^*} dx\right)^{\frac{p}{p^*}} \leq S^p\, \frac{1}{\beta}\left(\frac{\beta+p-1}{p}\right)^p \|f\|_{L^q(\Omega)} \left(\int_\Omega |v|^{\beta\, q'}\, dx\right)^{\frac{1}{q'}}.$$

Thanks to the choice $q > N/p$, we have

$$q' < \frac{N}{N-p} = \frac{p^*}{p},$$

which in turn entails that

$$(\beta+p-1)\,\frac{p^*}{p} > \beta\, q'.$$

Thus, the previous estimate gives an iterative scheme of reverse Hölder's inequalities. Moreover, if we set

$$\beta_0 = 1, \qquad \beta_{i+1} = (\beta_i + p - 1)\,\frac{p^*}{p\, q'}, \quad \text{for } i \in \mathbb{N} \setminus \{0\},$$

it is not difficult to see that this is a diverging sequence. Thus, this scheme should lead to the desired result after infinitely many iterations (*memento*: in the iteration process, pay attention to the constants!).

However, the proof above is not completely rigorous: observe that $\varphi = |v|^{\beta-1}\, v$ for $\beta > 1$ is not a feasible test function, because in general it does not even belong to $W^{1,p}(\Omega)$. In order to fix this issue, we can use a suitable modification of power functions, similarly as in the proof of Theorem 7.3.1. For example, we can define

$$f_{\beta,M}(t) = \begin{cases} \beta\, |t|^{\beta-1}, & \text{if } |t| \leq M, \\ \beta\, M^{\beta-1}, & \text{if } |t| > M, \end{cases}$$

and its primitive

$$F_{\beta,M}(t) = \int_0^t f_{\beta,M}(\tau)\, d\tau, \qquad t \in \mathbb{R},$$

for every $M > 0$. Then we can consider

$$\varphi = F_{\beta,M} \circ v.$$

Justify that this function belongs to $W_0^{1,p}(\Omega)$, for every $\beta \geq 1$ and $M > 0$. It will be useful to observe that

$$\begin{aligned}
\langle |\nabla v|^{p-2}\,\nabla v, \nabla(F_{\beta,M} \circ v)\rangle &= f_{\beta,M}(u)\,|\nabla u|^p \\
&= \beta \left(\frac{p}{\beta + p - 1}\right)^p \left(f_{\frac{\beta+p-1}{p},M}(u)\right)^p |\nabla u|^p \\
&= \beta \left(\frac{p}{\beta + p - 1}\right)^p \left|F'_{\frac{\beta+p-1}{p},M}(u)\right|^p |\nabla u|^p \\
&= \beta \left(\frac{p}{\beta + p - 1}\right)^p \left|\nabla\left(F_{\frac{\beta+p-1}{p},M} \circ u\right)\right|^p .
\end{aligned}$$

7.8.12 *Hints:* proceed as in the proof of Theorem 7.6.1, by using Nirenberg's difference quotient technique. One will need the following two pointwise inequalities:

$$\langle \nabla F_\alpha(z) - \nabla F_\alpha(w), z - w\rangle \geq \alpha^{\frac{p-2}{2}}\,|z - w|^2,$$

and

$$|\nabla F_\alpha(z) - \nabla F_\alpha(w)| \leq C_p \left((\alpha + |z|^2)^{\frac{p-2}{2}} + (\alpha + |w|^2)^{\frac{p-2}{2}}\right) |z - w|,$$

which hold for every $z, w \in \mathbb{R}^N$.

The first one follows from Proposition 1.6.2 and Remark 1.6.3. For the second one, we can proceed as follows: from the Fundamental Theorem of Calculus, we can write

$$\begin{aligned}
|\nabla F_\alpha(z) - \nabla F_\alpha(w)| &= \left|\int_0^1 \frac{d}{dt}\nabla F_\alpha(t\,z + (1-t)\,w)\,dt\right| \\
&= \left|\int_0^1 D^2 F_\alpha(t\,z + (1-t)\,w)\cdot(z - w)\,dt\right| \\
&\leq \int_0^1 |D^2 F_\alpha(t\,z + (1-t)\,w)\cdot(z - w)|\,dt.
\end{aligned}$$

By using the Cauchy-Schwarz inequality, we have

$$|D^2 F_\alpha(t\,z + (1-t)\,w)\cdot(z - w)| \leq \left(\sum_{i,j=1}^N \left|\frac{\partial^2 F_\alpha}{\partial z_i\,\partial z_j}(t\,z + (1-t)\,w)\right|^2\right)^{\frac{1}{2}} |z - w|.$$

Moreover, from Remark 1.6.3 we know that

$$D^2 F_\alpha(z) = (\alpha + |z|^2)^{\frac{p-2}{2}} \,\mathrm{Id}_N + (p-2)\, z \otimes z\, (\alpha + |z|^2)^{\frac{p-4}{2}},$$

which implies that

$$\left| \frac{\partial^2 F_\alpha}{\partial z_i\, \partial z_j}(t\, z + (1-t)\, w) \right| \leq (p-1)\, (\alpha + |t\, z + (1-t)\, w|^2)^{\frac{p-2}{2}}$$
$$\leq (p-1)\, (\alpha + t\, |z|^2 + (1-t)\, |w|^2)^{\frac{p-2}{2}}.$$

We thus obtain

$$|\nabla F_\alpha(z) - \nabla F_\alpha(w)| \leq (p-1) \left(\int_0^1 (\alpha + t\, |z|^2 + (1-t)\, |w|^2)^{\frac{p-2}{2}}\, dt \right) |z - w|.$$

Finally, we observe that:

- if $p \geq 4$, the function $\tau \mapsto \tau^{(p-2)/2}$ is convex and thus we get

$$(\alpha + t\, |z|^2 + (1-t)\, |w|^2)^{\frac{p-2}{2}} = \Big(t\, (\alpha + |z|^2) + (1-t)\, (\alpha + |w|^2) \Big)^{\frac{p-2}{2}}$$
$$\leq t\, (\alpha + |z|^2)^{\frac{p-2}{2}} + (1-t)\, (\alpha + |w|^2)^{\frac{p-2}{2}}.$$

  Accordingly, we obtain

$$|\nabla F_\alpha(z) - \nabla F_\alpha(w)| \leq (p-1) \left((\alpha + |z|^2)^{\frac{p-2}{2}} + (\alpha + |w|^2)^{\frac{p-2}{2}} \right) |z - w|;$$

- if $2 \leq p < 4$, from Problem 1.7.1 we get

$$(\alpha + t\, |z|^2 + (1-t)\, |w|^2)^{\frac{p-2}{2}} = \Big(t\, (\alpha + |z|^2) + (1-t)\, (\alpha + |w|^2) \Big)^{\frac{p-2}{2}}$$
$$\leq t^{\frac{p-2}{2}}\, (\alpha + |z|^2)^{\frac{p-2}{2}} + (1-t)^{\frac{p-2}{2}}\, (\alpha + |w|^2)^{\frac{p-2}{2}},$$

  and thus

$$|\nabla F_\alpha(z) - \nabla F_\alpha(w)| \leq \frac{2}{p}\, (p-1) \left((\alpha + |z|^2)^{\frac{p-2}{2}} + (\alpha + |w|^2)^{\frac{p-2}{2}} \right) |z - w|.$$

The result of this problem holds also for $1 < p < 2$: more precisely, we have $\nabla u \in W^{1,p}_{\mathrm{loc}}(\Omega)$. In this case the proof (still based on Nirenberg's technique) is slightly more complicated, see for example [6, Theorems 8.1].

Appendix A
Harmonic Functions in a Disk

We suppose to work in a disk D of radius $R > 0$, centered for simplicity at the origin $(0, 0)$. In this case, we can introduce the polar coordinates

$$x = \varrho \cos\vartheta, \qquad y = \varrho \sin\vartheta, \qquad 0 \le \varrho < R,\ 0 \le \vartheta \le 2\pi.$$

Thus, given a function $u : D \to \mathbb{R}$ of class $C^2(D)$, we want to recall the expression of its gradient and Laplacian in terms of the new coordinates ϱ and ϑ.

A.1 The Gradient in Polar Coordinates

We first use the chain rule for the function $u(x, y) = u(\varrho \cos\vartheta, \varrho \sin\vartheta)$, so to get

$$\frac{\partial u}{\partial \varrho} = \cos\vartheta\, \frac{\partial u}{\partial x} + \sin\vartheta\, \frac{\partial u}{\partial y}, \tag{A.1.1}$$

and

$$\frac{\partial u}{\partial \vartheta} = -\varrho \sin\vartheta\, \frac{\partial u}{\partial x} + \varrho \cos\vartheta\, \frac{\partial u}{\partial y}. \tag{A.1.2}$$

We now wish to invert these relations and write

$$\frac{\partial u}{\partial x} \text{ and } \frac{\partial u}{\partial y} \qquad \text{in terms of} \qquad \frac{\partial u}{\partial \varrho} \text{ and } \frac{\partial u}{\partial \vartheta}.$$

L. Brasco, *Handbook of Calculus of Variations for Absolute Beginners*,
La Matematica per il 3+2 163, https://doi.org/10.1007/978-3-031-87164-1

To this aim, for $0 < \varrho < R$ we multiply Eq. (A.1.1) by $\varrho \sin\vartheta$, multiply Eq. (A.1.2) by $\cos\vartheta$ and sum the two relevant equations. We get

$$\varrho \sin\vartheta \frac{\partial u}{\partial \varrho} + \cos\vartheta \frac{\partial u}{\partial \vartheta} = \cancel{\varrho \sin\vartheta \cos\vartheta \frac{\partial u}{\partial x}} + \varrho \sin^2\vartheta \frac{\partial u}{\partial y}$$
$$- \cancel{\varrho \cos\vartheta \sin\vartheta \frac{\partial u}{\partial x}} + \varrho \cos^2\vartheta \frac{\partial u}{\partial y}.$$

By recalling that $\cos^2\vartheta + \sin^2\vartheta = 1$, we obtain

$$\boxed{\frac{\partial u}{\partial y} = \sin\vartheta \frac{\partial u}{\partial \varrho} + \frac{\cos\vartheta}{\varrho} \frac{\partial u}{\partial \vartheta}}, \tag{A.1.3}$$

which is valid for $0 < \varrho < R$. In order to find $\partial u/\partial x$, we argue in a similar fashion: we multiply Eq. (A.1.1) by $\varrho \cos\vartheta$, multiply Eq. (A.1.2) by $-\sin\vartheta$ and the take the sum of the two resulting equations. We get

$$\varrho \cos\vartheta \frac{\partial u}{\partial \varrho} - \sin\vartheta \frac{\partial u}{\partial \vartheta} = \varrho \cos^2\vartheta \frac{\partial u}{\partial x} + \cancel{\varrho \sin\vartheta \cos\vartheta \frac{\partial u}{\partial y}}$$
$$+ \varrho \sin^2\vartheta \frac{\partial u}{\partial x} - \cancel{\varrho \cos\vartheta \sin\vartheta \frac{\partial u}{\partial y}}.$$

We use again that $\cos^2\vartheta + \sin^2\vartheta = 1$. We now get

$$\boxed{\frac{\partial u}{\partial x} = \cos\vartheta \frac{\partial u}{\partial \varrho} - \frac{\sin\vartheta}{\varrho} \frac{\partial u}{\partial \vartheta}}. \tag{A.1.4}$$

We also observe that (A.1.3) and (A.1.4) give

$$|\nabla u|^2 = \left(\frac{\partial u}{\partial x}\right)^2 + \left(\frac{\partial u}{\partial y}\right)^2 = \left(\frac{\partial u}{\partial \varrho}\right)^2 + \frac{1}{\varrho^2}\left(\frac{\partial u}{\partial \vartheta}\right)^2. \tag{A.1.5}$$

A.2 The Laplacian in Polar Coordinates

Let us now proceed to get the expression of the Laplacian

$$\Delta u = \frac{\partial^2 u}{\partial x^2} + \frac{\partial^2 u}{\partial y^2}.$$

By observing that

$$\frac{\partial^2 u}{\partial x^2} = \frac{\partial}{\partial x}\,\frac{\partial u}{\partial x} \qquad \text{and} \qquad \frac{\partial^2 u}{\partial y^2} = \frac{\partial}{\partial y}\,\frac{\partial u}{\partial y},$$

we need to iterate (A.1.3) and (A.1.4). Thus, we get

$$\begin{aligned}
\frac{\partial^2 u}{\partial x^2} = \frac{\partial}{\partial x}\,\frac{\partial u}{\partial x} &= \cos\vartheta\,\frac{\partial}{\partial\varrho}\left(\cos\vartheta\,\frac{\partial u}{\partial\varrho} - \frac{\sin\vartheta}{\varrho}\,\frac{\partial u}{\partial\vartheta}\right)\\
&\quad - \frac{\sin\vartheta}{\varrho}\,\frac{\partial}{\partial\vartheta}\left(\cos\vartheta\,\frac{\partial u}{\partial\varrho} - \frac{\sin\vartheta}{\varrho}\,\frac{\partial u}{\partial\vartheta}\right)\\
&= \cos^2\vartheta\,\frac{\partial^2 u}{\partial\varrho^2} - \cos\vartheta\,\sin\vartheta\,\frac{\partial}{\partial\varrho}\left(\frac{1}{\varrho}\,\frac{\partial u}{\partial\vartheta}\right)\\
&\quad - \frac{\sin\vartheta}{\varrho}\,\frac{\partial}{\partial\vartheta}\left(\cos\vartheta\,\frac{\partial u}{\partial\varrho}\right) + \frac{\sin\vartheta}{\varrho^2}\,\frac{\partial}{\partial\vartheta}\left(\sin\vartheta\,\frac{\partial u}{\partial\vartheta}\right)
\end{aligned}$$

and

$$\begin{aligned}
\frac{\partial^2 u}{\partial y^2} = \frac{\partial}{\partial y}\,\frac{\partial u}{\partial y} &= \sin\vartheta\,\frac{\partial}{\partial\varrho}\left(\sin\vartheta\,\frac{\partial u}{\partial\varrho} + \frac{\cos\vartheta}{\varrho}\,\frac{\partial u}{\partial\vartheta}\right)\\
&\quad + \frac{\cos\vartheta}{\varrho}\,\frac{\partial}{\partial\vartheta}\left(\sin\vartheta\,\frac{\partial u}{\partial\varrho} + \frac{\cos\vartheta}{\varrho}\,\frac{\partial u}{\partial\vartheta}\right)\\
&= \sin^2\vartheta\,\frac{\partial^2 u}{\partial\varrho^2} + \cos\vartheta\,\sin\vartheta\,\frac{\partial}{\partial\varrho}\left(\frac{1}{\varrho}\,\frac{\partial u}{\partial\vartheta}\right)\\
&\quad + \frac{\cos\vartheta}{\varrho}\,\frac{\partial}{\partial\vartheta}\left(\sin\vartheta\,\frac{\partial u}{\partial\varrho}\right) + \frac{\cos\vartheta}{\varrho^2}\,\frac{\partial}{\partial\vartheta}\left(\cos\vartheta\,\frac{\partial u}{\partial\vartheta}\right)
\end{aligned}$$

When we sum up the last two quantities, we get

$$\begin{aligned}
\Delta u = \frac{\partial^2 u}{\partial x^2} + \frac{\partial^2 u}{\partial y^2} &= \cos^2\vartheta\,\frac{\partial^2 u}{\partial\varrho^2} \cancel{- \cos\vartheta\,\sin\vartheta\,\frac{\partial}{\partial\varrho}\left(\frac{1}{\varrho}\,\frac{\partial u}{\partial\vartheta}\right)}\\
&\quad - \frac{\sin\vartheta}{\varrho}\,\frac{\partial}{\partial\vartheta}\left(\cos\vartheta\,\frac{\partial u}{\partial\varrho}\right) + \frac{\sin\vartheta}{\varrho^2}\,\frac{\partial}{\partial\vartheta}\left(\sin\vartheta\,\frac{\partial u}{\partial\vartheta}\right)\\
&\quad + \sin^2\vartheta\,\frac{\partial^2 u}{\partial\varrho^2} \cancel{+ \cos\vartheta\,\sin\vartheta\,\frac{\partial}{\partial\varrho}\left(\frac{1}{\varrho}\,\frac{\partial u}{\partial\vartheta}\right)}\\
&\quad + \frac{\cos\vartheta}{\varrho}\,\frac{\partial}{\partial\vartheta}\left(\sin\vartheta\,\frac{\partial u}{\partial\varrho}\right) + \frac{\cos\vartheta}{\varrho^2}\,\frac{\partial}{\partial\vartheta}\left(\cos\vartheta\,\frac{\partial u}{\partial\vartheta}\right),
\end{aligned}$$

that is, by using the fundamental trigonometric identity,

$$
\begin{aligned}
\Delta u = &= \frac{\partial^2 u}{\partial \varrho^2} \\
&- \frac{\sin\vartheta}{\varrho}\frac{\partial}{\partial\vartheta}\left(\cos\vartheta\,\frac{\partial u}{\partial\varrho}\right) + \frac{\sin\vartheta}{\varrho^2}\frac{\partial}{\partial\vartheta}\left(\sin\vartheta\,\frac{\partial u}{\partial\vartheta}\right) \\
&+ \frac{\cos\vartheta}{\varrho}\frac{\partial}{\partial\vartheta}\left(\sin\vartheta\,\frac{\partial u}{\partial\varrho}\right) + \frac{\cos\vartheta}{\varrho^2}\frac{\partial}{\partial\vartheta}\left(\cos\vartheta\,\frac{\partial u}{\partial\vartheta}\right),
\end{aligned}
$$

We are now left to compute the last derivatives: this yields

$$
\begin{aligned}
\Delta u = &\frac{\partial^2 u}{\partial \varrho^2} \\
&+ \frac{\sin^2\vartheta}{\varrho}\frac{\partial u}{\partial\varrho} - \cancel{\frac{\sin\vartheta\,\cos\vartheta}{\varrho}\frac{\partial^2 u}{\partial\varrho\,\partial\vartheta}} \\
&+ \cancel{\frac{\sin\vartheta\,\cos\vartheta}{\varrho^2}\frac{\partial u}{\partial\vartheta}} + \frac{\sin^2\vartheta}{\varrho^2}\frac{\partial^2 u}{\partial\vartheta^2} \\
&+ \frac{\cos^2\vartheta}{\varrho}\frac{\partial u}{\partial\varrho} + \cancel{\frac{\cos\vartheta\,\sin\vartheta}{\varrho}\frac{\partial^2 u}{\partial\varrho\,\partial\vartheta}} \\
&- \cancel{\frac{\cos\vartheta\,\sin\vartheta}{\varrho^2}\frac{\partial u}{\partial\vartheta}} + \frac{\cos^2\vartheta}{\varrho^2}\frac{\partial^2 u}{\partial\vartheta^2} \\
= &\frac{\partial^2 u}{\partial \varrho^2} + \frac{\cos^2\vartheta+\sin^2\vartheta}{\varrho}\frac{\partial u}{\partial\varrho} + \frac{\cos^2\vartheta+\sin^2\vartheta}{\varrho^2}\frac{\partial^2 u}{\partial\vartheta^2}.
\end{aligned}
$$

In conclusion, we get

$$
\boxed{\Delta u = \frac{\partial^2 u}{\partial\varrho^2} + \frac{1}{\varrho}\frac{\partial u}{\partial\varrho} + \frac{1}{\varrho^2}\frac{\partial^2 u}{\partial\vartheta^2}}. \tag{A.2.1}
$$

A.3 Circular Harmonics in the Plane

Let $n \in \mathbb{N}\setminus\{0\}$, we consider the functions defined in polar coordinates by

$$
u_n(\varrho,\vartheta) = \varrho^n\,\cos(n\,\vartheta) \qquad \text{and} \qquad v_n(\varrho,\vartheta) = \varrho^n\,\sin(n\,\vartheta).
$$

By using formula (A.2.1), it is easy to see that these are harmonic functions in $\mathbb{R}^2$. Indeed, we have

$$\begin{aligned}\Delta u_n &= \frac{\partial^2}{\partial \varrho^2}(\varrho^n \cos(n\,\vartheta)) + \frac{1}{\varrho}\,\frac{\partial}{\partial \varrho}(\varrho^n \cos(n\,\vartheta)) + \frac{1}{\varrho^2}\,\frac{\partial^2}{\partial \vartheta^2}(\varrho^n \cos(n\,\vartheta)) \\ &= n\,(n-1)\,\varrho^{n-2}\cos(n\,\vartheta) + n\,\varrho^{n-2}\cos(n\,\vartheta) - n^2\,\varrho^{n-2}\cos(n\,\vartheta) \\ &= (n^2 - n + n - n^2)\,\varrho^{n-2}\cos(n\,\vartheta) = 0.\end{aligned}$$

Similar computations apply to the case of v_n. We recall that by using polar coordinates in the complex plane $\mathbb{C}$, i.e.

$$z = \varrho\,(\cos\vartheta + \mathrm{i}\,\sin\vartheta),$$

then we know that

$$z^n = \varrho^n\,(\cos(n\,\vartheta) + \mathrm{i}\,\sin(n\,\vartheta)).$$

Thus, we can write the functions above as

$$u_n = \mathrm{Re}(z^n) \qquad \text{and} \qquad v_n = \mathrm{Im}(z^n),$$

where Re and Im denote the *real* and *imaginary parts* of a complex number, respectively.

This shows that (u_n, v_n) is a conjugate pair, for every $n \in \mathbb{N} \setminus \{0\}$, i.e. they satisfy the Cauchy-Riemann equations (see for example [39] for more details). We also observe that the relation above between u_n (or v_n) and z^n, permits to find u_n and v_n as functions of the standard cartesian variables (x, y). Indeed, we have

$$u_n(x, y) = \mathrm{Re}(z^n) = \mathrm{Re}((x + \mathrm{i}\,y)^n) = \mathrm{Re}\left(\sum_{k=0}^{n}\binom{n}{k}x^k\,(\mathrm{i}\,y)^{n-k}\right), \tag{A.3.1}$$

and

$$v_n(x, y) = \mathrm{Im}(z^n) = \mathrm{Im}((x + \mathrm{i}\,y)^n) = \mathrm{Im}\left(\sum_{k=0}^{n}\binom{n}{k}x^k\,(\mathrm{i}\,y)^{n-k}\right), \tag{A.3.2}$$

according to Newton's formula. Let us compute explicitly some of them:

- for $n = 1$ we have

$$u_1(x, y) = \mathrm{Re}(\mathrm{i}\,y + x) = x,$$

and

$$v_1(x, y) = \mathrm{Im}(\mathrm{i}\,y + x) = y;$$

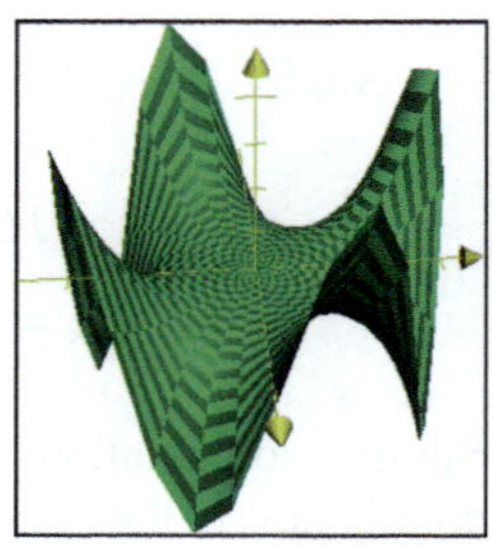

Fig. A.1 The graph of the circular harmonic $v_4(\varrho, \vartheta) = \varrho^4 \sin(4\vartheta)$

- for $n = 2$ we have

$$u_2(x, y) = \mathrm{Re}((\mathrm{i}\,y)^2 + 2\,x\,(\mathrm{i}\,y) + x^2) = x^2 - y^2,$$

and

$$v_2(x, y) = \mathrm{Im}((\mathrm{i}\,y)^2 + 2\,x\,(\mathrm{i}\,y) + x^2) = 2\,x\,y;$$

- for $n = 3$ we have

$$u_3(x, y) = \mathrm{Re}((\mathrm{i}\,y)^3 + 3\,x\,(\mathrm{i}\,y)^2 + 3\,x^2\,(\mathrm{i}\,y) + x^3) = x^3 - 3\,x\,y^2,$$

and

$$v_3(x, y) = \mathrm{Im}((\mathrm{i}\,y)^3 + 3\,x\,(\mathrm{i}\,y)^2 + 3\,x^2\,(\mathrm{i}\,y) + x^3) = 3\,x^2\,y - y^3;$$

- for $n = 4$ we have

$$u_4(x, y) = \mathrm{Re}((\mathrm{i}\,y)^4 + 4\,x\,(\mathrm{i}\,y)^3 + 6\,x^2\,(\mathrm{i}\,y)^2 + 4\,x^3\,(\mathrm{i}\,y) + x^4) = y^4 - 6\,x^2\,y^2 + x^4,$$

and

$$v_4(x, y) = \mathrm{Im}((\mathrm{i}\,y)^4 + 4\,x\,(\mathrm{i}\,y)^3 + 6\,x^2\,(\mathrm{i}\,y)^2 + 4\,x^3\,(\mathrm{i}\,y) + x^4) = -4\,x\,y^3 + 4\,x^3\,y.$$

The functions (u_n, v_n) are called *circular harmonics of order n* (see Fig. A.1).

Remark A.3.1 In light of (A.3.1) and (A.3.2), we observe that for every $n \in \mathbb{N}$, the functions u_n and v_n are harmonic n-homogeneous polynomials, as functions of the cartesian coordinates x, y. It can be shown that every harmonic n-homogeneous polynomial can be written as a linear combination of u_n and v_n. We refer for example to [51] for more details on the subject.

We finish this appendix with a simple result, which is sometimes useful: it permits to construct harmonic functions in the disk, in the form of a series.

Lemma A.3.2 *Let $\{\alpha_n\}_{n\in\mathbb{N}}, \{\beta_n\}_{n\in\mathbb{N}} \subseteq \mathbb{R}$ be two sequences such that*

$$\sum_{n=0}^{\infty}(|\alpha_n| + |\beta_n|) < +\infty \qquad \text{and} \qquad \sum_{n=0}^{\infty} n\,(\alpha_n^2 + \beta_n^2) < +\infty.$$

Then, for the function defined on the disk $D = \{(x, y) \in \mathbb{R}^2 : x^2 + y^2 < 1\}$ by

$$u(\varrho, \vartheta) = \alpha_0 + \sum_{n=1}^{\infty} \alpha_n\, \varrho^n \cos(n\,\vartheta) + \sum_{n=1}^{\infty} \beta_n\, \varrho^n \sin(n\,\vartheta),$$
$$\text{for } (\varrho, \vartheta) \in [0, 1) \times [0, 2\pi),$$

we have:

1. *$u \in C^2(D) \cap C^0(\overline{D})$;*
2. *u is harmonic in D;*
3. *$\nabla u \in L^2(D; \mathbb{R}^2)$ and we have*

$$\int_D |\nabla u|^2\, dx\, dy = \pi \sum_{n=1}^{\infty} n\,(\alpha_n^2 + \beta_n^2) < +\infty.$$

Proof We prove each point separately.

Proof of 1 We first observe that the assumption on the two sequences implies that the two series

$$\sum_{n=1}^{\infty} \alpha_n\, \varrho^n \cos(n\,\vartheta) \qquad \text{and} \qquad \sum_{n=1}^{\infty} \beta_n\, \varrho^n \sin(n\,\vartheta),$$

are totally converging on $[0, 1] \times [0, 2\pi]$. Thus, the function u is well-defined and continuous on $\overline{D}$. We also observe that the two series obtained by differentiating them term by term, i.e.

$$\sum_{n=1}^{\infty} n\,\alpha_n\, \varrho^{n-1} \cos(n\,\vartheta) + \sum_{n=1}^{\infty} n\,\beta_n\, \varrho^{n-1} \sin(n\,\vartheta), \tag{A.3.3}$$

and

$$-\sum_{n=1}^{\infty} n\,\alpha_n\, \varrho^n \sin(n\,\vartheta) + \sum_{n=1}^{\infty} n\,\beta_n\, \varrho^n \cos(n\,\vartheta),$$

are totally converging for $(\varrho, \vartheta) \in [0, \delta] \times [0, 2\pi]$, for every $0 < \delta < 1$. Thus, the two partial derivatives

$$\frac{\partial u}{\partial \varrho} \qquad \text{and} \qquad \frac{\partial u}{\partial \vartheta},$$

exist and are continuous for $(\varrho, \vartheta) \in [0, \delta] \times [0, 2\pi]$, for every $0 < \delta < 1$. Moreover, they coincide with the differentiated series obtained above, i.e.

$$\frac{\partial u}{\partial \varrho} = \sum_{n=1}^{\infty} n\, \alpha_n\, \varrho^{n-1} \cos(n\,\vartheta) + \sum_{n=1}^{\infty} n\, \beta_n\, \varrho^{n-1} \sin(n\,\vartheta),$$

and

$$\frac{\partial u}{\partial \vartheta} = -\sum_{n=1}^{\infty} n\, \alpha_n\, \varrho^{n} \sin(n\,\vartheta) + \sum_{n=1}^{\infty} n\, \beta_n\, \varrho^{n} \cos(n\,\vartheta).$$

By recalling that for $\varrho > 0$ we have from (A.1.4) and (A.1.3)

$$\begin{aligned}
\frac{\partial u}{\partial x} &= \cos\vartheta\, \frac{\partial u}{\partial \varrho} - \frac{\sin\vartheta}{\varrho}\, \frac{\partial u}{\partial \vartheta} \\
&= \sum_{n=1}^{\infty} n\, (\alpha_n \cos\vartheta - \beta_n \sin\vartheta)\, \varrho^{n-1} \cos(n\,\vartheta) \\
&+ \sum_{n=1}^{\infty} n\, (\alpha_n \sin\vartheta + \beta_n \cos\vartheta)\, \varrho^{n-1} \sin(n\,\vartheta),
\end{aligned} \tag{A.3.4}$$

and

$$\begin{aligned}
\frac{\partial u}{\partial y} &= \sin\vartheta\, \frac{\partial u}{\partial \varrho} + \frac{\cos\vartheta}{\varrho}\, \frac{\partial u}{\partial \vartheta} \\
&= \sum_{n=1}^{\infty} n\, (\alpha_n \sin\vartheta + \beta_n \cos\vartheta)\, \varrho^{n-1} \cos(n\,\vartheta) \\
&+ \sum_{n=1}^{\infty} n\, (\beta_n \sin\vartheta - \alpha_n \cos\vartheta)\, \varrho^{n-1} \sin(n\,\vartheta),
\end{aligned} \tag{A.3.5}$$

we get that the function u is C^1 on $\{(x, y) \in \mathbb{R}^2 : 0 < x^2 + y^2 \leq \delta^2\}$, for every $\delta < 1$. By arbitrariness of $0 < \delta < 1$, we then get that $u \in C^1(D \setminus \{0\})$. In order to prove that this is actually C^1 on the whole disk, we observe that for (x, y) converging to $(0, 0)$ we have

$$
\begin{aligned}
u(x,y) - u(0,0) &= \sum_{n=1}^{\infty} \alpha_n \varrho^n \cos(n\vartheta) + \sum_{n=1}^{\infty} \beta_n \varrho^n \sin(n\vartheta) \\
&= \alpha_1 \varrho \cos\vartheta + \beta_1 \varrho \sin\vartheta \\
&+ \sum_{n=2}^{\infty} \alpha_n \varrho^n \cos(n\vartheta) + \sum_{n=2}^{\infty} \beta_n \varrho^n \sin(n\vartheta) \\
&= \alpha_1 \varrho \cos\vartheta + \beta_1 \varrho \sin\vartheta + o(\varrho) \\
&= \alpha_1 x + \beta_1 y + o(\sqrt{x^2+y^2}).
\end{aligned}
$$

This shows that u is differentiable at the origin, with $\nabla u(0,0) = (\alpha_1, \beta_1)$. Moreover, by using the expression for the gradient obtained above, we have

$$
\begin{aligned}
\lim_{(x,y)\to(0,0)} \frac{\partial u}{\partial x}(x,y) &= \lim_{\varrho\to 0} \left[\sin\vartheta \, \frac{\partial u}{\partial \varrho} + \frac{\cos\vartheta}{\varrho} \, \frac{\partial u}{\partial \vartheta} \right] \\
&= (\alpha_1 \cos\vartheta - \cancel{\beta_1 \sin\vartheta}) \cos\vartheta + (\alpha_1 \sin\vartheta + \cancel{\beta_1 \cos\vartheta}) \sin\vartheta \\
&= \alpha_1 = \frac{\partial u}{\partial x}(0,0).
\end{aligned}
$$

Similarly, we can show that

$$
\lim_{(x,y)\to(0,0)} \frac{\partial u}{\partial y}(x,y) = \beta_1 = \frac{\partial u}{\partial x}(0,0).
$$

This shows that $u \in C^1(D)$. In order to complete the proof of the fact that $u \in C^2(D)$, we have to prove

$$
\frac{\partial u}{\partial x}, \frac{\partial u}{\partial y} \in C^1(D).
$$

This can be proved along the same lines as before, by starting from (A.3.4) and (A.3.5).

Proof of 2 We recall that the functions

$$
\varrho^n \cos(n\vartheta) \qquad \text{and} \qquad \varrho^n \sin(n\vartheta),
$$

are harmonic, as observed in Sect. A.3. In light of the previous discussion, we can compute the Laplacian of u by differentiating the series term by term and get the desired result.

Proof of 3 From point 1, we have in particular $|\nabla u| \in C^0(D)$. Thus, if set $D_\delta = \{(x,y) \in \mathbb{R}^2 : x^2 + y^2 \leq \delta^2\}$, we have

$$\int_{D_\delta} |\nabla u|^2 \,dx\,dy < +\infty, \qquad \text{for every } 0 < \delta < 1.$$

We fix $0 < \delta < 1$: by using polar coordinates, from formula (A.1.5) we have

$$\begin{aligned}\int_{D_\delta} |\nabla u|^2 \,dx\,dy &= \int_{[0,\delta]\times[0,2\pi]} \left[\left(\frac{\partial u}{\partial \varrho}\right)^2 + \frac{1}{\varrho^2}\left(\frac{\partial u}{\partial \vartheta}\right)^2\right] \varrho \,d\varrho \\ &= \int_0^\delta \left(\int_0^{2\pi} \left(\frac{\partial u}{\partial \varrho}\right)^2 d\vartheta\right) \varrho\,d\varrho \\ &\quad + \int_0^\delta \left(\int_0^{2\pi} \left(\frac{\partial u}{\partial \vartheta}\right)^2 d\vartheta\right) \frac{1}{\varrho}\,d\varrho.\end{aligned}$$

For every $0 < \varrho \leq \delta$ fixed, we notice that (A.3.3) can be regarded as the Fourier expansion of the function $\vartheta \to \partial u/\partial\varrho(\varrho, \vartheta)$. By the Plancherel identity for Fourier series, we then have

$$\int_0^{2\pi} \left(\frac{\partial u}{\partial \varrho}\right)^2 d\vartheta = \pi \sum_{n=1}^{\infty} n^2\,(\alpha_n^2 + \beta_n^2)\,\varrho^{2n-2}, \qquad \text{for every } 0 < \varrho \leq \delta.$$

With a similar argument, we also get

$$\int_0^{2\pi} \left(\frac{\partial u}{\partial \vartheta}\right)^2 d\vartheta = \pi \sum_{n=1}^{\infty} n^2\,(\alpha_n^2 + \beta_n^2)\,\varrho^{2n}, \qquad \text{for every } 0 < \varrho \leq \delta.$$

Both series are converging, thanks to the assumption on $\{\alpha_n\}_{n\in\mathbb{N}}$ and $\{\beta_n\}_{n\in\mathbb{N}}$. By multiplying the first term by ϱ, the second one by $1/\varrho$ and then summing up, we then obtain

$$\int_{D_\delta} |\nabla u|^2 \,dx\,dy = 2\pi \int_0^\delta \sum_{n=1}^{\infty} n^2\,(\alpha_n^2 + \beta_n^2)\,\varrho^{2n-1}\,d\varrho.$$

We can observe that the series

$$\sum_{n=1}^{k} n^2\,(\alpha_n^2 + \beta_n^2)\,\varrho^{2n-1},$$

is totally converging on $[0, \delta]$. Thus, we can exchange the series and the integral and get

$$\begin{aligned}\int_{D_\delta} |\nabla u|^2\, dx\, dy &= 2\pi \sum_{n=1}^{\infty} \int_0^{\delta} n^2\, (\alpha_n^2 + \beta_n^2)\, \varrho^{2n-1}\, d\varrho \\ &= 2\pi \sum_{n=1}^{\infty} n^2\, (\alpha_n^2 + \beta_n^2)\, \frac{\delta^{2n}}{2n} \\ &= \pi \sum_{n=1}^{\infty} n\, (\alpha_n^2 + \beta_n^2)\, \delta^{2n}.\end{aligned}$$

This holds for every $0 < \delta < 1$. By using the assumption that $\sum_{n=1}^{\infty} n\, (\alpha_n^2 + \beta_n^2) < +\infty$, we can conclude by taking the limit as δ goes to 0. □

Remark A.3.3 We refer to [8, page 6] for the example of a harmonic function of the type above, such that

$$\sum_{n=0}^{\infty} (|\alpha_n| + |\beta_n|) < +\infty \qquad \text{but} \qquad \sum_{n=0}^{\infty} n\, (\alpha_n^2 + \beta_n^2) = +\infty.$$

Appendix B
Compactness Results in L^p Spaces

B.1 Weak Compactness

We recall the following classical result, the proof can be found for example in [46, Theorem 2.18].

Theorem B.1.1 (Banach-Alaoglu) *Let $1 < p < \infty$ and let $\Omega \subseteq \mathbb{R}^N$ be a measurable set. For every $M > 0$, the set*

$$\mathcal{B}_M = \Big\{ u \in L^p(\Omega) \, : \, \|u\|_{L^p(\Omega)} \leq M \Big\},$$

is sequentially weakly compact, i.e. for every $\{u_n\}_{n\in\mathbb{N}} \subseteq \mathcal{B}_M$, there exists a subsequence weakly converging in $L^p(\Omega)$ to a function $u \in \mathcal{B}_M$.

The previous result admits a converse: this is a consequence of a more general result, valid in the setting of Banach spaces (see [20, Chapter 2]). We give an elementary proof, adapted from [88] (see also [46, Theorem 2.12]), directly for L^p spaces.

Proposition B.1.2 (Uniform Boundedness Principle) *Let $1 < p < \infty$ and let $\Omega \subseteq \mathbb{R}^N$ be a measurable set. If the sequence $\{u_n\}_{n\in\mathbb{N}} \subseteq L^p(\Omega)$ is weakly converging, then there exists a constant $M > 0$ such that*

$$\|u_n\|_{L^p(\Omega)} \leq M, \qquad \textit{for every } n \in \mathbb{N}.$$

Proof Without loss of generality, we can suppose that the sequence is weakly converging to 0. We argue by contradiction and assume that

$$\text{for every } M > 0, \text{ there exists } n_M \in \mathbb{N} \text{ such that } \|u_{n_M}\|_{L^p(\Omega)} > M.$$

In particular, by taking $M = 3^k$ there exists u_{n_k} such that

L. Brasco, *Handbook of Calculus of Variations for Absolute Beginners*,
La Matematica per il 3+2 163, https://doi.org/10.1007/978-3-031-87164-1

$$\|u_{n_k}\|_{L^p(\Omega)} > 3^k, \qquad \text{for every } k \in \mathbb{N}.$$

This subsequence is still weakly converging to 0. We then construct the new sequence

$$U_k = 3^k \frac{u_{n_k}}{\|u_{n_k}\|_{L^p(\Omega)}},$$

which is such that

$$\|U_k\|_{L^p(\Omega)} = 3^k, \qquad \text{for every } k \in \mathbb{N},$$

and for every $\varphi \in L^{p'}(\Omega)$

$$\lim_{k\to\infty} \left| \int_\Omega U_k \, \varphi \, dx \right| = \lim_{k\to\infty} \frac{3^k}{\|u_{n_k}\|_{L^p(\Omega)}} \left| \int_\Omega u_{n_k} \, \varphi \, dx \right| \le \lim_{k\to\infty} \left| \int_\Omega u_{n_k} \, \varphi \, dx \right| = 0.$$

By using the sequence $\{U_k\}_{k\in\mathbb{N}}$, we can now construct a suitable sequence in $L^{p'}(\Omega)$ as follows. We define

$$G_0 = 0 \qquad G_k = \omega_k \left(\frac{1}{3}\right)^{p\,k} |U_k|^{p-2} \, U_k + G_{k-1}, \ \text{for } k \in \mathbb{N} \setminus \{0\},$$

where the real number ω_k is either 1 or -1 and it is inductively defined as follows: we set

$$\omega_0 = 1,$$

and then for $k \ge 1$

$$\omega_k = \begin{cases} 1, & \text{if } \displaystyle\int_\Omega G_{k-1} \, U_k \, dx \ge 0, \\ -1, & \text{if } \displaystyle\int_\Omega G_{k-1} \, U_k \, dx < 0. \end{cases}$$

We notice that by construction we have

$$\begin{aligned} \left| \int_\Omega U_k \, G_k \, dx \right| &= \left| \omega_k \left(\frac{1}{3}\right)^{p\,k} \int_\Omega |U_k|^p \, dx + \int_\Omega U_k \, G_{k-1} \, dx \right| \\ &\ge \left(\frac{1}{3}\right)^{p\,k} \int_\Omega |U_k|^p \, dx = \left(\frac{1}{3}\right)^{p\,k} 3^{p\,k} = 1, \end{aligned} \tag{B.1.1}$$

thanks to the choice of ω_k. Observe that $\{G_k\}_{k\in\mathbb{N}}$ is a Cauchy sequence in $L^{p'}(\Omega)$: indeed, for $m > n$ we have by Minkowski's inequality

$$\begin{aligned}\|G_n - G_m\|_{L^{p'}(\Omega)} &\leq \sum_{k=n}^{m-1} \|G_k - G_{k+1}\|_{L^{p'}(\Omega)}\\ &= \sum_{k=n}^{m-1} \left(\frac{1}{3}\right)^{p\,(k+1)} \left\| |U_{k+1}|^{p-2}\, U_{k+1} \right\|_{L^{p'}(\Omega)} = \sum_{j=n+1}^{m} \left(\frac{1}{3}\right)^{j}.\end{aligned} \tag{B.1.2}$$

This gives the claimed property, by recalling that the geometric series with argument $1/3$ is convergent. By using the completeness of $L^{p'}(\Omega)$, the sequence $\{G_k\}_{k\in\mathbb{N}}$ strongly converges in $L^{p'}(\Omega)$. We call $G \in L^{p'}(\Omega)$ the limit function. By taking the limit as m goes to ∞ in (B.1.2), we get

$$\|G_n - G\|_{L^{p'}(\Omega)} \leq \sum_{j=n+1}^{\infty} \left(\frac{1}{3}\right)^{j} = \frac{1}{2}\,\frac{1}{3^n}.$$

By using this fact and Hölder's inequality, we obtain

$$\begin{aligned}\left|\int_\Omega U_k\, G\, dx\right| &\geq \left|\int_\Omega U_k\, G_k\, dx\right| - \left|\int_\Omega U_k\,(G - G_k)\, dx\right|\\ &\geq \left|\int_\Omega U_k\, G_k\, dx\right| - \|U_k\|_{L^p(\Omega)}\, \|G_k - G\|_{L^{p'}(\Omega)}\\ &\geq \left|\int_\Omega U_k\, G_k\, dx\right| - 3^k\, \frac{1}{2}\,\frac{1}{3^k}.\end{aligned}$$

We now recall (B.1.1), thus we get for every $k \in \mathbb{N}$

$$\left|\int_\Omega U_k\, G\, dx\right| \geq \frac{1}{2}.$$

By recalling that $\{U_k\}_{k\in\mathbb{N}}$ weakly converges to 0, we get a contradiction. □

Remark B.1.3 The previous two results hold also in the *vectorial case*, i.e. if $L^p(\Omega)$ is replaced by $L^p(\Omega;\mathbb{R}^k)$ with $k \geq 2$.

B.2 Strong Compactness

In Chap. 3, we needed a strong compactness result for L^p spaces, in order to prove the compact embeddings for Sobolev spaces. This is the classical *Riesz-Fréchet-*

Kolmogorov compactness theorem. For simplicity, we give here a weaker statement, which is however largely sufficient for our scopes. This is a consequence of [20, Theorem 4.26], to which we refer for the proof.

Theorem B.2.1 (Riesz-Fréchet-Kolmogorov) *Let $1 \le p < \infty$ and let $\{u_n\}_{n\in\mathbb{N}} \subseteq L^p(\mathbb{R}^N)$ be such that*

(H1) there exists $M > 0$ such that

$$\|u_n\|_{L^p(\mathbb{R}^N)} \le M, \qquad \text{for every } n \in \mathbb{N};$$

(H2) there exists a bounded measurable set $\Omega \subseteq \mathbb{R}^N$ such that

$$u_n = 0 \text{ a. e. on } \mathbb{R}^N \setminus \Omega, \qquad \text{for every } n \in \mathbb{N};$$

(H3) we have

$$\lim_{|h|\to 0} \left(\sup_{n\in\mathbb{N}} \int_{\mathbb{R}^N} |u_n(x+h) - u_n(x)|^p\, dx \right) = 0.$$

Then there exists $u \in L^p(\mathbb{R}^N)$ such that, up to a subsequence, we have

$$\lim_{n\to\infty} \|u_n - u\|_{L^p(\mathbb{R}^N)} = 0.$$

Remark B.2.2 Hypothesis (H2) can be weakened and replaced by the following:

(H2bis) *for every $\varepsilon > 0$, there exists $R_\varepsilon > 0$ such that*

$$\int_{\mathbb{R}^N \setminus B_{R_\varepsilon}(0)} |u_n|^p\, dx < \varepsilon, \qquad \text{for every } n \in \mathbb{N}.$$

We refer to [20, Corollary 4.27] for the proof.

Assumptions like (H2) or (H2bis) prevent a bounded sequence $\{u_n\}_{n\in\mathbb{N}}$ to lose some mass "at infinity". Without conditions of this kind, one can not hope to have strong compactness in L^p: as a simple counter-example, one may take the caracteristic function of the ball $B_R(0)$ and construct the sequence of "horizontal translations"

$$u_n(x) = 1_{B_R(0)}(x - n\,\mathbf{e}_1) = 1_{B_R(n\,\mathbf{e}_1)}(x), \qquad \text{for every } n \in \mathbb{N}.$$

It is not difficult to see that both conditions (H1) and (H3) of Theorem B.2.1 are verified. Indeed, we have

$$\|u_n\|_{L^p(\mathbb{R}^N)} = \|1_{B_R(n\,\mathbf{e}_1)}\|_{L^p(\mathbb{R}^N)} = |B_R(0)|^{\frac{1}{p}}, \qquad \text{for every } n \in \mathbb{N},$$

and by using the change of variables $x - n\,\mathbf{e}_1 = y$

$$\begin{aligned}\int_{\mathbb{R}^N} |u_n(x+h) - u_n(x)|^p\,dx &= \int_{\mathbb{R}^N} |1_{B_R(0)}(y+h) - 1_{B_R(0)}(y)|^p\,dy \\ &= \int_{\mathbb{R}^N} |1_{B_R(-h)}(y) - 1_{B_R(0)}(y)|^p\,dy \\ &= |B_R(-h)\Delta B_R(0)|, \qquad \text{for every } n \in \mathbb{N},\end{aligned}$$

where Δ stands for the symmetric difference of sets. The last quantity clearly converges to 0 as $|h|$ goes to 0.

On the other hand, the sequence $\{u_n\}_{n\in\mathbb{N}}$ can not converge strongly in $L^p(\mathbb{R}^N)$. We argue by contradiction and assume that it converges (possibly up to some subsequence) to a function $u \in L^p(\mathbb{R}^N)$. In particular, we would have

$$\|u\|_{L^p(\mathbb{R}^N)} = \lim_{n\to\infty} \|u_n\|_{L^p(\mathbb{R}^N)} = |B_R(0)|^{\frac{1}{p}},$$

and thus $u \not\equiv 0$. On the other hand, we would have weak convergence in $L^p(\mathbb{R}^N)$, as well. For every $\varphi \in C_0^\infty(\mathbb{R}^N)$, this implies that

$$\int_{\mathbb{R}^N} u\,\varphi\,dx = \lim_{n\to\infty} \int_{\mathbb{R}^N} u_n\,\varphi\,dx = \lim_{n\to\infty} \int_{B_R(n\,\mathbf{e}_1)} \varphi\,dx = 0,$$

thanks to the fact that φ is compactly supported. In light of Lemma 1.4.2, this would imply that u vanishes almost everywhere in $\mathbb{R}^N$, thus giving a contradiction.

B.3 Convex Combinations

Definition B.3.1 Let V be a vector space over $\mathbb{R}$ and let $\mathbf{v}_1, \ldots, \mathbf{v}_n \in V$. The *convex hull* of the family of vectors $\{\mathbf{v}_1, \ldots, \mathbf{v}_n\}$ is given by

$$\operatorname{conv}\Big(\{\mathbf{v}_1, \ldots, \mathbf{v}_n\}\Big) := \left\{\mathbf{v} \in V \,:\, \mathbf{v} = \sum_{i=1}^n \lambda_i\,\mathbf{v}_i \text{ for some } \lambda_i \ge 0 \text{ s. t. } \sum_{i=1}^n \lambda_i = 1\right\}.$$

By construction, this is a convex subset of V.

Theorem B.3.2 (Mazur's Lemma) *Let $1 < p < \infty$ and let $\Omega \subseteq \mathbb{R}^N$ be a measurable set. Let $\{f_n\}_{n\in\mathbb{N}} \subseteq L^p(\Omega)$ be a sequence, which weakly converges to $f \in L^p(\Omega)$. Then there exists a sequence $\{F_n\}_{n\in\mathbb{N}} \subseteq L^p(\Omega)$ such that:*

(i) for every $n \in \mathbb{N}$, the function F_n is a convex combination of $f_0, \ldots, f_n$, i.e. we have

$$F_n \in \mathrm{conv}\Big(\{f_0, \dots, f_n\}\Big), \qquad \textit{for every } n \in \mathbb{N}; \tag{B.3.1}$$

(ii) F_n *converges strongly in* $L^p(\Omega)$ *to* f.

Proof We give here an elementary proof, taken from [46, Theorem 2.13]. We define the *non-decreasing* sequence of convex sets $\{K_n\}_{n\in\mathbb{N}}$ given by

$$K_n = \mathrm{conv}\Big(\{f_0, \dots, f_n\}\Big).$$

We first observe that if there exists $n_0 \in \mathbb{N}$ such that $f \in K_{n_0}$, this means that

$$f = \sum_{i=0}^{n_0} \lambda_i f_i, \qquad \text{for some } \lambda_i \geq 0 \text{ such that } \sum_{i=0}^{n_0} \lambda_i = 1.$$

Thus, we get the conclusion of the statement by simply taking

$$F_n = \begin{cases} f_n, \text{ if } n \leq n_0, \\ \sum\limits_{i=0}^{n_0} \lambda_i f_i, \text{ if } n \geq n_0 + 1. \end{cases}$$

We now assume that

$$f \notin \bigcup_{n=0}^{\infty} K_n =: K. \tag{B.3.2}$$

We consider the closure in $L^p(\Omega)$ of K, denoted as usual by $\overline{K}$. We claim that $\overline{K}$ is a convex closed subset of $L^p(\Omega)$. The closedness is a direct consequence of its definition. In order to prove that $\overline{K}$ is convex, we take $g, h \in \overline{K}$ and we need to prove that

$$\lambda g + (1 - \lambda) h \in \overline{K}, \qquad \text{for every } \lambda \in [0, 1]. \tag{B.3.3}$$

By construction, there exist two sequences

$$\{g_k\}_{k\in\mathbb{N}} \subseteq K \qquad \text{and} \qquad \{h_k\}_{k\in\mathbb{N}} \subseteq K,$$

such that

$$\lim_{k\to\infty} \|g_k - g\|_{L^p(\Omega)} = 0 = \lim_{k\to\infty} \|h_k - h\|_{L^p(\Omega)}.$$

We then observe that for every $\lambda \in [0, 1]$

$$\lambda\, g_k + (1-\lambda)\, h_k \in K, \qquad \text{for every } k \in \mathbb{N},$$

and by Minkowski's inequality

$$\begin{aligned}&\lim_{k\to\infty} \left\| \Big(\lambda\, g_k + (1-\lambda)\, h_k\Big) - \Big(\lambda\, g + (1-\lambda)\, h\Big) \right\|_{L^p(\Omega)} \\ &\le \lambda \lim_{k\to\infty} \|g_k - g\|_{L^p(\Omega)} + (1-\lambda) \lim_{k\to\infty} \|h_k - h\|_{L^p(\Omega)} = 0.\end{aligned}$$

This shows that (B.3.3) holds true.

In order to conclude the proof, we prove at first that $f \in \overline{K}$: by definition this would show that f is an adherence point of K and thus there exists a sequence $\{G_n\}_{n\in\mathbb{N}} \subseteq K$ such that

$$\lim_{n\to\infty} \|G_n - f\|_{L^p(\Omega)} = 0.$$

We argue by contradiction and assume that $f \notin \overline{K}$. Since $\overline{K} \subseteq L^p(\Omega)$ is convex and closed, we thus know that there exists $\widetilde{f} \in \overline{K}$ such that

$$\min_{g\in\overline{K}} \|f - g\|_{L^p(\Omega)} = \|f - \widetilde{f}\|_{L^p(\Omega)} > 0. \tag{B.3.4}$$

We recall that the existence of $\widetilde{f}$ can be proved by using that, for $1 < p < \infty$, the space $L^p(\Omega)$ is uniformly convex (see [46, Lemma 2.8]). Moreover, the minimizer $\widetilde{f}$ satisfies the optimality condition

$$\int_\Omega (f - \widetilde{f})\, |f - \widetilde{f}|^{p-2}\, (g - \widetilde{f})\, dx \le 0, \qquad \text{for every } g \in \overline{K}. \tag{B.3.5}$$

Indeed, for every $g \in \overline{K}$ and every $t \in [0, 1]$, by convexity of $\overline{K}$ and minimality of $\widetilde{f}$ we have

$$\psi(t) := \int_\Omega |f - t\, g - (1-t)\, \widetilde{f}|^p\, dx \ge \int_\Omega |f - \widetilde{f}|^p\, dx = \psi(0).$$

This implies that

$$\lim_{t\to 0^+} \frac{\psi(t) - \psi(0)}{t} \ge 0.$$

The last limit can be computed by using the Dominated Convergence Theorem and we get (B.3.5). In particular, by taking $g = f_n$ in (B.3.5), using the hypothesis on the weak convergence and the fact that

$$(f - \widetilde{f})\, |f - \widetilde{f}|^{p-2} \in L^{p'}(\Omega),$$

we have

$$\begin{aligned}\|f-\widetilde{f}\|_{L^p(\Omega)}^p &= \int_\Omega (f-\widetilde{f})\,|f-\widetilde{f}|^{p-2}\,(f-\widetilde{f})\,dx \\ &= \lim_{n\to\infty}\int_\Omega (f-\widetilde{f})\,|f-\widetilde{f}|^{p-2}\,(f_n-\widetilde{f})\,dx \le 0.\end{aligned}$$

The latter clearly contradicts (B.3.4) and thus $f \in \overline{K}$.

We still need to prove that f is the strong limit in $L^p(\Omega)$ of a sequence $\{F_n\}_{n\in\mathbb{N}} \subseteq L^p(\Omega)$ with the property (B.3.1). To this aim, we take the sequence $\{G_n\}_{n\in\mathbb{N}} \subseteq K$ obtained above. By definition of K, this means that for every $n \in \mathbb{N}$ there exists $k_n \in \mathbb{N}$ such that

$$G_n \in K_{k_n} = \mathrm{conv}\Big(\{f_0, \dots, f_{k_n}\}\Big).$$

In light of the assumption (B.3.2), it is not difficult to see that the sequence of natural numbers $\{k_n\}_{n\in\mathbb{N}}$ diverges to $+\infty$. Thus, by observing that the positive sequence

$$\alpha_n = \min_{g\in K_n} \|g-f\|_{L^p(\Omega)},$$

is non-increasing (thanks to the monotonicity of the sets K_n), we get in particular that

$$\lim_{n\to\infty} \alpha_{k_n} = \lim_{n\to\infty} \alpha_n.$$

This means that

$$\lim_{n\to\infty} \min_{g\in K_{k_n}} \|g-f\|_{L^p(\Omega)} = \lim_{n\to\infty} \min_{g\in K_n} \|g-f\|_{L^p(\Omega)}.$$

On the other hand, we clearly have

$$\|G_n - f\|_{L^p(\Omega)} \ge \min_{g\in K_{k_n}} \|g-f\|_{L^p(\Omega)}..$$

By joining the last two equations and recalling that $\{G_n\}_{n\in\mathbb{N}}$ converges to f, we obtain

$$\lim_{n\to\infty} \min_{g\in K_n} \|g-f\|_{L^p(\Omega)} = 0.$$

If for every $n \in \mathbb{N}$ we take a function $F_n \in K_n$ attaining the last minimum, we eventually get the conclusion. □

Remark B.3.3 The previous result holds (and the proof is exactly the same) also in the *vectorial case*, i.e. if $L^p(\Omega)$ is replaced by $L^p(\Omega; \mathbb{R}^k)$ with $k \geq 2$. We leave the details to the reader.

Remark B.3.4 The previous result holds in the extremal cases $p = 1$ and $p = \infty$, as well. However, in these cases the previous proof does not apply and one should rather appeal to the *Hahn-Banach Theorem* in one of its geometric forms. See for example [20, Corollary 3.8] for a general statement of Mazur's Lemma, covering the cases $L^1(\Omega)$ and $L^\infty(\Omega)$, as well.

Appendix C
Homogeneous Sobolev Spaces

C.1 The Completion Space $\mathcal{D}_0^{1,p}$

We recall that the space $W_0^{1,p}(\Omega)$ is endowed with the norm

$$\|\varphi\|_{W^{1,p}(\Omega)} = \|\varphi\|_{L^p(\Omega)} + \|\nabla\varphi\|_{L^p(\Omega;\mathbb{R}^N)}, \qquad \varphi \in W_0^{1,p}(\Omega), \tag{C.1.1}$$

and it is actually a Banach space, see Theorem 3.7.4.

Lemma C.1.1 *Let $1 \le p \le \infty$ and let $\Omega \subseteq \mathbb{R}^N$ be an open set. Then*

$$\varphi \mapsto \|\nabla\varphi\|_{L^p(\Omega;\mathbb{R}^N)}, \tag{C.1.2}$$

is a norm on $W_0^{1,p}(\Omega)$.

Proof We have

$$\|\nabla(\lambda\, u)\|_{L^p(\Omega;\mathbb{R}^N)} = |\lambda|\, \|\nabla u\|_{L^p(\Omega;\mathbb{R}^N)}, \qquad \text{for every } u \in W_0^{1,p}(\Omega),\ \lambda \in \mathbb{R},$$

and

$$\|\nabla(u+v)\|_{L^p(\Omega;\mathbb{R}^N)} \le \|\nabla u\|_{L^p(\Omega;\mathbb{R}^N)} + \|\nabla v\|_{L^p(\Omega;\mathbb{R}^N)}, \quad \text{for every } u, v \in W_0^{1,p}(\Omega),$$

as a direct consequence of the properties of L^p norms. Moreover, by Proposition 3.7.11 we have

$$u \in W_0^{1,p}(\Omega) \text{ is such that } \|\nabla u\|_{L^p(\Omega;\mathbb{R}^N)} = 0 \qquad \Longleftrightarrow \qquad u \equiv 0 \text{ a. e. in } \Omega.$$

This concludes the proof. □

L. Brasco, *Handbook of Calculus of Variations for Absolute Beginners*,
La Matematica per il 3+2 163, https://doi.org/10.1007/978-3-031-87164-1

We have seen in Theorem 3.7.6 that, on an open set which is bounded in one direction, the reduced norm (C.1.2) is equivalent to the standard one (C.1.1) on $W_0^{1,p}(\Omega)$. This remains true even if Ω is unbounded in every direction, provided it has finite measure (see Problem 3.12.11).

However, for a general open set, this equivalence may fail to be true. Indeed, the space $W_0^{1,p}(\Omega)$ endowed with the new norm (C.1.2) *is not even complete*, in general. This is the content of the next result.

Proposition C.1.2 *Let* $1 \leq p \leq \infty$*, the set* $W_0^{1,p}(\mathbb{R}^N)$ *endowed with the norm* (C.1.2) *is not a Banach space. In particular, such a norm is not equivalent to* (C.1.1) *on* $W_0^{1,p}(\mathbb{R}^N)$.

Proof The proof is based on explicitly exhibiting a sequence $\{u_n\}_{n\in\mathbb{N}} \subseteq W_0^{1,p}(\mathbb{R}^N)$ with the following properties:

- it is a Cauchy sequence with respect to the reduced norm (C.1.2);
- it is not converging to a function $u \in W_0^{1,p}(\mathbb{R}^N)$, with respect to this norm.

We recall at first that we have

$$W_0^{1,p}(\mathbb{R}^N) = W^{1,p}(\mathbb{R}^N) \qquad \text{for } 1 \leq p < \infty,$$

by Proposition 3.7.10. We now divide the proof in four cases.

Case $1 \leq p < N$ We consider the function

$$\varphi(x) = \frac{1}{(1+|x|^2)^{\frac{\alpha}{2}}},$$

for an exponent α such that

$$\frac{N}{p} - 1 < \alpha \leq \frac{N}{p}. \tag{C.1.3}$$

We observe that $\varphi \in C^\infty(\mathbb{R}^N)$ and by using spherical coordinates

$$\begin{aligned}\int_{\mathbb{R}^N} |\nabla\varphi|^p\, dx &= |\alpha|^p \int_{\mathbb{R}^N} (1+|x|^2)^{-\frac{\alpha+2}{2}\, p}\, |x|^p\, dx \\ &= |\alpha|^p\, N\, \omega_N \int_0^{+\infty} (1+\varrho^2)^{-\frac{\alpha+2}{2}\, p}\, \varrho^{p+N-1}\, d\varrho.\end{aligned}$$

This integral converges, thanks to the choice of α. For every $n \in \mathbb{N}$, we take a cut-off function $\eta_n \in C_0^\infty(B_{n+1}(0))$ such that

$$0 \leq \eta_n \leq 1, \qquad \eta_n \equiv 1 \text{ on } B_n(0), \qquad |\nabla\eta_n(x)| \leq C.$$

Then we have

$$\varphi\,\eta_n \in C_0^\infty(B_{n+1}(0)) \subseteq C_0^\infty(\mathbb{R}^N) \subseteq W^{1,p}(\mathbb{R}^N),$$

and this is a Cauchy sequence with respect to the reduced norm (C.1.2). Indeed, we can show that

$$\lim_{n\to\infty} \|\nabla(\varphi\,\eta_n) - \nabla\varphi\|_{L^p(\mathbb{R}^N;\mathbb{R}^N)} = 0.$$

However, the limit function $\varphi \notin W^{1,p}(\mathbb{R}^N)$, since we have $\varphi \notin L^p(\mathbb{R}^N)$, thanks to the choice (C.1.3).

Case $p = 1 = N$ This is similar to the previous case. We consider the function

$$\varphi(x) = \frac{1}{\sqrt{1+|x|^2}}, \qquad x \in \mathbb{R}.$$

It is not difficult to see that

$$\varphi' \in L^1(\mathbb{R}) \qquad \text{but} \qquad \varphi \notin L^1(\mathbb{R}).$$

By proceeding as in the previous case, we can construct the required Cauchy sequence.

Case $p > N$ This is simpler. We take any non-negative function $\varphi \in C_0^\infty(\mathbb{R}^N)$ such that $\varphi(0) > 0$. Then we fix

$$0 < \alpha < 1 - \frac{N}{p},$$

a choice which is feasible, thanks to the fact that $p > N$. For every $n \in \mathbb{N} \setminus \{0\}$, we define the rescaled function

$$\varphi_n(x) = n^\alpha\,\varphi\left(\frac{x}{n}\right), \qquad \text{for every } x \in \mathbb{R}^N,$$

and compute the L^p norm of its gradient. This is given by

$$\|\nabla\varphi_n\|_{L^p(\mathbb{R}^N;\mathbb{R}^N)} = n^{\alpha-1+\frac{N}{p}}\,\|\nabla\varphi\|_{L^p(\mathbb{R}^N;\mathbb{R}^N)}.$$

Thanks to the choice of α, we have

$$\lim_{n\to\infty} \|\nabla\varphi_n\|_{L^p(\mathbb{R}^N;\mathbb{R}^N)} = 0,$$

and thus $\{\varphi_n\}_{n\in\mathbb{N}\setminus\{0\}}$ is certainly a Cauchy sequence, with respect to the reduced norm (C.1.2). On the other hand, by construction we have

$$\lim_{n\to\infty} \varphi_n(x) = +\infty, \qquad \text{for every } x \in \mathbb{R}^N,$$

thus such a sequence can not converge.

Case $p = N$ and $N \geq 2$ This borderline case is more delicate. We consider a sequence of truncated logarithms, similar to that of the solution of Problem 3.12.22. We take the following radially symmetric function for $n \in \mathbb{N} \setminus \{0\}$

$$u_n(x) = -\frac{1}{\log n} \log \frac{|x|}{n^2}, \qquad \text{for } n \leq |x| \leq n^2.$$

This is a C^1 function defined on the annulus

$$\overline{B_{n^2}(0)} \setminus B_n(0),$$

such that

$$u_n(x) = 1, \quad \text{for } |x| = n, \qquad u_n(x) = 0, \quad \text{for } |x| = n^2.$$

As in the solution of Problem 3.12.22, we "fill the hole" by considering the radially symmetric function

$$\psi_n(x) = 1 + \frac{n^2}{2\log n}\left(\frac{1}{n^2} - \frac{|x|^2}{n^4}\right), \qquad \text{for } |x| \leq n.$$

In this way, one can check that the function

$$\varphi_n(x) = \begin{cases} \psi_n(x), & \text{if } |x| \leq n, \\ u_n(x), & \text{if } n \leq |x| \leq n^2, \end{cases}$$

is C^1 on $B_{n^2}(0)$, with bounded gradient. Moreover, by construction this function vanishes on $\partial B_{n^2}(0)$. Thus, by Problem 3.12.15, we get that $\varphi_n \in W_0^{1,N}(B_{n^2}(0))$. We then consider its extension by 0 outside $B_{n^2}(0)$: we obtain a function, still denoted by φ_n, which belongs to $W_0^{1,N}(\mathbb{R}^N)$ (thanks to Lemma 3.7.9). We compute the L^N norm of its gradient: this is given by

$$\begin{aligned}
\int_{\mathbb{R}^N} |\nabla\varphi_n|^N \, dx &= \int_{B_{n^2}(0)} |\nabla\varphi_n|^N \, dx \\
&= \int_{B_n(0)} |\nabla\psi_n|^N \, dx + \int_{B_{n^2}(0)\setminus B_n(0)} |\nabla u_n|^N \, dx \\
&= \left(\frac{1}{n^2 \log n}\right)^N \int_{B_n(0)} |x|^N \, dx + \frac{1}{(\log n)^N} \int_{B_{n^2}(0)\setminus B_n(0)} \frac{1}{|x|^N} \, dx
\end{aligned}$$

$$= N\,\omega_N\left(\frac{1}{n^2\log n}\right)^N\int_0^n \varrho^{2N-1}\,d\varrho + \frac{N\,\omega_N}{(\log n)^N}\int_n^{n^2}\frac{1}{\varrho}\,d\varrho$$
$$= \frac{\omega_N}{2}\,\frac{1}{(\log n)^N} + \frac{N\,\omega_N}{(\log n)^{N-1}},$$

In particular, for $N \geq 2$ we have

$$\lim_{n\to\infty}\int_{\mathbb{R}^N}|\nabla\varphi_n|^N\,dx = 0.$$

As in the previous case, this shows that $\{\varphi_n\}_{n\in\mathbb{N}\setminus\{0\}}$ is a Cauchy sequence, with respect to the L^N norm of the gradient. By construction, we have that

$$\lim_{n\to\infty}\varphi_n(x) = \lim_{n\to\infty}\psi_n(x) = 1, \qquad \text{for every } x\in\mathbb{R}^N.$$

However, the function constantly equal to 1 can not belong to $W_0^{1,N}(\mathbb{R}^N)$ (this is not in $L^N(\mathbb{R}^N)$). Thus, the sequence $\{\varphi_n\}_{n\in\mathbb{N}}$ can not converge. This concludes the proof. □

The previous result naturally leads to the following

Definition C.1.3 Let $1 \leq p \leq \infty$ and let $\Omega \subseteq \mathbb{R}^N$ be an open set. We consider the set of all sequences $\{u_n\}_{n\in\mathbb{N}} \subseteq C_0^\infty(\Omega)$ which are Cauchy sequences with respect to the reduced norm (C.1.2). We introduce on this set the equivalence relation

$$\{u_n\}_{n\in\mathbb{N}} \sim \{v_n\}_{n\in\mathbb{N}} \qquad \text{if} \qquad \lim_{n\to\infty}\|\nabla u_n - \nabla v_n\|_{L^p(\Omega;\mathbb{R}^N)} = 0.$$

We then define $\mathcal{D}_0^{1,p}(\Omega)$ as the quotient space of the set of all these Cauchy sequences, with respect to such an equivalence relation. This is called the *completion* of $C_0^\infty(\Omega)$ with respect to the norm (C.1.2)

Observe that the elements of $\mathcal{D}_0^{1,p}(\Omega)$ are equivalence classes of sequences, by construction. For each of these equivalence classes

$$U = \Big[\{u_n\}_{n\in\mathbb{N}}\Big]_\sim \in \mathcal{D}_0^{1,p}(\Omega),$$

we define its norm in terms of a representative as follows

$$\|U\|_{\mathcal{D}_0^{1,p}(\Omega)} := \lim_{n\to\infty}\|\nabla u_n\|_{L^p(\Omega;\mathbb{R}^N)}.$$

It is easily seen that such a definition is independent of the representative $\{u_n\}_{n\in\mathbb{N}}$. With this construction, we have that

$$\left(\mathcal{D}_0^{1,p}(\Omega), \|\cdot\|_{\mathcal{D}_0^{1,p}(\Omega)}\right),$$

is a Banach space. The space $C_0^\infty(\Omega)$ can be naturally embedded into $\mathcal{D}_0^{1,p}(\Omega)$ by associating to each $u \in C_0^\infty(\Omega)$ the constant sequence

$$\{u_n\}_{n\in\mathbb{N}} \quad \text{such that} \quad u_n = u, \quad \text{for every } n \in \mathbb{N}.$$

By definition, this representation of $C_0^\infty(\Omega)$ is dense in $\mathcal{D}_0^{1,p}(\Omega)$.

C.2 Comparison with $W_0^{1,p}$

In light of Proposition C.1.2, we can not expect the completion space $\mathcal{D}_0^{1,p}(\Omega)$ to coincide with $W_0^{1,p}(\Omega)$. Nevertheless, the latter is contained in the former, in a suitable sense.

Lemma C.2.1 *Let $1 \le p \le \infty$ and let $\Omega \subseteq \mathbb{R}^N$ be an open set. Then we have*

$$W_0^{1,p}(\Omega) \subseteq \mathcal{D}_0^{1,p}(\Omega).$$

More precisely, if define the linear application $\mathcal{J} : W_0^{1,p}(\Omega) \to \mathcal{D}_0^{1,p}(\Omega)$ by

$$\mathcal{J}(u) = \Big[\{u_n\}_{n\in\mathbb{N}}\Big]_\sim, \qquad \text{for } \{u_n\}_{n\in\mathbb{N}} \subseteq C_0^\infty(\Omega) \text{ such that } \lim_{n\to\infty} \|u_n - u\|_{W^{1,p}(\Omega)} = 0,$$

we have that $\mathcal{J}$ is continuous, injective and such that

$$\|\mathcal{J}(u)\|_{\mathcal{D}_0^{1,p}(\Omega)} = \|\nabla u\|_{L^p(\Omega;\mathbb{R}^N)}, \qquad \text{for every } u \in W_0^{1,p}(\Omega). \tag{C.2.1}$$

Proof We first observe that $\mathcal{J}$ is well-defined. Indeed, by definition of $W_0^{1,p}(\Omega)$, for every $u \in W_0^{1,p}(\Omega)$ there exists $\{u_n\}_{n\in\mathbb{N}} \subseteq C_0^\infty(\Omega)$ such that

$$\lim_{n\to\infty} \|u_n - u\|_{W^{1,p}(\Omega)} = 0.$$

Moreover, if $\{v_n\}_{n\in\mathbb{N}} \subseteq C_0^\infty(\Omega)$ is any other sequence with the same property, both $\{u_n\}_{n\in\mathbb{N}}$ and $\{v_n\}_{n\in\mathbb{N}}$ are Cauchy sequences with respect to the reduced norm (C.1.2) and

$$\{u_n\}_{n\in\mathbb{N}} \sim \{v_n\}_{n\in\mathbb{N}},$$

thanks to the fact that they have the same limit. Thus, $\mathcal{J}(u)$ does not depend on the particular sequence chosen.

The linearity of $\mathcal{J}$ is straightforward. In order to prove (C.2.1) it is sufficient to observe that for every $u \in W_0^{1,p}(\Omega)$ we have by definition

$$\|\mathcal{J}(u)\|_{\mathcal{D}_0^{1,p}(\Omega)} = \lim_{n\to\infty} \|\nabla u_n\|_{L^p(\Omega;\mathbb{R}^N)} = \|\nabla u\|_{L^p(\Omega;\mathbb{R}^N)}.$$

Once we proved (C.2.1), both the continuity and the injectivity of $\mathcal{J}$ follow. □

Remark C.2.2 The previous result can be rephrased by saying that $W_0^{1,p}(\Omega)$ endowed with the norm (C.1.2) is linearly isometrically embedded into $\mathcal{D}_0^{1,p}(\Omega)$.

By Proposition C.1.2, we get that in general $\mathcal{J}$ is not surjective, i.e. for a general open set the two Sobolev spaces

$$W_0^{1,p}(\Omega) \qquad \text{and} \qquad \mathcal{D}_0^{1,p}(\Omega),$$

can not be identified. However, this becomes true, provided some suitable assumptions are added on the open set Ω.

Proposition C.2.3 *Let $1 \le p \le \infty$ and let $\Omega \subseteq \mathbb{R}^N$ be an open set with finite Lebesgue measure. Then we have*

$$\mathcal{D}_0^{1,p}(\Omega) = W_0^{1,p}(\Omega).$$

More precisely, by considering the linear application $\mathcal{J} : W_0^{1,p}(\Omega) \to \mathcal{D}_0^{1,p}(\Omega)$ defined in Lemma C.2.1, we get that this is surjective, as well.

Proof Let us take $U = [\{u_n\}_{n\in\mathbb{N}}]_\sim \in \mathcal{D}_0^{1,p}(\Omega)$. The assumption on Ω and Problem 3.12.11 entail that we have the Poincaré inequality

$$\|\varphi\|_{L^p(\Omega)} \le C\,|\Omega|^{\frac{1}{N}}\,\|\nabla\varphi\|_{L^p(\Omega;\mathbb{R}^N)}, \qquad \text{for every } \varphi \in C_0^\infty(\Omega),$$

at our disposal. This in particular permits to infer that

$$\|\nabla\varphi\|_{L^p(\Omega;\mathbb{R}^N)} \qquad \text{and} \qquad \|\varphi\|_{W^{1,p}(\Omega)},$$

are equivalent norms on $C_0^\infty(\Omega)$. Thus $\{u_n\}_{n\in\mathbb{N}}$ is a Cauchy sequence in $W_0^{1,p}(\Omega)$, as well. The latter being a Banach space (recall Theorem 3.7.4), we obtain that such a sequence converges in $W^{1,p}(\Omega)$ to $u \in W_0^{1,p}(\Omega)$. By definition of $\mathcal{J}$, this shows that

$$\mathcal{J}(u) = U = \Big[\{u_n\}_{n\in\mathbb{N}}\Big]_\sim.$$

This proves that $\mathcal{J}$ is surjective. In light of the other properties of $\mathcal{J}$ contained in Lemma C.2.1, we get the conclusion. □

Remark C.2.4 An inspection of the proof reveals that the previous result is still valid for every open set $\Omega \subseteq \mathbb{R}^N$ which admits the Poincaré inequality. This is the case, for example, of an open set bounded in one direction (see Theorem 3.7.6).

C.3 The Case $1 \le p < N$ on the Whole Space

When $1 \le p < N$ and $\Omega = \mathbb{R}^N$, as a consequence of the Sobolev inequality (see Theorem 3.6.1), we can obtain an explicit characterization of the homogeneous space $\mathcal{D}_0^{1,p}$. We start by considering an expedient Sobolev space: it will play a crucial role.

Proposition C.3.1 *For $1 \le p < N$, we define following vector space*

$$\dot{W}^{1,p}(\mathbb{R}^N) = \Big\{u \in L^{p^*}(\mathbb{R}^N) \,:\, \nabla u \in L^p(\mathbb{R}^N;\mathbb{R}^N)\Big\},$$

where as usual

$$p^* = \frac{N\,p}{N-p}.$$

We endow $\dot{W}^{1,p}(\mathbb{R}^N)$ with the reduced norm (C.1.2).

Then $\dot{W}^{1,p}(\mathbb{R}^N)$ is a Banach space, which contains $C_0^\infty(\mathbb{R}^N)$ as a dense subspace. In particular, for functions in $\dot{W}^{1,p}(\mathbb{R}^N)$ we still have the Sobolev inequality, i.e.

$$\left(\int_{\mathbb{R}^N} |u|^{p^*}\,dx\right)^{\frac{p}{p^*}} \le \mathcal{S}^p \int_{\mathbb{R}^N} |\nabla u|^p\,dx, \qquad \textit{for every } u \in \dot{W}^{1,p}(\mathbb{R}^N),$$

where $\mathcal{S} = \mathcal{S}(N,p) > 0$ is the same constant as in Theorem 3.6.1.

Proof We first observe that (C.1.2) is indeed a norm on this space. It is sufficient to check that

$$\text{if } u \in \dot{W}^{1,p}(\mathbb{R}^N) \text{ is such that } \|\nabla u\|_{L^p(\mathbb{R}^N;\mathbb{R}^N)} = 0, \text{ then } u \equiv 0,$$

the other properties being straightforward. This follows from Proposition 3.2.9 and the fact that the unique constant function contained in $L^{p^*}(\mathbb{R}^N)$ is the null one.

We then prove that $C_0^\infty(\mathbb{R}^N)$ is dense in $\dot{W}^{1,p}(\mathbb{R}^N)$. Let u be such that

$$u \in L^{p^*}(\mathbb{R}^N) \qquad \text{and} \qquad \nabla u \in L^p(\mathbb{R}^N;\mathbb{R}^N),$$

we need to show that there exists a sequence $\{u_n\}_{n\in\mathbb{N}} \subseteq C_0^\infty(\mathbb{R}^N)$ such that

$$\lim_{n\to\infty} \|\nabla u_n - \nabla u\|_{L^p(\mathbb{R}^N;\mathbb{R}^N)} = 0.$$

Such a sequence is constructed in a similar fashion as in the proof of Proposition 3.7.10. We first define

$$v_n = u * \rho_n,$$

where $\{\rho_n\}_{n\in\mathbb{N}\setminus\{0\}} \subseteq C_0^\infty(\mathbb{R}^N)$ is the sequence of standard mollifiers, with each ρ_n supported on the ball $\overline{B_{1/n}(0)}$. Then we take a family of cut-off functions $\{\eta_n\}_{n\in\mathbb{N}\setminus\{0\}} \subseteq C_0^\infty(\mathbb{R}^N)$ such that

$$0 \le \eta_n \le 1, \qquad \eta_n = 1 \text{ on } B_n(0), \qquad \eta_n = 0 \text{ on } \mathbb{R}^N \setminus B_{2n}(0),$$

and

$$|\nabla \eta_n| \le \frac{C}{n}.$$

Finally, we set $u_n = v_n\,\eta_n \in C_0^\infty(\mathbb{R}^N)$. Observe that

$$\begin{aligned}
&\int_{\mathbb{R}^N} |\nabla u_n - \nabla u|^p\,dx \\
&\quad = \int_{\mathbb{R}^N} |\eta_n\,\nabla v_n + v_n\,\nabla \eta_n - \nabla u|^p\,dx \\
&\quad \le 2^{p-1} \int_{\mathbb{R}^N} |\eta_n\,\nabla v_n - \nabla u|^p\,dx + 2^{p-1} \int_{\mathbb{R}^N} |v_n|^p\,|\nabla \eta_n|^p\,dx \\
&\quad \le 4^{p-1} \int_{\mathbb{R}^N} |\eta_n\,\nabla v_n - \eta_n\,\nabla u|^p\,dx + 4^{p-1} \int_{\mathbb{R}^N} |1-\eta_n|^p\,|\nabla u|^p\,dx \\
&\qquad + 4^{p-1} \int_{\mathbb{R}^N} |u - v_n|^p\,|\nabla \eta_n|^p\,dx + 4^{p-1} \int_{\mathbb{R}^N} |u|^p\,|\nabla \eta_n|^p\,dx.
\end{aligned}$$

By using the properties of η_n, we then get

$$\begin{aligned}
\int_{\mathbb{R}^N} |\nabla u_n - \nabla u|^p\,dx &\le 4^{p-1} \int_{\mathbb{R}^N} |\nabla v_n - \nabla u|^p\,dx + 4^{p-1} \int_{\mathbb{R}^N\setminus B_n(0)} |\nabla u|^p\,dx \\
&\quad + \frac{C}{n^p} \int_{B_{2n}(0)\setminus B_n(0)} |u - v_n|^p\,dx + \frac{C}{n^p} \int_{B_{2n}(0)\setminus B_n(0)} |u|^p\,dx.
\end{aligned}$$

We now show that the four integrals on the right-hand side converge to 0 as n goes to ∞. Indeed, since $v_n = u * \rho_n$ and $\nabla u \in L^p(\mathbb{R}^N; \mathbb{R}^N)$, by the properties of convolutions we have

$$\lim_{n\to\infty} \int_{\mathbb{R}^N} |\nabla v_n - \nabla u|^p \, dx = 0.$$

By using that $\nabla u \in L^p(\mathbb{R}^N; \mathbb{R}^N)$ and the Dominated Convergence Theorem, we have

$$\lim_{n\to\infty} \int_{\mathbb{R}^N \setminus B_n(0)} |\nabla u|^p \, dx = 0.$$

We are left with handling the last two integrals, which are the most delicate ones: indeed, observe that $u \notin L^p(\mathbb{R}^N)$, we only have $u \in L^{p^*}(\mathbb{R}^N)$. By using Hölder's inequality with exponents

$$\frac{p^*}{p} \quad \text{and} \quad \frac{p^*}{p^* - p},$$

we have

$$\begin{aligned}\frac{1}{n^p} \int_{B_{2n}(0)\setminus B_n(0)} |u - v_n|^p \, dx &\le \frac{1}{n^p} \left|B_{2n}(0) \setminus B_n(0)\right|^{1-\frac{p}{p^*}} \\ &\quad \times \left(\int_{B_{2n}(0)\setminus B_n(0)} |u - v_n|^{p^*} \, dx\right)^{\frac{p}{p^*}}.\end{aligned}$$

Observe that

$$\left|B_{2n}(0) \setminus B_n(0)\right|^{1-\frac{p}{p^*}} \le \left|B_{2n}(0)\right|^{1-\frac{p}{p^*}} = \left(\omega_N\,(2\,n)^N\right)^{1-\frac{p}{p^*}} = C\,n^p,$$

thanks to the definition of p^*. Here $C > 0$ is a constant depending on N and p, only. This permits to infer that

$$\begin{aligned}\frac{1}{n^p} \int_{B_{2n}(0)\setminus B_n(0)} |u - v_n|^p \, dx &\le C \left(\int_{B_{2n}(0)\setminus B_n(0)} |u - v_n|^{p^*} \, dx\right)^{\frac{p}{p^*}} \\ &\le C \left(\int_{\mathbb{R}^N} |u - v_n|^{p^*} \, dx\right)^{\frac{p}{p^*}}.\end{aligned}$$

The last integral converges to 0, thanks to the fact that $u \in L^{p^*}(\mathbb{R}^N)$ and the properties of convolutions. Finally, for the last integral we proceed similarly: we have

$$\frac{1}{n^p}\int_{B_{2n}(0)\setminus B_n(0)}|u|^p\,dx \le \frac{1}{n^p}\,|B_{2n}(0)\setminus B_n(0)|^{1-\frac{p}{p^*}}\left(\int_{B_{2n}(0)\setminus B_n(0)}|u|^{p^*}\,dx\right)^{\frac{p}{p^*}}.$$

As above, we obtain

$$\frac{1}{n^p}\int_{B_{2n}(0)\setminus B_n(0)}|u|^p\,dx \le C\left(\int_{B_{2n}(0)\setminus B_n(0)}|u|^{p^*}\,dx\right)^{\frac{p}{p^*}}.$$

The fact that $u \in L^{p^*}(\mathbb{R}^N)$ and the Dominated Convergence Theorem assure that

$$\lim_{n\to\infty}\left(\int_{B_{2n}(0)\setminus B_n(0)}|u|^{p^*}\,dx\right)^{\frac{p}{p^*}} = 0.$$

This finally proves the density of $C_0^\infty(\mathbb{R}^N)$ in $\dot{W}_0^{1,p}(\mathbb{R}^N)$.

The fact that the Sobolev inequality is still valid for $\dot{W}^{1,p}(\mathbb{R}^N)$ simply follows from Theorem 3.6.1 and the density of $C_0^\infty(\mathbb{R}^N)$. More precisely, let us take $u \in \dot{W}^{1,p}(\mathbb{R}^N)$. From the previous part of the proof, there exists a sequence $\{u_n\}_{n\in\mathbb{N}} \subseteq C_0^\infty(\mathbb{R}^N)$ such that the gradients converge in $L^p(\mathbb{R}^N;\mathbb{R}^N)$ to ∇u. In particular, we get that $\{\nabla u_n\}_{n\in\mathbb{N}\setminus\{0\}}$ is a Cauchy sequence in the Banach space $L^p(\mathbb{R}^N;\mathbb{R}^N)$. By using the Sobolev inequality for $u_m - u_m \in C_0^\infty(\mathbb{R}^N)$, we have

$$\left(\int_{\mathbb{R}^N}|u_n - u_m|^{p^*}\,dx\right)^{\frac{p}{p^*}} \le \mathcal{S}^p\int_{\mathbb{R}^N}|\nabla u_n - \nabla u_m|^p\,dx.$$

This shows that $\{u_n\}_{n\in\mathbb{N}}$ is a Cauchy sequence in the Banach space $L^{p^*}(\mathbb{R}^N)$. Thus, it must converge in this space to a limit function $v \in L^{p^*}(\mathbb{R}^N)$. We claim that

$$v = u, \qquad \text{a. e. in } \mathbb{R}^N. \tag{C.3.1}$$

Indeed, for every $\varphi \in C_0^\infty(\mathbb{R}^N)$ and every $i \in \{1,\ldots,N\}$, we have

$$\lim_{n\to\infty}\int_{\mathbb{R}^N}\varphi\,\frac{\partial u_n}{\partial x_i}\,dx = \int_{\mathbb{R}^N}\varphi\,\frac{\partial u}{\partial x_i}\,dx,$$

thanks to the strong convergence of the gradients. Also, by using the definition of weak partial derivatives

$$\lim_{n\to\infty}\int_{\mathbb{R}^N}\varphi\,\frac{\partial u_n}{\partial x_i}\,dx = -\lim_{n\to\infty}\int_{\mathbb{R}^N}\frac{\partial\varphi}{\partial x_i}\,u_n\,dx = -\int_{\mathbb{R}^N}\frac{\partial\varphi}{\partial x_i}\,v\,dx,$$

where we now used the strong convergence of $\{u_n\}_{n\in\mathbb{N}}$. In particular, we obtain

$$\int_{\mathbb{R}^N} \varphi \frac{\partial u}{\partial x_i}\,dx = -\int_{\mathbb{R}^N} \frac{\partial \varphi}{\partial x_i}\, v\,dx, \qquad \text{for every } \varphi \in C_0^\infty(\mathbb{R}^N),\ i \in \{1,\ldots,N\}.$$

This in turn implies that v has a weak gradient in $L^p(\mathbb{R}^N;\mathbb{R}^N)$ and

$$\nabla v = \nabla u, \qquad \text{a. e. in } \mathbb{R}^N.$$

By using Proposition 3.2.9, we can then infer that there exists a constant $C \in \mathbb{R}$ such that

$$v = u + C, \qquad \text{a. e. in } \mathbb{R}^N.$$

Since both u and v belong to $L^{p^*}(\mathbb{R}^N)$, the only possibility is that $C = 0$ and thus we get (C.3.1). In conclusion, we have obtained

$$\lim_{n\to\infty} \Big(\|u_n - u\|_{L^{p^*}(\mathbb{R}^N)} + \|\nabla u_n - \nabla u\|_{L^p(\mathbb{R}^N;\mathbb{R}^N)} \Big) = 0.$$

We can then apply Sobolev's inequality to each u_n and then pass to the limit as n goes to ∞. This finally gives that $u \in \dot{W}^{1,p}(\mathbb{R}^N)$ satisfies the same inequality, as well.

Finally, to prove that $\dot{W}^{1,p}(\mathbb{R}^N)$ is a Banach space, we take a Cauchy sequence $\{u_n\}_{n\in\mathbb{N}} \subseteq \dot{W}^{1,p}(\mathbb{R}^N)$. By relying on the Sobolev inequality, which holds in $\dot{W}^{1,p}(\mathbb{R}^N)$ as just shown, we get that this is a Cauchy sequence in $L^{p^*}(\mathbb{R}^N)$, as well. We can infer that ∇u_n converges in $L^p(\mathbb{R}^N;\mathbb{R}^N)$ to some $\phi \in L^p(\mathbb{R}^N;\mathbb{R}^N)$, while u_n converges in $L^{p^*}(\mathbb{R}^N)$ to some $u \in L^{p^*}(\mathbb{R}^N)$. By using the definition of weak gradient, it is easy to show that

$$\phi = \nabla u, \qquad \text{a. e. in } \mathbb{R}^N.$$

Thus, we get $u \in \dot{W}^{1,p}(\mathbb{R}^N)$, as desired. □

Thanks to the previous space $\dot{W}^{1,p}(\mathbb{R}^N)$, we can now give a concrete characterization of the completion space $\mathcal{D}_0^{1,p}(\mathbb{R}^N)$, in the case $1 \le p < N$.

Theorem C.3.2 (Characterization of $\mathcal{D}_0^{1,p}(\mathbb{R}^N)$) *Let $1 \le p < N$, then we have*

$$\mathcal{D}_0^{1,p}(\mathbb{R}^N) = \dot{W}^{1,p}(\mathbb{R}^N).$$

More precisely, if we define the linear application $\mathcal{J} : \dot{W}^{1,p}(\mathbb{R}^N) \to \mathcal{D}_0^{1,p}(\mathbb{R}^N)$ by

$$\mathcal{J}(u) = \Big[\{u_n\}_{n\in\mathbb{N}}\Big]_\sim, \qquad \begin{array}{l} \textit{for } \{u_n\}_{n\in\mathbb{N}} \subseteq C_0^\infty(\mathbb{R}^N) \textit{ such that} \\ \displaystyle\lim_{n\to\infty} \|\nabla u_n - \nabla u\|_{L^p(\mathbb{R}^N;\mathbb{R}^N)} = 0, \end{array}$$

we have that $\mathcal{J}$ is continuous, injective, surjective and such that

$$\|\mathcal{J}(u)\|_{\mathcal{D}_0^{1,p}(\mathbb{R}^N)} = \|\nabla u\|_{L^p(\mathbb{R}^N;\mathbb{R}^N)}, \qquad \textit{for every } u \in \dot{W}^{1,p}(\mathbb{R}^N). \tag{C.3.2}$$

Proof We first observe that $\mathcal{J}$ is well-defined, thanks to Proposition C.3.1. The identity C.3.2 can be proved exactly as in the proof of Lemma C.2.1. Accordingly, once we proved (C.3.2), both the continuity and the injectivity of $\mathcal{J}$ are straightforward.

Finally, in order to show that $\mathcal{J}$ is surjective, we proceed as in the proof of Proposition C.2.3: let us take $U = [\{u_n\}_{n\in\mathbb{N}}]_\sim \in \mathcal{D}_0^{1,p}(\mathbb{R}^N)$. In particular, $\{u_n\}_{n\in\mathbb{N}}$ is a Cauchy sequence in $\dot{W}^{1,p}(\mathbb{R}^N)$. The latter being a Banach space, we obtain that such a sequence converges in $\dot{W}^{1,p}(\mathbb{R}^N)$ to $u \in \dot{W}^{1,p}(\mathbb{R}^N)$. By definition of $\mathcal{J}$, this shows that

$$\mathcal{J}(u) = U = \Big[\{u_n\}_{n\in\mathbb{N}}\Big]_\sim.$$

This proves that $\mathcal{J}$ is surjective, as well. □

Remark C.3.3 The characterization of $\mathcal{D}_0^{1,p}(\mathbb{R}^N)$ in the case $p \geq N$ is more delicate: the interested reader is referred to the paper [67], on which this appendix is partly based. The case $p = N = 1$ is however exceptional: in this case, one can identify $\mathcal{D}_0^{1,1}(\mathbb{R})$ with the following Banach space

$$\dot{W}^{1,1}(\mathbb{R}) := \Big\{u \in C_b^0(\mathbb{R}) \,:\, u' \in L^1(\mathbb{R})\Big\}.$$

It is sufficient to proceed as in the proofs of Proposition C.3.1 and Theorem C.3.2, by using Theorem 3.6.6 which gives

$$\|u\|_{L^\infty(\mathbb{R})} \leq \frac{1}{2}\int_{\mathbb{R}} |u'|\,dx, \qquad \text{for every } u \in C_0^\infty(\mathbb{R}).$$

Appendix D
A Glimpse of Traces and Fractional Sobolev Spaces

D.1 The Trace Inequality in the Half-space

Let $1 \leq p < \infty$ and $0 < s < 1$, for a measurable function $u : \mathbb{R}^N \to \mathbb{R}$ we introduce the following *Sobolev–Slobodeckiĭ seminorm*

$$[u]_{W^{s,p}(\mathbb{R}^N)} = \left(\iint_{\mathbb{R}^N \times \mathbb{R}^N} \frac{|u(x) - u(y)|^p}{|x - y|^{N+s\,p}}\, dx\, dy \right)^{\frac{1}{p}}.$$

We can then introduce the *fractional Sobolev space* defined by

$$W^{s,p}(\mathbb{R}^N) = \left\{ u \in L^p(\mathbb{R}^N) \,:\, [u]_{W^{s,p}(\mathbb{R}^N)} < +\infty \right\}.$$

It is not difficult to see that this is a Banach space, when endowed with the norm

$$\|u\|_{W^{s,p}(\mathbb{R}^N)} = \|u\|_{L^p(\mathbb{R}^N)} + [u]_{W^{s,p}(\mathbb{R}^N)}.$$

An equivalent norm is given by the following

$$u \mapsto \left(\|u\|^p_{L^p(\mathbb{R}^N)} + [u]^p_{W^{s,p}(\mathbb{R}^N)} \right)^{\frac{1}{p}}.$$

In this appendix, we will work in the upper half-space $\mathbb{R}^N \times \mathbb{R}_+$, where $\mathbb{R}_+ = (0, +\infty)$. We will use the notation

$$(x, z) \in \mathbb{R}^N \times \mathbb{R}_+.$$

L. Brasco, *Handbook of Calculus of Variations for Absolute Beginners*, La Matematica per il 3+2 163, https://doi.org/10.1007/978-3-031-87164-1

Correspondingly, we will indicate

$$\nabla_x u = \left(\frac{\partial u}{\partial x_1}, \dots, \frac{\partial u}{\partial x_N}\right) \qquad \text{and} \qquad \partial_z u = \frac{\partial u}{\partial z},$$

for notational simplicity. Accordingly, we will also write $\nabla u = (\nabla_x u, \partial_z u)$.

Proposition D.1.1 (Trace Inequality) *Let* $1 \le p < \infty$ *and* $u \in C_0^\infty(\mathbb{R}^N \times \overline{\mathbb{R}_+})$. *Then we have*

$$\|u(\cdot, 0)\|^p_{L^p(\mathbb{R}^N)} \le p\, \|u\|^{p-1}_{L^p(\mathbb{R}^N\times\mathbb{R}_+)} \, \|\partial_z u\|_{L^p(\mathbb{R}^N\times\mathbb{R}_+)}, \tag{D.1.1}$$

In the case $1 < p < \infty$, *we also have*

$$\begin{aligned} \big[u(\cdot, 0)\big]^p_{W^{1-\frac{1}{p},p}(\mathbb{R}^N)} &\le C \iint_{\mathbb{R}^N\times\mathbb{R}_+} |\nabla_x u|^p \, dx\, dz \\ &\quad + C \left(\frac{p}{p-1}\right)^p \iint_{\mathbb{R}^N\times\mathbb{R}_+} |\partial_z u|^p \, dx\, dz, \end{aligned} \tag{D.1.2}$$

for a constant $C = C(N, p) > 0$.

Proof We first prove inequality (D.1.1). If $u \in C_0^\infty(\mathbb{R}^N \times \overline{\mathbb{R}_+})$, for every $x \in \mathbb{R}^N$ the function

$$z \mapsto |u(x, z)|^{p-1}\, u(x, z),$$

is C^1 and it identically vanishes for z large enough. We can then write

$$\begin{aligned} -|u(x, 0)|^{p-1}\, u(x, 0) &= \int_0^{+\infty} \partial_z \big(|u(x, z)|^{p-1}\, u(x, z)\big)\, dz \\ &= p \int_0^{+\infty} \partial_z u(x, z)\, |u(x, z)|^{p-1}\, dz. \end{aligned}$$

In particular, this implies that

$$|u(x, 0)|^p \le p \int_0^{+\infty} |\partial_z u(x, z)|\, |u(x, z)|^{p-1}\, dz.$$

The previous estimate is valid for every $x \in \mathbb{R}^N$: by integrating over $\mathbb{R}^N$, we obtain

$$\int_{\mathbb{R}^N} |u(x, 0)|^p\, dx \le p \iint_{\mathbb{R}^N\times\mathbb{R}_+} |\partial_z u(x, z)|\, |u(x, z)|^{p-1}\, dx\, dz.$$

It is now sufficient to use Hölder's inequality with exponents p and p' in order to get (D.1.1).

We assume $1 < p < \infty$ and come to the proof of (D.1.2), which is more sophisticated. We start by writing for every $x, h \in \mathbb{R}^N$

$$\begin{aligned}
u(x+h,0)-u(x,0) &= \Big(u(x+h,0)-u(x+h,|h|)\Big)\\
&\quad + \Big(u(x+h,|h|)-u(x,|h|)\Big)\\
&\quad + \Big(u(x,|h|)-u(x,0)\Big)\\
&= -\int_0^{|h|} \partial_z u(x+h,z)\,dz + \int_0^1 \langle \nabla_x u(x+z\,h,|h|), h\rangle\,dz\\
&\quad + \int_0^{|h|} \partial_z u(x,z)\,dz,
\end{aligned}$$

thanks to the Fundamental Theorem of Calculus. From this identity, we get

$$\begin{aligned}
|u(x+h,0)-u(x,0)| \le \int_0^{|h|} &|\partial_z u(x+h,z)|\,dz\\
&+ |h| \int_0^1 |\nabla_x u(x+z\,h,|h|)|\,dz\\
&+ \int_0^{|h|} |\partial_z u(x,z)|\,dz.
\end{aligned}$$

By taking the L^p norm in the x variable and using Minkowski's inequality, we thus get

$$\begin{aligned}
\|u(\cdot+h,0)-u(\cdot,0)\|_{L^p(\mathbb{R}^N)} \le \int_0^{|h|} &\|\partial_z u(\cdot+h,z)\|_{L^p(\mathbb{R}^N)}\,dz\\
&+ |h| \int_0^1 \|\nabla_x u(\cdot+z\,h,|h|)\|_{L^p(\mathbb{R}^N)}\,dz\\
&+ \int_0^{|h|} \|\partial_z u(\cdot,z)\|_{L^p(\mathbb{R}^N)}\,dz\\
&= 2\int_0^{|h|} \|\partial_z u(\cdot,z)\|_{L^p(\mathbb{R}^N)}\,dz\\
&+ |h|\,\|\nabla_x u(\cdot,|h|)\|_{L^p(\mathbb{R}^N)},
\end{aligned}$$

where we also used the invariance by translations of the $L^p(\mathbb{R}^N)$ norm. By observing that the Sobolev–Slobodeckiĭ seminorm can be rewritten as follows

$$
\begin{aligned}
[u(\cdot,0)]^p_{W^{1-\frac{1}{p},p}(\mathbb{R}^N)} &= \iint_{\mathbb{R}^N\times\mathbb{R}^N} \frac{|u(x,0)-u(y,0)|^p}{|x-y|^{N+p-1}}\,dx\,dy \\
&= \iint_{\mathbb{R}^N\times\mathbb{R}^N} \frac{|u(x+h,0)-u(x,0)|^p}{|h|^{N+p-1}}\,dx\,dh \\
&= \int_{\mathbb{R}^N} \|u(\cdot+h,0)-u(\cdot,0)\|^p_{L^p(\mathbb{R}^N)}\,\frac{dh}{|h|^{N+p-1}},
\end{aligned}
$$

we thus obtain

$$
\begin{aligned}
[u(\cdot,0)]^p_{W^{1-\frac{1}{p},p}(\mathbb{R}^N)} &\le 2^{2\,p-1}\int_{\mathbb{R}^N}\left(\int_0^{|h|}\|\partial_z u(\cdot,z)\|_{L^p(\mathbb{R}^N)}\,dz\right)^p\frac{dh}{|h|^{N+p-1}} \\
&\quad + 2^{p-1}\int_{\mathbb{R}^N}\|\nabla_x u(\cdot,|h|)\|^p_{L^p(\mathbb{R}^N)}\,\frac{dh}{|h|^{N-1}} =: \mathcal{I}_1+\mathcal{I}_2.
\end{aligned} \tag{D.1.3}
$$

We now estimate separately the two integrals. For the first one, by using spherical coordinates we have

$$
\mathcal{I}_1 = 2^{2\,p-1}\,N\,\omega_N\int_0^{+\infty}\frac{1}{\varrho^p}\left(\int_0^{\varrho}\|\partial_z u(\cdot,z)\|_{L^p(\mathbb{R}^N)}\,dz\right)^p d\varrho.
$$

Since $p>1$, we can use Hardy's inequality for the half-line (see Problem 3.12.8 with $N=1$)

$$
\left(\frac{p-1}{p}\right)^p\int_0^{+\infty}\frac{|\varphi(\varrho)|^p}{\varrho^p}\,d\varrho \le \int_0^{+\infty}|\varphi'(\varrho)|^p\,d\varrho,
$$

with the choice[1]

$$
\varphi(\varrho) = \int_0^{\varrho}\|\partial_z u(\cdot,z)\|_{L^p(\mathbb{R}^N)}\,dz.
$$

This yields

$$
\begin{aligned}
\mathcal{I}_1 &\le 2^{2\,p-1}\,N\,\omega_N\left(\frac{p}{p-1}\right)^p\int_0^{+\infty}\|\partial_\varrho u(\cdot,\varrho)\|^p_{L^p(\mathbb{R}^N)}\,d\varrho \\
&= 2^{2\,p-1}\,N\,\omega_N\left(\frac{p}{p-1}\right)^p\iint_{\mathbb{R}^N\times\mathbb{R}_+}|\partial_z u(x,z)|^p\,dx\,dz.
\end{aligned}
$$

In order to estimate $\mathcal{I}_2$, we use again spherical coordinates, so to directly get

[1] The reader should try to justify that such a function is feasible for Hardy's inequality, as a useful exercise.

$$\begin{aligned}\mathcal{I}_2 &= 2^{p-1}\, N\, \omega_N \int_0^{+\infty} \|\nabla_x u(\cdot,\varrho)\|^p_{L^p(\mathbb{R}^N)}\, d\varrho \\ &= 2^{p-1}\, N\, \omega_N \iint_{\mathbb{R}^N\times\mathbb{R}_+} |\nabla_x u(x,z)|^p\, dx\, dz.\end{aligned}$$

By inserting the last two informations in (D.1.3), we get the inequality (D.1.2). □

The following technical result will be useful, in order to define the trace operator.

Lemma D.1.2 *Let $1 \le p < \infty$, then $C_0^\infty(\mathbb{R}^N \times \overline{\mathbb{R}_+})$ is dense in $W^{1,p}(\mathbb{R}^N \times \mathbb{R}_+)$.*

Proof We need to show that for every $u \in W^{1,p}(\mathbb{R}^N \times \mathbb{R}_+)$ there exists a sequence $\{u_n\}_{n\in\mathbb{N}} \subseteq C_0^\infty(\mathbb{R}^N \times \overline{\mathbb{R}_+})$ such that

$$\lim_{n\to\infty} \|u_n - u\|_{W^{1,p}(\mathbb{R}^N\times\mathbb{R}_+)} = 0.$$

We split the proof in two parts, for ease of readability.

Part 1: Extension by Reflection For every $u \in W^{1,p}(\mathbb{R}^N \times \mathbb{R}_+)$, as in Problem 3.12.5 we define its *extension by even reflection*, given by

$$\widehat{u}(x,z) = \begin{cases} u(x,z), & \text{if } x \in \mathbb{R}^N,\ z \ge 0, \\ u(x,-z), & \text{if } x \in \mathbb{R}^N,\ z < 0. \end{cases}$$

We are going to show that $\widehat{u} \in W^{1,p}(\mathbb{R}^{N+1})$. The fact that $\widehat{u} \in L^p(\mathbb{R}^{N+1})$ is straightforward. We need to show that $\widehat{u}$ has a weak gradient which belongs to $L^p(\mathbb{R}^{N+1};\mathbb{R}^{N+1})$. We claim that

$$\nabla_x \widehat{u}(x,z) = \begin{cases} \nabla_x u(x,z), & \text{if } x \in \mathbb{R}^N,\ z \ge 0, \\ \nabla_x u(x,-z), & \text{if } x \in \mathbb{R}^N,\ z < 0, \end{cases} \tag{D.1.4}$$

and

$$\partial_z \widehat{u}(x,z) = \begin{cases} \partial_z u(x,z), & \text{if } x \in \mathbb{R}^N,\ z \ge 0, \\ -\partial_z u(x,-z), & \text{if } x \in \mathbb{R}^N,\ z < 0. \end{cases} \tag{D.1.5}$$

We take $\varphi \in C_0^\infty(\mathbb{R}^{N+1})$ and a cut-off function $\eta_\varepsilon \in C^\infty(\mathbb{R})$, such that

$$0 \le \eta_\varepsilon \le 1, \qquad \eta_\varepsilon \equiv 1 \text{ on } \mathbb{R}\setminus[-2\,\varepsilon, 2\,\varepsilon], \qquad \eta_\varepsilon \equiv 0 \text{ on } [-\varepsilon,\varepsilon],$$

with

$$|\eta_\varepsilon'(z)| \le \frac{C}{\varepsilon}.$$

Moreover, we take η_ε to be even, i.e. such that

$$\eta_\varepsilon(-z) = \eta_\varepsilon(z), \qquad \text{for every } z \in \mathbb{R}.$$

Then, for $i \in \{1, \dots, N\}$, we have by the Dominated Convergence Theorem

$$\iint_{\mathbb{R}^N \times \mathbb{R}} \frac{\partial \varphi}{\partial x_i} \widehat{u} \, dx\, dz = \lim_{\varepsilon \to 0^+} \iint_{\mathbb{R}^N \times \mathbb{R}} \frac{\partial \varphi}{\partial x_i} \eta_\varepsilon \widehat{u} \, dx\, dz. \tag{D.1.6}$$

By denoting $\mathbb{R}_- = (-\infty, 0)$, we rewrite the last integral as

$$\begin{aligned}
\iint_{\mathbb{R}^N \times \mathbb{R}} \frac{\partial \varphi}{\partial x_i}(x, z)\, \eta_\varepsilon(z)\, \widehat{u}(x, z)\, dx\, dz &= \iint_{\mathbb{R}^N \times \mathbb{R}_+} \frac{\partial \varphi}{\partial x_i}(x, z)\, \eta_\varepsilon(z)\, u(x, z)\, dx\, dz \\
&\quad + \iint_{\mathbb{R}^N \times \mathbb{R}_-} \frac{\partial \varphi}{\partial x_i}(x, z)\, \eta_\varepsilon(z)\, u(x, -z)\, dx\, dz \\
&= \iint_{\mathbb{R}^N \times \mathbb{R}_+} \frac{\partial \varphi}{\partial x_i}(x, z)\, \eta_\varepsilon(z)\, u(x, z)\, dx\, dz \\
&\quad + \iint_{\mathbb{R}^N \times \mathbb{R}_+} \frac{\partial \varphi}{\partial x_i}(x, -z)\, \eta_\varepsilon(-z)\, u(x, z)\, dx\, dz,
\end{aligned}$$

where we used a change of variable and the definition of $\widehat{u}$. We now observe that by construction $\varphi\, \eta_\varepsilon \in C_0^\infty(\mathbb{R}^N \times \mathbb{R}_+)$ and

$$\frac{\partial \varphi}{\partial x_i} \eta_\varepsilon = \frac{\partial}{\partial x_i}(\varphi\, \eta_\varepsilon),$$

since η_ε does not depend on x_i. Thus, by using that $u \in W^{1,p}(\mathbb{R}^N \times \mathbb{R}_+)$ and the definition of weak derivative, we get

$$\iint_{\mathbb{R}^N \times \mathbb{R}_+} \frac{\partial \varphi}{\partial x_i}(x, z)\, \eta_\varepsilon(z)\, u(x, z)\, dx\, dz = - \iint_{\mathbb{R}^N \times \mathbb{R}_+} \varphi(x, z)\, \eta_\varepsilon(z)\, \frac{\partial u}{\partial x_i}(x, z)\, dx\, dz,$$

and

$$\begin{aligned}
&\iint_{\mathbb{R}^N \times \mathbb{R}_+} \frac{\partial \varphi}{\partial x_i}(x, -z)\, \eta_\varepsilon(-z)\, u(x, z)\, dx\, dz \\
&\qquad = - \iint_{\mathbb{R}^N \times \mathbb{R}_+} \varphi(x, -z)\, \eta_\varepsilon(-z)\, \frac{\partial u}{\partial x_i}(x, z)\, dx\, dz.
\end{aligned}$$

From (D.1.6) we thus get

$$\iint_{\mathbb{R}^N \times \mathbb{R}} \frac{\partial \varphi}{\partial x_i}(x, z)\, \widehat{u}(x, z)\, dx\, dz = - \lim_{\varepsilon \to 0^+} \iint_{\mathbb{R}^N \times \mathbb{R}_+} \varphi(x, z)\, \eta_\varepsilon(z)\, \frac{\partial u}{\partial x_i}(x, z)\, dx\, dz$$

$$
\begin{aligned}
&- \lim_{\varepsilon\to 0^+} \iint_{\mathbb{R}^N\times\mathbb{R}_+} \varphi(x,-z)\,\eta_\varepsilon(-z)\,\frac{\partial u}{\partial x_i}(x,z)\,dx\,dz \\
&= - \iint_{\mathbb{R}^N\times\mathbb{R}_+} \varphi(x,z)\,\frac{\partial u}{\partial x_i}(x,z)\,dx\,dz \\
&- \iint_{\mathbb{R}^N\times\mathbb{R}_+} \varphi(x,-z)\,\frac{\partial u}{\partial x_i}(x,z)\,dx\,dz.
\end{aligned}
$$

By further making a change of variable in the last integral, we obtain

$$
\begin{aligned}
\iint_{\mathbb{R}^N\times\mathbb{R}} \frac{\partial\varphi}{\partial x_i}(x,z)\,\widehat{u}(x,z)\,dx\,dz = &- \iint_{\mathbb{R}^N\times\mathbb{R}_+} \varphi(x,z)\,\frac{\partial u}{\partial x_i}(x,z)\,dx\,dz \\
&- \iint_{\mathbb{R}^N\times\mathbb{R}_-} \varphi(x,z)\,\frac{\partial u}{\partial x_i}(x,-z)\,dx\,dz,
\end{aligned}
$$

which shows that $\widehat{u}$ has a weak partial derivative in $L^p(\mathbb{R}^{N+1})$, with respect to every $x_1,\ldots,x_N$. Moreover, formula (D.1.4) holds for these derivatives.

We are left with proving that $\widehat{u}$ has a weak partial derivative with respect to the variable z, as well. Again by the Dominated Convergence Theorem, we have

$$
\iint_{\mathbb{R}^N\times\mathbb{R}} \frac{\partial\varphi}{\partial z}\,\widehat{u}\,dx\,dz = \lim_{\varepsilon\to 0^+} \iint_{\mathbb{R}^N\times\mathbb{R}} \frac{\partial\varphi}{\partial z}\,\eta_\varepsilon\,\widehat{u}\,dx\,dz. \tag{D.1.7}
$$

As above, we can rewrite the last integral as

$$
\begin{aligned}
\iint_{\mathbb{R}^N\times\mathbb{R}} \frac{\partial\varphi}{\partial z}(x,z)\,\eta_\varepsilon(z)\,\widehat{u}(x,z)\,dx\,dz = &\iint_{\mathbb{R}^N\times\mathbb{R}_+} \frac{\partial\varphi}{\partial z}(x,z)\,\eta_\varepsilon(z)\,u(x,z)\,dx\,dz \\
&+ \iint_{\mathbb{R}^N\times\mathbb{R}_-} \frac{\partial\varphi}{\partial z}(x,z)\,\eta_\varepsilon(z)\,u(x,-z)\,dx\,dz \\
&= \iint_{\mathbb{R}^N\times\mathbb{R}_+} \frac{\partial(\varphi\,\eta_\varepsilon)}{\partial z}(x,z)\,u(x,z)\,dx\,dz \\
&- \iint_{\mathbb{R}^N\times\mathbb{R}_+} \varphi(x,z)\,\eta_\varepsilon'(z)\,u(x,z)\,dx\,dz \\
&+ \iint_{\mathbb{R}^N\times\mathbb{R}_-} \frac{\partial(\varphi\,\eta_\varepsilon)}{\partial z}(x,z)\,u(x,-z)\,dx\,dz \\
&- \iint_{\mathbb{R}^N\times\mathbb{R}_-} \varphi(x,z)\,\eta_\varepsilon'(z)\,u(x,-z)\,dx\,dz.
\end{aligned} \tag{D.1.8}
$$

Then we can use that, when restricted to both $\mathbb{R}^N \times \mathbb{R}_+$ and $\mathbb{R}^N \times \mathbb{R}_-$, the function $\varphi\,\eta_\varepsilon$ has compact support. By definition of weak derivative, we get

$$\begin{aligned}\iint_{\mathbb{R}^N\times\mathbb{R}_+} \frac{\partial(\varphi\,\eta_\varepsilon)}{\partial z}(x,z)\,u(x,z)\,dx\,dz + \iint_{\mathbb{R}^N\times\mathbb{R}_-} \frac{\partial(\varphi\,\eta_\varepsilon)}{\partial z}(x,z)\,u(x,-z)\,dx\,dz \\ = -\iint_{\mathbb{R}^N\times\mathbb{R}_+} \varphi(x,z)\,\eta_\varepsilon(z)\,\partial_z u(x,z)\,dx\,dz \\ -\iint_{\mathbb{R}^N\times\mathbb{R}_-} \varphi(x,z)\,\eta_\varepsilon(z)\,(-\partial_z u(x,-z))\,dx\,dz.\end{aligned}$$

Thus, up to now, from (D.1.7) and (D.1.8) we have obtained

$$\begin{aligned}&\iint_{\mathbb{R}^N\times\mathbb{R}} \frac{\partial\varphi}{\partial z}(x,z)\,\widehat{u}(x,z)\,dx\,dz \\ &= -\lim_{\varepsilon\to0^+}\iint_{\mathbb{R}^N\times\mathbb{R}_+} \varphi(x,z)\,\eta_\varepsilon(z)\,\partial_z u(x,z)\,dx\,dz \\ &\quad -\lim_{\varepsilon\to0^+}\iint_{\mathbb{R}^N\times\mathbb{R}_-} \varphi(x,z)\,\eta_\varepsilon(z)\,(-\partial_z u(x,-z))\,dx\,dz \\ &\quad -\lim_{\varepsilon\to0^+}\iint_{\mathbb{R}^N\times\mathbb{R}_+} \varphi(x,z)\,\eta_\varepsilon'(z)\,u(x,z)\,dx\,dz \\ &\quad -\lim_{\varepsilon\to0^+}\iint_{\mathbb{R}^N\times\mathbb{R}_-} \varphi(x,z)\,\eta_\varepsilon'(z)\,u(x,-z)\,dx\,dz \\ &= -\iint_{\mathbb{R}^N\times\mathbb{R}_+} \varphi(x,z)\,\partial_z u(x,z)\,dx\,dz \\ &\quad -\iint_{\mathbb{R}^N\times\mathbb{R}_-} \varphi(x,z)\,(-\partial_z u(x,-z))\,dx\,dz \\ &\quad -\lim_{\varepsilon\to0^+}\iint_{\mathbb{R}^N\times\mathbb{R}_+} \varphi(x,z)\,\eta_\varepsilon'(z)\,u(x,z)\,dx\,dz \\ &\quad -\lim_{\varepsilon\to0^+}\iint_{\mathbb{R}^N\times\mathbb{R}_-} \varphi(x,z)\,\eta_\varepsilon'(z)\,u(x,-z)\,dx\,dz.\end{aligned}$$

In order to conclude this part of the proof, we just need to show that the last two terms vanish. We make a change of variable in the last term and use that η_ε is even,[2] thus we get

[2] This implies that η_ε' is an odd function.

$$\begin{aligned}\lim_{\varepsilon\to 0^+}\iint_{\mathbb{R}^N\times\mathbb{R}_+} &\varphi(x,z)\,\eta_\varepsilon'(z)\,u(x,z)\,dx\,dz\\ &+\lim_{\varepsilon\to 0^+}\iint_{\mathbb{R}^N\times\mathbb{R}_-}\varphi(x,z)\,\eta_\varepsilon'(z)\,u(x,-z)\,dx\,dz\\ &=\lim_{\varepsilon\to 0^+}\iint_{\mathbb{R}^N\times\mathbb{R}_+}\Big(\varphi(x,z)-\varphi(x,-z)\Big)\,u(x,z)\,\eta_\varepsilon'(z)\,dx\,dz\\ &=\lim_{\varepsilon\to 0^+}\iint_{\mathbb{R}^N\times(0,2\,\varepsilon)}\Big(\varphi(x,z)-\varphi(x,-z)\Big)\,u(x,z)\,\eta_\varepsilon'(z)\,dx\,dz.\end{aligned}$$

We now observe that for every $(x,z)\in\mathbb{R}^N\times(0,2\,\varepsilon)$ we have

$$\Big|\varphi(x,z)-\varphi(x,-z)\Big|=\left|\int_{-z}^{z}\partial_z\varphi(x,\tau)\,d\tau\right|\le 2\,\|\partial_z\varphi\|_{L^\infty(\mathbb{R}^{N+1})}\,z\le 4\,\|\partial_z\varphi\|_{L^\infty}\,\varepsilon.$$

This, together with the fact that φ is compactly supported, permits to infer that the previous limit is 0. In conclusion, we obtain

$$\begin{aligned}\iint_{\mathbb{R}^N\times\mathbb{R}}\frac{\partial\varphi}{\partial z}(x,z)\,\widehat{u}(x,z)\,dx\,dz&=-\iint_{\mathbb{R}^N\times\mathbb{R}_+}\varphi(x,z)\,\partial_z u(x,z)\,dx\,dz\\ &\quad-\iint_{\mathbb{R}^N\times\mathbb{R}_-}\varphi(x,z)\,\left(-\partial_z u(x,-z)\right)\,dx\,dz,\end{aligned}$$

which shows that $\widehat{u}$ has a weak partial derivative in $L^p(\mathbb{R}^{N+1})$ with respect to z, as well. Moreover, formula (D.1.5) holds for this derivative.

Part 2: Approximation For every $u\in W^{1,p}(\mathbb{R}^N\times\mathbb{R}_+)$, we consider its extension $\widehat{u}\in W^{1,p}(\mathbb{R}^{N+1})$. By Proposition 3.7.10, we know that

$$W^{1,p}(\mathbb{R}^{N+1})=W^{1,p}_0(\mathbb{R}^{N+1}).$$

In other words, the space $C^\infty_0(\mathbb{R}^{N+1})$ is dense in $W^{1,p}(\mathbb{R}^{N+1})$. Thus, there exists a sequence $\{v_n\}_{n\in\mathbb{N}}\subseteq C^\infty_0(\mathbb{R}^{N+1})$ such that

$$\lim_{n\to\infty}\|v_n-\widehat{u}\|_{W^{1,p}(\mathbb{R}^{N+1})}=0.$$

We observe that

$$\|v_n-\widehat{u}\|_{W^{1,p}(\mathbb{R}^{N+1})}\ge\|v_n-\widehat{u}\|_{W^{1,p}(\mathbb{R}^N\times\mathbb{R}_+)}=\|v_n-u\|_{W^{1,p}(\mathbb{R}^N\times\mathbb{R}_+)},$$

thanks to the fact that $\widehat{u}$ coincides with u on $\mathbb{R}^N\times\mathbb{R}_+$, by construction. Thus, we have

$$\lim_{n\to\infty}\|v_n-u\|_{W^{1,p}(\mathbb{R}^N\times\mathbb{R}_+)}=0,$$

as well. By finally noticing that the restriction of v_n to $\mathbb{R}^N \times \overline{\mathbb{R}_+}$ is an element of $C_0^\infty(\mathbb{R}^N \times \overline{\mathbb{R}_+})$, we get the desired approximating sequence. □

D.2 The Trace Operator in the Half-space

The next result clarifies in which sense the space $W_0^{1,p}$ has to be intended as the space of Sobolev functions vanishing at the boundary. Without any attempt to give a complete discussion, we will confine ourselves to the case of the upper half-space.

Theorem D.2.1 (Trace Operator) *Let $1 < p < \infty$, then there exists a linear continuous operator*

$$\mathrm{tr} : W^{1,p}(\mathbb{R}^N \times \mathbb{R}_+) \to W^{1-\frac{1}{p},p}(\mathbb{R}^N),$$

with the following properties:

(1) for every $u \in C^0(\mathbb{R}^N \times \overline{\mathbb{R}_+}) \cap W^{1,p}(\mathbb{R}^N \times \mathbb{R}_+)$, the function $\mathrm{tr}(u)$ coincides with the restriction of u to the hyperplane $\{z = 0\}$;
(2) we have

$$W_0^{1,p}(\mathbb{R}^N \times \mathbb{R}_+) = \Big\{u \in W^{1,p}(\mathbb{R}^N \times \mathbb{R}_+) \, : \, \mathrm{tr}(u) = 0\Big\}.$$

Proof We first define the operator tr. By Lemma D.1.2, we know that for every $u \in W^{1,p}(\mathbb{R}^N \times \mathbb{R}_+)$ there exists a sequence $\{u_n\}_{n\in\mathbb{N}} \subseteq C_0^\infty(\mathbb{R}^N \times \overline{\mathbb{R}_+})$ such that

$$\lim_{n\to\infty} \|u_n - u\|_{W^{1,p}(\mathbb{R}^N\times\mathbb{R}_+)} = 0.$$

In particular, $\{u_n\}_{n\in\mathbb{N}}$ is a Cauchy sequence in $W^{1,p}(\mathbb{R}^N \times \mathbb{R}_+)$. By using this fact and the trace inequality of Proposition D.1.1, we get that

$$\{u_n(\cdot, 0)\}_{n\in\mathbb{N}},$$

is a Cauchy sequence in $W^{1-1/p,p}(\mathbb{R}^N)$. The latter being a Banach space, we can infer existence of a function $U \in W^{1-1/p,p}(\mathbb{R}^N)$, such that

$$\lim_{n\to\infty} \|u_n(\cdot, 0) - U\|_{W^{1-\frac{1}{p},p}(\mathbb{R}^N)} = 0.$$

We then define $\mathrm{tr}(u)$ by

$$\mathrm{tr}(u) := U.$$

By construction, this is a linear operator, taking values in the fractional Sobolev space $W^{1-1/p,p}(\mathbb{R}^N)$. It is not difficult to see that the definition is well-posed, i.e. it does not depend on the choice of the approximating sequence $\{u_n\}_{n\in\mathbb{N}}$. Indeed, if $\{v_n\}_{n\in\mathbb{N}} \subseteq C_0^\infty(\mathbb{R}^n \times \overline{\mathbb{R}_+})$ is another sequence such that

$$\lim_{n\to\infty} \|v_n - u\|_{W^{1,p}(\mathbb{R}^N\times\mathbb{R}_+)} = 0,$$

by the triangle inequality, we would get

$$\lim_{n\to\infty} \|u_n - v_n\|_{W^{1,p}(\mathbb{R}^N\times\mathbb{R}_+)} = 0,$$

as well. By applying the trace inequality of Proposition D.1.1 to $u_n - v_n$, we would obtain

$$\lim_{n\to\infty} \|u_n(\cdot,0) - v_n(\cdot,0)\|_{W^{1-\frac{1}{p},p}(\mathbb{R}^N)} = 0,$$

i.e. the limit function U obtained above is independent of the choice of the sequence.

We are left with proving that tr is continuous: by using again the trace inequality for a smooth approximating sequence $\{u_n\}_{n\in\mathbb{N}}$, we have

$$\begin{aligned}\|u_n(\cdot,0)\|_{W^{1-\frac{1}{p},p}(\mathbb{R}^N)} &= \|u_n(\cdot,0)\|_{L^p(\mathbb{R}^N)} + \Big[u_n(\cdot,0)\Big]_{W^{1-\frac{1}{p},p}(\mathbb{R}^N)}\\ &\le \widetilde{C}\,\|u_n\|_{W^{1,p}(\mathbb{R}^N\times\mathbb{R}_+)},\end{aligned}$$

for a constant $\widetilde{C} = \widetilde{C}(N,p) > 0$. By taking the limit as n goes to ∞ and using the definition of the trace operator, we get

$$\|\mathrm{tr}(u)\|_{W^{1-\frac{1}{p},p}(\mathbb{R}^N)} \le \widetilde{C}\,\|u\|_{W^{1,p}(\mathbb{R}^N\times\mathbb{R}_+)}.$$

Since tr is linear, this proves the required continuity property of the operator.

We now prove separately the claimed properties of the trace operator.

Proof of (1) Let $u \in C^0(\mathbb{R}^N\times\overline{\mathbb{R}_+})\cap W^{1,p}(\mathbb{R}^N\times\mathbb{R}_+)$, we take a sequence $\{u_n\}_{n\in\mathbb{N}} \subseteq C_0^\infty(\mathbb{R}^N \times \overline{\mathbb{R}_+})$ such that

$$\lim_{n\to\infty} \|u_n - u\|_{W^{1,p}(\mathbb{R}^N\times\mathbb{R}_+)} = 0. \tag{D.2.1}$$

By the Fundamental Theorem of Calculus and Hölder's inequality, for every $n \in \mathbb{N}$ and $z > 0$, we have

$$|u_n(x,z) - u_n(x,0)|^p = \left|\int_0^z \partial_\tau u_n(x,\tau)\,d\tau\right|^p \le z^{p-1}\int_0^z |\partial_\tau u_n(x,\tau)|^p\,d\tau.$$

By integrating with respect to $x \in \mathbb{R}^N$, we get for every $z > 0$

$$\begin{aligned}\int_{\mathbb{R}^N} |u_n(x,z) - u_n(x,0)|^p \, dx &\leq z^{p-1} \iint_{\mathbb{R}^N \times (0,z)} |\partial_\tau u_n|^p \, dx \, d\tau \\ &\leq z^{p-1} \iint_{\mathbb{R}^N \times \mathbb{R}_+} |\partial_\tau u_n|^p \, dx \, d\tau.\end{aligned} \tag{D.2.2}$$

From (D.2.1), we know in particular that

$$\lim_{n\to\infty} \iint_{\mathbb{R}^N \times \mathbb{R}_+} |u_n(x,z) - u(x,z)|^p \, dx \, dz = 0.$$

Thus, the Fubini-Tonelli Theorem entails that we must have

$$\lim_{n\to\infty} \int_{\mathbb{R}^N} |u_n(x,z) - u(x,z)|^p \, dx = 0, \qquad \text{for a. e. } z > 0.$$

By using this information, the definition of $\operatorname{tr}(u)$ and (D.2.1), we can pass to the limit in (D.2.2) and obtain

$$\int_{\mathbb{R}^N} |u(x,z) - \operatorname{tr}(u)(x)|^p \, dx \leq z^{p-1} \iint_{\mathbb{R}^N \times \mathbb{R}_+} |\partial_\tau u|^p \, dx \, d\tau, \qquad \text{for a. e. } z > 0.$$

By finally taking the limit as z goes to 0 and using the continuity assumption on u, we get from the previous estimate

$$u(\cdot, 0) = \operatorname{tr}(u),$$

as desired.

Proof of (2) For the second point, we follow the argument of [65, Theorem 5.1]. The proof of the inclusion

$$W_0^{1,p}(\mathbb{R}^N \times \mathbb{R}_+) \subseteq \Big\{ W^{1,p}(\mathbb{R}^N \times \mathbb{R}_+) \, : \, \operatorname{tr}(u) = 0 \Big\},$$

is easy. Indeed, if $u \in W_0^{1,p}(\mathbb{R}^N \times \mathbb{R}_+)$, by definition there exists a sequence $\{u_n\}_{n\in\mathbb{N}} \subseteq C_0^\infty(\mathbb{R}^N \times \mathbb{R}_+)$ such that

$$\lim_{n\to\infty} \|u_n - u\|_{W^{1,p}(\mathbb{R}^N \times \mathbb{R}_+)} = 0.$$

By observing that

$$u_n(x,0) = 0 \qquad \text{for every } x \in \mathbb{R}^N, \ n \in \mathbb{N},$$

and using the definition of the trace operator given at the beginning, we get that $\operatorname{tr}(u)$ is the function identically vanishing.

In order to complete the proof, we need to prove that for every $u \in W^{1,p}(\mathbb{R}^N \times \mathbb{R}_+)$ such that $\mathrm{tr}(u) = 0$, there exists a sequence $\{u_n\}_{n\in\mathbb{N}} \subseteq C_0^\infty(\mathbb{R}^N \times \mathbb{R}_+)$ such that

$$\lim_{n\to\infty} \|u_n - u\|_{W^{1,p}(\mathbb{R}^N\times\mathbb{R}_+)} = 0.$$

By assumption and the definition of tr, we know that there exists a sequence $\{v_n\}_{n\in\mathbb{N}} \subseteq C_0^\infty(\mathbb{R}^N \times \overline{\mathbb{R}_+})$ such that

$$\lim_{n\to\infty} \|v_n - u\|_{W^{1,p}(\mathbb{R}^N\times\mathbb{R}_+)} = 0,$$

and

$$\lim_{n\to\infty} \|v_n(\cdot,0)\|_{W^{1-\frac{1}{p},p}(\mathbb{R}^N)} = 0.$$

Without loss of generality, we can assume that

$$\|v_n - u\|_{W^{1,p}(\mathbb{R}^N\times\mathbb{R}_+)} \le \frac{1}{n}.$$

Thus, by using the continuity and linearity of the trace operator, together with (1), we get

$$\begin{aligned}\|v_n(\cdot,0)\|_{W^{1-\frac{1}{p},p}(\mathbb{R}^N)} &= \|v_n(\cdot,0) - \mathrm{tr}(u)\|_{W^{1-\frac{1}{p},p}(\mathbb{R}^N)}\\ &= \|\mathrm{tr}(v_n - u)\|_{W^{1-\frac{1}{p},p}(\mathbb{R}^N)} \le \widetilde{C}\,\|v_n - u\|_{W^{1,p}(\mathbb{R}^N\times\mathbb{R}_+)}.\end{aligned}$$

This yields

$$\|v_n(\cdot,0)\|_{W^{1-\frac{1}{p},p}(\mathbb{R}^N)} \le \frac{\widetilde{C}}{n}, \tag{D.2.3}$$

where $\widetilde{C} = \widetilde{C}(N,p) > 0$. We take a sequence of cut-off functions $\eta_n \in C^\infty(\mathbb{R}_+)$, such that

$$0 \le \eta_n \le 1, \qquad \eta_n \equiv 1 \text{ on } \left[\frac{2}{n}, +\infty\right), \qquad \eta_n \equiv 0 \text{ on } \left[0, \frac{1}{n}\right],$$

with

$$|\eta_n'(z)| \le C\,n, \qquad \text{for } z \in \mathbb{R}_+.$$

We then define $u_n = v_n\,\eta_n$, which belongs to $C_0^\infty(\mathbb{R}^N \times \mathbb{R}_+)$. It is not difficult to see that

$$\lim_{n\to\infty} \|u_n - u\|_{L^p(\mathbb{R}^N\times\mathbb{R}_+)} = 0,$$

and

$$\lim_{n\to\infty} \|\nabla_x u_n - \nabla_x u\|_{L^p(\mathbb{R}^N\times\mathbb{R}_+)} = 0.$$

In order to conclude, we need to prove that

$$\lim_{n\to\infty} \|\partial_z u_n - \partial_z u\|_{L^p(\mathbb{R}^N\times\mathbb{R}_+)} = 0.$$

We first observe that by the triangle inequality, we have

$$\|\partial_z u_n - \partial_z u\|_{L^p(\mathbb{R}^N\times\mathbb{R}_+)} \le \|\partial_z u_n - \partial_z v_n\|_{L^p(\mathbb{R}^N\times\mathbb{R}_+)} + \|\partial_z v_n - \partial_z u\|_{L^p(\mathbb{R}^N\times\mathbb{R}_+)}.$$

By hypothesis, the last term goes to zero as n goes to ∞, thus we are left with showing that

$$\lim_{n\to\infty} \|\partial_z u_n - \partial_z v_n\|_{L^p(\mathbb{R}^N\times\mathbb{R}_+)} = 0.$$

By using the definition of u_n and the properties of η_n, this term can be estimated as follows

$$\begin{aligned}
\|\partial_z u_n - \partial_z v_n\|^p_{L^p(\mathbb{R}^N\times\mathbb{R}_+)} &= \iint_{\mathbb{R}^N\times\mathbb{R}_+} \left|\partial_z v_n\, \eta_n + v_n\, \eta_n' - \partial_z v_n\right|^p\, dx\, dz \\
&\le 2^{p-1} \iint_{\mathbb{R}^N\times\mathbb{R}_+} |\partial_z v_n|^p\, |1-\eta_n|^p\, dx\, dz \\
&\quad + 2^{p-1} \iint_{\mathbb{R}^N\times\mathbb{R}_+} |v_n|^p\, |\eta_n'|^p\, dx\, dz \\
&\le 2^{p-1} \iint_{\mathbb{R}^N\times\left(0,\frac{2}{n}\right)} |\partial_z v_n|^p\, dx\, dz \\
&\quad + 2^{p-1}\, C\, n^p \iint_{\mathbb{R}^N\times\left(0,\frac{2}{n}\right)} |v_n|^p\, dx\, dz.
\end{aligned}$$

By using the convergence of v_n to u, we get

$$\begin{aligned}
\lim_{n\to\infty} \iint_{\mathbb{R}^N\times\left(0,\frac{2}{n}\right)} |\partial_z v_n|^p\, dx\, dz &\le 2^{p-1} \lim_{n\to\infty} \iint_{\mathbb{R}^N\times\left(0,\frac{2}{n}\right)} |\partial_z v_n - \partial_z u|^p\, dx\, dz \\
&\quad + 2^{p-1} \lim_{n\to\infty} \iint_{\mathbb{R}^N\times\left(0,\frac{2}{n}\right)} |\partial_z u|^p\, dx\, dz = 0.
\end{aligned} \tag{D.2.4}$$

We also used that $\partial_z u \in L^p(\mathbb{R}^N \times \mathbb{R}_+)$ and the Dominated Convergence Theorem. We still need to prove that

$$\lim_{n\to\infty} n^p \iint_{\mathbb{R}^N\times\left(0,\frac{2}{n}\right)} |v_n|^p \,dx\,dz = 0. \tag{D.2.5}$$

We rely once again on the Fundamental Theorem of Calculus. This gives

$$\begin{aligned}|v_n(x,z)| &= \left|\int_0^z \partial_z v_n(x,\tau)\,d\tau + v_n(x,0)\right|\\ &\le \int_0^z |\partial_z v_n(x,\tau)|\,d\tau + |v_n(x,0)|.\end{aligned}$$

We raise this estimate to the power p and use Jensen's inequality (Proposition 1.2.11), so to get

$$|v_n(x,z)|^p \le 2^{p-1}\,z^{p-1}\int_0^z |\partial_z v_n(x,\tau)|^p\,d\tau + 2^{p-1}\,|v_n(x,0)|^p.$$

We integrate with respect to z over the interval $(0, 2/n)$

$$\begin{aligned}\int_0^{\frac{2}{n}} |v_n(x,z)|^p\,dz &\le 2^{p-1}\int_0^{\frac{2}{n}}\left(z^{p-1}\int_0^z |\partial_z v_n(x,\tau)|^p\,d\tau\right)dz\\ &\quad + 2^{p-1}\,\frac{2}{n}\,|v_n(x,0)|^p\\ &\le 2^{p-1}\int_0^{\frac{2}{n}}\left(z^{p-1}\int_0^{\frac{2}{n}} |\partial_z v_n(x,\tau)|^p\,d\tau\right)dz\\ &\quad + 2^{p-1}\,\frac{2}{n}\,|v_n(x,0)|^p\\ &= \frac{2^{2\,p-1}}{p\,n^p}\int_0^{\frac{2}{n}} |\partial_z v_n(x,\tau)|^p\,d\tau + \frac{2^p}{n}\,|v_n(x,0)|^p.\end{aligned}$$

A further integration with respect to x leads to

$$\iint_{\mathbb{R}^N\times\left(0,\frac{2}{n}\right)} |v_n|^p\,dx\,dz \le \frac{C}{n^p}\iint_{\mathbb{R}^N\times\left(0,\frac{2}{n}\right)} |\partial_z v_n|^p\,dx\,dz + \frac{C}{n}\int_{\mathbb{R}^N} |v_n(x,0)|^p\,dx,$$

with $C = C(p) > 0$. If we multiply both sides by n^p and then use (D.2.3) for the last term, we get

$$n^p \iint_{\mathbb{R}^N\times\left(0,\frac{2}{n}\right)} |v_n|^p\,dx\,dz \le C\iint_{\mathbb{R}^N\times\left(0,\frac{2}{n}\right)} |\partial_z v_n|^p\,dx\,dz + \frac{C}{n},$$

possibly fo a different constant $C = C(N, p) > 0$. If we take into account (D.2.4), the validity (D.2.5) easily follows from the previous estimate. This concludes the proof. □

Remark D.2.2 Actually, one could prove that the trace operator constructed in Theorem D.2.1 is *surjective*. This means that the fractional Sobolev space $W^{1-1/p,p}(\mathbb{R}^N)$ *coincides* with the space of traces of functions in $W^{1,p}(\mathbb{R}^N \times \mathbb{R}_+)$, when $1 < p < \infty$. In the limit case $p = 1$, one can construct the linear and continuous trace operator

$$\mathrm{tr} : W^{1,1}(\mathbb{R}^N \times \mathbb{R}_+) \to L^1(\mathbb{R}^N),$$

and show again that this is surjective. These remarkable results are due to Gagliardo, see [74, Teorema 1.I & Teorema 1.II].

A similar remark applies more generally to $W^{s,p}(\mathbb{R}^N)$, for every $0 < s < 1$. These spaces coincide with the trace spaces of suitable *weighted* Sobolev spaces. The interested reader is referred to [50, Chapter 10], for example.

Remark D.2.3 We assumed throughout this appendix that $p < \infty$. The limit case $p = \infty$ is not particularly interesting: indeed, in this case we have

$$W^{1,\infty}(\mathbb{R}^N \times \mathbb{R}_+) = C^{0,1}(\mathbb{R}^N \times \overline{\mathbb{R}_+}),$$

thanks to Proposition 5.4.3. Accordingly, the trace operator would simply be given by

$$\begin{aligned} \mathrm{tr} : C^{0,1}(\mathbb{R}^N \times \overline{\mathbb{R}_+}) &\to C^{0,1}(\mathbb{R}^N) \\ u &\mapsto u(\cdot, 0). \end{aligned}$$

By Theorem 5.1.10, we have that this operator is surjective.

References

Some Books on Calculus of Variations

1. Angrisani, F., Ascione, G., Leone, C., Mantegazza, C.: Appunti di Calcolo delle Variazioni. Amazon Editore, Seattle (2019)
2. Buttazzo, G., Giaquinta, M., Hildebrandt, S.: One-dimensional Variational Problems. An Introduction. Oxford Lecture Series in Mathematics and Its Applications, vol. 15. The Clarendon Press/Oxford University Press, New York (1998)
3. Dacorogna, B.: Direct Methods in the Calculus of Variations. Applied Mathematical Sciences, vol. 78, 2nd edn. Springer, New York (2008)
4. Giaquinta, M., Hildebrandt, S.: Calculus of Variations. II. The Hamiltonian Formalism. Grundlehren der Mathematischen Wissenschaften, vol. 311. Springer-Verlag, Berlin (1996)
5. Giaquinta, M., Hildebrandt, S.: Calculus of Variations. I. The Lagrangian Formalism. Grundlehren der Mathematischen Wissenschaften, vol. 310. Springer-Verlag, Berlin (1996)
6. Giusti, E.: Direct Methods in the Calculus of Variations. World Scientific Publishing Co., Inc., River Edge (2003)
7. Goldstine, H.H.: A history of the calculus of variations from the 17th through the 19th century. Studies in the History of Mathematics and Physical Sciences, vol. 5. Springer-Verlag, New York-Berlin (1980)
8. Morrey, C.B.: Multiple Integrals in the Calculus of Variations. Classics in Mathematics, Reprint of the 1966 edn. Springer-Verlag, Berlin (2008)
9. Moser, J.: Selected Chapters in the Calculus of Variations. Lectures in Mathematics ETH Zürich. Birkhäuser Verlag, Basel (2003)
10. Santambrogio, F.: A Course in the Calculus of Variations – Optimization, Regularity, and Modeling. Universitext. Springer, Cham (2023)
11. Struwe, M.: Variational Methods. Applications to Nonlinear Partial Differential Equations and Hamiltonian Systems, 2nd edn. Ergebnisse der Mathematik und ihrer Grenzgebiete (3) [Results in Mathematics and Related Areas (3)], vol. 34. Springer-Verlag, Berlin (1996)
12. Talenti, G., Colesanti, A., Salani, P.: Un'introduzione al Calcolo delle Variazioni: teoria ed esercizi. Unione Matematica Italiana, Bologna (2016)

L. Brasco, *Handbook of Calculus of Variations for Absolute Beginners*,
La Matematica per il 3+2 163, https://doi.org/10.1007/978-3-031-87164-1

Other Books

13. Adams, R.A.: Sobolev Spaces. Pure and Applied Mathematics, vol. 65. Academic Press, New York-London (1975)
14. Adams, R.A., Fournier, J.J.F.: Sobolev Spaces. Pure and Applied Mathematics (Amst.), vol. 140, 2nd edn. Elsevier/Academic Press, Amsterdam (2003)
15. Ambrosio, L., Tilli, P.: Topics on Analysis in Metric Spaces. Oxford Lecture Series in Mathematics and Its Applications, vol. 25. Oxford University Press, Oxford (2004)
16. Ambrosio, L., Fusco, N., Pallara, D.: Functions of Bounded Variation and Free Discontinuity Problems. Oxford Mathematical Monographs. The Clarendon Press/Oxford University Press, New York (2000)
17. Beck, L.: Elliptic Regularity Theory. A First Course. Lecture Notes of the Unione Matematica Italiana, vol.19. Springer, ChamUnione Matematica Italiana, Bologna (2016)
18. Birman, M.S., Solomjak, M.Z.: Spectral theory of selfadjoint operators in Hilbert space. Mathematics and Its Applications (Soviet Series). D. Reidel Publishing Co., Dordrecht (1987). Translated from the 1980 Russian original by S. Khrushchev and V. Peller
19. Boothby, W.M.: An Introduction to Differentiable Manifolds and Riemannian Geometry. Pure and Applied Mathematics, vol. 120, 2nd edn. Academic Press, Inc., Orlando (1986)
20. Brezis, H.: Functional Analysis, Sobolev Spaces and Partial Differential Equations. Universitext. Springer, New York (2011)
21. Burago, Y.D., Zalgaller, V.A.: Geometric Inequalities. Grundlehren der mathematischen Wissenschaften, vol. 285 [Fundamental Principles of Mathematical Sciences]. Springer Series in Soviet Mathematics. Springer-Verlag, Berlin (1988). Translated from the Russian by A. B. Sosinskiĭ
22. Carlier, G.: Classical and Modern Optimization. Advanced Textbooks in Mathematics. World Scientific Publishing Co. Pte. Ltd., Hackensack (2022)
23. Davies, E.B.: Heat Kernels and Spectral Theory. Cambridge Tracts in Mathematics, vol. 92. Cambridge University Press, Cambridge (1989)
24. DiBenedetto, E.: Real Analysis. Birkhäuser Advanced Texts Basler Lehrbücher [Birkhäuser Advanced Texts: Basel Textbooks]. Birkhäuser Boston, Inc., Boston (2002)
25. do Carmo, M.P.: Differential Geometry of Curves and Surfaces. Prentice-Hall, Inc., Englewood Cliffs (1976). Translated from the Portuguese
26. Edmunds, D.E., Evans, W.D.: Fractional Sobolev Spaces and Inequalities. Cambridge Tracts in Mathematics, vol. 230. Cambridge University Press, Cambridge (2023)
27. Evans, L.C.: Partial Differential Equations, 2nd edn. Graduate Students Mathematics, vol. 19. American Mathematical Society, Providence (2010)
28. Evans, L.C., Gariepy, R.F.: Measure Theory and Fine Properties of Functions. Studies in Advanced Mathematics. CRC Press, Boca Raton (1992)
29. Gilbarg, D., Trudinger, N.S.: Elliptic Partial Differential Equations of Second Order. Classics in Mathematics. Springer-Verlag, Berlin (2001)
30. Giusti, E.: Minimal Surfaces and Functions of Bounded Variation. Monographs in Mathematics, vol. 80. Birkhäuser Verlag, Basel (1984)
31. Han, Q., Lin, F.: Elliptic Partial Differential Equations, 2nd edn. Courant Lecture Notes in Mathematics, vol. 1. Courant Institute of Mathematical Sciences/American Mathematical Society, Providence/New York (2011)
32. Hardy, G.H., Littlewood, J.E., Pólya, G.: Inequalities. Cambridge University Press, Cambridge (1934)
33. Heinonen, J.: Lectures on Lipschitz Analysis. Rep. Univ. Jyväskylä Dep. Math. Stat., vol. 100. University of Jyväskylä, Jyväskylä (2005)
34. Henrot, A.: Extremum Problems for Eigenvalues of Elliptic Operators. Frontiers in Mathematics. Birkhauser Verlag, Basel (2006)
35. Hewitt, E., Stromberg, K.: Real and Abstract Analysis. A Modern Treatment of the Theory of Functions of One Real Variable. Springer-Verlag, New York (1965)

36. Hiriart-Urruty, J.-B., Lemaréchal, C.: Fundamentals of Convex Analysis. Grundlehren Text Ed. Springer-Verlag, Berlin (2001)
37. Hislop, P.D., Sigal, I.M.: Introduction to Spectral Theory. With Applications to Schrödinger Operators. Applied Mathematics in Science, vol. 113. Springer-Verlag, New York (1996)
38. Jost, J.: Postmodern Analysis, 3rd edn. Universitext. Springer-Verlag, Berlin (2005)
39. Krantz, S.G.: Handbook of Complex Variables. Birkhäuser Boston, Inc., Boston (1999)
40. Kufner, A.: Weighted Sobolev Spaces. Wiley-Interscience Publication. John Wiley & Sons, Inc., New York (1985). Translated from the Czech
41. Kufner, A.: John, O., Fučík, S.: Function Spaces. Monographs and Textbooks on Mechanics of Solids and Fluids; Mechanics: Analysis. Noordhoff International Publishing/Academia, Leyden/Prague (1977)
42. Ladyzhenskaya, O.A.: The Mathematical Theory of Viscous Incompressible Flow, Revised English edition, translated from the Russian by Richard A. Silverman. Gordon and Breach Science Publishers, New York-London (1963)
43. Landau, L.D.: Lifšits, E.M.: Theory of Elasticity, 3rd edn. Pergamon Press, Oxford (1986)
44. Leoni, G.: A First Course in Sobolev Spaces, 2nd edn. Graduate Students Mathematics, vol. 181. American Mathematical Society, Providence (2017)
45. Leoni, G.: A First Course in Fractional Sobolev Spaces. Graduate Students Mathematics, vol. 229. American Mathematical Society, Providence (2023)
46. Lieb, E.H., Loss, M.: Analysis, 2nd edn. Graduate Students Mathematics, vol. 14. American Mathematical Society, Providence (2001)
47. Lindqvist, P.: Notes on the Stationary $p-$Laplace Equation. SpringerBriefs in Mathematics. Springer, Cham (2019)
48. Maggi, F.: Optimal Mass Transport on Euclidean Spaces. Cambridge Studies in Advanced Mathematics, vol. 207. Cambridge University Press, Cambridge (2023)
49. Maz'ya, V.: Sobolev Spaces. Springer Series in Soviet Mathematics. Springer-Verlag, Berlin (1985). Translated from the Russian by T. O. Shaposhnikova.
50. Maz'ya, V.: Sobolev Spaces with Applications to Elliptic Partial Differential Equations, Second, revised and augmented edition. Grundlehren der Mathematischen Wissenschaften [Fundamental Principles of Mathematical Sciences], vol. 342. Springer, Heidelberg (2011)
51. Müller, C.: Spherical Harmonics. Lecture Notes in Mathematics, vol. 17. Springer-Verlag, Berlin-New York (1966)
52. Opic, B., Kufner, A.: Hardy-type Inequalities. Pitman Research Notes in Mathematics Series, vol. 219. Longman Scientific & Technical, Harlow (1990)
53. Radó, T.: On the Problem of Plateau. Springer-Verlag, New York-Heidelberg (1971). Reprint
54. Rudin, W.: Functional Analysis. International Series in Pure and Applied Mathematics, 2nd edn. McGraw-Hill, Inc., New York (1991)
55. Schneider, R.: Convex Bodies: The Brunn-Minkowski Theory. Encyclopedia of Mathematics and its Applications, vol. 151, 2nd expanded edn. Cambridge University Press, Cambridge (2014)
56. Sobolev, S.L.: Some Applications of Functional Analysis in Mathematical Physics. Translations of Mathematical Monographs, vol. 90. American Mathematical Society, Providence (1991). Translated from the third Russian edition by Harold H. McFaden. With comments by V. P. Palamodov
57. Stein, E.M., Shakarchi, R.: Fourier Analysis. An Introduction. Princeton Lecture Notes in Analysis, vol. 1. Princeton University Press, Princeton (2003)
58. Vázquez, J.L.: The Porous Medium Equation. Mathematical Theory. Oxford Mathematical Monographs. The Clarendon Press/Oxford University Press, Oxford (2007)
59. Villani, C.: Topics in Optimal Transportation. Graduate Studies in Mathematics, vol. 58, American Mathematical Society, Providence (2003)
60. Willard, S.: General Topology. Addison-Wesley Publishing Co., Reading-London-Don Mills (1970)
61. Zeidler, E.: Applied Functional Analysis: Main Principles and Their Applications. Applied Mathematical Sciences, vol.109. Springer-Verlag, New York (1995)

62. Ziemer, W.P.: Weakly Differentiable Functions. Sobolev Spaces and Functions of Bounded Variation. Graduate Texts in Mathematics, vol. 120. Springer-Verlag, New York (1989)

Miscellaneous Articles

63. Aubin, T.: Problèmes isopérimétriques et espaces de Sobolev. J. Differ. Geom. **11**, 573–598 (1976)
64. Blåsjö, V.: The isoperimetric problem. Am. Math. Mon. **112**, 526–566 (2005)
65. Barros Neto, J.: Inhomogeneous boundary value problems in a half space. Ann. Scuola Norm. Sup. Pisa Cl. Sci. (3) **19**, 331–365 (1965)
66. Bousquet, P., Mariconda, C., Treu, G.: A survey on the non occurence of the Lavrentiev gap for convex, autonomous multiple integral scalar variational problems. Set-Valued Var. Anal. **23**, 55–68 (2015)
67. Brasco, L., Gómez-Castro, D., Vázquez, J.L.: Characterisation of homogeneous fractional Sobolev spaces. Calc. Var. Partial Differ. Equ. **60**, Paper No. 60, 40 pp. (2021)
68. Brezis, H., Oswald, L.: Remarks on sublinear elliptic equations. Nonlinear Anal. **10**, 55–64 (1986)
69. Buttazzo, G., Mintchev, M.: Brachistochrone curves in gravity fields. In: Remembering Franco Conti (Italian). Pubbl. Cent. Ric. Mat. Ennio Giorgi. Scuola Normale Superiore, Pisa (2004), pp. 49–58
70. Cinti, E.: Il problema isoperimetrico: una storia lunga 2000 anni. In: "Matematica, Cultura e Società" – Rivista dell'Unione Matematica Italiana, Serie 1 **4**, 95–106 (2019)
71. Clarke, F.: Continuity of solutions to a basic problem in the calculus of variations. Ann. Sc. Norm. Sup. Pisa Cl. Sci (5). **4**, 511–530 (2005)
72. De Giorgi, E.: Sulla differenziabilità e l'analiticità delle estremali degli integrali multipli regolari. Mem. Accad. Sci. Torino. Cl. Sci. Fis. Mat. Nat. (3) **3**, 25–43 (1957)
73. Fusco, N.: The classical isoperimetric theorem. Rend. Accad. Sci. Fis. Mat. Napoli (4) **71**, 63–107 (2004)
74. Gagliardo, E.: Caratterizzazioni delle tracce sulla frontiera relative ad alcune classi di funzioni in n variabili. Rend. Sem. Mat. Univ. Padova **27**, 284–305 (1957)
75. Gagliardo, E.: Proprietà di alcune classi di funzioni in più variabili. Ricerche Mat. **7**, 102–137 (1958)
76. Guo, Y., Seiringer, R.: On the mass concentration for Bose-Einstein condensates with attractive interactions. Lett. Math. Phys. **104**, 141–156 (2014)
77. Hartman, P.: On the bounded slope condition. Pac. J. Math. **18**, 495–511 (1966)
78. Hartman, P., Stampacchia, G.: On some non-linear elliptic differential-functional equations. Acta Math. **115**, 271–310 (1966)
79. Hile, G., Stanoyevitch, A.: Gradient bounds for harmonic functions Lipschitz on the boundary. Appl. Anal. **73**, 101–113 (1999)
80. Hynd, R., Seuffert, F.: Extremal functions for Morrey's inequality. Arch. Ration. Mech. Anal. **241**, 903–945 (2021)
81. Marcus, M., Mizel, V.J., Pinchover, Y.: On the best constant for Hardy's inequality in $\mathbb{R}^n$. Trans. Am. Math. Soc. **350**, 3237–3255 (1998)
82. Mingione, G.: Regularity of minima: an invitation to the dark side of the calculus of variations. Appl. Math. **51**, 355–426 (2006)
83. Miranda, M.: Un teorema di esistenza e unicità per il problema dell'area minima in n variabili. Ann. Scuola Norm. Sup. Pisa Cl. Sci. (3) **19**, 233–249 (1965)
84. Moser, J.: A new proof of De Giorgi's theorem concerning the regularity problem for elliptic differential equations. Commun. Pure Appl. Math. **13**, 457–468 (1960)
85. Nirenberg, L.: Remarks on strongly elliptic partial differential equations. Commun. Pure Appl. Math. **8**, 649–675 (1955)

86. Nirenberg, L.: On elliptic partial differential equations. Ann. Scuola Norm. Sup. Pisa Cl. Sci. (3) **13**, 115–162 (1959)
87. Sobolev, S.L.: On a theorem in functional analysis (Russian). Mat. Sb. **4**, 471–497 (1938) [Engl. transl.: Am. Math. Soc. Transl. (2) **34**, 39–68 (1963)]
88. Sokal, A.D.: A really simple elementary proof of the uniform boundedness theorem. Am. Math. Monthly **118**, 450–452 (2011)
89. Stampacchia, G.: On some regular multiple integral problems in the calculus of variations. Commun. Pure Appl. Math. **16**, 383–421 (1963)
90. Talenti, G.: Best constant in Sobolev inequality. Ann. Mat. Pura Appl. **110**, 353–372 (1976)
91. Talenti, G.: Inequalities in rearrangement invariant function spaces. In: Nonlinear Analysis, Function Spaces and Applications, vol. 5 (Prometheus Publishing House, Prague, 1994), pp. 177–230

Index

L. Brasco, *Handbook of Calculus of Variations for Absolute Beginners*, La Matematica per il 3+2 163, https://doi.org/10.1007/978-3-031-87164-1

GPSR Compliance

The European Union's (EU) General Product Safety Regulation (GPSR) is a set of rules that requires consumer products to be safe and our obligations to ensure this.

If you have any concerns about our products, you can contact us on ProductSafety@springernature.com

In case Publisher is established outside the EU, the EU authorized representative is:

Springer Nature Customer Service Center GmbH
Europaplatz 3
69115 Heidelberg, Germany

Batch number: 10365480

Printed by Printforce, the Netherlands